Mangroves: Ecology, Biodiversity and Management

Rajesh P. Rastogi • Mahendra Phulwaria •
Dharmendra K. Gupta
Editors

Mangroves: Ecology, Biodiversity and Management

Springer

Editors
Rajesh P. Rastogi
Research in Environment Division
Ministry of Environment, Forest and Climate Change
New Delhi, India

Mahendra Phulwaria
CS-I (Mangroves & Coral Reef) Division
Ministry of Environment, Forest and Climate Change
New Delhi, India

Dharmendra K. Gupta
Ministry of Environment, Forest and Climate Change
New Delhi, India

ISBN 978-981-16-2493-3 ISBN 978-981-16-2494-0 (eBook)
https://doi.org/10.1007/978-981-16-2494-0

This Springer imprint is published by the registered company Springer Nature Singapore Pte Ltd.
The registered company address is: 152 Beach Road, #21-01/04 Gateway East, Singapore 189721, Singapore

Preface

Mangroves are one of the most productive and biologically important blue-carbon ecosystems across the coastal intertidal zone, on the planet Earth. It develops very differently in various ecological settings; however, the expansion of mangroves is commonly found along the tropical and subtropical coastlines around the world. In the current scenario of serious environmental issues of global warming and climate change, mangrove forests may play a vital role in mitigating the rising greenhouse gas emissions due to their extreme capacity in carbon sequestration per unit area. It has been estimated that mangrove forests can store about four to five times more carbon than terrestrial rainforests. Moreover, mangroves play a key role in mitigating global climate change and defend against extreme natural disasters, such as onslaught of cyclones, floods, winds and tidal surges, to the human settlements along the coastal ecosystems.

Various flora and fauna inhabiting mangrove ecosystems have developed several discrete adaptations to withstand daily fluctuating environment with changes in salinity, water and oxygen content with the rise and fall of the sea-tide. Moreover, mangroves are rich in biological diversity of different taxonomic groups with great ecological and commercial importance. Mangroves yield large amounts of oysters, fish, prawns, crabs, fuel-wood, timbers, tannins and other natural products.

Owing to the utmost ecological as well as the economic importance of mangroves, their restoration and proper management are crucial. However, increasing anthropogenic activities and global climate change have raised worldwide concerns towards the conservation, survival and productivity of mangrove ecosystems. Therefore, effective conservation strategies and solid management action plan should be implemented at the global level for their sustainable use, ecological balance and development as a unique ecosystem intended for livelihood and healthy environment.

The proposed book will emphasize the emerging information on ecology of mangroves, with a special reference to their biodiversity and management.

This book has attempted to span the depth of mangrove's ecology starting with more general information such as types, importance and biogeography of the mangrove ecosystem in Chaps. 1 and 2. Mangrove is an important habitat for fauna, in relation to food availability, and as a nursery for several species during different stages of their life cycles, which has been discussed in Chap. 3 as feeding and breeding grounds of mangroves. Chapter 4 focuses on diverse environmental factors influencing the mangrove ecosystems. Chapter 5 highlights the energy flux in a mangrove ecosystem, which will help to understand the energy balance in a mangrove ecosystem. The information regarding nitrogen and phosphorus budget in mangrove ecosystems by exploring the storage, transformation and fluxes in mangrove sediment, biomass, atmosphere and tidal waters has been discussed in Chap. 6. Determination of carbon sinks and estimation of blue-carbon stock of mangrove ecosystem have been reviewed in Chaps. 7 and 8. Chapters 9 and 10 address the responses of mangroves to climate change in the Anthropocene and deep insight of mangroves in combating climate change, respectively. Chapter 11 describes the role of mangroves in pollution abatement. Chapter 12 presents the measurement and modelling of above-ground root systems as attributes of flow and wave attenuation function of mangroves. Chapter 13 explains the roles of mangrove as natural barrier to environmental risks and coastal protection. Chapters 14 and 15 all deal with structure and diversity of polychaetes in mangroves of the Indian coast, as well as structure, composition and diversity of plants in mangrove ecosystems in different locations of the world, respectively. Chapter 16 deals with the patters of livelihood of forest dependent dwellers in relation to the exploitation of resources at the fringe of Indian Sundarbans. The role of mangroves in sustainable tourism development and conserving it for future generations has been mentioned in Chap. 17. In Chap. 18, the role of mangroves in supporting the aquaculture industries has been discussed. Chapters 19 and 20 highlight the ecological valuation and ecosystems of mangroves along with the development of effective management action plans to promote the sustainability of mangrove forest reserves, respectively. Chapter 21 presents an overview of how the RS and GIS technologies are evolving in the context of their use for scientific and quantitative studies on mangroves.

I believe that this book will be helpful to a great extent for the academicians and researchers in the field of mangroves. Undoubtedly, the contents incorporated in this book can be used as a textbook by undergraduate and postgraduate students, teachers and researchers in the fields of mangroves ecology, biodiversity and management.

I thank Ms. Aakanksha Tyagi, Senior editor (Books), Springer Nature for her assistance in seeing it through to completion. I am sincerely grateful to the entire team of Springer for the coordination, support and implementation of this book project. Last but not least, I express my sincere gratitude to all the authors for their kind collaboration and scientific contributions towards the completion of this book, successfully.

New Delhi, India — Rajesh P. Rastogi
New Delhi, India — Mahendra Phulwaria
New Delhi, India — Dharmendra K. Gupta
March 2021

Contents

Editors and Contributors

About the Editors

Rajesh P. Rastogi is Scientist, at Ministry of Environment, Forest and Climate Change (MoEF&CC), New Delhi, India. He obtained his graduate and postgraduate degree from Patna University, Bihar and completed PhD in botany at Banaras Hindu University, Varanasi, India. He is the recipient of some national/international awards and fellowships. He was a visiting scientist at Friedrich Alexander University, Nuremberg, Germany and served as a Visiting Professor of biochemistry at Chulalongkorn University, Thailand. He has published several research/review papers in journals/books of international repute and also edited some books. His field of research includes freshwater/marine algae and cyanobacterial ecology, plant-physiology, photochemistry and photobiology and stress biology.

Mahendra Phulwaria is Scientist at Ministry of Environment, Forest and Climate Change (MoEF&CC), New Delhi, India. He received his graduation degree from Maharshi Dayanand Saraswati University and did postgraduation and PhD in botany from Jai Narain Vyas University, Rajasthan, India. His main field of research includes plant tissue culture (organogenesis, synthetic seed technology, in vitro conservation and micropropagation of woody and medicinal plants) and molecular biology (genetic diversity analysis using DNA barcoding). He was awarded with SERB Young Scientist by Department of Science and Technology, New Delhi and Post-Doctoral Fellowship by UGC, New Delhi. He is a Member of International Association of Plant Biotechnology, India. Chapter.

Dharmendra K. Gupta is Scientist 'F' and Director (S) of Hazardous Substance Management Division in the Ministry of Environment, Forest and Climate Change, Government of India at New Delhi, India and has already published more than 115 refereed research papers/review articles in peer-reviewed international journals and also has edited 17 books of Springer. His field of research includes abiotic stress by radionuclides/heavy metals and xenobiotics in plants, antioxidative system in plants and environmental pollution (radionuclides/heavy metals) remediation through plants (phytoremediation). According to Google Scholar, his total citation of research papers is more than 5000 and h-index is 36.

Contributors

Nur Hasmiza Abdul Hamid Faculty of Applied Sciences, Universiti Teknologi MARA (UiTM), Shah Alam, Malaysia
Institute for Biodiversity and Sustainable Development, Universiti Teknologi MARA (UiTM), Shah Alam, Malaysia

Yarina Ahmad Faculty of Administrative Science and Policy Studies, Universiti Teknologi MARA (UiTM), Shah Alam, Malaysia
Institute for Biodiversity and Sustainable Development, Universiti Teknologi MARA (UiTM), Shah Alam, Malaysia

Daniel M. Alongi Tropical Coastal & Mangrove Consultants, Pakenham, Australia

D. Arceo-Carranza Facultad de Ciencias, Unidad Multidisciplinaria de Docencia e Investigación Sisal, Universidad Nacional Autónoma de México, Sisal, Yucatán, Mexico

C. Yáñez Arenas Facultad de Ciencias, Unidad Multidisciplinaria de Docencia e Investigación Sisal, Sede Parque CientÚfico, Universidad Nacional Autónoma de México, Mérida, Mexico

Nazlin Asari Faculty of Applied Sciences, Universiti Teknologi MARA (UiTM), Shah Alam, Malaysia
Institute for Biodiversity and Sustainable Development, Universiti Teknologi MARA (UiTM), Shah Alam, Malaysia

Thangavel Balasubramanian Faculty of Marine Sciences, CAS in Marine Biology, Annamalai University, Parangipettai, India

X. Chiappa-Carrara Facultad de Ciencias, Unidad Multidisciplinaria de Docencia e Investigación Sisal, Universidad Nacional Autónoma de México, Sisal, Yucatán, Mexico
Escuela Nacional de Estudios Superiores Mérida, Universidad Nacional Autónoma de México, Mérida, Yucatán, Mexico

V. K. Dadhwal Indian Institute of Space Science and Technology (IIST), Trivandrum, India

Chandan Surabhi Das Department of Geography, Barasat Government College, Kolkata, India

Joanna C. Ellison School of Geography, Planning and Spatial Sciences, University of Tasmania, Launceston, TAS, Australia

Engku Azlin Rahayu Engku Ariff Faculty of Applied Sciences, Universiti Teknologi MARA (UiTM) Pahang Campus, Bandar Tun Razak Jengka, Malaysia

L. Gnanappazham Indian Institute of Space Science and Technology (IIST), Trivandrum, India

Tapan Kumar Jana Department of Marine Science, University of Calcutta, Kolkata, India

R. S. C. Jayaraj Rain Forest Research Institute, Jorhat, India

K. Kathiresan Centre of Advanced Study in Marine Biology, Annamalai University, Parangipettai, India

Sanjeev Kumar Physical Research Laboratory, Ahmedabad, India

Anupam Kumari Academy of Scientific and Innovative Research, (AcSIR), Ghaziabad, India
Division of Applied Phycology and Biotechnology, CSIR-Central Salt and Marine Chemicals Research Institute (CSIR-CSMCRI), Bhavnagar, Gujarat, India

R. Chávez López Facultad de Estudios Superiores Iztacala, Universidad Nacional Autónoma de México, Mexico City, Mexico

T. Mageswaran National Centre for Sustainable Coastal Management, MoEF & CC, Chennai, India

Sy Tuan Mai Mangrove Ecosystem Research Center, Hanoi National University of Education, Hanoi, Vietnam

Nurun Nadhirah Md Isa Faculty of Applied Sciences, Universiti Teknologi MARA (UiTM) Pahang Branch, Jengka, Pahang, Malaysia

Mohamad Danial Md Sabri Forestry and Environment Division, Forest Research Institute Malaysia, Kepong, Selangor, Malaysia

P. M. Mohan Department of Ocean Studies and Marine Biology, Pondicherry University, Port Blair, India

Nurul Atiqah Mohd Ayob Faculty of Applied Sciences, Universiti Teknologi MARA (UiTM), Shah Alam, Malaysia

Sandip Kumar Mukhopadhyay Department of Marine Science, University of Calcutta, Kolkata, India

P. Murugesan Faculty of Marine Sciences, Centre of Advanced Study in Marine Biology, Annamalai University, Parangipettai, India

Kazuo Nadaoka School of Environment and Society, Tokyo Institute of Technology, Tokyo, Japan

B. Nagarajan Institute of Forest Genetics and Tree Breeding, Coimbatore, India

Takashi Nakamura School of Environment and Society, Tokyo Institute of Technology, Tokyo, Japan

Thi Hong Hanh Nguyen Faculty of Environment, Hanoi University of Natural Resources and Environment, Hanoi, Vietnam

Ajith Nithin Faculty of Marine Sciences, CAS in Marine Biology, Annamalai University, Parangipettai, India

Hong Tinh Pham Faculty of Environment, Hanoi University of Natural Resources and Environment, Hanoi, Vietnam

Kumar Arun Prasad Department of Geography, Central University of Tamil Nadu, Thiruvarur, India

P. Ragavan Physical Research Laboratory, Ahmedabad, India

AL. Ramanathan School of Environmental Sciences, Jawaharlal Nehru University, New Delhi, India

T. S. Rana CSIR-National Botanical Research Institute, Rana Paratap Marg, Lucknow, India

Prabhat Ranjan Central Pollution Control Board, New Delhi, India

Karuna Rao School of Environmental Sciences, Jawaharlal Nehru University, New Delhi, India

Mangal S. Rathore Academy of Scientific and Innovative Research, (AcSIR), Ghaziabad, India
Division of Applied Phycology and Biotechnology, CSIR-Central Salt and Marine Chemicals Research Institute (CSIR-CSMCRI), Bhavnagar, Gujarat, India

Raghab Ray Atmosphere and Ocean Research Institute, The University of Tokyo, Kashiwa, Japan

Rene Rollon Institute of Environmental Science and Meteorology, College of Science, University of the Philippines, Quezon City, Philippines

V. Sachithanandam National Centre for Sustainable Coastal Management, MoEF & CC, Chennai, India

Ahmad Faris Seman Kamarulzaman Faculty of Applied Sciences, Universiti Teknologi MARA (UiTM) Pahang Campus, Bandar Tun Razak Jengka, Malaysia

Nur Hajar Zamah Shari Forestry and Environment Division, Forest Research Institute Malaysia, Kepong, Selangor, Malaysia

Arumugam Sundaramanickam Faculty of Marine Sciences, CAS in Marine Biology, Annamalai University, Parangipettai, India

Mohd Nazip Suratman Faculty of Applied Sciences and Institute for Biodiversity and Sustainable Development, UiTM, Shah Alam, Malaysia

Rempei Suwa Forestry Division, Japan International Research Center for Agricultural Sciences (JIRCAS), Tsukuba, Japan

Tengku Mohd Zarawie Tengku Hashim Faculty of Applied Sciences and Institute for Biodiversity and Sustainable Development, UiTM, Shah Alam, Malaysia

Masaya Yoshikai School of Environment and Society, Tokyo Institute of Technology, Tokyo, Japan

Mangroves: Types and Importance 1

K. Kathiresan

Abstract

Mangroves are of great ecological significance and economic importance. They are of different types—deltaic, estuarine, lagoon, and fringe mangroves—based on coastal location. The mangroves are of six functional types—fringe, riverine, basin, over-wash, scrub, and hammock. They are also of three types—river-dominated, tide-dominated, and interior mangroves—based on tidal range and sedimentation. In addition, there are six broad types of mangroves: large deltaic systems, tidal plains, composite plains, fringing barriers with lagoons, drowned bedrock valleys, and coral coasts. Mangroves are ecologically significant in protecting the coast from solar UV-B radiation, 'greenhouse' gases, cyclones, floods, sea level rise, wave action, and coastal soil erosion. They act as nutrient sinks, sediment traps, and nutrient source to support the food web in other coastal ecosystems. The mangroves are the most efficient in carbon sequestration and climate change mitigation. They provide feeding, breeding, and nursery grounds for many food fishes and wildlife animals. They protect other marine systems such as islands, coral reefs, seaweeds, and seagrass meadows. Mangroves are economically valuable in supplying the forestry and fishery products and also in serving as sites for developing a burgeoning eco-tourism. The mangroves are of great bioprospecting potential as a source of salt-tolerant genes, chemicals, and valuable products that can be used in medical, industrial, agricultural, and food sectors.

K. Kathiresan (✉)
Centre of Advanced Study in Marine Biology, Annamalai University, Parangipettai, Tamil Nadu, India

R. P. Rastogi et al. (eds.), *Mangroves: Ecology, Biodiversity and Management*,
https://doi.org/10.1007/978-981-16-2494-0_1

Keywords

Mangroves · Ecological role · Economic value · Bioprospecting · Coastal ecosystems

1.1 Introduction

Mangroves are highly productive and biologically diverse coastal habitats of tropical and subtropical regions of the world. They are the one place on earth where land, freshwater, and ocean mix. They are found in intertidal areas along sheltered shores and estuaries. The mangroves have diversified habitats, such as water bodies, forests, litter forest floor, mudflats and adjacent coral reefs and seagrass meadows; and this habitat diversity supports a wide variety of organisms. The mangrove habitats are chiefly colonized by flowering trees and shrubs, remarkably adapted to harsh coastal conditions, such as seawater, periodic inundation and exposure, wind, waves, strong currents, and anaerobic soil. There are no other groups of plants in the entire plant kingdom with such highly developed adaptations. The 'standing crop' of mangroves is greater than any other aquatic ecosystems on the earth (Duke et al. 1998; Kathiresan and Bingham 2001; Kathiresan and Qasim 2005; Spalding et al. 2010; Tomlinson 2016).

Mangroves occur in low-lying coastal plains where the topography is smooth, but not steep and the tidal amplitude is large. They have luxuriant growth in the alluvial soil substrates with fine-textured loose mud or silt, rich in humus. The mangrove plants find difficult to colonize the coastal zone with waves of high energy, and hence they normally establish themselves in sheltered shorelines (Kathiresan and Bingham 2001; Kathiresan and Qasim 2005).

Mangroves are the only tall tree forest system, located between land and sea in tropical and subtropical coasts. They are a rare forest type in the world with 80 tree and shrub species, occupying 13.8 million hectares in 118 countries and territories (Giri et al. 2011; Duke 2017). They are largely restricted to the latitudes between 32° N and 38° S. Growth and biomass production of the mangroves decrease with increasing latitudes and they are the highest around the equator region. The mangroves are often called as 'tidal forests', 'coastal woodlands', 'oceanic rain-forests' or 'blue carbon forest'. Unfortunately, long-term survival of the mangroves is at a great risk, and the ecosystem services offered by them may totally be lost in the world within the next 100 years (Duke et al. 2007). The present chapter deals with different types of mangroves, ecological significance, and economic importance, for better conservation and management.

1.2 Types of Mangroves

Mangroves and their environment are strongly interacting with each other. The mangroves influence chemical and physical conditions of their environment, which in turn influence growth and productivity of the mangroves. They are found

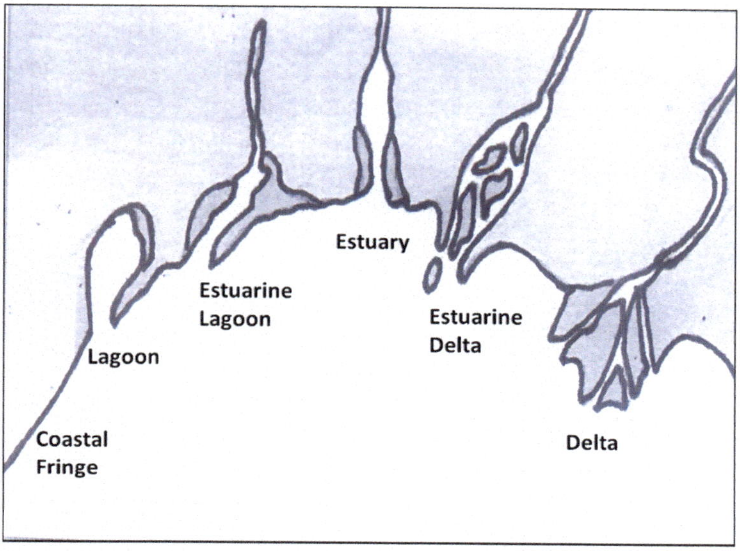

Fig. 1.1 Tropical coastal settings of mangrove forests

in a variety of tropical coastal settings such as the deltas, estuarine deltas, estuaries, estuarine lagoons, lagoons, and coastal fringes (Fig. 1.1) (Kjefve 1990). The deltas are formed by active deposition of sediment at the river mouth, which is colonized by mangroves. Estuaries are the sites of sediment deposition, which is colonized by mangroves. Coastal lagoons are formed behind sand spits and barrier islands; these sites are of less wave action and colonized by mangroves. Mangroves do occur on sediment substrates along the coastal fringes where wave energy is low.

There are six functional types of mangrove forests as shown in Fig. 1.2, namely, over-wash, fringe, riverine, basin, scrub (dwarf), and hammock forests (Lugo and Snedaker 1974; Woodroffe 1992). The last three types are the modified forms of the first three types. The six types can be summarized as follows:

1. **Over-wash mangrove forests**: These are small mangrove islands, frequently formed by tidal washings.
2. **Fringe mangrove forests**: These occur along the borders of protected shorelines and islands, influenced by daily tidal range. They are sensitive to erosion and long exposure to turbulent waves and tides.
3. **Riverine mangrove forests**: These are luxuriant patches of mangroves existing along rivers and creeks, which get flooded daily by the tides. Such forests are influenced by large amounts of freshwater and nutrients and thus making the system highly productive with tall trees.
4. **Basin mangrove forests**: These are stunted mangroves located along the interior side of the swamps and in drainage depressions.

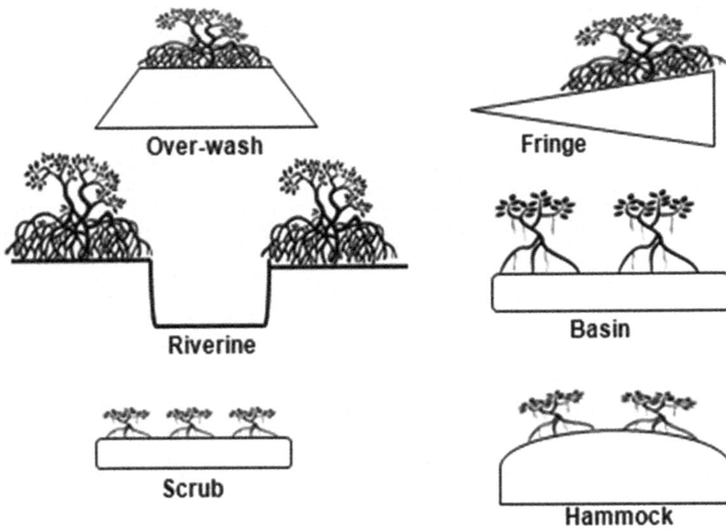

Fig. 1.2 Six functional types of mangrove forests

5. **Hammock mangrove forests**: These are similar to the basin type but occurring in more elevated sites than the four types given-above.
6. **Scrub mangroves**: These are dwarf mangrove forests occurring along flat coastal fringes.

The above classification is not providing any information on the physical processes that take place in mangrove forests. Considering the physical processes, another classification has been proposed with three types of mangrove forests: (1) river-dominated, (2) tide-dominated, and (3) interior mangrove forests (that have less influence of river/or tides) (Fig. 1.3) (Woodroffe 1992). The river-dominated mangroves have a strong out-welling, whereas the tide-dominated mangroves have bidirectional flux, and while the mangroves that are located in interior region are typical sinks for sediment nutrients.

There are six broad classes of mangrove settings, based on the tidal range and sedimentation:

1. **Large deltaic systems** (occurring in low tidal range and substrate with very fine sediments) (e.g. mangroves of Borneo, Sundarbans).
2. **Tidal plains** (where large mudflats are formed by alluvial sediments, reworked by tides, and then the mudflats are colonized by mangroves).
3. **Composite plains** (influenced by both tidal and alluvial conditions, e.g. lagoons formed behind wave-built barriers where mangroves grow).
4. **Fringing barriers with lagoons** (high wave energy conditions with sediments of fine sand and mud) (e.g. mangroves of the Philippines)

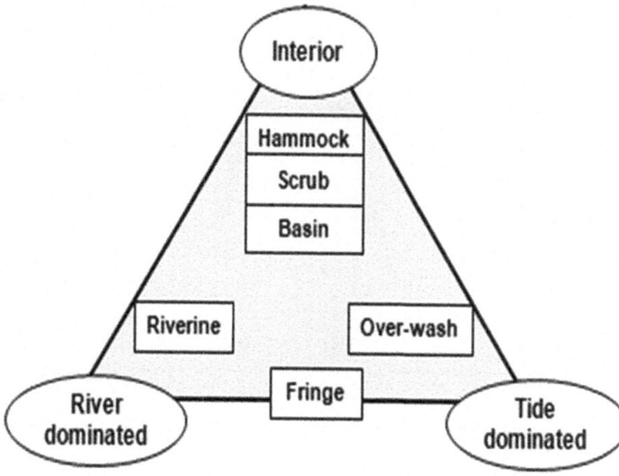

Fig. 1.3 Three Major types of mangrove forests

5. **Drowned bedrock valleys** (e.g. mangroves of Northern Vietnam or Eastern Malaysia)
6. **Coral coasts** (mangroves growing at the bottom of coral sand or in platform reefs) (e.g. mangroves of Indonesia, and Singapore).

Based on global distribution, the mangroves are of two types: 'Old world mangroves' and 'New world mangroves'. The 'Old world mangroves' is the place of origin for mangroves in the Eastern hemisphere, whereas the 'New world mangroves' is the place of relatively recent origin for mangroves in the Western hemisphere. The Eastern hemisphere is Indo-West Pacific region that includes East Africa, Indo-Malaysia, and Australasia. The Western hemisphere is Atlantic East Pacific region that includes West America, East America, and West Africa. Thus, there are six geographic regions of global mangroves in the world. The Eastern hemisphere has 57% of global mangrove area, while Western hemisphere has 43%. The Eastern hemisphere is rich in biodiversity with 63 mangrove species, while the Western hemisphere is poor in biodiversity with only 19 species. Mangroves have broader ranges along the warmer eastern coastlines of the Americas and Africa than along the cooler western coastlines (Fig. 1.4). This difference in distribution is attributed to the presence of warm and cold oceanic currents (Duke 1992).

Global mangrove habitats are inhabited by 80 tree and shrub species including 69 species and 11 hybrids, belonging to 32 genera under 17 families worldwide. Except one genus all the other genera are flowering plants (Tomlinson 2016; Duke et al. 1998; Duke 2017). Mangrove diversity is the highest around South East Asia in the old world mangroves. Some mangrove genera are specific to some regions: *Pelliciera*, *Conocarpus*, and *Laguncularia* are present only in the new world, whereas *Osbornia* and *Camptostemon* exist only in the old world. The mangrove fern *Acrostichum aureum* is the only species, common to both the new and old worlds. In the world, there are two places, richest in mangrove genetic diversity, and these are the Bhitarkanika (Odisha) in India and the Baimuru in Papua New Guinea.

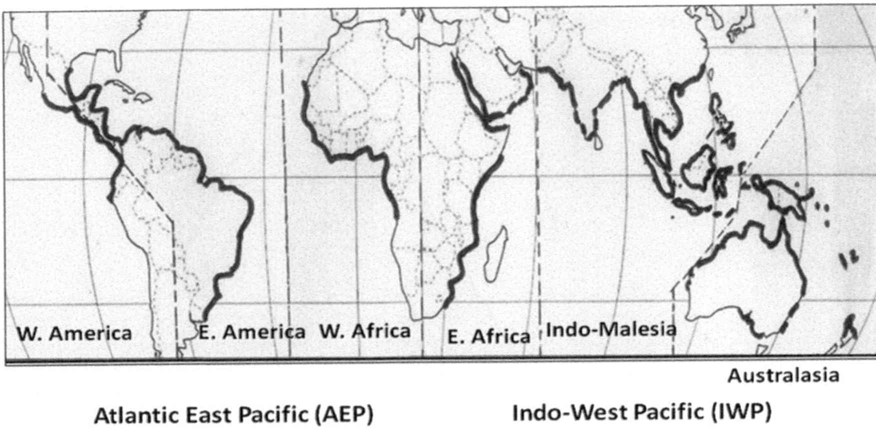

Fig. 1.4 Global distribution of mangroves with six geographic regions

Among the countries, Indonesia has the largest extent of mangrove cover. Sundarbans is the largest single block of mangrove forest in the world, and the second largest ones are in Niger Delta, northern Brazil, and southern Papua.

1.3 Mangrove Ecosystem Services

1.3.1 Protecting from Intense Sunlight and UV-Radiation

Mangroves are able to grow under intense sunlight and solar UV-B radiation. *Avicennia marina* grows under hot and arid conditions with high sunlight, while *Rhizophora* species are quite tolerant to solar UV-B radiation. The mangroves are rich in phenolic compounds that dissipate the excessive sunlight. The flavonoids accumulated in the mangrove leaves serve as UV-screen compounds. Hence, the mangroves are free from the deleterious effects of heat energy and UV-B radiation under-canopy environment (Moorthy and Kathiresan 1997a, b).

1.3.2 Reducing 'Greenhouse Gas' and Carbon Sequestration

The coastal vegetated habitats such as mangroves, salt marshes, and seagrasses are considered to be 'blue carbon ecosystems'. They are amongst the most significant carbon reservoir in the biosphere, and they play a great role in oceanic carbon cycle. The annual carbon burial by coastal vegetated habitats is 180 times greater than that in the deep sea sediments. The vegetated coastal habitats cover only less than 0.2% of the seafloor, but they contribute about 50% of the global burial of organic carbon in marine sediments (Duarte et al. 2005).

Among blue carbon ecosystems, the mangroves are efficient in removing atmospheric CO_2 through photosynthesis, and storing carbon in their biomass and soil substrates. By area, mangrove forests constitute only 0.1% of the world's plant biomes and only 0.7% of the global coastal zone, yet they contribute to the global carbon in a larger way. Mangrove net primary production equals 10% of total net primary production in the coastal zone (Alongi 2007).

Mangroves are the only blue carbon forest of the world. The mangrove forests account for only 1% (13.5 Gt.year^{-1}) of carbon sequestration by the world's forests, but they account for 14% of carbon sequestration by the global ocean (Alongi 2012). They supply more than 10% of the organic carbon, essential to the world oceans (Dittmar et al. 2006). Hence, the mangroves reduce the problems of 'greenhouse gases' and global warming (Kathiresan and Bingham 2001).

Mangrove forests are globally significant carbon sinks, storing carbon in a range of 455–856 mega-gram in one hectare of forest and 6.2–11.7 peta-gram in the world (Alongi 2020; Kauffman et al. 2020; Ouyang and Lee 2020). They are 10 times greater in carbon sequestration and four times efficient in carbon stocks than other tropical forests (McLeod et al. 2011). The mangroves also have larger carbon stock in tropics (895 $\pm$ 90 MgC ha^{-1}) than sub-tropics (547 $\pm$ 66 MgC ha^{-1}) (Sanders et al. 2016). The mangrove forests hold higher carbon stock in Asia-Pacific (1094 MgC ha^{-1}) than other regions: Latin America (939 MgC ha^{-1}), West Central Africa (799 MgC ha^{-1}), and Arabian/Oman Gulf (217 MgC ha^{-1}) (Kauffman and Bhomia 2017).

Mangroves are highly productive, storing large amounts of carbon in their soil system over a very long period of time due to high sedimentation rates and anoxic soils (Donato et al. 2011; Atwood et al. 2017; Alongi 2018) in contrast to other forest soils that store carbon only for a short time. The mangrove soil is reported to account for 76.5% of total carbon in ecosystem (Alongi 2014, 2020) and to store carbon in a range of 2.6–6.4 PgC in the top one metre of soil in the global mangroves (Jardine and Siikamaki 2014; Atwood et al. 2017; Sanderman et al. 2018). The mangrove soil is more efficient in carbon burial by 2.4-folds than salt marshes, by 5.2-folds than seagrasses, and by four-fold than in tropical forests (Duarte et al. 2005). In addition, mangrove root biomass is higher than other forest types. The biomass invested in mangrove roots is 40% of shoot, in contrast to 25% in upland forests. This higher root biomass helps to ensure stability in the soft substrates of mangrove environment.

Mangroves have high carbon sequestration, which is estimated to be 14.2 TgC. year^{-1} and the value per unit area is 1.71 $\pm$ 0.17 MgC ha^{-1} year^{-1} for global mangroves (Alongi 2018). The Indian Sundarbans has the capacity to sequester a total of 2.79 TgC in its natural forest area of 4264 km^2 (Ray and Jana 2017). The carbon sequestration is high at early stage of the forest. A 20-year old plantation of mangroves stores 11.6 kg m^{-2} of carbon with C burial rate of 580 g m^{-2} year^{-1} in Japan (Fujimoto 2000). In Malaysia, a 20 year old stand of *Rhizophora apiculata* mangrove forest is reported to store 7.14 MgC ha^{-1} year^{-1}. The rate of carbon sequestered in mangrove mud is estimated to be 1.5 MgC ha^{-1} year^{-1}. Each hectare of mangrove sediment contains 700 Mg carbon per metre depth (Ong 1993; Ong et al. 1995). However, species-wise variations do occur. *Avicennia marina* performs better to display 75% higher rate of carbon sequestration than that in *Rhizophora mucronata* (Kathiresan et al. 2013a). Total carbon is 98.2% higher in natural mangroves and 41.8% in planted mangroves than that in non-mangrove soil (Kathiresan et al. 2014).

Mangrove loss disturbs the carbon stocks resulting in emission of greenhouse gas (Alongi 2012). A loss of about 35% of the world's mangroves has resulted in a net loss of 3.8×10^{14}g C stored as mangrove biomass (Cebrain 2002). Mangrove deforestation in the world generates emissions of 0.02–0.12 pico grams of carbon per year. This is as much as around 10% of emissions from global deforestation of all forests (Spalding et al. 2010). Carbon storage is reduced in disturbed mangroves than that in intact mangroves: by four folds lower in degraded or deforested forests, three folds lower in the forests, impacted by domestic or aquaculture effluents, and two folds lower in the forests, affected by storms and flood (Perez et al. 2018). Thus failing to preserve mangrove forests can cause considerable carbon emissions and thus global warming. Therefore, mangrove restoration can be a novel counter-measure for global warming issue.

1.3.3 Protecting from Cyclone and Storms

Mangrove forests save the coastal people against cyclones and storms. The 'super-cyclone' that occurred on the 29th October 1999 with a wind speed of 310 km/h along the Odisha coast in India caused heavy damage in the coastal area that was devoid of mangroves. But, there was practically no damage occurred in the coastal areas with dense mangrove forest. This super-cyclone killed over 10,000 people and caused a heavy loss of livestock and property. This loss would have been avoided, had the mangrove forests been intact. The protection value of one hectare of intact mangroves is reported to be 8700 US dollars, as against the value of 5000 US dollars fetched for one hectare of cleared land. Yet another example is the cyclone 'Nargis' that hit the coast of Myanmar on the third May, 2008 killed over 30,000 people and heavy loss of properties only in the areas that were devoid of mangroves. The beneficial effect of mangroves was also recorded during the cyclones especially 'Aila'—2009, 'Ockhi'—2017, 'Gaja'—2018 along the east coast of India. The mangroves act as a defence force against natural calamity and hence protecting mangroves as storm buffers generates more value to society.

1.3.4 Protecting from Giant Waves

Mangrove forests protect the coast against strong wave actions. This was evident during the 26th December 2004 tsunami that occurred in the Indian Ocean area, which killed 3 million people in Asian and African countries and caused a loss of 6 billion US dollars in 13 countries. However, mangroves mitigated the deleterious impact of the tsunami waves and protected the shoreline against damage (Kathiresan

Fig. 1.5 The 2004 tsunami broken a long boat jetty in to pieces, while mangroves intact without damage in Parangipettai, south east India

and Rajendran 2005; Danielsen et al. 2005). However, the 2004 tsunami affected mangroves in some places, especially in Andaman and Nicobar Islands. In Andaman, the tsunami caused considerable change in the mangrove stands of the islands; where *Avicennia marina* and *Sonneratia alba* were not generally affected, while *Rhizophora* species got affected due to continuous submergence by seawater as a result of the tsunami waves. In South Andaman, 30–80% of mangrove stands got affected due to natural elevation of land; however, in middle Andaman and North Andaman, mangroves were not affected (Dam Roy and Krishnan 2005; Ramachandran et al. 2005).

The role of mangroves in reducing the sea-waves is well-known. A hydraulic experiment has proved that mangroves are more effective for reduction of wave damages than concrete seawall structures such as wave dissipating block, breakwater rock, and houses (Harada et al. 2002) (Fig. 1.5). Another study has proved that six-year-old mangrove forest of 1.5 km width reduces the sea-waves by 20 times, from 1 m high waves at the open sea to 0.05 m at the coast (Mazda et al. 1997). The reduction of wave amplitude and energy by tree vegetation has also been proved by measurements of wave forces and modeling of fluid dynamics (Massel et al. 1999). According to an analytical model, 30 trees from 10 m^2 in a 100 m wide belt can reduce the maximum tsunami flow pressure by more than 90%, if the wave height is less than 4–5 m (Hiraishi and Harada 2003). As per our observation, the mangroves can provide protection against tsunami in the situation, where the height of mangrove forest (with >25 trees/10 m^2) is higher than the tsunami wave height (Kathiresan and Rajendran 2005). Therefore, conserving or restoring mangroves will save the coast from future events of natural disasters such as tsunami.

1.3.5 Controlling Flood Damage

Mangroves protect coastal systems against floods that are often caused by tidal waves or heavy rainfall in association with storms. The mangroves are able to control the flood due to the root system, which has a larger spread out area and also ability to promote sedimentation.

Global mangroves provide flood protection benefits exceeding 65 billion US dollars per year. If mangroves were lost, annually 15 million more people would be affected by flood across the world. The greatest economic benefits are recorded with USA, China, India, and Mexico. The economic benefits in terms of people protected are found in Vietnam, India, and Bangladesh. Many 20-km coastal stretches receive more than 250 million US dollars annually in flood protection benefits from mangroves (Pelayo et al. 2020). The mangroves protect more than 150,000 people from flooding every year in Abidjan and Lagos in West Africa, Mumbai and Karachi in South Asia, Wenzhou in East Asia, and Cebu and Denpasar in South east Asia. The mangroves provide annually over 500 million US dollars in avoiding the property loss in cities such as Miami in the USA and Cancun in Mexico. However, the mangroves are beneficial not only to urban areas but also to less populated coastal floodplains (Pelayo et al. 2020).

Besides flood control, the mangroves do prevent the entry of seawater inland and protect the underground water systems, which are a source of drinking water supply to coastal population (Kathiresan 2018). In addition, the mangroves reduce the salinity of the groundwater. This is evident by the fact that there is a very sharp decline in salt concentrations of groundwater at the interface between salt flats and mangroves (Ridd and Sam 1996).

1.3.6 Preventing Coastal Soil Erosion

Mangroves reduce the wave action and prevent the coastal erosion. The reduction of waves increases with density of vegetation. In a tall mangrove forest, the rate of wave reduction is as large as 20% in a distance of 100 metre (Mazda et al. 1997). Mangrove forests are 'live seawall'-like natural formations and are very cost-effective as compared to the concrete seawall and other structures for coastal protection against erosion (Harada et al. 2002). The mangrove forest with 100 m width is proved to protect the sea dyke, lying behind the forest, for more than 50 years, in contrast to rock fencing that protects the sea dyke for only 5 years. This is because of the fact that the rock fencing is not long resistant to wave damage as compared to mangrove forest, as proved in the Red River Delta, Vietnam. The cost of mangrove planting is 1.1 million US dollars, but it has helped to avoid the maintenance cost of 7.3 million US dollars per year for the sea dyke. However, mangrove deforestation causes coastal erosion, as proved in the Gulf of Kachchh and other regions (World Disaster Report 2002).

1.3.7 Trapping Coastal Sediments

Mangroves are capable of trapping sediment, and thus acting as sinks for the suspended sediments (Woodroffe 1992; Wolanski et al. 1992; Wolanski 1995; Furukawa et al. 1997). The mangroves catch sediments by their complex aerial root systems and thus they function as land expanders. Generally annual sedimentation rate ranges between 1 and 8 mm, in the mangrove areas (Bird and Barson 1977). A contrary view is also proposed that the mangrove forests are the result, and not the cause of sedimentation in coastal areas, but they accelerate the sedimentation process. This depends largely on the exchange process taking place between mangroves and the adjoining coastal areas (Woodroffe 1992).

The mechanism of sediment trapping in mangrove habitats is shown in Fig. 1.6 (Furukawa et al. 1997; Kathiresan 2003). The mangroves inhibit tidal flows, due to the friction force that is provided by the trees with complex aerial root structures. The soil particles are carried in suspension by the incoming tide, whereas the soil particles are left behind in the mangrove swamps by the outgoing tides. Thus, the particles settle down in the forests during the low tide, when the turbulence gets reduced and the water velocity becomes low and sluggish to carry the soil particles back to the sea. In contrast, the particles are held in suspension during the high tide, when the turbulence is high.

Density of mangrove species and their complexity of root systems are the most important factors that determine the sedimentation process (Kathiresan 2003). The sedimentary process also varies in different types of mangrove forests: riverine, basin, and fringe types. The process falls in decreasing order: Riverine > basin > fringe (Ewel et al. 1998). The river-dominated system receives

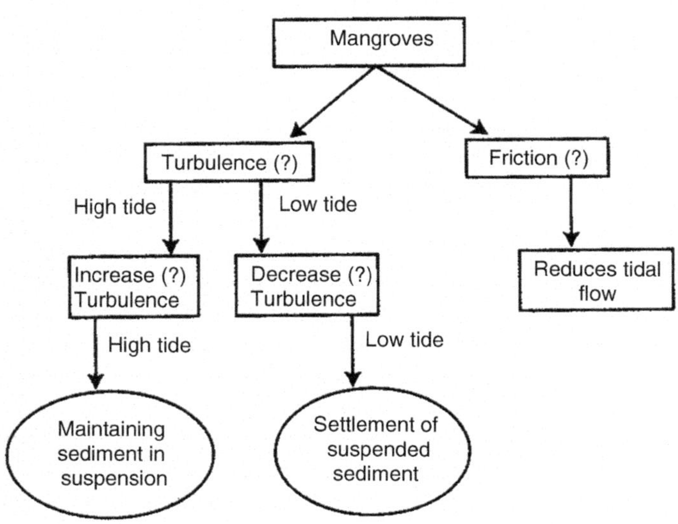

Fig. 1.6 Mechanism of sedimentation as induced by mangroves

sediment supply from the upland areas. The tide-dominated fringe system contains abundant sediment from the sea, but the sedimentation gets disturbed by the tides. The interior basin mangroves are the sinks for sediments (Woodroffe 1992).

1.3.8 Deepening of Creeks

Water circulation through mangrove forests is important for keeping creeks deeper. The water movement in tidal creek is different from surrounding mangrove swamps. This is because of the fact that the tidal creek is deep, while the mangrove swamp is a shallow system, colonized with vegetation. The cause of water movement is the tide. The flow of water during the low tide is much greater than that of the high tide. For example, the riverine mangroves produce asymmetrical tidal currents which are stronger at the low tide than at the high tide (Medeiros and Kjerfve 1993). The fast low tide tends to flush out the material from the mangrove swamp area and maintains the depth of creeks. When the area of forest swamp is reduced, the speed of the low tide is reduced, and the creeks get clogged up. This is commonly observed in some Southeast Asian countries where deforestation of mangroves has reduced the navigability of the canals and river mouths (Wolanski et al. 1992).

1.3.9 Trapping & Recycling of Nutrients

Mangrove soil serves as a 'sink' for retaining nutrients. This depends on the soil characteristics and water flow patterns of the mangrove habitats. Mangrove soil, algae, microbes, and physical processes absorb large amounts of organic and heavy metal pollutants (Wong et al. 1995). Therefore, mangroves especially *Avicennia marina* are capable of surviving in the areas that are dumped with heavy organic wastes and toxic heavy metals. Ammonium nitrogen is rapidly assimilated by bacteria and benthic algae present in the mangrove soil and hence, export of ammonium nitrogen is largely prevented. The loss of nitrogen and phosphorus is also significantly prevented due to reduced flow of water in the severely damaged mangrove areas, as reported in the North Queensland (Kaly et al. 1997).

Mangrove soil is highly efficient in absorbing and holding heavy metals such as Fe, Zn, Cr, Pb, Cd, Mn, and Cu, thereby preventing the metal pollution in coastal environments. In the mangrove ecosystem, heavy metals are mostly trapped down in the soil, but only $<1\%$ of all these metals are present in mangrove vegetation (Silva et al. 1990). This trend of higher levels of heavy metals in soil than vegetation is due to (1) low availability of metals to the plants from sediments, (2) exclusion of the metals by the mangrove plant itself, and (3) physiological adaptations that prevent accumulation of metals inside the plants. Oxygen exuded by the underground roots promotes the formation of iron plaques that adhere to the root surfaces and prevent the trace metals from entering the root cells. The metals are precipitated in the form of stable metal sulphides under anoxic conditions and the sulphides are buried deep into the mangrove soil. This process decreases the bioavailability of trace metals to the plants from the mangrove soil. Mercury does not form sulphides, and hence it is immobilized in organic complexes in mangrove soil. Disturbances may cause the mangrove soil to lose its metal binding capacity, resulting in the mobilization of metals. The degrading mangroves then shift their site from 'sink' to 'source' of heavy metals (Lacerda 1998; Kathiresan and Bingham 2001).

In addition to retaining nutrients, the mangrove systems also help in recycling of nutrients especially carbon, nitrogen, and sulphur and making the nutrients available in assimilable forms to other organisms. It is noteworthy that mangrove ecosystem is the only biotic system that most efficiently recycles sulphur.

1.3.10 Litter Decomposition & Nutrient Enrichment

Mangroves produce large amounts of litter in the form of falling leaves, branches, and other debris. These litters are subjected to leaching of nutrients and microbial colonization, which produce high levels of dissolved organic matter and the recycling of nutrients both in the mangrove and in adjacent habitats. The nutrients can potentially enrich the coastal sea and, ultimately supporting the fishery resources (Lee et al. 1990; Benner et al. 1990; Chale 1993).

Microbes play a vital role in decomposition of mangrove litter and recycling of nutrients. During early decomposition process, potassium and carbohydrates are quickly leached out in a very short time. Tannins, in contrast, leach out very slowly and the high level of tannin reduces the colonization of bacterial populations in the initial period of decomposition. However, tannins are degraded by fungi which are the earlier colonizers on the mangrove litter. Once the tannins are leached out and fungally decomposed, the bacterial populations rapidly increase (Steinke et al. 1990, 1993; Rajendran 1997; Rajendran and Kathiresan 1999). The N_2-fixing azotobacters are one of the important groups in decomposing litter (Chale 1993; Wafar et al. 1997), and their activities increase the content of protein nitrogen by 2–3 times in the litter after 1 month of decomposition (Fig. 1.7; Rajendran 1997). This protein-rich detritus in turn, attracts shrimp, crabs, and fish.

1.3.11 Supporting Fishes and Wildlife

Mangroves are important for fish production by serving as nursery, feeding, and breeding grounds for crabs, prawns, mollusks, and finfish. The calm waters provide habitat for fishes, while the mangrove aerial roots, tree trunks and forest floor support oysters, snails, barnacles, crabs, and other invertebrates. The muddy or sandy sediments of mangroves are inhabited by a variety of epibenthic, infaunal, and meiofaunal invertebrates. Nearly 80% of the fish catches are directly or indirectly dependent on mangrove and other coastal ecosystems worldwide (Kjerfve and Macintosh 1997; Kathiresan 2000; Kathiresan and Rajendran 2002). The mangroves also support a variety of wildlife such as the Bengal tiger, dolphins, monitor water lizard, estuarine crocodile, deer, sea turtles, monkeys, wild pigs, snakes, fishing cats,

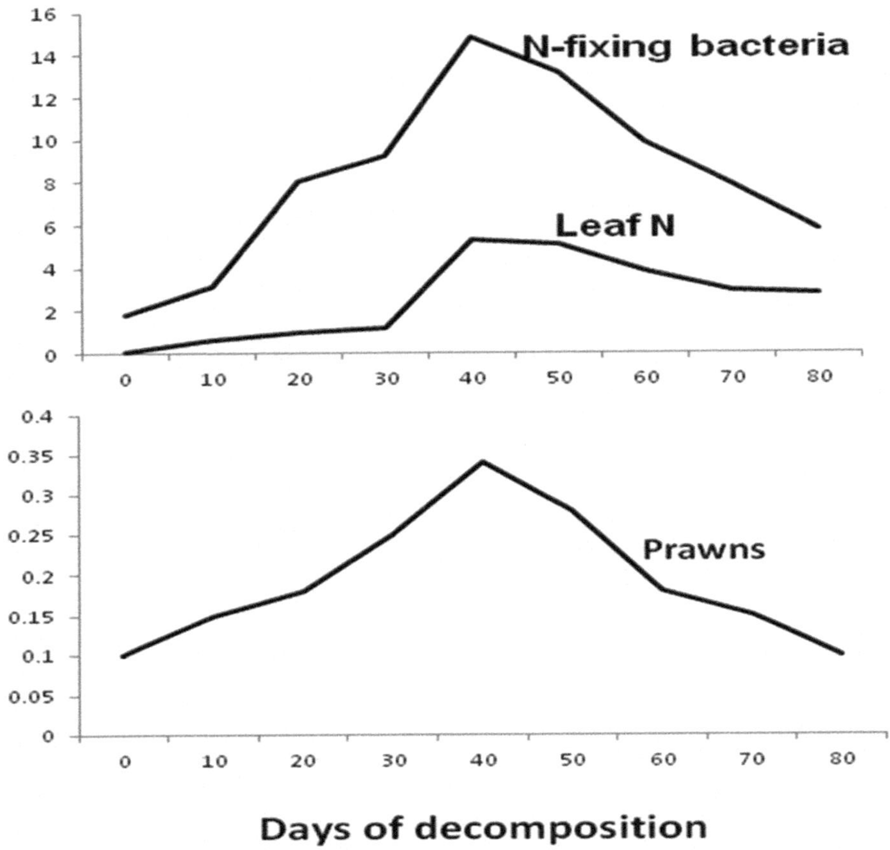

Fig. 1.7 Changes in the nitrogen–fixing azotobacter counts (1×10^4 g^{-1} leaf tissue), the total nitrogen content (% of leaf tissue) and the juvenile prawns (no. $haul^{-1}$) along with decomposing senescent leaves of *Avicennia marina*

insects, and birds. Mangrove leaves, flowers, and fruits are fed in fresh condition by insects, mollusks, crabs, and mammals. Ants protect mangrove trees from insect grazers. A variety of species including bees, bats, and birds such as humming birds, sun-birds, and honey eaters facilitate pollination. Many bird species use mangroves as nesting or roosting grounds, including terrestrial and marine species (Kathiresan and Bingham 2001; Kathiresan and Qasim 2005).

1.3.12 Supporting Coastal Food Web

Mangrove habitats contribute to complex food webs and energy transfers between terrestrial and marine systems (Fig. 1.8). The mangroves produce decomposing

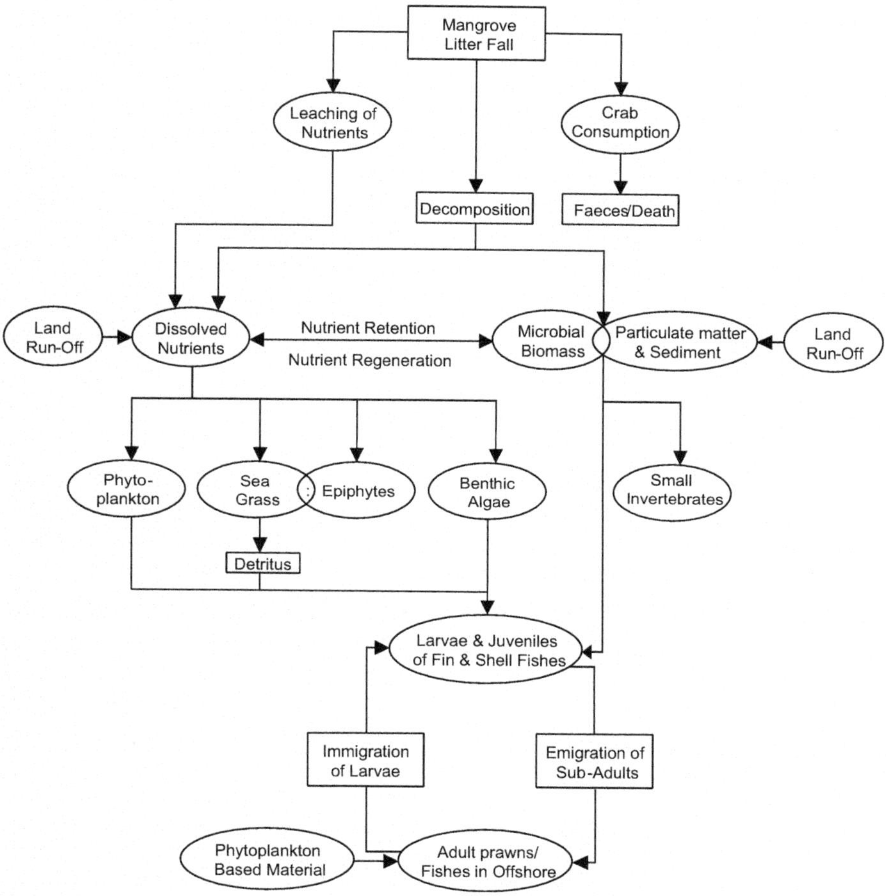

Fig. 1.8 A food web in a mangrove and coastal offshore ecosystems

organic matter, known as 'detritus' which serves as nutritious food to innumerable finfish and shellfish especially juveniles. These detritus–feeding fishes are preyed upon by larger carnivorous organisms. In addition to detritus, the mangrove litter decomposition provides essential nutrients to boost the growth of microorganisms and phytoplankton which are fed by zooplankton and fishes in mangroves and offshore regions (Marshall 1994; Robertson and Alongi 1995; Alongi et al. 1992; Twilley et al. 1992; Hemminga et al. 1995; Alongi 1998).

Mangroves are detritus-based system. The juvenile fishes feed directly on mangrove detritus, small detritivorous invertebrates and on benthic microalgae growing in the mangrove system. The sub-adult fishes migrate from mangroves into sea that is predominant with plankton-based materials, while the juveniles immigrate from sea into the mangrove system that is predominant with detritus-based materials. Thus the life cycle of fishes is completed with the help of mangroves and other coastal systems.

Mangrove plants are the key primary producers, in addition to algae, benthic algae, and phytoplankton in the mangrove forests. These forests also receive considerable quantities of external organic material that are transported from upstream or offshore ecosystems. About 10% of net productivity of mangroves is incorporated within local sediments, while 50–60% is consumed or decomposed, and 30% is exported to offshore regions. The nutrient export also supports productivity in adjacent waters including benthic and pelagic systems of other coastal ecosystems such as seagrasses and coral reefs (Kathiresan and Qasim 2005).

1.3.13 Protecting Other Coastal Systems

Mangroves provide protection to other coastal systems such as coral reefs, seagrasses, and seaweeds. The mangroves are preventing the coastal soil erosion, supplying the clean water after trapping soil particles, and providing the nutrient-rich water through litter decomposition. However, when the mangroves are removed, the loose sediment makes water turbid and not allowing required light for primary production, thereby destroying the associated systems (Fig. 1.9).

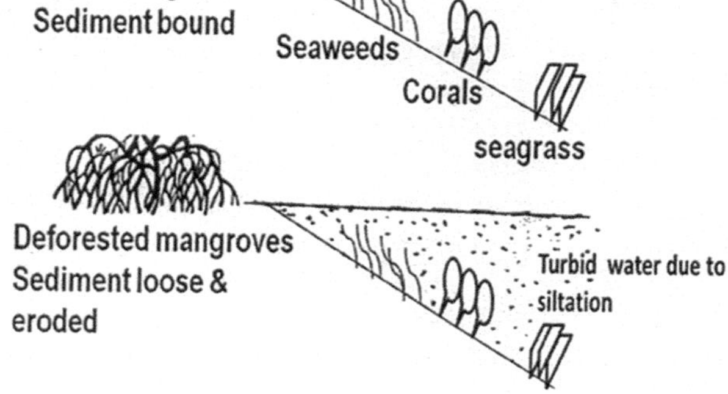

Fig. 1.9 Influence of intact or deforested mangroves on seaweed, coral and seagrass ecosystems

1.4 Uses of Mangroves

Mangroves are of great economic value, providing the ecosystem services worth of at least 1.6 billion US dollars per year, and supporting the coastal livelihoods worldwide (Costanza et al. 1997). Ecosystem economic value is estimated at 91000 US dollars for one hectare of mangrove forest, which is greater than other marine ecosystems: seagrasses, deep sea, coastal plankton, and tidal marsh (Macreadie et al. 2019). The mangroves are known to support over 70 direct human activities, ranging from fuel-wood collection to fisheries (Dixon 1989; Lucy 2006). The economic value is placed in a range of 2000–9990 US dollars per hectare per year, and this is much greater than that of coral reefs, continental shelves or the open sea (Costanza et al. 1997; Spalding et al. 2010). One hectare of mangroves is estimated to store 794 tons of CO_2 equivalent for the carbon credit value of Rs. 168 per ton of CO_2 storage, to protect 243 people from flood, to support the commercial fish stock of 25.3 million individuals, to yield the annual fish catch of 5.7 tons with economic gain of Rs. 4.3 lakh (Kathiresan and Rajendran 2002; Thomas and Mark 2018).

1.4.1 Firewood and Wood Products

Mangrove twigs are used for making charcoal and firewood, due to high calorific value. Charcoal is an energy product largely derived from *Rhizophora* species in Thailand, Malaysia, Vietnam, and Indonesia. One ton of mangrove firewood is equivalent to 5 tons of coal, and it burns producing high heat without generating smoke. In Thailand, about 90% of the felled timber is used for charcoal production. From Thailand and Indonesia, mangrove charcoal is exported to Singapore, Malaysia, Hong Kong, and other Asian countries (Kathiresan 2015).

Mangrove wood is termite-resistant and durable due to its high content of tannin, and hence the mangroves are used as timber. Pneumatophores are used to make bottle stoppers and floats. Mangrove palm (*Nypa*) leaves are used to thatch roofs, mats, and baskets (Fig. 1.10). Its sap of young inflorescence is used for sugar production, alcohol distillation, and vinegar production. Its soft endosperm of fruits is edible and widely used in Thailand, Indonesia, and Philippines.

Mangrove wood chips are a priced commodity for export in Indonesia and East Malaysia. The stalks and fibres are processed into cellulose, paper, and artificial silk (rayon) and supplied from Indonesia to Japan and Taiwan, for cellulose industrial use. Japan has established paper mills and chipboard factories in Kalmantan and Sumatra of Indonesia (Kathiresan 2015).

Tannin is extracted from the bark of mangroves belonging to the family Rhizophoraceae and is used for tanning leather and fishing nets in India, Malaysia, and Pakistan. Extracts of *Bruguiera parviflora, B. gymnorrhiza, Rhizophora apiculata, R. mucronata,* and *Ceriops tagal* are used for tanning fishing nets and fish traps in northern Australia.

Fig. 1.10 *Nypa fruticans*, an economically valuable and freshwater loving mangrove palm

1.4.2 Honey Collection

Mangrove forests are a rich source of honey, facilitating apiculture activities. Indian Sundarbans provide employment to 2000 people engaged in extracting 111 tons of honey annually and this accounts for about 90% of honey production among the mangroves of India (Krishnamurthy 1990). In Bangladesh, 185 tons of honey and 44.4 tons of wax are harvested each year in the western part of the mangrove forest (Siddiqi 1997). A bulk of honey comes from *Ceriops* species and *Excoecaria agallocha*, while the honey of best quality is produced from mangroves such as *Aegialitis rotundifolia* and *Cynometra ramiflora*.

1.4.3 Mangrove Foliage as Fodder

Mangroves especially *Avicennia* are nutritive fodders to buffaloes, sheep, goats, and camels (Fig. 1.11). It is believed that the cattle feeding on mangroves yield highly nutritious milk. Camel grazing is very common in India, Pakistan, Persian Gulf region, and Indonesia (Qasim 1998). Over 16,000 camels are herded into the mangroves of Indus delta of Pakistan (Vannucci 2002). The camel herding is one of the activities practiced by the pastoral communities known as 'Maldharis' in Gujarat, India. The maldharis are shifting along with their camels to far away areas in search of fodder for their cattle. The degradation and non-access of mangroves has critically impacted their livelihoods.

Fig. 1.11 Goat grazing on the leaves of *Avicennia marina*

1.4.4 Fisheries and Livelihood

Mangroves are known for fish production, facilitating the fishing and aquaculture activities. Upto 80% of global fish catch is dependent on mangroves, thereby ensuring the food security of coastal people (Ellison 2008). About 40,000 fishers get an annual yield of about 540 million seeds of *Penaeus monodon* for aquaculture, in the Sundarbans mangroves of West Bengal (Chaudhuri and Choudhury 1994). One hectare of mangroves can yield 767 kg of wild fish and crustaceans which can provide the annual revenue of about 11,300 US dollars, which is on par with the most profitable intensive shrimp farming (Primavera 1991). The mangrove-rich area provides about 70-times more catches of fish and economic gain than the mangrove-poor area does (Kathiresan and Rajendran 2002). Shells of mangrove mollusks are also collected for the manufacture of lime. According to coastal people, 'No mangroves, no prawns; no mangroves, no crabs; no forest on land, either no fish or only fewer fish in the sea'.

1.4.5 Eco-tourism Development

Mangroves are the attractive sites for developing a burgeoning eco-tourism especially in Southeast Asian countries. This is due to rich natural treasures associated with the mangroves. In Malaysia, night tourism is well-developed in mangroves by using fire-flies. In India, mangroves are increasingly attractive due to the presence of (1) world's largest nesting site for the Olive Ridley turtle in Gahirmatha coast of Odisha; (2) seagrass meadows associated with the seacow (Dugong); (3) coral reefs associated with the most beautiful ornamental fishes; (4) intertidal mudflats teeming with millions of the migratory and residential birds; and, (5) dense mangrove forest colonized with Royal Bengal tigers.

1.4.6 Environmental Risk Reduction

The potential of mangroves in removal of atmospheric CO_2 is remarkable in mitigating the impacts of global warming and climate change. For example, Indian mangrove forests can remove 96 million tons of atmospheric CO_2 everyday which is equivalent to the carbon credit value of 386 million US dollars in the international market (Kathiresan 2018). The mangroves save coastal human lives, and prevent heavy economic loss of properties during extreme weather conditions and natural calamities such as flood, storm surges, cyclone, and tsunami (Kathiresan and Rajendran 2005). One hectare of land with mangroves has 2 times higher cyclone-protection cost than the selling price of mangrove 'cleared' land (Das 2004). Mangroves protect groundwater aquifers from seepage of seawater, thereby ensuring the water security of coastal population. The mangroves remove the solid and wastewater pollution, and the heavy metals are buried deep into the mangrove soil thereby avoiding their toxic pollution effects (Kathiresan 2018).

1.4.7 Traditional Medicinal Value

Mangrove extracts have been traditionally used as medicine. *Bruguiera* species are used for reducing blood pressures and *Excoecaria agallocha* for the treatment of leprosy and epilepsy. Roots and stems of *Derris trifoliata* are used as insecticides and for narcotising fishes, whereas *Acanthus ilicifolius* is used in the treatment of rheumatic disorders. Seeds of *Xylocarpus* species have antidiarrhoeal properties, while barks of *Rhizophora* species have astringent, antidiarrhoea, and antiemetic activities. *Avicennia* species have tonic effect, whereas *Ceriops* species produce hemostatic activity. Tender leaves of *Acrostichum* are used as a vegetable. A beverage is prepared from the fruits of *Sonneratia* species. The traditional knowledge deserves scientific validation. Several mangroves are in clinical trials towards drug developments. One such is *Rhizophora racemosa* in the treatment of type 2 diabetes (Tsabang et al. 2016).

1.4.8 Herbal Tea from Mangroves

The mangrove plants are rich in polyphenols that are essential ingredients for tea-making. Catechin is a predominant group of polyphenols involved in enzymatic oxidation by polyphenol oxidase during fermentation. This process results in the synthesis of two major compounds of tea, namely theaflavins (flavouring chemicals) and thearubigins (nerve stimulant chemicals). Our laboratory has developed a protocol for making the tea from the mangrove plant, *Ceriops decandra*, with the beverage qualities similar to the conventional tea (Kathiresan 1995; Kathiresan and Pandian 1991, 1993). Further, polyphenols such as tannins are commercially important plant products. Gallotannins can be used in leather, medical, pharmaceutical, food, and beverage industries (Veera Ravi and Kathiresan 1990).

1.4.9 Antimicrobial and Antioxidant Activities

Mangrove extracts can inhibit the bacterial growth. The lignin extracted from *Ceriops decandra* is reported to significantly protect the mice from lethal infection of *E. coli.* This antibacterial activity is due to the antioxidant property of the lignins (Sakagami et al. 1998). In general, mangroves display a strong antioxidant activity (Table 1.1) due to the presence of high amounts of phenolics. The antioxidants remove radical oxygen that otherwise damages cellular biomolecules during disease incidence.

1.4.10 Anticancer Activity of Mangroves

Mangroves are traditionally used to treat the cancers and tumours (Kathiresan et al. 2006b). Earlier reports have revealed the anticancer activities of *Avicennia africana, A. nitida, Bruguiera exeristata,* and *B. parviflora* (Bandaranayake 1998, 2002). We have proved that black tea from *Ceriops decandra* prevents the incidence of chemically induced oral cancer in the animal model. The carcinogenic effect of DMBA is proved in animals by well-developed squamous cell carcinoma, along with well-defined epithelial and keratin pearls in the connective tissue with cellular pleomorphism. However, the tumour-bearing animals when treated with mangrove tea, exhibit no tumour but only hyperplasia. In addition, the hair loss is shown to be caused in the animals-bearing tumour whereas the hair loss is prevented in the animals treated with mangrove tea (Sithranga Boopathy and Kathiresan 2010; Sithranga Boopathy et al. 2011a, b). The cancer is a serious issue with growing threat of global warming. In this regard, we have experimentally proved that the oral cancer increases with increasing temperature (Kathiresan and Sithrangaboopathy 2008). We have found that the callus extract of *Excoecaria agallocha* and *Acanthus*

Table 1.1 Antioxidant property of mangroves

Mangrove extract	% of DPPH radical scavenging
Acanthus ilicifolius (leaf)	99.25
Excoecaria agallocha (leaf)	98.59
Avicennia marina (leaf)	75.82
Avicennia marina (flower)	57.97
Rhizophora annamalayana (leaf)	16.57
Rhizophora mucronata (leaf)	74.8
Rhizophora mucronata (flower)	45.97

ilicifolius suppresses the lung cancer induced by a chemical carcinogen (benzo (a)pyrene) in albino mice, and the plant completely cures the cancer within 16 weeks after the treatment (Singh and Kathiresan 2013, 2014).

1.4.11 Anti-Viral Activities

Polyphenols have the property of precipitating protein. Hence, the polyphenol-rich mangroves are believed to inactivate viral proteins and pathogenic viral activity. Our laboratory has reported that mangroves inhibit the viruses that cause human and animal diseases such as human immunodeficiency virus (HIV), hepatitis-B-virus, Newcastle disease virus, vaccinia virus, Semiliki forest virus, and encephalomyocarditis virus (Premanathan et al. 1992, 1993, 1994a, b, 1995, 1996). The members of the family Rhizophoraceae are most inhibitory to the viruses tested (Premanathan et al. 1992). Bioactive compounds that are responsible for anti-HIV are lignins and acid sugars (glucose, galactose, uronic acid, arabinose, and galactosamine). The sugar molecules protect the host cells from the virus-induced cytopathogenicity and they block the viral antigens and their expression, thereby completely inhibiting viral binding to the cells (Premanathan et al. 1999).

Mangroves contain limonoids with anti-viral activities. Seeds of *Xylocarpus moluccensis* possess anti-viral activities against pandemic influenza A virus (Li et al. 2015).The bioactive limonoid is krishnolide-A, effective against the human immunodeficiency virus (HIV) (Zhang et al. 2017) and influenza A virus (Ren et al. 2018). Thus the mangroves are promising for the treatment of incurable viral diseases.

1.4.12 Anti-Diabetic Activity

A diabetic is abnormal condition of a high level of post-prandial glucose, insufficient insulin, and low insulin action in humans. We have proved that the mangrove extract of *Ceriops decandra* at a concentration of 120 mg/kg exhibits promising anti-diabetic activity on par with the standard drug, glibenclamide (Nabeel et al. 2010). This deserves further clinical studies and drug development.

Other mangroves reported to have anti-diabetic activity are *Sonneratia alba, Rhizophora mangle, R. apiculata, R. mucronata, Avicennia marina, A. officinalis, Lumnitzera littorea, L. racemosa, Bruguiera cylindrica, Kandelia candel, Xylocarpus moluccensis, Nypa fruticans,* and *Aegiceras corniculatum*. Bioactive compounds responsible for the anti-diabetic activity are complex polysaccharides, epicatechin, flavonoids, terpenoids, saponin, tannins, quercetin, stigmasterol, and corosolic acid (Gajula et al. 2020).

1.4.13 Anticoagulant Activity of Mangroves

Mangrove extract prolongs the time taken for blood clotting due to the presence of anticoagulant polysaccharides. The activity increases with increasing concentrations of the extract from 100 to 1000 g/mL and also with increasing level of sulphate in the extracts. We have reported the highest anticoagulant activity in *Avicennia marina*, followed by *Rhizophora* species, *Excoecaria agallocha*, and *Aegiceras corniculatum* (Kathiresan et al. 2006a).

1.4.14 Neuro-Protective Activity

Mangroves are efficient in inhibiting the acetyl cholinesterase enzyme (AChE), resulting in neuro-protection. This enzyme is responsible for damaging the central nervous system thereby causing disorders such as Alzheimer disease. The mangroves with anti-AChE are *Rhizophora lamarckii* (leaf), *Avicennia officinalis* (leaf), *Ceriops tagal* (leaf, root), *Sonneratia alba* (leaf, bark), *Nypa fruiticans,* and *R. mucronata* (leaf) (Ravikiran et al. 2020).

1.4.15 Bioactive Nanoparticle Synthesis

Our laboratory has reported for the first time the ability of mangroves in synthesizing the nanoparticles. Among the mangrove extracts, *Xylocarpus mekongensis* is efficient, with the highest production of silver nanoparticles and others (gold, calcium, copper, magnesium, zinc, and lead). Gallic acid is mainly responsible for the nanoparticle synthesis, whereas glucose is required for stabilizing the nanoparticles synthesized (Asmathunisha and Kathiresan 2013, 2018; Asmathunisha 2013). It is also proved that the callus tissue of *Sesuvium portulacastrum* produces silver nanoparticles better than intact plant tissue does (Asmathunisha et al. 2010).

The nanoparticles possess different biological activities. The silver nanoparticles, synthesized by leaf extract of *Rhizophora mucronata Avicennia marina,* and *Acanthus ilicifolius* exhibit mosquito larvicidal activity against *Aedes aegypti* and *Culex quinquefasciatus* (Gnanadesigan et al. 2012). The silver nanoparticles, synthesized by coastal stand of *Prosopis chilensis* have strong antibacterial activity in controlling the vibriosis, a common shrimp disease (Kathiresan et al. 2013b) and inhibiting the human pathogenic bacteria (Asmathunisha et al. 2010). The silver nanoparticles are reportedly promising in stabilizing the cotton fabrics and making them odour resistant, preserving the apple fruits, purifying the drinking water from microbial contaminants, detoxifying the carcinogenic ethidium bromide, and in controlling the cancer cells (e.g. Kathiresan 2020).

Antimicrobial activity is well-known for the silver nanoparticles, synthesized by mangroves (*Rhizophora mucronata, Ceriops decandra, C. tagal, Excoecaria agallocha, Heritiera fomes,* and *Sonneratia apetala*). The silver nanoparticles inhibit bacteria through damaging the structure and function of bacterial cell wall (Das and

Thatoi 2020). Antioxidant activity is reported for the silver/zinc nanoparticles, synthesized by mangroves (*Avicennia officinalis, Excoecaria agallocha, Heritiera fomes, Xylocarpus granatum,* and *Sonneratia apetala*). This activity is due to free radical scavenging (Das and Thatoi 2020). Anti-diabetic activity is also reported for the silver/zinc nanoparticles, synthesized by mangroves (*Avicennia officinalis, Heritiera fomes, Xylocarpus granatum,* and *Sonneratia apetala*). This activity is due to inhibition of α-amylase and α-glucosidase (Das and Thatoi 2020). Anti-inflammatory activity is also reported for the silver nanoparticles, synthesized by mangroves (*Avicennia officinalis, Heritiera fomes, Xylocarpus granatum,* and *Sonneratia apetala*). This activity is due to inhibition of protein denaturation (Das and Thatoi 2020).

1.4.16 Mosquito Repellents and Larvicides

Mangrove extracts are effective in killing larvae of mosquito vectors for dengue fever, filariasis, and malarial diseases, transmitted by *Aedes aegypti, Culex tritaeniorhynchus,* and *Anopheles stephensi,* respectively (Thangam and Kathiresan 1988, 1989, 1991, 1992a, b, 1993a, 1994). We have shown that the extract from stilt root of *Rhizophora apiculata* is much effective for mosquito larvicidal activity at low concentrations (17–26 ppm). This activity is due to pyrethrum derivative as a bioactive compound (Thangam and Kathiresan 1997). The mangrove extract exhibits repellent activity against *Aedes aegypti* when it is applied as a paste on human skin, and the mangrove extract also shows smoke repellency and produces lethal effect on *Culex quinquefasciatus* and *Aedes aegypti* (Thangam and Kathiresan 1992a, b, 1993b).

1.4.17 Lead Molecules for Drug Development

Our computer-based drug discovery has identified lead molecules from mangroves to develop drugs. The compounds are scalaradial for cervical cancer (Senthilraja and Kathiresan 2011), triterpenoid for anti-breast cancer (Senthilraja et al. 2011), dinitrophenylhydrazone for anti-oral cancer, heptadecanoic acid for anti-skin cancer (Saravanakumar et al. 2012), stigmasterol as antimalarial (Senthilraja et al. 2012), and against white spot syndrome virus (Sahu et al. 2012). These compounds are being tested towards the development of drugs.

1.4.18 Valuable Genes from Mangroves

Mangroves are rich in genes that confer resistance to environmental stressors. M.S. Swaminathan Research Foundation has isolated a salt-tolerant gene (AM 244) from the mangrove species *Avicennia marina,* and introduced it into a paddy crop (Pusa Basmati and IR64) via Agrobacterium. The salt-tolerant paddy variety may be cultivated along the coast using saltwater in the future.

1.5 Concluding Remarks

Mangrove habitats are of different types that vary in productivity and ecosystem services. Based on coastal location, they are of three types: deltaic, estuarine, lagoon, and fringe. Functionally, they are of six types: fringe, riverine, basin, over-wash, scrub, and hammock. According to tidal range and sedimentation, the mangroves are of three types: river-dominated, tide-dominated, and interior. In addition, there are of six broad types of mangroves: large delta, tidal plains, composite plains, fringing barriers with lagoons, drowned bedrock valleys, and coral coasts.

Mangroves are promising for bioprospecting due to the presence of structurally new chemicals to overcome harsh coastal conditions, and valuable genes for tolerating wind, salinity, and flood (Kathiresan and Ravikumar 2010; Kathiresan 2010, 2015). Bioprospecting the mangroves is likely to develop patents, which can provide revenue and employment opportunities. The economic benefits are many folds greater than the cost of plantation or conservation. There is a greater potential for raising mangroves as 'cash crops' for their possible applications in food, medical, agricultural and industrial sectors in the future.

References

Alongi DM (1998) Coastal Ecosystems Processes. CRC Press, Boca Raton, FL, USA.

Alongi DM (2007) Mangrove forests of Papua, In: Marshall AJ, Beehler B.M, (eds.) Ecology of Papua, Part 1, Periplus Editions (HK) Ltd., pp. 824–857.

Alongi DM (2012) Carbon sequestering in mangrove forests. Carbon Manage 3: 313–322

Alongi DM (2014) Carbon cycling and storage in mangrove forests. Ann Rev Mar Sci 6: 195–219.

Alongi DM (2018) (ed.) Mangrove Forests: In blue Carbon, Springer, Switzerland, pp. 23–36.

Alongi DM (2020) Global Significance of Mangrove Blue Carbon in Climate Change Mitigation. Sci 2(3): 57. https://doi.org/10.3390/sci2030057.

Alongi DM, Boto KG, Robertson AI (1992) In: Robertson AI, Alongi DM, (eds.) Tropical Mangrove Ecosystems, American Geophysical Union, Washington, D.C, pp. 251–292.

Asmathunisha N (2013) Mangrove-Based Nanoparticles: Synthesis, Characterization andApplications (Ph.D. thesis). CAS in Marine Biology, Annamalai University, India, 261 pp.

Asmathunisha N, Kathiresan K (2013) A review on biosynthesis of nanoparticles by marine organisms. Colloids Surf B: Biointerfaces 103: 283–287.

Asmathunisha N, Kathiresan K (2018) Seasonal variation changes in biosynthesis of nanoparticles by mangroves. Inter J Pharmacol Biosci 9B: 302–310.

Asmathunisha N, Kathiresan K, Anburaj R, Nabeel MA (2010) Synthesis of antimicrobial silver nanoparticles by callus and leaf extracts from saltmarsh plant, Sesuvium portulacastrum L. Colloids Surf B: Biointerfaces 79: 488–493.

Atwood TB, Connolly RM, Almahasheer H, Carnell PE, Duarte CM, Lewis CJE, Irigoien X, Kelleway JJ, Lavery PS, Macreadie PI, Serrano O (2017) Global patterns in mangrove soil carbon stocks and losses. Nat Clim Change 7: 523.

Bandaranayake WM (1998) Traditional and medicinal uses of mangroves. Mangroves Salt Marshes 2: 133–148.

Bandaranayake WM (2002) Bioactivities, bioactive compounds and chemical constituents of mangrove plants. Wetl Ecol Manage 10: 421–452.
Benner R, Hatcher PG, Hedges JI (1990) Early diagenesis of mangrove leaves in a tropical estuary; bulk chemical characterization using solid-state (super 13) C NMR and elemental analyses. Geochim Cosmochim Acta 54 (7): 2003–2013.
Bird ECF, Barson MM (1977) Measurement of physiographic changes on mangrove-fringed estuaries coastlines. Mar Res Indonesia 18: 73–80.
Cebrain J (2002) Variability and control of carbon consumption, export, and accumulation in marine communities. Limnol Ocean 47: 11–22.
Chale FMM (1993) Degradation of mangrove leaf litter under aerobic conditions. Hydrobiologia 257 (3): 177–183.
Chaudhuri AB, Choudhury A (1994) Mangroves of the Sundarbans Vol. I: India. IUCN, Bangkok, Thailand, 247 pp.
Costanza R, D'Arge R, De Groot R, Farber S, Grasso M, Hannon B, Linnberg K, Naeema S, O'Neill RV, Parvelo J, Raskin RG, Sutton P, Van den Belt M (1997) The value of the world's ecosystem services and natural capital. Nature 387: 253–260.
Dam Roy S, Krishnan P (2005) Mangrove stands of Andaman vis-à-vis tsunami. Curr Sci 89: 1800–1804.
Danielsen F, Sorensen MK, Olwig MF, Selvam V, Parish F, Burgess ND, Hiralshi T, Karunagaran VM, Rasmussen MS, Hansen LB, Quarto A, Suryadiputra N (2005) The Asian Tsunami: a protective role for coastal vegetation. Science 310: 643.
Das, Saudamini (2004) Mangroves: A natural defense against cyclones: An investigation from Orissa, India. SANDEE Policy Brief, No. 24–07.
Das SK, Thatoi H (2020) Mangrove plant mediated green synthesis of nanoparticles and their pharmaceutical applications: an overview In: Patra JK, Mishra RR, Thatoi H (eds.) Biotechnological Utilization of Mangrove Resources Academic Press—an imprint of Elsevier. Pp. 355–369.
Dittmar T, Hertkorn N, Kattner G, Lara RJ (2006) Mangroves, a major source of dissolved organic carbon to the oceans. Global Biogeochem Cycles 20: 1 GB101210, doi: https://doi.org/10.1029/2005GB002570.
Dixon JA (1989) The value of mangrove ecosystems. Tropical Coastal Area Management Newsletter 4: 5–8.
Donato DC, Kauffman JB, Murdiyarso D, Kurnianto S, Stidham M, Kanninen M (2011) Mangroves among the most carbon-rich forests in the tropics. Nat Geosci 4: 293–297.
Duarte CM, Borum J, Short FT, Walker DI (2005) Seagrass Ecosystems: Their Global Status and Prospects. In: Polunin NVC (ed) Aquatic Ecosystems: Trends and Global Prospects, Cambridge Univ. Press.
Duke NC (1992) Mangrove floristics and biogeography. In: Robertson AI, Alongi DM (eds.) Coastal and Estuarine Studies: Tropical Mangrove Ecosystems. American Geophysical Union, Washington DC, USA: 63–100.
Duke NC (2017) Mangrove Floristics and Biogeography Revisited: Further Deductions from Biodiversity Hot Spots, Ancestral Discontinuities, and Common Evolutionary Processes Chapter 2 V.H. Rivera-Monroy et al. (eds.), Mangrove Ecosystems: A Global Biogeographic Perspective, Springer International Publishing AG pp. 17–53. https://doi.org/10.1007/978-3-319-62206-4_2.
Duke NC, Ball MC, Ellison JC (1998) Factors influencing biodiversity and distributional gradients in mangroves. Glob Ecol Biogeogr Lett 7: 27–47.
Duke NC, Meynecke J-O, Dittmann S, Ellison AM, Anger K, Berger U, Cannicci S, Diele K, Ewel KC, Field CD, Koedam N, Lee SY, Marchand C, Nordhaus I, Dahdouh-Guebas F (2007) A world without mangroves? Science 317: 41–42.
Ellison AM (2008) Managing mangroves with benthic biodiversity in mind: moving beyond roving banditry. J Sea Res 59: 2–15.

Ewel, K, Twilley R, Ong J (1998) Different kinds of mangrove forests provide different goods and services. Glob Ecol Biogeogr Lett 7: 83–94.
Fujimoto K (2000) Below-ground carbon sequestration of mangrove forests in the Asia-Pacific region. Proceedings of Asia-Pacific Cooperation on Research for Conservation of Mangroves, Okinawa, Japan, pp. 87–96.
Furukawa K, Wolanski E, Mueller H (1997) Currents and sediment transport in mangrove forests. Estuar Coast Shelf Sci 44 (3): 301–310.
Gajula H, Kumar V, Vijendra PD, Rajashekar J, Sannabommaji T, Basappa G (2020) Secondary metabolites from mangrove plants and their biological activities. In: Biotechnological Utilization of Mangrove Resources (Eds. J.K. Patra, R.R. Mishra & H. Thatoi). Academic Press—an imprint of Elsevier. pp. 117–134.
Giri C, Ochieng E, Tieszen LL, Zhu Z, Singh A, Loveland T, Masekand J, Duke N (2011) Status and distribution of mangrove forests of the world using earth observation satellite data. Global Ecol Biogeogr 20: 154–159.
Gnanadesigan M, Anand M, Ravikumar S, Maruthupandy M, Syed Ali M, Vijayakumar V, et al., (2012) Antibacterial potential of biosynthesized silver nanoparticles using *Avicennia marina* mangrove plant. Appl Nanosci 2: 143–147.
Harada K, Imamura F, Hiraishi T (2002) Experimental study on the effect in reducing Tsunami by the coastal permeable structures. Final Proc. Int. Offshore Polar Eng. Conf., USA, pp. 652–658.
Hiraishi T, Harada K (2003) Green belt Tsunami Prevention in South-Pacific Region. Greenbelt tsunami prevention in South Pacific region. Report of the port and airport research institute 42: 1–23.
Hemminga MA, Gwada P, Slim FJ, De Koeyer P, Kazungu J (1995) Leaf production and nutrient contents of the seagrass *Thalassodendron ciliatum* in the proximity of a mangrove forest (Gazi Bay, Kenya). Aquat Bot 50 (2): 159–170.
Jardine SL, Siikamaki JV (2014) A global predictive model of carbon in mangrove soils. Environ Res Lett 9: 104013.
Kaly UL, Eugelink G, Robertson AI (1997) Soil conditions in damaged North Queensland mangroves. Estuaries 20(2): 291–300.
Kathiresan K (1995) Studies on tea from mangrove leaves. Environ Ecol 13(2): 321–323
Kathiresan K (2000) A review of studies on Pichavaram mangrove, southeast India. Hydrobiologia 430: 185–205.
Kathiresan K (2003) How do mangrove forests induce sedimentation?. Revista de Biol Trop 51(2): 355–360.
Kathiresan K (2010) Importance of mangrove forests of India. J Coast Environ 1(1): 11–26.
Kathiresan K (2015) Bioprospecting potential of mangrove resources. In: Kathiresan K, Ajmal Khan S (eds.) International Training Course on 'Mangrove Biodiversity and Ecosystems'Course Manual. Annamalai University (Centre of Advanced Study in Marine Biology), Parangipettai, India, 753 pp.
Kathiresan K (2018) Mangrove Forests of India. Curr Sci 114: 976–981.
Kathiresan K (2020) Bioprospecting potential of mangrove resources. In: Patra JK, Mishra RR, Thatoi H (eds.) Biotechnological Utilization of Mangrove Resources. Academic Press—an imprint of Elsevier. Pp.225–242.
Kathiresan K, Ravindran VS, Muruganandham A (2006a) Mangrove extracts prevent the blood coagulate. Indian J Biotechnol 5: 252–254.
Kathiresan K, Sithranga BN, Kavitha S (2006b) Coastal vegetation an under explored source of anticancer drugs. Indian J Nat Prod Rad 5(2): 115–119.
Kathiresan K, Bingham BL (2001) Biology of mangroves and mangrove ecosystems. Adv Mar Biol 40: 81–251.
Kathiresan K, Pandian M (1991) Effect of UV on quality of black tea from *Ceriops decandra*. Sci Cult 57 (3&4): 93–95.
Kathiresan K, Pandian M (1993) Effect of UV on black tea constituents of mangrove leaves. Sci Cult 59 (7–10): 61–63.

Kathiresan K, Qasim SZ (2005) Biodiversity of Mangrove Ecosystems. Hindustan Publishing Corporation, New Delhi, 251 pp.
Kathiresan K, Ravikumar S (2010) Marine pharmacology: an overview. Mar Pharmacol 1: 1–37.
Kathiresan K, Rajendran N (2002) Fishery resources and economic gain in three mangrove areas on the southeast coast of India. Fish Manage Ecol 49(5): 277–283.
Kathiresan K, Rajendran N (2005) Coastal mangrove forests mitigated tsunami. Estuar coast Shelf Sci 65: 601–606.
Kathiresan K, Sithrangaboopathy N (2008) Temperature effect on chemical induced carcinogenesis in hamster cheek pouch. Environ Toxic Pharmacol 26(2): 147–149.
Kathiresan K, Anburaj R, Gomathi V, Saravanakumar K (2013a) Carbon sequestration potential of *Rhizophora mucronata* and *Avicennia marina* as influenced by age, season, growth and sediment characteristics in southeast coast of India. J Coastal Conserv 17: 397–408.
Kathiresan K, Gomathi V, Anburaj R, Saravanakumar K (2014) Impact of mangrove vegetation on seasonal carbon burial and other sediment characteristics in the Vellar-Coleroon estuary, India. J For Res 25: 787–794.
Kathiresan K, Nabeel MA, Gayathridevi M, Asmathunisha N, Gopalakrishnan A (2013b) Synthesis of silver nanoparticles by coastal plant *Prosopis chilensis* (L.) and their efficacy in controlling vibriosis in shrimp *Penaeus monodon*. Appl Nanosci 3: 65–73.
Kauffman JB, Bhomia RK (2017) Ecosystem carbon stocks of mangroves across broad environmental gradients in West Central Africa: Global and regional comparisons. PLoS ONE 12, 30187749.
Kauffman JB, Adame MF, Arifanti VB, Schile-Beers LM, Bernardino AF, Bhomia RK, Donato DC, Feller IC, Ferreira TO, Jesus Garcia MC, MacKenzie RA, Megonigal JP, Murdiyarso D, Simpson L, Hernandez Trejo H (2020). Total ecosystem carbon stocks of mangroves across broad global environmental and physical gradients. Ecol Monogr e01405. https://doi.org/10.1002/ecm.1405.
Kjefve B (1990) Manual for investigation of hydrological processes in mangrove ecosystems. UNESCO/UNDP Regional Project, 79 pp.
Kjerfve B, Macintosh DJ (1997) In: Kjerfve B, Lacerda LD, Diop, EHM (eds.) Mangrove Ecosystem Studies in Latin America and Africa. UNESCO, Paris, pp. 1–7.
Krishnamurthy K (1990) The apiary of mangroves. In: Whigham DF, Dykyjoya D, Hejny S, (eds.) Wetland Ecology and management: Case studies pp. 135–140. Kluwer Academic press, Netherlands.
Lacerda LD (1998) Trace metals biogeochemistry and diffuse pollution in mangrove ecosystems. ISME Mangrove Ecosystems Occasional Papers 2: 1–61.
Lee HL, Seleena P, Winn Z (1990) *Bacillus thuringiensis* serotype H-14 isolated from mangrove swamp soil in Malaysia. Mosquito Borne Diseases Bulletin 7(4): 134–135.
Li W, Jiang Z, Shen L, Pedpradab P, Bruhn T, Wu J, et al. (2015) Antiviral limonoids including khayanolides from the Trang mangrove plant *Xylocarpus moluccensis*. J Nat Prod 78 (7): 1570–1578.
Lucy E (2006) Counting mangrove ecosystems as economic components of Asia's coastal infrastructure. Proceedings of International Conference and Exhibition on Mangroves of Indian and Western Pacific Oceans (ICEMAN 2006), Aug. 21–24, 2006 Kuala Lumpur. Pp. 1–14
Lugo AE, Snedaker SC (1974) The ecology of mangroves. Ann Rev Ecol Syst 5 : 39–64.
Macreadie PI, Anton A, Raven JA, Beaumont N, Connolly RM, Friess DA, Kelleway JJ, Kennedy H, Kuwae T, Lavery PS, Lovelock CE, Smale DA, Apostolaki ET, Atwood TB, Baldock J, Bianchi TS, Chmura GL, Eyre BD, Fourqurean JW, Hall-Spencer JM, Huxham M, Hendriks IE, Krause-Jensen D, Laffoley D, Luisetti T, Marbà N, Masque P, McGlathery KJ, Megonigal JP, Murdiyarso D, Russell BD, Santos R, Serrano O, Silliman BR, Watanabe K, Duarte CM, 2019. The future of Blue Carbon science. Nat Commun 10: 3998. doi: https://doi.org/10.1038/s41467-019-11693-w.
Marshall N (1994) Mangrove conservation in relation to overall environmental considerations. Hydrobiologia 285 (1–3): 303–309.

Massel SR, Furukawa K, Brinkman RM (1999) Surface wave propagation in mangrove forests. Fluid Dyn Res 24 (4): 219–249.
Mazda Y, Magi M, Kogo M, Hong PN (1997) Mangrove on coastal protection from waves in the Tong King Delta, Vietnam. Mangr Salt Marsh 1: 127–135.
Mcleod E, Chmura GL, Bouillon S, Salm R, Bjork M, Duarte M, Lovelock CE, Schlesinger HW, Silliman BR (2011) A blueprint for blue carbon: toward an improved understanding of the role of vegetated coastal habitats in sequestering CO_2. Ecol Environ 9 (10): 552–560.
Medeiros CQ, Kjerfve B (1993) Hydrology of a tropical estuarine system: Itamaraca, Brazil. Estuar Coast Shelf Sci 36: 495–515.
Moorthy P, Kathiresan K (1997a) Photosynthetic pigments in tropical mangroves: Impacts of seasonal flux on UV-B radiation and other environmental attributes. Bot Mar 40: 341–349.
Moorthy P, Kathiresan K (1997b) Influence of UV-B radiation on photosynthetic and biochemical characteristics of a mangrove *Rhizophora apiculata* Blume. Photosynthetica 34(3): 465–471.
Nabeel MA, Kathiresan K, Manivannan S (2010) Antidiabetic activity of the mangrove species *Ceriops decandra* in alloxan-induced diabetic rate. J Diabetes 2: 97–103.
Ong JE (1993) Mangroves—a carbon source and sink. Chemosphere 27: 1097–1107.
Ong JE, Gong WK, Clough BF (1995) Structure and productivity of a 20-year-old stand of *Rhizophora apiculata* Bl. mangrove forest. J Biogeogr 22: 417–424.
Ouyang X, Lee SY (2020) Improved estimates on global carbon stock and carbon pools in tidal wetlands. Nat Commun 11: 317.
Pelayo M, Losada IJ, Torres-Ortega S, Narayan S, Beck MW (2020) The global flood protection benefits of mangroves. Sci Rep 10: 1–1.
Perez A, Libardoni BG, Sanders CJ (2018) Factors influencing organic carbon accumulation in mangrove ecosystems. Biol Lett 14: 20180237.
Premanathan M, Chandra K, Bajpai SK, Kathiresan K (1992) A survey of some Indian marine plants of antiviral activity. Bot Mar 35: 321–324.
Premanathan M, Kathiresan K, Chandra K, Bajpai SK (1993) Antiviral activity of marine plants against New castle disease virus. Trop Biomed 10: 31–33.
Premanathan M, Kathiresan K, Chandra K (1994a) Antiviral activity of marine and coastal plants from India. Int J Pharmacogn 32: 330–336.
Premanathan M, Kathiresan K, Chandra K (1994b) In vitro anti-vaccinia virus activity of some marine plants. Indian J Med Res 99: 236–238.
Premanathan M, Kathiresan K, Chandra K (1995) Antiviral evaluation of some marine plants against Semliki forest virus. Int J Pharmacogn 33 (1): 75–77.
Premanathan M, Nakashima H, Kathiresan K, Rajendran N, Yamamoto N (1996) *In vitro* anti human immunodeficiency virus activity of mangrove plants. Indian J Med Res 130: 276–279.
Premanathan M, Arakaki R, Izumi H, Kathiresan K, Nakano M, Yamamoto N (1999) Antiviral properties of a mangrove plant, *Rhizophora apiculata* Blume, against human immunodeficiency virus. Antivir Res 44 (2): 113–122.
Primavera JH (1991) Marine shrimp culture: Principles and Practices 23: 701–728.
Qasim SZ (1998) Mangroves, In: *Glimpses of the Indian Ocean*, (University Press, Hyderabad), pp 123–129.
Rajendran N (1997) Studies on mangrove—associated prawn seed resources of the Pichavaram, Southeast coast of India. Ph.D thesis. Annamalai University, India.
Rajendran N, Kathiresan K (1999) Biochemical changes in decomposing leaves of mangroves. Chem Ecol 17: 91–102.
Ramachandran S, Anitha S, Balamurugan V, Dharanirajan K, Vendhan KE, Divien MIP, Senthil Vel A, Sjjahad Hussain I, Udayaraj A (2005) Ecological impact of tsunami on Nicobar Islands (Camorta, Katchal, Nancowry and Trinkat). Curr Sci 89: 195–200.
Ray R, Jana TK (2017) Carbon sequestration by mangrove forest: One approach for managing carbon dioxide emission from coal-based power plant. Atmospheric environment 171: 149–154.
Ravikiran T, Anand S, Ramachandregowda S, Kariyappa AS, Dundaiah B, Gopinath MM (2020) Neuroprotective effects of mangrove plants. In: Patra JK, Mishra RR, Thatoi H (eds.)

Biotechnological Utilization of Mangrove. Resources Academic Press—an imprint of Elsevier. pp. 261–273.
Ren JL, Zou XP, Li WS, Shen L, Wu J (2018). Limonoids containing a C1-O-C29 moiety: isolation, structural modification and antiviral activity. Mar Drugs 16(11): 434.
Ridd PV, Sam R (1996) Profiling groundwater salt concentrations in mangrove swamps and tropical salt flats. Estuar Coast Shelf Sci 43(5): 627–635.
Robertson AI, Alongi DM (1995) Role of riverine mangrove forests in organic carbon export to the tropical coastal ocean; a preliminary mass balance for the Fly Delta (Papua New Guinea). Geo-Marine Letters 15(3–4): 134–139.
Sakagami H, Kasdhimata M, Toguchi M, Satoh K, Odanaka Y, Ida Y, et al. (1998) Radical modulation activity of lignins from a mangrove plant *Ceriops decandra* (Griff.) Ding. Hou. In Vivo 12: 327–332.
Sahu SK, Kathiresan K, Singh R, Senthilraja P (2012) Molecular docking analyses of *Avicennia marina* derived phytochemicals against white spot syndrome virus (WSSV) envelope protein-VP28. Bioinformation 8 (18): 897–900.
Sanders CJ, Mather DT, Tait DR, Williams D, Holloway C, Sippo JZ, Santos IR (2016) Are global mangrove carbon stocks driven by rainfall? J Geophys Res Biogeosci 121: 2600–2609.
Sanderman J, Hengl T, Fiske G, Solvik K, Adame MF, Benson L, et al. (2018) A global map of mangrove forest soil carbon at 30 m spatial resolution. Environ Res Lett 13: 055002.
Saravanakumar K, Sunil KS, Kathiresan K (2012) In-silico studies on fungal metabolites against breast cancer protein (BRCA1). Asian J Trop Biomed B488 (3): 1–3.
Senthilraja P, Kathiresan K (2011) Computational selection of compounds derived from mangrove ecosystem for anti-cervical cancer activity. J Recent Sci Res 2 (4): 93–98.
Senthilraja P, Kathiresan K, Saravanakumar K (2011) Comparative analysis of bioethanol production by different strains of immobilized marine yeast. J Yeast Fungal Res 2 (8): 113–116.
Senthilraja P, Sahu SK, Kathiresan K (2012) Potential of mangrove derived compounds against dihydrofolate reductase: an in-silico docking study. J Comput Biol Bioinforma Res 4: 23–27.
Singh CR, Kathiresan K (2013) Anticancer efficiency of root tissue and root callus of *Acanthus ilicifolius* L., on Benzo pyrene induced pulmonary carcinoma in Musmusculaus. World J Pharm Pharm Sci 2: 5271–5283.
Singh CR, Kathiresan K (2014) In vitro cytotoxicity effect of mangroves against non-small cell lung carcinoma A549 and NCI-H522. World J Pharm Pharm Sci 3: 1067–1078.
Boopathy NS, Kathiresan K (2010) Effect of mangrove species *Ceriops decandra* on hair loss in the hamster rats induced with oral cancer. Zool Surv India pp. 447–454.
Boopathy NS, Kathiresan K, Manivannan S, Jeon YJ (2011a) Effect of mangrove tea extract from *Ceriops decandra* (Griff.) Ding Hou. on salivary bacterial flora of DMBA induced Hamster buccal pouch carcinoma. Indian J Microbiol 51(3): 338–344.
Boopathy NS, Kathiresan K, Jeon YJ (2011b) Effect of mangrove tea extract from *Ceriops decandra* (Griff.) on haematology and biochemical changes in DMBA induced Hamster buccal pouch carcinogenesis. J Environ Toxicol Pharmacol 32: 193–200.
Siddiqi NA (1997) Management of Resources in the Sunderbans Mangroves of Bangladesh. *International News letter of coastal Management—Intercoast Network.* Special edition 1: 22–23.
Silva CAR, Lacerda LD, Rezende CE (1990) Heavy metal reservoirs in a red mangrove forest. Biotropica 22: 339–345.
Spalding M, Kainuma M, Collins L (2010) World Atlas of Mangroves. Earthscan, London, Washington DC. Pp. 319.
Steinke TD, Barnabas AD, Somaru R (1990) Structural changes and associated microbial activity accompanying decomposition of mangrove leaves in Mgeni estuary (South Africa). South Afr J Bot 56(1): 39–48.
Steinke TD, Holland AJ, Singh Y (1993) Leaching losses during decomposition of mangrove leaf litter. South Afr J Bot 59: 21–25.

Tomlinson PB (2016) The botany of mangroves. Second Edition. Cambridge University Press, Cambridge, p 418.
Thangam TS, Kathiresan K (1988) Toxic effect of mangrove plant extracts on mosquito larvae *Anopheles stephensi* L. Curr Sci 47 (16): 914–915.
Thangam TS, Kathiresan K (1989) Larvicidal effect of marine plant extracts on mosquito *Culex tritaeniorhynchus*. J Mar Biol Assoc India 31 (1–2): 306–307.
Thangam T, Kathiresan K (1991) Mosquito larvicidal activity of marine plants with synthetic insecticides. Bot Mar 34: 537–539.
Thangam TS, Kathiresan K (1992a) Smoke repellency and killing effect of marine plants against *Culex quinquefasciatus*. Trop Biomed 9: 35–38.
Thangam TS, Kathiresan K (1992b) Mosquito larvicidal activity of mangrove plant extract against *Aedesa egypti*. Int Pest Control 34 (4): 116–119.
Thangam TS, Kathiresan K (1993a) Mosquito larvicidal activity of mangrove plants extract against *Aedes aegypti* and *Culex quinquefasciatus*. Int Pest Control 35: 94–95.
Thangam TS, Kathiresan K (1993b) Repellency of marine plant extracts against *Aedes aegypti*. Int J Pharmacogn 31: 321–323.
Thangam TS, Kathiresan K (1994) Mosquito larvicidal activity of *Rhizophora apiculata*. Int J Pharmacogn 32: 33–36.
Thangam TS, Kathiresan K (1997) Mosquito larvicidal activity of mangrove plant extracts and synergistic activity of *R. apiculata* with pyrethrum against *Culex quinquefasciatus*. Int J Pharmacogn 35 (1): 69–71.
Thomas W, Mark S (2018) Technical report on Mangrove Restoration potential IUCN, University of Cambridge, the Nature Conservancy, p. 33.
Tsabang N, Ngah N, Estella FT, Agbor GA (2016) Herbal medicine and treatment of diabetes in Africa: case study in Cameroon. Diabetes Case Rep 1 (112): 2.
Twilley RR, Chen R, Hargis T (1992) Carbon sinks in mangroves and their implication to carbon budget of tropical ecosystems. Water Air and Soil Pollut 64: 265–288.
Vannucci M (2002) Indo-west pacific mangroves. In: "Mangrove ecosystems" (Lacerda L.D ed.) pp. 123-215. Springer-Verlag, Berlin.
Veera Ravi A, Kathiresan K (1990) Seasonal variations in gallotannin from mangroves. Indian J Mar Sci 19: 224–225.
Wafar S, Untawale AG, Wafar M (1997) Litter fall and energy flux in a mangrove ecosystem. Estuar Coast Shelf Sci 44: 111–124.
Wolanski E (1995) Transport of sediment in mangrove swamps. Hydrobiologia 295: 31–42.
Wolanski E, Mazda Y, Ridd P (1992) Mangrove hydrodynamics. In: Robertson AI, Alongi DM (eds.) Coastal and Estuarine Studies: Tropical Mangrove Ecosystems. American Geophysical Union, Washington, DC., USA, pp. 43–62.
Wong YS, Lam CY, Che SH, Li XR, Tam NFY (1995) Effect of wastewater discharge on nutrient contamination of mangrove soil and plants. Hydrobiologia 295: 243–254.
Woodroffe C (1992) Mangrove sediments and geomorphology. In : Robertson AI, Alongi DM (eds.) Coastal and Estuarine Studies: Tropical Mangrove Ecosystem. American Geophysical Union, Washington DC., USA, pp. 7–41.
World Disasters Report (2002) Focus on Reducing Risk. Geneva: IFRC.
Zhang Q, Satyanandamurty T, Shen L, Wu J (2017) Krishnolides A-D: New 2-Ketokhayanolides from the Krishna mangrove, *Xylocarpus moluccensis*. Mar Drugs 15 (11): 333.

Biogeography of the Mangrove Ecosystem: Floristics, Population Structure, and Conservation Strategies

2

P. Ragavan, K. Kathiresan, Sanjeev Kumar, B. Nagarajan, R. S. C. Jayaraj, P. M. Mohan, V. Sachithanandam, T. Mageswaran, and T. S. Rana

Abstract

Biogeography refers to reconstruction in the patterns of distribution of biological diversity and to identify the processes that are responsible for those distributions over the time. Despite the better understanding of global patterns of mangrove distribution using improved exploration techniques, satellite cartography, or the use of geographic information systems (GIS), the causal factors and processes underlying such patterns are still debated. The biogeography of mangroves has been widely discussed based on two alternative biogeographic processes, viz., dispersal and vicariance. However, in recent decades, human developmental activities, climate change and extreme natural events have affected the distribution patterns of species in both the terrestrial and marine environments.

P. Ragavan (✉) · S. Kumar
Physical Research Laboratory, Ahmedabad, India

K. Kathiresan
Centre of Advanced Study in Marine Biology, Faculty of Marine Science, UGC-BSR Faculty Fellow, Annamalai University, Parangipettai, Tamil Nadu, India

B. Nagarajan
Institute of Forest Genetics and Tree Breeding, Coimbatore, Tamil Nadu, India

R. S. C. Jayaraj
Rain Forest Research Institute, Jorhat, Assam, India

P. M. Mohan
Department of Ocean Studies and Marine Biology, Pondicherry University, Port Blair, Andaman and Nicobar Islands, India

V. Sachithanandam · T. Mageswaran
National Centre for Sustainable Coastal Management, MoEF & CC, Chennai, India

T. S. Rana
CSIR-National Botanical Research Institute, Rana Paratap Marg, Lucknow, India

R. P. Rastogi et al. (eds.), *Mangroves: Ecology, Biodiversity and Management*,
https://doi.org/10.1007/978-981-16-2494-0_2

Expansion of mangroves towards the poles and recent massive mangrove dieback in the Gulf of Carpentaria, Australia, for instance, is considered an impact of climate change and extreme events, respectively, whereas degradation of ecological health of existing mangroves are attributed to unsustainable human developmental activities. In addition, changes in distributional limits of the mangroves species, genetic discontinuity of population of widely distributed mangroves resulted from the recent molecular floristic studies, and recognition of the role of Pleistocene sea-level fluctuations, ocean currents, geomorphology, nutrient cycling, and hydrology of estuary in shaping the present distribution and population structure of mangrove species warrant a further comprehensive account of the biogeography of mangroves. Understanding the changes in plant species composition, distribution and the underlying processes are imperative for conservation and management of threatened mangrove habitats, where plants are the founder species which create habitats, modulate ecosystem functions, and support entire ecological communities. Considering these facts, the biogeography of the global mangroves has been examined in detail, and threats to the mangroves of various regions are also briefly reviewed. In addition, given the ecological and economic values of mangroves, the effectiveness of existing conservation measures are evaluated, and the safeguards needed to maintain the ecological health of the existing mangroves is highlighted.

Keywords

Biogeography · Conservation · Mangroves · Population Structure

2.1 Introduction

Mangroves are phylogenetically unrelated group of plants that evolved mechanisms to adapt, disperse, and establish in intertidal areas of tropical and subtropical coast around the world (Tomlinson 2016). Despite the pan-tropical distribution, the mangroves are species-poor in comparison with other tropical ecosystems (Duke 2017). However, it supports rich biological diversity and offers valuable ecological services to mankind like sediment filter, carbon sink, breeding ground for coastal fauna and coastal defence against storm surges and tsunami (Lee et al. 2014). In the past, it has been presumed that the water buyout propagules and seeds of the mangrove species facilitate the long distance dispersal, therefore the resultant population across the region are homogenous (Duke et al. 1998a, b; Maguire et al. 2000). In contrast, recent studies on phylogeography of widely distributed mangrove species revealed the existences of strong genetic differentiation among the population (Guo et al. 2018a). High genetic differentiation indicates the low gene flow between the populations and suggesting the restricted seed dispersal. Recent studies have provided sufficient evidences that the homogenization driven by long distance dispersal is weak due to historical barriers (continental drift and vicariant) and contemporary barriers (vast oceanic surface distance, sea surface currents, and geomorphology) (Guo et al. 2018a; Wee et al. 2020). In addition, in the last few decades, many mangrove forests are rapidly getting small, fragmented, and lost due to human developmental activities, climate change, and extreme natural events (Feller et al. 2017; Thomas et al. 2017; Bryan-Brown et al. 2020). Population size reduction and fragmentation cause a loss of genetic diversity and lead to the loss of adaptability to environmental changes and subsequently accelerate the extinction risk. Hence, understanding the changes in plant species composition, distribution pattern, and population genetic structure of mangrove species is imperative for conservation and management. Albeit, the biogeography of mangrove has been dealt by several workers earlier as well, recently Saenger et al. (2019) has provided a detailed account on biogeography of IWP mangroves considering the taxonomical studies carried out in the last two decades (Sheue et al. 2003; Sheue et al. 2005; Sheue et al. 2009a, b, 2010; Duke 2010; Cooper et al. 2016; Ragavan et al. 2016; Ono et al. 2016). Furthermore, in recent times, several studies have been carried out to understand the phylogeography of mangrove species based on molecular data (Wee et al. 2015; Yang et al. 2016; Guo et al. 2016; Yang et al. 2017; Guo et al. 2018a, b, c). Phylogeography is a fundamental component of a core modern biogeography (Lomolino et al. 2006) as well as an ever-expanding bridge between biogeography and related disciplines (Riddle et al. 2008). Hence, the present text discussed biogeography of global mangrove species with respect to their phylogeographical pattern, and also highlighted the role of the Indian subcontinent in diversification of Indo-West Pacific (IWP) mangroves. In addition, given the ecological and economic values of mangroves, the effectiveness of existing conservation measures are evaluated, and betterment needed to safeguard the ecological health of the existing mangroves is highlighted.

2.2 Mangrove Floristics of the World and Distribution Pattern

Globally, the mangrove realm is largely confined to sheltered tropical and subtropical coastlines within the latitudes of around 32°N and 38°S and covers an area of 137,600 km^2 spanning in 118 countries and territories (Bunting et al. 2018; Fig. 2.1). Mangroves are sensitive to freezing and chilling temperatures; therefore, they are most common in the tropics and subtropics (Duke 2017). The defined threshold limit of mangroves is mean winter sea surface temperature of at least 20 °C (Duke et al. 1998a). However, the mangroves are now expanding into the temperate regions of multiple continents due to the reduced frequency and intensity of extreme freeze events caused by climate change (Saintilan et al. 2014). About 96% of global mangroves are confined between the Tropic of cancer (23.4°N) and Tropic of Capricorn (23.4°S) (Bunting et al. 2018). Longitudinally there are two main centres of mangroves, viz., Atlantic East Pacific (AEP) and Indo-West Pacific (IWP). Mangroves of IWP represent 56% of global mangroves whereas AEP mangroves represent 44% (Bunting et al. 2018). South-east Asia alone represents 32% of global mangroves. Detailed global mangroves statistics are given in Table 2.1.

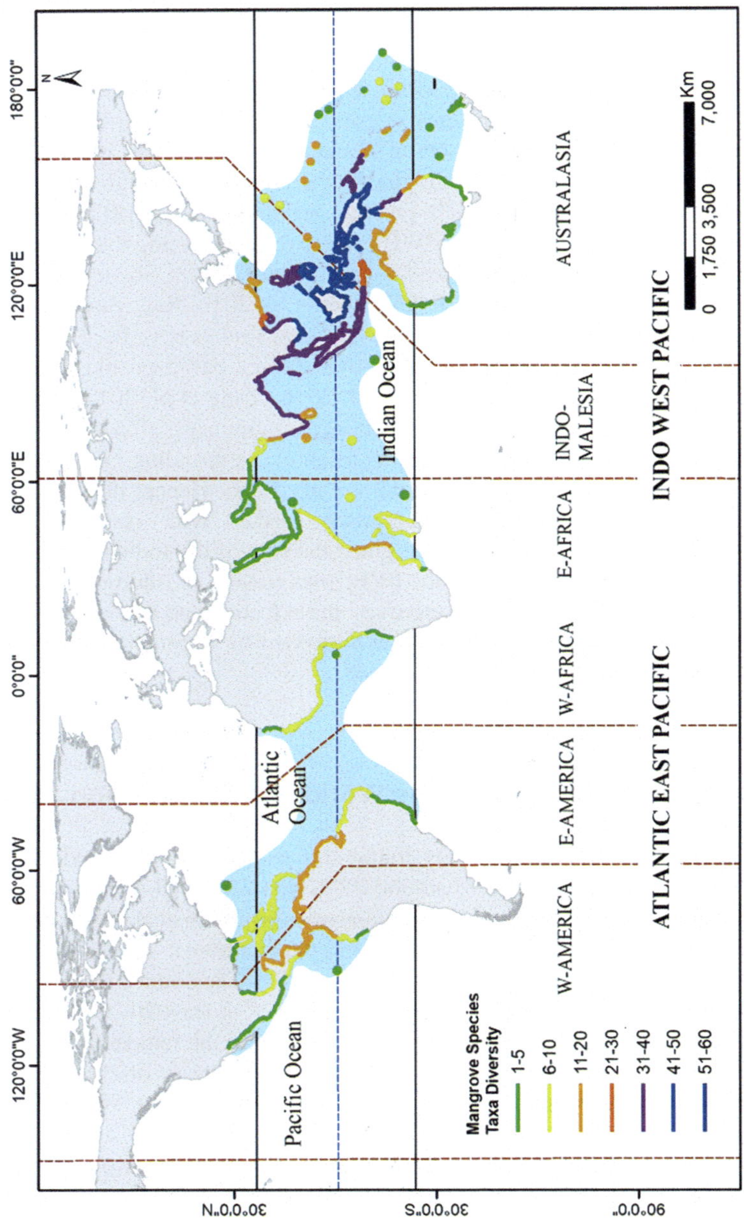

Fig. 2.1 Map showing distribution and species richness of global mangroves (reproduced based on Duke 2017)

Table 2.1 Statistics of global mangroves (net loss and gain, protected, restorable and degraded areas)

Region	Area (km^2) in 1996	Area (km^2) in 2016	Loss (km^2)	Gain (km^2)	Area of mangrove protected (km^2) in 2016	Proportion of Area protected (%)	Area restorable (km^2) in 2016	Proportion of original mangrove areas restorable (%)	Extent of highly restorable mangrove areas (km^2)	Area of degraded mangrove areas (km^2) in 2016	Proportion of mangrove degraded (%)
Australia & new Zealand	10,332	10,037	370	74	4553	45.4	350.9	3.3	328.6	54.6	0.5
East & southern Africa	7630	7329	424	122	3112	42.5	412.0	5.3	407.0	133.0	1.6
East Asia	159	159	12	13	21	12.9	7.0	4.0	6.5	2.6	1.8
Middle East	334	319	19	4	100	31.3	11.4	3.3	7.9	2.7	0.8
North & Central America & the Caribbean	22,702	21,072	2196	566	12,411	58.9	2277.2	9.6	1636.3	140.2	0.7
Pacific Islands	6410	6327	146	63	563	8.9	166.6	2.6	147.1	5.0	0.1
South America	19,632	19,063	1106	537	13,649	71.6	1068.2	5.2	794.9	92.6	0.5
South Asia	8701	8492	435	226	5428	63.9	352.7	3.9	279.7	32.4	0.4
Southeast Asia	46,789	44,060	3308	579	8769	19.9	3037.1	6.4	2591.2	847.0	1.9

(continued)

Table 2.1 (continued)

Region	Area (km^2) in 1996	Area (km^2) in 2016	Loss (km^2)	Gain (km^2)	Area of mangrove protected (km^2) in 2016	Proportion of Area protected (%)	Area restorable (km^2) in 2016	Proportion of original mangrove areas restorable (%)	Extent of highly restorable mangrove areas (km^2)	Area of degraded mangrove areas (km^2) in 2016	Proportion of mangrove degraded (%)
West & Central Africa	20,107	19,857	422	171	5317	26.8	437.1	2.1	430.5	78.5	0.4
Total	**142,795**	**136,714**	**8437**	**2356**	**53,923**	**39.4**	**8120.0**	**5.5**	**6629.9**	**1388.6**	**1.0**

Source: Worthington and Spalding (2019)

At present, the updated list of true mangrove species of the world consists of 85 species (including 14 natural hybrids) in 30 genera from 17 families worldwide (Tables 2.2 and 2.3). Despite similar areas (Saenger et al. 1983; Ellison et al. 1999) and essentially identical physical conditions of the mangrove environments (Chapman 1976; Duke 1992), the IWP region harbors 72 mangrove species (including 13 natural hybrids), whereas the AEP region has 15 species, which include one natural hybrid (Tables 2.2 and 2.3). *Pelliciera benthamii* is a new mangrove species discovered from AEP, recently (Duke 2020). Natural hybrids are not discussed here because they are usually of limited distribution, occurring within the range of the putative parental species, and they do not offer a great deal of biogeographic information. Only two species (*Acrostichum aureum, Rhizophora samoensis*) and three genera (*Acrostichum, Avicennia,* and *Rhizophora*) have a common distribution in both the IWP and the AEP. Except for the family Tetrameristaceae (= Pelliceriaceae), all the mangrove families of AEP are represented in IWP. Difference in species richness and distinct species composition between the IWP and AEP is attributed to complete isolation caused by the closure of Tethys seaway and regional differences in the rate of origin of new mangrove lineages (Duke 2017). Indo-West Pacific has a continuous presence of large areas of the continental shelf with islands scattered among shallow tropical seas resulted by a more complex geological history of tectonic movements, and these geomorphic features are favourable for a high rate of origin of new mangrove lineages in IWP compared to the AEP.

Mangroves of IWP are often divided into three sub-regions, viz., East Africa, Indo-Malesia, and Australasia. East African mangroves represent 10 mangroves species, whereas Indo-Malesia and Australasia are represented by 55 and 43 species, respectively. Species present in East African bioregions are widely distributed across IWP. However, significant distributional discontinuity exists between Indo-Malesia and Australasia. Both the regions share 39 mangrove species, but 16 mangrove species are exclusive to Indo-Malesia, these are *Acanthus volubilis, Acanthus xiamenensis, Phoenix paludosa, Excoecaria indica, Sonneratia apetala, Sonneratia griffithii, Brownlowia tersa, Camptostemon philippinensis, Heritiera fomes, Aglaia cucullata, Aegiceras floridum, Aegialitis rotundifolia, Ceriops decandra, Ceriops zippeliana, Kandelia candel,* and *Kandelia obovata.* Whereas, only 4 mangrove species, viz., *Avicennia integra, Diospyros littorea, Excoecaria ovalis,* and *Rhizophora samoensis* are restricted to Australasia region. Within Indo-Malesia, distributional discontinuity also exists between South Asia (also known as Indian Subcontinent) and South-east Asia. All the 38 mangrove species present in South Asian mangroves are common between both the regions, whereas 17 species are restricted to South-east Asia, and these are *Acanthus xiamenensis, Avicennia rumphiana, Sonneratia griffithii, Sonneratia lanceolata, Sonneratia ovata, Brownlowia argentea, Camptostemon philippinensis, Camptostemon schultzii, Aegiceras floridum, Osbornia octodonta, Aegialitis annulata, Bruguiera exaristata, Ceriops australis, Ceriops pseudodecandra, Ceriops zippeliana, Kandelia obovata,* and *Rhizophora stylosa.* Six species are endemic to Southeast Asia, and they are *Acanthus xiamenensis, Sonneratia griffithii, Camptostemon philippinensis, Aegiceras floridum, Ceriops zippeliana,* and *Kandelia obovata.* Ten species are

Table 2.2 Mangrove floristics of the World (* indicates presence of mangrove species)

Family	Genus	Species	Indo-West Pacific			Atlantic East Pacific		
			East Africa	Indo-Malesia	Australasia	West America	East America	West Africa
1. Acanthaceae	*1. Acanthus*	*1. Acanthus ebracteatus*		*	*			
		2. Acanthus ilicifolius		*	*			
		3. Acanthus volubilis		*				
		4. Acanthus xiamenensis		*				
	2. Avicennia	*5. Avicennia alba*		*	*			
		6. Avicennia bicolor[a]				*	*	
		7. Avicennia germinans[a]				*	*	*
		8. Avicennia schaueriana[a]				*	*	
		9. Avicennia integra			*			
		10. Avicennia marina	*	*	*			
		11. Avicennia officinalis		*	*			
		12. Avicennia rumphiana		*	*			
2. Arecaceae	*3. Nypa*	*13. Nypa fruticans*		*	*			
	4. Phoenix	*14. Phoenix paludosa*		*				
3. Bignoniaceae	*5. Dolichandrone*	*15. Dolichandrone spathacea*		*	*			
	6. Tabebuia	*16. Tabebuia palustris*[a]				*	*	
4. Combretaceae	*7. Conocarpus*	*17. Conocarpus erectus*[a]				*	*	*
	8. Laguncularia	*18. Laguncularia racemosa*[a]				*	*	*
	9. Lumnitzera	*19. Lumnitzera littorea*		*	*			
		20. Lumnitzera racemosa	*	*	*			
5. Ebenaceae	*10. Diospyros*	*21. Diospyros littorea*			*			
6. Euphorbiaceae	*11. Excoecaria*	*22. Excoecaria agallocha*		*	*			

		23. Excoecaria indica		*				
		24. Excoecaria ovalis			*			
7. Fabaceae	*12 Cynometra*	*25. Cynometra iripa*		*	*			
	13.Mora	*26. Mora oleifera*[a]				*	*	
8. Lythraceae	*14. Pemphis*	*27. Pemphis acidula*	*	*	*			
	15. Sonneratia	*28. Sonneratia alba*	*	*	*			
		29. Sonneratia apetala		*				
		30.Sonneratia caseolaris		*	*			
		31. Sonneratia griffithii		*				
		32. Sonneratia lanceolata		*	*			
		33. Sonneratia ovata		*	*			
9. Malvaceae	*16. Brownlowia*	*34. Brownlowia argentea*		*	*			
		35. Brownlowia tersa		*				
	17. Camptostemon	*36. Camptostemon philippinensis*		*				
		37. Camptostemon schultzii		*	*			
	18. Heritiera	*38. Heritiera fomes*		*				
		39. Heritiera littoralis	*	*	*			
10. Meliaceae	*19. Aglaia*	*40. Aglaia cucullata*		*				
	20. *Xylocarpus*	*41. Xylocarpus granatum*	*	*	*			
		42. Xylocarpus moluccensis		*	*			
11 Primulaceae	21 *Aegiceras*	*43. Aegiceras corniculatum*		*	*			
		44. Aegiceras floridum		*				
12. Myrtaceae	22. *Osbornia*	*45. Osbornia octodonta*		*	*			
13. Plumbaginaceae	23.*Aegialitis*	*46. Aegialitis annulata*		*	*			
		47. Aegialitis rotundifolia		*				

(continued)

Table 2.2 (continued)

Family	Genus	Species	Indo-West Pacific			Atlantic East Pacific		
			East Africa	Indo-Malesia	Australasia	West America	East America	West Africa
14.Pteridaceae	24. *Acrostichum*	*48. Acrostichum aureum*[b]	*	*	*	*	*	*
		49. Acrostichum danaeifolium[a]				*	*	*
		50. Acrostichum speciosum		*	*			
15.. Rhizophoraceae	25. *Bruguiera*	*51. Bruguiera cylindrica*		*	*			
		52. Bruguiera exaristata		*	*			
		53. Bruguiera gymnorhiza	*	*	*			
		54. Bruguiera parviflora		*	*			
		55. Bruguiera sexangula		*	*			
	26. *Ceriops*	*56. Ceriops australis*		*	*			
		57. Ceriops decandra		*				
		58. Ceriops pseudodecandra		*	*			
		59. Ceriops tagal	*	*	*			
		60. Ceriops zippeliana		*				
	27. *Kandelia*	*61. Kandelia candel*		*				
		62. Kandelia obovata		*				
	28. *Rhizophora*	*63. Rhizophora apiculata*		*	*			
		64. Rhizophora mangle[a]				*	*	*
		65. Rhizophora samoensis[b]			*	*	*	
		66. Rhizophora mucronata	*	*	*			
		67. Rhizophora racemosa[a]				*	*	*

		68. Rhizophora stylosa		*	*			
16. Rubiaceae	29. *Scyphiphora*	*69. Scyphiphora hydrophylacea*		*	*			
17. Tetrameristaceae	30. *Pelliciera*	*70. Pelliciera rhizophorae*[a]				*	*	
		71. Pelliciera benthamii[a]				*	*	
Total			**10**	**55**	**43**	**14**	**14**	**7**

[a]Present in Atlantic East Pacific mangroves
[b]Present in both Atlantic East Pacific and Indo-West Pacific mangroves

Table 2.3 Natural hybrids of global mangroves

	References
Natural hybrids (Parental Species)	
1. Lumnitzera × rosea (L. littorea × L. racemosa)	Tomlinson et al. (1978)
2. Sonneratia × gulngai (S. alba × S. caseolaris)	Duke (1984)
3. Sonneratia × hainanensis (S. alba × S. ovata)	Ko (1985), Wang et al. (1999)
4. Sonneratia × urama (S. alba × S. lanceolata)	Duke (1994)
5. Sonneratia × zhongcairongii (Sonneratia alba × S. apetala)	Xie et al. (2020), Zhong et al. (2020)
6. Bruguiera × rhynchopetala (B. gymnorhiza × B. sexangula)	Ge (2001) and Sun and Lo (2011)
7. Bruguiera × hainesii (B. gymnorhiza × B. cylindrica	Ono et al.(2016)
8. Bruguiera × dungarra (B. exaristata and B. gymnorhiza)	Duke and Kudo (2018)
9. [a]Rhizophora × harrisonii (R. mangle × R. racemosa)	Tomlinson and Womersley (1976)
10. Rhizophora × lamarckii (R. apiculata × R. stylosa)	Duke and Bunt (1979)
11. Rhizophora × annamalayana (R. apiculata × R. mucronata)	Kathiresan (1995)
12. Rhizophora × mohanii (R. mucronata × R. stylosa)	Ragavan et al. (2015)
13. Rhizophora × selala (R. stylosa × R. mangle)	Tomlinson (1978)
14. Rhizophora × tomlinsonii (R. apiculata × R. mangle)	Duke (2010)
Unnamed natural hybrids	
1. Avicennia marina × A. rumphiana	Huang et al. (2014)
2. A. germinans × A. bicolor	Mori et al. (2015)
3. Sonneratia alba × S. griffithii	Qiu et al. (2008)
4. Acrostichum aureum × A. speciosum	Tsai et al. (2012)
5. Ceriops tagal × C. australis	Zhang et al. (2013)

[a]Hybrid confined to the Atlantic East Pacific only

not found in Australasia but are shared between South-east Asia and South Asia, and these are *Acanthus volubilis*, *Phoenix paludosa*, *Excoecaria indica*, *Sonneratia apetala*, *Brownlowia tersa*, *Heritiera fomes*, *Aglaia cucullata*, *Aegialitis rotundifolia*, *Ceriops decandra*, and *Kandelia candel*. Mangroves of AEP are divided into three sub-regions, viz., West America, East America, and West Africa. All the known AEP mangrove species are present in West and East American sub-regions, whereas only 8 mangrove species are present in the West African region. In general, longitudinally mangrove species richness increases towards east, whereas latitudinally the species richness increases towards equator.

2.3 Phylogeography of Extant Mangrove Species

Recent phylogeographical studies revealed the low variation within populations, but high differentiation among populations of widely distributed mangrove species (Yan et al. 2016; Guo et al. 2018a), which is in contrast to the expectation that long-lived woody species maintain more variation within species and within populations but have less variation among populations (Hamrick et al. 1992). The major cause of the low genetic diversity within populations and within species, especially in marginal populations of the species' distribution ranges are repeated founder effects or bottleneck events associated with geo-climatic history in IWP region and genetic drift in small and geographically isolated populations (Yan et al. 2016). Furthermore, these studies reveal the role of barriers of gene flow, viz., geographic distance, glacial vicariance (land masses), and ocean currents in shaping the distribution pattern of extant mangrove species. The genetic differentiation with respect to geographical distance is minimal in most of the mangrove species due to their potential of long distance dispersal, whereas the levels of gene flow between populations in most of the mangrove species are determined by land barriers, vast oceanic surface distance, and directionality and dynamic interactions of wind and current, acting alone or in synergy (Lo et al. 2014; Wee et al. 2014; Wee et al. 2020).

In the Indo-West Pacific (IWP) region, genetic differentiation between populations has been reported in many mangrove species on both sides of the Malay Peninsula (=Sunda shelf) and of the Wallaces. The Malay Peninsula is a land barrier associated with the genetic break between the Indian Ocean region (referring to the coasts located on the west of the Malay Peninsula) and Southeast Asian region (referring to the coasts located on the east of the Malay Peninsula). This distinctive genetic break across the Malay Peninsula has been observed in many mangrove species, including *Lumnitzera racemosa* (Su et al. 2006; Li et al. 2016), *Lumnitzera littorea* (Su et al. 2007), *Ceriops tagal* (Ge and Sun 2001; Liao et al. 2007), *Ceriops decandra* (Tan et al. 2005; Huang et al. 2008), *Sonneratia alba* (Wee et al. 2017; Yang et al. 2017), *Bruguiera gymnorhiza* (Minobe et al. 2010; Urashi et al. 2013; Ono 2016), *Rhizophora apiculata* (Inomata et al. 2009; Wee et al. 2014; Yan et al. 2016), *Xylocarpus granatum* (Tomizawa et al. 2017; Guo et al. 2018c), *Excoecaria agallocha* (Zhang et al. 2008; Guo et al. 2018b), *Xylocarpus moluccensis* (Guo et al. 2018c), *Acanthus ilicifolius* (Guo et al. 2020) and *Heritiera littoralis* (Banerjee et al. 2020). During the glacial periods, falls in sea levels caused the Malay Peninsula, Sumatra, and Java to become connected and form the Sunda Land, which halted the exchange of seawater between the Indian Ocean and the South China Sea (Wyrtki 1961; Wang et al. 1995). This glacial isolation resulted in genetic divergence between populations from these two oceans. Despite the prominent gene flow barrier, recent studies identified the Malay peninsula as a corridor for genetic exchange between the oceanic regions, especially during interglacial periods of the Pleistocene when the land shelves submerged (Wee et al. 2014; Yang et al. 2017). A study on the contemporary gene flow across the Malay Peninsula has revealed that species with higher dispersal potential (*Bruguiera gymnorhiza* and *Rhizophora mucronata*) exhibit much higher proportion of recent inter-population migration along the Malacca Strait than the species with lower dispersal potential (*Avicennia alba* and *Sonneratia alba*) (Wee et al. 2020). Thus, the genetic differentiation in *Rhizophora mucronata* is found only at the edge of the Andaman Sea and the Strait of Malacca (Wee et al. 2014).

The genetic break across Wallaces has been observed in *Ceriops tagal* (Huang et al. 2008), *Lumnitzera racemosa* (Li et al. 2016; Su et al. 2006), *L. littorea* (Su et al. 2007), *Rhizophora stylosa* (Wee et al. 2014; Yan et al. 2017), *Rhizophora apiculata* (Guo et al., 2016; Yan et al. 2016), *Sonneratia caseolaris* (Yang et al. 2016), *S. alba* (Yang et al. 2017), *Xylocarpus granatum* (Tomizawa et al. 2017; Guo et al. 2018c), and *Heritiera littoralis* (Banerjee et al. 2020). Plate motions of the Indo-Australian Archipelago, the emergence of the Sahul shelf during glacial time and Ocean currents have been identified as barriers for gene flow between Southeast Asia and Australasia/Oceania. However, it is seldom considered to be the strongest genetic break except for *R. apiculata* and *R. stylosa*. Genetic differentiation is distinct between South-East Asian (Japan, Vietnam, Philippine, and Indonesia) and Oceanian (Fiji, Vanuatu, and New Caledonia) in the populations of *Rhizophora stylosa* (Wee et al. 2015), and *Rhizophora apiculata* (Guo et al. 2016). As the locations of these genetic breaks correspond to the boundaries of oceanic currents, Wee et al. (2015) suggested that oceanic circulation patterns might have acted as "cryptic barriers". In contrast, Banerjee et al. (2020) and Guo et al. (2018c) reported only a weak genetic differentiation between South-east Asian and Australasia populations of *Heritiera littoralis* and *Xylocarpus moluccensis*, respectively, and noted the existences of considerable gene flow between these two regions being mediated by the Indonesian throughflow which moves from the Celebes Sea through the Makassar Strait to the Timor Sea.

In addition, several studies have also reported the genetic structure within each oceanic region. For instance, within Oceania, a strong genetic differentiation is found in the populations of *Sonneratia alba* and *Rhizophora apiculata* between Australia and South-West Pacific Islands (Wee et al. 2017; Guo et al. 2016). Within South China Sea, the genetic differentiation is recorded between mainland and Island coast populations of *Sonneratia alba* (Wee et al. 2017) and Northern and Southern SCS populations of *Heritiera littoralis* (Banerjee et al. 2020). Similarly, within the Indian Ocean, genetic differentiation occurs between the Arabian Sea and Bay of Bengal populations of *Bruguiera gymnorhiza* (Urashi et al. 2013) and between African and Asian populations of *B. gymnorhiza* (Ono 2016). Oceanic currents and the vastness of the Ocean/sea are often attributed to the genetic differentiation within the oceanic region. In Atlantic East Pacific (AEP) genetic differentiation is observed between populations of the Atlantic and Pacific Oceans in all studied mangrove species and it is attributed to vicariance following the final closure of the Central American Isthmus (Takayama et al. 2013; Cerón-Souza et al. 2015).

2.4 Origin and Diversification of Mangroves Based on the West Tethyan Hotspot

Mangroves evolved from terrestrial rather than marine habitats. This is proved by the presence of mangrove pollen fossils in India below marine foraminiferan assemblages (lower deposits of estuarine environments) suggesting the mangrove evolution from a non-marine habitat and adapted to an estuarine habitat (Srivastava and Binda 1991). There are two hypotheses available for the origin of mangroves, viz., centre-of-origin and vicariance hypothesis. According to the centre-of-origin hypothesis (van Steenis 1962; Chapman 1975, 1976), the entire mangrove taxa first appeared in the Indo-West Pacific and subsequently dispersed to other regions. According to the vicariance hypothesis (McCoy and Heck 1976; Mepham 1983), all mangroves originated around the Tethys Sea, and continental drift then isolated the flora in different regions of the earth. In addition, Specht (1981) proposed an Australasian centre-of-origin hypothesis mainly based on pollen observations made by Churchill (1973). However, vicariance hypothesis has been widely accepted by many subsequent researchers (Tomlinson 1986; Duke 1995; Duke et al. 1998a; Ellison et al. 1999; Kathiresan and Bingham 2001). In the past, three possible routes have been suggested to explain the modern distribution of mangroves, viz., (1) Eastward dispersal—initial diversification in IWP and dispersal eastward across Pacific into AEP, (2) Westward dispersal—initial diversification in IWP and dispersal westward through Tethys seaways into the Atlantic, (3) Connection between the IWP and AEP regions around the southern tip of African continent (Ellison et al. 1999; Srivastava and Prasad 2019). All the three routes cannot be dismissed, although there is no adequate fossil evidence to support it. Further, all these hypothetical routes emphasize the initial colonization occurred primarily in Southeast Asia/Malaysia and dispersed to other regions; poor dispersal abilities and the closure of the Tethys connection to the Atlantic is often attributed to the restriction of most mangrove taxa to the IWP. Due to the presence of high species richness and the prevalence of equable climatic conditions since the end of the Cretaceous, the South-East Asia is believed to be the place of origin for angiosperms which first acquired the mangrove habitat, from where most contemporary mangrove genera originated. However, the oldest South-East Asian mangrove fossils are often less older than equivalent ones from distant sites (Mepham 1983). For instance, fossil records of *Rhizophora* and *Sonneratia* are known from the Paleocene of India and France, but it appears in Borneo during Middle Eocene and Lower Miocene, respectively (Ellison et al. 1999). Thus a consistent hypothesis accounting for the distribution of modern mangrove species is far from complete.

The combined evidence from fossil and molecular studies revealed that the biodiversity hotspots occurred at different places through time and there have been at least three marine biodiversity hotspots during the past 50 million years (Renema et al. 2008). Recent paleobiogeographical studies based on the re-examination of fossil evidences revealed that mangroves were at their prime during the geological period of Eocene rather than in the Late Cretaceous (Srivastava and Prasad 2019). Particularly, the records of mangrove fossils of recognized mangrove genera like

Nypa, Acrostichum, Rhizophora, Avicennia, Pelliciera, Sonneratia, Bruguiera, Ceriops, Heritiera, and *Aegiceras* were present around the Tethyan shoreline during Eocene at the convergence zone between the African and European continents (Ellison et al. 1999; Srivastava and Prasad 2019). This supports the existences of Eocene West Tethyan and Arabian hotspots prior to Southeast Asian hotspot. Thus, it is apparent that rich biodiversity of South-east Asia is not a unique feature, but is the latest manifestation of a pattern that was present in the past (Renema et al. 2008). Further, it is also plausible that the initial colonization of mangroves with high species richness occurred in west tethyan hotspot and dispersed west to the Americas, as well as south and east, along the coasts of Asia, East Africa, Malagasy, and Greater India. Subsequent continental drift resulted current distribution pattern of mangroves.

2.4.1 Founder Effect Speciation- Causes for Genotypically and Phenotypically Distinct AEP Population

Birth and death of successive hotspots are attributed to environmental changes caused by the geological events such as Plate tectonics (Renema et al. 2008). Duke (2017) recognized the role of four geological events in shaping the current distribution pattern of mangroves, and they are (1) the separation of Africa and South America (~100 mya) and progressive opening of the South Atlantic Ocean,(2) the closure of the Tethys Sea between Africa and Eurasia 25–35 mya, (3) the separation of India (70–75 mya) and Australia (~50 mya) with their subsequent collisions, and (4) the opening of the North Atlantic (from ~60 mya onwards). Of these, except event 3, others are responsible for the formation of genotypically and phenotypically distinct AEP population. Collision of Africa and Eurasia had resulted in contraction of West tethyan hotspot and subsequently terminated the tethyan seaway. As a result, the populations of AEP remain isolated from parental populations and experience the founder event along the coast of West Africa and South and North America. Subsequent widening of the Atlantic, lack of gene flow and regional climate change are attributed to the strong genetic drift in the founder population leading to the evolution of genotypically and phenotypically distinct AEP population through founder effect speciation. Lack of speciation in AEP species can be attributed to the prevalence of dry tropical climate of the New World tropics since Miocene (Mepham 1983). Wet climates support terrestrial-mangrove transition, but a marked drying trend in the AEP region as evident from the extinction of *Nypa.* The range contraction of *Pelliciera* since the Miocene also suggested the lack of suitable environmental conditions to support the terrestrial-mangrove transition. Whereas in IWP, the continuous presence of extensive wet habitat since the end of the Cretaceous favours the terrestrial-mangrove transition. Thus, diversity continues to increase in IWP through invasion of new clades and autochthonous production of new taxa. Predictably, many more nonexclusive taxa of trees invade mangrove habitat from terrestrial habitats in the IWP region than in the AEP region.

2.4.2 Indian Subcontinent View of Diversification of IWP Mangroves

In IWP, the collision of India and Australia with Asia would be the key tectonic movement (Duke 2017). Further, this collision altered the geomorphology of South-East Asia into massive fragmentation of landmasses with shallow shelf areas, which are highly suitable for colonization of mangroves. Indian continent is the first one, separated from Gondowana around 90–80 mya and move northward towards Asia in tropical latitudes about 40 mya and collision with Asia began around 40 mya. This complete isolation of India for 40 mya and vast continental shelf area, particularly the Northern and North-eastern parts, around the Indian Island would be highly suitable for terrestrial-mangrove transition/colonization and also for speciation. In the past, India is seldom considered as part of tethyan shore, but actually Indian subcontinent was an important part of the ancient Tethyan shore line (particularly North and North-eastern parts) as wells as the part of Eocene west Tethyan and Arabian hotspot (Renema et al. 2008). So it is more obvious that the Indian subcontinent could be refugia of ancestral population of IWP rather than that of South-East Asia. For instance, when Western tethyan hotspot matured in the late Eocene, collision of India with Asia began, which would lead to the creation of numerous shallow marine platforms in the Northern and North-eastern India. Further the rifting of the Mediterranean also would create shallow areas in the Middle East. Due to this contraction of west tethyan hotspot, the population of west tethyan hotspot migrated toward east and colonized in adjacent suitable habitats, viz., Arabian hotspot and Indian coast. This is evident by the occurrence of mangrove fossils since late cretaceous period (Ellison et al. 1999), the rich mangrove fossils in India during Eocene (Srivastava and Prasad 2019) and the presence of more than 60% of extant species in Indian subcontinent. However, the role of the Indian subcontinent in the current distribution pattern of extant mangrove species has been overlooked in most of earlier bio-geographical distribution. Furthermore, recent studies on phylogeography of widely distributed mangroves of IWP, mostly excluded the sampling from Indian subcontinent, therefore, it is very important to highlight the significance of the Indian Subcontinent in the current distribution pattern of mangroves in IWP.

The mangroves of IWP consist of 59 mangrove species, which excludes 13 natural hybrids. India subcontinents host 38 mangrove species, whereas South-East Asia and Australasia host 55 and 43 species, respectively. Out of the 59 mangrove species of IWP, 21 species are not known from the Indian subcontinent, of which, many species are recently diverged from their sister/closely related species (Table 2.4). Recent phylogeographical studies revealed that the divergence time of these species is within the last 3 Myrs, and cycles of isolation interspersed by episodes of gene flow during Pleistocene had resulted in this speciation in South-East Asia—a boundary between the Indian Ocean and Pacific Ocean (He et al. 2019). Further, recent molecular studies have revealed that the gene flow between the Indian and Pacific Ocean is asymmetrical and eastward gene flow from North Indian Ocean to South China Sea and Australia is stronger than that in the opposite direction (Guo

Table 2.4 Recently diverged mangrove species and their close relatives

Species	Remark
1. Acanthus xiamenensis	Restricted distribution. Moreover identity is uncertain
2. Avicennia integra	Recently diverged from *A. officinalis*
3. Avicennia rumphiana	Recently diverged from *A. alba*
4. Diospyros littorea	Independent terrestrial-mangrove transition (endemic to Australia)
5. Excoecaria ovalis	Subspecies of *E. agallocha*
6. Sonneratia griffithii	Though not found in SA. Possibilities for its occurrence are high. Known from Bruma
7. Sonneratia lanceolata	Recently diverged from *S. caseolaris*
8. Sonneratia ovata	Recently diverged from *S. caseolaris*
9. Brownlowia argentea	Recently diverged from *B. tersa*
10. Camptostemon philippinensis	Independent terrestrial-mangrove transition (endemic to South-East Asia)
11. Camptostemon schultzii	Independent terrestrial-mangrove transition (known in South-East Asia and Australia)
12. Aegiceras floridum	Recently diverged from *A. corniculatum*
13. Osbornia octodonta	Independent terrestrial-mangrove transition (known in South-East Asia and Australia)
14. Aegialitis annulata	Recently diverged from *A. rotundifolia*
15. Bruguiera exaristata	Recently diverged from *B. gymnorhiza*
16. Ceriops australis	Recently diverged from *C. tagal*
17. Ceriops pseudodecandra	Recently diverged from *C. decandra*
18. Ceriops zippeliana	Recently diverged from *C. decandra*
19. Kandelia obovata	Recently diverged from *K. candel*
20. Rhizophora samoensis[c]	Known in west pacific islands. Identity is uncertain
21. R. stylosa	Recently diverged from *R. mucronata*

et al. 2016; Li et al. 2016). Particularly in summer, due to the effects of the strong south-west monsoon and the absence of the north-east trades, strong current flows from west to east, which facilitates the west to east migration rather than that in opposite direction. Further, summer monsoon current completely obliterates the north equatorial current and hence, there is also no counter-equatorial current, and makes eastward migration more prominent.

All mangrove species of the Indian subcontinent, except *Lumnitzera littorea*, are known from Sundarbans (considering both Indian and Bangladesh). In the past, the formation of Bengal Basin was viewed as a result of the collision of India and Eurasia, and the formation Sundarban delta is considered most recent. If the Sundarban is of most recent origin, how does it host about 60% of extant mangrove species of IWP. Are they dispersed from South-East Asia? This chance for dispersal from South-East Asia to Sundarban is very rare considering the very rare long distance dispersal of mangroves species and barrier of Sundaland during Pleistocene climate oscillation. Though the Bengal basin is widely considered as Foreland basin formed via continent–continent collision, there is speculation that the formation of Bengal basin is not a recent one, and it began during the cretaceous period (Alam et al. 2003), viz., formation of the Bengal basin started in the Jurassic with initiation of rifting of the Pangaea and completed by the Miocene with docking of eastern India with Burma Plate. Collision of India with Eurasia resulted in expansion of Bengal basin due to huge sediment deposition from Himalayan Rivers. The presence of rich mangrove fossils of Bengal basin during Eocene supports the occurrence of ancestral populations along the Indian coast (Venkatachala and Rawat 1972; Mathur and Chowdhary 1977; Kar 1985; Thanikaimoni 1987; Mathur 1984). India was the part of West Tethyan and Arabian hotspot (Renema et al. 2008), and hence, it is more plausible that Indian subcontinent could be the refugia of ancestral population of IWP mangroves. Climate /geo-morphological changes that resulted from the collision of India with Eurasia would be attributed to the extinction of few mangroves (e.g. *Nypa*) along west coast. However, formation of deltas along the east coast would have enhanced the colonization of ancestral populations. Water buoyant propagules of mangroves are suitable for long-shore dispersal, especially of viviparous species rather than transoceanic dispersal. Thus the formation of contiguous coastline from India to SEA and seasonally directed eastward monsoon current as well as East Indian Coastal current would have facilitated dispersal from the Indian subcontinent to SEA prior to the Pleistocene. The collision of India and Australia with Asia resulted in massive fragmentation of landmasses and shallow continental shelf areas which would have facilitated the rapid colonization in SEA. Further, cycles of isolation interspersed by episodes of gene flow during Pleistocene sea-level fluctuation resulted in speciation event in SEA (He et al. 2019).

Collision of Australia with Asia began 10 mya and it separated from Antarctica around 40 mya. So Australia spent more time in the temperate zone before its collision with Asia. Thus, it is widely accepted that Australia had gained floristic diversity when it came close to the SEA. The Sahul shelf formed due to collision of Australia with SEA prevents gene flow between SEA and Australia during glacial periods; but when the collision began, Australia was close to SEA, which might have facilitated migration of mangrove species from SEA to Australia. It is particularly true with Indonesia, where current flow was strong enough to facilitate gene flow. Recent phylogeographical studies also reported the weak genetic differentiation of many mangrove species due to the presence of gene flow between South-East Asia and Australia facilitated by Indonesian throughflow current. Phylogeographical studies also revealed the close genetic similarity of east African population of *Bruguiera gymnorhiza, Sonneratia alba*, and *Xylocarpus granatum* with south-east Asian populations and with *Rhizophora mucronata* with Australian population. Further, a strong gene flow is found to be possible from Southeast Asia/Australia to East Africa. Hence, the ancestors of African populations might have emigrated from Southeast Asian/Australian populations, via major ocean currents (South equatorial currents) (Lo et al. 2014). There are seven hypothetical dispersal routes have been proposed for IWP mangroves (Duke 2017). Of which, IWP south-east and IWP North-East routes seem to be more prominent considering the existing ocean current pattern as supported by many phylogeographical studies. Based on above

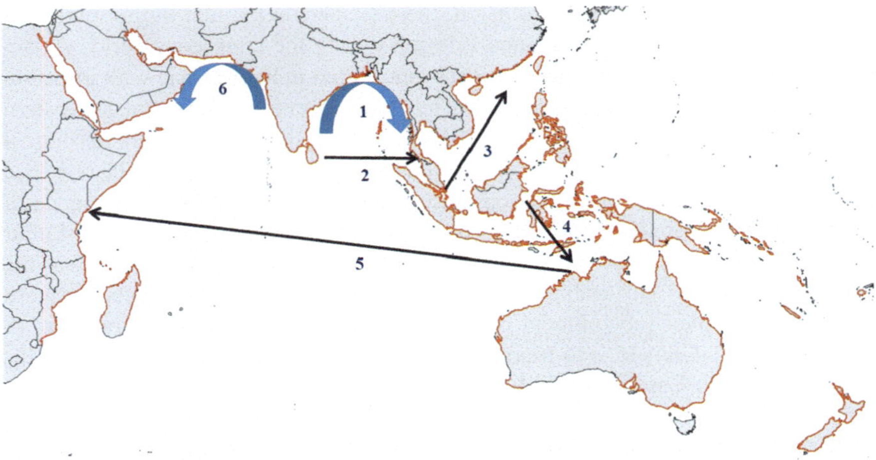

Fig. 2.2 Map showing dispersal route based on Indian Subcontinent view (*1*) East Indian Coastal current, (*2*) Summer monsoon current (*3*) South China Sea Warm Current (*4*) Indonesia throughflow current (*5*) South equatorial current (*6*) West Indian Coastal Current

speculation, more plausible dispersal routes in IWP are proposed here (Fig. 2.2). However, it needs to be verified in future in the present context of very little fossil evidences and lack of phylogeographical information on mangrove species of Indian subcontinent and Middle East.

2.5 Conservation and Management

Despite the slowdown in average rate of mangrove loss in recent decade, still mangroves experience an annual loss of 0.2–0.7% between 2000 and 2012, and remain the most threatened ecosystems of the world (Hamilton and Casey 2016). Anthropogenic activities and climate change consequences are the potential threat for mangroves. Overexploitation, land use change, hydrological alteration, and environmental pollution are the contemporary anthropogenic factors affecting mangrove distributions, whereas sea-level rise, increasing frequency of natural calamities such as drought, storm, cyclones, tsunami are the climate related threat factors. Recently Goldberg et al. (2020) estimated that about 3363 km^2 (2.1%) of global mangrove areas was lost between 2000 and 2016, at an average annual rate of 0.13%. Anthropogenic causes are accounted for 62% of total mangrove loss, whereas natural causes are attributed to 38% of total losses. Habitat conversion for rice, shrimp, and oil palm cultivation remains the primary global driver of mangrove loss, which represents 47% of global losses from 2000 to 2016; land reclamation for non-productive conversions and human settlements represents 12 and 3% of global losses, respectively. Shoreline erosion resulted from the increasing sea-level rise represent the second highest as well as primary natural causes of global mangrove losses, which represent 27% of global mangrove loss from 2000 to 2016; while extreme weather events (EWE) contribute to 11% of losses (Goldberg et al. 2020). In order to reverse the mangrove loss, various legislative measures and massive rehabilitation/restoration efforts have been undertaken in the last three decades. Declaration of protected areas is a major legislative measure in many mangrove nations. Worldwide, there are about 2500 protected areas that include mangrove forests within their boundaries, which represent over 39% (around 54,000 km^2; Table 2.1) of world's remaining mangroves; and, 34 countries have placed more than half of their mangroves under such protection (Worthington and Spalding 2019; IUCN and UNEP-WCMC 2019). An extensive coverage of mangroves under protected areas represents a strong positive trend in coastal conservation However, Southeast Asia- a hotspot of mangroves in terms of area cover, species diversity, and deforestation rate—has only 20% area under protection (Table 2.1). Protected areas prevent some drivers of degradation, such as unsustainable timber extraction. However, other drivers of degradation, such as upstream water abstraction or changes to sediment supplies, cannot be influenced when they occur beyond boundaries of the protected area (Worthington and Spalding 2019). So, considering the hydrological and ecosystem connectivity is imperative to minimize the ecological degradation within the protected areas.

In the last three decades, massive efforts were globally made to restore the degraded mangrove areas. Incorporation of mangroves into engineered hard coastal defence structures, monoculture plantations, and "ecological mangrove restoration" (EMR) approaches are common methods of mangrove restoration (Ellison et al. 2020). However, despite existing guidance for successful restoration efforts (Lewis et al. 2019), most of the efforts are not successful (Lee et al. 2019) and successful cases are rare (e.g. Stubbs and Saenger 2002; Lewis et al. 2005; Bosire et al. 2008; Rey et al. 2012; Dale et al. 2014; Brown et al. 2014a, b; Begam et al. 2017). Unsuccessful restoration is a waste of time, money, and human resources. Further, the habitat characteristics altered for the purpose of mangrove restoration render the natural corridors unsuitable for natural migration of mangroves to cope-up with the impacts of climate change. Further, functionality of restored areas is rarely monitored to fully ascertain the restoration success based on faunal diversity, vegetation structure (e.g., basal area, species diversity), and function (e.g., net primary productivity, carbon storage, resilience) (López-Portillo et al. 2017). The costs of mangrove restoration are around US$3000 per hectare (Bayraktarov et al. 2016), however, the cost is high over US$100,000 per hectare when large-scale engineering to restore hydrology, combined with high staffing (Worthington and Spalding 2019). So, mangrove restoration is expensive for many developing nations' especially South-East Asian countries. Thus, investments in large-scale planting must be the last option. Rehabilitation of abandoned aquaculture areas should be directed to the hydrological correction (Lewis et al. 2016) to facilitate natural recruitment of mangroves and also improving soil fertility. Recently, Ellison et al. (2020) proposed the adaptive management of restored mangroves areas—a *structured, iterative process of "learning-by-doing" and decision-making in the face either of continuous change (environmental, social, cultural, or political) or uncertainty*, through regular monitoring of key indicators of the objectives and goals of a restoration project and identifying the clear triggers or decision-points for appropriate intervention and action if the objectives or goals are not being met.

As a result of last three decadal conservation measures, human-driven mangrove loss decreased significantly since 2000. For instance, the area of mangroves converted by direct human intervention has declined by 73% over the 16-year period (viz., from 1186 km^2 between 2000 and 2005 to 314 km^2 between 2010 and 2016) and during the same period, the total loss of mangroves area is attributed to natural causes declined by 60% from 624 km^2 between 2000 and 2005 to 249 km^2 between 2010 and 2016 (Goldberg et al. 2020). The decreasing rate of natural loss (3.75% per year) was significantly lower than that of anthropogenic loss (4.56% per year); however, the relative contribution of natural drivers to global mangrove losses increased by 10% over the 16-year period (2000–2016) (Goldberg et al. 2020). Shoreline erosion and extreme weather events are the key drivers of natural losses of mangroves. So it is certain that the emergence of natural drivers as the primary causes of modern mangrove loss in upcoming decades. For example, the recent massive mangrove dieback in the Gulf of Carpentaria, Australia, for instance, is considered to be an impact of climate change-induced extreme weather events (Duke et al. 2017; Lovelock et al. 2017).

Despite being threatened by human developmental activities and climate change, the mangrove forests are highly resilient ecosystems that have the potential to adapt and adjust to changing conditions (Woodroffe et al. 2016). This is evident by the recent expansion of mangroves towards the polar regions in response to increasing minimum winter temperature (Saintilan et al. 2014) as well expansion and contraction in response to temperature fluctuations and Pleistocene sea-level drop and rise in the long past (Ludt and Roacha 2014). Vertical adjustment and horizontal movement across the landscape are the processes that govern the responses of mangrove forest to sea-level rise (Woodroffe et al. 2016; Krauss et al. 2014; Lovelock et al. 2015). Furthermore, the observed higher expression of diversity than the genetic diversity and smaller genome size warrants the significant evolutionary potential of mangroves (Lira-Medeiros et al. 2010; Wee et al. 2018; Lyu et al. 2018). So enhancing the adaptive potential of mangroves is the need of the hour to cope-up with the climate change. Since, widely distributed mangroves species exhibit low genetic diversity and high genetic differentiation among the populations across their distribution ranges, it is imperative to delineate the Conservation Units, including Evolutionarily Significant Units (ESUs) and Management Units (MUs), which warrant management as a separate unit (Frankham et al. 2010) Thus, use of genetic information in conservation is crucial for its long-term effectiveness to preserve the adaptive and evolutionary potentials of ecosystem/species (Hoffmann and Sgro 2011). However, little has been translated into on-the-ground conservation (Wee et al. 2018). Further, structure and functions of mangrove are highly site-specific, and they even differ significantly along the same coast. The mangrove ecosystem responses to climate change processes are expected to be greatly influenced by plant-mediated processes. So the site-specific and species-specific knowledge of mangroves is highly imperative to enhance the sustainability of mangroves.

Acknowledgement The author (P.R.) is thankful to the Department of Space, Government of India for funding under Post-Doctoral Fellowship at Physical Research Laboratory and the authorities of the Physical Research Laboratory for providing facilities.

References

Alam M, Alam MM, Curray JR, Chowdhury ALR, Gani MR (2003) An overview of the sedimentary geology of the Bengal Basin in relation to the regional tectonic framework and basin-fill history. Sediment Geol 155: 179–208.

Banerjee AK, Guo W, Qiao S, Li W, Xing F, Lin Y, Hou Z, Li S, Liu Y, Huang Y (2020) Land masses and oceanic currents drive population structure of *Heritiera littoralis*, a widespread mangrove in the Indo-West Pacific. Ecol Evol 00:1–15. doi: https://doi.org/10.1002/ece3.6460.

Bayraktarov E, Saunders MI, Abdullah S, Mills M, Beher J, Possingham HP, Mumby PJ, Lovelock CE. 2016. The cost and feasibility of marine coastal restoration. Ecol Appl 264: 1055–1074.

Begam MM, Sutradhar T, Chowdhury R, Mukherjee C, Basak SK, Ray K (2017) Native salt-tolerant grass species for habitat restoration, their acclimation and contribution to improving edaphic conditions: a study from a degraded mangrove in the Indian Sundarbans. Hydrobiologia 803: 373–387.

Bosire JO, Dahdouh-Guebas F, Walton M, Crona BI, Lewis RR, Field C, Koedam N (2008) Functionality of restored mangroves: a review. Aquat Bot 89(2): 251–259.

Brown B, Fadilla R, Nurdin Y, Soulsby I, Ahmad R (2014b) Community based ecological mangrove rehabilitation (CBEMR) in Indonesia. SAPIENS 7: 53–64.

Brown B, Yuniati W, Ahmad R, Soulsby I (2014a) Observations of natural recruitment and human attempts at mangrove rehabilitation after seismic (tsunami and earthquake) events in Simulue Island and Singkil Lagoon, Acheh, Indonesia. In Santiago-Fandino V, Kontar YA, Kaneda Y (eds), *Post-Tsunami Hazard Reconstruction and Restoration*. Springer, New York: 311–327.

Bryan-Brown DN, Connolly RM, Richards DR, Adame F, Friess DA, Brown CJ (2020) Global trends in mangrove forest fragmentation. Sci Rep 10: 7117.

Bunting P, Rosenqvist A, Lucas RM, Rebelo LM, Hilarides L, Thomas N, Hardy A, Itoh T, Shimada M, Finlayson CM (2018) The Global Mangrove Watch-A New 2010 Global Baseline of Mangrove Extent. Remote Sens 10: 1669.

Cerón-Souza I, Gonzalez EG, Schwarzbach AE, Salas-Leiva DE, Rivera-Ocasio E, Toro-Perea N, Bermingham E, McMillan WO (2015) Contrasting demographic history and gene flow patterns of two mangrove species on either side of the central American isthmus. Ecol Evol 5: 3486–3499.

Chapman VJ (1975) Mangrove Biogeography. In: Proceedings of the International Symposium on Biology and Management of Mangroves, eds. Walsh GE, Snedaker SC, Teas HJ, University of Florida Press, Gainesville. pp. 3–22.

Chapman VJ (1976) Mangrove vegetation. J. Cramer, Lehre, Germany, 447 pp.

Churchill DM (1973) The ecological significance of tropical mangroves in the early Teniary floras of southern Australia. Geological Society of Australia-Special Publication 4: 79-86.

Cooper WE, Kudo H, Duke NC (2016) *Bruguiera hainesii* C.G.Rogers (Rhizophoraceae), an endangered species recently discovered in Australia. Austrobaileya 9:481–488.

Dale PER Knight JM, Dwyer PG (2014) Mangrove rehabilitation: a review focusing on ecological and institutional issues. Wetl Ecol Manag 22: 587–604.

Duke NC (1984) A mangrove hybrid, *Sonneratia* × *gulngai* (Sonneratiaceae) from north-eastern Australia. Austrobaileya 2: 103–105.

Duke NC (1992) Mangrove floristics and biogeography. In: Robertson AI, Alongi DM (eds) Tropical mangrove ecosystems. American Geophysical Union, Washington DC, USA, pp. 63–100

Duke NC (1994) A Mangrove Hybrid, Sonneratia x urama (Sonneratiaceae) from Northern Australia and Southern New Guinea. Aust Syst Bot 7:521–526.

Duke NC (1995) Genetic diversity, distributional barriers and rafting continents—more thoughts on the evolution of mangroves. Hydrobiologia 295: 167-181.

Duke NC (2010) Overlap of eastern and western mangroves in the South-western Pacific: hybridization of all three *Rhizophora* (Rhizophoraceae) combinations in New Caledonia. Blumea 55:171–188.

Duke NC (2017) Mangrove floristics and biogeography revisited: further deductions from biodiversity hot spots, ancestral discontinuities, and common evolutionary processes. In: Rivera-Monroy VH, Lee SY, Kristensen E, Twilley RR (e) (eds) Mangrove ecosystems: a global biogeographic perspective. Springer, Cham, pp. 17–53.

Duke NC (2020)Asystematic revision of the vulnerable mangrove genus Pelliciera (Tetrameristaceae) in equatorialAmerica. Blumea 65(2):107–120. https://doi.org/10.3767/blumea.2020.65.02.04

Duke NC, Ball MC, Ellison JC (1998a) Factors influencing biodiversity and distributional gradients in mangroves. Glob Ecol Biogeogr 7(1):27–47.

Duke NC, Benzie JAH, Goodall JA, Ballment ER (1998b) Genetic structure and evolution of species in the mangrove genus *Avicennia* (Avicenniaceae) in the Indo-West Pacific. Evolution 52(6):1612–1626.

Duke NC, Bunt JS (1979) The genus *Rhizophora* (Rhizophoraceae) in north-eastern Australia. Aust J Bot 27: 657–678.

Duke NC, Kovacs JM, Griffith A, Preece L, Hill DJ, van Oosterzee P, Mackenzie J, Morning HS, Burrows D (2017) Large-scale dieback of mangroves in Australia's Gulf of Carpentaria: a severe ecosystem response, coincidental with an unusually extreme weather event. Mar Freshw Res doi:https://doi.org/10.1071/MF16322.

Duke NC, Kudo H (2018) *Bruguiera* × *dungarra*, a new hybrid between mangrove species B. exaristata and *B. gymnorhiza (Rhizophoraceae) recently discovered in north-east Australia.* Blumea 63*:279–285*

Ellison AM, Farnsworth EJ, Merkt RE (1999) Origins of mangrove ecosystems and the mangrove biodiversity anomaly. Glob Ecol Biogeogr 8:95–115. https://doi.org/10.1046/j.1466-822X.1999.00126.x

Ellison AM, Felson AJ, Friess DA (2020) Mangrove Rehabilitation and Restoration as Experimental Adaptive Management. Front Mar Sci 7: 327. doi: https://doi.org/10.3389/fmars.2020.00327.

Feller IC, Friess DA, Krauss KW, Lewis III RR (2017) The state of the world's mangroves in the 21st century under climate change. Hydrobiologia 803: 1–12.

Frankham DA, Ballou R, Briscoe JD (2010) Introduction to conservation genetics, 2nd ed. Cambridge University Press, Cambridge.

Ge XJ (2001) Genetic diversity and conservation genetics of mangrove species in South China and Hong Kong. Hong Kong: University of Hong Kong.

Ge XJ, Sun M (2001) Population genetic structure of *Ceriops tagal* (Rhizophoraceae) in Thailand and China. Wetl Ecol Manag 9: 213–219.

Goldberg L, Lagomasino D, Thomas N, Fatoyinbo T (2020) Global declines in human-driven mangrove loss. Global Change Biol 00:1–12. doi: https://doi.org/10.1111/gcb.15275.

Guo Z, Huang YJ, Chen Y, Duke NC, Zhong C, Shi S (2016) Genetic discontinuities in a dominant mangrove *Rhizophora apiculata* (Rhizophoraceae) in the Indo-Malesian region. J Biogeogr 43 (9): 1856–1868.

Guo Z, Li X, He Z, Yang Y, Wang W, Zhong C, Greenberg AJ, Wu C-I, Duke NC, Shi S (2018a) Extremely low genetic diversity across mangrove taxa reflects past sea level changes and hints at poor future responses. Global Change Biol 24:1741–1748.

Guo W, Ng WL, Wu H, Li W, Zhang L, Qiao S, et al. (2018b). Chloroplast phylogeography of a widely distributed mangrove species, *Excoecaria agallocha*, in the Indo-West Pacific region. Hydrobiologia 807: 333–347. doi: https://doi.org/10.1007/s10750-017-3409-7.

Guo Z, Guo W, Wu H, Fang X, Ng WL, et al. (2018c) Differing phylogeographic patterns within the Indo-West Pacific mangrove genus *Xylocarpus* (Meliaceae). J Biogeogr 45: 676–689.

Guo W, Banerjee AK, Ng WL, Yuan Y, Li W and Huang Y (2020) Chloroplast DNA phylogeography of the Holly mangrove Acanthus ilicifolius in the Indo-West Pacific. Hydrobiologia 847:3591–3608.

Hamilton SE, Casey D (2016) Creation of a high spatio-temporal resolution global database of continuous mangrove forest covers for the 21st century (CGMFC-21). Glob Ecol Biogeogr 25 (6): 729–738.

Hamrick JL, Godt MJW, Sherman-Broyles SL (1992) Factors influencing levels of genetic diversity in woody plant species. New For 6: 95–124. https://doi.org/10.1007/BF00120641.

He Z, Li X, Yang M, et al. (2019) Speciation with gene flow via cycles of isolation and migration: insights from multiple mangrove taxa. Natl. Sci. Rev. 6: 275–88.

Hoffmann AA, Sgro CM (2011) Climate change and evolutionary adaptation. Nature 470: 479–485.

Huang L, Li X, Huang V, Shi S, Zhou R (2014) Molecular evidence for natural hybridization in the mangrove genus *Avicennia*. Pak J Bot 46: 1577–1584.

Huang Y, Tan F, Su G, Deng S, He H, Shi S (2008) Population genetic structure of three tree species in the mangrove genus *Ceriops* (Rhizophoraceae) from the Indo West Pacific. Genetica 133: 47–56. doi: https://doi.org/10.1007/s10709-007-9182-1.

Inomata N, Wang X-R, Changtragoon S, Szmidt AE (2009) Levels and patterns of DNA variation in two sympatric mangrove species, *Rhizophora apiculata* and *R. mucronata* from Thailand. Genes Genet Syst 84: 277–286. doi: https://doi.org/10.1266/ggs.84.277.

IUCN and UNEP-WCMC (2019) www.protectedplanet.net.

Kar RK (1985) The fossil floras of Kachchh-IV. Tertiary Palynostratigraphy. Palaeobotanist 34: 1-280.

Kathiresan K (1995) *Rhizophora annamalayana*: a new species of mangroves. Environ Ecol 13: 240–241.

Kathiresan K, Bingham BL (2001) Biology of mangroves and mangrove ecosystems. Adv Mar Biol 40:81–251.

Ko WC (1985) Notes on the genus *Sonneratia* (Sonneratiaceae) in S. E. Asia. Acta Phytotax Sin 23: 311–314.

Krauss KW, McKee KL, Lovelock CE, Cahoon DR, Saintilan N, Reef R, Chen L (2014) How mangrove forests adjust to rising sea level. New Phytol 202: 19–34.

Lee SY, Hamilton S, Barbier EB, Primavera J, Lewis III RR (2019) Better restoration policies are needed to conserve mangrove ecosystems. Nat Ecol Evol 3: 870–872.

Lee SY, Primavera JH, Dahdouh-Guebas F, McKee K, Bosire JO, Cannicci S, Diele K, Fromard F, Koedam N, Marchand C, Mendelssohn I, Mukherjee N, Record S (2014) Ecological role and services of tropical mangrove ecosystems: a reassessment. Glob Ecol Biogeogr 23: 726–743.

Lewis RR, Brown BM, Flynn LL (2019) "Methods and criteria for successful mangrove forest rehabilitation," *In Coastal Wetlands: An Integrated Ecosystem Approach*, 2nd Edn, eds G. Perillo ME, Wolanski E, Cahoon DR, Hopkinson CS (Amsterdam: Elsevier), pp. 863–887. doi: https://doi.org/10.1016/B978-0-444-63893-9.00024-1.

Lewis RR, Hodgson AB, Mauseth GS (2005) Project facilitates the natural reseeding of mangrove forests (Florida). Ecol Restor 23: 276–277.

Lewis RR, Milbrandt EC, Brown B, Krauss KW, Rovai AS, Beever JW, Flynn LL (2016) Stress in mangrove forests: early detection and pre-emptive rehabilitation are essential for future successful worldwide mangrove forest management. Mar Pollut Bull 109: 764–771.

Li J, Yang Y, Chen Q, Fang L, He Z, Guo W, Qiao S, Wang Z, Guo M, Zhong C, Zhou R, Shi S (2016) Pronounced genetic differentiation and recent secondary contact in the mangrove tree *Lumnitzera racemosa* revealed by population genomic analyses. Sci Rep 6: 29486. doi: https://doi.org/10.1038/srep29486.

Liao PC, Havanond S, Huang S (2007) Phylogeography of *Ceriops tagal* (Rhizophoraceae) in Southeast Asia: The land barrier of the Malay Peninsula has caused population differentiation between the Indian Ocean and South China Sea. Conserv Genet 8: 89–98. doi: https://doi.org/10.1007/s10592-006-9151-8.

Lira-Medeiros CF, Parisod C, Fernandes RA, Mata CS, Cardoso MA, Ferreira PCG (2010) Epigenetic variation in mangrove plants occurring in contrasting natural environment. PLoS ONE 5: e10326.

Lo EY, Duke NC, Sun M (2014) Phylogeographic pattern of *Rhizophora* (Rhizophoraceae) reveals the importance of both vicariance and long-distance oceanic dispersal to modern mangrove distribution. BMC Evol Biol 14: 83. doi:https://doi.org/10.1186/1471-2148-14-83.

Lomolino MV, Riddle BR, Brown JH (2006) Biogeography, 3rd edn. Sinauer, Sunderland, MA.

López-Portillo J, Lewis RR, Saenger P, Rovai A, Koedam N, Dahdouh-Guebas F, Agraz-Hernández C, Rivera-Monroy VH (2017) Mangrove forest restoration and rehabilitation. In: Rivera-Monroy VH, Lee SY, Kristensen E, Twilley RR (Eds.), *Mangrove Ecosystems: A Global Biogeographic Perspective*. Springer, New York, New York, USA, pp. 301–345

Lovelock CE, Cahoon DR, Friess DA, Guntenspergen GR, Krauss KW, Reef R, Rogers K, Saunders ML, Sidik F, Swales A, Saintilan N, Thuyen AX, Triet T (2015) The vulnerability of Indo-Pacific mangrove forests to sea-level rise. Nature 526: 559–563.

Lovelock CE, Feller IC, Reef R, Hickey S, Ball MC (2017) Mangrove dieback during fluctuating sea levels. Sci Rep 7: 1680.

Ludt WB, Rocha LA (2014) Shifting seas: the impacts of Pleistocene sea-level fluctuations on the evolution of tropical marine taxa. J Biogeogr 42(1): 25–38.

Lyu H, He Z, Wu C-I, Shi S (2018) Convergent adaptive evolution in marginal environments: unloading transposable elements as a common strategy among mangrove genomes. New Phytol 217: 428–438.

Maguire TL, Saenger P, Baverstock P, Henry R (2000) Microsatellite analysis of genetic structure in the mangrove species *Avicennia marina* (Forsk.) Vierh. (Avicenniaceae). Mol Ecol 9:1853–1862.

Mathur YK (1984) Cenozoic palynofossils, vegetation, ecology and climate of the North and North-western Sub-Himalayan region. In R. O. Whyte (Ed.), The Evolution of the East Asian Environment, 2 (pp. 504–551). Hong Kong, China: Centre of Asian Studies, Univ. Hong Kong.

Mathur YK, Chowdhary LR (1977) Palaeoecology of the Kalol Formation, Cambay Basin, India. In B. S. Venkatachala, & V. V. Sastri (Eds.), (pp. 164–178). Proceedings of the 4th Colloquium on Indian Micropalaeontology and Stratigraphy, Dehradun, 1974–75. Dehradun, Uttarakhand: Institute of Petroleum Exploration, Oil and Natural Gas Commission.

McCoy ED, Heck KL Jr (1976). Biogeography of corals, seagrasses, and mangroves: An alternative to the center of origin concept. Syst Zool 25:201–210. doi: https://doi.org/10.2307/2412488.

Mepham RH (1983) Mangrove floras of the southern continents. Part 1. The geographical origin of Indo-Pacific mangrove genera and the development and present status of the Australian mangroves. South Afr J Bot 2: 1–8.

Minobe S, Fukui S, Saiki R, Kajita T, Changtragoon S, Shukor NAA, Latiff A, Ramesh BR, Koizumi O, Yamazaki T (2010) Highly differentiated population structure of a Mangrove species, *Bruguiera gymnorhiza* (Rhizophoraceae) revealed by one nuclear Gap Cp and one chloroplast intergenic spacer trnF–trnL. Conserv Genet 11: 301–310.

Mori GM, Zucchi MI, Sampaio I, Souza AP (2015) Species distribution and introgressive hybridization of two *Avicennia* species from the Western Hemisphere unveiled by phylogeographic patterns. BMC Evol Biol 15: 1–15.

Ono J (2016) Conservation genetic study of mangrove plant genus *Bruguiera*. Ph D thesis, Chiba University.

Ono J, Yong JWH, Takayama K, Saleh MNB, Wee AKS, Asakawa T, Yllano OB, Salmo SG, Meenakshisundaram SH, Watano Y, Webb EL, Kajita T (2016) *Bruguiera hainesii*, a critically endangered mangrove species, is a hybrid between *B. cylindrical* and *B. gymnorhiza* (Rhizophoraceae). Conserv Genet 17:1137–1144. doi: https://doi.org/10.1007/s10592-016-0849-y.

Qiu S, Zhou RC, Li YQ, Havanond S, Jaengjai C, Shi SH (2008) Molecular evidence for natural hybridization between *Sonneratia alba* and *S. griffithii*. J Syst Evol 46: 391–395.

Ragavan P, Saxena A, Jayaraj RSC, Ravichandran K, Saravanan S (2015). *Rhizophora × mohanii*: a putative hybrid between *Rhizophora mucronata* and *Rhizophora stylosa* from mangroves of the Andaman and Nicobar Islands, India. ISME/GLOMIS Electronic Journal 13: 3–7.

Ragavan R, Saxena A, Jayaraj RSC, Mohan PM, Pavichandran K, Saravanan S, Vijayaraghavan A (2016) A review of the mangrove floristics of India. Taiwania 61:224–242.

Renema W, Bellwood DR, Braga JC, Bromfield K, Hall R et al (2008) Hopping hotspots: global shifts in marine biodiversity. Science 321:654–657.

Rey JR, Carlson DB, Brockmeyer RE (2012) Coastal wetland management in Florida: environmental concerns and human health. Wetl Ecol Manag 20: 197–211.

Riddle BR, Dawson MN, Hadly EA, Hafner DJ, Hickerson MJ, Mantooth SJ, Yoder AD (2008) The role of molecular genetics in sculpting the future of integrative biogeography. Prog Phys Geogr 32: 173–202.

Saenger P, Hegerl EJ, Davie JDS (1983) Global Status of Mangrove Ecosystems. The Environmentalist 3 (Supplement): 1–88.

Saenger P, Ragavan P, Sheue CR, López-Portillo J, Yong JWH, Mageswaran T (2019) Mangrove Biogeography of the Indo-Pacific. In Sabkha Ecosystems Volume VI: Asia/Pacific, Gul B, Böer B, Khan, MA, Cluesener-Godt M, Hameed A (Eds.) Chapter 23, pp. 379–400.

Saintilan N, Wilson NC, Rogers K, Rajkaran A, Krauss KW (2014). Mangrove expansion and salt marsh decline at mangrove poleward limits. Glob Change Biol 20: 147–157.

Sheue CR, Liu HY, Tsai CC, Rashid SMA, Yong JWH, Yang YP (2009b) On the morphology and molecular basis of segregation of two species *Ceriops zippeliana* Blume and *C. decandra* (Griff.) Ding Hou (Rhizophoraceae) from southeastern Asia. Blumea 54:220–227

Sheue CR, Liu HY, Tsai CC, Yang YP (2010) Comparison of *Ceriops pseudodecandra sp. nov.* (Rhizophoraceae), a new mangrove species in Australasia, with related species. Bot Stud 51:237–248.

Sheue CR, Liu HY, Yong JWH (2003) *Kandelia obovata* (Rhizophoraceae), a new mangrove species from Eastern Asia. Taxon 52:287–294

Sheue CR, Yang YP, Liu HY, Chou FS, Chang SC, Saenger P, Mangion CP, Wightman G, Yong JWH, Tsai CC (2009a) Reevaluating the taxonomic status of *Ceriops australis* (Rhizophoraceae) based on morphological and molecular evidence. Bot Stud 50:89–100

Sheue CR, Yong JWH, Yang YP (2005) The *Bruguiera* (Rhizophoraceae) species in the mangroves of Singapore, especially on the new record and the rediscovery. Taiwania 50:251–260

Specht RL (1981) Biogeography of halophytic angiosperms (saltmarsh, mangrove and sea-grass). In A. Keast (ed.), Ecological Biogeography of Australia. W. Junk, The Hague.

Srivastava J, Prasad V (2019) Evolution and paleobiogeography of mangroves. Mar. Ecol. 00: e12571. doi:https://doi.org/10.1111/maec.12571.

Srivastava SK, Binda PL (1991) Depositional history of the early eocene shumaysi formation, Saudi Arabia. Palynology 15: 47–61.

Stubbs BJ, Saenger P (2002). The application of forestry principles to the design, execution and evaluation of mangrove restoration projects. Bois et Foréts des Tropiques 56: 5–21.

Su G, Huang Y, Tan F, Ni X, Tang T, Shi S (2007) Conservation genetics of *Lumnitzera littorea* (Combretaceae), an endangered mangrove, from the Indo-West Pacific. Mar Biol 150: 321–328. doi: https://doi.org/10.1007/s00227-006-0357-6.

Su GH, Huang YL, Tan FX, Ni XW, Tang T, Shi SH (2006) Genetic variation in *Lumnitzera racemosa*, a mangrove species from the Indo-West Pacific. Aquat Bot 84: 341–346. https://doi.org/10.1016/j.aquabot.2006.01.001.

Sun M, Lo EY (2011) Genomic markers reveal introgressive hybridization in the Indo-West Pacific mangroves: a case study. PLoS One 6: e19671.

Takayama K, Tamura M, Tateishi Y, Webb EL, Kajita T (2013) Strong genetic structure over the American continents and transoceanic dispersal in the mangrove genus *Rhizophora* (Rhizophoraceae) revealed by broad-scale nuclear and chloroplast DNA analysis. Am J Bot 100: 1191–1201.

Tan F, Huang Y, Ge X, Su G, Ni X, Shi S (2005) Population genetic structure and conservation implications of *Ceriops decandra* in Malay Peninsula and North Australia. Aquat Bot 81:175–188. https://doi.org/10.1016/j.aquabot.2004.11.004.

Thanikaimoni G (1987) *Mangrove palynology*. Tr Sec. Sci Tech. Institute of French Pondichery 24: 100.

Thomas N, Lucas R, Bunting P, Hardy A, Rosenqvist A, Simard M (2017) Distribution and drivers of global mangrove forest change, 1996–2010. PLoS ONE 12(6): e0179302. doi: https://doi.org/10.1371/journal.pone.0179302.

Tomizawa Y, Tsuda Y, Saleh M, Wee A, Takayama K, Yamamoto T, et al. (2017) Genetic structure and population demographic history of a widespread mangrove plant *Xylocarpus granatum* J. Koenig across the Indo-West Pacific Region. Forests 8: 480. doi: https://doi.org/10.3390/f8120480.

Tomlinson BP, Womersley JG (1976) A *Rhizophora* new to Queensland and New Guinea, with notes relevant to the genus. Contrib Herb Aust 19: 1–10.

Tomlinson PB (1978) Rhizophora in Australasia—some clarification of taxonomy and distribution. J Arnold Arbor 59: 156–169.

Tomlinson PB (1986) The Botany of Mangroves. Cambridge University Press, Cambridge, 413 pp.

Tomlinson PB (2016) The botany of mangroves, 2nd edn. Cambridge University Press, Cambridge.

Tomlinson PB, Bunt JS, Primack RB, Duke NC (1978) *Lumnitzera* × *rosea* (Combretaceae)—its status and floral morphology. J Arnold Arbor 59: 342–351.

Tsai CC, Li SJ, Su YY, Yong JWH, Saenger P, Chesson P, Das S, Wightman G, Yang YP, Liu HY, Sheue CR (2012) Molecular phylogeny and evidence for natural hybridization and historical introgression between *Ceriops* species (Rhizophoraceae). Biochem Syst Ecol 43: 178–191.

Urashi C, Teshima KM, Minobe S, Koizumi O, Inomata N (2013) Inference of evolutionary history of a widely distributed mangrove species, *Bruguiera gymnorrhiza*, in the Indo-West Pacific region. Ecol Evol 3: 2251–2261.

Van Steenis CGGJ (1962) The distribution of mangrove plant genera and its significance for Palaeogeography. Nederl. Akad. Von Wetenschappen 65:164–169.

Venkatachala BS, Rawat MS (1972) Palynology of the Tertiary sediments in the Cauvery basin 1. Palaeocene-Eocene palynoflora from the subsurface. In Proceedings of the Seminar on Palaeopalynology and Indian Stratigraphy, Calcutta (pp. 292–335).

Wang PX, Wang LJ, Bian YH, Jian ZM (1995) Late Quaternary paleoceanography of the South China Sea: surface circulation and carbonate cycles. Mar Geol 127:145–165.

Wang RJ, Chen ZY, Chen EY, Zheng XR (1999) Two hybrids of the genus *Sonneratia* (Sonneratiaceae) from China. Guihaia 19: 199–204.

Wee A, Teo J, Chua J, Takayama K, Asakawa T, Meenakshisundaram S, et al. (2017) Vicariance and oceanic barriers drive contemporary genetic structure of widespread mangrove species *Sonneratia alba* J. Sm in the Indo-West Pacific. Forests 8: 483. doi: https://doi.org/10.3390/f8120483.

Wee AKS, Mori GM, Lira CF, Núñez-Farfán J, Takayama K, Faulks L, Shi S, Tsuda Y, Suyama Y, Yamamoto T, Iwasaki T, Nagano Y, Wang Z, Watanabe S, Kajita T (2018) The integration and application of genomic information in mangrove conservation. Conserv Biol 33: 206–209.

Wee AKS, Noreen AME, Ono J, et al. (2020) Genetic structures across a biogeographical barrier reflect dispersal potential of four Southeast Asian mangrove plant species. J Biogeogr 00: 1–14. doi: https://doi.org/10.1111/jbi.13813.

Wee AKS, Takayama K, Asakawa T, Thompson B, Sungkaew S, Tung NX et al. (2014) Oceanic currents, not land masses, maintain the genetic structure of the mangrove *Rhizophora mucronata* Lam. (Rhizophoraceae) in Southeast Asia J Biogeogr 41: 954–964.

Wee AKS, Takayama K, Chua JL, Asakawa T et al. (2015) Genetic differentiation and phylogeography of partially sympatric species complex *Rhizophora mucronata* Lam. and *R. stylosa* Griff. using SSR markers. BMC Evol Biol 15: 57. https://doi.org/10.1186/s12862-015-0331-3.

Woodroffe CD, Rogers K, McKee KL, Lovelock CE, Mendelssohn IA, Saintilan N (2016) Mangrove sedimentation and response to relative sea-level rise. Ann Rev Mar Sci 8: 243–266.
Worthington T, Spalding M (2019) Mangrove restoration potential: A global map highlighting a critical opportunity. 34pp.
Wyrtki K (1961) Scientific results of marine investigations of the South China Sea and the Gulf of Thailand 1959–1961. NAGA report 2. pp. 164–169.
Xie W, Zhong G, Li X, Guo Z, Shi S (2020) Hybridization with natives augments the threats of introduced species in *Sonneratia* mangroves. Aquat Bot 160: 103166.
Yan Y-B, Duke NC, Sun M (2016) Comparative Analysis of the Pattern of Population Genetic Diversity in Three Indo-West Pacific *Rhizophora* Mangrove Species. Front Plant Sci 7: 1434. doi: https://doi.org/10.3389/fpls.2016.01434.
Yang Y, Duke NC, Peng F, Li J, Yang S, Zhong C, Zhou R, Shi S (2016) Ancient Geographical Barriers Drive Differentiation among *Sonneratia caseolaris* Populations and Recent Divergence from *S. lanceolata*. Front Plant Sci 7:1618. doi: https://doi.org/10.3389/fpls.2016.01618.
Yang Y, Li J, Yang S, Li X, Fang LU, Zhong C, et al. (2017) Effects of Pleistocene sea-level fluctuations on mangrove population dynamics: A lesson from *Sonneratia alba*. BMC Evol Biol 17: 22. doi: https://doi.org/10.1186/s12862-016-0849-z.
Zhang R, Liu T, Wu W, Li Y, Chao L, Huang L, Huang Y, Shi S, Zhou R (2013) Molecular evidence for natural hybridization in the mangrove fern genus *Acrostichum*. BMC Plant Biol 13: 74.
Zhang ZH, Zhou RC, Tang T, Huang YL, Zhong Y, Shi SH (2008) Genetic variation in central and peripheral populations of *Excoecaria agallocha* from Indo-West Pacific. Aquat Bot 89: 57–62.
Zhong C, Li D, Zhang Y (2020) Description of a new natural *Sonneratia* hybrid from Hainan Island, China. PhytoKeys 154: 1–9.

Mangroves as Feeding and Breeding Grounds

3

D. Arceo-Carranza, X. Chiappa-Carrara, R. Chávez López, and C. Yáñez Arenas

Abstract

Mangroves are considered as ecosystems that provide shelter, food and breeding grounds for many groups of inhabiting fauna. Much of the fauna present are organisms in different stages of their life cycle, mostly juveniles. The three-dimensional structure of the mangrove roots and the combination of the aquatic and terrestrial environments are factors that bring together a great diversity. Such diversity within mangrove sites includes aquatic and terrestrial vertebrates and invertebrates such as fish, amphibians, reptiles, mammals and birds. Present within the fauna are representatives of different trophic guilds that perform key functions in the ecosystem, such as pollination, seed dispersal and nutrient recirculation. The food produced by the ecosystem is based on the production of detritus caused by leaf litter and its decomposition, where transformation of energy and accumulation of biomass for higher trophic levels begins with the

D. Arceo-Carranza (✉)
Facultad de Ciencias, Unidad Multidisciplinaria de Docencia e Investigación Sisal, Universidad Nacional Autónoma de México, Sisal, Yucatán, Mexico
e-mail: darceo@ciencias.unam.mx

X. Chiappa-Carrara
Facultad de Ciencias, Unidad Multidisciplinaria de Docencia e Investigación Sisal, Universidad Nacional Autónoma de México, Sisal, Yucatán, Mexico

Escuela Nacional de Estudios Superiores Mérida, Universidad Nacional Autónoma de México, Mérida, Yucatán, Mexico

R. Chávez López
Facultad de Estudios Superiores Iztacala, Universidad Nacional Autónoma de México, Mexico City, Mexico

C. Yáñez Arenas
Facultad de Ciencias, Unidad Multidisciplinaria de Docencia e Investigación Sisal, Sede Parque Científico, Universidad Nacional Autónoma de México, Mérida, Mexico

R. P. Rastogi et al. (eds.), *Mangroves: Ecology, Biodiversity and Management*,
https://doi.org/10.1007/978-981-16-2494-0_3

invertebrates. With regard to breeding activity, many fish families spawn in the mangrove roots (e.g. Fundulidae), or nest in the canopy (herons and cormorants) that provides protection against predators and food for juvenile organisms. Undoubtedly, mangroves function as feeding and breeding grounds that are essential in maintaining populations of marine organisms, especially fish. Many fish species that grow in the mangroves are important for fisheries. Unequivocally, maintaining these breeding sites for marine and terrestrial fauna is crucial to the general functioning of adjacent ecosystems.

Keywords

Guilds · Nursery · Life stage · Juvenile · Refuge · Trophic ecology

3.1 Introduction

Coasts throughout the world are comprised of ecosystems formed by vegetation types where environmental adaptations are necessary for life in an environment with constant changes in hydrological characteristics, prominently, salinity and flooding. These halophytes from the mangrove ecosystems nurture a great diversity worldwide. The niche diversity of this ecosystem occurs at the aquatic, sediment, roots, trunk, and foliage level, serving as a habitat and food source to many animal species. Therefore, mangrove fauna is diverse due to the geographic, hydrological and climatic characteristics as well as ecological processes that occur in these ecosystems. This chapter will discuss important mangrove characteristics that make this ecosystem an important habitat for fauna, in relation to food availability, and as a nursery for several species during different stages of their life cycles, making mangroves an essential habitat.

3.2 Mangroves, a Habitat for Fauna

Mangrove ecosystems are considered as a type of coastal wetland dominated by woody plants, that are distributed on particular latitudinal gradients (30°N to 37°S), tidal height (<1 m to >4 m), with particular geomorphology (from continental river systems to oceanic islands), various sediment composition (from organic peat to alluvial zones), to types of climate (from temperate warm, arid zones to humid tropics), various nutrient availability (from oligotrophic to eutrophic levels), and with soils that are principally anaerobic. Their success is the result of highly developed morphological and physiological adaptations.

All these characteristics categorize mangroves as biocomplex systems (Feller et al. 2010) due to the processes that occur at the individual, population and ecosystem level, as well as their natural biotic and abiotic interactions. Mangroves have been characterized over time as being very productive sites that hold great diversity, this is because they are located on the borders between marine and terrestrial ecosystems. In this coastal ecotone, we can find fauna with distinctive features, those of saltwater and freshwater origins, or fauna that adapt and tolerate fluctuating environmental conditions (e.g. salinity, flooding, pH and dissolved oxygen concentration in sediments). Inhabiting the mangrove is local fauna known as permanent residents, found at birth, growing, reproducing and dying within this ecosystem. Also, there is fauna that can be considered as temporary visitors, inhabiting mangroves for short periods, such as migratory bird species (e.g. herons and pelicans), fish (e.g. grunts and snappers), marine invertebrates (e.g. shrimp) and amphibians, to mention a few. These species use mangroves to reproduce, spawn or as resting sites (Kathiresan and Bingham 2001; Nagelkerken et al. 2008; De Dios et al. 2019). Occasional or facultative visitors can be found in mangroves due to its high food source, but these species do not require mangroves to fulfill their life cycle. These systems are considered relicts for endemic species or for species with a need for protection (Luther and Greenberg 2009). Ecologically, mangroves are crucial for estuarine ecosystems, and are one of the most utilized and threatened habitats due to anthropogenic activities, as increased degradation in the last 30 years has caused a 35% loss of mangrove surface (Barbier et al. 2011). The most deteriorated ecosystem functions are the nursery areas (consequently, causing the collapse of some coastal fisheries), filtration and purification (services provided by detritus feeders, e.g. bivalves, barnacles and sponges), decreased water quality, loss of biodiversity and their role in protecting the coastline (Worm et al. 2006).

The connectivity of this ecosystem is a feature resulting from the interactions between the mainland and geomorphology, hydrology, weather and tidal systems, structural characteristics, accessibility to fauna, and the borders of the mangrove forests; altogether this stimulates the exchange between the ecosystem limits (Nagelkerken et al. 2008). This connectivity not only contributes to the economic value and ecosystem services, but also increases its vulnerability to natural and human disturbances (Alongi 2008).

3.2.1 Canopy

Found in the aerial section of mangroves, comprising a considerable portion of the fauna, are the insects, either as permanent residents or temporary visitors, their presence is important given that they carry out different functions in this dynamic ecosystem. Due to their mobility, many insect species are temporary visitors of mangroves, and thus create links to other habitats they inhabit (Balakrishnan et al. 2014). The mangrove canopy sustains terrestrial fauna, which as in other ecosystems is dominated by insects, but also includes spiders, crabs and members of all terrestrial invertebrates. However, only a few endemic species are reported (Luther and Greenberg 2009). Brooks and Bell (2001) indicate that many species are subgroups of terrestrial fauna that scatter in the mangrove by flying, swimming or even floating on debris. Accordingly, species that feed on flowers and fruits such as frugivores, folivores and palynivores can be found (e.g. butterflies, bees, crickets, bats, rodents, birds, among others). Although the presence of dragonflies, saddles and termites is common (Balakrishnan et al. 2018), a considerable number of studies have focused on bees, ants and mosquitos, and as per the former, honey produced in mangroves is valued in places such as India, the Caribbean, Florida and Brazil (Fernandes et al. 2011).

In Panama, Adams (1994) demonstrated that four species of ants divided the resources of the mangrove canopy by marking territories with a combination of tactile and chemical pheromone signals, whereas other animals (e.g. birds, reptiles) use the upper part of mangroves for resting or as a refuge site (Gopal and Chauhan 2006). In Mexico, De Dios et al. (2019) reported five species of marine migratory birds that nest in the mangrove canopy of the Yucatan Peninsula.

3.2.2 Trunks and Branches

Terrestrial insects that inhabit mangroves face severe environmental conditions such as insolation, high temperatures and high probabilities of desiccation. Nonetheless, insects as well as other arthropods deal with these adversities through nocturnal behaviour or by living inside plants. Moths and beetles in Belize have been found to dig tunnels through mangroves, a habitat modification used by more than 70 species of ants, spiders, mites, moths, cockroaches, termites and scorpions (Feller 2002). Insects and spiders also temporarily use trunks and branches in flooding events, while other species (e.g. isopods, amphipods and myriapods) show affinity for trees in the intertidal zone to avoid desiccation (Feller and Mathis 1997). The feeding habits of some organisms such as herbivorous insects can harm the mangrove vegetation through drilling of the bark and wood, they also cause defoliation, and seedlings can be vulnerable when growing next to adult trees (Jenoh et al. 2016).

One of the reasons mangroves are disregarded and unpopular is due to them harbouring large populations of mosquitos; holes inside tree barks made by insects and crab burrowers are ideal sites, particularly in species of *Avicennia* spp. (Ismail et al. 2018). The insects that inhabit mangroves can be grouped as cryptic and endophytic species, e.g. miners, borers and gall creators. In the intertidal zone substrates such as trunks, aerial roots and muddy saline plains are used by some benthonic organisms.

3.2.3 Roots

Considering that mangroves are surrounded by sand or lime sediments, with roots, branches and trunks, converts them into ecological islands that are an attractive habitat for a variety of epifaunal communities (Guerra-Castro et al. 2011). This type of biota can include an assemblage of different invertebrates, such as sponges, hydroids, anemones, polychaetes, bivalves, barnacles, bryozoans and ascidians; some of these organisms show morphological adaptations to mangrove life, including a display of different morphology when inhabiting roots (Díaz et al. 2004).

A basic function of the mangrove roots at the water–sediment interface is to recycle nutrients, these chemical compounds are carried by freshwater currents and tidal circulation patterns, also included, are inorganic particles, organic matter, sediments and other contaminants.

Found associated with the roots is a community of various bacteria, that are involved in the decomposition of organic matter. These bacterial communities occupy different niches and are important elements for the mangrove to function, since they control all the chemical aspects. For instance, sulphate reducing bacteria *Desulfovibrio*, *Desulfotomaculum*, *Desulfosarcina* and *Desulfococcus* are the first participants in the decomposition processes. These bacteria control the dynamics of the ions, iron, phosphorous and sulphur, which contribute to soil formation and vegetation patterns (Varon-Lopez et al. 2014). Methanogenic cyanobacteria are temporally abundant in *Avicennia* species and are critical for recycling nitrogen, and as components of the mangrove microbiota, as they provide an important nitrogen source (Arai et al. 2016). The function of aerial roots is to influence the tidal flow rate, which determines the sedimentation and particle retention rates. The variation in water flow and height of the water column affects the availability of food particles for filtering organisms such as zooplankton (Srikanth et al. 2016).

In the entangled roots below water are spaces or tunnels shaped by the tidal level and root system, where a zooplankton community can be found with abundances that can be very high in densities of up to 10,000 organisms/m^3 and 623 mg/m^3 of biomass, surpassing records obtained from marine waters (Saravanakumar et al. 2007). Another important group that finds refuge and food between the roots is the fish community, mostly juvenile marine fish that use this habitat for a short period of their life cycle.

3.2.4 Sediments

The basic structure and functional attributes of the mangrove ecosystem are usually represented by models that show the main external energy sources and the pressures that affect them (Lugo and Snedaker 1974; Mukherjee et al. 2014). In these representations, the ecosystem is divided into two compartments. The first, groups the structures found above the ground, and the second includes, what remains flooded for most of the time (i.e. roots and sediments) with aerobic and anaerobic processes occurring as well. In these models a high primary productivity is recognized for mangrove ecosystems, with a continuous nutrient supply essential for plant growth. Besides availability, nutrient requirements are possible due to efficient capture, absorption and nutrient recycling systems (Woodroffe 1992; Alongi et al. 1993; Alongi 1994; Kristensen et al. 1994).

Mangrove systems offer a wide range of benthonic substrates for the infauna (Hsieh 1995), from poorly consolidated sludge, sand mixed with detritus from the mangrove itself or neighbouring ecosystems, to hard substrates such as roots and other supporting structures (Alongi 1989; Day et al. 2013). The complexity of this habitat enables the assumption that mangroves host a great diversity of communities (Eller and Grassle 1992).

Three main carbon sources available for invertebrates are recognized in the surface layers of the sediment: mangrove leaf litter, plant detritus from neighbouring systems and microphytobenthos. However, reports by some authors (Bouillon et al. 2002; Oakes et al. 2010) show that mangroves provide for carbon assimilation which is important to a limited number of species. The production of local and imported microphytobenthos is the most important carbon source of the invertebrate benthonic communities within the intertidal mangroves. Nonetheless, most studies conclude that mangroves are the carbon source that sustains the dominant communities, both in numbers and biomass (e.g. crabs and gastropods) (Wells 1984; Camilleri 1992).

These species play an important role in leaf litter degradation, unlike other components of the meiofauna which are normally found low in abundance in the mangrove sediments (Alongi 1987). Small-sized infauna from mangrove sediments is not well studied, even though these organisms might play an important role in the food web, being an important food source for fish and crustaceans that use mangroves as a nursery area (Daniel and Robertson 1990; Dittmann 2000).

Several factors associated with hydrodynamics are responsible for the presence and abundance of the infauna in mangrove sediments. Other biological factors such as competition with epifauna, predation, food quality and the chemical defence from mangroves (Alongi 1989) should be studied, in order to explain spatial distribution patterns and species composition. Recently, taxonomic diversity in water and mangrove sediment has been explored using next generation sequencing technology, as well as, isotope techniques to demonstrate the rich diversity of bacteria consortia, diverse in metabolic functions (Pascal et al. 2016).

3.2.5 The Aquatic Environment

The hydrological characteristics in mangroves vary according to season; these changes can follow a daily cycle relative to the tides, or stationary, according to precipitation input during the rainy season. The main change is observed in the salinity gradient, and as a result, fauna distributes itself according to osmoregulatory capacity.

A great variety of marine invertebrates in juvenile phases (zooplankton) can be found in the water that floods the mangrove, some of these species return to the sea as adults to continue forming part of zooplankton (e.g. Mysidacea) or turn towards a marine benthic life (e.g. shrimps, gastropods), whereas others such as encrusting species (e.g. oysters and barnacles) settle in estuarine waters to grow and develop (Guerra-Castro et al. 2011). The zones permanently flooded are used by some fish species as a refuge during daylight, and move towards other sites during the night to feed (Kimirei et al. 2011; Ramírez-Martínez et al. 2016).

Turbid waters high in quantities of suspended solids are another feature that offers protection and food for juvenile species, with notable migrations to these waters observed in crustaceans and distinct fish species, according to ecological hypotheses mentioned by various authors (Dorenbosch et al. 2004; Wang et al. 2009; Arceo-Carranza et al. 2016). This migratory behaviour by some species of fish, and use of the mangrove during early, or juvenile life stages is known as estuarine dependence (Nagelkerken et al. 2001; Dorenbosch et al. 2004; Able 2005; Kimirei et al. 2011); some examples are *Ocyurus chrysurus*, *Haemulon flavolineatum*, *Gerres cinereus*, *Lagodon rhomboides* and *Elops saurus* (Nagelkerken et al. 2001; Able 2005). This behaviour can be facultative or obligatory, depending on the biology of the species, or due to temporality and hydrological characteristics of the system.

3.3 Mangrove as Food Resources

Mangroves are one of the most productive ecosystems worldwide (Robertson and Alongi 1992). Their high productivity is reflected on its diverse food resources, such as leaves, flowers, seeds and propagules, and a high detritus production generated by leaf litter decomposition. Mangrove ecosystems are constantly defoliating their trees, and thus, detritus production occurs throughout the year. Due to this amount of resources, it is possible to find diverse trophic guilds within mangroves. Among the primary consumers, we can find palynivores, herbivores, frugivores and folivores, as well as detritivores and omnivores that obtain their energy from the system, which is transferred to carnivores and ichthyophages at higher levels. Overall, these food resources form the base of the food chain in mangrove ecosystems. According to Azam et al. (1983), an essential element in mangrove productivity is the *microbial loop* defined as a group of microorganisms that transform particle and dissolved organic matter into food for higher trophic levels. Therefore, bacteria, viruses, fungi and protozoans play an important role in nutrient assimilation and detritus production, which is the base of the food web responsible in supporting secondary productivity for mangroves (Pascal et al. 2016).

3.3.1 Primary Production

The organic matter produced by mangroves has been used to estimate its primary production, which is directly correlated to biomass, i.e. density of trees and canopy regulated by environmental variables such as precipitation, temperature and nutrient availability (Vitousek and Sanford 1986).

Primary production in mangroves comprised of biomass (organic matter) generated by mangrove trees (including roots and pneumatophores), seedlings and periphyton (a biofilm of organisms contained in a polysaccharide matrix). Mangrove productivity is magnified by the microbial associations occurring in the sediments, resulting in recycling and biogeochemical processes for better nutrient use within the ecosystem. Bacterial communities help to sustain mangroves, as well as in productivity, based on three mechanisms: (1) mineralization of organic matter under anaerobic and microaerophilic conditions mostly by sulphur-reducing bacteria;

(2) increase in the rate of nitrogen fixation and (3) bacterial symbiosis at the rhizosphere that provide nutrients and regulate growth substances (Holguin and Bashan 2007).

Benthonic macrofauna such as crabs (e.g. Grapsid and Sesarma) also have an important role in the leaf litter dynamics, in the cycling of nutrients, energy flow, and consequently, in the productivity of the ecosystem (Smith III et al. 1991). Predatory interactions between herbivores and non-herbivores (e.g. interaction between isopods and barnacles) have a positive influence on the ecosystem's productivity, since predation between species controls populations of organisms that can feed on propagules, causing a decrease in production and growth of mangrove roots (Perry 1988).

3.3.2 Detritus

Detritus production in mangrove ecosystems is the main energy source and most detritus in mangroves come from leaf litter. Detritus is defined as all types of biogenic material at various decomposition stages by microbes (McLusky and Elliott 2004). Some authors have demonstrated transfer of carbon to the meiofauna and macroinvertebrates (Oakes et al. 2010); both fauna are important links between producer and consumer levels in this system, as they contribute by adding nitrogen waste to the sediments, intervene with organic matter decomposition, and serve as prey for predatory invertebrates, fish and birds. Different authors agree that detritus is a key element in the diet of many mangrove species (Nagelkerken et al. 2008; Oakes et al. 2010; Pascal et al. 2016). Holguin and Bashan (2007) mentioned that from 120 mangrove species, one-third feeds on detritus, serving as a link to energetic interactions in the ecosystem; some species are penaeid shrimp, insect larvae, nematodes, oysters, polychaetes, isopods, bivalves and some fish families (e.g. Mugilidae). Additionally, some authors (Odum and Heald 1972) have noted that detritus is a food source that is exported to other neighbouring ecosystems such as sea grasses and coral reefs through suspended particles.

3.3.3 Secondary Production

Mangroves are sites with high secondary productivity due to the growth of inhabiting fauna. Although mangrove productivity is higher on coastal and riverine forests, many species are associated to subtidal habitats. Island mangroves with distinguished clear water and fewer salinity shifts have groups of species rich in sessile epibionts (e.g. algae, sponges, tunicates and anemones), where numerous parasitic and mutualistic associations have been detected (Guerra-Castro et al. 2011). Species diversity in mangroves can also be explained through habitat varieties created by the heterogenous structures of these plants. Even though these ecosystems are recognized with low species richness, studies have shown the importance of interactions between fauna and vegetation for forest development, productivity and structural complexity (Nagelkerken et al. 2008).

Increase in growth and biomass of aquatic organisms is one of the hypotheses why mangroves are considered nursery habitats (Beck et al. 2001). This has been recorded by various authors in marine species; for instance, Lutjanidae and Haemulidae fish families feed, increase size and weight in mangroves during juvenile stages, reaching to adult phases that eventually migrate to, and inhabit coral reefs (Dorenbosch et al. 2004; Kimirei et al. 2011). This was analysed by Lara-Domínguez and Yáñez-Arancibia (1999) in mangroves of the Gulf of Mexico, where different uses were evaluated for distinct lengths and abundance in juveniles (*Eucinostomus gula* and *Orthopristis chrysoptera*) and adult species (*Arius melanopus* and *Lutjanus synagris*) for different habitats; and these contribute to the biomass generated by the ecosystem.

3.4 Fauna Diversity and Habitat Utilization

3.4.1 Terrestrial Invertebrates

Arthropods are the most diverse groups on the planet; so for mangroves, insects and spiders are found in all structural levels of this ecosystem, from aerial roots to, pneumatophores, trunks and canopies, hence, different guilds of these organisms can be found at various trophic levels. Despite their great diversity and many functions, little attention has been given to this fauna, and their role in mangrove ecosystems.

Although some studies have revealed that mangroves are important sites for feeding and refuge of terrestrial invertebrates (Clay and Andersen 1996; Musyafa et al. 2020). Nagelkerken et al. (2008) mention three main trophic groups within the insect community: (1) herbivore insects that feed on leaves, flowers, seeds and any other plant part (e.g. crickets, butterflies and bees); (2) saprophagous and saproxylic invertebrates that feed on dead or decaying organic matter (e.g. flies and beetles) and (3) parasitic or predatory insects (e.g. dragonflies, bedbugs and spiders).

Many terrestrial invertebrates are involved in key processes such as pollination and nutrient recycling, in particular, detritivore organisms inhabiting sediments (e.g. springtails, woodlouse). These organisms are an essential part of the food web as they are the main food source for birds, fish and amphibians of the mangrove ecosystems (Rajpar and Zakaria 2014).

Palynivore insects are classified within the herbivore group, and besides feeding in the mangrove ecosystems, they are vital in maintaining the genetic variability of mangroves. Worldwide, insects have been recorded in these ecosystems; e.g. Diptera, Hymenoptera and Lepidoptera (Sánchez-Nuñez and Mancera-Pineda 2012), with up to 25 species in *Rhizophora mangle*, *Avicennia germinans* and *Laguncularia racemosa* in mangroves from the Caribbean.

Veenakumari et al. (1997) found 276 species of insects in mangroves off islands of the Indian Ocean, of which 197 were herbivores, 43 parasitic, and 36 predatory species. A similar composition in diversity and abundance was reported for mangroves from Thailand by Murphy (1990).

In Zanzibar, Mchenga and Ali (2013) reported 103 species of orders in which hymenopterans were the most abundant. In Indonesian mangroves, Musyafa et al. (2020) recorded 10 insect orders, where coleopterans were the most abundant. In New Zealand, Dencer-Brown et al. (2020) recorded a total of four classes of arthropods, where spiders were the most abundant.

In the south of the Gulf of Mexico, García-Martínez et al. (2019) recorded 11 orders of arthropods within a mangrove system formed by *A. germinans* and *R. mangle*, where spiders and ants from the Crematogaster genus were the most relevant in number and abundance. In Brazil, 22 ant species have been reported, the most common genus being *Camponotus* and *Solenopsis* (Gissel 2010); in Australia, 16 species have been registered (Clay and Andersen 1996), including reports of *Polyrhachis sokolova*, an endemic species of the area that are capable of building their nests in sludge. Terrestrial arthropods are an important group in terms of energy transfer, with links to different groups of terrestrial and aquatic vertebrates, whether these arthropods obtain energy directly from detritus, or by adding it through herbivory. As observed, fish predation on the aquatic larval phase of mosquitoes serves as a natural control, with some studies showing low mosquito densities when fish densities are high (Griffin and Knight 2012).

3.4.2 Aquatic Invertebrates

Mangroves are important habitats for aquatic invertebrate populations such as crustaceans, molluscs and worms (Zakaria and Rajpar 2015). Crustaceans such as shrimps, crabs and prawns contribute to the recycling of organic matter through bioturbation of mangrove sediments, but other groups such as barnacles can cause considerable damage to plants since they grow on aerial roots and pneumatophores (e.g. Euraphia, Elminius and Hexaminius). Apparently, only desiccation and an increase in temperature can limit the colonization of barnacle larvae on the root systems (Ross and Underwood 1997).

Isopods known as a burrowing crustacean group (e.g. *Sphaeroma terebrans* and *S. peruvianum*) have also damaged mangroves of the Atlantic, Caribbean and east Pacific regions, as juveniles and adults embed themselves in roots and stems, affecting root growth (Brooks and Bell 2005; Davidson et al. 2016). There are other crustaceans that temporarily inhabit mangrove waters during a certain phase of their life cycle, for example, the Caribbean spiny lobster (*Panulirus argus*) that uses the mangrove habitat as nursery areas. Lobsters, shrimps and prawns migrate from the mangrove once they have grown, with adults remaining in the mangrove vegetation, only if it is adjacent to a coral reef. However, if several species are present, then, they will tend to distribute themselves differently (Hing et al. 2014).

Crabs occupy a great diversity of habitats within the complex mangrove structure. Some species have obligatory relationships with mangroves while others are facultative, allowing them to occupy seagrass beds and algae located within the mangrove system (Ravichandran et al. 2007). Six of the 30 Brachyura families of crabs are frequent colonizers of mangroves; the most common families are Mictyridae, Grapsidae, Geocarcinidae, Portunidae, Ocypodidae and Xanthidae, with around 127 species. The Ocypodidae family (8 of the 19 genera are the most numerous of at least 80 species) and the Grapsidae family (the Sesarma genus has 60 species inhabiting mangroves) where 30 species are found in the Indo-Malaysian region, 16 on the coasts of Africa, 14 in Australia and 5 in tropical America (Diele et al. 2010).

Among the Anomura crustaceans, hermit crabs from the Diogenidae family (e.g. *Clibanarius* spp. and *Coenobita* spp.) form numerous groups as they climb and rest on the mangrove roots during the immersion period (Teoh et al. 2014). Crab behaviour in mangroves varies, with nocturnal activity for some, perhaps avoiding high temperatures or predators (e.g. *Coenobita rugosus* and *C. cavipes*), whereas others are active all day (Barnes 2001). Mangrove crabs are divided into groups according to their feeding mode; *Uca* and *Macrophthalmus* spp. are detritivores; *Scylla serrata* are opportunistic diggers, and *Thalamita crenata* are predators that occupy the marine edges of the mangrove, as they feed on slowly moving bivalves and crustaceans (Cannicci et al. 2008). Herbivore crabs are another group of consumers that feed directly on mangrove leaf litter, for instance, in the diet of Sesarmidae crabs from Thailand up to 82% of leaf litter was found to be consumed (Diele et al. 2010).

These crustaceans may be greatly important for leaf degradation and transport of organic matter to other marine and estuarine habitats (Ravichandran et al. 2006; Andreetta et al. 2014). Some studies have indicated that crabs play a key role in consuming mangrove propagules. Crabs from the Grapsidae family in Malaysia and Australia consume 95% of the post-dispersion propagules (Sousa et al. 2007), such herbivore influences explaining the pattern of distribution and composition of mangroves in the intertidal zone; at the same time determining the assemblages of associated fauna, and successive stages of vegetation (Lindquist et al. 2009).

Molluscs are considered a dominant aquatic group in the mangrove fauna. They are an important link for the transfer of energy from organic matter produced there to secondary consumers such as invertebrates, mammals, fish and birds (Alfaro 2006). The nature of the mollusc community is strongly influenced by physical conditions; they live inside, above the sediment, and are firmly attached to the roots.

In mangroves from China, the density and biomass of 52 mollusc species were higher at high tide, decreased with depth, and abundance was more related to an increase in salinity (Printrakoon et al. 2008). Mollusc fauna associated with mangroves is mostly composed of bivalves and gastropods, but other groups, more marine based, are also found in mangroves, such as nudibranchs, chitons and scaphopods (Kathiresan and Bingham 2001).

Mangrove molluscs can be divided into three groups: (1) native to the habitat (e.g. *Cerithidea*, *Terebralia* and *Nerita*); (2) facultative molluscs (e.g. *Littoraria* and *Crassostrea*) and (3) migratory molluscs (e.g. *Nerita* and *Clypeomorus*) (Irma and Sofyatuddin 2012). Molluscs occupy all level of the food web, but detritus and filter feeders predominate (Cannicci et al. 2008). Intertidal zones have been documented to function as critical habitats of some organisms (Krauss et al. 2008), since they facilitate aggregation of other species that form an association, such as *Rhizophora mangle* and the *Crassostrea virginica* oyster (Aquino-Thomas and Proffitt 2014).

3.4.3 Amphibians

Amphibians play an important role in the food web of terrestrial and aquatic ecosystems. Due to their predatory behaviour they act as primary and secondary consumers of insects, some of which are crop pests or disease vectors (Behangana 2004). Additionally, many amphibian species are considered as biomarkers, since changes in their population are observed when disturbances occur in their habitat (Welsh Jr and Ollivier 1998). However, amphibian diversity in mangroves is comparatively low relative to other vertebrate groups (Alfred and Ramakrishna 2004).

Most amphibians have a limited osmoregulation capacity and are particularly sensitive to saltwater, which generally restricts their presence in brackish and salt environments (Gomez-Mestre et al. 2004). Approximately, 144 amphibian species capable of tolerating saltwater (between 0.5 and 32 ppt) and brackish conditions have been reported (Wu and Kam 2009; Hopkins and Brodie Jr 2015). However, only 26 amphibian species have been globally recorded in mangrove ecosystems (Rog et al. 2017). Among the species recorded in these ecosystems, only adult stages have been observed, thus, it could be assumed that they inhabit mangroves only in certain seasons (Hopkins and Brodie Jr 2015). The facultative dependence on mangroves shown by these species is probably due to them searching for, food resources, reproduction sites, dispersion pathways between primary habitats, and/or as a temporary refuge against biotic (e.g. predators, competition), abiotic (e.g. extreme temperatures, desiccation) or anthropogenic stress (Rog et al. 2017).

Amphibians can tolerate salinity conditions typical of mangrove ecosystems. They are predators in these environments feeding on small organisms such as insects (e.g. beetles, bees, ants, termites and crickets), snails, small frogs, shrimp and fish (Rajpar and Zakaria 2014). Frogs from the *Fejervarya* genus (located in southeast Asia) (Yodthong et al. 2019) have successfully colonized mangroves and are closely linked to them due to their high tolerance to brackish waters (up to 35 ppt) and show specialized feeding on crabs (Wright et al. 2004; Hopkins and Brodie Jr 2015). Other authors have observed larvae of these frog species, including reports of egg laying in these ecosystems (Uchiyama et al. 1990). The commonly known cane toad (*Rhinella marina*) is another species linked to these habitats (Rajpar and Zakaria 2014); the larval presence of this species in mangroves suggests a similar pattern regarding salinity tolerance. In other species of frogs, larvae and egg laying have been observed such as for *Smilisca baudinii*, *Trachycephalus typhonius* (both from Costa Rica; Sasa et al. 2009) and *Leptodactylus macrosternum* (in mangroves from Ceara, northeast Brazil; Ferreira et al. 2019). However, the latter of these species has been reported for the rainy season, and authors speculate that low salinity concentrations, characterized by this season, allow for their presence (Ferreira et al. 2019).

The presence of larva and/or egg laying suggests that some individuals, or populations of the species mentioned above, complete their life cycle within mangroves; species that have records in other ecosystems, such that they are regarded as facultative inhabitants of mangroves (Rog et al. 2017). On the contrary, the distribution of the Caribbean robber frog (*Eleutherodactylus caribe*) is strictly restricted to mangroves. This species was first documented for a marsh dominated by red mangroves (*Rhizophora mangle*) on the western side of the Tiburon peninsula in Haiti (10 males, 4 females and 2 juveniles). Frogs were observed on trees of *R. mangle* that were completely flooded by brackish water, with no epiphytes, or other elements where this species could potentially lay eggs. Authors of this result concluded that mangroves could be the preferred habitat for *E. caribe*, since other wetland communities without mangroves, also analysed, had no sightings of *E. caribe*. To date, this species, worldwide, is considered the only endemic amphibian species, or with a restricted distribution to mangrove ecosystems (Luther and Greenberg 2009).

3.4.4 Reptiles

Reptiles are an essential component in mangrove ecosystems, as their unique life history and diverse role in food webs are as, prey, predators, herbivores, seed dispersers and commensal species (Raxworthy et al. 2008). This group constitutes a large percentage of the fauna biomass, and their presence is vital for the proper functioning of many ecological processes (Campbell and Campbell 2000). They also serve as biomarkers of environmental health (Read 1998; Raxworthy et al. 2008) due to their sensitivity to habitat loss and degradation, contamination, disease and climate change (Todd et al. 2010). The wide diversity of shapes, and the nature of reptiles, has enabled them to conquer all continents (expect Antarctica) and to inhabit a variety of diverse environments (Roll et al. 2017), including saltwater and brackish environments (Schmidt-Nielsen and Fange 1958), such as mangroves (Voris and Murphy 2012).

Currently 118 reptile species have been reported in these ecosystems, majority as predators, since more than 90% are carnivores (Rog et al. 2017). Reptiles have successfully colonized mangroves, temporally inhabiting, or relying on their resources; these species come from independent evolutionary lines: freshwater turtles, sea turtles, crocodiles, marine and terrestrial lizards and snakes (Nagelkerken et al. 2008).

Some turtle species are known to rely on estuaries and other brackish environments although little is known on their specific requirements. For example, the northern river terrapin, *Batagur baska* (distributed in central and southwest Asia), and the painted terrapin *Batagur borneoensis* (distributed in Thailand and west Malaysia and Indonesia), both species classified as critically endangered (IUCN 2020), feed mainly on riverside vegetation including mangrove fruits

(Blanco et al. 1991). In other areas, various studies have documented the fidelity of sea turtles to feeding areas within mangroves (Limpus and Limpus 2000; Godley et al. 2002). Green sea turtles (*Chelonia mydas*) have a pelagic distribution during the first 3 to 5 years, thereafter, recruiting occurs up to sexual maturity in coastal waters, where they occupy a series of habitats for development, including the mangroves (Makowski et al. 2006). These distribution changes coincide with ontogenetic diet shifts from omnivore to herbivore, as they consume fruits, cotyledons and propagules from *Avicennia marina* (Pendoley and Fitzpatrick 1999; Limpus and Limpus 2000). Similarly, mangroves can indirectly provide food for turtles, since algae, also a food source, grow on the mangrove roots, trunks and pneumatophores. *Rhizophora mangle* has been reported important for foraging, and the development of Kemp turtles (*Lepidochelys kempii*) (Schmid 2000). Finally, the diamondback terrapin (*Malaclemys terrapin rhizophorarum*), which is native to the eastern and southern saltmarshes of the USA, is considered an endemic species to mangroves (Luther and Greenberg 2009). It mainly feeds on decapod crustaceans, bivalves, gastropods, fish and the typical vegetation of this habitat (Tucker et al. 1995; Butler et al. 2012).

Regarding crocodiles, most of the fish and marine animals they feed on use mangroves for their reproduction and nesting sites. Notwithstanding, its importance as a habitat varies depending on the species. The saltwater crocodile *Crocodylus porosus* is closely associated to mangroves, and lays eggs in the vegetation of adjacent ecosystems (Webb et al. 1977; Magnusson 1980). A study from Sri Lanka reported that the population decrease in this species was correlated to an increase in coastal runoff due to mangrove tree cuttings, and thus destruction of nests and eggs (Santiapillai and de Silva 2001). According to Santiapillai and de Silva (2001), prop roots of *Rhizophora* spp. provide an important structural shelter for *C. porosus* juveniles. Other crocodile species are closely linked to mangroves (Table 3.1).

Many lizards from geckos to iguanas inhabit intertidal mangroves. Some are terrestrial species that enter the ecosystem in an opportunistic manner to access the resources, while others such as monitor lizards are semi-aquatic with a distribution that is more restricted to these environments (Nagelkerken et al. 2008). The littoral whiptail-skink (*Emoia atrocostata*) from Southeast Asia and the Pacific is frequently observed in mangrove habitats feeding on crabs, fish and insects (Voris and Murphy 2012). The oriental garden lizard (*Calotes versicolor*), widely distributed in Asia, is also a common species of these ecosystems (Ghosh 2011).

Finally, within the reptile group, are the snakes. Snakes are by far the most successful and diverse reptiles in marine and brackish habitats, with several species found living in the mangroves (Voris and Murphy 2012). For instance, 19 species of mangrove snakes have been reported in Nigeria (Luiselli and Akani 2002). To date, at least 45 species associated with these ecosystems have been globally documented (Rog et al. 2017); some only access mangroves occasionally, while others are regular inhabitants. According to Luther and Greenberg (2009), worldwide, approximately 11 snake species are restricted to mangroves. Some species distinctive of North America's mangroves are *Nerodia clarkii* and *Agkistrodon piscivorus conanti* (Voris

Table 3.1 Main vertebrate species and their protection category (IUCN 2020) in mangrove ecosystems

Class	Common name	Specie	Conservation status IUCN	Reference
Mammals	Vordermann's Pipistrelle Northern Pipistrelle Proboscis Monkey Pygmy three-toed Sloth Garrido's Hutia Cabrera's Hutia Bengal Tiger Fishing Cat Sumatran Tiger Leopard Temminck's Red Colobus Zanzibar Red Colobus Colombian White-faced capuchin Balabac Mouse Deer Smooth coated Otter Asian small clawed otter Caribbean manatee	*Hypsugo vordermanni* *Pipistrellus westralis* *Nasalis larvatus* *Bradypus pygmaeus* *Mysateles garridoi* *Mesocapromys angelcabrerai* *Panthera tigris tigris* *Prionailurus viverrinus* *Panthera tigris sumatrae* *Panthera pardus* *Procolobus badius temminck* *Procolobus kirkii* *Cebus capucinus* *Tragulus nigricans* *Lutrogale perspicillata* *Aonyx cinereus* *Trichechus manatus*	DD LC EN CR CR EN VU CR VU EN EN VU EN VU VU EN	Luther and Greenberg (2009), Barlow et al. (2011), Rajpar and Zakaria (2014), Adhya et al. (2011), Giesen et al. (2007), Nowak (2012), Galat-Luong and Galat (2005) and Nowak and Lee (2011)
Birds	Mangrove cuckoo Mangrove Finch Mangrove Hummingbird Mangrove Warbler Mangrove Vireo Reddish Egret Chinese Egret Lesser Adjutant Spotted Greenshank	*Coccyzus minor* *Camarhynchus heliobates* *Amazilia buocardi* *Setophaga petechia bryanti* *Vireo pallens* *Egretta rufescens* *Egretta eulophotes* *Leptoptilos javanicus* *Tringa guttifer* *Phoenicopterus ruber* *Halcyon senegaloides*	LC CR LC NT VU VU EN LC LC LC	Polidoro et al. (2010), Buelow and Sheaves (2015), Canales-Delgadillo et al. (2019), Zakaria and Rajpar (2015) and Rajpar and Zakaria (2014)

(continued)

Table 3.1 (continued)

Class	Common name	Specie	Conservation status IUCN	Reference
	American flamingo Mangrove kingfisher White-Bellied Sea-eagle	*Haliaeetus leucogaster*		
Amphibians	Mangrove frog Crab eating frog Cuban treefrog Cane toad	*Eleutherodactylus caribe* *Fejervarya cancrivora* *Osteopilus septentrionalis* *Rhinella marina*	CR LC LC LC	Luther and Greenberg (2009), Nagelkerken et al. (2008) and Rajpar and Zakaria (2014)
Reptiles	Mangrove diamondback Terrapin Northern River Terrapin Mugger Nile crocodile American crocodile Common caiman North-western Mangrove Sea Snake Richardson's Mangrove Snake King Cobra Rusty monitor Mangrove Pit Viper	*Malaclemys terrapin rhizophorarum* *Batagur baska* *Crocodylus palustris* *Crocodylus niloticus* *Crocodylus acutus* *Caiman crocodilus* *Ephalophis greyae* *Myron rischardsonii* *Ophiophagus Hannah* *Varanus semiremex* *Trimeresurus purpureomaculatus*	VU CR VU LC VU LC LC LC VU LC LC	Luther and Greenberg (2009), Rajpar and Zakaria (2014), Macintosh and Ashton (2002) and Nagelkerken et al. (2008)
Fish	Mayan Cichlid Silver Jenny Bay Anchovy Flathead Mullet Common Snook Yucatan Killifish Giant Killifish Grey Snapper Great Barracuda Rainbow Parrotfish	*Mayaheros urophthalmus* *Eucinostomus gula* *Anchoa mitchilli* *Mugil cephalus* *Centropomus undecimalis* *Fundulus persimils* *Fundulus grandissimus* *Lutjanus griseus* *Sphyraena barracuda* *Scarus guacamaia* *Lutjanus apodus*	LC LC LC LC LC EN VU LC LC NT LC LC LC LC LC	Arceo-Carranza and Vega-Cendejas (2009), Arceo-Carranza et al. (2016), Faunce and Serafy (2006), Huxam et al. (2004), Igulu et al. (2013), Verweij and Nagelkerken (2007), Zakaria and Rajpar (2015), McDonald et al. (2009) and Able (2005)

(continued)

Table 3.1 (continued)

Class	Common name	Specie	Conservation status IUCN	Reference
	Schoolmaster Snapper Yellowfin Mojarra Whipfin Mojarra French Grunt Hardhead Sea Catfish Checkered Puffer Mud Skipper Silver croaker Tarpon Northern ladyfish Sail fin Molly Yucatan pupfish	*Gerres cinereus* *Gerres* *filamentosus* *Haemulon* *flavolineatum* *Ariopsis felis* *Sphoeroides* *testudineus* *Boleophthalmus* *dussumieri* *Bairdiella* *chrysoura* *Megalops* *atlanticus* *Elops saurus* *Poecilia velifera* *Cyprinodon* *artifrons*	LC LC LC VU LC VU LC	

Source: DD = data deficient; LC = least concern; NT = near threatened; VU = vulnerable; EN = endangered; CR = critically endangered

and Murphy 2012). The latter of these species has been observed below the herons and cormorants searching for food, at the same time, they nest in the intertidal zone (Wharton 1966). In the Neotropics (a region with few studies on mangrove associated fauna) some snake genera have been identified (e.g. *Helicops*, *Hydrops*, *Liophis* and *Tretanorhinus*, inhabiting mangroves, although occasionally (Voris and Murphy 2012). In Africa, the *Grayia smythii* and *Crotaphopeltis hotamboeia* species have been documented for brackish waters, mangroves and freshwater (Luiselli and Akani 2002). The relative frequency in which the African rock python (*Python sebae*) has been reported in mangroves, in comparison to other habitats, suggests this habitat could act as an important refuge or dispersal corridor for this species (Nagelkerken et al. 2008). In Asia, the Indian python (*Python molurus*) and king cobra (*Ophiophagus hannah*) are two examples of terrestrial snakes that commonly enter the mangroves for foraging (Macintosh and Ashton 2002). In Australasia, some sea snakes (Hydrophiidae family) access the mangroves during high tide, while other species such as the Grey's mud snake (*Ephalophis greyae*) have conserved its mode of terrestrial movement, migrating to dry mangrove substrate for fish foraging, during low tide (Storr et al. 2002). Other snakes from this region depend on mangrove trees as a physical structure in the habitat, species such as *Myron richardsonii* (Guinea et al. 2004) and *Boiga dendrophila* (Norhayati et al. 2009). The dog-faced water snake (*Cerberus rynchops*) (Lim et al. 2001), little file snake (*Acrochordus granulatus*) (Gorman et al. 1981) and mangrove Pit viper

(*Trimeresurus purpureomaculatus*) (Lim et al. 2001) are equally common in the mangroves of Australasia (Rajpar and Zakaria 2014).

3.4.5 Mammals

Worldwide, mammals are a key component of biodiversity since they perform roles that are vital for the proper functioning of most ecosystems, such as controlling the population of species at lower trophic levels (McLaren and Peterson 1994); they are also seed dispersers (Asquith et al. 1997), some large-size species are landscape modifiers (Sinclair 2003), and many are top predators on land (Van Valkenburgh 1999) and aquatic ecosystems (Bowen 1997). Moreover, various mammals are considered as flagship species, because they are a great attraction to the general public, and thus helps the creation and maintenance of natural protected areas (Câmara and Oliveira 2012).

In mangrove forests, mammals are the terrestrial vertebrates with the highest species recorded (~320 species) (Rog et al. 2017), although only a few mammals are exclusive to mangroves, since these ecosystems have extreme conditions that only a few species can tolerate (Vanucci 2001).

Most mammal species reported in mangrove ecosystems temporarily live in them by means of opportunistic circumstances, or may change their distribution so as to shelter in mangroves, due to the destruction of their terrestrial habitats by humans (Hogarth 2015). Due to widespread deforestation in the tropics, mangroves and other wetland ecosystems that are not easily accessible to human populations have become an important refuge for the persistence of various populations of threatened mammal species (Nowak 2012). In West Africa, human disturbance has had a strong impact on predators like the spotted hyena (*Crocuta crocuta*) and African civet (*Civettictis civetta*), species that have been observed hunting for the Patas monkey (*Erythrocebus patas*), harnessed bushbuck (*Tragelaphus scriptus*) and warthogs (*Phacochoerus africanus*) in mangroves with more frequency than in their preferred habitats (Galat-Luong and Galat 2007).

Similarly, poaching has caused the disappearance of majority of the Javan rhinoceros populations (*Rhinoceros sondaicus*) from their preferred habitats; this species has currently found refuge in mangrove ecosystems (Macintosh and Ashton 2002).

Mangroves are also fundamental for the conservation of wild cat species (Table 3.1). Primates are frequent inhabitants of mangroves; in West Africa, (south of Senegal) guenons (*Cercopithecus*) feed on fiddler crabs (*Uca tangeri*) and fruits, flowers and young leaves of *Rhizophora*. The Macaques monkeys (*Macaca*) from southwest Asia feed in the mud looking for crabs and bivalves; although this has been documented to have a negative impact on mangrove restoration projects, since Macaques tend to uproot the *Rhizophora* propagules. The Surilis monkeys (*Presbytis*) also in Southeast Asia feed mainly on mangrove leaves and fruits. Lastly, the Proboscis monkeys (*Nasalis larvatus*) feed on large quantities of leaves and fruits per day (Macintosh and Ashton 2002). The importance of mangrove ecosystems for wild cats and primates of Africa and Asia has been documented by Nowak (2012). The data shows mangroves being used by 39 primates and 2 felines (species and subspecies) from Africa, and 29 primates and 18 felines from Asia.

Even though mangroves provide multiple benefits to several mammalian species, mammals also return important services in these ecosystems. Besides being important links in the food web, some species are key in transporting nutrients from terrestrial to marine ecosystems through the use of mangroves (Reef et al. 2014). In Australia, the role of herbivorous mammals and their nutrient input to the mangroves were evaluated. On the one side, the relationship and presence of the red kangaroo (*Macropus rufus*) and that of the Wallaroo (*Macropus robustus*) showed high nutrient concentrations coming from terrestrial environments. On the other hand, a positive relationship was observed between the presence of the black fruit bat (*Pteropus alecto*) and nutrition levels, as well as for growth in mangrove trees. A colony of thousands of *P. Alecto* was observed resting on the canopies of *Rhizophora stylosa*, *Ceriops tagal* and *Lumnitzera rosea* mangroves, making inland trips for fruit and nectar. Bats defecate where they rest, thus providing and transporting nutrients coming from inland mangroves (Reef et al. 2014).

Mangroves around the world are of great importance to several mammal species with aquatic lifestyles, such as dugongs, porpoises, dolphins and some whales. In Sundarbans, Bangladesh, the South Asian River dolphin (*Platanista gangetica*) and Irrawaddy dolphin (*Orcaella brevirostris*) are noticeable inhabitants of mangrove ecosystems. Also found, are the Sea otters, another mammalian group that use the mangrove to feed. In Southeast Asia, aquiculturist considered them as pests because they steal their produce (Macintosh and Ashton 2002).

3.4.6 Birds

Most mangroves have experienced a loss in connectivity and decrease in habitat heterogeneity, which has reduced the diversity in fauna, including land and aquatic migratory bird populations (Mohd-Azlan et al. 2015; Amir 2018). Additionally, there are opportunistic bird species that use the canopy and the aerial root system to shelter from predators, for reproduction, or as feeding sites (Naranjo 1997). Mangroves together with neighbouring environments offer the resources for nest building, including a great diversity of potential prey. Although mangroves are an important habitat for birds, few endemic species are found in these ecosystems (Table 3.1); majority of the birds found in the mangroves are also found in other habitats.

Species composition of birds has been used to indicate mangrove health (Behrouzi-Rad 2014) and to evaluate the impacts of climate change and coastal development (Ogden et al. 2014). In addition, this allows each site to be characterized due to the similarity between sites being relatively low, which suggests that physical characteristics such as habitat heterogeneity and vegetation, including the flooding surface of each zone, allows for the presence of particular bird groups.

Other species can use mangroves when their preferred habitats are unavailable or because mangrove trees provide an additional habitat to bird populations that would mainly be found in neighbouring habitats. Nonetheless, fewer mangrove bird species are observed in highly disturbed sites as compared to those less disturbed (Mohd-Taib et al. 2020). Results from Mohd-Azlan and Lawes (2011) show that the number of specialized species depend on the mangrove patch area, which suggests that these species are limited by resource availability.

The presence of aquatic birds in mangrove ecosystems, mainly piscivore guild, has been used as a status indicator of habitat conservation. These birds are found at the top of the food chain, and are susceptible to environment and habitat changes. These groups reflect the conditions of the terrestrial environment, as well as that of the aquatic environment since they rely on both environments for feeding, refuge and reproduction (Catterall et al. 2012; Zakaria and Rajpar 2015; De Dios et al. 2019). Observations from Canales-Delgadillo et al. (2019) indicate that bird abundance in tropical mangroves is more influenced by habitat condition (i.e. hydroperiod and forest structure), rather than water quality variables. Although without disregard, environmental characteristics possibly have an influence over resource availability due to its effects on primary producers, as well as, the presence of benthic communities, fish and crustaceans (that bird species feed on).

Birds mainly congregate on the soft bottom substrates found close to the borders of the mangrove forests; this has been linked to their feeding behaviour. Sites far from the canopy provide better feeding areas for wading birds (Curado et al. 2013). Furthermore, open and semi-open areas are probably chosen to reduce the risk of predation and increases feeding efficiency, since this type of substrate favours the presence of aquatic invertebrates (Pomeroy 2006; Chacin et al. 2015).

The dynamics of bird populations depend on multiple environmental factors and food availability (Goodsell 1990; Halse et al. 1993; Murkin et al. 1997); it is one of the reasons why piscivorous birds are distributed according to prey availability (Kerekes et al. 1997). Another factor of great importance is the vegetation structure, where there is the likelihood that it can be used as a habitat for roosting, as a refuge, and nesting sites (Zakaria and Rajpar 2015).

Some species of aquatic birds, mostly the Suliformes and Ciconiiformes orders, choose wetlands for their nesting sites because it provides the necessary resources for reproduction, nesting and breeding (Cairns and Kerekes 2000). The location of colonies depends on the accessibility to nearby sites with the available food to satisfy nutritional requirements of both parents and chicks (Buckley and Buckley 1980; Cairns and Kerekes 2000). The coastal wetlands boarded by mangrove forests on the northern Yucatan coast host colonies of aquatic birds with different feeding strategies and whose success depends on the water depth. The shallowest area of the wetland is particularly used by *Platalea ajaja* adults, the intertidal area is used by *Ardea alba*, *Egretta rufescens*, *E. thula* and *E. tricolor* adults, the deepest area is used by *Phalacrocorax brasilianus* adults, and the peripheral area is used by *Cochlearius cochlearius* which displays a foraging bout where it attacks prey starting from trunks or branches, unlike wading birds that feed directly in the water.

Overall, leg length acts as a limiting factor in resource partitioning. Birds with longer legs have access to deeper areas, and thus have more fishing sites than smaller birds that may be restricted to fishing areas. Due to their wide trophic plasticity, herons and cormorants feed on a wide variety of prey, such as fish, crustaceans, insects, amphibians, reptiles and other birds (Ramo and Busto 1993; Kushlan and Hancock 2005; Nelson 2005); however, they are mostly considered piscivorous because their diet heavily depends on fish (Kneib 1982; Britton and Moser 1982; Barquete et al. 2008; Ẑydelist and Kontautas 2008). Therefore, the permanency of a colony is dependent on sites that provide food high in nutritional quality, food such as fish (Kushlan et al. 2002). These birds form a colony in an islet on the northern coast of Yucatan to reproduce. Robles-Toral (2019) mentioned that mangrove forests are exploited by aquatic bird colonies for nest building, because the complexity of the vegetation structure protects nests from inclement weather and predators, and showing a significant correlation between the abundance of nests of the colony and vegetation cover. Thus, trees with greater coverage provide more shelter from predators and inclement weather.

3.4.7 Fish

Mangroves and other shallow aquatic habitats are sites that have recorded densities of large numbers of juvenile species, including invertebrates and fish groups, observed higher in numbers than in nearby areas without vegetation. These observations gave rise to the hypothesis that mangroves act as nurseries for many species that migrate to other habitats once they become adults (Lee et al. 2014).

Beck et al. (2001) proposed three reasons to explain the great abundance of crustaceans and fish juveniles within the mangroves: (1) food availability, (2) reduced predation due to the presence of microhabitats found in shallow waters, and turbidity which is high for nearby habitats that lack vegetation, and (3) the complex physical structure created by the aerial and submerged mangrove roots. These factors could act in synergy to favour the presence of juvenile organisms, given that mangroves offer optimal conditions for their growth and survival. The structure of mangrove plants creates a canopy that reduces light penetration and the presence of fine sediments that increase turbidity, decreasing the interaction between predator and prey (Lee 2008).

These hypotheses, without a doubt are important; the relative importance of each depends on the specific conditions of each system, the type of mangrove and the species present *t* (Pittman et al. 2004; Lugendo et al. 2006). Recent studies using stable isotopes have shown that carbon comes from other primary producers such as phytoplankton, microphytobenthos, macroalgae and epiphytes, which are necessary so as to maintain trophic levels (Bouillon et al. 2008; Nagelkerken et al. 2008). Carbon from leaf litter has been shown to enter the marine food web through serranid and snapper fish species that visit mangroves to feed on crabs through a process called "*short-circuit*". Sheaves and Molony (2000) noted that a considerate amount of mangrove production incorporated in Sesarmidae crabs is exported to neighbouring ecosystems once fish complete their incursion of the mangroves. The low frequency in fish consumption by serranid and snapper species in Australian estuaries (Sheaves and Molony 2000) supports theories that a reduction in predatory pressure over ichthyofauna adds value to tropical mangroves as nursing sites for many fish species.

For decades, studies have been done to describe the diet of fish inhabiting mangrove ecosystems, all in an effort to understand trophic relationships and thus expand on the biology of species, their ecological role in aquatic systems and the factors that regulate their periodic presence and absence (Braga et al. 2012). Notably, tidal cycles have been shown to promote fish movement although there are existing differences imposed by the amplitude of the tide. In areas with small amplitudes, mangrove ecosystems seem to function mainly as habitat shelters (Kimirei et al. 2013), however, in sites with greater tidal amplitudes, mangroves are used as feeding sites during high tides by various fish species (Unsworth et al. 2007; Pülmanns et al. 2018; Loera-Pérez et al. 2020). However, these observed differences in fish feeding patterns suggest that it is important to consider other factors, such as local flooding regimes, time of flooding and the geomorphological context of the basin (Unsworth et al. 2007; Pülmanns et al. 2018; Loera-Pérez et al. 2020).

Unequivocally, it is difficult to conclude on the importance of mangroves as feeding sites, considering the logistical dilemmas in obtaining information from areas difficult to access, or with spatial differences in the structure of the landscape at different scales, and with uneven geomorphologies. Additionally, Igulu et al. (2013) show that there are other sources of uncertainties related with: (1) terminology, given that some studies inappropriately use the concepts “mangrove microhabitat” and “mangroves”, (2) studies in tidal regimen, since these have been done in areas with distinct tidal amplitudes and sampled at specific times of the tide, (3) the nature of the species studied, leading to intrinsic variations of results depending on the bioecological features of each species, (4) the spatial-temporal variability of a dynamic system, that are mangroves, occurs at distinct scales that promotes changes in the presence and abundance of fish species that use these habitats, and (5) the study methods, which, even though, after decades of efforts in analysing stomach contents and nutritional value of prey, it has not yet been standardized, making it difficult to compare between species in distinctive sites, and within a specific time frame, further reducing a potential meta-analysis (Amundsen and Sánchez-Hernández 2019; Fonteles de Vasconcelos Filho et al. 2019).

Moreover, several fish species show ontogenic migrations between habitats, with shifts in their diet, possibly due to patterns in prey distribution (Ramírez-Martínez et al. 2016). The feeding activities shown by sexually immature individuals found in the mangroves prove vital for this stage of their life cycle. However, it has been shown that available prey in these systems do not always have a high nutritional value, but not all fish migrations between mangroves and nearby ecosystems are related to food (Igulu et al. 2013), hence, these habitats are important refuges thanks to the architecture of mangrove roots. Thereby, understanding the processes and linkages of ecological interactions between neighbouring coastal ecosystems is fundamental. The importance of documenting the spatial and temporal variability of fish communities in coastal ecosystems, particularly in mangroves, has been highlighted by Nagelkerken et al. (2008). Nonetheless, environmental dynamics imposes at particular stages are distinctive to time and space attempting to establish the relationships between the variables that determine fish distribution (Bonilla-Gómez et al. 2013). Techniques in direct observations, based on underwater video monitoring, reveal that only a small number of species frequent to these different tropical estuary habitats, actually make extensive use of mangroves (Sheaves et al. 2016), suggesting a re-evaluation of patterns in which fish use habitat.

An inventory of fish biodiversity and studies on their life history allows the comprehension of the complexity and functioning of coastal food webs, and the importance of connectivity between different habitats. It has been shown that the reduction and fragmentation of mangrove and seagrass habitats, as well as loss of connectivity between ecosystems, negatively affect productivity. There are studies that have addressed the consequences of fisheries and ecosystem services would suffer if these socio-environmentally fragile mangrove systems are affected. Conservation efforts should be directed towards protecting habitat patches instead of single habitats, or small portions of an ecosystem since it is known that species move throughout the borders of these ecosystems.

Studies comprising large extensions of mangroves done for coastal wetlands in the southern Gulf of Mexico show the presence of piscivorous fish, benefitting from the richness and abundance of fish prey (Arceo-Carranza and Chiappa-Carrara 2017). Icthyophagous species show significant values in diet overlap, which, under conditions of low prey abundance, may trigger competition, and hence, add a predatory pressure to juveniles of various marine fish species using the mangroves as a nursery, and for feeding.

Migrations of organisms with sufficient swimming capacity to move between adjacent coastal systems are common, and occur at different time scales. Many fish use the mangrove systems differentially, resulting in species composition, abundance and feeding variability throughout a 24 hour cycle, and annual cycle (Arceo-Carranza et al. 2013). Consequently, many authors have indicated that these migrations occur to maximize feeding opportunities, or reduce predation risk, but without specifying the relative value each process has for each species (Zárate-Hernández et al. 2012; Kruse et al. 2016).

Globally, the loss of mangrove coverage is an important subject, due to the functions and services it provides. Coastal development and tourism infrastructure are the main impacts disrupting water flow, and causing deforestation. Throughout this chapter, descriptions have been provided on the diversity and functions displayed by the mangrove fauna, including the role of indicator species in determining the status of habitat health (e.g. amphibians, birds and fish), necessary to determine changes resulting from the effects of impacts on the mangroves. In recent years, different groups of fauna have been evaluated, some to a certain degree of impact, or fauna in the process of restoration, with the objective of proving ecological hypotheses, as it pertains to the functions mangroves offer, i.e. for feeding and as nurseries (Arceo-Carranza et al. 2016; García-Martínez et al. 2019; Hernández Mendoza 2020).

The important role of mangrove ecosystems merits a global and monumental response, with efforts geared towards emphasizing the significance of these ecosystems for both terrestrial and aquatic fauna.

References

Able KW (2005) A re-examination of fish estuarine dependence: Evidence for connectivity between estuarine and ocean habitats. Estuar Coast Shelf Sci 64:5–17.

Adams ES (1994) Territory defense by the ant *Azteca trigona*: Maintenance of an arboreal ant mosaic. Oecol 97: 202–208.

Adhya T, Dey P, Das U, Hazra P (2011) Status survey of Fishing Cat (*Prionailurus viverrinus*) in Howrah and Hooghly, West Bengal. Intermediate report submitted to the small grants programme, WWF, India. WWF. India.

Alfaro AC (2006) Benthic macro-invertebrate community composition within a mangrove seagrass estuary in northern New Zealand. Estuar Coast Shelf Sci 66:97–110.

Alfred JRB, Ramakrishna (2004) Faunal Resources in Mangrove Ecosystem. Envis Forestry Bull 4:24–31.

Alongi DM (1987) Intertidal zonation and seasonality of meiobenthos in tropical mangrove estuaries. Mar Biol 95: 447–458.

Alongi DM (1989) Benthic processes across mixed terrigenous-carbonate sedimentary facies on the central great barrier reef continental shelf. Cont Shelf Res 9(7): 629–663.

Alongi DM, Christoffersen P, Tirendi F (1993) The influence of forest type on microbial-nutrient relationships in tropical mangrove sediments. J Exp Mar Biol Ecol 171: 201–223.

Alongi DM (1994) The role of bacteria in nutrient recycling in tropical mangrove and other coastal benthic ecosystems. Hydrobiology 285: 19–32.

Alongi D (2008) Mangrove forests: resilience, protection from tsunamis, and responses to global climate change. Estuar Coast Shelf Sci 76:1–13. https://doi.org/10.1016/j.ecss.2007.08.024.

Amir AA (2018) Mitigate risk for Malaysia's mangroves. Science 359(6382): 1342–1343

Amundsen PA, Sánchez-Hernández J (2019) Feeding studies take guts–critical review and recommendations of methods for stomach contents analysis in fish. J Fish Biol 95: 1364–1373.

Andreetta A, Fusi M, Cameldi I, Cimo F, Carnicelli S, Cannicci S (2014) Mangrove carbon sink. Do burrowing crabs contribute to sediment carbon storage? Evidence from a Kenyan mangrove system. J Sea Res 85:524–533. https://doi.org/10.1016/j.seares.2013.08.010.

Arai H, Yoshioka R, Hanazawa S, Minh VQ, Tuan VQ, Tinh TK, Phu TQ, Jha CS, Reddy SR, Dadhwal VK, Mano M, Inubushi K (2016) Function of the methanogenic community in mangrove soils as influenced by the chemical properties of the hydrosphere. Soil Sci Plant Nut 62(2):150–163. https://doi.org/10.1080/00380768.2016.1165598.

Arceo-Carranza D, Vega-Cendejas ME (2009) Spatial and temporal characterization of fish assemblages in a tropical coastal system influenced by freshwater inputs: northwestern Yucatan peninsula. Rev Biol Trop57(1-2):89–103.

Arceo-Carranza D, Vega-Cendejas ME, Hernández de Santillana M (2013) Day and night trophic variations of dominant fish species in a lagoon influenced by freshwater seeps. J Fish Biol 82:54–68. https://doi.org/10.1111/j.1095-8649.2012.03463.x.

Arceo-Carranza D, Gamboa E, Teutli-Hernández C, Badillo-Alemán M, Herrera Silveira JA (2016) Los peces como indicador de restauración de áreas de manglar en la costa norte de Yucatán. Rev Mex Biodiv 87:489–496. https://doi.org/10.1016/j.rmb.2016.03.001.

Arceo-Carranza D, Chiappa-Carrara X (2017) Feeding ecology of juvenile marine fish in a shallow coastal lagoon of southeastern Mexico. Lat Am J Aquat Res 43(4): 621–631.

Aquino-Thomas JS, Proffitt E (2014) Oysters *Crassostrea virginica* on red mangrove *Rhizophora mangle* prop roots: Facilitation of one foundation species by another. Mar Ecol Prog Ser. https://doi.org/10.3354/meps10742.

Asquith NM, Wright SJ, Clauss MJ (1997) Does mammal community composition control recruitment in neotropical forests? Evidence from Panama. Ecology 78:941–946.

Azam F, Fenchel T, Field JG, Gray JS, Meyer-Reil, Thingstad F (1983) The ecological role of water-column microbes in the sea. Mar Ecol Prog Ser 10: 257–263.

Balakrishnan S, Srinivasan M, Mohanraj J (2014) Diversity of some insect fauna in different coastal habitats of Tamil Nadu, southeast coast of India. J Asia-Pacific Biodiv. https://doi.org/10.1016/j.japb.2014.10.010

Balakrishnan S, Srinivasan M, Santhanam P (2018) Insect Fauna of Pitchavaram and Parangipettai Mangroves of Southeast Coast of India. Proc Zool Soc. https://doi.org/10.1007/s12595-016-0182-y.

Barbier EB, Hacker SD, Kennedy C, Koch EW, Stier AC, Silliman BR (2011) The value of estuarine and coastal ecosystem services. Ecol Monog 81:169–193.

Barlow AC, Smith JL, Ahmad IU, Hossain AN, Rahman M, Howlader A (2011) Female tiger Panthera tigris home range size in the Bangladesh Sundarbans: the value of this mangrove ecosystem for the species' conservation. Oryx 45:125–128.

Barnes DKA (2001) Hermit crabs, humans and Mozambique mangroves. Afr J Ecol. https://doi.org/10.1046/j.1365-2028.2001.00304.x

Barquete V, Bugoni L, Vooeren CM (2008) Diet of neotropic cormorant (*Phalacrocorax brasilianus*) in an estuarine environment. Mar Biol 153(3):431–443. https://doi.org/10.1007/s00227-007-0824-8.

Beck MW, Heck KL Jr, Able KW, Childers DL, Eggleston DB, Gillanders BM, Halpern B, Hays CG, Hoshino K, Minello TJ, Orth RJ, Sheridan PF, Weinstein MP (2001) The identification, conservation, and management of estuarine and marine nurseries for fish and invertebrates. BioScience 51: 633–641.

Behangana M. (2004) The diversity and status of amphibians and reptiles in the Kyoga Lake Basin. Afr J Ecol 42:51–56. https://doi.org/10.1111/j.1365-2028.2004.00461.x.

Behrouzi-Rad T (2014) Population dynamic and species diversity of wintering waterbirds in mangroves wetland (Persian Gulf) in 1983 and 2013. Int J Mar Sci 4(63):1–7. https://doi.org/10.5376/ijms.2014.04.63.

Blanco S, Behler J, Kostel F (1991) Propagation of the Batagurine Turtles *Batagur baska* and *Callagur borneoensis* at the Bronx Zoo. In: Beaman KR, Caporaso F, McKeown S, Graff MD (ed) Proceedings of the 1st international symposium on turtles and tortoises: conservation and captive husbandry. pp. 63–65.

Bonilla-Gómez JL, Badillo M, López K, Gallardo A, Galindo C, Arceo D, Chiappa-Carrara X (2013) Environmental influences on the abundance of dominant fishes in a very shallow tropical coastal lagoon in Northwestern Yucatan Peninsula, Mexico. J Marine Sci Res Dev 3:118. https://doi.org/10.4172/2155-9910.1000118.

Bouillon S, Koedam N, Raman AV, Dehairs F (2002) Primary producers sustaining macro-invertebrate communities in intertidal mangrove forests. Oecol 130: 441–448.

Bouillon S, Connolly R, Lee SY (2008) Organic matter exchange and cycling in mangrove ecosystems: recent insights from stable isotope studies. J Sea Res 59: 44–58.

Bowen W (1997) Role of marine mammals in aquatic ecosystems. Mar Ecol Prog Ser 158: 267–274.

Britton RH, Moser ME (1982) Size specific predation by herons and its effect on the sex-ratio of natural populations fish Gambusia affinis Baird and Girard. Oecol 53: 14–151.

Braga RR, Bornatowski H, Vitule JRS (2012) Feeding ecology of fishes: an overview of worldwide publications. Rev Fish Biol Fish 22: 915–929.

Brooks RA, Bell SS (2001) Colonization of a dynamic substrate: factors influencing recruitment of the wood-boring isopod, *Sphaeroma terebrans*, onto red mangrove, *Rhizophora mangle* prop roots. Oecol 127:522–532.

Brooks RA, Bell SS (2005) The distribution and abundance of *Sphaeroma terebrans*, a wood-boring isopod of red mangrove (*Rhizophora mangle*) habitat within Tampa Bay. Bull Mar Sci. 76:27–46.

Buckley FG, Buckley PA (1980) Habitat selection and marine birds. In: Burger J, Olla BL, Winn HE (eds) Behavior of marine animals, Plenum Press, New York. pp. 69–112.

Buelow C, Sheaves M (2015) A birds-eye view of biological connectivity in mangrove systems. Estuar Coast Shelf Sci 152: 33–43. https://doi.org/10.1016/j.ecss.2014.10.014.

Butler JA, Heinrich GL, Mitchell ML (2012) Diet of the Carolina diamondback terrapin (*Malaclemys terrapin centrata*) in northeastern Florida. Chelonian Conserv Bi 11:124–128.

Cairns DK, Kerekes JJ (2000) Fish harvest by common loons and common mergansers in Kejimkujik National Park, Nova Scotia, Canada, as estimated by bioenergetic modelling. In: Comín FA, Herrera-Silveira JA, Ramírez-Ramírez J (eds): Limnology and aquatic birds. Monitoring, modelling and management, Proceedings of the 2nd International Symposium on Limnology and Aquatic Birds, Universidad Autónoma de Yucatán, Mérida, México. pp. 125–135.

Câmara E, Oliveira L (2012). Mammals of Serra do Cipó National Park, southeastern Brazil. Check List 8:355–359. https://doi.org/10.15560/8.3.355.

Camilleri JC (1992) Leaf-litter processing by invertebrates in a mangrove forest in Queensland. Mar Biol 114: 139–145.

Campbell KR, Campbell TS (2000) Lizard contaminant data for ecological risk assessment. In: de Voogt P (ed) Reviews of Environmental Contamination and Toxicology. Springer, Netherlands. pp. 39–116.

Canales-Delgadillo JC, Perez-Ceballos R, Zaldivar-Jimenez MA, Merino-Ibarra M, Cardoza G, Cardoso-Mohedano J (2019) The effect of mangrove restoration on avian assemblages of a coastal lagoon in southern Mexico. PeerJ 7:e7493. https://doi.org/10.7717/peerj.7493.

Cannicci S, Burrows D, Fratini S, Smith TJ III, Offenberg J, Dahdouh-Guebas F (2008) Faunal impact on vegetation structure and ecosystem function in mangrove forests: a review. Aquat Bot 89:186–200.

Catterall CP, Freeman AND, Kanowski J, Freebody K (2012) Can active restoration of tropical rainforest rescue biodiversity? A case with bird community indicators. Biol Conserv 146(1): 53–61. https://doi.org/10.1016/j.biocon.2011.10.033.

Chacin DH, Giery ST, Yeager LA, Layman CA, Brian-Langerhans R (2015) Does hydrological fragmentation affect coastal bird communities? A study from Abaco Island, The Bahamas. Wetl Ecol Manag 23: 551–557. https://doi.org/10.1007/s11273-014-9389-8.

Clay RE, Andersen AN (1996) Ant fauna of a mangrove community in the Australian seasonal tropics, with particular reference to zonation. Aust J Zool 44:521–533.

Curado G, Figueroa E, Sanchez MI, Castillo JM (2013) Avian communities in *Spartina maritima* restored and non-restored salt marshes. Bird Study 60: 185–194.

Daniel PA, Robertson AI (1990) Epibenthos of mangrove waterways and open embayments: community structure and the relationship between exported mangrove detritus and epifaunal standing stocks. Estuar Coast Shelf Sci 31:599–619.

Davidson TM, Ruiz GM, Torchin ME (2016) Boring crustaceans shape the land–sea interface in brackish Caribbean mangroves. Ecosphere. https://doi.org/10.1002/ecs2.1430.

Day JW, Crump BC, Kemp WM, Yáñez-Arancibia A (2013) Estuarine ecology. 2nd ed. John Wiley and Sons, New Jersey.

De Dios Arcos C, Badillo-Alemán M, Arceo-Carranza D, Chiappa-Carrara X (2019) Feeding ecology of the waterbirds in a tropical mangrove in the southeast Gulf of Mexico. Stud Neotrop Fauna Environ 55(1): 1–9. https://doi.org/10.1080/01650521.2019.1682232.

Dencer-Brown AM, Alfaro AC, Bourgeois C, Sharma S (2020) The secret lives of mangroves: Exploring New Zealand's urban mangroves with integrated biodiversity assessments. Ocean Coast Manag 191:1–16. https://doi.org/10.1016/j.ocecoaman.2020.105185.

Díaz MC, Smith KP, Rützler K (2004) Sponge species richness and abundance as indicators of mangrove epibenthic community health. Atoll Res Bull. https://doi.org/10.5479/si.00775630.518.1.

Diele K., Koch V., Abrunhosa F, Darlan JB, Simith (2010) The Brachyuran Crab Community of the Caeté Estuary, North Brazil: Species Richness, Zonation and Abundance. In: Saint-Paul U,

Schneider H (eds) Mangrove Dynamics and Management in North Brazil. Ecological studies vol 211. Springer, Berlin.
Dittmann S (2000) Zonation of benthic communities in a tropical tidal flat of north-east Australia. J Sea Res 43: 33–51.
Dorenbosch M, Van Riel MC, Nagelkerken I, Van der Velde G (2004) The relationship of reef fish densities to the proximity of mangrove and seagrass nurseries. Estuar Coast Shelf Sci 60: 37–48.
Eller RJ, Grassle JF (1992) Patterns of species diversity in the deep-sea as a function of sediment particle size. Nature 360: 576–578.
Faunce CH, Serafy JE (2006) Mangroves as fish habitat: 50 years of field studies. Mar Ecol Prog Ser 318:1–18.
Feller IC, Mathis WN (1997) Primary herbivory by wood-boring insects along an architectural gradient of *Rhizophora mangle*. Biotropica 29: 440–451.
Feller IC (2002) The role of herbivory by wood-boring insects in mangrove ecosystems in Belize. Oikos 97:167–176.
Feller IC, Lovelock CE, Berger U, McKee KL, Joye SB, Ball MC (2010) Biocomplexity in Mangrove ecosystems. Annu Rev Mar Sci 2:395–417.
Fernandes C, Fernandes CP, Barth MO (2011) Pollen analysis of honey and beebread derived from Brazilian mangroves. Braz J Bot. https://doi.org/10.1590/S0100-84042012000100009.
Ferreira AC, Cascon P, Matthews-Cascon H (2019) Occurrence and egg-laying of Leptodactylus macrosternum Miranda-Ribeiro, 1926 in mangrove habitat in Ceará, Northeast Brazil. Herpetol Notes 12:865–868.
Fonteles de Vasconcelos Filho JI, Camargo Maia R, Salles R (2019) Desempenho dos diferentes métodos de amostragem para caracterização da ictiofauna associada ao manguezal da praia de Arpoeiras em Acaraú, Ceará. Arq Ciên Mar 52(1):81–98.
Galat-Luong A, Galat G (2005) Conservation and survival adaptations of Temminck's red colobus (*Procolobus badius temmincki*), in Senegal. Internat J Primat 26:585–603.
Galat-Luong A, Galat G (2007). Influence of anthropization on the distribution of large wildlife: the mangroves, a refuge environment. In: Fournier A, Sinsin B, Mensah GA (eds) Quelles aires protégées pour l'Afrique de l'Ouest? Conservation de la biodiversité et développement, IRD Editions, Paris 568–569.
García-Martínez G, Arceo-Carranza D, Teutli Hernández C, Flores Rivero MA (2019) Diversidad de artrópodos terrestres asociados a un ecosistema de manglar en estado conservado y en proceso de restauración dentro de la ciénega de Progreso, Yucatán In: Molina Moreira (ed) Manglares de América. Ed Compas Pp 83–93.
Ghosh D (2011) Mangroves. Resonance. 16:47–60.
Giesen W, Wulffraat S, Zieren M, Scholten L (2007). Mangrove guidebook for Southeast Asia. Wageningen, FAO and Wetlands International.
Gissel NM (2010) Ants (Hymenoptera: Formicidae) of mangrove and other regularly inundated habitats: Life in physiological extreme. Myrmecol News 14:113–121.
Godley B, Richardson S, Broderick A, Coyne M, Glen F, Hays G (2002) Long-term satellite telemetry of the movements and habitat utilisation by green turtles in the Mediterranean. Ecography 25:352–362.
Gomez-Mestre I, Tejedo M, Ramayo E, Estepa J (2004) Developmental alterations and osmoregulatory physiology of a larval anuran under osmotic stress. Physiol Biochem Zool 77:267–274.
Goodsell JT (1990) Distribution of waterbird broods relative to wetland salinity and pH in southwestern Australia. Aust Wildl Res 17: 219–229.
Gopal B, Chauhan M (2006) Biodiversity and Its Conservation in the Sundarban Mangrove Ecosystem. Aquat Sci 68:338–354. https://doi.org/10.1007/s00027-006-0868-8.
Gorman GC, Licht P, McCollum F (1981) Annual reproductive patterns in three species of marine snakes from the central Philippines. J Herpetol 335–354.
Griffin LF, Knight JM (2012) A review of the role of fish as biological control agents of disease vector mosquitoes in mangrove forests: reducing human health risks while reducing environmental risk. Wetl Ecol Manag. https://doi.org/10.1007/s11273-012-9248-4.

Guerra-Castro E, Cruz Motta JJ, Conde JE (2011) Cuantificación de la diversidad de especies incrustantes asociadas a las raíces de *Rhizophora mangle* L. en el Parque Nacional Laguna de la Restinga. Interciencia. 36(12): 923–930.

Guinea M, Limpus C, Whiting S (2004) Marine snakes. In: Description of key species groups in the northern planning area. National Oceans Office Hobart, Australia.

Halse SA, Williams MR, Jaensch RP, Lane JAK (1993) Wetland characteristics and waterbird use of wetlands in south-western Australia. Wildl Res 20: 103–125.

Hernández Mendoza LC (2020) Ictiofauna como indicador de recuperación de las funciones ecológicas de una zona de manglar en la Reserva de la Biosfera Sian Ka'an, Quintana Roo, México. MSc thesis, Universidad Nacional Autónoma de México, México.

Hing MH, Chong VC, Sasekumar A (2014) Distribution and burrow morphology of three sympatric species of *Thalassina* mud lobsters in relation to environmental parameters on a Malayan mangrove shore. J Sea R. 95: 75–83. https://doi.org/10.1016/j.seares.2014.10.006.

Hogarth PJ (2015) The biology of mangroves and seagrasses. Oxford University Press.

Holguin G, Bashan Y (2007) La importancia de los manglares y su microbiología para el sostenimiento de las pesquerías costeras. In R. Ferrera Cerrato, A. Alarcón (eds) Microbiología agrícola: hongos, bacterias, micro y macrofauna, control biológico, planta organismo. Trillas, México, México. p 239–252.

Hopkins GR, Brodie Jr ED (2015) Occurrence of amphibians in saline habitats: a review and evolutionary perspective. Herpetol Monogr 29:1–27.

Hsieh HL (1995) *Laonome albicingillum*, a new fan worm species (Polychaeta: Sabellidae: Sabellinae) from Taiwan. Proc Biol Soc Wash 108(1):130–131.

Huxam M, Kimani E, Augley J (2004) Mangrove fish: a comparison of community structure between forested and cleared habitats. Estuar Coast Shelf Sci. 60:637–647.

Igulu MM, Nagelkerken I, van der Velde G, Mgaya YD (2013) Mangrove fish production is largely fuelled by external food sources: a stable isotope analysis of fishes at the individual, species, and community levels from across the globe. Ecosystems 16:1336–1352.

Irma D, Sofyatuddin K (2012) Diversity of Gastropods and Bivalves in mangrove ecosystem rehabilitation areas in Aceh Besar and Banda Aceh districts, Indonesia. AACL Bioflux 5 (2):55–59.

Ismail T, Kassim A, Rahman A, Yahya AK, Webb CE (2018) Day biting habits of mosquitoes associated with mangrove forests in Kedah, Malaysia. Trop Med Inf Dis. https://doi.org/10.3390/tropicalmed3030077.

IUCN 2020. *The IUCN Red List of Threatened Species. Version 2020-2* https://www.iucnredlist.org. Downloaded on 09 July 2020.

Jenoh EM, Robert EM, Lehmann I, Kioko E, Bosire JO, Ngisiange N, Dahdouh-Guebas F, Koedam N (2016) Wide ranging insect infestation of the pioneer mangrove *Sonneratia alba* by two insect species along the Kenyan Coast. PloS One. https://doi.org/10.1371/journal.pone.0154849.

Kathiresan K, Bingham BL (2001) Biology of Mangroves and Mangrove Ecosystems. Adv Mar Biol 40:81–251.

Kerekes JJ, Duggan M, Tordon R, Boros G, Bronkhorst M (1997) Abundance and distribution of fish-eating birds in Kejimkujik National Park, Canada (1988–1994). In: Faragó S, Kerekes JJ (eds) Limnology and waterfowl. Monitoring, modelling and management, Wetlands International Publication. p 211–227.

Kimirei IA, Nagelkerken I, Griffioen B, Wagner C, Mgaya YD (2011) Ontogenetic habitat use by mangrove/seagrass-associated coral reef fishes shows flexibility in time and space. Estuar Coast Shelf Sci 92:47–58.

Kimirei IA, Nagelkerken I, Mgaya YD, Huijbers CM (2013) The mangrove nursery paradigm revisited: otolith stable isotopes support nursery-to-reef movements by Indo-Pacific fishes. PLoS ONE 8(6): e66320 https://doi.org/10.1371/journal.pone.0066320.

Kneib RT (1982) The effects of predation by wading birds (Ardeidae) and blue crabs (*Callinectes sapidus*) on the population size structure of the common mummichog *Fundulus heteroclitus*. Estuar Coast Shelf Sci 14:159–165.

Krauss KW, Lovelock CE, McKee KL, Lopez-Hoffman L, Ewe SML, Sousa WP (2008) Environmental driver in mangrove establishment and early development: a review. Aquat Bot 89:105−127.

Kristensen E, King GM, Holmer M, Banta GT, Jensen MH, Hansen K, Bussarawit N (1994) Sulfate reduction, acetate turnover and carbon metabolism in sediments of Ao Nam Bor mangrove, Phuket, Thailand. Mar Ecol Prog Ser 109:245–255.

Kruse M, Taylor M, Muhando CA, Reuter H (2016) Lunar, diel, and tidal changes in fish assemblages in an East African marine reserve. Reg Stud Mar Sci 3:49–57.

Kushlan JA, Hancock JA (2005) The Herons. Oxford University Press, Oxford.

Kushlan JA, Steinkamp MJ, Parsons KC, Capp J, Acosta-Cruz M, Coulter M, Davidson I, Dickson L, Edelson N, Elliot R, Erwin RM, Hatch S, Kress S, Milko R, Miller S, Mills K, Paul R, Phillips R, Saliva JE, Sydeman B, Trapp J, Wheeler J, Wohl K (2002) Waterbird conservation for the Americas: the North American waterbird conservation plan, Version 1, Waterbird Conservation for the Americas Washington, DC.

Lara-Domínguez AL, Yáñez-Arancibia A (1999) Productividad secundaria, utilización del hábitat y estructura trófica. In: Yáñez-Arancibia A, Lara-Domínguez AL (eds.). Ecosistemas de Manglar en América Tropical. Instituto de Ecología A.C. México, UICN/ORMA, Costa Rica, NOAA/NMFS Silver Spring MD, USA p. 153–166.

Lee SY (2008) Mangrove macrobenthos: assemblages, services, and linkages. J Sea Res 59(1–2): 16–29.

Lee SY, Primavera JH, Dahdouh-Guebas F, McKee K, Bosire JO, Cannicci S, Diele K, Fromard F, Koedam N, Marchand C, Mendelssohn I, Mukherjee N, Record S (2014) Reassessment of mangrove ecosystem services. Global Ecol Biogeogr 23:726–743.

Lim KK, Murphy D, Morgany T, Sviasothi N, Ng P, Soong B, Tan H, Tan H, Tan T (2001) A guide to the mangroves of Singapore. Raffles Museum of Biodiversity Research, National University of Singapore, and Singapore Science Centre, Singapore, 1.

Limpus CJ, Limpus DJ (2000) Mangroves in the diet of Chelonia mydas in Queensland, Australia. Mar Turt Newsl 89:13–15.

Lindquist ES, Krauss KW, Green PT, O'Dowd DJ, Sherman PM, Smith TJ (2009) Land crabs as key drivers in tropical coastal forest recruitment. Biol Rev. https://doi.org/10.1111/j.1469-185X.2008.00070.x.

Loera-Pérez J, Hernández-Stefanoni J, Chiappa-Carrara X (2020) How do spatial and environmental factors affect the fish community structure in seasonally flooded karst systems? Lat Am J Aquat Res 48(2):268–279.

Lugendo BR, Nagelkerken I, Van Der Velde G, Mgaya YD (2006) The importance of mangroves, mud and sand flats, and seagrass beds as feeding areas for juvenile fishes in Chwaka Bay, Zanzibar: gut content and stable isotope analyses. J Fish Biol 69(6): 1639–1661.

Lugo AE, Snedaker SC (1974) The Ecology of Mangroves. Annu Rev Ecol Syst 5: 39–64.

Luiselli L, Akani GC (2002) An investigation into the composition, complexity and functioning of snake communities in the mangroves of south-eastern Nigeria. Afr J Ecol 40, 220–227.

Luther D, Greenberg R (2009) Mangroves: a global perspective on the evolution and conservation of their terrestrial vertebrates. BioScience 59:602–612.

Macintosh D, Ashton E (2002) A Review of Mangrove Biodiversity Conservation and Management. Centre for Tropical Ecosystems Research. University of Aarhus, Denmark.

Magnusson WE (1980) Habitat Required for Nesting by Crocodylus porosus (Reptilia: Crocodilidae). Wildl Res 7:149–156.

Makowski C, Seminoff JA, Salmon M (2006) Home range and habitat use of juvenile Atlantic green turtles (*Chelonia mydas* L.) on shallow reef habitats in Palm Beach, Florida, USA. Mar Biol 148:1167–1179.

MacDonald J, Shahrestani S, Weis J (2009) Behavior and space utilization of two common fishes within Caribbean mangroves: implications for the protective function of mangrove habitats. Estuar Coast Shelf Sci 84(2):195–201.

McLusky DS, Elliott M. (2004) The Estuarine Ecosystem. Ecology, threats and management. Third edition. Oxford University Press. New York, USA.

Mchenga ISS, Ali AI (2013). Macro-fauna communities in tropical mangrove forest of Zanzibar island, Tanzania. GJBB 2(1):260–266.

McLaren BE, Peterson RO (1994) Wolves, moose, and tree rings on Isle Royale. Science 266:1555–1558.

Mohd-Azlan J, Lawes MJ (2011) The effect of the surrounding landscape matrix on mangrove bird community assembly in north Australia. Biol Conserv 144(9):2134–2141.

Mohd-Azlan J, Noske RA, Lawes MJ (2015) The role of habitat heterogeneity in structuring mangrove bird assemblages. Diversity 7:118–136.

Mohd-Taib FS, Mohd-Saleh W, Asyikha R, Mansor MS, Ahmad-Mustapha M, Mustafa-Bakray NA, Mod-Husin S, Md-Shukor A, Amat-Darbis ND, Sulaiman N (2020) Effects of anthropogenic disturbance on the species assemblages of birds in the back mangrove forests. Wetlands Ecol Manage 28:479–494.

Mukherjee N, Sutherland WJ, Khan MNI, Berger U, Schmitz N, Dahdouh-Guebas F, Koedam N (2014) Using expert knowledge and modeling to define mangrove composition, functioning, and threats and estimate time frame for recovery. Ecol Evol 4(11):2247–2262.

Murkin HR, Murkinand EJ, Ball JP (1997) Avian habitat selection and prairie wetland dynamics: A 10-year experiment. Ecol Appl 7:1144–1159.

Murphy DH (1990) The recognition of some insects associated with mangroves in Thailand. In: Mangrove Ecosystem Occasional Paper 7. UNDP /UNESCO, New Delhi.

Musyafa B, Hardiwinoto S, Ahbudin A (2020) Changes in insect biodiversity on rehabilitation sites in the southern coastal areas of Java Island, Indonesia. Biodiversitas 21(1):1–7.

Nagelkerken I, Kleijnen S, Klop T, Van den Brand CJ, Cocheret de la Moriniere E, Van der Velde G (2001) Dependence of Caribbean reef fishes on mangroves and seagrass beds as nursery habitats: a comparison of fish faunas between bays with and without mangroves/seagrass beds. Mar Ecol Prog Series 214:225–235.

Nagelkerken I, Blaber S, Bouillon S, Green P, Haywood M, Kirton L, Meynecke J-O, Pawlik J, Penrose H, Sasekumar A (2008) The habitat function of mangroves for terrestrial and marine fauna: a review. Aquat Bot 89:155–185.

Naranjo L G (1997) A note on the birds of the Pacific mangroves of Colombia. In: Kjerfve B, de La Cerda LD, Diop HS (eds) Mangrove ecosystem studies in Latin America and Africa, UNESCO–International Society for Mangrove Ecosystems, Forest Service Department of Agriculture, Paris p 64–70.

Nelson B (2005) Pelicans, cormorants and their relatives: Pelecanidae, Sulidae, Phalacrocoracidae, Anhingidae, Phaethontidae. Oxford University Press, Oxford.

Norhayati A, Shukor M, Juliana S, Wa WJ (2009) Mangrove flora and fauna of Klang islands mangrove forest reserves, Selangor, Malaysia. MJS 28:275–288.

Nowak K (2012) Mangrove and peat swamp forests: refuge habitats for primates and felids. Folia Primatol 83:361–376.

Nowak K, Lee PC (2011) Demographic structure of Zanzibar red colobus populations in unprotected coral rag and mangrove forests. Internat J Primatol 32:24–45.

Oakes JM, Connolly RM, Revill AT (2010) Isotope enrichment in mangrove forests separates microphytobenthos and detritus as carbon sources for animals. Limnol Oceanogr. https://doi.org/10.4319/lo.2010.55.1.0393.

Odum WE, Heald EJ (1972) Trophic Analyses of an Estuarine Mangrove Community. Bull Mar Sci 22(3):671–738.

Ogden JC, Baldwin JD, Bass OL, Browder JA, Cook MI, Frederick PC, Frezza PE, Galvez RA, Hodgson AB, Meyer KD, Oberhofer LD, Paul AF, Fletcher PJ, Davis SM, Lorenz JJ (2014) Waterbirds as indicators of ecosystem health in the coastal marine habitats of Southern Florida: 2. Conceptual ecological models. Ecol indic 44:128–147.

Pascal PY, Gros O, Boschker HTS (2016) Temporal fluctuations in the trophic role of large benthic sulfur bacteria. Food webs 7:20–28.

Pendoley K, Fitzpatrick J (1999) Browsing of mangroves by green turtles in Western Australia. Mar Turtl Newsl 84:10.

Perry D (1988) Effects of associated fauna on growth and productivity in the red mangrove. Ecology 69(4):1064–1075.

Pittman SJ, McAlpine CA, Pittman KM (2004) Linking fish and prawns to their environment: a hierarchical landscape approach. Mar Ecol Prog Ser 283:233–254.

Polidoro BA, Carpenter KE, Collins L, Duke NC, Ellison AM, Ellison JC, Farnsworth EJ, Fernando ES, Kathiresan K, Koedam NE, Livingstone SR, Miyagi T, Moore GE, Nam VN, Ong JE, Primavera JH, Salmo SG III, Sanciangco JC, Sukardjo S, Wang Y, Yong JWH (2010) The Loss of Species: Mangrove Extinction Risk and Geographic Areas of Global Concern. PLoS ONE 5 (4): e10095. https://doi.org/10.1371/journal.pone.0010095.

Pomeroy AC (2006) Tradeoffs between food abundance and predation danger in spatial usage of a stopover site by western sandpipers, *Calidris mauri*. Oikos 112: 629–637.

Printrakoon C, Wells FE, Chitramvong Y (2008) Distribution of mollusks in mangrove at six sites in the upper gulf of Thailand. Raffles Bull Zool 18:247–257.

Pülmanns N, Castellanos-Galindo GA, Krumme U (2018) Tidal-diel patterns in feeding and abundance of armed snook *Centropomus armatus* from macrotidal mangrove creeks of the tropical eastern Pacific Ocean. J Fish Biol 93: 850–859.

Rajpar MN, Zakaria M (2014) Mangrove fauna of Asia. In: Faridah-Hanum I, Latiff A, Hakeem K, Ozturk M (eds). Mangrove Ecosystems of Asia. Springer, New York, p. 153–197.

Ramírez-Martínez GA, Castellanos-Galindo GA, Krumme U (2016) Tidal and diel patterns in abundance and feeding of a marine-estuarine-dependent fish from macrotidal mangrove creeks in the Tropical Eastern Pacific (Colombia). Estuaries Coasts 39:1249–1261.

Ramo C, Busto B (1993) Resource use by herons in Yucatan wetland during the breeding season. Wilson Bull 105(4):573–586.

Ravichandran S, Kannupandi T, Kathiresan K (2006) Mangrove leaf litter processing by sesarmid crabs. Cey J Sci (Bio Sci) 35:107–114.

Ravichandran S, Anthonisamy T, Kannupandi S, Balasubramanian T (2007) Habitat Preference of crabs in Pichavaram mangrove environment, Southeast coast of India. J Fish Aq Sci 2:47–55.

Raxworthy C, Pearson R, Zimkus B, Reddy S, Deo A, Nussbaum R, Ingram C (2008) Continental speciation in the tropics: contrasting biogeographic patterns of divergence in the Uroplatus leaf-tailed gecko radiation of Madagascar. J Zool 275:423–440.

Read J (1998) Are geckos useful bioindicators of air pollution? Oecol 114:180–187.

Reef R, Feller IC, Lovelock CE (2014) Mammalian herbivores in Australia transport nutrients from terrestrial to marine ecosystems via mangroves. J Trop Ecol 30:179–188.

Robertson AI, Alongi DM (1992) Tropical mangrove ecosystems, American Geophysical Union, Washington DC.

Robles-Toral PJ (2019) Uso de hábitat de una colonia de aves acuáticas (Aves: Ardeidae, Threskiornithidae y Phalacrocoracidae) en la Reserva Estatal de Ciénegas y Manglares de la Costa Norte de Yucatán, México. MSc thesis, Universidad Nacional Autónoma de México, México.

Rog SM, Clarke RH, Cook CN (2017) More than marine: revealing the critical importance of mangrove ecosystems for terrestrial vertebrates. Divers Distrib 23:221–230.

Roll U, Feldman A, Novosolov M, Allison A, Bauer AM, Bernard R, Böhm M, Castro-Herrera F, Chirio L, Collen B (2017) The global distribution of tetrapods reveals a need for targeted reptile conservation. Nat Ecol Evol 1:1677–1682.

Ross PM, Underwood AJ (1997) The distribution and abundance of barnacles in a mangrove forest. Aust J Ecol 22:37–47.

Sasa M, Chaves GA, Patrick LD (2009) Marine reptiles and amphibians. In: Wehrtmann IS, Cortés J (eds) Marine Biodiversity of Costa Rica, Central America. Springer, p. 459–468.

Santiapillai C, de Silva M (2001) Status, distribution and conservation of crocodiles in Sri Lanka. Biol Conserv 97:305–318.

Sánchez-Nuñez DA, Mancera-Pineda JE (2012) Pollination and fruit set in the main neotropical mangrove species from the southwestern Caribbean. Aquat bot 103:60–65.
Saravanakumar A, Rajkumar M, Serebiah JS, Thivakaran GA (2007) Abundance and seasonal variations of zooplankton in the arid zone mangroves of gulf of Kachchh-Gujarat, westcoast of India. Pak J Biol Sci. https://doi.org/10.3923/pjbs.2007.3525.3532.
Schmid JR (2000) Activity patterns and habitat associations of Kemp's ridley turtles, Lepidochelys kempi, in the coastal waters of the Cedar Keys, Florida. PhD Thesis University of Florida.
Schmidt-Nielsen K, Fange R (1958) Salt glands in marine reptiles. Nature 182:783–785.
Sinclair A (2003) The role of mammals as ecosystem landscapers. Alces 39:161–176.
Sheaves M, Johnston R, Baker R (2016) Use of mangroves by fish: new insights from in-forest videos. Mar Ecol Prog Ser 549: 167–182.
Sheaves M, Molony B (2000) Short-circuit in the mangrove food chain. Mar Ecol Prog Ser 199:97–109.
Sousa WP, Kennedy PG, Mitchell BJ, Ordonez BM (2007) Supply-side ecology in mangroves: Do propagule dispersal and seedling establishment explain forest structure? Ecol Monogr. 77:53–76.
Smith III TJ, Boto KG, Frusher SD, Giddins RL (1991) Keystone species and mangrove forest dynamics: the influence of burrowing by crabs on soil nutrient status and forest productivity. Estuar Coast Shelf Sci 33:419–432.
Srikanth S, Yamauchi LSK, Chen Z (2016) Mangrove root: adaptations and ecological importance. Trees. https://doi.org/10.1007/s00468-015-1233-0.
Storr GM, Smith LA, Johnstone RE (2002) Snakes of Western Australia. Western Australian Museum.
Teoh HW, Hussein MAS, Chong VC (2014) Influence of habitat heterogeneity on the assemblages and shell use of hermit crabs (Anomura: Diogenidae). Zool Stud https://doi.org/10.1186/s40555-014-0067-6.
Todd BD, Willson JD, Gibbons JW (2010) The global status of reptiles and causes of their decline. In: Sparling D, Linder G, Bishop C, Krest S (eds.) Ecotoxicology of amphibians and reptiles, CRC Press Boca Raton, Fl, p. 47–67.
Tucker AD, Fitzsimmons NN, Gibbons JW (1995) Resource partitioning by the estuarine turtle Malaclemys terrapin: trophic, spatial, and temporal foraging constraints. Herpetologica 167–181.
Uchiyama M, Murakami T, Yoshizawa H (1990) Notes on the development of the crab-eating frog, *Rana cancrivora*. Zool Sci 7:73–78.
Unsworth R, Bell J, Smith D (2007) Tidal fish connectivity of reef and sea grass habitats in the Indo-Pacific. J Mar Biol Ass UK 87(5):1287–1296.
Van Valkenburgh B (1999) Major patterns in the history of carnivorous mammals. Annu Rev Earth Planet Sci 27:463–493.
Vanucci M (2001) What is so special about mangroves? Braz J Biol 61:599–603.
Varon-Lopez M, Dias AC, Fasanella CC (2014) Sulphur-oxidizing and sulphate-reducing communities in Brazilian mangrove sediments. Environ Microbiol. https://doi.org/10.1111/1462-2920.12237.
Veenakumari K, Mohanraj P, Bandyopadhyay AK (1997) Insect herbivores and their natural enemies in the mangals of the Andaman and Nicobar Islands. J Nat Hist 31:1105–1126.
Verweij M, Nagelkerken I (2007) Short and long-term movement and site fidelity of juvenile Haemulidae in back reef habitats of a Caribbean embayment. Hydrobiologia 592(1): 257–270.
Vitousek PM, Sanford Jr RL (1986) Nutrient cycling in moist tropical forest. Ann. Rev. Ecol. Syst 17: 137–167.
Voris HK, Murphy JC (2012) Sampling marine and estuarial reptiles. In: McDiarmid RW, Foster MS, Guyer C, Gibbons JW, Chernoff N (eds). Reptile biodiversity: standard methods for inventory and monitoring. University of California Press, Berkeley, 192–196.
Wang M, Huang Z, Shi F, Wang W (2009) Are vegetated areas of mangroves attractive to juvenile and small fish? Estuar Coast Shelf Sci 85:208–216.

Webb GJ, Messel H, Magnusson W (1977) The nesting of Crocodylus porosus in Arnhem Land, northern Australia. Copeia 238–249.

Wells FE (1984) Comparative distribution of macromolluscs and macrocrustaceans in a north-western Australian mangrove system. Aust J Mar Freshw Res 35:591–596.

Welsh Jr HH, Ollivier LM (1998) Stream amphibians as indicators of ecosystem stress: a case study from California's redwoods. Ecol Appl 8:1118–1132.

Wharton CH (1966) Reproduction and growth in the cottonmouths, *Agkistrodon piscivorus* Lacépede, of Cedar Keys, Florida. Copeia 149–161.

Woodroffe CD (1992) Mangrove sediments and geomorphology. In: Robertson AI, Alongi DM (eds) Tropical mangrove ecosystems, American Geophysical Union, Washington, DC. p 7–41.

Worm B, Barbier EB, Beaumont N, Duffy JE, Folke C, Halpern BS, Jackson J B, Lotze HK, Micheli F, Palumbi SR, Sala E, Selkoe KA, Stachowicz JJ, Watson R (2006) Impacts of biodiversity loss on ocean ecosystem services. Science. https://doi.org/10.1126/SCIENCE.1132294.

Wright P, Anderson P, Weng L, Frick N, Wong WP, Ip YK (2004) The crab-eating frog, *Rana cancrivora*, up-regulates hepatic carbamoyl phosphate synthetase I activity and tissue osmolyte levels in response to increased salinity. J Exp Zool A Comp Exp Biol 301:559–568.

Wu CS, Kam YC (2009) Effects of salinity on the survival, growth, development, and metamorphosis of *Fejervarya limnocharis* tadpoles living in brackish water. Zool Sci 26:476–482.

Yodthong S, Stuart BL, Aowphol A (2019) Species delimitation of crab-eating frogs (*Fejervarya cancrivora* complex) clarifies taxonomy and geographic distributions in mainland Southeast Asia. ZooKeys 883:119–153.

Zakaria M, Rajpar MN (2015) Assessing the fauna diversity of Marudu Bay mangrove forest, Sabah, Malaysia, for future conservation. Diversity 7:137–148. https://doi.org/10.3390/d7020137.

Zárate-Hernández R, Castillo-Rivera M, Sanvicente-Añorve L, Ortiz-Burgos S (2012) Spatial, diel, and seasonal changes in the fish community structure of a Mexican tropical estuary. Cienc Mar 38(4): 665–676.

Žydelist R, Kontautas A (2008) Piscivorous birds as top predators and fishery competitors in the lagoon ecosystem. Hydrobiologia 611: 45–54.

Factors Influencing Mangrove Ecosystems 4

Joanna C. Ellison

Abstract

Mangroves occur in coastal settings of estuaries, deltas, lagoons, open coasts and oceanic low islands. In these settings, mangrove attributes are influenced by physical factors of temperature, coastal typology, ocean currents and land barriers, wave action and sediment supply, river catchment discharge and sediment yield, and tidal range and inundation frequencies. Factors of gradients and tidal ranges control the lateral extent of mangroves through inundation frequency, and factors influencing accretion rates in the context of relative sea level change can shift or eliminate mangrove extents over time. Mangroves are however resilient systems within steady state equilibrium, that allows recovery from minor perturbations. Factors influencing mangroves can however exceed tipping points of tolerance, bringing a sudden change in ecosystem function and breakdown of equilibrium. Stressors that may cause critical reduction of mangrove resilience are the impacts from humans, climate becoming significantly drier, increased inundation, reduced sedimentation supply, and relative sea level rise. Rehabilitation can be successful if ecological guidance on mangrove restoration is followed, particularly topographic positioning with respect to tidal inundation frequency factors. Understanding of the physical factors that influence mangrove ecosystems that contribute to variation in processes, that result in spatial and temporal differences in mangrove attributes, is essential to effective management.

J. C. Ellison (✉)
School of Geography, Planning and Spatial Sciences, University of Tasmania, Launceston, TAS, Australia
e-mail: Joanna.Ellison@utas.edu.au

R. P. Rastogi et al. (eds.), *Mangroves: Ecology, Biodiversity and Management*,
https://doi.org/10.1007/978-981-16-2494-0_4

Keywords

Tidal inundation · Coastal typology · Oceanic currents · Temperature · Rainfall · Tipping points

4.1 Introduction

Mangrove forests occur most extensively on low latitude, low energy, sedimentary shorelines, between intertidal elevations. Global extends are greatest in Asia (38.7%), of a global 137,600 km^2, with 20.3% in Central America and the Caribbean, 20% in Africa, 11.9% in Oceania, 8.4% in North America and 0.7% in European Territories (Bunting et al. 2018). The IUCN Redlist recognises 70 species of mangroves in 17 families (Polidoro et al. 2010), the majority of which grow in mid to high intertidal positions indicating the stress of deeper water. Mangrove trees are also more diverse in Asia/SW Pacific (Spalding et al. 2010). Adaptations of mangroves to their intertidal, anaerobic, salty habitat are four main types: aerial roots, waxy leaves, salt regulation and viviparity, which is the germination of seeds before they drop into the low-oxygen salty water (Hutchings and Saenger 1987; Tomlinson 2016).

Adapting general concepts from Odum (1972), mangrove ecosystems include all of the organisms in the mangrove area interacting with the physical environment, so that a flow of energy leads to exchange of materials between the living and non-living parts of the system. The interactions with the physical environment are the factors influencing mangroves. Ecosystems are a dynamic network consisting of organisms, and interconnection with their environment and the definition includes abiotic factors which include those factors that limit distributions (Putman and Warren 1984).

Mangrove ecosystems therefore comprise mangrove trees, and floral associates, particularly at the landward edge, as well as fauna linked through food-chains, in a physical setting of sunlight, rainfall, tidal waters and sediments. The mangrove trees are the structuring species group creating habitats variety for other biotas, and the mangroves can biogeomorphically influence processes to become self-maintaining systems (Lugo and Snedaker 1974; Ellison 2019). Factors influence the spatial ranges of species, and particularly include the climatic factor of temperature, and physical factors that in intertidal systems include tidal inundation and hydroperiod (Sasmito et al. 2020), salinity, and position in the estuary (Valiela et al. 2018). Dispersal capability of propagules, restricted availability of mangrove habitats and adverse oceanic currents are major limiting factors of mangroves (Saenger et al. 2019).

Factors, attributes, characteristics and processes are all nouns that are applied to the analysis of mangrove systems, but each has different meanings (Lincoln et al. 1982; Macmillan 2020) (Fig. 4.1). Factors are an element or circumstance contributing to a result, and other words for the same definition are component, constituent or contributing element. A process is rather the series of actions or procedures or mechanisms that cause change, such as through explanatory verbs that may apply to

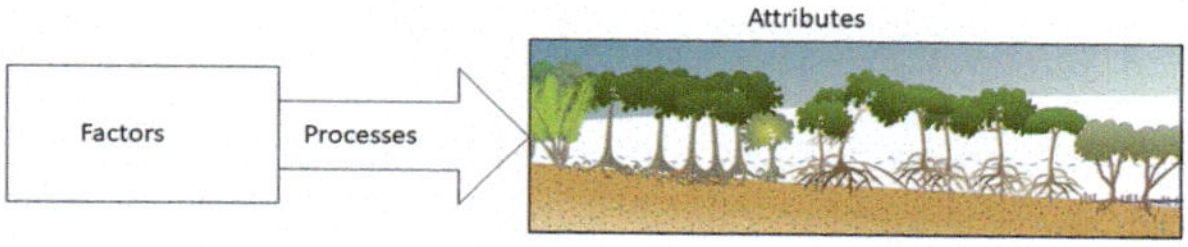

Fig. 4.1 Relationships between mangrove attributes and influencing factors and processes

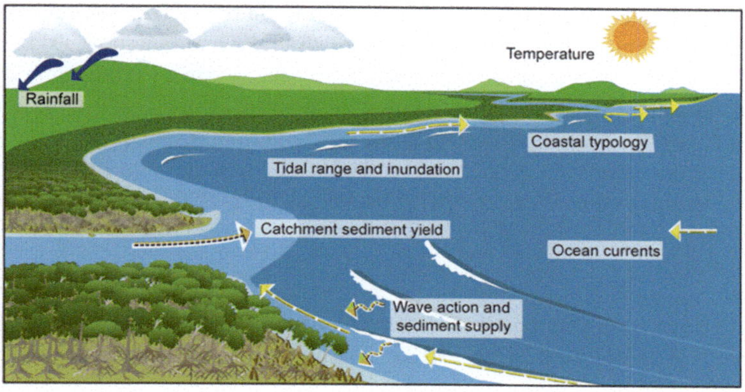

Fig. 4.2 Schematic illustration of the factors influencing mangroves (Source: adapted from Ellison 2019)

the mangroves of respire, photosynthesise, grow, breed, propagate, accumulate, metabolise, break down, transfer, adsorb, compress, or store. An attribute is a characteristic, quality, property so is a descriptive term for results of factors influencing processes. Hence the nouns all have associated meanings that help scientists understand how ecosystems such as mangroves work (Fig. 4.1). This chapter reviews physical factors that influence mangrove ecosystems that contribute to variation in processes to result in spatial and temporal differences in mangrove attributes.

In the well cited article "Factors influencing biodiversity and distributional gradients in mangroves", Duke et al. (1998) review the influences of environmental factors prevalent at global, regional, estuarine and intertidal scales. Environmental factors of influence on mangrove distributions are climate, hydrology, geomorphology and water conditions. Climatic factors include temperature, rainfall and storms, and hydrological factors include tides, currents and sea level. Geomorphological factors include sediments, catchment size and slope, and water condition factors include salinity, nutrients, oxygen and pH (Duke et al. 1998). The extent of influencing abiotic factors on mangroves are the climate (temperature and rainfall), tidal inundation, salinity fluctuations, exposure to wind and waves, sea level, currents variations and sedimentation (Duke 2016, 2017; Horstman et al. 2018). These have notable and profound influences on the distributions, functioning and well-being of mangrove habitats, along with human disturbance (Duke 2016; Saenger et al. 2019). Limiting abiotic factors controlling mangrove attributes are schematically illustrated in Fig. 4.2.

4.2 Climate: Temperature and Rainfall

Mangrove global extents are usually attributed to temperature limits of sea surface temperatures, based on mangrove distributions extending into higher latitudes with prevalent warmer currents (Tomlinson 2016). Temperature influences both photosynthetic and respiratory processes, controlling a large number of internal processes such as salt regulation, excretion and root respiration (Hutchings and Saenger 1987). Mangroves are tolerant of high temperature stresses, but mangrove photosynthesis is limited by low temperature at higher latitudes (Steinke and Naidoo 1991). Coastal distributions in the sub-tropics are influenced by the origin of influencing oceanic currents indicated by the 20 °C isotherm, with frost incidence also a limiting factor (Duke et al. 1998). *Avicennia marina* is the most tolerant of low temperatures in the southern hemisphere, where the highest latitudes for mangrove distributions are reached (Hutchings and Saenger 1987; Duke et al. 1998). Climatic factors primarily result in mangrove latitudinal expansion and contraction (Van der Stocken et al. 2019), particularly temperature (Saenger et al. 2019).

Winter air temperature extremes are an especially important factor controlling mangrove limits along the coasts of eastern North America, western Gulf of Mexico and eastern Asia (Osland et al. 2017). Frequency of freeze events controls the distribution, canopy height and coverage of *Avicennia germinans* at the species northern limits in Louisiana, USA, showing the limiting factor of winter temperature extremes (Osland et al. 2020). Freeze events cause mangrove mortality, leaf damage, while warmer conditions favour more abundant and tall mangroves. Concomitant abiotic factors are inundation, salinity and surface elevation gradients. However, for mangroves to establish in productive marshes at high latitudes an entry point is permitted by an external disturbance such as drought, fire, or storm damage causing marsh dieback (Osland et al. 2020). Coastal wetlands in northeast Florida have shown shifts between mangroves and saltmarshes at least 6 times in the last two centuries due to decadal scale fluctuations in frequency and intensity of extreme cold events (Cavanaugh et al. 2019). Northern extents of mangroves in the Gulf of Mexico and Florida approach 30°N. Indicators of vegetation photosynthetic activity and leaf area index at these mangrove limits are strongly controlled by climate factors, with significant relationships with annual minimum temperature (Cavanaugh et al. 2018). The influences on mangrove distributions of warm ocean currents are evident from the occurrence of mangroves in Bermuda at 32°N (Ellison 1997).

At latitudinal extremes such as New Zealand, mangrove diversity is restricted by factors such as temperature and day length, causing mean tree height to reduce with increasing latitude (Horstman et al. 2018). Temperature stress causes reduced net primary productivity resulting in smaller tree sizes, leading to stunted or dwarfed tree architecture at higher latitudes. For example, at the northern latitudinal limit of mangroves in the Atlantic at Bermuda 32°N, *Rhizophora mangle* reaches a maximum height of 6 m and *Avicennia germinans* 10 m (Ellison 1997), with strong seasonal patterns in productivity in a mean annual temperature range of 17–29 °C.

This contrasts with close to the equator in central Africa, where mangroves including these species can exceed 40 m in height (Ajonina et al. 2014).

As well as temperature limits on mangrove extents, along arid or semi-arid high latitude coastlines, increase or decrease in rainfall can lead to contraction and expansion. This is demonstrated from western North America, western Gulf of Mexico and Western Australia, though west central Africa lacks data to confirm the trend (Osland et al. 2017). The northern most arid mangrove margin of this region at 20°N showed dwarf growth of *Avicennia germinans* at average height of 1.2 m, and poor regeneration rates (Otero et al. 2016), distributed among halophytic vegetation.

Reduced rainfall and humidity cause reduction in mangrove diversity, photosynthesis, productivity and growth rates (Waycott et al. 2011). In low rainfall areas, high evaporation and soil salinity in higher intertidal mangrove zones cause stunting of mangroves or replacement by hypersaline sand flats. During drought periods, extensive mangrove mortality of such areas can occur (Duke et al. 2017; Lovelock et al. 2017). In the Gulf of Carpentaria, Australia, more than 74 km^2 of mangroves died during El Niño related drought, high temperatures and lower sea level (Duke et al. 2017; Duke 2020), mostly along the higher elevation zones. Similar losses have been analysed from Kakadu, northern Australia (Ashbridge et al. 2019) this affecting the landward *Avicennia marina* zone, and Western Australia (Lovelock et al. 2017) where 20–30% increase in soil salinisation occurred.

4.3 Ocean Currents and Land Barriers

The limiting factor of oceanic currents and land barriers on dispersal of mangrove propagules has influenced the species presence in different parts of the world, and the biodiversity complements of mangrove areas (Duke et al. 1998; Spalding et al. 2010). Tertiary records of mangrove occurrence do not show clear trends of mangrove migration with climate change, rather distributions developed over time are due to biogeographic factors and habitat availability (Ellison 2008). Biogeographic factors include oceanic connectivity, such as the western end of the Tethys Sea became isolated with the enclosure of the Mediterranean by the collision of Africa and southern Asia around 18 million years ago (Saenger 1998). At that time the pantropical mangrove flora became disjunct and subsequently developed the different species groups of the S.E. Asian and American centres of diversity (Ellison et al. 1999).

The Himalayan uplift and establishment of the Asian summer monsoon towards the Late Neogene further affected the coastal dynamics affecting mangrove distributions of the Indian subcontinent (Srivastava and Prasad 2019). Around three million years ago the Panama Gap closed with the collision of North and South America, to disjunct the mangroves on either side of the American land masses (Saenger 1998).

Dispersal capability and the barriers to oceanic connection isolated the American biodiversity centre from the Indo-West Pacific, demonstrating the combination of

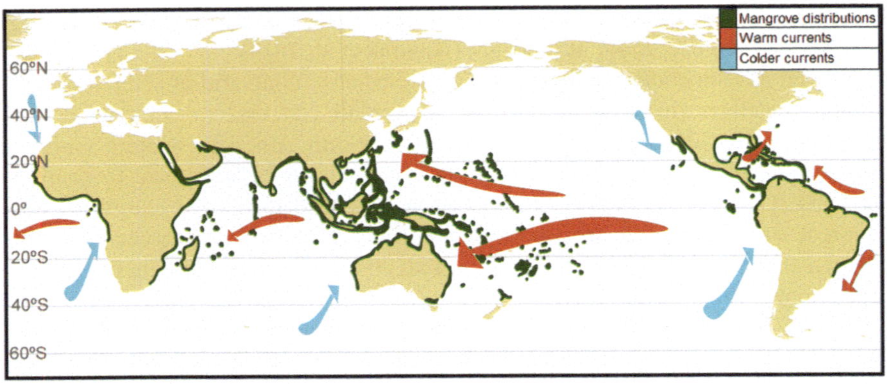

Fig. 4.3 World distributions of mangroves, and warm and cold currents that influence these distributions. Source: Mangrove distributions are adapted from Spalding et al. (2010)

these other factors influencing mangrove distributions. The most widely distributed genus in the Pacific Islands, *Rhizophora* shows high genetic variation across islands indicating that long-distance dispersal of propagules is rare (Yan et al. 2016). Other factors such as wind and water currents, estuarine geomorphology and coastal topographies likely played important roles in facilitating or blocking the mangrove gene flow (Yan et al. 2016). The distributions of mangroves across the world, and influences of warm and cold currents and land barriers are illustrated in Fig. 4.3.

4.4 Coastal Typology Factors

Factors of catchment area and geomorphic setting are regional scale influences on mangroves with geomorphological factors, including sediments and catchment size (Duke et al. 1998). Geomorphology and coastal topographies are important in facilitating or blocking the mangrove gene flow (Yan et al. 2016). At climate extremes, lack of suitable habitats within the dispersal range of the propagules can limit mangrove ranges (de Lange and De Lange 1994). These factors influence mangroves through variation of sediment type and supply, and mangrove habitats. Table 4.1 shows the prevalent categories of mangrove typological settings, each illustrated by a location in Fig. 4.4.

Drainage basin size, gradients and land uses influence fluvial runoff, solutes and sediment yield. Suspended sediment in river discharge allows coastal progradation and mangrove substrate accretion, and if abundant can cause mangroves to become opportunistic (Ellison 2019). Catchment geology and land uses influence sediment yield, with climate determining weathering rates and types, rainfall characteristics influencing runoff variability, vegetation and soil cover influencing slope hydrology, and river channel erodibility (Tedford and Ellison 2018). Sediment yield can be increased by catchment vegetation clearance to increase overland flow, riparian vegetation disturbance that increases river bank erodibility, and increase in flow

Table 4.1 Coastal typologies of mangroves

Typology	Deltas and estuaries, river dominated distributaries	Estuarine tidal creeks, strong tidal currents	Wave-dominated berm or barrier, sheltering low energy lagoons	Oceanic karst or small islands, lacking fluvial discharge
Mangrove locations	Seaward edge and river distributaries	Tidal creek margins and elongated islands	Inside low energy lagoons	Fringing or basin
Sediment supply type	Catchment discharged allochthonous sediment, flood deposits	Catchment allochthonous sediment reworked by tidal currents	Sand barriers, autochthonous mangrove fringed lagoons	Autochthonous mangrove basins and fringes
Examples shown in Fig. 4.4	North Rufiji delta, Tanzania	Fly delta, Papua New Guinea	Pichavaram, NE Cauvery River Delta, India	Pipon Island, Queensland, Australia

Fig. 4.4 Examples of coastal typologies of mangroves, imagery is adapted from Google Earth Pro 2020. (**a**) North Rufiji delta, Tanzania. (**b**) Fly delta, Papua New Guinea. (**c**) Pichavaram mangroves, Eastern India. (**d**) Pipon Island, Queensland, Australia

velocities caused by drainage of wetlands and channel straightening. These factors influence mangrove settings in estuarine, deltaic and riverine typologies (Table 4.1), to cause coastal progradation (Fig. 4.4A and B) that can be offset by relative sea level rise.

Wave-dominated coastlines are concentrated in the mid-latitudes, and feature intermittently closed lagoons (McSweeney et al. 2017). The strength of wave processes relative to tidal processes builds shore-parallel barriers (such as a spit or berm) which allow the sheltered habitats that mangroves favour as demonstrated in Fig. 4.4C. Analysis of Australian estuaries showed mangroves are most extensive in tide-dominated estuaries and deltas, and smallest in wave-dominated estuaries (Mahoney and Bishop 2018). The area increased with tidal rage, estuary entrance width and catchment area. By contrast, mangroves of New Zealand are mostly in wave-dominated settings, or river dominated, or a combination of both (Horstman et al. 2018).

Oceanic marine dominated mangrove settings have no fluvial sediment supply, and freshwater is limited to rainfall and groundwater outflow at low tide. This reduces mangrove biodiversity, with inner mangrove zone species of riverine habitats such as *Nypa fruticans* absent, and favours full salinity tolerant species such as *Rhizophora stylosa* in the Pacific and *Rhizophora mangle* in the Caribbean. With the rising sea level these systems can accumulate autochthonous sediment deterministically (Ellison 2019), sustaining low gradients across the mangroves to allow extensive areas even in microtidal ranges.

4.5 Tidal Range and Inundation Regimes

Tidal range varies across the mangrove world from microtidal in the Caribbean to macrotidal in locations such as North West Australia and eastern Africa, and for mangroves tidal range is a limiting factor in spatial extent (Ellison 2015). Accommodation space for the mangrove habitat is only from mean sea level to high tide levels, which acting against the intertidal gradient limits the spatial expansion of mangrove areas. This does not however mean that microtidal areas have narrow mangroves, as the geomorphic setting is a moderator (Table 4.1). Deterministic mangroves can maintain their own habitat in the tidal range by net accretion during slow relative sea level rise (Ellison 2019), and with substrate surfaces controlled by root mat development can achieve very low gradient surfaces in microtidal areas to allow expansive mangroves.

Micro-topography influences the distribution of mangroves, controlling the critical periods of tidal inundation and air exposure that governs the health of the forest (Watson 1928; Kjerfve 1990, Friess 2017). Flooding depths, durations and frequencies are critical factors in the survival of each mangrove species (Lewis et al. 2019), resulting in zones of different species, each with preferred habitats (Fig. 4.5). In the first research on this subject, Watson derived inundation classes for Port Klang, Malaysia, recording mangroves across a 2.13 m range (Friess 2017) within a 5.5 m spring tidal range (Port Klang Malaysia Marine Information Handbook 2010). This research commenced a mangrove paradigm relating ground surface elevation to frequency of flooding that result in species zonation (Friess 2017; Ellison 2019).

Fig. 4.5 Typical zonation by elevation of mangroves, with graph showing amount of time each zone has the ground surface in the air (*in yellow*) and under water (*in blue*)

Mangrove species are adapted to tidal inundation with aerial roots architectures; these include pneumatophores in *Avicennia* and *Sonneratia*, stilt roots in *Rhizophora*, and knee roots in *Bruguiera* and *Lumnitzera* species (Tomlinson 2016). These are illustrated in Fig. 4.5 from right to left. Each species has specific preferences for timeframes and depths of inundation as controlled by the level of tidal waters relative to the roots and their substrate, which results in species zonation by micro-elevation. Figure 4.5 adapts a standard tidal curve from Maritime Safety Queensland (2019: 118) to demonstrate in the base graph the proportion of a tidal cycle that mangroves at different elevation zones have their substrate in the air and underwater. This is 50% at the seaward edge at mean sea level, and increased time in the air towards the high tide margins, the gradient controlling oxygen availability and salinity.

Mangroves can experience decline and mortality when their inundation patterns are altered, and this is part of their vulnerability to rising sea levels (Ellison 2015). Inundation changes caused by direct human influences that alter hydrological exchange show clear evidence of this sensitivity. These are well documented from the decades before mangroves became better appreciated and valued, as they have become this century.

Hydrological alteration is a major cause of loss of mangroves in the Indo-West Pacific (Saenger et al. 2019), indicating the importance of the inundation regime as a key factor for mangrove vitality. Evidence is also well recorded from American longitudes, with a mortality of 15 hectares of *Avicennia germinans* and *Laguncularia racemosa* occurred in Puerto Rico when dredging caused an abrupt change to the hydrological regime, resulting in permanent flooding with cause of death attributed to reduced soil oxygen (Jimenez et al. 1985). Mortality of *Avicennia germinans* at Clam Bay, Florida occurred following blockage of tidal channels to cause excess inundation by accumulated flood water (Turner and Lewis 1997).

Blockage of tidal creeks in NW Australia caused death of 6 hectares of *Avicennia marina* and *Ceriops tagal* (Gordon 1988). Construction of an evaporator pond caused permanent ponding and mass mortality to about 11 km^2 of mangroves

(Gordon 1988). In Florida, impoundment of mangroves for mosquito control provided several early case studies of prolonged inundation causing mangrove mortality (Brockmeyer et al. 1997). *Avicennia germinans* trees are killed if their short pneumatophores are submerged after just weeks. The inundation was of sudden onset, and then sustained to cause mortality of *Rhizophora mangle* and *Avicennia germinans* at Indian River, East Florida as a result of 4 months flooding of 0.3–0.45 m depth (Harrington and Harrington 1982).

Following closure of the river mouth, water levels in the Kosi Estuary, Natal rose by about 0.3 m over the succeeding 140 days (Forbes and Cyrus 1992). Mass mortality to mangroves occurred, primarily *Avicennia marina*, with the bark of all trees rotted around 0.2 m above the mud level. Mangroves on higher elevations, mostly *Lumnitzera racemosa* survived.

Human-caused changes to hydrological exchange in Columbia resulted in hyper-salinisation of soils, resulting in mortality of 30,000 hectares of mangroves (Elster et al. 1999). At Clarence Estuary in NSW, structural flood mitigation works caused loss of tidal fluctuations in mangrove areas (Pollard and Hannan 1994). This caused near total loss of mangroves above the flood gates, from reduced tidal inundation.

Natural causes of sustained inundation have also been recorded to have caused mangrove mortality, further indicating the importance of this factor to system tolerance. River flooding causing inundation and sediment deposition on *Bruguiera* caused 42 hectares of mortality of in Java (Soerianegara 1968). Inundation of 2.5 m above the mean summer level falling slowly to 1.7–1.5 m two weeks later, and returning to typical levels by 2 months later caused mortalities to *Bruguiera gymnorrhiza* (Forbes and Cyrus 1992).

Flooding after enclosure by a natural sand bar of 50 hectares of *Avicennia* mangroves in Brunei-Darussalam, Borneo, brought water depths of 0.5 m above tidal high water for 8 weeks before the floodwater was released (Choy and Booth 1994). *Avicennia marina* suffered substantial losses of shrub-sized plants, while most trees >5 m survived. Necrosis (death of tissue) occurred to bark and surface wood tissues on mangroves stems below flood level, and death of pneumatophores of trees occurred.

Experimental flooding resulted in the lower ability of leaves to conduct water, and increase in stomatal closing in *Bruguiera gymnorrhiza*, leading to reduced rates of photosynthesis (Naidoo 1983). When lenticels of aerial roots become inundated, oxygen concentrations in the plant fall dramatically (Scholander et al. 1955). If inundation is sustained, low-oxygen conditions and mortalities follow.

Mangrove seedlings are thought to be more susceptible to death from excess inundation than adult trees. During sustained flooding of about 30 cm, all seedlings of *Avicennia germinans* and *Laguncularia racemosa* died within 1 month (Elster et al. 1999). Inundation effects on mangrove seedling growth have been demonstrated experimentally (Ellison and Farnsworth 1997), showing that normal tidal fluctuation around a 16 cm raised water level slowed growth at the sapling stage, and by 2.5 years age plants were 10–20% smaller than control plants.

Hence, while mangroves are tolerant of tidal inundation, inundation above normal for prolonged periods brings mangrove tree mortality, and decline is observed during less extreme flood events. This is due to reduced capability of gas exchange in the root systems (Jimenez et al. 1985). These case studies indicate the importance of the inundation factor (Fig. 4.5) in limiting mangroves, bringing a tipping point to mangrove system equilibrium.

4.6 Wave Action

Exposure to wave action is a limit to mangrove distributions (Duke 2017), and on wave-dominated coastlines mangroves are limited to sheltered lagoons inside barriers (Table 4.1, Fig. 4.4C). Wave energy is higher outside of the equatorial latitudes for mangrove distributions, as prevalent winds are stronger. Swell waves are generated by the frictional effect of wind as it passes over the ocean surface, where swell consists of long low waves generated in a major wave generation area due to the action of prevailing/storm winds on the water. Wave dispersion causes regularly spaced successions of waves to migrate from the source zone. As waves move towards land, wave velocity decreases and wave height increases, and offshore gradient moderates wave energy in approach to shore. Tsunami waves are a rare series of large waves of extremely long wavelength generated by undersea earthquakes or disturbance near the coast or in the ocean. Wave energy limits mangrove distributions, and large waves can cause damage and mortality. Friction of intertidal root systems and intertidal mud banks trapped under mangroves forces storm waves to spill not plunge in break. Energy is dissipated as the waves traverse the mangrove fringe.

An important value of mangroves to human communities is the protection the forest provides to land from waves and tidal surges, during cyclones and other storms. The dense foliage of mangroves combined with friction effects of aerial roots provides some facility in reducing wave power, including reducing damage from extreme events such as tsunami waves (Mazda et al. 2007). At high tide in a *Rhizophora* forest, there is a 50% decline in wave energy within 150 m of the seaward edge (Brinkman et al. 2007). The degree of protection obviously depends on the size of the waves, and the maturity and density of the forest (Yanagisawa et al. 2009).

Following the 2004 Asian tsunami it was found that, in some places, human deaths and loss of property were reduced by the presence of coastal vegetation shielding coastal villages (Dahdouh-Guebas et al. 2005; Katharesan and Rajendran 2005). However, a tsunami wave of over 6 m resulted in mangroves being mostly destroyed (Yanagisawa et al. 2009), with total destruction occurring from larger tsunamis (Koh et al. 2018).

Evidence of damage to mangroves from high energy events is available from many accounts over the last several decades, such as Cyclone Nargis impacting Myanmar in 2008, with winds of up to 215 km/h in the area of mangroves for about 12 hours, and most mangroves were damaged (Aung et al. 2011). Most defoliation and uprooting occurred to taller plantations, and *Avicennia marina* was found to be the most resistant to defoliation. After Hurricane Mitch hit Honduras in 1998, after 2 days of winds of around 200 km/h, nearly complete defoliation of mangroves occurred, and taller trees uprooted or broken. Of about 311 hectares of mangrove forests in Guanaja, only 11 hectares (3%) survived (Cahoon et al. 2003), leading to rapid sediment elevation loss as peat collapsed.

In moderate events, mangroves have an important role in protecting coasts during storm and tsunami events, primarily by reduction of the energy of waves (Massel et al. 1999; Dahdouh-Guebas et al. 2005; Katharesan and Rajendran 2005; Danielson et al. 2005; Vermaat and Thampanya 2006). During a typhoon wave height can be reduced by 20% on a mangrove shore of 100 m width (Mazda et al. 1997), and a similar tract of mangroves could reduce tsunami wave height by 50% (Hirashi and Harada 2003). The survival rate of mangroves to tsunami impacts is greater where there is a higher stem diameter (Yanagisawa et al. 2009).

Where waves are combined with wind action such as in severe storms, damage to mangroves has been shown to be severe in many case studies. Mangroves prefer habitats of low energy shorelines, sheltered from strong wave action. Mangroves can be impacted or damaged by high energy events such as storms or tsunamis. Consistent wave energy is a factor influencing the natural occurrence of mangroves on shorelines.

4.7 Tipping Points

Mangroves are complex systems influenced by physical factors of temperature, frost, wave power, tides, currents, sea level, sediment supply and geomorphic setting. These control attributes such as biodiversity, net productivity and sedimentation rates (Fig. 4.2). Such resilient systems are in dynamic equilibrium as shown by (1) in Fig. 4.6, where a small horizontal mark is a period of steady time (Ritter et al. 2002), such as represented by observations a scientist or student may make on visiting a mangrove area. Over time within that equilibrium state, ups and downs occur related to perturbations or El Niño cycles, these not so great as to break out of the long-term equilibrium, the system recovers. For example, storms physically impact mangroves, and recovery can then occur by local recruitment. Large waves cause erosion, and deposition and mangrove extent adjust. Pests or human cutting may have some impact, but the system recovers robustness and younger trees grow up in light gaps caused. The equilibrium steady state is represented in Fig. 4.6 by the regular up and down waves, such as at (1).

Tipping points have conceptual roots in the palaeoecological literature, with major geological time periods categorised by extinction events caused by major perturbations (Alvarez et al. 1984). After mangroves first appeared in the Late Palaeocene in the Asian/Pacific region, during warmer climates than today (Saenger 1998), they extended up to 50° north and south latitude (Ellison 2008). As temperatures fell from the end of the Eocene to the Pliocene, mangrove latitudinal

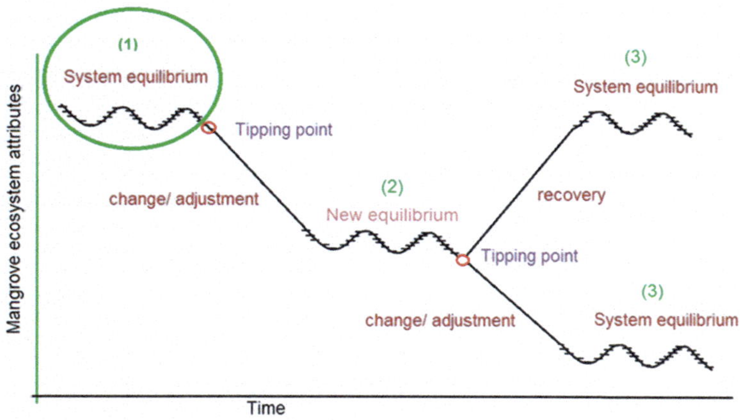

Fig. 4.6 Conceptual graph showing system equilibrium states of mangrove ecosystems over time, and tipping points leading to change and adjustment

ranges reduced, each location reaching a tipping point followed by the loss from the record. Tipping points are also evident in relation to sea level rise, of large mangrove expanses in the mid-Holocene that were replaced by offshore communities such as seagrass or lagoons (Ellison 2008, 2019). The actual causes of these changes by factors of salinity, inundation during mid-Holocene events, or temperature fall or frost in the Late Tertiary can only be inferred from stratigraphic records. They can however be understood through concurrent ecological system studies (Otero et al. 2016; Cavanaugh et al. 2019; Osland et al. 2020).

The interplay of growth factors promoted by beneficial factors balanced against stress or reduction factors regulates population size and hence ecosystem balance. Stable or balanced ecosystems have become a concept behind successfully managed ecosystems in a sustainable manner. Physical factors change to bring habitat stress or disturbance, to which the ecosystem may respond with resilience recovery or progress towards a tipping point of change to an altered system. A tipping point is any influence that causes a threshold change (Brunsden and Thornes 1979; van Belzen et al. 2017), such that regulatory processes that maintained the system equilibrium are so sufficiently altered that it cannot be recovered. For ecosystems a phenomenon of critical slowing down has been recognised (Scheffer et al. 2009; Van Belzen et al. 2017; Eslami-Andergoli et al. 2015) which can give early warning of an imminent tipping point. It is a state of low resilience, which for mangroves could be measured within dimensions of ecosystem robustness, magnitude of stressors and effectiveness of management actions (Ong and Ellison 2021).

Stressors that may cause critical reduction of mangrove resilience are impacts from humans, climate becoming significantly drier, increased inundation, reduced net sedimentation and relative sea level rise. The combination of low sediment supply and relative sea level rise indicates how factors may apply in combination to cause a tipping point that the system resilience may endure if only one were applying. A notable factor in mangrove recovery from oil spill impacts is the prior condition and resilience of the system before impact (Duke 2016). Ecosystem robustness indicators of critical reduction are mangrove condition and mortality, spatial loss, low net accretion or reduced adjacent ecosystem resilience such as the co-benefits of adjacent ecosystems such as coral reefs or seagrass are reduced. Stressors and reduced robustness apply in parallel, such as a critical threshold for future mangrove vulnerability is net mangrove accretion rates relative to rates of relative sea level rise (Ellison 2019). Management actions that may lead to a critical reduction of mangrove resilience are poor mangrove protection legislation or lack of enforcement, local community management in need of capacity building, and poor stakeholder involvement in mangrove conservation. These indicators of mangrove resilience can each be monitored and evaluated (Ong and Ellison 2021), to help potentially avoid a tipping point for the mangrove system.

A tipping point (Fig. 4.6) is a sudden change in ecosystem function, a breakdown of equilibrium. Change or adjustment after a tipping point is schematically shown as a line in Fig. 4.6, but could be a crash or jagged decline, with loss of the previous stable equilibrium. It is followed by a time of change/adjustment which may reach a new stable equilibrium. After impoundment, mangrove areas suffered mortality to become aquatic lagoons (Gordon 1988; Brockmeyer et al. 1997). Sea level rise caused conversion of mangrove areas to shallow bays (Ellison and Zouh 2012; Kemp et al. 2019). Mangrove areas are frequently cleared and transformed into aquaculture ponds (Spalding et al. 2010; Oh et al. 2017), such as to allow shrimp farming in Indonesia (Koh et al. 2018). Water exclusion such as by road construction causes mangrove mortality owing to hyper-salinisation (Elster et al. 1999), and subsequently mangroves have reduced, ecosystems have altered (de Klerk 2016; Gómez et al. 2017). These alternative systems become the new equilibrium for the location, as shown by (2) in Fig. 4.6. From here, a tipping point may further change the system to a different equilibrium, which may be rehabilitation back to a replanted mangrove area, these shown by (3) in Fig. 4.6. Rehabilitation can be successful if ecological guidance on mangrove restoration is followed (Lewis and Brown 2014; Lewis et al. 2019), particularly topographic positioning with respect to tidal inundation frequency factors (Ellison 2020).

4.8 Conclusions

Mangroves are sensitive to even minor changes in influencing factors (Fig. 4.1), such as altered drainage patterns, saltwater intrusion, accretion or erosion in response to changes in sea level (Ellison 2009). Each factor influences mangroves simultaneously (Fig. 4.2), though with varying intensity relative to coastal setting (Table 4.1), worldwide location (Fig. 4.3) and over time such as with climate change. The response of mangroves to these factors is observed through variations in the composition and relative abundance of plant species within the mangrove habitat (Blasco et al. 1996; Ellison 2005). Although the responses may be gradual, particularly in undisturbed systems, the alterations in coverage and composition of species can be used to assess the effects of climate change and other environmental impacts on mangrove habitats. This can be demonstrated through palaeo-environmental reconstruction (Ellison 2019), spatial monitoring (Bunting et al. 2018) or mangrove ecosystem monitoring (Ellison et al. 2012).

Understanding of the physical factors controlling mangrove ecosystems is essential to effective management. Factors of temperature, ocean currents and land barriers influence the spatial range of species, and rainfall can be a limiting factor in drier areas. Tidal ranges control vertical mangrove extents, across intertidal gradients and hydroperiod controls the locations of different mangrove zones. Coastal typology influences sediment supply, and the relative balance between wave, tide and river processes. Change in factors can bring a tipping point of loss of ecosystem equilibrium, followed by a change and adjustment to a different system which may be undesired to management goals. The rehabilitation investment may be able to return mangrove ecosystem balance, if the factors for mangrove successful growth are managed.

References

Ajonina GN, Kairo J, Grimsditch G, Sembres T, Chuyong G, Diyouke E (2014) Assessment of mangrove carbon stocks in Cameroon, Gabon, the Republic of Congo (RoC) and the Democratic Republic of Congo (DRC) including their potential for reducing emissions from deforestation and forest degradation (REDD+). In: Diop S, Barusseau J-P, Descamps C (eds) The land/ ocean interactions in the coastal zone of West and Central Africa, Springer, Cham, p. 177–189.

Alvarez W, Alvarez LW, Asaro F, Michel HV (1984) The end of the Cretaceous: sharp boundary or gradual transition? Science 223(4641):1183–1186.

Ashbridge EF, Bartolo R, Finlayson CM, Lucas RM, Rogers K, Woodroffe CD (2019) Assessing the distribution and drivers of mangrove dieback in Kakadu National Park, northern Australia. Estuar Coast Shelf Sci 228. https://doi.org/10.1016/j.ecss.2019.106353

Aung TY, Than MM, Katsuhiro O, Yukira M (2011) Assessing the status of three mangrove species restored by the local community in the cyclone-affected area of the Ayeyarwady Delta, Myanmar. Wetl Ecol Manage 19(2):195–208.

Blasco F, Saenger P, Janodet E (1996) Mangroves as indicators of coastal change. Catena 27:167–178.

Brinkman RM, Massel SR, Ridd PV, Furukawa K (2007) Surface wave attenuation in mangrove forests. Proc 13th Austral Coastal Ocean Engineer Conf 2:941–949.

Brockmeyer RE Jr, Rey JR, Virnstein RW, Gilmore RG, Ernest L (1997) Rehabilitation of impounded estuarine wetlands by hydrologic reconnection to the Indian River Lagoon, Florida (USA). Wetl Ecol Manage 4:93–109.

Brunsden D, Thornes JB (1979) Landscape sensitivity and change. Trans Inst Brit Geogr 4 (4):463–484.

Bunting P, Rosenqvist A, Lucas RM, Rebelo LM, Hilarides L, Thomas N, Hardy A, Itoh T, Shimada M, Finlayson CM (2018) The global mangrove watch-a new 2010 global baseline of mangrove extent. Remote Sens 10:1669. https://doi.org/10.3390/rs10101669

Cavanaugh KC, Osland MJ, Bardou R, Hinojosa-Arango G, López-Vivas JM, Parker JD, Rovai AS (2018) Sensitivity of mangrove range limits to climate variability. Global Ecol Biogeogr 27 (8):925–935.

Cavanaugh KC, Dangremond EM, Doughty CL, Williams AP, Parker JD, Hayes MA, Rodriguez W, Feller IC (2019) Climate-driven regime shifts in a mangrove–salt marsh ecotone over the past 250 years. Proc Nat Acad Sci 116(43):21602–21608.

Cahoon DR, Hensel P, Rybczyk J, McKee KL, Proffitt CE, Perez BC (2003) Mass tree mortality leads to mangrove peat collapse at Bay Islands, Honduras after Hurricane Mitch. J Ecol 91 (6):1093–1105.

Choy SC, Booth WE (1994) Prolonged inundation and ecological changes in an *Avicennia* mangrove: implications for conservation and management. Hydrobiol 285(1):237–247.

Dahdouh-Guebas F, Jayatissa LP, Di Nitto D, Bosire JO, Lo Seen D, Koedam N (2005) How effective were mangroves as a defence against the recent tsunami? Curr Biol 15:R443–447.

Danielson F, Soerensen M, Olwig M, Selvam V, Parish F, Burgess N, Hiraishi T, Karunagaran V, Rasmussen M, Hansen L, Quarto A, Nyoman S (2005) The Asian tsunami: A protective role for coastal vegetation. Science 310:643.

de Klerk J (2016) The slow death of the Ciénaga Grande de Santa Marta. The City Paper (Bogata) 22 September 2016. https://thecitypaperbogota.com/features/the-slow-death-of-the-cienaga-grande-de-santa-marta/14235.

de Lange WP, De Lange PJ (1994) An appraisal of factors controlling the latitudinal distribution of mangrove (*Avicennia marina* var. *resinifera*) in New Zealand. J Coast Res 10(3):539–548.

Duke NC (2020) Mangrove harbingers of coastal degradation seen in their responses to global climate change coupled with ever-increasing human pressures. CHEC Hum Ecol 30:19–23.

Duke NC (2016) Oil spill impacts on mangroves: recommendations for operational planning and action based on a global review. Mar Pollut Bull 109(2):700–715.

Duke NC (2017) Mangrove floristics and biogeography revisited: further deductions from biodiversity hot spots, ancestral discontinuities, and common evolutionary processes. In: Rivera-Monroy VH, Lee SY, Kristensen E, Twilley RR (eds) Mangrove ecosystems: A global biogeographic perspective, Springer, Cham, p. 17–53.

Duke NC, Kovacs JM, Griffiths AD, Preece L, Hill DJ, Van Oosterzee P, Mackenzie J, Morning HS, Burrows D (2017) Large-scale dieback of mangroves in Australia's Gulf of Carpentaria: a severe ecosystem response, coincidental with an unusually extreme weather event. Marine Freshw Res 68(10):1816–1829.

Duke NC, Ball MC, Ellison JC (1998) Factors influencing biodiversity and distributional gradients in mangroves. Global Ecol Biogeog Let 7:27–47

Ellison AM, Farnsworth EJ (1997) Simulated sea level change alters anatomy, physiology, growth, and reproduction of red mangrove (*Rhizophora mangle* L.). Oecologia 112(4):435–446.

Ellison AM, Farnsworth EJ, Merkt RE (1999) Origins of mangrove ecosystems and the mangrove biodiversity anomaly. Global Ecol. Biogeogr. 8, 95–115.

Ellison JC (2020) Mangrove ecosystem-based adaptation: Advice on improved success. CHEC Hum Ecol 30:37–40.

Ellison JC (2019) Biogeomorphology of mangroves. In: Wolanski E, Cahoon D, Perillo GME (eds) Coastal wetlands: An ecosystem integrated approach, Elsevier Science, Amsterdam, p. 687–715.

Ellison JC (2005) Holocene palynology and sea-level change in two estuaries in Southern Irian Jaya. Palaeogeogr Palaeoclimatol Palaeoecol 220:291–309.

Ellison JC (2009) Geomorphology and sedimentology of mangrove swamps. In: Wolanski E, Cahoon D, Perillo GME (eds) Coastal wetlands: an ecosystem integrated approach. Elsevier Science, Amsterdam, p 564–591.

Ellison JC (2015) Vulnerability assessment of mangroves to climate change and sea-level rise impacts. Wetl Ecol Manage 23:115–137.

Ellison JC (2008) Long-term retrospection on mangrove development using sediment cores and pollen analysis. Aquat Bot 89:93–104.

Ellison JC (1997). Mangrove community characteristics and litter production in Bermuda. In: Kjerfve B, Lacerda LD, Diop ES (eds) Mangrove forests of the Latin America and Africa regions, UNESCO, Paris, p. 5–17.

Ellison JC, Jungblut V, Anderson P, Slaven C (2012) Manual for mangrove monitoring in the Pacific Islands Region. Secretariat of the Pacific Regional Environment Program, Apia, Samoa.

Ellison JC, Zouh I (2012) Vulnerability to climate change of mangroves: Assessment from Cameroon, Central Africa. Biology 1:617–638.

Elster C, Perdomo L, Schnetter M-L (1999) Impact of ecological factors on the regeneration of mangroves in the Cienaga Grande de Santa Marta, Colombia. Hydrobiol 413:35–46.

Eslami-Andergoli L, Dale PE, Knight JM, McCallum H. (2015) Approaching tipping points: a focussed review of indicators and relevance to managing intertidal ecosystems. Wetl Ecol Manage 23(5):791–802.

Forbes AT, Cyrus DP (1992) Impact of a major cyclone on southeast African estuarine lake system. Neth J Sea Res 30:265–272.

Friess DA (2017) JG Watson, inundation classes, and their influence on paradigms in mangrove forest ecology. Wetl 37(4):603–613.

Gómez JF, Byrne ML, Hamilton J, Isla F (2017) Historical coastal evolution and dune vegetation in isla salamanca national park, colombia. J Coast Res 33(3):632–641.

Google Earth Pro (2020) Imagery attributed to CNES/Airbus, Terra Metrics, and Maxar Technologies, used following attribution guidelines https://www.google.com/permissions/geoguidelines/attr-guide/

Gordon DM (1988) Disturbance to mangroves in tropical-arid Western Australia: hypersalinity and restricted tidal exchange as factors leading to mortality. J Arid Environ 15:117–145.

Harrington, RW, Harrington ES (1982) Effects on fishes and their forage organisms of impounding a Florida salt marsh to prevent breeding by salt marsh mosquitos. Bull Mar Sci 32:523–531.

Hirashi T, Harada K (2003) Greenbelt tsunami prevention in South-Pacific region. Rep Port Airport Res Inst 42:3–25.

Horstman EM, Lundquist CJ, Bryan KR, Bulmer RH, Mullarney JC, Stokes DJ (2018) The dynamics of expanding mangroves in New Zealand. In: Makowski C, Finkl CW (eds) Threats to mangrove forests, Springer, Cham, p. 23–51.

Hutchings P, Saenger P (1987) Ecology of Mangroves. University of Queensland Press, St Lucia.

Jimenez JA, Lugo AE, Cintron G (1985) Tree mortality in mangrove forests. Biotrop 17:177–185.

Katharesan K, Rajendran N (2005) Coastal mangrove forests mitigates tsunami. Estuar Coast Shelf Sci 65:601–606.

Kemp AC, Vane CH, Khan NS, Ellison JC, Engelhart SE, Horton BP, Nikitina D, Smith SR, Rodrigues LJ, Moyer RP (2019) Testing the utility of geochemical proxies to reconstruct holocene coastal environments and relative sea level: a case study from Hungry Bay, Bermuda. Open Quat 5(5):1–8.

Kjerfve B (1990) Manual for investigation of hydrological processes in mangrove ecosystems. Baruch Institute for Marine Biology and Coastal Research, University of South Carolina.

Koh H, Teh SY, Kh'ng XY, Raja Barizan RS (2018) Mangrove forests: protection against and resilience to coastal disturbances. J Trop For Sci 30(5):446–460.

Lewis RR, Brown BM, Flynn LL (2019) Methods and criteria for successful mangrove forest rehabilitation. In: Wolanski E, Cahoon D, Perillo GME (eds) Coastal wetlands: An ecosystem integrated approach, Elsevier Science, Amsterdam, p 863–887.

Lewis RR, Brown B (2014) Ecological mangrove rehabilitation-a field manual for practitioners. Mangrove Action Project Indonesia, Blue Forests, Canadian International Development Agency and OXFAM, Gatineau, Quebec, Canada. http://mangroverestoration.com/.

Lincoln R, Boxshall G, Clark P (1982) Dictionary of ecology, evolution and systematics. Cambridge University Press, Cambridge.

Lovelock CE, Feller IC, Reef R, Hickey S, Ball MC (2017) Mangrove dieback during fluctuating sea levels. Sci Rep 7(1):1–8.

Lugo AE, Snedaker SC (1974) The ecology of mangroves. Ann Rev Ecol Syst 5(1):39–64.

Macmillan dictionary (2020) https://www.macmillandictionary.com/thesaurus-category/british/biological-processes.

Mahoney PC, Bishop MJ (2018) Are geomorphological typologies for estuaries also useful for classifying their ecosystems? Aquat Conserv 28(5):1200–1208.

Maritime Safety Queensland (2019) Queensland tide tables standard port tide times 2020. Department of Transport and Main Roads, State of Queensland.

Mazda Y, Magi M, Kogo M, Hong PN (1997) Mangroves as a coastal protection from waves in the Tong King delta, Vietnam. Mangr Salt Marsh 1(2):127–135.

Mazda Y, Parish F, Danielsen F, Imamura F (2007) Hydraulic functions of mangroves in relation to tsunamis. Mangr Sci 4(5):57–67.

McSweeney SL, Kennedy DM, Rutherfurd ID, Stout JC (2017) Intermittently closed/open lakes and lagoons: Their global distribution and boundary conditions. Geomorph 292:142–152.

Naidoo G (1983) Effects of flooding on leaf water potential and stomatal resistance in *Bruguiera gymnorrhiza*. New Phytol 93:369–373.

Odum EP (1972) Ecosystem theory in relation to man. In: Wiens JA (ed) Ecosystem structure and function, Oregon State University Press, Corvallis, p. 11–24.

Oh RRY, Friess DA, Brown BM (2017) The role of surface elevation in the rehabilitation of abandoned aquaculture ponds to mangrove forests, Sulawesi, Indonesia. Ecol Engineer 100:325–334.

Ong WJ, Ellison JC (2021) A framework for the quantitative assessment of mangrove resilience. In: Friess D, Sidik F (eds) Dynamic sedimentary environments of mangrove coasts, Elsevier, Amsterdam, p. 513–538.

Osland MJ, Day RH, Hall CT, Brumfield MD, Dugas JL, Jones WR (2017) Mangrove expansion and contraction at a poleward range limit: climate extremes and land-ocean temperature gradients. Ecology 98:125–137.

Osland MJ, Day RH, Michot TC (2020) Frequency of extreme freeze events controls the distribution and structure of black mangroves (*Avicennia germinans*) near their northern range limit in coastal Louisiana. Divers Distrib 00:1–17. https://doi.org/10.1111/ddi.13119

Otero V, Quisthoudt K, Koedam N, Dahdouh-Guebas F (2016) Mangroves at their limits: Detection and area estimation of mangroves along the Sahara Desert Coast. Remote Sens 8:512. https://doi.org/10.3390/rs8060512

Polidoro BA, Carpenter KE, Collins L, Duke NC, Ellison AM, Ellison JC, Farnsworth EJ, Fernando ES, Kathiresan K, Koedam NE, Livingstone SR, Miyagi T, Moore GE, Nam VN, Primavera JH, Salmo III SG, Sanciangco JC, Sukardjo S, Wang Y, Yong JWH (2010) The loss of species: Mangrove extinction risk and geographic areas of global concern. PLoS One 5(4): e10095. https://doi.org/10.1371/journal.pone.0010095

Pollard DA, Hannan JC (1994) The ecological effects of structural flood mitigation works on fish habitats and fish communities in the lower Clarence River system of south-eastern Australia. Estuar 17:427–461.

Port Klang Malaysia Marine Information Handbook (2010) https://www.pka.gov.my/phocadownloadpap/information/marine/MARINE%20HANDBOOK.pdf.

Putman R, Warren SD (1984) Principles of ecology. Chapman and Hall, London.

Ritter DF, Kochel C, Miller JR (2002) Process geomorphology. McGraw Hill, Dubuque.

Saenger P (1998) Mangrove vegetation: an evolutionary perspective. Mar Freshw Res 49:277–286.

Saenger P, Ragavan P, Sheue CR, López-Portillo J, Yong JWH, Mageswaran, T (2019) Mangrove biogeography of the Indo-Pacific. In: Gul B, Böer B, Khan M, Clüsener-Godt M, Hameed A (eds) Sabkha ecosystems, Springer, Cham, p. 379–400.

Sasmito SD, Kuzyakov Y, Lubis AA, Murdiyarso D, Hutley LB, Bachri S, Friess DA, Martius C, Borchard N (2020) Organic carbon burial and sources in soils of coastal mudflat and mangrove ecosystems. Catena 187:104414.

Scheffer M, Bascompte J, Brock WA, Brovkin V, Carpenter SR, Dakos V, Held H, Van Nes EH, Rietkerk M, Sugihara G (2009) Early-warning signals for critical transitions. Nature 461 (7260):53–59.

Scholander PF, van Dam L, Scholander SI (1955) Gas exchange in roots of mangroves. Amer J Bot 42(1):92–98.

Soerianegara I (1968) The causes of mortality of *Bruguiera* trees in the mangrove forest near Tjilatjap, Central Java. Rimba Ind 13:1–11.

Spalding M, Kainuma M, Collins L (2010) World Atlas of Mangroves. Earthscan, London and Washington, DC.
Srivastava J, Prasad V (2019) Evolution and paleobiogeography of mangroves. Mar Ecol 40(6): e12571.
Steinke TD, Naidoo Y (1991) Respiration and net photosynthesis of cotyledons during establishment and early growth of propagules of the mangrove, *Avicennia marina*, at three temperatures. S Afr J Bot 57:171–174.
Tedford, M, Ellison, JC (2018) Analysis of river rehabilitation success, Pipers River, Tasmania. Ecol Indic 91:350–358.
Tomlinson PB (2016) The botany of mangroves. Cambridge University Press, Cambridge.
Turner RE, Lewis RR (1997) Hydrologic restoration of coastal wetlands. Wetl Ecol Manage 4:65–72.
Valiela I, Pascual J, Giblin A, Barth-Jensen C, Martinetto P, Otter M, Stone T, Tucker J, Bartholomew M, Viana IG (2018) External and local controls on land-sea coupling assessed by stable isotopic signatures of mangrove producers in estuaries of Pacific Panama. Mar Env Res 137:133–144.
van Belzen J, van de Koppel J, Kirwan M, van der Wal D, Herman PMJ, Dakos V, Kefi S, Scheffer M, Guntenspergen GR, Bouma TJ (2017) Vegetation recovery in tidal marshes reveals critical slowing down under increased inundation. Nat Commun 8:15811. https://doi.org/10.1038/ncomms15811.
Van der Stocken T, Wee AK, De Ryck DJ, Vanschoenwinkel B, Friess DA, Dahdouh-Guebas F, Simard M, Koedam N, Webb EL (2019) A general framework for propagule dispersal in mangroves. Biol Rev 94(4):1547–1575.
Vermaat JE, Thampanya U (2006) Mangroves mitigate tsunami damage: A further response. Estuar Coast Shelf Sci 69:1–3.
Watson JG (1928) Mangrove forests of the Malay Peninsula. Malay For Rec 6:1–275.
Waycott M, McKenzie LJ, Mellors JE, Ellison JC, Sheaves MT, Collier C, Schwarz AM, Webb A, Johnson JE, Payri CE (2011) Vulnerability of mangroves, seagrasses and intertidal flats in the tropical Pacific to climate change. In: Bell JD, Johnson JE, Hobday AJ (eds) Vulnerability of fisheries and aquaculture in the Pacific to climate change, Secretariat of the Pacific Community, Noumea, p. 297–368.
Yan YB, Duke NC, Sun M (2016) Comparative analysis of the pattern of population genetic diversity in three Indo-west Pacific *Rhizophora* mangrove species. Front Plant Sci 30(7):1434.
Yanagisawa H, Koshimura S, Goto K, Miyagi T, Imamura F, Ruangrassamee A, Tanavud C (2009) The reduction effects of mangrove forest on a tsunami based on field surveys at Pakarang Cape, Thailand and numerical analysis. Estuar Coast Shelf Sci 81:27–37.

Energy Flux in Mangrove Ecosystems

5

Engku Azlin Rahayu Engku Ariff,
Ahmad Faris Seman Kamarulzaman, and Mohd Nazip Suratman

Abstract

The energy balance at the Earth's land surface requires that the energy gained from net radiation be balanced by the fluxes of sensible and latent heat to the atmosphere and the storage of heat in the soil. Latent and sensible heat are crucial variables in ecology, hydrology and meteorology because they give influence to a climate that can be used to determine environmental parameters which alter mass and energy exchange between the soil and the atmosphere. These energy fluxes are a primary determinant of surface climate. The annual energy balance at the land surface differs geographically depending on the incoming solar radiation and soil water availability. Thus, provides key insight into processes such as photosynthesis and respiration. Throughout the days and years, energy flux varies depending on the diurnal and annual cycles of solar radiation and also soil water availability. The various terms in the energy budget (net radiation, sensible heat flux, latent heat flux and soil heat flux) are illustrated for different climate zones and for various vegetation types. In this review, the energy flux in mangrove ecosystem in the trophic level is highlighted. We propose that integrating view point from community and ecosystem ecology in mangrove by quantifying energy fluxes from solar energy to producers, consumers and decomposers which can provide vital information for understanding the connections between the

E. A. R. Engku Ariff · A. F. Seman Kamarulzaman
Faculty of Applied Sciences, Universiti Teknologi MARA (UiTM) Pahang Campus, Bandar Tun Razak Jengka, Malaysia

M. N. Suratman (✉)
Faculty of Applied Sciences and Institute for Biodiversity and Sustainable Development, UiTM, Shah Alam, Malaysia
e-mail: nazip@uitm.edu.my

R. P. Rastogi et al. (eds.), *Mangroves: Ecology, Biodiversity and Management*,
https://doi.org/10.1007/978-981-16-2494-0_5

diversity of complex multitrophic systems as well as multiple ecosystem functions in mangrove ecosystem.

Keywords

Mangroves · Energy flux · Photosynthesis · Respiration · Consumer · Decomposer

5.1 Introduction

Mangrove is referring to the trees, large shrubs, ferns and palm that grow in the intertidal coastal zone of the tropics and subtropics regions of the world between approximately 30° N and 30° S latitude (Spalding 2010; Giri et al. 2011). There is a range of 50 to 73 different species of mangrove all over the world as the exact number of species is still under discussion due to different classification systems (FAO 2007; Spalding 2010).

Mangrove is considered as vascular halophytes plant as it is able to withstand or tolerate the high salt condition (Flowers and Colmer 2015). The plant can impressively grow in the harsh environment of oxygen poor intertidal coastal area as the plant has developed the convergent adaptation in its morphology, physiology, which includes the roots, leaves and stem anatomy and reproductive strategies. The plant has an aerial rooting system which functions in respiration during high tide. It also functions as anchorage and for nutrient uptake in the muddy soil (FAO 2007).

The plant adapted the high salt soil by preventing or reducing the Na+ from the outside of the root from entering the xylem through apoplastic route by having a wider Casparian strip and by producing a compatible solute as osmoprotectants (Naskar and Palit 2015). Some of the adaptation to reduce salt content in the mangrove is species-specific such as *Xylocarpus* and *Excoecaria* channel the salt into the senescent leaves, while *Aegialitis*, *Aegiceras* and *Avicennia* secrete out the salt actively using special glands through the leaves (Spalding 2010). Despite preventing the entering and secreting out the salt from the plant, mangroves also have succulent leaves with thick cuticle wax and large vacuole to preserve water (Naskar and Palit 2015).

All mangroves take advantage on the water tides to disperse their seeds. However, mangrove did not undergo seed dormant stage as the plant reproduces through vivipary where the seeds produced immediately germinate into a seedling. The seedling will fall off from the parent tree and float until it sinks, or the tides are low to take root and lodge in the soil (Spalding 2010; Feller 2018). The different adaptation of the mangrove towards the regular environmental changes and different intertidal condition has led to different diversity of mangrove plant which can be categorized into five types of mangrove forest, namely fringe, basin, riverine, overwash and dwarf (Spalding 2010; Feller 2018). Each of the categories is resulted from different ecosystems and will support an incredible diversity of creatures, including some species unique to mangrove forests.

In 1993, the world total area of mangroves estimated from 91 countries was 19.89 million hectares (Fisher and Spalding 1993) and further estimated by Giri et al.

(2011) using the data in the year of 2000 from 118 countries as 13.77 million hectares. The estimate shows a reduction up to 30.7% of the mangrove's world total area in just within seven years period. The reduction of the mangroves will also leave an impact towards energy flux between the biodiversity in its ecosystem. The unique geomorphic characteristic of mangrove grows where the land, ocean and freshwater overlap is a great habitat or hunting ground for a wide range of organisms (Spalding 2010). The distribution of organisms may vary according to the regions of the world, salinity and tidal level.

The ecosystems of mangrove may involve the sustainability of some endangered animals such as proboscis monkey (*Nasalis larvatus*) which is endemic to mangrove forests of Sabah and Sarawak states of Malaysia, the pygmy three-toed sloth (*Bradypus pygmaeus*), in mangrove forests at the off coast of Panama, and Bengal tigers (*Panthera tigris*) in the largest mangrove forests in Sundarban of Bangladesh. The same is true with the endangered mangrove hummingbirds (*Amazilia boucardi*) that feed on the sweet nectar of the rare and vulnerable Pacific mangroves and only grow in about a dozen patchy forests from Nicaragua to Ecuador.

Mangroves not only provide habitat, but also as spawning grounds and nurseries which make the diversity ranging from young to adult organisms. The diversity of organism includes reptiles, amphibian, mammals and birds. However, major parts of mangrove food web are driven by detritus (Spalding 2010). The dry leaves that fall from the tree are decomposed by detritivore and saprophyte which recycle the energy flux to be re-used by the plants. This chapter aimed at highlighting the energy fluxes in mangrove ecosystems in order to enhance our understanding of the magnitude and changes of their regulatory mechanisms, which will help to understand the mangrove ecosystem energy balance.

5.2 Energy Fluxes Within and Over Mangroves

Active regulation of water, energy and carbon fluxes between forests and the atmosphere has been conducted over the past few years, in order to understand both forest functioning and the role of forests as sinks or sources. Similar to all the other forest ecosystems, mangrove trees received powerful energy sources from the sun that supplied the energy to the earth that gave the strongest influence on other environmental factors.

There are several energy fluxes processes that occur between Earth's surface and overlying atmospheres, which include (1) Thermal conduction of heat energy within the ground, (2) Absorption and emission of 'natural' electromagnetic radiation by the surface, (3) Evaporation of water stored in the soil or condensation of atmospheric water vapour onto the surface and (4) Turbulent transfer of heat energy towards or away from the surface within the atmosphere.

Therefore, each of these processes is correlated with an energy flux density (SI unit: $J\ m^{-2}\ s^{-1}$) that can be referred to as the rate of energy transferred perpendicular to a surface of the unit area. The SI unit for the energy flux density is also equivalent to Wm^{-2}.

5.3 Short-wave Radiation and Long-wave Radiation

The earth's surface is affected by two main types of radiations, namely short-wave (280–2800 nm) and long-wave (2800–100,000 nm) components. Short-wave radiation comprises higher energy about 85% as compared to long-wave radiation at the waveband from 0.1 to 5.0 μm, which includes ultraviolet (0.1–0.4 μm), visible light (0.4–0.7 μm) and near-infrared (0.7–5.0 μm) spectral regions. Furthermore, short-wave radiation is the radiation received from the sun and is composed of both the beams (or direct) and diffuse radiations where beam radiation reaches the earth's surface directly from the sun after travelling through space, while diffuse radiation is subjected to interference from any interspatial matter (Kumar et al. 1997). Short-wave radiation is the energy source that drives evaporation, transpiration, photosynthesis and many other important processes linked to agricultural systems.

The remaining 15% is long-wave radiation, which falls into the range of approximately 4–50 μm and reaches the surface of the Earth via contributions from the atmosphere and the sun's spectrum (Kiehl and Trenberth 1997; Wild et al. 2013). The majority of long-wave radiation falling on the surface of the Earth is from the atmosphere which is emitted by the gasses, especially water vapour and CO_2, present in the atmosphere.

5.4 Energy Balance

The temperature of the Earth–atmosphere system basically depends on its energy balance on the way of absorption of solar energy and distribution between different levels of the atmosphere and the ground. Thus, the temperature of any plant organ such as leaf, stem and root depends on the balance between incoming energy and also energy loss. 'Steady state' is where the rate of energy absorption exactly balances the rate of energy loss and the temperature of the absorbing tissue stays constant. The short-wave radiation entering the canopy and meet a leaf or other elements in the canopy (e.g. root, branch, part of stem) is absorbed. A variety of processes involve, such as evaporation, transpiration, heat conduction and photosynthesis. Some of the energy that absorbed goes to heating the surface and another some is emitted back to the atmosphere as long-wave radiation or transported as sensible heat.

The global energy balance considers the energy flows within the climate system and their exchanges with outer space. The energy balance of a surface layer of finite depth and unit horizontal area can be written as:

$$\frac{dQ}{dt} = Rn - H - LE - G - Qp$$

Where Q is the total heat energy stored in the surface layer and Rn is the net surface irradiance (commonly referred to as the net radiation flux, Wm^{-2}). The Rn represents the gain of energy by the surface from radiation and it is a positive number when it is towards the surface. *G* is the ground heat flux (W^{-2}) and it is the loss of energy by heat conduction through the lower boundary. The *G* is a positive number when it is directed away from the surface into the ground. The value at the surface is denoted as G_0. *H* is the sensible heat flux (Wm^{-2}) and represents the loss of energy from the surface through heat transfer to the atmosphere. The *H* is positive when directed away from the surface into the atmosphere. *LE* is the latent heat flux which represents a loss of energy from the surface due to evaporation, whereas *Qp* represents the energy used during photosynthesis, (Wm^{-2}).

Obviously, the largest terms of the fluxes are *Rn*, *H* and *LE* and energy used in photosynthesis accounts only a few percentages of net radiation. Net radiation, *Rn* which involves division of energy into sensible and latent heat fluxes strongly depends on the surface characteristics, vegetation functioning and weather conditions.

5.5 Photosynthesis in Mangroves

In life, about 5% of energy reaching a leaf surface is absorbed by photosynthetic pigments, primarily chlorophyll, and around 2% of this absorbed energy is converted into chemical energy. The first stage of reactions involved in the driving force of photosynthesis is called the light dependent reaction stage.

When the reaction centre of the chlorophyll a molecules reaches an excited state, it loses its electron entirely. The losses of the electron from chlorophyll a is a crucial for the light energy to be converted into chemical energy within the plant (Malkin and Fork 1981; Kühlbrandt et al. 1994; Scott 2008). The capture of electron and transduction of energy occurs between photosystem I (PSI; found mainly in the intergranal lamellae) and photosystem II (PSII) is an extremely fast reaction of approximately 200 to 500 picoseconds and almost exclusively independent of temperature (Whitmarsh and Govindjee 1995). Both PSI and PSII are two membrane-protein complexes linked by the complexes of antennae.

Besides PSI and PSII, there are two other large membrane-protein complexes called Cytochrome b6 and ATP synthase in thylakoid membranes that are involved in light independent reaction of photosynthesis (Moore et al. 1998; Nelson and Ben-Shem 2004; Dekker and Boekema 2005). The light energy absorbed by photosynthetic pigments is first captured in transient forms of chemical energy, namely (ATP), that along with NADPH, facilitates the enzymatic fixation and reduction of atmospheric CO_2 into sugars in the chloroplasts of the leaves. Of the chemical energy and mass contained in these sugars, approximately 50% ends up in biomass, primarily cellulose in tree stems, and the other half is utilized in the growth and maintenance respiration.

Compared to other plants, the light response curve of mangroves is similar with steep linear increase up to ≈300 to 400 μmol photons m^{-2} s^{-1} after which saturation is reached. It has been shown that during low light level, mangrove photosynthesis reaches saturation due to decrease in stomatal conductance and intercellular CO_2 concentrations (Cheeseman 1994; Cheeseman and Lovelock 2004). Among the mangrove species, the rate of photosynthesis varies widely depending on the regulatory factors such as soil salinity, vapour pressure deficit between leaf and surrounding which are air and light intensity (Lovelock and Ball 2002).

5.6 Mangrove Respiration

Respiration in plant refers to as the energy produced by oxidation of organic compound resulted from energy stored during photosynthesis in living cells, which required for growth and maintenance of tissues, absorption of mineral nutrients and translocation of organic and inorganic materials (Hoque et al. 2010). Mangroves can survive in various conditions of coastline ranging from the extremely arid coast of the Persian Gulf to the west coast of Asia, where the soil types ranging from heavy consolidated clay to coral rubble and organic peats with different level of salinity (Clough 1992). Different conditions of the coastline may give different impacts on energy flux, gross productivity, net productivity and primary productivity of mangrove forests especially the availability of oxygen for the respiration.

Respirations by the plant were estimated to consume approximately 30–70% of the total carbon fixed in photosynthesis (Hoque et al. 2010). It is widely stated that mangroves can survive to grow in flooded anoxic soils. It is due to the mechanism of the plants which are able to get the access towards oxygen for respiration from the air through aerial rooting systems (Feller 2018). However, the root respiration can still be affected by several other factors such as salinity and temperature. High level of salinity was reported to increase leaf respiration due to the higher energy needed in secreting out the salt or transporting the oxygen to the submerged root (Ye et al. 2005) but the root respiration decreases due to the limitation of oxygen supply which resulted with a higher cost of metabolic activity (Burchett et al. 1989).

The changes in temperature can also affect the respiratory enzyme through kinetic response (Hoque et al. 2010), where the respiration reduces in warm summer and the respiration increases in cool winter, as more energy is needed in colder temperature to maintain the metabolic activity. Mangrove forests have higher capacity in fixing the carbon compared to terrestrial forest and producing a higher value of *Qp* but at the same time would have high energy as the metabolic rate used in respiration of the mangrove forest could be affected by several factors when it grows in different places. Hence, the different net energy resulted from the different localization of mangrove forest may provide different energy flux toward the consumer.

5.7 Consumers in Mangrove Ecosystems

Each ecosystem, including mangrove has its own dynamicity which develops from a tangle of thousand other species of plants, animals, fungi and bacteria. The mangrove itself is the engineer of the ecosystems by providing and maintaining the physical structure of the habitat and becomes budget energy as key primary producer (Spalding 2010; Sahu and Kathiresan 2019). Other than mangrove itself, there are various types of animals that involve in the energy flux of mangrove ecosystem, including vertebrates (birds, reptiles, amphibians, mammals, fishes) and invertebrates (insects, crabs, prawns).

Birds which live in mangrove ecosystem can be divided into several types, namely aerial feeder, surface forager and foliage gleaner. The birds are important as pollinators and as a control agent on insect pest population. In addition, insect also was preyed by a few species of mangrove frogs. The existence of birds and frogs enables the energy flux from insect pests as first consumer to its predator as the next consumer. Reptiles that inhabited the mangroves including snakes, crocodiles and alligators also involve in energy flux as they use mangrove areas as the hunting ground. They become a predator of different animals, such as birds, fishes, mammals and other reptiles where the energy will be further passed from the prey.

Compared to birds and reptiles, mammals are more significant component in energy flux of mangrove ecosystems as it functions as a major food source for a variety of animals. Mammals can become the prey for crocodiles, snakes and even for bigger mammals such as Bengal tigers (Sandilyan and Kathiresan 2012). The mammals also can function as seed dispersal agent when frugivore mammals such as monkeys, bats and squirrels eat the fruits and disperse the seeds.

Other important players in energy flux of mangrove ecosystems are the aquatic invertebrates, especially the detritivores such as crabs and bivalves (Spalding 2010). The major part of herbivores on mangrove trees belong to detritivore which devours the fallen mangrove leaves as it is more palatable than the fresh mangrove leaves which protected by thick wax and many secondary metabolites such as tannin (Spalding 2010). They break down the majority of leaf litter into detritus that acts as fertilizer for mangrove trees. The energy flux is returned back to mangroves as the mangrove uses its own degraded leaves as a nutrient to grow. The crabs also help to increase the root respiration by developing a honeycomb of complex tunnel in the mangrove soil, which leads to increase of dissolved oxygen in the soils.

Aside from the organisms mentioned above, there are also important organisms that play a critical role in balancing the final stage of energy flux in the mangrove ecosystems which are often overlooked. These include the microscopic life such as fungi, protist, bacteria and archaea, which functioned as decomposer (Faridah-Hanum et al. 2014).

5.8 Decomposer of Mangrove Ecosystems

Mangroves are considered as detritus-based ecosystems where the decomposers are very important in cycling the nutrients efficiently from the litter of mangrove that are sedimented in the soil and absorbed by the mangrove trees (Nazim et al. 2013). Nutrient cycling begins when the litter accumulated are subjected to microbial degradation. There is a critical dependence of mangrove towards decomposer as it releases the nutrient from the decomposition of litter near to the root and maintaining the growth of the mangroves (Spalding 2010). Unlike the herbivory and predation that occur in the mangrove ecosystem, the cycle of nutrient provided by decomposers causes the energy flux to stay unchanged (Potapov et al. 2019). In addition, the highly productive mangrove ecosystem can provide a high amount and continuous amount of litter which consist of fallen leaves, branches and other debris (Nazim et al. 2013). However, the productivity can vary according to various climatic condition of mangroves forest in various places.

Different localization and conditions of mangroves can also affect the rate of decomposition. Low intertidal zone, low latitude, high feeding activities of invertebrate, leaves with lower tannin and leaves that easier to sink show faster rate of decomposition (Nazim et al. 2013). As the bacteria and fungi are the organisms that contribute to the decomposition of mangrove material, high temperature is reported to result with higher fungi growth rate and lead to a faster decomposition. Besides, the biodiversity of decomposer and detritivores can also be affected by the quantity and quality of detritus throughout the decomposition process. Until 2012, a sum of 120 fungi species or also called as manglicolous fungi have been recorded from 29 different mangrove forests around the world (Sandilyan and Kathiresan 2012).

The ability of decomposer to provide nutrients to the mangrove forests to increase its growth to become a greater habitat for various organisms has brought the importance of decomposers in mangrove ecosystem to the extent where it can physically alter the habitat. The alteration of the habitat may affect the transfer efficiency of energy through different trophic level of consumer and stabilize the energy flux in mangrove ecosystems (Roy et al. 2008). Complex interactions of these decomposers maintain the harmony of different biogeochemical processes and sustain the nutritional status and ecological balance of mangroves (Thatoi et al. 2012).

5.9 Conclusions

An exchange of energy between the land surface and the atmosphere is an important process in an ecosystem. In the case of the mangrove forests, this process affects the forest ecosystem which includes photosynthesis, respiration, plant growth, water transport and many other processes. Understanding on the latent and sensible heat fluxes is fundamental for mangrove ecological analysis. In this chapter, the energy flux in the multitrophic level of mangrove ecosystems from solar energy to the producers, consumers and decomposers is discussed. Based on previous studies, some pertinent examples of processes were highlighted on how energy from solar is transferred and lost along the multitrophic level which is expanding previous findings of ecological scale and complexity. It is crucial to understand the impact of environmental change factors such as climate change, species invasion, nutrient deposition at the trophic level and the study of energy effects. In addition, the regulatory mechanism along with the magnitude and changes of energy flux in mangrove forests is very important for understanding the climatological processes at the local, global and regional levels.

References

Burchett MD, Clarke CJ, Field CD, Pulkownik A (1989) Growth and respiration in two mangrove species at a range of salinities. Physiol Planta 75(2): 299–303.

Cheeseman JM (1994) Depressions of photosynthesis in mangrove canopies. In NR Baker, JR Bowyer, eds, Photoinhibition of Photosynthesis: From molecular mechanisms to the field. Bios Scientific Publishers, Oxford, UK, p 377–389.

Cheeseman JM, Lovelock CE (2004) Photosynthetic characteristics of dwarf and fringe *Rhizophora mangle* in a Belizean mangrove. Plant Cell Environ 27: 769–780.

Clough BF (1992) Primary productivity and growth of mangrove forests. Tropical mangrove ecosystems. American Geophysical Union, Washington DC complex by electron crystallography. Nature 367: 614–621.

Dekker JP, Boekema EJ (2005) Supramolecular organization of thylakoid membrane proteins in green plants. Biochim Biophys Acta 1706: 12–39.

FAO (2007) The World's Mangroves 1980–2005, FAO Forestry Paper 153 Rome, Forest Resources Division, FAO. p 77.

Faridah-Hanum I, Latiff A, Hakeem KR, Ozturk M (Eds.) (2014). Mangrove ecosystems of Asia: Status, challenges and management strategies. London/New York/Heidelberg: Springer. https://doi.org/10.1007/978-1-4614-8582-7.

Feller I (2018) Mangroves. Retrieved from https://ocean.si.edu/ocean-life/plants-algae/mangroves.

Fisher P, Spalding MD (1993) Protected areas with mangrove habitat. Draft Report World Conservation Centre, Cambridge, UK, p 60.

Flowers TJ, Colmer TD (2015) Plant salt tolerance: Adaptations in halophytes. Ann Bot 115(3): 327–331.

Giri C, Ochieng E, Tieszen LL, Zhu Z, Singh A, Loveland T, Duke N (2011) Status and distribution of mangrove forests of the world using earth observation satellite data. Glob Ecol Biogeogr 20 (1): 154–159.

Hoque AR, Sharma S, Suwa R, Mori S, Hagihara A (2010) Seasonal variation in the size-dependent respiration of mangroves Kandelia obovata. Mar Ecol Prog Ser 404: 31–37.

Kiehl J, Trenberth K (1997) Earth's Annual Global Mean Energy Budget. Bull Amer Meteor Soc 78: 197–208.

Kühlbrandt W, Wang DN, Fujiyoshi Y (1994) Atomic model of plant light-harvesting complex by electron crystallography. Nature 367(6464): 614–621.

Kumar L, Skidmore AK, Knowles E (1997) Modelling topographic variation in solar radiation in a GIS environment. Int J Geogr Inf Sci 11(5): 475–497.

Lovelock CE, Ball MC (2002) Influence of salinity on photosynthesis of halophytes. In: Lauchli A, Luttge U (eds) Salinity: Environment plants molecules. Kluwer Academic, New York, p 315–339.

Malkin S, Fork DC (1981) Photosynthetic units of sun and shade plants. Plant Physiol 67: 580–583.

Moore R, Clark WD, Vodopich DS (1998) Botany. McGraw-Hill Companies, Inc.

Naskar S, Palit PK (2015) Anatomical and physiological adaptations of mangroves. Wetl Ecol Manag 23(3): 357–370.

Nazim K, Ahmed M, Shaukat SS, Khan MU (2013) Seasonal variation of litter accumulation and putrefaction with reference to decomposers in a mangrove forest in Karachi, Pakistan. Turk J Bot 37(4): 735–743.

Nelson N, Ben-Shem A (2004) The complex architecture of oxygenic photosynthesis. Nat Rev Mol Cell Biol 5: 971–982.

Potapov AM, Klarner B, Sandmann D, Widyastuti R, Scheu S (2019) Linking size spectrum, energy flux and trophic multifunctionality in soil food webs of tropical land-use systems. J Anim Eco 88(12): 1845–1859.

Roy M, Mandal S, Ray S (2008) Detrital ontogenic model including decomposer diversity. Ecol Model 215: 200–206.

Sahu SK, Kathiresan K (2019) The age and species composition of mangrove forest directly influence the net primary productivity and carbon sequestration potential. Biocatal Agric Biotechnol 20: 101235.

Sandilyan S, Kathiresan K (2012) Mangrove conservation: A global perspective. Biodivers Conserv 21(14): 3523–3542.

Scott P (2008) Physiology and behaviour of plants. John Wiley & Son, Ltd, England.

Spalding M (2010) World atlas of mangroves. Routledge.

Thatoi H, Behera BC, Mishra RR, Dutta SK (2012) Biodiversity and biotechnological potential of microorganisms from mangrove ecosystems: A review. Ann Microbiol 63(1): 1–19.

Whitmarsh J, Govindjee (1995) Photosynthesis. Encyclopedia of Applied Physics, VCH Publishers, 13, p 513–532.

Wild M, Folini D, Schär C, Loeb N, Dutton EG, König-Langlo G (2013) The global energy balance from a surface perspective. Clim Dyn 40: 3107–3134.

Ye Y, Tam NFY, Lu CY, Wong YS (2005) Effects of salinity on germination, seedling growth and physiology of three salt-secreting mangrove species. Aquati Bot 83(3): 193–205.

Nitrogen and Phosphorus Budget in Mangrove Ecosystem

6

Raghab Ray, Sandip Kumar Mukhopadhyay, and Tapan Kumar Jana

Abstract

Mangrove biogeochemistry focuses on large, slow moving reservoirs (atmosphere-biosphere-geosphere-hydrosphere), and their smaller but active exchanges of elements within the reservoirs, mainly driven by biogeochemical activities in the geosphere or sediment. Other than carbon (C), nitrogen (N) and phosphorus (P) involve in such active cycling of elements and play essential role in the biogeochemical processing of organic matter and primary productivity of mangrove ecosystems. Despite being highly productive, mangroves maintain very low-nutrient level, which has been emphasized earlier in this chapter. Cycling of both N and P in mangrove ecosystem has been discussed, followed by their budget calculations. World mangroves generally act as a sink for essential elements, storing majority into the live biomass and sediment reservoir. A major fraction of the nutrient is conserved and recycled within the organic structure of the ecosystem. Finally, world's largest deltaic mangrove, the Sundarbans is recognized as one model site for developing regional N and P budget in view of the box model approach.

Keywords

Nitrogen · Phosphorus · Organic matter · Budget · Box model · Mangroves

R. Ray (✉)
Atmosphere and Ocean Research Institute, The University of Tokyo, Kashiwa, Japan
e-mail: raghab.ray@aori.u-tokyo.ac.jp

S. K. Mukhopadhyay · T. K. Jana
Department of Marine Science, University of Calcutta, Kolkata, India

R. P. Rastogi et al. (eds.), *Mangroves: Ecology, Biodiversity and Management*,
https://doi.org/10.1007/978-981-16-2494-0_6

6.1 Introduction

There are several excellent reviews and books available on comprehensive understanding of nitrogen (N) and phosphorus (P) cycling in the marine ecosystem (Fenchel and Blackburn 1979). In this chapter, we summarize N, P cycling in mangrove ecosystems by examining the storage, transformation, and fluxes in mangrove sediment, biomass, atmosphere, and tidal waters. Specific to mangrove ecosystems, such synoptic overview comprising both nutrients is very rare except the pioneering one by Alongi et al. (1992), but that too lacked P budget. However, before going into the details of mangrove N and P dynamics, an initial background knowledge on the state of nutrients in various water systems is essential.

Over-enrichment of nutrients or "eutrophication" in lakes, reservoirs, estuaries, and rivers are widespread all over the world and the severity is increasing. Although the mechanisms of water eutrophication are not fully understood, yet excessive nutrient loading via discharge of domestic wastes and non-point pollution from agricultural practices and urban development into surface water system are considered to be among the major factors. There is a long history of such human-induced eutrophication in the coastal waters. The negative feedback of eutrophication has been very apparent in rivers and estuaries for centuries, as being historically documented in a detailed landscape paintings by the Dutch artist Salomon van Ruysdael (ca. 1648) depicting the eutrophic waterways of the Netherlands as early as the seventeenth century (Fig. 6.1). Key nutrients of concern are nitrogen and phosphorus because the supply rates of these nutrients most often control or "limit" aquatic plant primary production and biomass formation (Paerl 2009). From individual freshwater basin and coastal zone studies, e.g. the Baltic region, Mississippi River, Gulf of Mexico, North Sea, Northern Adriatic, the Black sea (Wulff and Stigebrandt 1989; IGBP 1997), it was observed that coastal eutrophication was a consequence of elevated levels of waterborne N and P.

Downstream estuarine and coastal waters are physically and biogeochemically distinct from the freshwater ecosystems and, as a result, their responses to nutrient inputs and eutrophication may contrast to those observed in the freshwater ecosystems (Smith 2003; Bianchi 2007). Interactive effects of bathymetry (basin morphology), hydrology (upstream discharge and tidal mixing), collectively as "hydrodynamics" are the physical controls, whereas minerals deposition derived from watershed geological (erosional) and biological (plant production, microbial, and higher trophic level cycling) processes are the geochemical drivers of nutrient enrichment to the downstream estuaries and sea.

Coastal vegetations such as mangroves, salt marshes, sea grasses are the efficient repositories of such nutrients derived in abundance from the external sources. Mangroves, in particular, are often limited in N and P in order to aid their growth and sustenance (Alongi 2011). Despite nutrient limitation, how mangroves maintain high productivity is often considered as a "paradox," and that unique feature has been discussed in the next section. Because mangroves are the central focus of this chapter, other coastal vegetations like seagrass or tidal marshes are excluded from

Fig. 6.1 "River landscape," a painting by Salomon van Ruysdael (ca. 1648) depicting the eutrophic waterways of the Netherlands in the seventeenth century. Note the agricultural and human sources of nutrients and scums (presumably algal blooms) present on the water surface (image copyright permitted by Hans Paerl, UNC-Chapel Hill)

discussing despite their importance in nutrient retention capacity and role in ecosystem functioning.

6.2 Mangroves High Productivity in Low-Nutrient Environment

Mangroves are swamped forests, occurring mostly in the tropics and subtropics spanning between about 35° north latitude and about 40° south latitude. They provide many ecosystem services to the coastal communities such as attenuating inflow of flooded water and acting as a speed-breaker against storm surges, providing enough resources for fishery and other aquatic lives. They could provide important ecosystem service in recycling different forms of nutrients, thereby protecting the coastal ecosystems from negative impacts of eutrophication and atmospheric pollution. This typical functioning of mangrove ecosystems in maintaining high productivity and nutrient recycling has particular importance for understanding biogeochemistry of the adjacent marine ecosystems.

Mangrove are highly productive ecosystems with global carbon stock of 956 Mg C ha^{-1}, and sequestration rate of 174 g C m^{-2} yr^{-1} (Donato et al. 2011). Among the wetland vegetations such as rainforests, peat swamps, salt marshes, and seagrass meadows, mangrove C stock are much higher (Alongi 2014), and sequester at a faster rate than any other coastal habitats, e.g. sea grasses (138 g C m^{-2} yr^{-1}, Fourqurean et al. 2012) highlighting their importance as most proficient C fixers on earth (Donato et al. 2011). Despite such high productivity, mangroves are often limited by the availability of micro- and macronutrients specially N and P, mainly due to high nutrient use efficiency for developing their cell walls (Holguin et al. 2001; Lovelock et al. 2005; Reef et al. 2010). Nutrient availability depends on multiple environmental factors such as salinity, temperature, redox reactions between dissolved organic and inorganic constituents in interstitial water (or pore water), root uptake, efficiency of metabolic processes, etc. (Alongi 2013). For instance, in highly reducing mangrove sediment, it is observed that metal sulfide complexes readily bind to organic nutrients, and limit the amount of nutrients available for plant uptake (Alongi 2009). Nitrogen or phosphorus limitation also depends on tidal regime, like in the Caribbean, fringing mangroves are N-limited, but permanently flooded forests or those deep within the islands are P-limited (Feller et al. 2002).

Such low-nutrient availability in mangroves implies that in order to satisfy nutrient demands, recycling rates (or turnover time) should be very fast and transfers of nutrients efficient within the geosphere and mangrove vegetation. Nutrients are regenerated by mangrove litter decomposition (Holguin et al. 2001). It is also seen that increased nutrient delivery to the mangroves via anthropogenic nutrient loading, can have negative consequences for mangrove forests and their capacity for retention of nutrients may be limited (Reef et al. 2010). To explain such a wide range of mangrove ecosystem properties (like nutrient recycling, transfer efficiency, and conservation), elemental stoichiometry of major macronutrients, i.e. C, N, and P have become an efficient tool in biogeochemical and ecological studies (Scharler et al. 2015).

6.2.1 Implication of C:N:P

Marine phytoplankton has an average C:N:P ratio of 106:16:1 (Redfield 1934). In contrast, terrestrial plants are characterized by C:P ratios ranging from 300 to 1300 and C:N ratios ranging from 10 to 100 for soft tissue, and C:P ratios greater than 1300 and C:N ratios ranging from 100 to 1000 for woody tissue (Hedges et al. 1986; Ruttemberg and Goñi 1997). For the mangroves, the C:N:P stoichiometry differs strongly between the living and non-living compartments, such as concentrations in mangrove tissues are very low (*Avicennia marina* leaves: N $<$ 1.5%, P $<$ 0.1%, Ray et al. 2017) and can differ significantly between locations. In the Red Sea, where growth of mangroves is stunted, C:N:P of *Avicennia marina* leaves was reported to be 1918:36:1 (Almahasheer et al. 2016), that is slightly different than the taller *A. marina* in the Sundarbans (1200:40:1; Ray et al. 2017) or old growth *Rhizophora mangle* in Belize (1547:47:1; Scharler et al. 2015). Furthermore, a comparative results of stoichiometric C:N:P between the mangroves in Belize and Sundarbans showed statistically significant differences between plant tissues, microbial mat, and sediment ($p < 0.005$) (Table 6.1). Both mangrove settings are P-limited which is

Table 6.1 Comparison of elemental ratio between fringe *Rhizophora mangle L.* (red mangrove) in Belize (Scharler et al. 2015), and *Avicennia marina* (black mangroves) in the Indian Sundarbans (Ray et al. 2017)

Twin Cay, Belize				Sundarbans, India			
Molar ratio	Leaf	Litter	Microbial mat	Leaf	Litter	Root	Sediment
C:N	46.7	58.4	21.7	25.3	34	38.4	11.4
C:P	1570.2	4018.5	480	930	364	900	42.8
N:P	33.6	68.8	22.1	36.8	11	11.7	3.7
C:N:P	1569:47:1	4018:69:1	479:22:1	1200:40:1	940:36:1	1162:26:1	183:8:01

often a common feature for such marine environment. Primary producers like plants show higher C:N, C:P than the decomposers or heterotrophs, and overall C:N:P decreases with increasing trophic level (Scharler et al. 2015). It is therefore believed that nutrients are initially immobilized from the environment by the decomposers, and as decomposition proceeds, this difference decreases further, as also evident from a decreasing ratio in the sediment (C:N = 11.4, C:P = 42.8). When they become lower than the critical level (C:N = 24.5, C:P = 681–979), net release of nutrients occur (Moore et al. 2006; Parton et al. 2007). For the Sundarbans mangroves, Ray et al. (2017) calculated carbon-use efficiency of decomposers to be 0.86–1.0 (when sediment C:N = 10.6–13) and supported the hypothesis that decomposers adapt to low sediment C:N conditions by enhancing their carbon-use efficiency during decomposition processes in the sediment.

6.2.2 Transfer and Conservation of N, P

Most mangrove trees are evergreen with sclerophyllous leaves and high root/shoot biomass ratios (Komiyama et al. 2000). Root/shoot ratios in mangrove vegetation have been observed to be an order of magnitude higher than the tropical terrestrial forests. High root biomass in mangroves, especially the abundance of fine roots (Poungparn et al. 2016), is conducive to nutrient uptake from low-nutrient soils. Nutrient transfer efficiencies (TE) from one trophic level to the next, and recycling rates (calculated as Finn's cycling index or FCI, Finn 1980) are, in general, highest for P (29.3–45.3%, 15.4–84%, respectively), followed by N (14.5–28%, 12.4–16.8) and C (7.4–9.5%, 6.5–10.6%) and such difference could be even higher for the dwarf mangroves like in the Middle-East (Scharler et al. 2015) (Table 6.2). Benthic faunal activities of larger sizes (crabs, gastropods) and smaller sizes (leaf litter fauna) are among the key players in such fast recycling process of the limiting nutrients (Kristensen and Alongi 2006), and contribute to mangrove C budget (Andreetta et al. 2014). Indeed, slow recycling and turnover rate of C (several years in the sediment) result into their conservation within the biomass pool and sustenance of high productivity despite thriving in nutrient-poor conditions (months of 1–2 years

Table 6.2 Finn cycling index (FCI or % recycled), transfer efficiency (NTE) for three mangrove forest zones and nutrients (C, N, P) in Belize

Settings	Elements	FCI%	TE%
Fringe Mangroves	C	10.6	7.8
	N	12.4	14.5
	P	51	34.1
Transition Mangroves	C	9.3	7.4
	N	15	15.6
	P	15.4	29.3
Dwarf Mangroves	C	6.5	9.5
	N	16.8	28.2
	P	84.3	45.3

Data Source: Scharler et al. (2015)

of N, P turnover rates, Alongi et al. 2005). Depending on nature of nutrient recycling within the forest, mangroves may serve as either a source or sink of nutrient to the adjacent coastal waters (Twilley and Day 1999).

6.3 Mangroves as Source or Sink of Nutrient

Export of nutrients to the adjacent coastal environment have been well recognized by a number of studies (Dittmar and Lara 2001; Leopold et al. 2016). This idea came from the "Outwelling Theory" by Odum and Heald (1975) that initially was put forward for the salt marshes showing their ability to sustain high productivity of the adjacent water systems through a supply of nutrients and organic matter (OM). Later this theory has been supported by the recent studies that suggest export of nutrients/OM to be an important feature of the mangrove forests (Lee 1995). Furthermore, porewater seepage has been suggested as one of the most important processes in the outwelling of C, N, and P (Dittmar et al. 2006; Gleeson et al. 2013; Sippo et al. 2017). The outwelling of nutrients would occur only if export from porewater is higher than the consumption by the benthic community and trees (Dittmar and Lara 2001). Moreover, the capacity of mangroves to release or retain nutrients depends on geomorphological settings, such as the current velocity, frequency of inundation, and topographic elevation (Adame and Lovelock 2011). Especially at low tide, water body in the mangrove creek is enriched with high concentrations of ammonium (NH_4^+), phosphate (PO_4^{3-}), and dissolved organic C and nitrogen (DOC, DON) due to the strong influence by groundwater or pore water flow (Lara and Dittmar 1999). Mangrove types can also cause changes in the source/sink pattern, e.g. it is seen that fringe forests primarily act as sink for dissolved inorganic N (DIN) and a source of DON, while basin forests may exhibit the reverse trend (Rivera Monroy et al. 1995). Mangrove forests can also import particulate nutrients and C associated with suspended sediment and organic debris. The capacity of mangroves to import nutrients and C has been proposed to be an important mechanism in maintaining the health of the adjacent seagrass communities (Valiela and Cole 2002). Adame et al. (2010) showed that geomorphological setting could determine mangroves to retain nutrients with some riverine site receiving more nutrients than the tidal sites.

With regard to biogeochemical flux estimates, DIN exchange between the mangrove estuaries and coastal waters can be highly variable ranging from an export flux of 357.14 mmol m^{-2} yr^{-1} to the coastal waters (Caete River, Brazil; Dittmar and Lara 2001) to an import of 2621 mmol m^{-2} yr^{-1} to the mangrove basin (Taylor river, US, Davis et al. 2001), resulting in a net global export of 42.8 $\pm$ 117 g N m^{-2} yr^{-1} (Adame and Lovelock 2011). Dissolved inorganic P (DIP) exchange ranges from an export of 20.72 mmol m^{-2} $year^{-1}$ (Sundarbans, India; Ray et al. 2017) to an import of 45.2 mmol m^{-2} yr^{-1} (Taylor River, US; Davis et al. 2001) resulting in a net global import of 1.0 $\pm$ 11 mmol m^{-2} yr^{-1}. In the world's largest deltaic mangroves, the Sundarbans, net N and P export was estimated to be 264 mmol m^{-2} yr^{-1} and 188 mmol m^{-2} yr^{-1} (Indian part only; Ray et al. 2014, 2017). Mean N and P exchange as particulate matter were reported to be 94.3 $\pm$ 99.3 and

124 ± 285 mmol m^{-2} yr^{-1}, respectively (Adame and Lovelock 2011) with import rate cited for the SE Asian mangroves. Results of N and P concentrations (μmol L^{-1}) and their exchange fluxes (mmol m^{-2} yr^{-1}) in various mangrove locations are shown in Table 6.3.

6.4 Mangrove N Cycle

6.4.1 Nitrogen Stock in Biomass and Sediment

Mangroves uptake N via atmospheric and belowground sources leading to its storage within the different compartments of the live plant. Allometric regression equations are generally used as a non-destructive method to estimate mangrove above ground and below ground biomass (AGB and BGB) (Kauffman and Donato 2012) which can be converted into C and N stocks. However, compared to C, very few studies provided direct estimates of N stocks in mangrove biomass. AGB-N data (Mg N ha^{-1}) are sparsely available from the Oceania and Asian countries such as New Zealand (15.4; Bulmer et al. 2016), Australia (12.2, Alongi et al. 2003), Indian Sundarbans (1.28, Ray et al. 2014), Japan (35, Khan et al. 2007), Micronesia (56, Fujimoto et al. 1999). Very few below ground allometric functions exist for the mangroves due to hard labor needed for extracting mangrove roots and careful sieving (Komiyama et al. 2000). The N stock in root biomass or BGB ranges from 0.08 to 0.69 Mg N ha^{-1} (northern Australia: ~0.08–0.3; New Zealand: 0.69 ± 0.17; Sundarbans: 0.36 ± 0.03; Alongi et al. 2003; Bulmer et al. 2016; Ray et al. 2014). Similarly very few direct measurements of sediment N stock range from 0.04 to 24 Mg N ha^{-1} (up to1.2 m depth from the surface) with maximum observed for the Micronesian mangroves (20–24, Fujimoto et al. 1999) and minimum for the Indian Sundarbans. Such differences in N stock in sediment generally arise from different sampling depths considered for stock estimation, external sources of N (such as anthropogenic input), supply of OM and decomposition rates. The AGB-N tends to be 1.4 times as large as that in the BGB, and the sediment N stock is 3.3 times as large as the biomass N stock (Purvaja et al. 2008). Therefore, global mean of the ecosystem N stock (AGB + BGB + Sediment) in mangroves is calculated to be ~20 Mg N ha^{-1}. Global N cycle comprising stocks and major biogeochemical fluxes in mangrove ecosystems are summarized schematically in Fig. 6.2.

6.4.2 Nitrogen Transformation Processes and Fluxes

6.4.2.1 Nitrogen Fixation

The mangrove biogeochemistry of plant and sediment-derived nutrients (mainly N, P) focus on the large, slow moving chemical reservoirs and their smaller but active exchange or cycling driven by biogeochemical activities in the mangrove reservoirs. In particular, N cycle in mangrove benthic system is mediated predominantly by microbial rather than chemical processes (Alongi et al. 1992). Biological N fixation

Table 6.3 Average N and P concentrations (μmol L^{-1}) and their net exchange fluxes (mmol m^{-2} yr^{-1}) of their dissolved inorganic and particulate forms in various mangrove locations, negative and positive fluxes meaning export and import, respectively

Location	Mangrove settings	N conc. μmol L^{-1}	P conc. μmol L^{-1}	Net N flux mmol m^{-2} yr^{-1}	Net P flux mmol m^{-2} yr^{-1}	References
Dissolved inorganic fraction						
Klong Ngao, Thailand	River dominated	0.43	0.13	−35.00	−0.19	Wattayakorn et al. 1990
Lobos bay, Mexico	Tide dominated	24.14	1.55	−128.57	5.81	Sánchez-Carrillo et al. 2009
Caete River, Brazil	Tide dominated	15.00	0.52	−357.14	−6.13	Dittmar and Lara 2001
Sundarbans, India	Tide dominated	20.00	0.65	−264.00	−20.72	Ray et al. 2017
Conn creek, Australia	Tide dominated	0.21	0.00	−85.71	−19.68	Ayukai et al. 1998
Hinchinbrook, Australia	Tide dominated	0.07	0.10	114.29	10.97	Boto and Wellington 1988, Ayukai et al. 1998
Taylor River, USA	Carbonate setting	6.00	0.10	2621.43	45.16	Davis et al. 2001
S. Everglades, USA	Carbonate setting	6.00	0.10	−2.14	−0.03	Sutula et al. 2003
Particulate fraction						
Tapi River, Thailand	River dominated	–	–	3592.8	60	Wattayakorn et al. 2001
Red River, Vietnam	River dominated	–	–	8071.4	435	Wösten et al. 2003

Data Source: Review by Adame and Lovelock (2011)

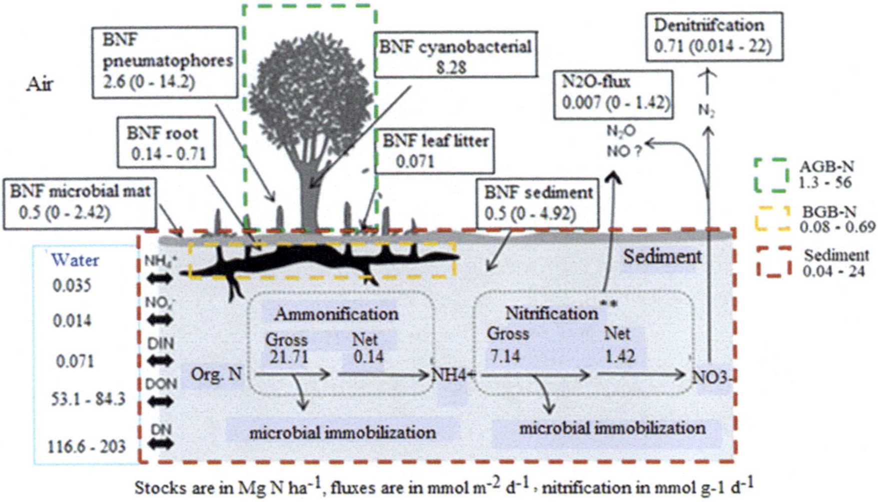

Fig. 6.2 Global mangrove N cycle. N stocks are given in Mg N ha^{-1}, and biogeochemical fluxes of N are presented in mmol m^{-2} d^{-1} except for nitrification rates** given as mmol g^{-1} d^{-1}. BNF: biological nitrogen fixation, AGB: above ground biomass, BGB: below ground biomass. Data are taken from 123 mangrove sites comprising areas in the Atlantic Ocean, Caribbean Sea, Gulf of Mexico, and Indo-Pacific coasts (Alongi et al. 1992; Purvaja et al. 2008; Reis et al. 2017). Image has been modified from the global syntheses by Reis et al. (2017)

(BNF) is the reduction of nitrogen gas (N_2) to NH_4^+ carried out by the Eubacteria and Archaea that possess a required enzyme, nitrogenase (termed as diazotrophs). The BNF has been detected in mangrove stands associated with plant roots, sediments (free-living), microbial mats, leaf litter, pneumatophores, and cyanobacterial crusts (free-living). Low rates of BNF in mangrove sediments were reported than those in seagrass and salt marsh communities with rates varying from 0 to 4.9 mmol m^{-2} d^{-1}. It has often been suggested that high DOC present in mangrove sediment pore water could limit BNF contrast to sea grasses which stimulate N fixation otherwise, suggesting that the N-fixing communities of seagrass and mangroves may be dominated by different bacterial groups. Sengupta and Chaudhuri (1991) isolated diazotrophic bacteria associated with root samples of several mangrove species from the Indian Sundarbans. It was observed that regardless of mangrove species, root samples from tidally inundated mangroves sustained greater BNF rates than the samples from occasionally inundated or drier sites, attributing to the presence of a larger number of diazotrophs belonging to a greater number of O_2 response groups in the tidally inundated mangrove sites. Despite the dominance of variable groups of microorganisms during N fixation (e.g., heterotrophic bacteria in sediment and roots, cyanobacteria in pneumatophores, mixture of both in leaf litter), mangrove sediment, roots, pneumatophores, litter debris, and cyanobacterial mats tend to show similar BNF rates, attesting to their insignificantly different contributions towards the total N input in mangrove benthic system ($H = 2.84$, $df = 3{,}51$, $p = 0.416$; Reis et al. 2017).

6.4.2.2 Nitrogen Mineralization

N mineralization is the microbial mediated process by which organic N is converted to inorganic N forms through a series of reactions like NH_4^+ production or ammonification, and NH_4^+ oxidation to nitrite (NO_2^-) and nitrate (NO_3^-) or nitrification. Nitrifying bacteria or chemolithotrophic organisms motivate nitrification process (e.g. species of the genera *Nitrosomonas*, *Nitrosococcus*, *Nitrobacter,* and *Nitrococcus*). In mangrove N cycle, high magnitude of difference between the gross and net ammonification (NH_4^+ production minus NH_4^+ utilization by microbes) implies efficient microbial immobilization of NH_4^+ that may constrain net ammonification rates in sediment. Chen and Twilley (1999) found a very strong positive correlation of net ammonification rates with P availability in mangrove stands in Florida, suggesting a P-limitation of microbial activities. Nitrification rates, on other hand, could be limited by many factors, such as the NH_4^+ availability and microbial immobilization, absence of root activity in rather anoxic non-vegetated tidal flat (Kristensen et al. 1998), Mn and OC availability (Krishnan and LokaBharathi 2009). Generally slow rates of nitrification in mangrove forests are associated with extensive uptake of NH_4^+ by the mangrove below ground part and microbes (Purvaja et al. 2008).

6.4.2.3 Nitrogen Removal

The microbial mediated N-removing pathway in mangrove sediment is important not only because it acts to mitigate N-loading but also because it means a loss of fixed nitrogen from such an ecosystem where N is frequently a limiting nutrient (Holguin et al. 2001). The process, called denitrification is one of such mechanisms that involves dissimilatory reduction of nitrate ion to nitrous oxide (N_20) and N_2 by the denitrifying bacteria (e.g. some species of Serratia, Pseudomonas). In reducing environment like mangrove sediment, denitrification step is energetically second most favorable during the diagenesis of organic matter. In this step, NO_3^- is utilized as an alternative electron source of oxygen within few centimeters of top-soil. Global mean denitrification is higher than the BNF in sediment (0.7 versus 0.5 mmol m^{-2} d^{-1}, Fig. 6.2) that generally suggests the N_2 source of mangrove sediment to the atmosphere. However, in most regional studies in pristine conditions, denitrification is slower than the N mineralization, that is about 15% of total N input to mangrove soils is denitrified. In other coastal ecosystems, the percentage of N lost via denitrification ranges from 20% to 70% (Seitzinger 1988). That is due to the low NO_3^- level in sediment and pore water, whereas mangrove forests receiving large NO_3^- discharges from sewage treatment plants show relatively high denitrification rates of 1 to 2-order of magnitude higher than that usually observed (Nedwell 1975; Corredor and Morell 1994).

N removal as N_2O is one of the key N transformation processes in the mangrove benthic environment. Nitrous oxide, a major greenhouse gas, is produced as a by-product between the microbial pathways of denitrification and nitrification.

Many studies have highlighted the impact of biogenic activities and seasonal changes on N_2O emissions from mangrove sediment (Corredor and Morell 1994; Allen et al. 2007; Chen et al. 2012; Chauhan et al. 2015). Lower N_2O fluxes in natural crab-bioturbated areas are observed in Brazilian mangroves (0.007–4.5 mmol m^{-2} d^{-1}) due to constant soil oxidation by macrofauna, whereas higher N_2O fluxes (0.01–0.08 mmol m^{-2} d^{-1}) in crab-exclusion mangrove areas are due to wet/anaerobic soil conditions that favor denitrification (Otero et al. 2020). Global mean N_20 emission flux from mangrove sediment is ~0.007 mmol m^{-1} d^{-1} (Fig. 6.2).

Anaerobic ammonium oxidation (anammox) is another important microbial mediated N removal mechanism that oxidizes NH_4^+ anaerobically coupled to NO_2^- reduction to N_2 (Dalsgaard et al. 2005). The presence of anammox activity in mangrove sediment was noted for the first time in the north-eastern Australia where rates of anammox were low (<2μmol N_2 m^{-2} d^{-1}) compared to other marine environments (Meyer et al. 2005). Later in Vietnam, Amano et al. (2011) measured anammox rate on sediment volume basis (0–0.7 nmol N_2 cm^{-3} h^{-1}) that corresponded to only ~2% of denitrification. Anammox reactions are believed to be inhibited by the soluble tannins or sulfides in the interstitial water (Alongi 2014).

Net tidal exchange represents the largest loss of N via mangrove waterways. This pattern is consistent with the idea that tidal pumping and pore water seepage transports high dissolved concentrations of nutrients into adjacent waters after the hydrostatic pressure gradually declines towards low tide ("outwelling concept" discussed before). Studies that quantified dissolved N fluxes between the tidal creek or estuarine waters in mangrove areas and coastal or ocean waters indicated that mangroves can act as a source of dissolved N for the adjacent water bodies (Adame and Lovelock 2011). A study by Rivera Monroy et al. (1995) in the fringe mangrove forest in Mexico observed net import of NH_4^+ and NO_3^- to the tidal creek and basin forest, and net export of DON to the sea. Hence fringe forests might primarily act as sink for DIN and a source of DON, while basin forests may exhibit the opposite pattern. Higher DIN fluxes were reported for the dwarf mangroves in Everglades (Davis et al. 2001).

6.4.3 Mangrove N Budget

Only two complete N budgets exist for the mangroves, one is the Missionary Bay mangroves of Hinchinbrook Island in Australia, and the second is the Sundarbans mangrove, India. The former was based on the research done by D.M Alongi and others in early 90s (Alongi et al. 1992), and latter was a recent one by Ray et al. (2014). In this section, N budget will be discussed based on the Bay ecosystem which is in balance considering many extrapolations and systematic and relative errors involved in a large number of individual measurements made over time (synthesized by Alongi 2013). For the Sundarbans, the comprehensive mass budget was more regional but robust, and would be discussed in detail later in this chapter.

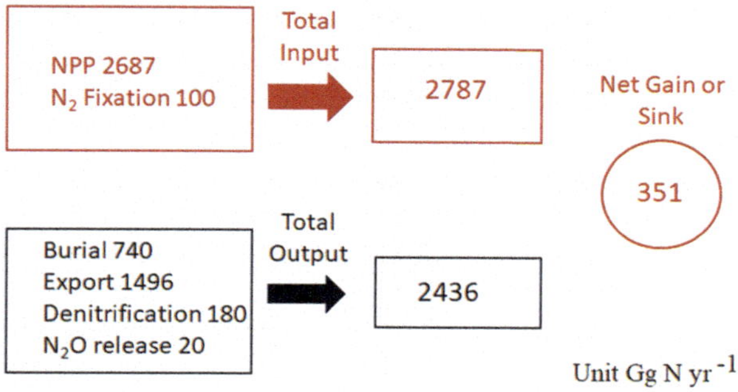

Fig. 6.3 A mass-based global N budget in mangrove ecosystem, assuming total area as15.2 million ha (Spalding et al. 2010). Annual flux data are retrieved from Alongi (2013)

The existing mangrove N budget shows two main N inputs to the ecosystem and four N outputs from the ecosystem (Fig. 6.3). The mass balance indicates that 2687 Gg N yr^{-1} is required to sustain global mangrove net primary production (combining wood, litter, and root) and only ~5% of the total N input is contributed by N_2 fixation. Less than 10% of mangrove N is lost via denitrification and N_2O emissions, while the majority of N loss occurs through tidal export (~60%). N fluxes are typically well balanced with unaccounted sink of only 351 Gg N yr^{-1} which is very small compared to the total inputs, outputs, and sources of error. Most of the allochthonous N (mainly tidally imported or anthropogenic) is efficiently recycled via plant-soil-microbe pathways and 75% of mangrove N is either stored within the organic structure (~1000–1300 Gg N yr^{-1} in leaf litter plus root, 740 Gg N yr^{-1} as sediment burial) or exported to the sea (1496 Gg N yr^{-1}). In the long run, N status as a source or sink is dependent on the balance between inputs and outputs of nutrient, and the biogeochemical coupling between different reservoirs of a mangrove ecosystem.

6.5 Mangrove P Cycle

The basic distinction between the N and P cycle is the absence of gaseous phases in latter, which makes P cycle relatively simple in nature, although the relationship between microbial activities and changes in P geochemistry can be highly complex and difficult to measure. Furthermore, despite rapid flowing of dissolved P through plants and animals, the processes governing their movement through the soil or sea are very slow and make the P cycle overall one of the slowest biogeochemical cycles (Oelkers 2008). General aspects of P cycling in estuarine and marine environments can be found in the paper by Nixon (1980).

6.5.1 Storage and Bioavailability of Phosphorus

In mangrove ecosystem, major pools of P are live biomass (AGB and BGB) and sediment. Mangrove sediments act as a sink for P with high retention capacity (Tam and Wong 1996). For instance, in Australia, mangrove sediment have been reported to have adsorption maxima in the range 8.1–22.6 mmol P kg^{-1} dry wt of sediment, that is ~50% of the total concentration in dry sediment (Clough et al. 1983). It has been estimated that up to 88% of the forest P pool is retained within the system in tropical mangroves. Furthermore, total P in sediment tends to be not easily influenced by the degradation or restoration of the wetlands, owing to its more conservative cycling process than those of C and N (due to the lack of exchange with the atmosphere).

In the domain of pH that is relevant to most mangrove soils (i.e. generally between 5 and 7.5), $H_2PO_4^-$ and HPO_4^{2-} are the dominant orthophosphate ions (Lindsay 1979). Although organic P is the major fraction, phosphate-P represents the largest potential pool of plant-available, soluble reactive form (Boto and Wellington 1988). Differing from soil total P concentration, plant-available soil-P plays an important role in controlling mangrove species distributions even though comparatively few data exist on this topic. It has been seen that invasion of mangrove associates (*Spartina alterniflora*) heavily decreases plant-available P, but exhibited only a slight influence on the sediment total P (Feng et al. 2018). Occurrences of phosphatase enzyme in sediment and phosphate-solubilizing bacteria associated with mangrove roots serve an important role in providing enough phosphate to plant biomass (Das et al. 2014).

6.5.2 Input and Output of Phosphorus

Biogeochemical input fluxes of P within the mangrove ecosystem is driven by various processes (Pomeroy et al. 1965), such as

Input
- Atmospheric (dry and wet deposition),
- Mangrove (canopy nutrient transfer and litter fall),
- Mineralization from soil,
- Anthropogenic sources (sewage, agriculture, aquaculture, etc.).

Output
- Mangrove plant assimilation,
- Microbial uptake,
- Uptake by macro-feeder,
- Tidal exchange
- Soil immobilization/sedimentation.

A comprehensive overview of P cycle has been provided for the Malaysia peninsular mangroves which has very impressive forest coverage globally (~3.7%). These specific mangroves located in Merbok stand for an ideal site for P cycle where above mentioned input and output sources of P are present, and anthropogenic inputs of nutrient are heavy due to intensive aquaculture practices (Fig. 6.4). However, for

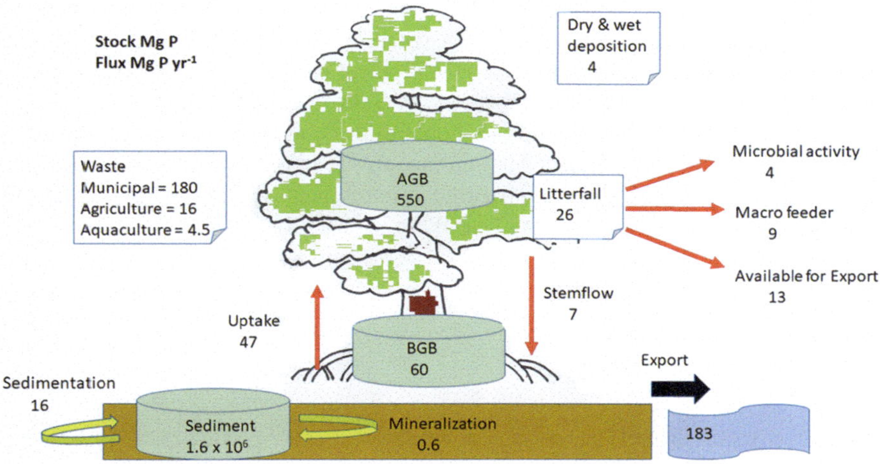

Fig. 6.4 Schematic representation of P cycle with stock and fluxes in a anthropogenic mangrove environment in Malaysia. Phosphorus stock (in Mg) present in biomass (AGB and BGB) is order of magnitude lower than the sediment. Various biogeochemical fluxes are given in Mg P yr^{-1}. (Data Source: Yeok 2002; Gong and Ong 1990)

more pristine mangroves, Indian Sundarbans is chosen as model site in this chapter to revisit the P budget (Ray et al. 2017, discussed in Sect. 6.6.3.3).

Primary producers mainly depend upon the internal input of P which is mobilized from the sediments. Deposition of P in the sediment and below ground biomass takes place through litterfall and streamflow, respectively. However, remobilization of P is not very smooth in such sediment conditions due to their occurrence in geochemically protected forms, i.e. P either associated to Ca, Fe or Al-hydroxides or can be adsorbed onto mineral surfaces or protected within the mineral matrices or present in organic compounds (Ruttenberg 1992). Chemical speciation of P largely controls biogeochemistry of this element. Studies of S, Fe, and P dynamics in wetlands indicate a strong sulfide/reactive Fe dependency controlling the P solubility under reduced conditions. For instance, in North Brazil mangroves, it was found that the speciation of P with Fe/Al (P-Fe/Al) was the main chemical bound species in the sediment (0.35 ± 0.09 to 0.56 ± 0.26 mg g^{-1}) compared to Ca-bound P (P-Ca) (0.03 ± 0.01 mg g^{-1}) (Ursula 2007). Sedimentation and subsequent immobilization of P within the geosphere is a greater flux term than its mineralization.

Despite temporary sink of P in the mangrove estuarine sediments, river water-driven point sources like agricultural/aquacultural run-off and wastewater discharges are always a dominant source of P to their estuary and coastal zone (Fig. 6.4). Direct runoff containing dissolved PO_4^{3-} is also very significant, particularly when rainfall follows the application of fertilizers in upland (Kleinman et al. 2009). During estuarine exchange, the pore- and groundwater P can be leached by rainfall, tidal inundation or drainage (Dittmar and Lara 2001).

Phosphate enters the atmosphere from a variety of sources such as continental-derived dust, sea spray, and plant pollen (Mahowald et al. 2005). Among these external atmospheric sources, dry and wet deposits in aerosol system also make a significant contribution to the total input flux. The sea-air exchange of aerosol particles plays an important role in the global biogeochemical cycle of phosphorus (Graham and Duce 1982). Dry deposition mangroves in the Indian Sundarbans was estimated to be higher than the marine environment in the North Island of New Zealand (0.12 to 27 versus 0.14 μg P cm^{-2} yr^{-1}; Ray et al. 2017; Chen et al. 1985).

6.5.3 Mangrove P Budget

Despite the importance of P cycle in maintaining mangrove productivity, a complete ecosystem scale budget is still very rare. A holistic scientific approach on an overall P budget is needed to elucidate the relative importance of various pools as a source or sink. Based on the results of P fluxes from the Merbok mangroves in Malaysia, a budget has been calculated (Fig. 6.5).

Analysis of data from various sources, balancing the input, output, and fluxes between the major pools of P leads to a conclusion that the mangrove ecological role as P source/sink has been greatly masked by the huge input of P from human activities. It is evident from the budget that municipal sewage loads in the Merbok catchment is the largest input flux of total P (>85% of the total TP input). Agriculture runoff contains high dissolved PO_4^{3-} concentration (~20-fold higher than the levels in more pristine region) that has been increased in recent decades when inputs of anthropogenic nutrients into the coastal sea have resulted in eutrophication, modifying aquatic food webs and causing severe hypoxic events in the coastal environment (review by Ramesh et al. 2015).

In the budget, tidal export is the largest loss term despite over-estimation of P output from tidal exchange due to the inherent variability of the tidal system on both temporal and spatial scales (Yeok 2002). According to the author, this is "unrealistically high" flux, but still makes a point given the riverine mangroves, unlike tide-dominated settings, always act as a net source of P to the estuary and sea despite their

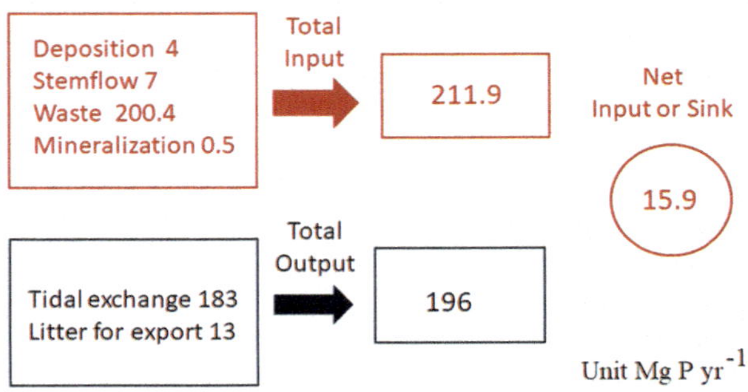

Fig. 6.5 A mass-based P budget in the anthropogenic mangrove ecosystem in Malaysia

temporary storage. A similar trend for the riverine Sundarbans is highlighted in Sect. 6.6.3.3.

Overall budget for the anthropogenically perturbed mangrove locations in Malaysia results into a net P gain of 15.9 Mg P yr^{-1}. Growing human population and industrializations around Asia-Pacific settings post an immediate major concern for the overall health of the mangrove ecosystem, but still these they manage to conserve P at net flux basis, and assign as a potential solution in hosting man-made nutrients through conservation mechanism.

6.6 Nutrient Budget in the Sundarbans

6.6.1 Overview of the Mangrove System

Our focus here is the Sundarbans, a UNESCO heritage site and the largest deltaic mangrove ecosystem in the world. The Sundarbans is bounded by 21°32′–22°40′ N latitude and 88°05′–89°51′ E longitude covering an area of around 10,000 km^2, of which approximately 60% lies in Bangladesh and 40% in India. The Harinbhanga River forms a natural demarcation, separating the Bangladeshi and Indian Sundarbans (Fig. 6.6). The mangrove ecosystem is characterized by high biodiversity, monsoonal rains, flooding, delta formation, tidal influences, exposure to super cyclones (Mandal and Hosaka 2020).

This unique ecosystem hosts a large number of flora and fauna. The forest is particularly rich in floral biodiversity, such as *Avicennia alba, Avicennia marina, Avicennia officinalis* are the predominant ones followed by *Ceriops decandra, Excoecaria agallocha, Bruguiera gymnorrhiza, Aegialitis rotundifolia, Sonneratia apetala, Aegiceras corniculatum, Xylocarpus granatum, Heritiera fomes* and mangrove associates like *Porteresia coarctata, Phoenix paludosa, Acanthus ilicifolius.* Average canopy height rarely exceeds 10 m.

Most of the rivers at the Sundarbans biosphere region flow north to south and are influenced by the tides from the Bay of Bengal. The main estuaries from west to east are Hooghly, Saptamukhi, Thakuran, Matla, Bidya, Ajmalmari, Bidyadhari, Gosaba, Kalindi, and Raimangal. These rivers, apart from the Hooghly, have no direct connection with the Ganges. Therefore, eastern part of Indian Sundarbans is more of tide-dominated settings over river influence.

Geologically the area is a result of extensive fluvio-marine deposits of the river Ganges and Bay of Bengal and the character of the sediment is silty clay with composition of quartzo-feldspathic minerals (quartz, albite, microcline) contributed from the eroded rocks of acidic composition of the drainage basin.

Climate in the region is characterized by the southwest monsoon (June–September), north east monsoon or post-monsoon (October–January) and pre-monsoon (February–May); 70–80% of annual rain fall occurs during the summer monsoon (South west monsoon), resulting in high river discharge (2952 and 11897 m^3s^{-1}), which gradually diminishes to 900–1500 m^3s^{-1} during non-monsoonal months (Mukhopadhyay et al. 2006).

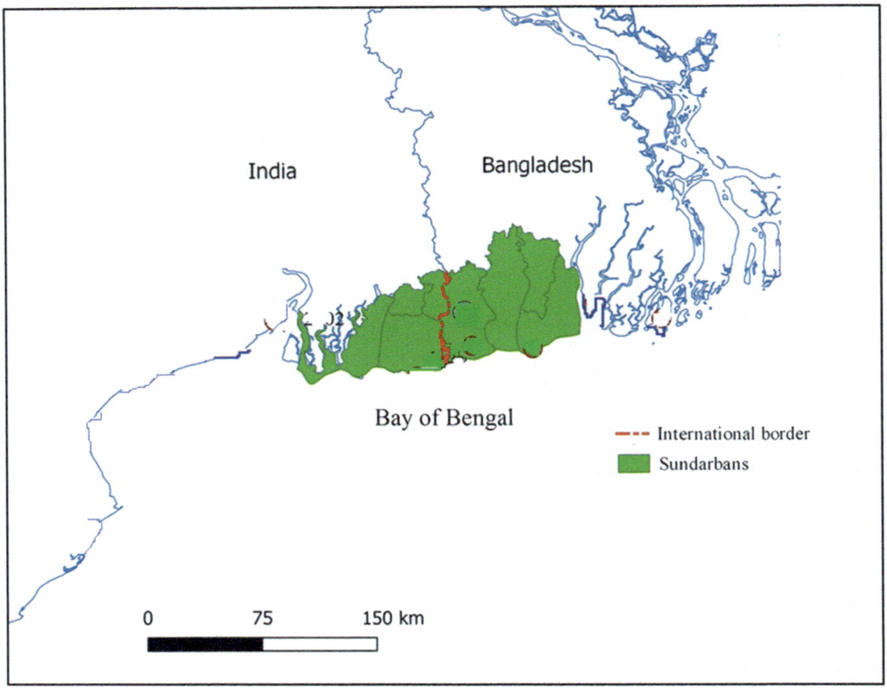

Fig. 6.6 Map of the Sundarbans mangrove covering India and Bangladesh (modified from Mandal and Hosaka 2020)

6.6.2 State of Nutrients in the Sundarbans

Mangrove dominated estuaries in the Sundarbans are the main sources of nutrients to the coastal water of Bay of Bengal, with significantly higher concentrations during the monsoonal run-off periods (53% for DIN, 31% for DIP) compared to the non-monsoon seasons (Mukhopadhyay et al. 2006). Such monsoonal enhancement of nutrient loads is due to the anthropogenic sources derived from upland aquaculture farms, waste discharge from Industry from adjacent Haldia port and domestic sewage discharge-points of the Kolkata mega city. Despite these anthropogenic impacts, a large section of the mangroves are typically pristine where human interferences are minimal in the protected areas (especially the tiger reserve forest in the east). High litter fall and its degradation and re-mineralization are the major biogenic sources of N and P in the Sundarbans where significant positive correlation was observed between OC and nutrients indicating ($R^2 = 0.80$; $p < 0.05$) sediment in situ processes to control N, P dynamics (Ray et al. 2014, 2017). In Pichavaram mangroves of southern India, high concentration of total N and its weak correlation with OC was observed owing to ex situ sources (Bala Krishna Prasad and Ramanathan 2008).

Concentrations of dissolved and sedimentary N and P in the Sundarbans are comparable with the global mangroves and estuaries (Table 6.4). Both nitrate and phosphate concentrations in the mangrove estuaries remained almost stable varying between ~14 and 20 μmol L^{-1} for DIN and between 0.5 and 1.0 μmol L^{-1} for dissolved PO_4^{3-} during the period between 1990 and 2011 (Ghosh et al. 1992; Mukhopadhyay et al. 2006; Chowdhury et al. 2012; Nandy et al. 2017). These concentrations are generally found to be higher during the neap tide than the spring tide. Monsoonal run-off increase significant amount of nutrients concentration (53% for dissolved inorganic nitrate, 31% for dissolved inorganic phosphate) compared to the pre-monsoon from riverine and estuaries surroundings of the Sundarbans (Mukhopadhyay et al. 2006).

6.6.3 Nutrient Budget in the Sundarbans

Standard allometric models have been developed by Ray et al. (2011) for estimating AGB and BGB in the Sundrabans. Plant N and P stock and their accrual to live biomass were estimated from their concentration in dry biomass stock of different plant parts (leaf+wood+root) and its monthly increment in the AGB and BGB during the study period between 2009 and 2011 (details about model and method, refer to Ray et al. 2014, 2017)

6.6.3.1 Box Model Approach

The basic characteristics of carbon and nutrient cycle are often described in terms of the content in various reservoirs and the fluxes between them (Lerman et al. 1975). Box model approach has been used for budgeting various biogeochemical processes in the ocean (Frost and Franzen 1992) and estuaries (Mukhopadhyay et al. 2006). Box models are representations of a system where quantities of materials are depicted as uniform within each box, and the flux between them is shown with arrows depending on their net concentrations (Rodhe and Bjorkstrom 1979). Wide classes of natural processes like radioactive decay, many forms of chemical decomposition, advective transport increase in a rate proportional to the number of molecules available and in many cases the increase could be smaller than proportional; for example, carbon-dependent photosynthesis in the sea is limited by nitrogen and phosphorus. Despite mass-based box model approach being only indicative, but it is very much instructive in pinpointing the magnitude of sinks and sources of elements. The implications of such model could also be used to guide management decisions with respect to a global carbon sequestration program and nutrient state (enrichment or limitation) in the marine ecosystem.

6.6.3.2 Nitrogen Budget

A schematic box model based N cycle has been developed for the Sundarbans considering multiple fluxes and stocks associated to the cycle (Fig. 6.7). Mangrove ecosystem N stock (i.e. total biomass and sediment up to 60 cm depth) is 720 Gg and most of it is in the biomass pool (98%). Major fraction of the available N in the

Table 6.4 Comparison of N, P concentrations in dissolved and sediment forms of world's selective mangrove locations

Mangrove location	Dissolved $NO_3^- + NO_2^-$ μmol L^{-1}	Dissolved NH_4^+ μmol L^{-1}	Sedimentary N μmol g^{-1}	Dissolved PO_4^{3-} μmol L^{-1}	Sedimentary P μmol g^{-1}	Reference
Sundarbans, India	12–15	2–5	134–225	0.5–1	0.07–0.3	Ray et al. (2017)
Amazon, Brazil	2–3	5–30	n.d.	1–4	n.d.	Dittmar and Lara (2001)
SE Brazil	0.5–5.5	n.d.	207	0.1–1.1	13	Sanders et al. (2014)
French Guiana	1.5–3.5	2.5–6	50–450	0.25–0.8	1–10	Ray et al. (unpublished)
Vietnam	6.8–11.4	2–8	n.d.	1.2–3.1	n.d.	Taillardat et al. (2019)
Malaysia	4.5	2.5	190–415	0.2–0.8	n.d.	Tanaka and Choo (2001)

n.d: no data available

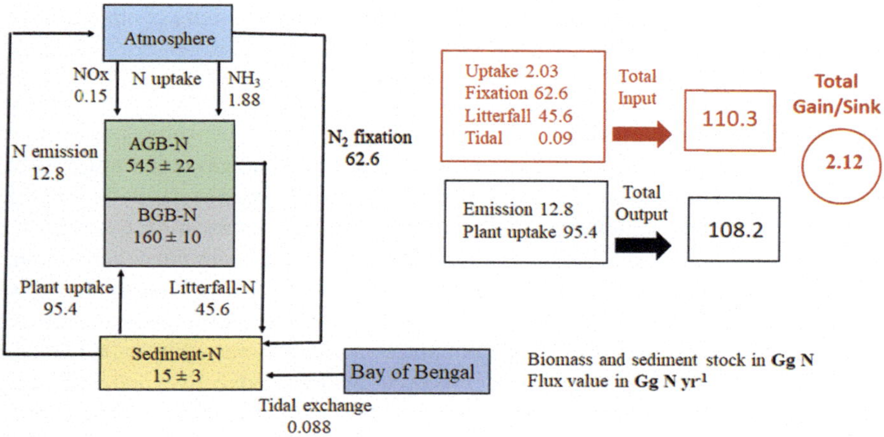

Fig. 6.7 Box model of N cycle in the left side, and mass-based budget (right side) calculating net sink/source of the Sundarbans. Stock are in Gg N (colored boxes, mean ± SD) and fluxes in Gg N yr^{-1} (solid arrow). (Data Source: Ray et al. 2014)

Sundarbans mangrove ecosystem is at all times in the biomass and is recycled within the organic structure of the biosphere-geosphere system (short N residence time ~58 days in sediment; Ray et al. 2014). Biological mechanisms help to retain and conserve N in the forest.

The major losses or sources of N from the sediment system occur via plant assimilation and, to a lesser extent, by N emission (probably denitrification). N emission appears to be very low compared with N_2 fixation for the Sundarbans, although in situ experiments of these fluxes were not performed except for the bioassay experiments of fixation (12 nmol cm^{-2} hr^{-1}), but it appears to follow same trend of the Missionary Bay mangroves (0–18 nmol cm^{-2} hr^{-1}, Alongi 2009). However, this is in contrast to the global N budget where N loss via denitrification was a greater flux component than the fixation (Fig. 6.3), revealing diverse geomorphic features of the mangrove sediment conditions from location to location.

After N_2 fixation, litter N input is the second largest input flux to the mangrove sediment. Net uptake of both NO_x and NH_3 by the Sundarbans mangrove forest was observed, but they altogether could account for only 2% of N required for mangrove net production. After summing up all those input fluxes, total N gain by the sediment is estimated to be 110.3 Gg N yr^{-1}. Total output flux or removal of N is 108.2 Gg N yr^{-1}, resulting into a net sink of 2.12 Gg N yr^{-1}. Considering extrapolations made from the measurements over a small area of such a huge ecosystem and also unaccounted sources (like sedimentation, volatilization, river run-off, etc.), the input and output sums are very close with the net loss well within the range of probable error (~5% difference). The net N sink is 1.9% of the total N input flux, that is in line with the global percentage (12.6%, refers to Fig. 6.3). Therefore, potentiality of such budget for the Sundarbans should be accepted with confidence, and included in the existing global data so that the more comprehensive N budget for the mangroves could be achieved.

6.6.3.3 Phosphorus Budget

Box model representation of P cycle in the Sundarbans mangrove is presented in Fig. 6.8. There are six coupled reservoirs of P in the model (Ray et al. 2017): P in the form of aerosol in the atmosphere, P in the form of organic matter in AGB and BGB, P in the form of organic and inorganic matter in soil, land estuary, and ocean. Fundamental processes involved in the model are: (1) dust aerosol deposition from the atmosphere and aerosol emission from the forest and surrounding mangrove water, (2) litter input from AGB, (3) Root uptake as BGB from sediment, (4) breakdown of plant litter to inorganic P followed by their advective and diffusive exchange between sediment and water, (5) deposition of particulate P on mud floor, (6) export of P from the mangrove system to the coastal water.

Forest biomass and sediment P stock (up to 60 cm depth) was 49.67 Gg P, out of which 97% was in the biomass pool. The residence time of P (= total mass of P in reservoir/rate of P removal from reservoir) incorporated into the sediment is very short (51 days) compared to the mangrove biomas (7.9 years; Ray et al. 2017) suggesting conservation of P in the biological reservoir and its rapid recycling within the biosphere-geosphere system.

Total P input fluxes to the sediment after summing up all sources, i.e., litter fall (6.1), sedimentation of particulate matter (0.8), and net atmospheric deposition (9.1) is 16.06 in Gg P yr^{-1}. The total output fluxes of P combining plant uptake (7.4), advective transport (3.6), and export (3.7) are 14.7 Gg P yr^{-1}. The net forest

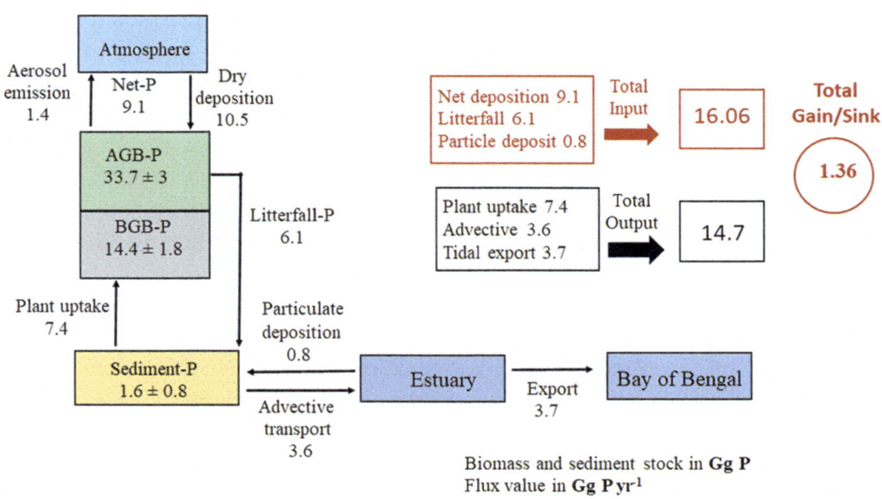

Fig. 6.8 Box model of P cycle in the left side, and mass-based budget (right side) calculating net sink/source of the Sundarbans. Stock are in Gg P (colored boxes, mean ± SD) and fluxes in Gg P yr^{-1} (solid arrow). (Data Source: Ray et al. 2017)

deposition of P from atmospheric sources is the largest input flux (dry deposition–emission = 10.5−1.4 = 9.1 Gg P yr^{-1}) that also compares well with the P deposition of long range African dust reported for the Amazon forest (6–37 Gg P yr^{-1}; Yu et al. 2015). Plant uptake accounted for 50.3% of the total output or P removal from the sediment. It was found that in contrast to N, concentration of P in pore water was roughly double the tidal water value and loss of P through sediment pores characterized by advective dispersal was about sixty two-fold as large as that of dispersal due to molecular diffusion.

The budget results into a net P gain or sink as 1.36 Gg P yr^{-1} that corresponds to 8.4% of the total input flux. This percentage in the pristine Sundarbans is very similar to the perturbed mangroves in Malaysia (7.4%, Fig. 6.5), suggesting an excellent agreement of budget formulation with high confidence, and consideration of these data in the global budget.

6.7 Conclusion and Perspective

With the rapidly changing world caused by human-induced pressures, atmospheric temperature, CO_2, and sea level rise (collectively as Climate Change), balancing the budgets of nutrients in coastal and mangrove environment are subject to hamper. However, at present, there are still substantial limitations on the impacts of global change on mangroves and their relationships with the nutrients. Although there are only few articulations available on pros and cons of global changes to nutrient dynamics in mangrove ecosystems, experimental approaches to test the effects of such relationship are largely descriptive than empirical.

In this context, an excellent review by Alongi (2018) is available for further discussion, where Author described in detail about how sources or sinks of N and P in mangroves could be influenced by the global alterations, such as land-use change, intensification of tropical storms and flooding, elevated CO_2 and temperature in the atmosphere. Author hinted that the effects of land-use change and the resulting eutrophication would lead to changes in rates and pathways of nutrient transformation processes that in turn affect rates of net primary productivity and survival of specific mangrove species. Recent frequency of intense storms and flooding could result in pulses of freshwater and sediment loads to the downstream coastal water, and relieve nutrient limitation in the mangroves by the excessive upstream discharge. While monsoon could lower the rate of denitrification (Fernandes et al. 2013), flooding has been suggested to release plant-available phosphate, hence affecting nutrient limitation and availability in mangroves (Mendoza et al. 2012). Temperature increases are likely to result in faster cycling of N and P transformation processes because microbial growth and rates of transformation processes are closely and positively linked to changes in temperature (Alongi 1988). Increased CO_2 concentration has shown positive feedback on terrestrial plant productivity (known as "CO_2 fertilization"; Norby et al. 2005), but the consequences of negative feedback on atmospheric CO_2 is uncertain because of the expectation that feedbacks through N and P cycles would reduce CO_2 fertilization effect (Thornton et al. 2007). For example, C cycle model predicted 1.17- and 2-fold increase of C storage in live biomass and sediment, respectively, in response to the hypothetical atmospheric CO_2 increment from 364 to 580 ppmv (Ray et al. 2013). Under such high CO_2 condition, enhanced storage of N and P in the long-lived reservoirs can significantly reduce their bioavailable fractions in the tight N and P budget of the Sundarbans, and that could eventually induce a negative feedback on mangrove productivity increase in high CO_2 world.

All these arguments and assumptions on the relationship between mangrove nutrients and global changes can be ascertained through more results from regional and global surveys, and also by comparing time series data over the decades. A better resolution mangrove N and P budget can only be achieved after that.

Acknowledgement Authors sincerely thank Sundarbans Biosphere Reserve and Divisional forest office, South 24 Parganas, Govt. of West Bengal, for giving permissions to carry out the consistent mangrove research. Authors are indebted to Prof. Hans Paerl (UNC-Chapel Hill) for providing the image used in Fig. 6.1.

References

Adame MF, Neil D, Wright SF, Lovelock CE (2010) Sedimentation within and among mangrove forests along a gradient of geomorphological settings. Estuar Coast Shelf Sc 86:21–30.

Adame ME, Lovelock CE (2011) Carbon and nutrient exchange of mangrove forests with the coastal ocean. Hydrobiol 663:23–50.

Almahasheer H,Duarte CM, Irigoien X (2016) Nutrient Limitation in Central RedSea Mangroves. Front. Mar.Sci.3:271. https://doi.org/10.3389/fmars.2016.00271

Alongi DM (1988) Bacterial production and microbial biomass in tropical mangrove sediments. Microb Ecol 15:59–79.

Alongi DM, Boto KG, Robertson AI (1992) Nitrogen and phosphorus cycles. In: Tropical Mangrove Ecosystems, obertson, A.I. and D.M. Alongi, eds., Coastal & Estuarine Studies No. 41, American Geophysical Union, Washington D.C., pp. 251–292.

Alongi D, Clough B, Dixon P, Tirendi F (2003) Nutrient partitioning and storage in arid-zone forests of the mangroves Rhizophora stylosa and Avicennia marina. Trees 17:51–60.

Alongi, D.M., B.F. Clough and A.I. Robertson (2005a) Nutrientuse efficiency in mangrove forests along the arid tropical coast of Western Australia. Aquatic botany, 82: 121–131

Alongi, D.M. (2009) The energetics of mangrove forests. Springer, Dordrecht

Alongi, D.M. (2013) Cycling and Global Fluxes of Nitrogen in Mangroves. Global Environmental Research, 17/2013: 173–182

Alongi DM (2014) Carbon cycling and storage in mangrove forests. Ann Rev Mar Sci 6:195–219.

Alongi DM (2018) Impact of Global Change on Nutrient Dynamics in Mangrove Forests. Forests 9:596; https://doi.org/10.3390/f9100596

Alongi DM (2011) Early growth responses of mangroves to different rates of nitrogen and phosphorus supply. J Exp Mar Biol Ecol 397:85–93.

Allen DE, Dalal RC, Rennenberg H, Meyer RL, Reeves S, Schmidt S (2007) Spatial and temporal variation of nitrous oxide and methane flux between subtropical mangrove sediments and the atmosphere. Soil Biol Biochem 39:622–631.

Amano, T, Yoshinaga, I., Yamagishi, T. et al., 2011. Contribution of Anammox Bacteria to Benthic Nitrogen Cycling in a Mangrove Forest and Shrimp Ponds, Haiphong, Vietnam. Microbes and Environment, 26, 1–6. https://doi.org/10.1264/jsme2.ME10150

Andreetta A, Fusi M, Cameldi I, Cimò F, Carnicelli S, Cannicci S (2014) Mangrove carbon sink. Do burrowing crabs contribute to sediment carbon storage? Evidence from a Kenyan mangrove system. J Sea Res 85:524–533.

Ayukai T, Miller D, Wolanski E, Spagnol S (1998) Fluxes of nutrients and dissolved and particulate organic carbon in two mangrove creeks in northeastern Australia. Mangroves and Salt Marshes 2:223–230.

Bianchi T (2007) Biogeochemistry of estuaries, 700. Oxford: Oxford University Press.

Bala Krishna Prasad M, Ramanathan AI (2008). Sedimentary nutrient dynamics in tropical estuarine mangrove ecosystem. Estuar Coast Shelf Sci 80:60–66.

Boto KG, Wellington JT (1988) Seasonal variation in concentrations and fluxes of dissolved organic and inorganic materials in a tropical, tidally-dominated, mangrove waterway. Mar Ecol Prog Ser 50:151–160.

Bulmer RH, Schwendenmann L, Lundquist CJ (2016) Carbon and Nitrogen Stocks and Below-Ground Allometry in Temperate Mangroves. Front Mar Sci 3:150. https://doi.org/10.3389/fmars.2016.00150.

Corredor JE, Morell JM (1994) Nitrate depuration of secondary sewage effluents in mangrove sediments. Estuaries 17:295–300.

Chen GC, Tam NFY, Ye Y (2012) Spatial and seasonal variations of atmospheric N_2O and CO_2 fluxes from a subtropical mangrove swamp and their relationships with soil characteristics. Soil Biol Biochem 48:175–181.

Chen L, Arimoto R, Duce RA (1985) The sources and forms of phosphorus in marine aerosol particles and rain from Northern New Zealand. Atmos Environ 19:779–787.

Chen R, Twilley RR (1999) A simulation model of organic matter and nutrient accumulation in mangrove wetland soils. Biogeochem 44:9–118.

Chauhan R, Datta A, Ramanathan A, Adhya TK (2015) Factors influencing spatio-temporal variation of methane and nitrous oxide emission from a tropical mangrove of eastern coast of India. Atmos Environ 107:95–106.

Chowdhury C, Majumder N, Ray R, Jana TK (2012) Inter-annual variation in some genera of diatom and zooplankton in a mangrove ecosystem. Biodiv Conserv 21:2029–2043

Clough BF, Boto KG, Attiwill PM (1983) Mangrove and sewage: A re-evaluation. In Biology and Ecology of Mangroves (ed. Teas, H. J.), Tasks for Vegetation Science Series, Dr W. Junk Publishers, Lancaster, UK, pp. 151–162.

Dalsgaard T, Thamdrup B, Canfield DE (2005) Anaerobic ammonium oxidation (anammox) in the marine environment. Res Microbiol 156:457–464.

Das S, De TK, Jana TK (2014) Vertical profile of phosphatase activity in the Sundarban mangrove forest, North East Coast of Bay of Bengal, India. Geomicrobiol J 31:716–725.

Davis SE, Childers DL, Day JWJ, Rudnick D, Sklar F (2001) Nutrient dynamic in vegetated and unvegetated areas of a southern Everglades mangrove creek. Estuar Coast Shelf Sci 52:753–776.

Dittmar T, Lara RJ (2001) Do mangroves rather than rivers provide nutrients to coastal environments south of the Amazon River? Evidence from long-term flux measurements. Mar Ecol Prog Ser 213:67–77.

Dittmar T, Hertkorn N, Kattner G, Lara RJ (2006) Mangroves, a major source of dissolved organic carbon to the oceans. Glob Biogeochem Cy 20:GB1012. https://doi.org/10.1029/2005GB002570.

Donato DC, Kauffman JB, Murdiyarso D, Kurnianto S, Stidham M, Kanninen M (2011) Mangroves among the most carbon-rich forests in the tropics. Nature Geosci 4:293–297.

Feller I, McKee K, Whigham D, O'Neill J (2002) Nitrogen vs. phosphorus limitation across an ecotonal gradient in a mangrove forest. Biogeochem 62:145–175.

Fenchel T, Blackburn TH (1979) Bacteria and Mineral Cycling, Academic Press, London, p. 225.

Feng W, Wu F, He Z, Song F, Zhu Y, Giesy JP, Wang Y, Qin N, Zhang C, Chen H, Sun F (2018) Simulated bioavailability of phosphorus from aquatic macrophytes and phytoplankton by aqueous suspension and incubation with alkaline phosphatase. Sci Total Environ 616:1431–1439.

Fernandes SO, Gonsalves, M-J, Michotey VD, Bonin P, Loka Bharathi PA (2013) Denitrification activity is closely linked to the total ambient Fe concentration in mangrove sediments of Goa, India. Estuar Coast Shelf Sci 131:64–74.

Finn JT (1980) Flow analysis of models of the Hubbard Brook ecosystem. Ecology 61:562–571.

Fourqurean JW, Duarte CM, Kennedy H, Marbà N, Holmer M, Mateo MA, Apostolaki ET, Kendrick GA, Krause-Jensen D, McGlathery KJ, Serrano O (2012) Seagrass ecosystems as a globally significant carbon stock. Nature Geosci 5:505–509.

Frost BW, Franzen NC (1992) Grazing and iron limitation in the control of phytoplankton stock and nutrient concentration: A chemostat analogue of the Pacific equatorial upwelling zone. Mar Ecol Prog Ser 83:291–303.

Fujimoto K, Imaya A, Tabuchi R, Kuramoto S, Utsugi H, Murofushi T (1999) Below ground carbon storage of Micronesian mangrove forests. Ecol Res 14:409–413.

Ghosh SK, De TK, Chowdhury A, Jana TK (1992) Distributions of nutrients in estuarine waters of Hooghly River. Trop Ecol 23:72–77.

Gong WK, Ong JE (1990) Plant biomass & nutrient flux in a managed mangrove forest in Malaysia. Estuar Coast Shelf Sci 31:519–530.

Graham WF, Duce RA (1982) The atmospheric transport of phosphorus to the western North Atlantic. Atmos Environ 16:1089–1097.

Hedges JI, Clark WA, Quay PD, Richey JE, Devol AH, Santos U de M (1986) Compositions and fluxes of particulate organic material in the Amazon River. Limnol Oceanogr. 31:717–738.

Holguin G, Vazquez P, Bashan Y (2001) The role of sediment microorganisms in the productivity, conservation, and rehabilitation of mangrove ecosystems: An overview. Biol Fertil Soils 33:265–278.

IGBP (1997) Modelling the transport and transformation of terrestrial materials to fresh water and coastal ecosystem, Stockholm work shop Report No. 39.

Kauffman JB, Donato DC (2012) Protocols for the measurement, monitoring and reporting of structure, biomass and carbon stocks in mangrove forests Working Paper 86, CIFOR, Bogor, Indonesia.

Khan MNI, Suwa R, Hagihara A (2007) Carbon and nitrogen pools in a mangrove stand of Kandelia obovata (S., L.) Wetlands Ecol Manage (2009) 17:585–599 597 123 Yong: vertical distributionin the soil-vegetation system. Wetlands Ecol Manage 15:141–153

Kleinman PJA, Sharpley AN, Saporito LS, Buda AR, Bryant RB (2009) Application of manure to no-till soils: Phosphorus losses by sub-surface and surface pathways. Nutr Cycl Agroecosyst 84:215–227.

Komiyama A, Havanond S, Srisawatt W, Mochida Y, Fujimoto K, Ohnishi T, Ishihara S, Miyagi T (2000) Top/root biomass ratio of a secondary mangrove (Ceriops tagal (Perr.) C. B. Rob.) forest. Forest Ecol Manag 139:127–134.

Kristensen E, Jensen MH, Banta GT, Hansen K, Holmer M, King GM (1998) Transformation and transport of inorganic nitrogen in sediments of a Southeast Asian mangrove forest. Aquat Microb Ecol 15:165–175.

Kristensen E, Alongi DM (2006) Control by fiddler crabs (Uca vocans) and plant roots (Avicennia marina) on carbon, iron, and sulfur biogeochemistry in mangrove sediment. Limnol Oceanogr 51:1557–1571.

Krishnan KP, LokaBharathi PA (2009) Organic carbon and iron modulate nitrification rates in mangrove swamps of Goa, south west coast of India. Estuar Coast Shelf Sci 84:419–426.

Lara RJ, Dittmar T (1999) Nutrient dynamics in a mangrove creek (North Brazil) during the dry season. Mangroves Salt Marshes 3:185–195.

Lee SY (1995) Mangrove outwelling: A review. Hydrobiol 295:203–212.

Lerman A, MacKenzie FT, Garrels RM (1975) Modeling of geochemical cycles: Phosphorus as an example. Geological Society of America 142:205–218.

Leopold A, Marchand C, Deborde J, Allenbach M (2016) Water Biogeochemistry of a Mangrove-Dominated Estuary under a Semi-Arid Climate (New Caledonia). Estuar Coast 40:773–791.

Lindsay WL (1979) Chemical equilibria in soils. John Wiley and Sons: New York, USA. p. 449.

Lovelock CE, Feller IC, Mc Kee KL, Thompson R (2005) Variation in mangrove forest structure and sediment characteristics in Bolas del Toro, Panama. Caribb J of Sci 41:456–464.

Mahowald NM, Artaxo P, Baker AR, Jickells TD, Okin GS, Randerson JT, Townsend AR (2005) Impacts of biomass burning emissions and land use change on Amazonian atmospheric phosphorus cycling and deposition, Glob Biogeochem Cy 19, GB4030. https://doi.org/10.1029/2005GB002541.

Mandal, M.S.H. & Hosaka, T. Assessing cyclone disturbances (1988–2016) in the Sundarbans mangrove forests using Landsat and Google Earth Engine. Nat. Hazards 102, 133–150 (2020)

Mendoza UN, da Cruz CC, Menezes MP, Lara RJ (2012) Flooding effects on phosphorus dynamics in an Amazonian mangrove forest, Northern Brazil. Plant Soil 353:107–121.

Meyer RL, Risgaard-Petersen N, Allen DE (2005) Correlation between anammox activity and microscale distribution of nitrite in a subtropical mangrove sediment. Appl Environ Microbiol 71:6142–6149.

Moore TR, Trofymow JA, Prescott CE, Fyles J, Titus BD (2006) Patterns of carbon, nitrogen and phosphorus dynamics in decomposing foliar litter in Canadian forests. Ecosystems 9:46–62.

Mukhopadhyay SK, Biswas H, De TK, Jana TK (2006) Fluxes of nutrients from the tropical River Hooghly at the land–ocean boundary of Sundarbans, NE Coast of Bay of Bengal, India. J Mar Syst 62:9–21.

Nandy T, Mandal S, Deb S, Ghosh M, Nath T, Chatterjee M (2017) Short-term variations in surface water properties in the Sundarban Estuarine System, India. Sustain. Water Resour Manag 4:559–566.

Nedwell DB (1975) Inorganic nitrogen-metabolism in a eutrophicated tropical mangrove estuary. Water Res 9:221–231.

Nixon SW (1980) Between coastal marshes and coastal waters – A review of twenty years of speculation and research on the role of salt marshesin estuarine productivity and water diversity. In: Hamilton, R. and MacDonald, K.B. (Eds), Estuarine and Wetland Processes, pp. 437–525, Plenum, New York.

Norby RJ, Delucia EH, Gielen B, Calfapietra C, Giardina CP, et al. (2005) Forest response to elevated CO2 is conserved across a broad range of productivity. Proc Natur Ac Sci 102:18052–18056.

Odum WE, Heald EJ (1975) The detritus-band food web on an estuarine mangrove community. In: Cromin, L.E. (Ed.), Estuarine Research. Academic Press, New York (265–286 pp).

Oelkers EH (2008) Phosphate mineral reactivity: From global cycles to sustainable development. Mineralogical Magazine 72: 337–340.

Otero XL, Araújo JMC Jr, Barcellos D, Queiroz HM, Romero DJ, Nóbrega GN, Siqueira Neto M, Ferreira TO (2020) Crab Bioturbation and Seasonality Control Nitrous Oxide Emissions in Semiarid Mangrove Forests (Ceará, Brazil). Appl Sci 10:2215.

Paerl HW (2009) Controlling Eutrophication along the Freshwater–Marine Continuum: Dual Nutrient (N and P) Reductions are Essential. Estuar Coast 32:593–601.

Parton W, Silver WL, Burke IC, Grassens L, Harmon ME, Currie WS, King JY, Adair EC, Brandt LA, Hart SC, Fasth B (2007) Global-scale similarities in nitrogen release patterns during long-term decomposition. Science 315:361–364.

Pomeroy LR, Smith EE, Grant CM (1965) The exchange of phosphate between estuarine water & sediments. Limnol Oceanogr 10:167–172.

Poungparn S, Charoenphonphakdi T, Sangtiean T et al. (2016) Fine root production in three zones of secondary mangrove forest in eastern Thailand. Trees 30:467–474.

Purvaja R, Ramesh R, Ray AK, Rixen T (2008) Nitrogen cycling: A review of the processes, transformations and fluxes in coastal ecosystems. Current Sc. 94:1419–1438.

Ramesh R, Robin RS, Purvaja R (2015) An inventory on the phosphorus flux of major Indian rivers. Current Sci 108 (10):1294–1299.

Ray R, Ganguly D, Chowdhury C, Dey M, Das S, Dutta MK, Mandal SK, Majumder N, De TK, Mukhopadhyay, SK, Jana, TK (2011) Carbon sequestration and annual increase of carbon stock in a mangrove forest. Atmos Environ 45:5016–5024.

Ray R, Chowdhury C, Majumdar N, Dutta MK, Mukhopadhyay SK, Jana TK (2013) Improved model calculation of atmospheric CO_2 increment in affecting carbon stock of tropical mangrove forest. Tellus B: Chem Phys Meteorol 65:18981.

Ray R, Majumder N, Das S, Chowdhury C, Jana TK (2014) Biogeochemical cycle of nitrogen in a tropical mangrove ecosystem, east coast of India. Mar Chem 167:33–43.

Ray R, Majumder N, Chowdhury C, Das S, Jana TK (2017). Phosphorus budget of the Sundarban mangrove ecosystem: Box model approach. Estuar Coast 41:1036–1049.

Redfield AC (1934) On the proportions of organic derivatives in sea water and their relation to the composition of plankton. In: Laboratory LS-F (ed) James Johnstone memorial volume. University Press, Liverpool, pp. 176–192.

Reef R, Feller IC, Lovelock CE (2010) Nutrition of mangroves. Tree Physiology 30:1148–1160.

Reis, CRG, Nardoto, GB, Oliveira, RS. 2017. Global overview on nitrogen dynamics in mangroves and consequences of increasing nitrogen availability for these systems. Plant Soil (2017) 410:1–19. https://doi.org/10.1007/s11104-016-3123-7

Rivera Monroy VH, Day JW, Twilley RR, Vera Herrera F, Coronado Molina C (1995) Flux of nitrogen and sediment in a fringe mangrove forest in Terminos Lagoon, Mexico. Estuar Coast Shelf Sci 40:139–160.

Rodhe H, Bjorkstrom A (1979) Some consequence of nonproportionality between fluxes and reservoir contents in natural systems. Tellus 31:269–278.

Ruttenberg KC (1992) Development of a sequential extraction method for different forms of phosphorus in marine sediments. Limnol Oceanogr 37:1460–1482.

Ruttemberg KC, Goñi MA (1997) Phosphorus distribution, C:N:P ratios, and 13 Coc in artic, temperate, and tropical coastal sediments: Tools or characterizing bulk sedimentary organic matter. Mar Geol 139:123–145.

Sánchez-Carrillo S, Sánchez-Andrés R, Alatorre LC, Angeler DG, Arreola-Lizárraga JA, Álvarez-Cobelas M (2009) Nutrient fluxes in a semi-arid microtidal mangrove wetland in the Gulf of California. Estuar Coast Shelf Sci 82: 654–662.

Sanders, C. J., et al. (2014), Elevated rates of organic carbon, nitrogen, and phosphorus accumulation in a highly impacted mangrove wetland, Geophys. Res. Lett., 41, 2475–2480. https://doi.org/10.1002/2014GL059789.

Scharler UM, Ulanowicz RE, Fogel ML et al. (2015) Variable nutrient stoichiometry (carbon: nitrogen:phosphorus) across trophic levels determines community and ecosystem properties in an oligotrophic mangrove system. Oecologia 179:863–876.

Seitzinger SP (1988) Denitrification in freshwater and coastal ecosystems: Ecological and geochemical significance. Limnol Oceanogr 33:702–724.

Smith VH (2003) Eutrophication of freshwater and coastal marine ecosystems: A global problem. Environ Sci Pollut Res Inter. 10:126–139.

Sengupta A, Chaudhuri S (1991) Ecology of heterotrophic dinitrogen fixation in the rhizosphere of mangrove plant community at the Ganges River estuary in India. Oecologia 87:560–564.

Sippo JZ, Maher DT, Tait DR, Ruiz-Halpern S, Sanders CJ, Santos IR (2017) Mangrove outwelling is a significant source of oceanic exchangeable organic carbon. Limnol Oceanog Lett 2:1–8.

Spalding, M., Kainuma, M., Collins, L., 2010. World atlas of Mangroves. Earthscan.

Sutula MA, Perez ABC, Reyes E, Childers DL, Davis S, Day JWJ, Rudnick D, Sklar F (2003). Factors affecting spatial and temporal variability in material exchange between the Southern Everglades wetlands and Florida Bay (USA). Estuar Coast Shelf Sci 57:757–781.

Taillardat P, Ziegler AD, Friess DA, Widory D, David F et al (2019). Assessing nutrient dynamics in mangrove porewater and adjacent tidal creek using nitrate dual-stable isotopes: a new approach to challenge the outwelling hypothesis? Mar. Chem. 214:103662. https://doi.org/10.1016/j.marchem.2019.103662

Tanaka K, Choo P-S (2001) Influences of Nutrient Outwelling from the Mangrove Swamp on the Distribution of Phytoplankton in the Matang Mangrove Estuary, Malaysia. J Oceanograph 56:69–78. https://doi.org/10.1023/A:1011114608536

Tam NFY, Wong YS (1996) Retention of wastewater-borne nitrogen and phosphorus in mangrove soils. Environ Technol 17:851–859.

Thornton PE, Lamarque JF, Rosenbloom NA, Mahowald NM (2007) Influence of carbon-nitrogen cycle coupling on land model response to CO_2 fertilization and climate variability. Glob Biogeochem Cy 21:GB4018.

Twilley RR, Day JW Jr. (1999) The productivity and nutrient cycling of mangrove ecosystems, pp. 127–152. In: A. Yáñez-Arancibia y A. L. LaraDomínguez (eds.). Ecosistemas de Manglar en América Tropical. Instituto de Ecología A.C. México, UICN/ORMA, Costa Rica, NOAA/NMFS Silver Spring MD USA, p. 380.

Ursula MNM (2007) Dynamics of phosphorus and sulphur in a mangrove forest in Bragança, North Brazil. Dissertation thesis, Zentrum für Marine Tropenökologie, der Universität Bremen.

Valiela I, Cole M (2002) Comparative Evidence that Salt Marshes and Mangroves May Protect Seagrass Meadows from Land-derived Nitrogen Loads. Ecosystems 5:92–102.

Wattayakorn, G., E. Wolanski & K. Björn, 1990. Mixing, trapping and outwelling in the Klong Ngao Mangrove Swamp, Thailand. Estuarine, Coastal and Shelf Science 31: 667–688.

Wattayakorn G, Prapong P, Noichareon D (2001) Biogeochemical budgets and processes in Bandon Bay, Suratthani, Thailand. J Sea Res 46:133–142.

Wösten JH, Willigen MP, Tri NH, Lien TV, Smith SV (2003). Nutrient dynamics in mangrove areas of the Red River estuary in Vietnam. Estuar Coast Shelf Sci 57:65–72.

Wulff F, Stigebrandt A (1989) A time dependent budget model for nutrients in the Baltic Sea. Glob Biogeochem Cy 3:53–78.

Yeok FS (2002) Phosphorus Budget in a Mangrove Ecosystem. Ph.D thesis, Universiti Sains Malaysia.

Yu HB, Chin M, Bian HS, Yuan TL et al. (2015) Quantification of trans-Atlantic dust transport from seven-year (2007–2013) record of CALIPSO lidar measurements. Remote Sensing Environ 159:232–249.

Mangroves as a Carbon Sink/Stocks

7

Tengku Mohd Zarawie Tengku Hashim and Mohd Nazip Suratman

Abstract

Mangroves are recognized as ecosystem that grow and dominate the coastal areas of tropical and sub-tropical regions across the world. The high adaptability properties of these halophytic trees enable them to thrive in harsh conditions such as the intertidal zones. They not only provide ecological and socio-economic support, but also play pivotal role in ecosystem function, especially in offsetting an excess of carbon from the atmosphere. Recently, the global climate change scenario has generated interest in understanding the carbon storage of mangroves. Despite the crucial roles provided by mangroves, the ecosystem has degraded at an alarming rate mainly due to climate change and anthropogenic activities. The existence of mangroves in the coastal areas where they are considered as the most biogeochemically active area makes them potential to store/sink a large amount of carbon. The ability of mangroves to sink excessive carbon is reported to be more superior from other terrestrial forests, and this could hold the key component in mitigating global climate change. However, there is still uncertainty in quantifying the biomass and characterizing carbon dynamics in mangroves. Therefore, it is important to understand the functions of mangroves in reducing the impact of climate change. Moreover, an understanding the productivity of mangroves such as biomass, primary productivity and carbon accumulation could have a significant impacts to this uncertainty. In this chapter, recent advancements on the determination of mangroves carbon sinks are highlighted. Apart from that, this paper also reviews on future challenges that are faced by the mangroves to maintain their status as a blue carbon area.

T. M. Z. Tengku Hashim · M. N. Suratman (✉)
Faculty of Applied Sciences and Institute for Biodiversity and Sustainable Development, UiTM, Shah Alam, Malaysia
e-mail: nazip@uitm.edu.my

R. P. Rastogi et al. (eds.), *Mangroves: Ecology, Biodiversity and Management*,
https://doi.org/10.1007/978-981-16-2494-0_7

Keywords

Mangroves · Carbon sink · Carbon stocks · Biomass · Climate change

7.1 Introduction

Since ancient time the word mangrove is believed to originate from Malay languages mangi-mangi which mean above soil. Mangroves can be defined as an assortment of salt tolerant plants such as trees, palms, shrubs and ferns that form a community and flourish within transitional or intertidal zones of coastal, estuary and riverine areas of tropical and sub-tropical regions across the globe. There is unanimity over the fact that this halophytic ecosystem is architecturally much simpler compared to the terrestrial forest, usually harbouring trees, shrubs, palms and scare ground ferns with height generally exceeding one half-metre and can easily be spotted across the coastlines (Duke 2011). These evergreen trees are a true ecotones, where it commonly found on mudflats and banks and can be easily identified by stands with rooted in salty sediments where the area are frequently submerged by daily ocean tides.

Mangroves develop numerous special adaptation capabilities to facilitate their survival against the harsh condition of the coastal climate environment. Constant inundation by frequent tidal action has transformed sediment in the area to become soft and muddy, the muddy condition makes mangroves to possess a bottom-heavy form that not only produce a high amount of biomass, but also to ensure tree can stand still, grow straight and strong in those kinds of condition (Naidoo 2016). According to Suratman (2008), to encounter the anoxic condition due to the waterlogged soils, mangroves adapt by having an aerial roots to aid them in gas exchange, water uptake and give support to older trees. Certain species of mangroves such as *Avicennia* and *Sonneratia* developed a special root system known as pneumatophores that act as a medium to overcome the low amount of oxygen (Naidoo 2016). This special root like pencil is equipped with lenticels that can be found a few centimetres above the anaerobic soil, enabling the diffusion of oxygen. One of the distinguish features uniquely possess by the mangroves that enable them to successfully inhabit the coastal area is their unique reproductive trait known as viviparous embryos. The mass production of viviparous propagules for species such as *Bruguiera, Rhizophora* and *Ceriops* can maintain and produce a mass number of individual trees by enabling the seed to germinate and mature into seedling while still attached to the mother tree before descending into the ground floor (Kathiresan and Bingham 2001; Naidoo 2016).

Despite their existence in the restricted coastal zone, mangroves are considered as a natural treasure in the coastal area as it plays a vital role in supporting socio-economics needs such as providing timber products, home to many important commercial fisheries resources and suitable area for aquaculture activity (Hutchison et al. 2014; Abdul Aziz et al. 2015; Venkatachalam et al. 2018). Mangroves also prove essentials in its ability to provide crucial ecological functions that effecting both upland communities and oceanic resources. For example, mangroves act as a first line of defence in anticipating a storm surge and tsunami while shielding the coastal community (Ahmadun et al. 2020). The presence of mangroves in coastal area can be termed as a natural ecosystem engineer in reducing coastal erosion and provide soil stabilization by binding sedimentation with their complex root system (Horstman et al. 2015; Gracia et al. 2018). Another ecological benefit that offered by the mangroves is by providing sound, suitable and safe nursery ground for many high commercial aquatic inhabitants such as bream, snapper, barramundi, grouper, banana prawns and mangroves mud crabs (Hutchison et al. 2014; Nanjo et al. 2014). Among all of the benefits provided by the mangroves, one particular important trait that possesses by this marine community that given less attention and always underestimates compare to upland forests is their ability to sink excessive carbon from the atmosphere. Mangroves have an enormous capacity for carbon storage and considered as earth's blue carbon sinks (Donato et al. 2011; Kauffman and Donato 2012).

Although mangroves are considered to be one of the highly productive biotopes, have vibrant, rich and endemic biodiversity while offered so many benefits both ecological and socio-economic, this ecosystem continue to experience losses at the highest degree. A previous study has shown that mangroves are the most threatened ecosystems in the whole world that caused by the calamity of global climate change and become more susceptible when uncontrollable anthropogenic activity intertwined (Ahmed and Glaser 2016; Das and Mandal 2016; Richards and Friess 2016). While mangroves receive a constant threat from the global climate change factors such as sea level rises, storms and tsunamis from the past decade, the rapid anthropogenic activities seem to be the new menace that could be the catalyst to the destruction of mangroves worldwide. According to Alongi (2012), mangroves around the globe experiencing high degradation about 1–3% annually, which are driven by dense human population and poverty in the coastal area. Furthermore, persistent hunger for more advance civilization, the human race has pushed the mangroves to the brink of extinction as demand for aquaculture farming, human settlement, illegal logging, agricultural activities and land development loomed large in the coastal area.

The uncertainty of the global climate change factors nowadays that have moved permanently outside the range of historical variation has come to the point that needs to be given serious defining by the human race. From shifting weather patterns to escalating combustion of fossil fuel has resulted increase in the concentration of carbon dioxide (CO_2) in the atmosphere suggest that the impact of climate change is global in scope and unprecedented in scale. The fluxes of the CO_2 concentration will cause perturbation in the global CO_2 reservoir and can have a significant impact on the global carbon cycle and sequestration (Le Quéré et al. 2017). Furthermore, the changes of carbon storage in the land and ocean reservoirs in response to increasing atmospheric CO_2 can be an additional fuel that could accelerate the global climate change scenario to a whole new level. Recent concern regarding climate change and increasing CO_2 in the atmosphere has generated interest among the researchers in the capabilities of mangroves to sink the excessive carbon concentration in the atmosphere. Despite account, only 2.4% of tropical forests, the capacity of mangroves to store carbon are four times greater than most other tropical forests around the world (Donato et al. 2011). Furthermore, the rapid carbon cycle, sediment and organic material that takes place between land and seas at the coastal area provide an opportunity for carbon sequestration potential in mangroves (Hashim et al. 2015). With the great hype surrounding their capacity in sinking excessive carbon, this fringe coastal community might hold the possible answer in ameliorate the impact of climate change.

This chapter aims to provide insights regarding the potential role of mangroves as a medium for sinking excessive carbon from the atmosphere. The perspective function of this halophytic plant in the global carbon cycle and mitigating the global climate change is underlined. Additionally, recent advancements on the determination of mangroves carbon sinks will be further discussed.

7.2 Ecosystem Services by Mangroves

7.2.1 Ecological Role of Mangroves

Mangroves form one of the unique wetland ecosystems that said to be most productive and biodiverse on the Earth, which comprise both living and non-living things. The unique adaptation ability of the mangroves not only enables them to cope with the harsh condition of the coastal area, but also provide a tremendous ecosystem function for the organism that lives in the area. For ages, mangroves serve as a frontline protector in terms of their position relative to many coastal hazards. They provide ecological support in buffering the impact of storms surge, wave activities and tsunamis. The existence of mangroves that possess complex root structure, width of mangroves zone and density in the coastal area acts as a blockage and reflects part of the wave current backward to offshore will reduce the destructive impact of the storm surge and tsunami (Zhang et al. 2012; Gracia et al. 2018). According to Krauss et al. (2009), during the Hurricanes Charley in 2004 and Wilma in 2004, the impact of these hurricanes when come in contact with mangroves, reduce the storm surge height and help in reducing the destruction to the coastal area and human society. Furthermore, the study also found out that the destruction by Hurricane Wilma could extend more than 70% further inland without the mangroves in the coastal area. Another study in Andaman Island during the catastrophic tsunami in 2004 indicated that existence of mangroves in the coastal area helped to reduce the damage caused by the tsunami only to 7%, meanwhile for the area that mangroves have degraded from the coastal area, it was estimated that the area received 80% to 100% of the damage (Dahdouh-Guebas and Koedam 2006).

As a coastal engineer in terms of reducing the soil erosion unique root systems of mangroves in limiting sediment exposure to wave energy and binding the sediment help to consolidate soil that will reduce the effect of erosion and in turns promoting soil stabilization (Horstman et al. 2015; Gracia et al. 2018). According to Thampanya et al. (2006), Southern Thailand which had lost almost 50% of its total mangrove since 1961 had promoted shorelines erosion by 0.01 to 0.32 km^2/year from 1967 to 1998, whereas the coastal area that dominated by the mangroves has experienced less erosion. Cabral et al. (2017) in their study found out that the rapid clearing of mangroves has resulted in 10% of the Mozambique coastal area is being highly exposed to erosion.

One of the major functions of the mangroves to the aquatic habitat is by providing sound and suitable place to life and survives. The unique structure of mangroves such as the complex prop roots and canopy shades that exists from the combination of branches and leaves not only form a strategic hideout spot from predators, but also provide shelter for the aquatic organism from predation thus create more microhabitat availability and increase the amount of food (Hutchison et al. 2014). Mangroves play a crucial role as the basis of the food chains that support a wide range of marine habitats in the coastal area. The high level of primary productivity from the mangrove vegetation such as litter, branches and trunks and other primary producers are important in establishing a complex food web (Hutchison et al. 2014).

7.2.2 Socio-Economic Role of Mangroves

The most tangible products from mangroves that can be commercialized are timbers. With a characteristic of mangroves that often grow as an almost pure stand in the coastal area made the mangroves suitable to be harvested for its timber. Timbers extracted from mangroves are used for many commercial goods such as charcoal production, poles for construction, industrial tannin, dye and Nipah products. Charcoal production is the most popular industry from mangrove trees, especially *R. apiculata* to produce high quality charcoal (Kridiborworn et al. 2012). Mangroves are also sources of poles for export and local house buildings purposed. As indicated by Ong and Gong (2013), poles from the mangroves especially from *Rhizophora* and *Bruguiera* make excellent materials for house pillars.

The roles of mangroves in providing breeding, feeding and nursery ground for many high values of commercial fish, shrimp and crab can be valuable assets for the commercial fishing industry. The availability of marine fish species such as the bream, snapper, barramundi, grouper, banana prawns that spend some part of their life cycle in the mangroves ecosystem can contribute to sources of income to the coastal households (Hutchison et al. 2014). According to Mohammad Abdullah et al. (2016), in Sundarbans India, the availability of rich resources of marine and fish species are fully utilized by the people in the area for their source of livelihoods. A study conducted in the coastal area of Madagascar found that 87% of adults in the area work in fisheries industry where most of the men employed in fishing and woman actively gleaning along the shorelines (Barnes-Mauthe et al. 2013).

The existence of mangroves throughout the coastal areas of the Gulf of California provided significant contribution to related fish species and total fisheries catch in the range of 10–32% (Aburto-Oropeza et al. 2008). In a study, an estimate annual values per sq. km of mangroves to fisheries was valued around USD 0.14 million to USD 6.1 million for offshore prawn and for USD 34 to USD 2.7 million for inshore coastal fisheries (Christensen et al. 2008). Das (2017) mentioned that the increased of mangroves in the coastal area boost the annual catchment of commercial fisheries and contributed to 15% of the total annual landing of the India, which saw the annual monetary gained of USD 0.57 billion.

7.3 Destruction of Mangroves

Despite the potential roles of mangroves to give benefit to the environment and surrounding coastal community, this ecosystem has degraded rapidly as compared to the upland forests throughout the decade. Table 1 reveals global extent of previous total world mangroves area estimate. With a constant deforestation rate of 1% annually (Alongi and Mukhopadhyay 2015), mangrove forests experience downward pattern in terms of global coverage from 18,100,000 ha in 1997 (Spalding et al. 1997) to 8,349,500 ha in 2016 (Hamilton and Casey 2016) (Table 7.1). Constant losses of dense mangroves throughout the globe to a smaller and fragmented area prove to have a devastating impact to the mangroves environment. This factor will result in mangroves long essential ecosystem service and survival decline tremendously and at great risk and this prompted the idea that mangroves may be wiped out from its existence in the near future.

As mentioned before, the loss of mangroves around the globe is due to the impact of global climate change and human encroachment activity. The significant losses of

Table 7.1 Global extent of previous total world mangroves area estimate

References	Number of countries included	Estimated total area (ha)
FAO (1981)[a]	51	15,642,673
FAO (1994)	65	16,221,000
Groombridge (1992)	56	16,500,000
ITTO/ISME (1993)[b]	87	19,847,861
Fisher and Spalding (1993)	54	12,429,115
Spalding et al. (1997)	91	18,100,077
Aizpuru and Blasco (2000)[a]	112[c]	17,075,600
FAO (2003)	112	14,653,000
FAO (2007)	124	15,231,000
Spalding et al. (2010)	123	15,236,100
Giri et al. (2011)	118	13,776,000
Hamilton and Casey (2016)	105	8,349,500

Sources: FAO (2003, 2007), Hamilton and Casey (2016) and Giri et al. (2011)

[a]Except for FAO (1981) and Aizpuru and Blasco (2000), the reference year is the year of the publications in which the estimate is cited, not the weighted average of all the national area estimates

[b]Combined figure from three publications by Clough (1993), Diop (1993), and Lacerda and Diop (1993)

[c]New estimates were provided for 21 countries, and for the remaining countries the study relied on Spalding et al. (1997)

mangroves due to these factors are reported higher in Southeast Asia region as compared to any part of the world. According to Hamilton and Casey (2016), five countries in the Southeast regions out of ten in the world are accounted to be the most countries in the world that experienced losses of its mangrove areas, where Indonesia top the ranks (6240 ha), followed by Malaysia (2020 ha), Myanmar (1960 ha), Thailand (390 ha) and Philippines (220 ha).

7.3.1 Anthropogenic Activities

Globally, Asian region representing the largest mangrove forests where 42% of world totals mangrove areas are located in this region. However, rapid development that occurs in the regions in recent decade has resulted mangrove forests in Asia experienced 30% of reduction since 1980, which is believed to be the highest rate of global mangrove area loss (Hamilton and Casey 2016). During the period between the years 2000 and 2012, Southeast Asia has lost more than 100,000 ha of mangroves at an average of 0.18% yearly where Myanmar is believed to be the highest nation that lost their mangroves in recent decade (Richards and Friess 2016). Meanwhile, Indonesia has stripped almost 74,900 ha or 3.11% of its mangrove area largely in the Provinces of Kalimantan Timur and Kalimantan Selatan, largely due to the land development, overexploitation of resources and aquaculture expansion (Hamilton and Casey 2016).

According to Truong and Do (2018), a total of 72,825 ha of mangroves in Mekong River Delta, Vietnam was cleared during 1980–1995 with an annual rate of 4855 per ha for shrimp aquaculture. Malaysia also faced losses of its mangroves, according to Hamdan et al. (2012), it was estimated that the rate of loss of mangrove in Malaysia to be about 1% or 1282 ha/year. Furthermore, according to Romañach et al. (2018), mangrove areas in Malaysia have lost 17% of its coverage in the coastal area to 570,516 ha from 695,000 ha in the 1970s. For the last decade, the exponential growth of aquaculture activities and rapid coastal development are becoming synonymous with the mangrove forest in Malaysia as it becomes the primary drivers that contribute to the loss of mangroves. In Kedah, Malaysia alone a total of 1041 ha of mangroves permanent reserve forests was converted into an aquaculture site in order to satisfy the demand for the aquaculture products (Ahmad and Mohammad 2005). With the nation gunning towards the status to become the number one exporter of oil palm product, oil palm crops expansion emerged as a new contender in Malaysia for the destruction of mangroves. According Richards and Friess (2016), it was estimated that 38% of mangrove loss in Malaysia was resulted from rapid expansion of oil palm plantation.

The loss of mangroves around the world not only can be seen in Asia continent, but also reported in other various continents. In the African continent, it was estimated that about 13.8% equates to 500,000 ha of mangrove forest has disappeared over the last 25 year with major losses are reported to be occurred in the Democratic Republic of the Congo, Gabon, Sierra Leone, Guinea-Bissau and Senegal (FAO 2007). According to FAO (2007), over the past two decades, the

North and Central America has lost almost 700,000 ha of their mangrove forests, meanwhile in the Caribbean, approximately 24% of mangrove area has degraded for the past quarter-century due to the land development. In South America, 90% of the mangroves are found in five countries which include Columbia, the Bolivarian Republic of Venezuela, Ecuador, Suriname and Brazil. During the periods 1980s and 1990s, 250,000 ha of the mangroves have lost all over region due to idea that considers the mangrove is unproductive ecosystem, where Brazil alone in the past 25 years has lost at least 50,000 ha (around 4%) of its total mangroves (Ferreira and Lacerda 2016).

7.3.2 Climate Change

One of unprecedented events in this decade that can have a significant impact to the survival of mangroves is the global climate change. The changes in climate condition at the coastal area such as an increase in sea level rise, storminess/tsunami and precipitation not only threaten the mangrove ecosystem, but it also disturbs their valuable ecosystem and socio-economic functions. One of the major threats that triggered by the changing of the global climate to the coastal ecosystem is the sea level rise. According to Church and White (2011), it was estimated that over the past decades the global sea level rise has risen by 3.2 mm per year due to the constant rising of heat content and continues melting of land ice. With the current trend of sea level rise, Nicholls et al. (2011) predicted that by 2100 it could rise to more than 4 m per year. The increase in water column resulted from sea level rise will eventually promote constant coastal flooding thus resulted in the high occurrence of submerged of mangroves under the seawater. The inability of mangroves to commensurate with this situation will may cause death to the mangroves due to its sensitivity for long inundation duration and frequency in the water (Ward et al. 2016). According to Rahman et al. (2011), Sundarbans, India has lost almost 17,000 ha mangroves since 1970 that largely been impacted by sea level rise. Using spatial analysis, Ellison and Zouh (2012) during the period 1957–2007 found that over two third of the shoreline edge of mangroves in Cameron suffered dieback and 89% of Mangrove Island located offshore was destroyed due to sea level rise. A study conducted in Guanxi, China using the Sea Level Affecting Marshes Model (SLAMM), indicated that an increase in sea level rise (2.9 mm per year), has resulted loss of mangrove habitat by 9.3%, 9.6% and 18.2% in 2005, 2050 and 2100, respectively (Li et al. 2015).

The continuous increase of temperature in the atmosphere resulted by the global warming phenomenon is likely to increase in tropical cyclone and storms activities. Coastal vegetation, especially the mangroves that inhabit the coastal area are vulnerable to the extreme weather event such as a storm, cyclone and tsunami that will result in loss of the mangroves. According to Long et al. (2016), the impact of Typhoon Haiyan that struck the Philippines coast has removed about staggering 8568 ha or 3.5% of the Philippines total mangrove area. Meanwhile, similar pattern of destruction was reported during the catastrophic Cyclone Sidr in Bangladesh. Trees were bent and removed from the soil, large number of trees were uprooted and overturn and the bark and twig were broken (Tanaka 2007). The catastrophic tsunami in 2004 in the Indian Ocean has resulted mangroves loss in South Andaman Island of India between 3825–10,000 ha followed by Aceh province, Indonesia (300–750 ha) and Andaman Coast (306 ha) (Alongi 2008).

The global climate change is predicted to have tremendous influence in terms of precipitation rates mainly through runoff and it was estimated that rainfall will increase 25% more by 2050 (Gilman et al. 2008). Variable precipitation will be further complicated by changes in temperature, influence both evaporation and transpiration rate (Ward et al. 2016). As mentioned by Gilman et al. (2008), changes in rainfall pattern have the potential influence on the growth, distribution and potential extent of the mangroves. The distribution of the world's mangroves showed that they are productive forests with complex structure exist in an area that receive a high amount of rainfall and high runoff as compared to the area that have a low amount of rainfall and runoff inputs (Kumara et al. 2010). In the drier condition, the decreased precipitation and increased evaporation will result in an increase in soil salinity that creates saline flats, which eventually will reduce the survival rate of seedling and growth rate and consequently resultant to mangrove loss (Ward et al. 2016). A study conducted in the Gulf of Carpentaria suggested that drier condition increased the salinities in the soil and resulted in destruction of *A. marina* species (Conacher et al. 1996).

7.4 Mangroves as a Carbon Sink

Since the beginning of the pre-industrial era, fossil fuels extraction as a source of energy, combustion and transportation fuel, triggered a rise of CO_2 concentration from approximately 277 ppm in 1750 to 402 ppm in 2016 (Le Quéré et al. 2017). The radiative properties and excessive of CO_2 in the atmosphere will influence the direction of the earth's global climate condition. With the increasing trend of greenhouse gasses, mangroves are disproportionately important component for the ecosystem as a medium in the global carbon cycle. Although mangroves represent only a small fraction of the tropical forests, their unique location in the coastal area might be crucial toward carbon biogeochemistry. Furthermore, this halophytic plant's capacity to sink carbon is believed to be among the best as compared to the tropical forests (Donato et al. 2011). According to Simard et al. (2019), mangroves were estimated to contribute to approximately 10–15% of total global carbon storage in the coastal regions. According to Suratman (2008), the carbon cycle that occurs in the coastal community was influenced by environmental factors such as temperature and precipitation while for the rate of carbon cycle is determined by the primary productivity and decomposition.

Mangroves stored carbon that is acquired from the photosynthesis process in stems, leaves, roots and branches as a part of their biomass. In terms of primary production in the coastal area, mangroves are believed to have higher and rapid carbon production than their adjacent marine primary producer ecosystem (Duarte et al. 2005). However, the rate of photosynthesis in the mangroves is varied according to species composition. In a study conducted in Tamil Nadu, India, it was found that *E. agallocha* has the highest net canopy photosynthesis (21.65 gC/m^2/ day) followed by *R. apiculata* (21.05 gC/m^2/ day), while B. cylindrical (15.99 gC/m^2/ day) has the lowest canopy photosynthesis rate (Sahu and Kathiresan 2019).

Mangroves are considered highly productive forests. Dead parts of the mangroves that fall to the floor do not completely decompose. This material slowly decomposes into a much simpler chemical substance that producing CO_2, water and energy (Fernando and Bandeira 2009). The breaking down of this organic component is a part of the coastal community's unique condition that comprise of waterlogged soil condition, flora and fauna not only create a unique food web in the mangroves area but also responsible for the carbon cycle process. Furthermore, the strategic location of the mangroves in the coastal area, not only enable some of the excess carbon transfer and subsidize to the adjacent ecosystems such as the ocean and beach ecosystems, but also stored the majority of organic carbon in the soil (Bouillon et al. 2008). The potential for storage of carbon in this ecosystem may be an important carbon sink candidate in order to combat the climate change scenario.

With evidence from the previous literature, mangroves may hold as one of the important candidates for carbon cycle that always been underestimated and overlooked. With the capability of mangroves in removing CO_2 from the atmosphere and stored it as a part of their plant materials and in soils, mangroves is undoubted contain the largest blue carbon in the coastal area. Therefore, to better understanding the dynamic of carbon sinks of this halophytic ecosystem, it is crucial to know the productivity of mangrove forests mainly in terms of biomass and primary productivity.

7.4.1 Biomass

Biomass can broadly be defined as weight or mass of its total above ground living organic matter in trees and usually expressed in a unit of a metric ton which can be divided into two different parts, i.e., the above ground biomass (AGB) and below ground biomass (BGB) (Walker et al. 2011). Mangroves can be categorized as an ecosystem that has higher biomass as compared to terrestrial forests. Several studies have been conducted around the world to quantify the biomass in mangroves. A study conducted in the Matang Mangrove Reserve, Malaysia which was dominated by *R. apiculata* species to be 480 Mg/ha (Putz and Chan 1986). Meanwhile, Hashim et al. (2015), in their study to estimate the AGB in a dominant *R. apiculata* forest in Merbok Mangrove Reserve, Malaysia found out that the AGB was 179 Mg/ha. Although the estimation was done in an area dominated by the same type of species, the AGB reading might be different due to the size and age of the mangroves in the area. A plant species such as *R. apiculata* can do photosynthesis and rapidly store carbon as a part of its biomass for almost 20 years during its lifespan and when the process level is off the storage of AGB and carbon does not decline (Alongi 2012).

The mangroves that exist at lower latitudes have higher biomass as compared to mangroves in higher latitudes. For example, a study that was conducted in Thailand found that the AGB of the riverine mangrove forests to be 449 Mg/ha (Jachowski et al. 2013). Similar results were also found by Kauffman et al. (2011), the estimation of AGB in Micronesia yielded about 514 Mg/ha. Meanwhile, in high latitude, the AGB estimation in Biscayne National Park, Florida, USA estimated that for fringe mangrove areas the AGB is about 56 Mg/ha while for dwarf mangroves 22 Mg/ha (Ross et al. 2001). Komiyama et al. (2008) highlighted that low latitudes area have greater AGB as compared to the temperate area. The influence of different climates might explain the variation of mangroves AGB between lower and higher latitudes. For lower latitude area, the amount of precipitation received and favourable climate condition enables the mangroves to grow bigger and have mature stands which contribute to higher AGB. Meanwhile different climate conditions, frequency of storm, precipitation and temperature affect the growth of mangroves that grow as a dwarf mangroves and yield lower AGB estimation. Distribution of mangroves that largely in Asia region and having a suitable climatic condition could be the most important area for carbon sinks.

To quantify the BGB in mangroves is not an easy task, with little literature discussed in estimating it. Even though the mangroves BGB component is considered higher than their AGB and important criteria for the estimation of mangroves biomass (Kauffman et al. 2011; Alongi 2012), the challenging task in sampling this type of biomass such as labour intensive, time consuming and costly makes it less popular among researchers. For example, Mackey (1993) estimated the BGB for *A. marina* in the secondary forests of Australia was 121.0 Mg/ha. Meanwhile, another study conducted in Australia for the same species in a primary forests indicated that the BGB was about 147.3 Mg/ha while the AGB were 144.5 Mg/ha (Briggs 1977). Other study conducted in Western and Central Sundarbans Mangroves, India, estimated that BGB for a natural forest consisting of *S. apetala, A. alba* and *Excoecaria agallocha* are 32.84 Mg/ha and 27.46 Mg/ha, respectively (Banerjee et al. 2013). While a study in Bangladesh to quantify the biomass and net primary productivity of mangroves recorded that the mean value of BGB was 84.2 Mg/ha with *A. officinalis* contributed the highest in terms of BGB reading (Kamruzzaman et al. 2017).

Even though carbon is stored in various parts such as stems, shoots, roots and down woods, soils are considered the most important parts where 50% of total carbon stocks are stored (Kauffman and Donato 2012). It was estimated that mangroves are carbon rich ecosystem with a total combination of tree and soil carbon containing up to 1023 Mg/ha in the tropics, which are higher than any other wetland ecosystems (Kauffman et al. 2011). In order to survive in the harsh coastal climate, mangroves possess a bottom-heavy structure where much of their biomass allocated into the root systems (Komiyama et al. 2008). According to Reef et al. (2010), mangroves invest more fixed carbon than any plant for adaptation purposes, such as to maximize water uptake, increase in stability and to transport oxygen. As a result of constant carbon fixation in the root system, it is fair to assume that mangroves store a disproportionate of their carbon underground. Furthermore, the amount of carbon that store in the soil increased with forest age. A study conducted in the Matang Mangrove Reserve, Malaysia suggested three-fold of soil carbon stock ranging from 385.2–545.0 Mg/ha (Adame et al. 2018).

7.4.2 Primary Productivity

The primary productivity occurs in mangroves through the photosynthesis process. The conversion of atmospheric CO_2 into organic compound is essential to producing new mangroves parts (stems, leaves, branches and root tissue) and maintaining the existing tissues (Alongi 2009). The main reason for the study of primary productivity in mangroves is to evaluate their carbon stocks. However, due to the mangroves that exists in the coastal area, physical and chemical factor such as solar radiation, salinity, fresh water sources, tides, soil type and temperature influence this process (Twilley et al. 1992, 2017). According to Bouillon et al. (2008), it was estimated that the global mangroves primary productivity is about 218 TgC yr.$^{-1}$. In a study conducted in the Sundarbans Mangrove forest, Bangladesh, the primary production was estimated to be 17.2 Mg/ha yr.$^{-1}$ (Kamruzzaman et al. 2017). The study indicated that mangroves location in the oligohaline zone that frequently flushes by tidal action might influence the results of primary productivity. Furthermore, it was concluded that mangroves in an oligohaline zone might have high productivity as compared to the mangroves that exist in other ecological zone (Kamruzzaman et al. 2017).

In another study, Ross et al. (2001) reported that the net primary production for fringe and dwarf mangroves in Florida, USA was 26.1 Mg/ha yr.$^{-1}$ and 8.1 Mg/ha yr.$^{-1}$, respectively. The study concludes that the high primary production in fringe mangroves is due to the rapid development of woody tissue compared to dwarf mangroves which slightly slower. In a study conducted in Tamil Nadu, India, it was reported that age plays an important role in mangroves primary productivity (Sahu and Kathiresan 2019). The studies estimated that for young mangrove trees in the area, the primary productivity was 30.80 Mg/ha yr.$^{-1}$ while for mature stand it was 17.04 Mg/ha yr.$^{-1}$. Another study conducted by Putz and Chan (1986) in estimating the tree growth and productivity in mature *R. apiculata* stands in Matang Mangrove Forests, Malaysia, found that the net primary productivity was about 17.7 Mg/ha yr.$^{-1}$. The best possible explanation for this trend might be when mangroves reach its maturity, their ability to sequester carbon declines, as for young mangroves the rate of sequestration increased.

7.5 Recent Advancements in Mangrove Carbon Sinks Studies

The exponential growth of remote sensing technology in recent decades has driven many forest ecologists and stakeholders to incorporate the usage of imagery data for many forest applications, especially in biomass monitoring in mangroves. Even though the popular traditional method such as destructive and allometric functions can yield better and accurate biomass results, it can measure only to small scale areas and cannot be applied to a larger area. Moreover, the treacherous mangrove areas that are almost impossible to access, time consuming, labour intensive and can be costly made the traditional method less relevance in today inventorying and monitoring purposes for biomass and carbon study. Therefore, remote sensing technology provides better spatially explicit, and can be efficiently combined forest biomass estimates and has the potential to give input at the large range of spatial and temporal scales (Galidaki et al. 2017).

Over recent decades, many literatures and publication have grown and applied the relationships between remote sensing parameters and mangroves attributes in determining the mangrove biomass. A relationship between the biomass and multiimagery satellite wavelength data was established during the study conducted in a mangrove replanting site in Jiulong River Estuary, China (Wang et al. 2018). The study found out that there is a significant correlation between AGB and Landsat bands 3 and 4. This study is an agreement with an earlier study that was conducted by Wu et al. (2016), where they conclude that the individual Landsat 8 band showed a promising correlation with AGB and carbon stocks where shortwave infra-red was the highest correlated (−0.57). A similar finding was also reported in the study in estimating the biomass and carbon stocks of the Matang Mangrove Forests Malaysia that utilized the Landsat TM and SPOT 5 data, where it was found that there are significant relationships between the vegetation indices and the forest variables (Hamdan et al. 2013). This concludes that satellite image data can have enormous potential in determining the biomass and carbon stocks in the mangroves which eventually help in understanding the potential of carbon sinks in mangroves.

A medium resolution sensor such as the Landsat can prove to be a good medium in estimating the potential of carbon sinks in mangroves. A study that was conducted in the Merbok Mangrove Reserves, Malaysia using Landsat 8 data to predict the carbon stocks of mangroves found out that the estimation in the range of 16.88 Mg/ha to 138.20 Mg/ha with R^2 0.56 and RMSE 22.24 Mg/ha (Hashim et al. 2020). A study conducted in West Kalimantan, Indonesia in their attempt to study the standing biomass of mangroves in the area using Landsat 8 data found out that the biomass in the region was 45 Mg/ha to 100 Mg/ha (Yusandi et al. 2018). A recent study conducted in the Matang Mangrove Forests Malaysia using the SPOT-5 sensor for AGB modelling estimated that AGB ranged between 33.65 and 437.46 Mg/ha with an average of 133.97 Mg/ha while for carbon stocks 16.86–218.73 Mg/ha (Muhd-Ekhzarizal et al. 2018). A study to estimate the mangroves AGB using a traditional field data collection and Unmanned Aerial Vehicle (UAV) method in two separate areas which are categorized as productive and protective area were done in the Matang Mangrove Forests, Malaysia (Otero et al. 2018). The study reported that for productive area yielded an estimate of AGB 217 Mg/ha for UAV and 238 for field inventory. Meanwhile, in the protective area, the estimate of AGB was 210 Mg/ha using UAV and 147 Mg/ha for field inventory.

According to Hamdan et al. (2013), during the study of carbon stocks in the Matang Mangrove Forests, Malaysia using Landsat 5 TM and SPOT-5 with different year interval found out that carbon stocks in the area ranged from 1.03 to 263.65 Mg/ha for year 1991 and 1.01–259.68 Mg/ha for the year 2011. Meanwhile Hamdan et al. (2014) predicted that using L-band ALOS PALSAR sensor in estimating the value of AGB in the Matang Mangrove Forests, Malaysia was ranging from 2.98 to 378.32 Mg/ha meanwhile the carbon stocks was ranging from 1.49 to 189.16 Mg/ha. Another study that conducted in a mangrove plantation area in North Vietnam to estimate the AGB of mangroves using a combination of ALOS-2 PALSAR-2, Sentinel-2A data and machine learning approach yielded AGB in the range of 36.22 Mg/ha to 230.14 Mg/ha with an average of 87.67 Mg/ha (Pham et al. 2018). Furthermore, the study concluded that the combination of satellite data and machine learning approach can be useful tools in estimating the mangroves AGB. Biomass in the mangroves can also be determined using the Multifrequency Radar data. The unique ability of radar to penetrate the vegetation canopies and to interact with tree parts and underlying water surface make radar an interesting tool to gather three-dimensional information in mangrove forests (Proisy et al. 2003).

7.6 Conclusions

The main objectives of this chapter were to review the role of mangroves as a carbon sink ecosystem as it might be the potential answer to mitigate the global climate change scenario. Studies from the past indicated that mangrove ecosystems could contribute to important blue carbon sinks in the coastal area as the carbon stored as part of their AGB and BGB. Furthermore, the potential of high primary productivity in young mangroves stands while enormous storage of carbon in mature stands suggested this ecosystem have a role to play in the global carbon budget.

The destruction of mangrove ecosystems due to the impacts of the uncertainty of global climate change and encroachment by human activity are supposed to be the main catalyst for the destruction of mangroves worldwide. This trend is expected to increase in the near future and put more stress on mangrove ecosystems as coastal development, human population and aquaculture industry expend further into the mangrove areas. The destruction of mangroves could turn the mangroves to become a source of carbon that might further accelerate global climate change. Therefore, strategic mitigation such as reducing forest degradation, reforestation and sustainable management of existing mangrove can increase their capacity in the global carbon cycle. Furthermore, the usage of remote sensing technology in monitoring and studying the mangroves carbon sinks and ecosystem dynamics can be a crucial solution to maintain their status as the largest blue carbon reservoir in the coastal area.

References

Abdul Aziz A, Dargusch P, Phinn S, Ward A (2015) Using REDD+ to balance timber production with conservation objectives in a mangrove forest in Malaysia. Ecol Econ 120: 108–116.

Aburto-Oropeza O, Ezcurra E, Danemann G, Valdez V, Murray J, Sala E (2008) Mangroves in the Gulf of California increase fishery yields. Proc Nat Acad Sci 105: 10456–10459.

Adame MF, Zakaria RM, Fry B, Chong VC, Then YHA, Brown CJ, Lee SY (2018) Loss and recovery of carbon and nitrogen after mangrove clearing. Ocean Coast Manag 161: 117–126.
Ahmad Y, Mohammad A (2005) Management and conservation of mangrove: Kedah experience. sustainable management of Matang Mangroves: 100 years and beyond. Forestry Department Peninsular Malaysia, p. 81–90.
Ahmadun FL-R, Wong MMR, Mat SA (2020) Consequences of the 2004 Indian Ocean Tsunami in Malaysia. Safety Science 121: 619–631.
Ahmed N, Glaser M (2016) Coastal aquaculture, mangrove deforestation and blue carbon emissions: Is REDD+ a solution? Mar Pol 66: 58–66.
Aizpuru M, Blasco F (2000) Global assessment of cover change of the mangrove forests using satellite imagery at medium to high resolution. In: EEC research project no 15017-1999-05 FIED ISP FR. Joint Research Center, Ispra.
Alongi DM (2009) The energetics of mangrove forests, Springer Science & Business Media.
Alongi DM (2008) Mangrove forests: Resilience, protection from tsunamis, and responses to global climate change. Estuar Coas Shelf Sci 76: 1–13.
Alongi DM (2012) Carbon sequestration in mangrove forests. Carbon Manag 3: 313–322.
Alongi DM, Mukhopadhyay SK (2015) Contribution of mangroves to coastal carbon cycling in low latitude seas. Agric For Meteorol 213: 266–272.
Banerjee K, Sengupta K, Raha A, Mitra A (2013) Salinity based allometric equations for biomass estimation of Sundarban mangroves. Biomass Bioenergy 56: 382–391.
Barnes-Mauthe M, Oleson KL, Zafindrasilivonona B (2013) The total economic value of small-scale fisheries with a characterization of post-landing trends: An application in Madagascar with global relevance. Fish Res 147: 175–185.
Bouillon S, Borges AV, Castañeda Moya E, Diele K, Dittmar T, Duke NC, Kristensen E, Lee SY, Marchand C, Middelburg JJ (2008) Mangrove production and carbon sinks: A revision of global budget estimates. Global biogeochemical cycles 22.
Briggs S (1977) Estimates of biomass in a temperate mangrove community. Austr J Ecol 2: 369–373.
Cabral P, Augusto G, Akande A, Costa A, Amade N, Niquisse S, Atumane A, Cuna A, Kazemi K, Mlucasse R, Santha R (2017) Assessing Mozambique's exposure to coastal climate hazards and erosion. Int J Disaster Risk Reduct 23: 45–52.
Christensen SM, Tarp P, Hjortsø CN (2008) Mangrove forest management planning in coastal buffer and conservation zones, Vietnam: A multimethodological approach incorporating multiple stakeholders. Ocean Coast Manag 51: 712–726.
Church JA, White NJ (2011) Sea-level rise from the late 19th to the early 21st century. Surv Geophys 32: 585–602.
Clough B (1993) The economic and environmental values of mangrove forests and their present state of conservation in the South-East Asia/Pacific Region. Mangrove Ecosystems Technical Reports. International Society for Mangrove Ecosystems, Okinawa, Japan.
Conacher C, O'brien C, Horrocks J, Kenyon R (1996) Litter production and accumulation in stressed mangrove communities in the Embley River estuary, North-eastern Gulf of Carpentaria, Australia. Mar Freshw Res 47: 737–743.
Dahdouh-Guebas F, Koedam N (2006) Coastal vegetation and the Asian tsunami. Science 311: 37–38.
Das CS, Mandal RN (2016) Coastal people and mangroves ecosystem resources vis-à-vis management strategies in Indian Sundarban. Ocean Coast Manag 134: 1–10.
Das S (2017) Ecological restoration and livelihood: Contribution of planted mangroves as nursery and habitat for artisanal and commercial fishery. World Dev 94: 492–502.
Diop E (1993) Conservation and sustainable utilization of mangrove forests in Latin America and Africa Regions. Part II. Africa. Mangrove Ecosystems Technical Reports No. 3, ITTO/ISME Project PD114/90. Nishihara, Japan, International Society for Mangrove Ecosystems (ISME) & International Tropical Timber Organization (ITTO).

Donato DC, Kauffman JB, Murdiyarso D, Kurnianto S, Stidham M, Kanninen M (2011) Mangroves among the most carbon-rich forests in the tropics. Nat Geosci 4: 293.
Duarte CM, Middelburg JJ, Caraco N (2005) Major role of marine vegetation on the oceanic carbon cycle. Biogeosciences 2: 1–8.
Duke NC (2011) Mangroves. In: HOPLEY, D. (ed.) Encyclopedia of Modern Coral Reefs: Structure, Form and Process. Dordrecht: Springer Netherlands.
Ellison J, Zouh I (2012) Vulnerability to climate change of mangroves: Assessment from Cameroon, Central Africa. Biology 1: 617–638.
FAO (1981) Tropical forest resources assessment project. Forest resources of tropical Africa. Part II: Country Briefs. FAO, UNEP.
FAO (1994) *Mangrove forest management guidelines*. FAO Forestry Paper 117, Rome, 319 pp.
FAO (2003) FAO's database on mangrove area estimates. Forest Resources Assessment Working Paper No. 62. Forest Resources Division, FAO, Rome (Unpublished).
FAO (2007) The world's mangroves 1980–2005. FAO Rome, Italy.
Fernando SM, Bandeira SO (2009) Litter fall and decomposition of mangrove species Avicennia marina and Rhizophora mucronata in Maputo Bay, Mozambique. West Indian Ocean J Mar Sci 8:139–145.
Ferreira AC, Lacerda LD (2016) Degradation and conservation of Brazilian mangroves, status and perspectives. Ocean Coast Manag 125: 38–46.
Fisher P, Spalding M (1993) Protected areas with mangrove habitat. Unpublished report to WCMC, Cambridge, UK.
Galidaki G, Zianis D, Gitas I, Radoglou K, Karathanassi V, Tsakiri-Strati M, Woodhouse I, Mallinis G (2017) Vegetation biomass estimation with remote sensing: Focus on forest and other wooded land over the Mediterranean ecosystem. Int J Remote Sens 38: 1940–1966.
Gilman EL, Ellison J, Duke NC, Field C (2008) Threats to mangroves from climate change and adaptation options: A review. Aquat Bot 89: 237–250.
Giri C, Ochieng E, Tieszen LL, Zhu Z, Singh A, Loveland TR, Masek J, Duke NC (2011) Status and distribution of mangrove forests of the world using earth observation satellite data. Glob Ecol Biogeogr 20: 154–159.
Gracia A, Rangel-Buitrago N, Oakley JA, Williams AT (2018) Use of ecosystems in coastal erosion management. Ocean Coast Manag 156: 277–289.
Groombridge B (1992) Global biodiversity: Status of the earth's living resources. Chapman & Hall.
Hamdan O, Aziz HK, Hasmadi IM (2014) L-band ALOS PALSAR for biomass estimation of Matang Mangroves, Malaysia. Remote Sens Environ 155: 69–78.
Hamdan O, Khairunnisa M, Ammar A, Hasmadi IM, Aziz HK (2013) Mangrove carbon stock assessment by optical satellite imagery. J Trop For Sci 25: 554–565.
Hamdan O, Khali Aziz H, Shamsudin I, Raja Barizan R (2012) Status of mangroves in Peninsular Malaysia. Forest Research Institute Malaysia, Kepong.
Hamilton SE, Casey D (2016) Creation of a high spatiotemporal resolution global database of continuous mangrove forest cover for the 21st century (CGMFC21). Glob Ecol Biogeogr 25: 729–738.
Hashim T, Suratman M, Jaafar J, Hasmadi IM, Abu F (2015) Field assessment of above ground biomass (agb) of mangrove stand in Merbok, Malaysia. Malaysian Appl Biol 44: 81–86.
Hashim T, Suratman M, Singh H, Jaafar J, Bakar A (2020) Predictive model of mangroves carbon stocks in Kedah, Malaysia using remote sensing. IOP Conference Series: Earth and Environmental Science, 2020. IOP Publishing, 012033.
Horstman EM, Dohmen-Janssen CM, Bouma TJ, Hulscher SJMH (2015) Tidal-scale flow routing and sedimentation in mangrove forests: Combining field data and numerical modelling. Geomorphology 228: 244–262.
Hutchison J, Spalding M, Ermgassen P (2014) The role of mangroves in fisheries enhancement. The Nature Conservancy and Wetlands International: 54.
Itto/Isme (1993) Project PD71/89.Rev.1. International Society for Mangrove Ecosystem (ISME), Okinawa, Japan.

Jachowski NRA, Quak MSY, Friess DA, Duangnamon D, Webb EL, Ziegler AD (2013) Mangrove biomass estimation in Southwest Thailand using machine learning. Appl Geogr 45: 311–321.
Kamruzzaman M, Ahmed S, Osawa A (2017) Biomass and net primary productivity of mangrove communities along the Oligohaline zone of Sundarbans, Bangladesh. Forest Ecosys 4: 1–9.
Kathiresan K, Bingham BL (2001) Biology of mangroves and mangrove ecosystems.
Kauffman JB, Donato DC (2012) Protocols for the measurement, monitoring and reporting of structure, biomass, and carbon stocks in mangrove forests. Working Paper 86, CIFOR, Bogor.
Kauffman JB, Heider C, Cole TG, Dwire KA, Donato DC (2011) Ecosystem carbon stocks of Micronesian mangrove forests. Wetlands 31: 343–352.
Komiyama A, Ong JE, Poungparn S (2008) Allometry, biomass, and productivity of mangrove forests: A review. Aquat Bot 89: 128–137.
Krauss KW, Doyle TW, Doyle TJ, Swarzenski CM, From AS, Day RH, Conner WH (2009) Water level observations in mangrove swamps during two hurricanes in Florida. Wetlands 29: 142–149.
Kridiborworn P, Chidthaisong A, Yuttitham M, Tripetchkul S (2012) Carbon sequestration by mangrove forest planted specifically for charcoal production in Yeesarn, Samut Songkram. J Sustain Energ Environ 3: 87–92.
Kumara M, Jayatissa L, Krauss K, Phillips D, Huxham M (2010) High mangrove density enhances surface accretion, surface elevation change, and tree survival in coastal areas susceptible to sea-level rise. Oecologia 164: 545–553.
Lacerda LDD, Diop E. (1993) Published. Conservation and sustainable utilization of mangrove forests in Latin America and Africa regions. Workshop on Conservation and Sustainable Utilization of Mangrove Forests in Latin America and Africa Regions., 1993. International Society for Mangrove Ecosystems.
Le Quéré C, Andrew RM, Friedlingstein P, Sitch S, Pongratz J, Manning AC, Korsbakken JI, Peters GP, Canadell JG, Jackson RB (2017) Global carbon budget 2017. Earth System Science Data Discussions: 1–79.
Li S, Meng X, Ge Z, Zhang L (2015) Evaluation of the threat from sea-level rise to the mangrove ecosystems in Tieshangang Bay, southern China. Ocean Coast Manag 109: 1–8.
Long J, Giri C, Primavera J, Trivedi M (2016) Damage and recovery assessment of the Philippines' mangroves following Super Typhoon Haiyan. Mar Poll Bullet 109: 734–743.
Mackey A (1993) Biomass of the mangrove Avicennia marina (Forsk.) Vierh. near Brisbane, south-eastern Queensland. Mar Freshwater Res 44: 721–725.
Mohammad Abdullah AN, Stacey N, Garnett ST, Myers B (2016) Economic dependence on mangrove forest resources for livelihoods in the Sundarbans, Bangladesh. For Policy Econ 64: 15–24.
Muhd-Ekhzarizal M, Mohd-Hasmadi I, Hamdan O, Mohamad-Roslan M, Noor-Shaila S (2018) Estimation of aboveground biomass in mangrove forests using vegetation indices from SPOT-5 image. J Trop For Sci 30: 224–233.
Naidoo G (2016) The mangroves of South Africa: An ecophysiological review. South Afr J Bot 107: 101–113.
Nanjo K, Kohno H, Nakamura Y, Horinouchi M, Sano M (2014) Effects of mangrove structure on fish distribution patterns and predation risks. J Exp Mar Biol Ecol 461: 216–225.
Nicholls RJ, Marinova N, Lowe JA, Brown S, Vellinga P, De Gusmão D, Hinkel J, Tol RSJ (2011) Sea-level rise and its possible impacts given a 'beyond 4°C world' in the twenty-first century. Philos Trans A Math Phys Eng Sci 369: 161–181.
Ong JE, Gong WK (2013) Structure, function and management of mangrove ecosystems, International Society for Mangrove Ecosystems.
Otero V, Van De Kerchove R, Satyanarayana B, Martínez-Espinosa C, Fisol MaB, Ibrahim MRB, Sulong I, Mohd-Lokman H, Lucas R, Dahdouh-Guebas F (2018) Managing mangrove forests from the sky: Forest inventory using field data and Unmanned Aerial Vehicle (UAV) imagery in the Matang Mangrove Forest Reserve, Peninsular Malaysia. For Ecol Manag 411: 35–45.

Pham TD, Yoshino K, Le NN, Bui DT (2018) Estimating aboveground biomass of a mangrove plantation on the Northern coast of Vietnam using machine learning techniques with an integration of ALOS-2 PALSAR-2 and Sentinel-2A data. Int J Remote Sens 39: 7761–7788.

Proisy C, Mitchell A, Lucas R, Fromard F, Mougin E. (2003) Estimation of Mangrove Biomass using Multifrequency Radar Data. Application to Mangroves of French Guiana and Northern Australia. Proceedings of the Mangrove 2003 Conference, 2003, p. 20–24.

Putz FE, Chan HT (1986) Tree growth, dynamics, and productivity in a mature mangrove forest in Malaysia. For Ecol Manag 17: 211–230.

Rahman AF, Dragoni D, El-Masri B (2011) Response of the Sundarbans coastline to sea level rise and decreased sediment flow: A remote sensing assessment. Remote Sens Environ 115: 3121–3128.

Reef R, Feller IC, Lovelock CE (2010) Nutrition of mangroves. Tree Physiol 30: 1148–1160.

Richards DR, Friess DA (2016) Rates and drivers of mangrove deforestation in Southeast Asia, 2000–2012. Proc Nat Acad Sci 113: 344–349.

Romañach SS, Deangelis DL, Koh HL, Li Y, Teh SY, Raja Barizan RS, Zhai L (2018) Conservation and restoration of mangroves: Global status, perspectives, and prognosis. Ocean Coast Manag 154: 72–82.

Ross MS, Ruiz PL, Telesnicki GJ, Meeder JF (2001) Estimating above-ground biomass and production in mangrove communities of Biscayne National Park, Florida (USA). Wet Ecol Manag 9: 27–37.

Sahu SK, Kathiresan K (2019) The age and species composition of mangrove forest directly influence the net primary productivity and carbon sequestration potential. Biocatal Agric Biotechnol 20: 101235.

Simard M, Fatoyinbo L, Smetanka C, Rivera-Monroy VH, Castañeda-Moya E, Thomas N, Van Der Stocken T (2019) Mangrove canopy height globally related to precipitation, temperature and cyclone frequency. Nat Geosci 12: 40–45.

Spalding M, Blasco F, Field C (1997) World mangrove Atlas.

Spalding M, Kainuma M, Collins L (2010) World Atlas of Mangroves, Taylor & Francis Group.

Suratman MN (2008) Carbon sequestration potential of mangroves in Southeast Asia. Managing forest ecosystems: The challenge of climate change. Springer.

Tanaka K (2007) Effectiveness and limitation of the coastal vegetation for storm surge disaster mitigation. Investigation report on the storm surge disaster by Cyclone Sidr.

Thampanya U, Vermaat J, Sinsakul S, Panapitukkul N (2006) Coastal erosion and mangrove progradation of Southern Thailand. Estuar Coast Shelf Sci 68: 75–85.

Truong TD, Do LH (2018) Mangrove forests and aquaculture in the Mekong River Delta. Land Use Policy 73: 20–28.

Twilley R, Chen R, Hargis T (1992) Carbon sinks in mangroves and their implications to carbon budget of tropical coastal ecosystems. Water Air Soil Pollut 64: 265–288.

Twilley RR, Castañeda-Moya E, Rivera-Monroy VH, Rovai A (2017) Productivity and carbon dynamics in mangrove wetlands. Mangrove ecosystems: A global biogeographic perspective. Springer.

Venkatachalam S, Kathiresan K, Krishnamoorthy I, Narayanasamy R (2018) Survival and growth of fish (Lates calcarifer) under integrated mangrove-aquaculture and open-aquaculture systems. Aquac Rep 9: 18–24.

Walker W, Baccini A, Nepstad M, Horning N, Knight DE, Braun BA (2011) Field guide for forest biomass and carbon estimation. Version 1.0. Woods Hole Research Center, Falmouth, Massachusetts, USA.

Wang M, Cao W, Guan Q, Wu G, Wang F (2018) Assessing changes of mangrove forest in a coastal region of southeast China using multi-temporal satellite images. Estuar Coast Shelf Sci 207: 283–292.

Ward RD, Friess DA, Day RH, Mackenzie RA (2016) Impacts of climate change on mangrove ecosystems: A region by region overview. Ecosystem Health and Sustainability 2(4): e01211.
Wu C, Shen H, Wang K, Shen A, Deng J, Gan M (2016) Landsat imagery-based above ground biomass estimation and change investigation related to human activities. Sustainability 8: 159.
Yusandi S, Jaya INS, Mulia F (2018) Biomass estimation model for mangrove forest using medium-resolution imageries in BSN Co Ltd concession area, West Kalimantan. Int J Remote Sens Earth Sci 15: 37–50.
Zhang K, Liu H, Li Y, Xu H, Shen J, Rhome J, Smith TJ (2012) The role of mangroves in attenuating storm surges. Estuar Coast Shelf Sci 102: 11–23.

Estimation of Blue Carbon Stock of Mangrove Ecosystem and Its Dynamics in Relation to Hydrogeomorphic Settings and Land Use-land Cover

8

Karuna Rao, Prabhat Ranjan, and AL. Ramanathan

Abstract

This chapter reviews blue carbon stock of different mangrove ecosystems across globe through published literature. It also tries to evaluate its dynamics with different land use and land cover changes. The study reveals that mangroves have a high potential to store carbon compared to other terrestrial and coastal ecosystems. Indian Sundarbans stores 160–360 tC/ha based on salinity and vegetation types. The emission of carbon from the degradation of above-ground biomass in Indian Sundarbans was 427,242 tons between 1975 and 2013. Deforestation of Bangladesh Sundarbans causes loss of 8500, 1800, 670, 290, 133, and 104 hectares of mangroves along Chakaria, Naf river estuary and offshore Island, Naf river, Maiskhali Island, Jaliardwip Island, and Matabar Island, respectively. There is a rapid decline in plantations (-58.2%), mangrove swamps (-49.3%), and mangrove forests (-21.3%) during 2000–2017 due to their conversion to aquaculture farms in Bangladesh. Along Indian coastlines, Andhra Pradesh is the most affected area and shows significant decrease in paddy fields due to their conversion to aquaculture farms between 1980–81 and 2000–01. It also indicates massive increase in shrimp production from 1990-2017. Bangladesh shows a dramatic rise in shrimp production from 56,569 to 75,274 tons from 2010-2011 to 2014-2015. Destruction of mangroves releases carbon dioxide in the atmosphere and can reverse mangroves' role from a sink to source. Since aquaculture farming helps in high revenue generation but negatively affects the coastal ecosystem, it is

K. Rao · A. Ramanathan (✉)
School of Environmental Sciences, Jawaharlal Nehru University, New Delhi, India

P. Ranjan
Central Pollution Control Board, New Delhi, India

R. P. Rastogi et al. (eds.), *Mangroves: Ecology, Biodiversity and Management*,
https://doi.org/10.1007/978-981-16-2494-0_8

essential to maintain a balance. Hence Integrated Multi-Trophic Aquaculture (IMTA) has been introduced along with mangrove forest restoration, and REDD+.

Keywords

Mangroves · Deforestation · Aquaculture · Shrimp farming · Land use land cover change

8.1 Introduction

The carbon stored and sequestered by the coastal ecosystems from the atmosphere and oceans in the organic-rich sediments or in the form of biomass could be termed as "Blue Carbon." These ecosystems include various tidal wetlands like salt marshes, seagrasses, and mangroves. Among them, mangroves have a significant capacity to store carbon compared to seagrasses and salt marshes. Mangroves contain almost 1,023 Mg C/ha, and its soil has been estimated to contribute around 49-98% of total stored carbon in estuarine ecosystems (Donato et al. 2011). Due to its high potential to store carbon, this ecosystem plays a significant role in offsetting the increased atmospheric carbon dioxide (CO_2) (Mcleod et al. 2011; Siikamäki et al. 2012), one of the main challenges at a present scenario.

Disturbances created in the mangroves and coastal systems by anthropogenic activities, especially fossil fuel combustion, release a high amount of carbon in the atmosphere (in the form of methane (CH_4), carbon dioxide (CO_2), or other species of carbon). Other major factors include land use activities; mainly deforestation accounts for almost 8-20% of all global greenhouse gas emissions (GHGs) (van der Werf et al. 2009) and releases carbon stored in the living and dead biomass and deep sediments. Loss of this vegetated ecosystem may introduce a considerable amount of carbon dioxide in the atmosphere known as "pulse" release. It may have the largest and most instant effect on GHGs release and is estimated to be 50 times greater than the annual net carbon sequestration rate (Eong 1993; McLeod et al. 2011). The release of carbon dioxide is due to destabilization or exposure of the mangrove sediments, thereby, increasing the rate of microbial activities, which in turn increases the emission of GHGs in a significant amount to the water column or the atmosphere (Eong 1993; Sjöling et al. 2005; Kristensen et al. 2008; Granek and Ruttenberg 2008; Strangmann et al. 2008; Sweetman et al. 2010). Carbon emission from mangroves is relatively unacknowledged or ignored in most climate change mitigation policies. Various studies reveal that land clearing has reduced sediment carbon and increased CO_2 emission to a large extent, e.g. the potential emission of carbon dioxide from the global loss of mangrove vegetation biomass and near-surface carbon stock (a few meters from surface soil) is around 1492 $MgCO_2$/ha (Pendleton et al. 2012). Land clearing in Panamanian mangrove has reduced sediment carbon by 50% within eight years (Granek and Ruttenberg 2008). The mean potential carbon dioxide emission from the degradation of above-ground biomass in Indian Sundarban mangroves was 1567.98 ± 551.69 Gg between 1975 and 2013

(Akhand et al. 2017). Hence factors like land clearing and deforestation and degradation of these coastal ecosystems significantly impact very large pools of previously sequestered carbon.

Keeping in mind the importance of mangroves in capturing carbon, the review from the various available literature has been carried out to characterize the sedimentary carbon stock of different mangroves and its potential emission due to several anthropogenic disturbances like land use land cover changes which include mangrove degradation, deforestation, agriculture, and aquaculture activities. In this study, an attempt has been made 1) to assess the carbon stock of various mangroves across the world 2) to estimate the potential carbon loss as a result of various anthropogenic disturbances. This assessment might support the conservation, preservation, and management of existing carbon stock in mangrove forests under immense pressure and threat.

8.2 Significance of Mangrove in Storing Carbon Over Other Ecosystems

8.2.1 Mangroves and Other Forest Ecosystems

Soil carbon pool within 1m depth from mangroves and other forest ecosystems like deciduous and evergreen needle leaf forest, permanent wetlands, open shrublands, mixed forest, grasslands open shrublands, evergreen and deciduous broadleaf forest, croplands, savannah, and closed shrublands were studied by Sanderman et al. 2018. Upon comparison with them, mangroves found to store the maximum amount of carbon (361 ton/ha) than others, as shown in (Fig. 8.1). This signifies that mangrove ecosystems are the most efficient ecosystem in storing and fixing carbon in their sediments and biomass (Kristensen et al. 2008; Donato et al. 2011).

8.2.2 Mangroves and Tidal Marshes and Seagrasses

In a study by Mcleod et al. 2011, carbon burial rates of mangroves, tidal marshes, and seagrasses were compared. It is found that mangroves and tidal marshes have approximately similar carbon burial rates of 226 g C/m^2/yr and 218 g C/m^2/yr, respectively, stating that they have almost the same potential to capture carbon (Fig. 8.2). At the same time, seagrass shows 138 gC/m^2/yr of burial rate. This indicates the significance of mangroves in capturing carbon over other blue carbon ecosystems.

In another study, in Victoria, Southeast Australia, Lewis et al. 2018 estimated the stock in the top 30cm of sediments across the station, namely Glenelg, Corangamite, Port Phillip Western port ecosystem, West, and East Gippsland. The lowest carbon stock (170 ha) in Glenelg might be due to the lower organic carbon value relative to other regions. In this study, Lewis revealed that carbon stored by mangrove ecosystem was 65.6±4.17 Mg C/ha, and tidal marshes were 87.1±4.90 MgC/ha

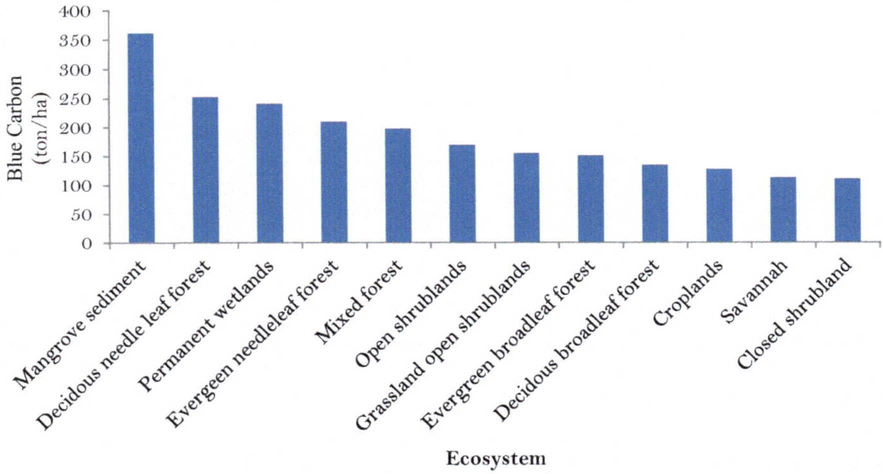

Fig. 8.1 Soil carbon pool in 1 m depth, indicating the significance of mangrove over other forest ecosystems in sequestering carbon (Source: Sanderman et al. 2018)

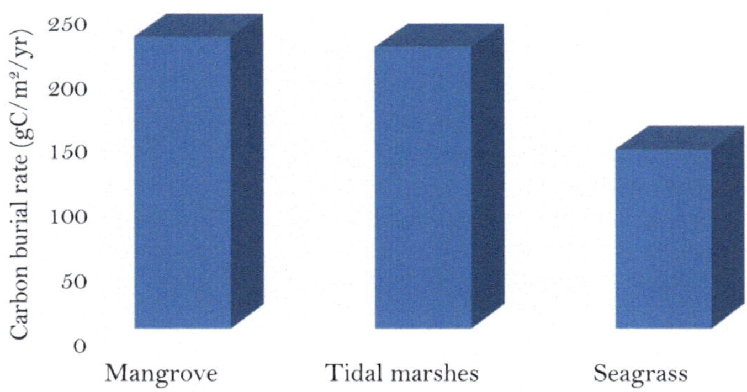

Fig. 8.2 Carbon burial rate of various blue carbon ecosystems (Data from Mcleod et al. 2011, and references therein)

(Average±SD) which were not significantly different. Overall, tidal marshes were shown to have the most extensive carbon stock due to their high average value of organic carbon and a large area covered by them (Fig. 8.3). Seagrasses showed the lowest average organic carbon stock (24.3±1.82 Mg C/ha) of the three ecosystems. Further, the study done by Mcleod et al. 2011 and Lewis et al. 2018 proposed that the potential carbon losses lead to the potential losses of monetary values related to these ecosystems, especially when they are vanishing at a faster rate.

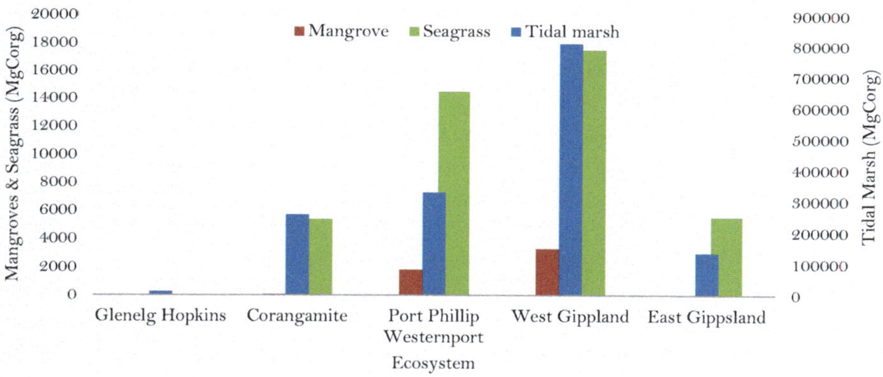

Fig. 8.3 Sedimentary carbon stock (in the top 30 cm) across various catchments areas of Victoria, southeast Australia (Source: Lewis et al. 2018)

8.3 Blue Carbon Sequestration

Blue carbon sequestration is around 53 million tons annually around the globe, out of which, approximately 16 million tons (i.e., about 30%) are sequestered by mangroves alone (Siikamäki et al. 2012). On a global scale, mangrove stores approximately 44.6 TgC/yr with a mean sequestration value of 210 g CO_2 m^{-2} yr^{-1}. The value of the carbon stock ranged from 441.76 ± 120.76 to 1267.00 ± 872.72 t C ha^{-1} with a global average value of 78.0 ± 64.5 t C ha^{-1} and carbon sequestration rate of 2.9 ± 2.2 t C ha^{-1} $year^{-1}$ (Murdiyarso et al. 2015; Estrada and Soares 2017). Hamilton and Friess 2018 estimated that global carbon stock in a mangrove in 2012 is around 4.19±0.62 Pg, out of which 2.96±0.53 Pg is captured in the soil and 1.23±0.06 Pg in standing and living biomass.

The sedimentary carbon stock across different mangrove settings in the Indonesian archipelago is depicted in Table 8.1. It can be observed that marine mangroves

Table 8.1 Sedimentary carbon stock across different mangrove settings in the Indonesian archipelago

Mangroves	Mg $ha^{-1}m^{-1}$	References
Seaward oriented marine mangroves	354-377	Kauffman et al. 2011
Interior marine mangroves	380-424	Kauffman et al. 2011
Landward marine mangroves	480-503	Kauffman et al. 2011
Yap	465	Donato et al. 2012
Palau	465	Donato et al. 2012
Marine mangroves	542	Murdiyarso et al. 2015
Tanjung Puting National Park	1059.2	Murdiyarso et al. 2009
Segara Anakan	571	Murdiyarso et al. 2009
Bunaken National Park	822.1	Murdiyarso et al. 2009

stored more carbon than estuarine mangroves, which could be due to their differences in organic carbon content and bulk densities in their soils. The bulk densities and carbon content of estuarine mangroves are around 0.33–0.80 g cm^{-3} and 11 to 85 mg SOC g^{-1}, and in marine mangrove soil, it is 0.18 0.27 g cm^{-3} and 170 to 260 mg SOC g^{-1}, respectively. The bulk density of estuarine soil is about four times higher than the marine soils, while the carbon concentration of marine mangroves is up to 25 times more than the estuarine mangroves. This reveals that marine mangroves have higher soil carbon stock despite low bulk density, making soil organic carbon the determining factor of soil carbon stock.

Mangroves are considered to be a crucial carbon sink. When these ecosystems are disturbed directly or indirectly, either by natural or anthropogenic means, these stored carbons get disturbed and released in the atmosphere. Thus, they act as a sink in natural conditions but as sources in the degraded and disturbed conditions. Since the last few decades, various researches have been carried out to reduce or minimize the emission of CO_2 and maximize the sink capacity of mangrove forests. Hence, they may provide an essential contribution to low-cost mitigation techniques for climate change.

8.4 Indian Mangrove Ecosystems

In South Asian mangroves, India occupies the second largest position (3400 km^2) after Bangladesh. Total mangrove in India covers 0.15% of the country's land, and around 3% of the global mangrove cover; around them, 8% of global mangrove area belongs to Asia's mangroves (Sahu et al. 2015). The carbon pool in Indian mangrove ecosystems and its partitioning into above-ground and below ground is represented in (Fig. 8.4). The overall assessment reveals that maximum carbon is stored in Sundarban, followed by Bhitarkanika, Kadalundi, Mahanadi, and Thalassery, etc. Sundarban has higher stock as it has the most extensive coverage area than any other mangroves. The carbon stock of Thalassery estuarine wetland was 153.64 t C ha^{-1} which was approximately equivalent to 536.86 t CO_2 ha^{-1}. The area covered by Thalassery mangroves was about 5.8 ha. Thus, this mangrove's potential carbon sequestration is 891.11 t C, and the amount of equivalent CO_2 is 3270.37 t CO_2 (Vinod et al. 2019). Further, total area of Mahanadi and Bhitarkanika mangrove (141,589+672) is considered to be 142,261 km^2. The equivalent mean CO_2 is estimated to be 455.47 $\pm$ 110.56 tones, which comes out to be approximately 64.80 TgC (Banerjee et al. 2020). The variation in carbon stock across different mangroves might be due to differences in the extent of mangrove cover, geomorphic settings, land use land cover patterns, different hydrodynamic conditions, etc. These studies support the notion that mangrove forests act as an essential carbon sink and the necessity to conserve and protect these critical ecosystems in light of climate change mitigation.

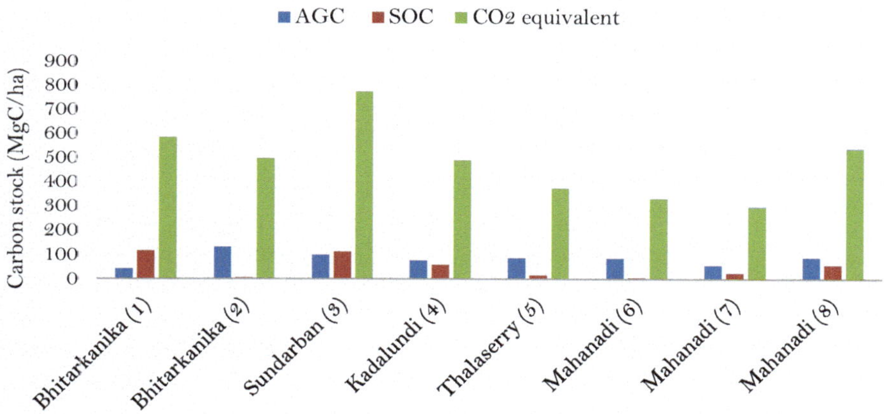

Fig. 8.4 Ecosystem C pools in mangrove ecosystems representing ABC, BGC, and their CO_2 equivalents across various mangrove ecosystems (1. Bhomia et al. 2016; 2. Banerjee et al. 2020; 3. Rahman 2015; 4. Vinod et al. 2018; 5. Vinod et al. 2019; 6. Banerjee et al. 2020; 7. Sahu et al. 2016; 8. Sahu et al. 2016)

8.4.1 Carbon Stored in Sundarban Mangrove (Bangladesh and India)

Sundarbans mangrove is the largest mangrove cover globally (occupying around 4,000 sq km in India and 6,017 sq km in Bangladesh). Sundarbans forest, nationally or internationally, is of great importance for its biodiversity and environmental services (Seidensticker and Hai 1983; Iftekhar and Saenger 2008). Hence, the assessment of carbon sequestration in Sundarbans mangrove (both Bangladesh and Indian part) is of immense importance. Degradation of above-ground biomass in Indian Sundarbans mangrove leads to the potential carbon emission of about 1567.98 Gg during 1975 to 2013, which is equivalent to US$64.29 million (Akhand et al. 2017). The carbon stored in Sundarbans mangrove forest is about 160–360 t/ha based on vegetation types and salinity (Rahman 2015), which for Bangladesh adds approximately 70–158 million tons of carbon stock (Table 8.2). Humanitywatch et al. 2011 estimated the cost of US$1.87 billion in the international market for about 56 million tons of carbon captured in the Bangladesh Sundarbans.

8.5 Threats to Mangrove Ecosystems

According to Saenger 2002 and Alongi 2002, the coast of Andaman has decreased by 79% between 1961 and 1989 due to anthropogenic activities like agriculture, aquaculture, and various other land use and land cover changes. A similar significant loss of mangroves has been observed in Southern Thailand due to extensive shrimp farming between 1975 and 1993 (CORIN 1995). The destruction of mangroves around the globe disturbs the large stock of previously sequestered carbon (956 Mg C ha^{-1}) at the present rate of 1% annually. It causes an additional emission of around

Table 8.2 Carbon sequestration and stock in Bangladesh mangrove ecosystems

Mangroves	Carbon sequestration (annually)	References
Blue C sequestration rate	1.15-1.39 t/ha	Siikamäki et al. 2012; Nellemann et al. 2009
Bangladesh	0.56 million tons	Chowdhury et al. 2015
Bangladesh, Sundarban	56 million tons	Humanitywatch et al. 2011
Bangladesh, Sundarban	91 million tons (36 million tons AG, 55 million tons BG)	Chanda et al. 2016
Bangladesh Sundarban C storage	70-158 million tons	Ahmed et al. 2017
Indian Sundarban C stock	160-360 t/ha	Rahman 2015
Indian Sundarban C stock	212.5-312.5 t/ha	Donato et al. 2011

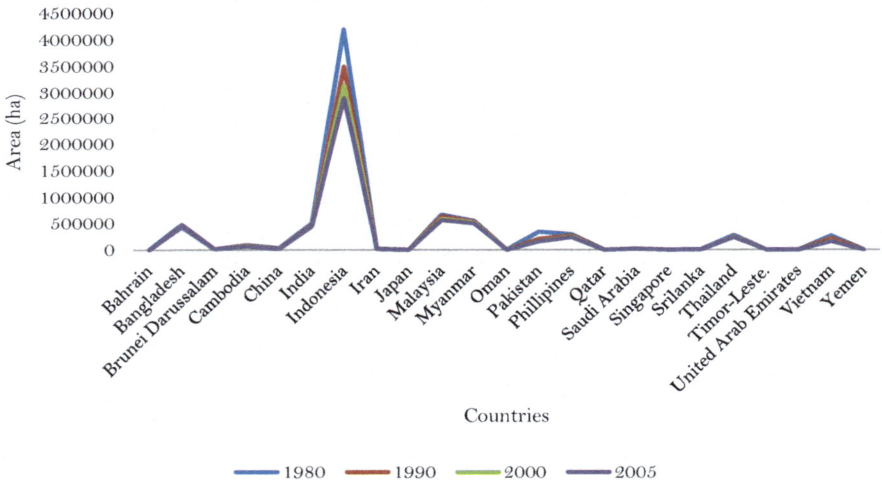

Fig. 8.5 South Asian Mangroves area changes in the year 1980, 1990, 2000, and 2005 obtained from FAO-forest resources assessments

133 Tg C yr^{-1} to the atmosphere. On the global scale, mangroves' destruction contributes to 10% of global carbon dioxide release from the deforestation (Alongi and Mukhopadhyay 2015). The severity and extent of impacts on mangrove carbon emission may range from small scale (harvesting of trees for fuelwood) (Malik et al. 2015) to industrial scale (timber harvesting) (Sillanpää et al. 2017). Mangrove forest degradation for the conversion to aquacultural ponds is the largest threat to Southeast Asian mangroves (Richards and Friess 2016). The changes in the extent of the South East Asian mangroves area by FAO, Forest Resources Assessments in the years 1980, 1990, 2000, and 2005 have been shown in Fig. 8.5 and reveal that there is a decrease in mangrove areas as results of various land use land cover changes.

8.5.1 Degradation and Deforestation

A recent estimate by Sanderman et al. 2018 reveals that mangrove degradation leads to the annual emission of soil carbon of around 2.0-8.1 $TgCyr^{-1}$. If all the mangroves across the world are destroyed, and it is assumed that 95% of all mangrove carbon gets oxidized, then the loss of CO_2 would be approximately 30.2 Pg CO_2 equivalents, which would be equivalent to around 6.5 years of carbon emissions from the loss of the global forests (Alongi 2018; Kennedy et al. 2014). The adverse effect of deforestation and degradation of the mangrove ecosystem leads to the emission of carbon and other greenhouse gas (GHGs) like methane and nitrous oxide, which may lead to anthropogenic climate change (Ahmed et al. 2013) in the long run, if not mitigated with time. Fig. 8.6 represents the mangrove area that underwent deforestation in Sundarban, Bangladesh. The 8500 ha of mangroves of Chakaria, Sundarban has been deforested for shrimp farming (Hossain et al. 2001; Shahid and Islam 2002). In the same way, 1800, 670, 290, 133, and 104 ha of mangroves along Naf and offshore Island, Naf river (Shahid and Islam 2002; Hossain 2001), Maiskhali Island, Jaliardwip Island, and Matabar Island has been destroyed for the shrimp cultivation.

8.5.2 Land Use and Land Cover Changes (LULC)

Land use and land cover changes comprise the conversion of mangrove forest lands to aquaculture, agriculture, upstream dams, forest cutting, exploitation, industrial use, dredging urban development, etc. (Short and Wyllie-Echeverria 1996; Valiela et al. 2001; Duke et al. 2007; Giri et al. 2008; Waycott et al. 2009; McLeod et al. 2011). LULC result in a loss of around 25-50% of the total global area since the last 50-100 years (McLeod et al. 2011). Loss of forested land continues and even

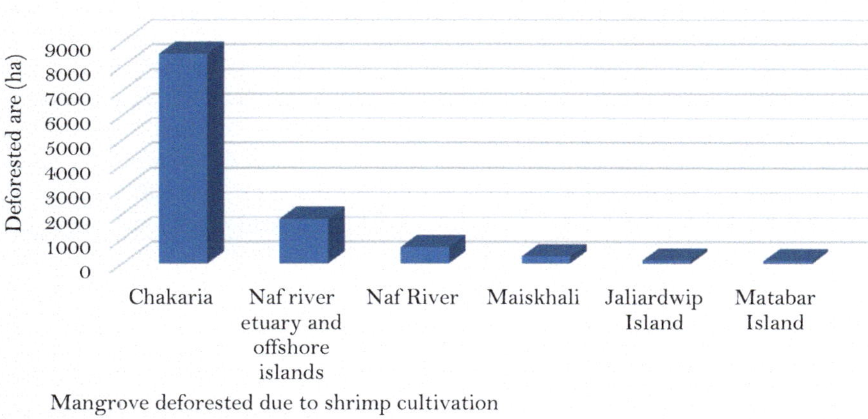

Fig. 8.6 Area of mangrove deforested as a result of shrimp cultivation in Sundarbans, Bangladesh mangroves (Ahmed et al. 2017)

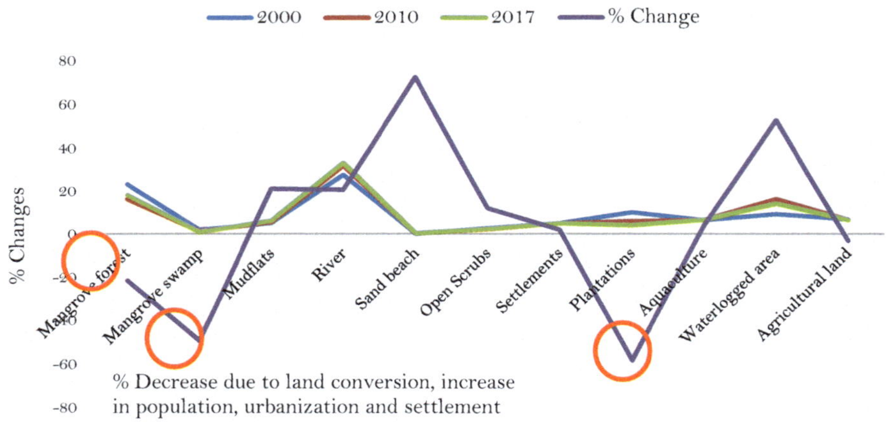

Fig. 8.7 Land use and land cover changes from 2000 to 2017 in the mangrove forest of Sundarban (Source: Thakur et al. 2020)

increasing in recent days depending upon the type of forest and ecosystems, resulting in the loss of approximately 8000 km^2 land each year (Valiela et al. 2001; Alongi 2002; Duarte et al. 2004; Bridgham et al. 2006; FAO 2016; Duarte et al. 2008; Spalding 2010; Mcleod et al. 2011). If land use and land cover change persist at the same rate, then the estimated loss of tidal marsh and seagrass ecosystem would be around 30-40% (IPCC 2007) and 100% of the mangroves in next 100 years.

Thakur et al. 2020, in his study in Sundarban mangroves, identified eleven LULC classes and detected various negative and positive changes as they have been exposed to multiple developmental activities. Different classes include Mangrove forest, mangrove swamp, Mudflats, river, sand beach, open scrubs, settlements, plantations, aquaculture, waterlogged areas, and agricultural land. Among all these classes, the percentage of land cover increases in mudflats (21.3%), sand beaches (72.5%), and the waterlogged regions (52.7%) during these 17 years (Fig. 8.7). On the other hand, decline in the land cover of mangrove forests (-21.3%), plantations (-58.2%), and mangrove swamps (-49.3%) has been observed due to the unrestricted destruction of a vast cover of open mangrove forests to meet the demand of land to accommodate the continuously rising population (Dutta et al. 2014). Some parts of the mangrove forests were also converted to mudflats, waterlogged areas, aquaculture, and agricultural ponds, and human settlements. This highlights that mangrove swamps, mangrove forests, and plantation decreased due to forest degradation, growing urbanization, land conversion, increase in settlements, etc. This indicates that human activities affected the mangroves on a large scale in a very short period while natural factors (like a cyclone, coastal flooding, rainfall, and temperature) affect them on a small level and with a slow pace (Banerjee et al. 2012; Ghosh et al. 2015). Thus, these mangroves act as carbon storehouses under natural conditions, but rapid human activities may shift it from sink to source of carbon.

8.5.3 Aquaculture

Aquaculture is a near-shore and on-shore saltwater culture and brackish water fish, including shellfish. It has increased over the last three decades, with a mean rate of 8.6% annually. In 2012, the production of fish touched 66.6 million tons globally, and the contribution of Asia in this production was 88.4%. There is a rapid expansion in aquaculture industries between the 1980s and1990s on a global scale. Conversion of mangroves to aquaculture leads to the removal of 60% of the soil organic carbon and 85% of the living biomass carbon stock. A similar observation is seen in the Dominican Republic (Kauffman et al. 2014). Since the last 50 years, approximately one-third of mangroves in the whole world have been lost due to the conversion of the mangrove land to the aquacultural and agricultural fields (Alongi 2002). Conversion of mangrove areas at such high rates may lead to the global carbon emission of about 0.12 Pg C/yr, which is around 10% of the total carbon emission from deforestation (Donato et al. 2011). In Asia, the rank of countries in aquaculture production follows an order of China, Indonesia, India, Vietnam, the Philippines, and Bangladesh (FAO 2016). Bangladesh is one of the leading aquaculture industries in the world, with around 2.06 million tons of production in 2014-2015 (FRSS 2016). Table 8.3 represents the loss of global mangroves due to the increase in coastal aquaculture activities, which accounts for the economic loss of 3.78-17.01 billion.

In India, Andhra Pradesh is the most affected coastline in terms of aquaculture. Significant changes in the distribution of land use patterns have been observed between 1980 and 81 and 2000-01. The area occupied by aquaculture is around 40.11%, 38.16%, 29.82%, 28.81%, 21.56%, 5.86%, and 29.58% in Bhimavaram Mandal, Kalla Mandal, Akiveedu Mandal, Mogalthuru Mandal, Narasapuram Mandal, and Palakol Mandal, respectively, in West Godavari district, Andhra Pradesh. The land under paddy shows decreases in percentage from 78.21% to 38.71% in Bhimavaram Mandal, 77.52% to 51.89% in Akiveedu Mandal, 73.32% to 43.27% in Kalla Mandal, 60.05% to 37.25% in Mogalthuru Mandal, 70.28% to 64.62% in Palakol Mandal, 62.28% to 40.62% in Narasapuram Mandal (Dorababu 2013). The decrease in paddy land is because of the conversion of these paddy lands to aquaculture during 1980-81 and 2000-01 (Fig. 8.8).

Table 8.3 Loss of global mangrove forests through coastal aquaculture

Features	Information	Reference
Mangroves area lost to aquaculture (ha) (other than shrimp)	0.49 million	Valiela et al. 2001
Total mangrove area lost to coastal aquaculture (ha)	1.89 million	Valiela et al. 2001
Mangrove area lost to coastal aquaculture (%)	52	Valiela et al. 2001
The total economic value of mangrove loss to aquaculture (US$/yr)	3.78-17.01 billion	Ahmed and Glaser 2016

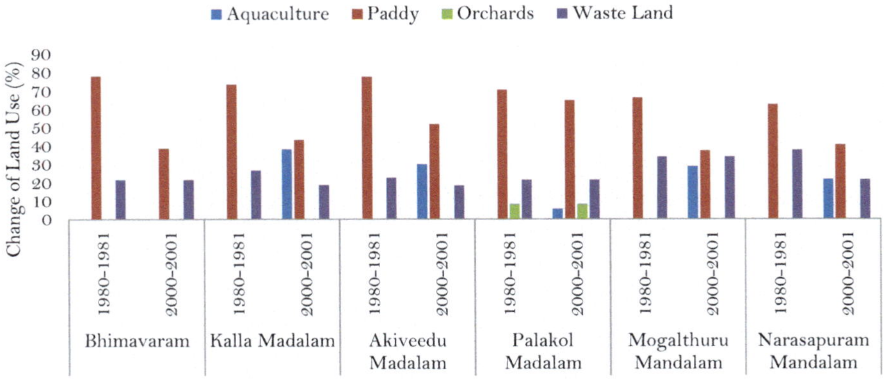

Fig. 8.8 Distribution of Land Use during 1980-81 and 2000-01 in West Godavari, Andhra Pradesh (Source: Dorababu 2013)

8.5.4 Shrimp Farming

Currently, shrimp farming is among the crucial sectors of the country's economy. Global mangrove loss resulting from shrimp farming accounted for 1.4 million ha (38%) (Valiela et al. 2001). In India, the duration between 1993-2000 was the golden period for shrimp production, and during this time, India became the fifth-largest country of global shrimp production (Briggs et al. 2004). As a result, many shrimp farms took over a huge part of Indian coastlines. Initially, these industries were supported by various government agencies (CIBA 2009), but later on, neither the central government nor state government took an interest in developing uniform practices needed to run the industries for a long run (Jong 1989; Jana and Jana 2003; Puthucherril 2016). Later on, various new advanced technological developments were adopted on the international level in Thailand, Taiwan, and China, etc. (Kongkeo 1997) and India also benefitted from this advancement, and its business became flourishes along many Indian coastlines. Among these coastlines, Andhra Pradesh was at the forefront, followed by West Bengal, Kerala, Orissa, Karnataka, Maharashtra, Tamil Nadu, Gujarat, and Goa. Fig. 8.9 represents an overview of the massive increase in the area of shrimp farming from the 1990s until 2017 in all the mangroves across Indian coastlines.

Bangladesh has also become a multimillion-dollar industry in shrimp farming in global markets, particularly the United States of America (USA) and the European Union (EU). In 2014-15, Bangladesh knew to export 44,278t of prawn, and shrimp valued US$ 142million and US$364 million, respectively (FRSS 2016). Due to their high export value, shrimp has been referred to as "white gold" in Bangladesh (Ahmed et al. 2013). In Bangladesh, shrimp production has dramatically increased from 56,569 t to 75,274 t from 2010-2011 to 2014-2015.

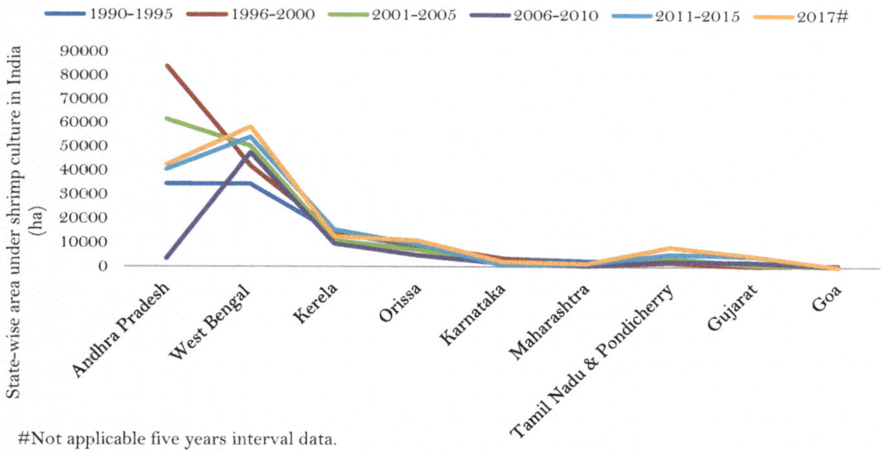

Fig. 8.9 State-wise massive increase in the area of shrimp farming from the 1990s until 2017 in India (Source: MPEDA 2018)

8.6 Emission of the Carbon Dioxide as Significant Greenhouse Gas (GHGs) Concern

The concentration of CO_2 is increasing continuously at a higher pace. Its value was 278 ppm in 1750, which rises to 390 ppm in 2011, 407.4 ppm in 2018 and would likely to increase between 467 and 555 ppm by the year 2050 (Table 8.4) (IPCC 2007; Anderson et al. 2009; Stocker et al. 2013; Lindsey 2018) and can increase the temperature which could melt the polar ice and sea level may rise to 5m (Detwiler and Hall 1988; IUFRO 2009). Among GHGs, CO_2 comprises 77% of its composition (Devi et al. 2012), and hence it is crucial to study its emission and sink.

8.6.1 Carbon Stock and Related Potential Carbon dioxide Emission of Various Mangroves Across the Globe

Bhomia et al. 2016 studied the top 30 cm of mangrove soil and revealed that the total amount of equivalent carbon dioxide (CO_2e) releases when soil carbon is oxidized would be 21.0×10^5 to 23.2×10^5 Mg CO_2e for an area of 155–183 km^2 in Bhitarkanika mangroves. This is comparable to the amount of CO_2 emitted by fossil

Table 8.4 Concentration of carbon dioxide from 1750 to 2018

Year	CO_2 concentration
1750	278 ppm
2011	390 ppm
2018	407.4
2050	467-555 ppm

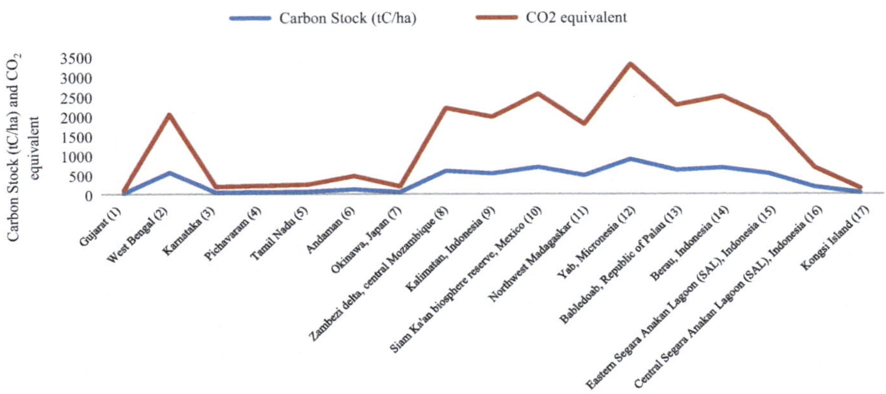

Fig. 8.10 The carbon stock and equivalent CO_2 emission of many mangroves across the world. (1. Pandey and Pandey 2013; 2. Ray et al. 2011; 3. Suresh et al. 2013; 4. ISFR 2019; 5. Kathiresan et al. 2013; 6. Mall 1991; 7. Khan et al. 2007; 8. Bosire et al. 2012; 9. Murdiyarso et al. 2010; 10. Adame et al. 2013; 11. Jones et al. 2014; 12. Kauffman et al. 2011; 13. Kauffman and Donato 2012; 14. Kusumaningtyas et al. 2019; 15. Kusumaningtyas et al. 2019; 16. Kusumaningtyas et al. 2019; 17. Kusumaningtyas et al. 2019)

fuel (oil) combustion in Nepal in 2005 (IEA 2013). Similarly, various carbon stocks have been compared across the world. CO_2 equivalent has been calculated by multiplying the carbon stock by 3.67 as one ton of carbon is equal to 3.67 tons of CO_2 and represents an equivalent amount of carbon lost from long-lived pool such as sediments. Fig. 8.10 depicts the comparison of carbon stock and equivalent CO_2 across many mangroves around the world. The maximum carbon stored and maximum equivalent CO_2 emission is observed in Yap (Micronesia), Sian Ka'an biosphere reserve (Mexico), Berau (Indonesia), Zambezi Delta (Central Mozambique), West Bengal (India), Kalimantan (Indonesia), and Eastern Segara Anakan Lagoon (SAL) (Indonesia). This suggests that since mangrove has a high capacity to sequester carbon, it has the same ability to emit carbon in the atmosphere, if it gets disturbed.

8.6.2 Estimated Carbon dioxide Emission Based on Ecosystem Loss Since European Settlement

A similar observation has been reflected in Victorian coastlines (Southeast Australia) where Lewis et al. 2018 represent that its total mean carbon stock is 4970 Mg C in the top 30 cm of soils and estimated the potential losses of carbon stock since European Settlements. The estimates of carbon loss are based on 50-90% remineralization of organic carbon. Hence, they may underestimate the loss of actual organic carbon deep down a meter or more (Pendleton et al. 2012). The total carbon stock in the Inlets, French Island, Nooramunga Coast, Western Port, Shallow Inlet, and Corner Inlet is 114, 328, 1094, 7256, 9054, and 11,972 MgC, respectively. The

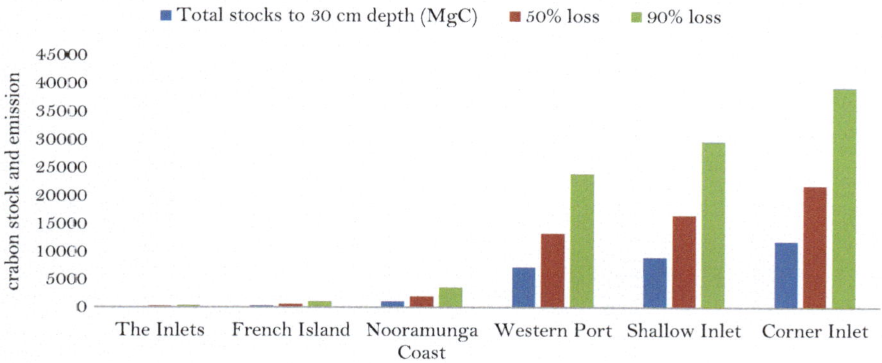

Fig. 8.11 Estimated Sediment carbon emissions based on ecosystem loss since European Settlement along Victorian coastlines (Lewis et al. 2018)

carbon loss from 50% remineralization in the Inlets, French Island, Nooramunga Coast, Western Port, Shallow Inlet, and Corner Inlet is 210, 602, 2008, 13314, 16614, and 21968 MgC, respectively, and from 90% mineralization is 378, 1083, 3614, 23996, 29905 and 39543 MgC, respectively (Fig. 8.11). The largest estimated carbon stock losses could be observed in Western Port, Shallow Inlet, Corner Inlet, totalling to approximately 95% of the estimated stock.

8.6.3 Carbon dioxide Emission from Mangrove Conversion to Shrimp Farming

The rapid increase in aquaculture and shrimp farming industries has adversely affected the mangroves of several countries, including China, India, Brazil, Mexico, Bangladesh, Myanmar, Indonesia, Sri Lanka, Thailand, Vietnam, and the Philippines (FAO 2007; UNEP 2014). Various adverse environmental impacts have been overlooked over substantial economic benefits. In Puttalam Lagoon, Sri Lanka, the mangroves conversion to shrimp ponds might cause a loss of 1,91,584 t of total carbon between 1990 and 2012, which is about 75.5% of the total carbon loss (Bournazel et al. 2015). In Ecuador mangroves, about 80% of the living carbon lost (7.01 million t) resulted from the replacement of the mangroves by shrimp cultivation (Hamilton and Lovette 2015)

Kauffman et al. 2014 estimated that about 1,036,971 Mg C is lost due to the mangrove conversion to cultivated shrimp ponds, salt ponds, and other uses. Donato et al. 2011 estimated that mangrove stores about 20 PgC globally, which is about 2.5 times of annual carbon dioxide annually. Hence, land use and land cover type have a significant effect on total carbon storage and sequestration rates in the coastal ecosystem (Guo and Gifford 2002). Figure 8.12 represents the potential carbon

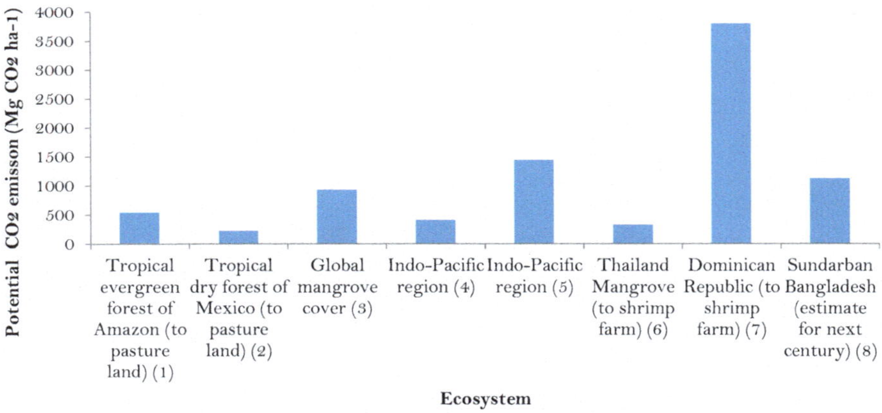

Fig. 8.12 Potential CO_2 emission from mangrove conversion to different land use land cover pattern. (1 Kauffman et al. 2003; 2. Kauffman et al. 2009; 3. Pendleton et al. 2012; 4. Donato et al. 2011; 5. Donato et al. 2011; 6. Yee 2010; 7. Kauffman et al. 2014; 8. Chanda et al. 2016)

dioxide loss due to the conversion of mangroves to shrimp farming across the globe, where the Dominican Republic shows the maximum carbon dioxide per hectare.

8.7 Adaptation Strategies to Balance the Land Use Land Cover and Ecosystem

It is difficult to control the spreading of aquaculture farming as a large population depends on them for their livelihood, especially the poor inhabitants living across the coastlines. The productivity of mangroves decreases following 5-10 years of conversion of mangroves to aquaculture and when the ponds are abandoned (Bosma et al. 2012; Cameron et al. 2019). Some of the adaptation strategies have been taken into account to run the aquaculture industries without much affecting the ecology and environment of coastal ecosystems. Some of the strategies are as follows:

8.7.1 IMTA (Integrated Multi-Trophic Aquaculture)

This is the procedure of producing diverse species of shellfish and finfish in integrated farming with the seaweeds from various trophic levels. IMTA increases profitability and productivity via recycling and reusing of nutrients. IMTA is based on the principle of co-cultivation of different organic and inorganic extractive species and fed fish (Troell et al. 2009; Chopin et al. 2010; Chopin et al. 2012). This is also called as "greening of aquaculture."

Benefits of IMTA-

1) Environmental friendly.
2) Hinder intensive fishing.
3) It could minimize the emission of blue carbon, thereby enhancing sequestration and storage of blue carbon.
4) It increases the biodiversity (Clements and Chopin 2017) and resilience of the coastal ecosystems (Worm et al. 2006; Levin and Lubchenco 2008).
5) IMTA may not be disturbed by natural calamities like sea-level rise, flood, and an increase in water temperature.
6) Can acclimatize to a wide range of water salinity as molluscs, seaweeds, and shrimp can tolerate variation in salinity.
7) Cultivation of seaweed in IMTA could play a crucial role in the sequestration of blue carbon via photosynthesis (Chung et al. 2013).
8) Seaweeds in IMTA keep the water cool and clean due to the absorption of toxic materials, pollutants, and sediments (Chung et al. 2013).

8.7.2 Restoration of Mangroves

Mangroves restoration would reimburse for mangrove loss by various aquaculture and shrimp farming. Regeneration of mangrove includes plantations. There are two main types of mangrove restoration: 1) mangrove restoration following natural disturbances and 2) mangrove restoration following anthropogenic degradation (Biswas et al. 2009).

8.7.3 REDD+ (Reducing Emissions from Deforestation and Forest Degradation)

The REDD+ program facilitates the afforestation, reforestation, and mangroves restoration and thus is wholly invested in the conservation of mangroves (UNEP 2014; Beymer-Farris and Bassett 2012; Olander et al. 2012). The REDD+ method is appropriate for mitigating the emission of greenhouse gas. It plays a significant role in minimizing anthropogenic emissions, thus helping in climate change mitigation. REDD+ if applied worldwide, can avoid almost 2.5 billion tons of carbon dioxide emission annually (Overmars et al. 2014). The combined approach of IMTA, mangrove restoration, and REDD+ could bring a broad range of economic, social, and environmental benefits.

8.8 Conclusion

The concentration of carbon dioxide is increasing continuously as a result of various anthropogenic activities. Hence, it is essential to recognize and acknowledge the ecosystems which can store carbon. The ability of coastal ecosystems like seagrasses, tidal marshes, and mangroves to store carbon is massive. This study assessed the various carbon stocks in the mangroves across the globe and the effect of anthropogenic disturbances on them. Indonesia has maximum stored carbon (especially in its marine settings) among Asia. In India, Sundarbans shows a significant amount of stored carbon. The disturbances include various land use land cover changes which comprise forest degradation and deforestation, aquaculture, and shrimp farming. This study reveals that mangrove's forest land shows a significant decrease in their area due to their conversion to aquaculture and shrimp farming. Aquaculture and shrimp farming are the fastest growing food industry in the world and show million-dollar profits in the global market. Bangladesh coast (In Bangladesh) and the Andhra Pradesh coast (in India) are the most affecting the coastal ecosystem in terms of shrimp farming and aquaculture.

On the one hand, these industries give a boost to the country's economy; on the other hand, it adversely affects the mangroves ecosystem by releasing previously sequestered carbon in the form of carbon dioxide in the atmosphere. This reveals that anthropogenic disturbances can turn the mangroves from carbon sink to source. Since aquaculture farming helps in high revenue generation but negatively affects the coastal ecosystem, it is essential to maintain a balance. Hence greening of aquaculture has been introduced (like IMTA) along with mangrove forest restoration, and REDD+.

References

Adame MF, Kauffman JB, Medina I, Gamboa JN, Torres O, Caamal JP, Reza M, Herrera-Silveira JA (2013) Carbon stocks of tropical coastal wetlands within the karstic landscape of the Mexican Caribbean. PloS one 8(2): p.e56569.

Ahmed N, Glaser M (2016) Coastal aquaculture, mangrove deforestation and blue carbon emissions: is REDD+ a solution?. Mar Policy 66:58-66.

Ahmed N, Cheung WW, Thompson S, Glaser M (2017) Solutions to blue carbon emissions: Shrimp cultivation, mangrove deforestation and climate change in coastal Bangladesh. Mar Policy 82:68-75.

Ahmed N, Occhipinti-Ambrogi A, Muir JF (2013) The impact of climate change on prawn postlarvae fishing in coastal Bangladesh: socioeconomic and ecological perspectives. Mar Policy 39:224-233.

Akhand A, Mukhopadhyay A, Chanda A, Mukherjee S, Das A, Das S, Hazra S, Mitra D, Choudhury SB, Rao KH (2017) Potential CO_2 emission due to loss of above ground biomass from the Indian Sundarban mangroves during the last four decades. J Indian Soc Remote Sens 45(1):147-154.

Alongi DM (2002) Present state and future of the world's mangrove forests. Environ Conserv 29 (3):331-349.

Alongi DM, Mukhopadhyay SK (2015) Contribution of mangroves to coastal carbon cycling in low latitude seas. Agric For Meteorol 213:266-272.

Alongi DM (2018) Mangrove forests. In Blue Carbon (pp. 23-36). Springer, Cham.

Anderson K, Starkey R, Bows A (2009) Defining dangerous climate change - a call for consistency. Policy 14(2):103-110.

Banerjee K, Roy Chowdhury M, Sengupta K, Sett S, Mitra A (2012) Influence of anthropogenic and natural factors on the mangrove soil of Indian Sundarbans wetland. Arch Environ Sci 6:80-91.

Banerjee K, Sahoo CK, Bal G, Mallik K, Paul R, Mitra A (2020) High blue carbon stock in mangrove forests of Eastern India. Trop Ecol 61:150-167.

Beymer-Farris BA, Bassett TJ (2012) The REDD menace: Resurgent protectionism in Tanzania's mangrove forests. Glob Environ Change 22(2):332-341.

Bhomia RK, MacKenzie RA, Murdiyarso D, Sasmito SD, Purbopuspito J (2016) Impacts of land use on Indian mangrove forest carbon stocks: Implications for conservation and management. Ecol Appl 26(5):1396-1408.

Biswas SR, Mallik AU, Choudhury JK, Nishat A (2009) A unified framework for the restoration of Southeast Asian mangroves - bridging ecology, society and economics. Wet Ecol Manage17 (4):365-383.

Bosma R, Sidik AS, van Zwieten P, Aditya A, Visser L (2012) Challenges of a transition to a sustainably managed shrimp culture agro-ecosystem in the Mahakam delta, East Kalimantan, Indonesia. Wet Ecol Manag 20(2):89-99.

Bosire JO, Bandeira S, Rafael J (2012) Coastal climate change mitigation and adaptation through REDD+ carbon programs in mangroves in Mozambique: Pilot in the Zambezi Delta. Determination of carbon stocks through localized allometric equations component, WWF.

Bournazel J, Kumara, MP, Jayatissa LP, Viergever K, Morel V Huxham M (2015) The impacts of shrimp farming on land-use and carbon storage around Puttalam lagoon, Sri Lanka. Ocean Coast Manage 113:18-28.

Bridgham SD, Megonigal JP, Keller JK, Bliss NB, Trettin C (2006) The carbon balance of North American wetlands. Wetlands 26(4):889-916.

Briggs M, Funge-Smith S, Subasinghe R, Phillips M. Introductions and movement of Penaeus vannamei and Penaeus stylirostris in Asia and the Pacific. RAP publication, 10(2004):92.

Cameron C, Hutley LB, Friess DA, Brown B (2019) High greenhouse gas emissions mitigation benefits from mangrove rehabilitation in Sulawesi, Indonesia. Eco Ser 40:101035.

Chanda A, Mukhopadhyay A, Ghosh T, Akhand A, Mondal P, Ghosh S, Mukherjee S, Wolf J, Lázár AN, Rahman MM, Salehin M (2016) Blue carbon stock of the Bangladesh Sundarban mangroves: what could be the scenario after a century?. Wetlands 36(6):1033-1045.

Chopin T, Troell M, Reid GK, Knowler D, Robinson SMC, Neori A, Buschmann AH, Pang SJ, Fang J (2010) Integrated multi-trophic aquaculture (IMTA)-A responsible practice providing diversified seafood products while rendering biomitigating services through its extractive components. Abstracts. Aquaculture Europe.

Chopin T, Cooper JA, Reid G, Cross S, Moore, C (2012) Open-water integrated multi-trophic aquaculture: environmental biomitigation and economic diversification of fed aquaculture by extractive aquaculture. Rev Aquac 4(4):209-220.

CIBA (2009) Training manual on Better management practices in shrimp farming. CIBA Special Publications No.44. Madras, India, p. 133.

Chowdhury SR, Hossain MS, Sharifuzzaman SM, Sarker S (2015) Blue carbon in the coastal ecosystems of Bangladesh. Project Document, Support to Bangladesh on Climate Change Negotiation and Knowledge Management on Various Streams of UNFCCC Process Project, funded by DFID and Danida, implemented by IUCN Bangladesh Country Office.

Chung IK, Oak JH, Lee JA, Shin JA, Kim JG, Park KS (2013) Installing kelp forests/seaweed beds for mitigation and adaptation against global warming: Korean Project Overview. ICES J Mar Sci 70(5):1038-1044.

Clements JC, Chopin T (2017) Ocean acidification and marine aquaculture in North America: potential impacts and mitigation strategies. Rev Aqu 9(4):326-341.

Coastal Resource Institute (CORIN) (1995) The effect of aquaculture on agricultural land and coastal environment. Mimeo, Prince of Songkla University, Songkhla, Thailand.
Detwiler RP, Hall CAS (1988) Tropical forests and the global carbon cycle. Science 239:42–47.
Devi B, Bhardwaj DR, Panwar P, Pal S, Gupta NK, Thakur CL (2012) Carbon allocation, sequestration and carbon dioxide mitigation under plantation forests of north western Himalaya, India. Ann Forest Res 56(1):123–135.
Donato DC, Kauffman JB, Murdiyarso D, Kurnianto S, Stidham M, Kanninen M (2011) Mangroves among the most carbon-rich forests in the tropics. Nat Geosci 4(5):293-297.
Donato DC, Kauffman JB, Mackenzie RA, Ainsworth A, Pfleeger AZ (2012) Whole-island carbon stocks in the tropical Pacific: Implications for mangrove conservation and upland restoration. J Environ Manage 97:89–96.
Dorababu KK (2013) Impact of aquaculture on land use patterns, environment and economy: A case study of west Godavari district, Andhra Pradesh, India. Int J Curr Res 5(7):1993–1996
Duarte CM, Middelburg JJ, Caraco N (2004) Major role of marine vegetation on the oceanic carbon cycle. Biogeosciences 2:1–8
Duarte CM, Dennison WC, Orth RJ, Carruthers TJ (2008) The charisma of coastal ecosystems: addressing the imbalance. Estuar Coast 31(2):233-238.
Duke NC, Meynecke JO, Dittmann S, Ellison AM, Anger K, Berger U, Cannicci S, Diele K, Ewel KC, Field CD, Koedam N (2007) A world without mangroves?. Science 317(5834):41–42.
Dutta D, Das PK, Paul S, Sharma JR Dhadwal VK (2014) Spatio-temporal assessment of ecological disturbance and its intensity in the mangrove forest using MODIS derived disturbance index. International Archives of the Photogrammetry, Remote Sensing and Spatial Information Sciences 8:555–559.
Eong OJ (1993) Mangroves-a carbon source and sink. Chemosphere 27(6):1097–1107.
Estrada GC, Soares ML (2017) Global patterns of aboveground carbon stock and sequestration in mangroves. An Acad Bras Ciênc 89(2):973–989.
FAO, The State of World Fisheries and Aquaculture (2016) Contributing to Food Security and Nutrition for All, Food and Agriculture Organization of the United Nations, Rome.
FAO, Food and Agriculture Organization (FAO) of the United Nations (2007) The World's Mangroves 1980–2005. FAO, Rome.
FRSS, Fisheries Statistical Report of Bangladesh, Fisheries Resources Survey System, Department of Fisheries, Bangladesh, 2016.
Ghosh A, Schmidt S, Fickert T, Nüsser M (2015) The Indian Sundarban mangrove forests: history, utilization, conservation strategies and local perception. Diversity 7(2):149-169.
Giri C, Zhu Z, Tieszen LL, Singh A, Gillette S, Kelmelis JA (2008) Mangrove forest distributions and dynamics (1975–2005) of the tsunami-affected region of Asia. J Biogeograph 35 (3):519–528.
Granek E, Ruttenberg BI (2008) Changes in biotic and abiotic processes following mangrove clearing. Estuar Coast Mar Sci 80(4):555-562.
Guo LB, Gifford RM (2002) Soil carbon stocks and land use change: a meta-analysis. Glob Change Biol 8(4):345-360.
Hamilton SE, Friess DA (2018) Global carbon stocks and potential emissions due to mangrove deforestation from 2000 to 2012. Nat Clim Change 8(3):240-244.
Hamilton SE, Lovette J (2015) Ecuador's mangrove forest carbon stocks: A spatiotemporal analysis of living carbon holdings and their depletion since the advent of commercial aquaculture. PloS one 10(3):e0118880.
Hossain M, Lin CK, Hussain MZ (2001) Goodbye Chakaria Sundarban: the oldest mangrove forest. Wet Sci Prac18(3):19-22.
Hossain MS (2001) Biological aspects of the coastal and marine environment of Bangladesh. Ocean Coast Manage 44(3-4):261-282.
Humanity watch Carbon Trading, the Sundarbans and Climate Justice, Campaign Paper, Equity and Justice Working Group-Bangladesh, Humanitywatch, 2011.

IEA (International Energy Agency) (2013) Definition. Paris, France: Total Primary Energy Supply. http://www.iea.org/stats/defs/Tpes.asp; accessed on 24 June 2013.

Iftekhar MS, Saenger P (2008) Vegetation dynamics in the Bangladesh Sundarbans mangroves: a review of forest inventories. Wet Ecol Manage 16(4):291-312.

India State of Forest Report (ISFR) (2019).

Intergovernmental Panel on Climate Change (IPCC) (2007) Climate change 2007: Assessment Report. IPCC, Valencia.

IUFRO (2009) Adaptation of forests and people to climate change. A global assessment report. IUFRO World Series 22:224

Jana BB, Jana S (2003) The potential and sustainability of aquaculture in India. J Appl Aquacul 13 (3-4):283-316.

Jones TG, Ratsimba HR, Ravaoarinorotsihoarana L, Cripps G, Bey A (2014) Ecological variability and carbon stock estimates of mangrove ecosystems in north-western Madagascar. Forests 5 (1):177-205.

Jong JD (1989) Aquaculture in India. Rijksdienst Voor Ondernemend. 2:1117–1152.

Levin SA, Lubchenco J (2008) Resilience, robustness, and marine ecosystem-based management. Bioscience 58(1):27-32.

Lindsey R (2018) Climate change: atmospheric carbon dioxide. National Oceanographic and Atmospheric Administration, News & Features. August.

Kathiresan K, Anburaj R, Gomathi V, Saravanakumar K (2013) Carbon sequestration potential of Rhizophora mucronata and Avicennia marina as influenced by age, season, growth and sediment characteristics in southeast coast of India. J Coast Conserv 17(3):397-408.

Kauffman JB, Donato DC (2012) Protocols for the measurement, monitoring and reporting of structure, biomass, and carbon stocks in mangrove forests (pp. 50-p). Bogor, Indonesia: CIFOR.

Kauffman JB, Heider C, Cole TG, Dwire KA, Donato DC (2011) Ecosystem carbon stocks of Micronesian mangrove forests. Wetlands 31(2):343-352.

Kauffman JB, Heider C, Norfolk J, Payton F (2014) Carbon stocks of intact mangroves and carbon emissions arising from their conversion in the Dominican Republic. Ecol Appl 24(3):518-527.

Kauffman JB, Hughes RF, Heider C (2009) Dynamics of C and nutrient pools associated with land conversion and abandonment in Neotropical landscapes. Ecol Appl 19:1211-1222.

Kauffman JB, Steele MD, Cummings DL, Jaramillo VJ (2003) Biomass dynamics associated with deforestation, fire, and, conversion to cattle pasture in a Mexican tropical dry forest. Forest Ecol Manage 176(1-3):1-12.

Kennedy H. et al., (2014) In Supplement to the 2006 IPCC Guidelines for National Greenhouse Gas Inventories: Wetlands (eds Hiraishi, T. et al.), Intergovernmental Panel on Climate Change, Gland, Switzerland.

Khan MNI, Suwa R, Hagihara A (2007) Carbon and nitrogen pools in a mangrove stand of Kandelia obovata (S., L.) Yong: vertical distribution in the soil–vegetation system. Wet Ecol Manage 15 (2):141-153.

Kongkeo H (1997) Comparison of intensive shrimp farming systems in Indonesia, Philippines, Taiwan and Thailand. Aquacul Res 28(10):789-796.

Kristensen E, Bouillon S, Dittmar T, Marchand C (2008) Organic carbon dynamics in mangrove ecosystems: a review. Aquat Bot 89(2):201-219.

Kusumaningtyas MA, Hutahaean AA, Fischer HW, Pérez-Mayo M, Ransby D, Jennerjahn TC (2019) Variability in the organic carbon stocks, sources, and accumulation rates of Indonesian mangrove ecosystems. Estuar Coast Mar Sci 218:310-323.

Lewis CJE, Carnell PE, Sanderman J, Baldock JA, Macreadie PI (2018) Variability and vulnerability of coastal 'blue carbon'stocks: a case study from southeast Australia. Ecosystems 21(2):263-279.

Malik A, Fensholt R, Mertz O (2015) Mangrove exploitation effects on biodiversity and ecosystem services. Biodivers Conserv 24(14):3543-3557.

Mall LP (1991) Study of biomass, litter fall, litter decomposition and soil respiration in monogenic mangrove and mixed mangrove forests of Andaman Islands. Trop Ecol 32:144-152.

Mcleod E, Chmura GL, Bouillon S, Salm R, Björk M, Duarte CM, Lovelock CE, Schlesinger WH, Silliman BR (2011) A blueprint for blue carbon: toward an improved understanding of the role of vegetated coastal habitats in sequestering CO_2. Front Ecol Environ 9(10):552-560.

MPEDA (2018) Annual Report 2017–18. The Marine Products Export Development Authority. Cochin, India, p. 59.

Murdiyarso D, Donato D, Kauffman JB, Kurnianto S, Stidham M, Kanninen M (2009) Carbon storage in mangrove and peatland ecosystems: A preliminary account from plots in Indonesia. Working paper 48. Bogor Banat, Indonesia: Center Internat Forest Res 35:1-35.

Murdiyarso D, Hergoualc'h K, Verchot LV (2010) Opportunities for reducing greenhouse gas emissions in tropical peatlands. Proc Nat Acad Sci 107(46):19655-19660.

Murdiyarso D, Purbopuspito J, Kauffman JB, Warren MW, Sasmito SD, Donato DC, Manuri S, Krisnawati H, Taberimadon S, Kurnianto S (2015) The potential of Indonesian mangrove forests for global climate change mitigation. Nat Clim Change 5(12):1089-1092.

Nellemann C, Corcoran E, Duarte CM, Valdes L, DeYoung C, et al., (2009) Blue Carbon. A Rapid Response Assessment. United Nations Environment Programme, GRID-Arendal website. www.grida.no. Accessed 2011 Nov 11.

Olander LP, Galik CS, Kissinger GA (2012) Operationalizing REDD+: Scope of reduced emissions from deforestation and forest degradation. Curr Opin Environ Sustain 4(6):661-669.

Overmars KP, Stehfest E, Tabeau A, van Meijl H, Beltrán AM, Kram T (2014) Estimating the opportunity costs of reducing carbon dioxide emissions via avoided deforestation, using integrated assessment modelling. Land Use Policy 41:45-60.

Pandey CN, Pandey R (2013) Carbon sequestration in mangroves of Gujarat, India. Internat J Bot Res 3(2):57-70.

Pendleton L, Donato DC, Murray BC, Crooks S, Jenkins WA, Sifleet S, Craft C, Fourqurean JW, Kauffman JB, Marbà N, Megonigal P (2012) Estimating global "blue carbon" emissions from conversion and degradation of vegetated coastal ecosystems. PloS One 7(9):e43542.

Puthucherril TG (2016) Sustainable aquaculture in India: looking back to think ahead. In Aquaculture Law and Policy. Edward Elgar Publishing.

Rahman MM (2015) Carbon and nitrogen dynamics and carbon sequestration in soils under different residue management. The Agriculturists 12(2):48-55.

Ray R, Ganguly D, Chowdhury C, Dey M, Das S, Dutta MK, Mandal SK, Majumder N, De TK, Mukhopadhyay SK, Jana TK (2011) Carbon sequestration and annual increase of carbon stock in a mangrove forest. Atmos Environ 45(28):5016-5024.

Richards DR, Friess DA (2016) Rates and drivers of mangrove deforestation in Southeast Asia, 2000–2012. Proc Nat Acad Sci 113(2):344-349.

Saenger P (2002) Mangrove ecology, silviculture and conservation. Springer Science & Business Media.

Sahu SC, Kumar M, Ravindranath NH (2016) Carbon stocks in natural and planted mangrove forests of Mahanadi Mangrove Wetland, East Coast of India. Curr Sci 110:2253-2260.

Sahu SC, Suresh HS, Murthy IK, Ravindranath NH (2015) Mangrove area assessment in India: implications of loss of mangroves. J Earth Sci Climat Change 6(5):1.

Sanderman J, Hengl T, Fiske G, Solvik K, Adame MF, Benson L, Bukoski JJ, Carnell P, Cifuentes-Jara M, Donato D, Duncan C (2018) A global map of mangrove forest soil carbon at 30 m spatial resolution. Environ Res Lett 13(5):055002.

Seidensticker J, Hai MA (1983) The Sundarbans wildlife management plan: conservation in the Bangladesh coastal zone.

Shahid MA, Islam J (2002) March. Impact of denudation of mangrove forest due to shrimp farming on the coastal environment in Bangladesh. In Technical proceedings of BAU-NURAD workshop on environment and socio-economic impacts of shrimp farming in Bangladesh 5:67-75.

Short FT, Wyllie-Echeverria S (1996) Natural and human-induced disturbance of seagrasses. Environ Conserv 23:17-27.

Siikamäki J, Sanchirico JN, Jardine SL (2012) Global economic potential for reducing carbon dioxide emissions from mangrove loss. Proc Nat Acad Sci 109(36):14369-14374.

Sillanpää M, Vantellingen J, Friess DA (2017) Vegetation regeneration in a sustainably harvested mangrove forest in West Papua, Indonesia. Forest Ecol Manage 390:137-146.

Sjöling S, Mohammed SM, Lyimo TJ, Kyaruzi JJ (2005) Benthic bacterial diversity and nutrient processes in mangroves: impact of deforestation. Estuar Coast Shelf Sci 63(3):397-406.

Spalding M (2010) World atlas of mangroves. Routledge.

Stocker TF, Qin D, Plattner GK, Tignor M, Allen SK, Boschung J, Nauels A, Xia Y, Bex V, Midgley PM (2013) IPCC, 2013. Climate change.

Suresh B, Manjappa S, Puttaiah ET (2013) Dynamics of phytoplankton succession in Tungabhadra river near Harihar, Karnataka (India). J Microbiol Antimicrob 5(7):65-71.

Sweetman AK, Middelburg JJ, Berle AM, Bernardino AF, Schander C, Demopoulos AWJ, Smith CR (2010) Impacts of exotic mangrove forests and mangrove deforestation on carbon remineralization and ecosystem functioning in marine sediments. Biogeosciences 7(7).

Strangmann A, Bashan Y, Giani L (2008) Methane in pristine and impaired mangrove soils and its possible effect on establishment of mangrove seedlings. Biol Fert Soils 44(3):511.

Thakur S, Maity D, Mondal I, Basumatary G, Ghosh PB, Das P, De TK (2020) Assessment of changes in land use, land cover, and land surface temperature in the mangrove forest of Sundarbans, northeast coast of India. Environ Develop Sustain 1-27.

Troell M, Joyce A, Chopin T, Neori A, Buschmann AH, Fang JG (2009) Ecological engineering in aquaculture-potential for integrated multi-trophic aquaculture (IMTA) in marine offshore systems. Aquaculture 297(1-4):1-9.

UNEP (2014) The importance of mangroves to people: A call to action. Cambridge: UNEP World Conservation Monitoring Centre.

Valiela I, Bowen JL, York JK (2001) Mangrove Forests: One of the World's Threatened Major Tropical Environments: At least 35% of the area of mangrove forests has been lost in the past two decades, losses that exceed those for tropical rain forests and coral reefs, two other well-known threatened environments. Bioscience 51(10):807-815.

Van der Werf GR, Morton DC, DeFries RS, Olivier JG, Kasibhatla PS, Jackson RB, Collatz GJ, Randerson JT (2009) CO_2 emissions from forest loss. Nat Geosci 2(11):737-738.

Vinod K, Anasu Koya A, Kunhikoya VA, Shilpa PG, Asokan PK, Zacharia PU, Joshi KK (2018) Biomass and carbon stocks in mangrove stands of Kadalundi estuarine wetland, south-west coast of India. Ind J Fisher 65(2):89-99.

Vinod K, Asokan PK, Zacharia PU, Ansar CP, Vijayan G, Anasukoya A, Kunhi Koya VA, Nikhiljith M (2019) Assessment of Biomass and Carbon Stocks in Mangroves of Thalassery Estuarine Wetland of Kerala, South West Coast of India. J Coast Res 86(SI):209-217.

Waycott M, Duarte CM, Carruthers TJ, Orth RJ, Dennison WC, Olyarnik S, Calladine A, Fourqurean JW, Heck KL, Hughes AR, Kendrick GA (2009) Accelerating loss of seagrasses across the globe threatens coastal ecosystems. Proc Nat Acad Sci 106(30):12377-12381.

Worm B, Barbier EB, Beaumont N, Duffy JE, Folke C, Halpern BS, Jackson JB, Lotze, H.K., Micheli, F., Palumbi, S.R. and Sala, E., (2006) Impacts of biodiversity loss on ocean ecosystem services. Science 314(5800):787-790.

Yee S (2010) REDD and BLUE carbon: carbon payments for mangrove conservation.

Responses of Mangrove Ecosystems to Climate Change in the Anthropocene

9

Daniel M. Alongi

Abstract

Mangrove ecosystems are well-adapted to living in a dynamic and harsh environment and have survived catastrophic climate events since their first appearance 66 Mya along Tethys Sea shores. There have been past episodes of localized extinction due mostly to abrupt, rapid rises in sea-level. Living at the edge between land and sea, mangrove ecosystems are inclined to be resilient to environmental disturbance, such as highly variable changes in salinity, solar insolation, temperature, tidal inundation, freshwater inputs, and sea-level. Based on current knowledge of mangrove responses to environmental change, mangroves in subsiding river deltas, on low Pacific and Caribbean islands, and in the arid tropics are likely to decline in area, biodiversity, forest structure, and function into the Anthropocene. Mangrove forests will continue to expand latitudinally along the Gulf of Mexico, the east coast of Florida, the coasts of South Africa, eastern Australia, northern New Zealand, south eastern Brazil, and China. Regions that experience tropical storms or other extreme events, as in northern Australia, the Gulf of Mexico, the east India coast, Florida, the Philippines, south China coast, and the northern Caribbean, mangroves will likely suffer increased damage, destruction, and loss. The east coast of Sumatra, north coast of Java, Sulawesi, and southern Vietnam are likely to lose mangroves because of their low tidal range. Mangroves in other parts of Southeast Asia may response positively to increased rainfall and warmer temperatures, as on the west coast of Peninsular Malaysia and the southwest coast of Thailand. Mangroves will likely respond positively or exhibit no significant change along the tropical west and north coasts of South America and the Caribbean and Pacific

D. M. Alongi (✉)
Tropical Coastal & Mangrove Consultants, Pakenham, VIC, Australia

R. P. Rastogi et al. (eds.), *Mangroves: Ecology, Biodiversity and Management*,
https://doi.org/10.1007/978-981-16-2494-0_9

coasts of Central America. Globally, mangroves are likely to survive, perhaps in altered states, into the next century.

Keywords

Anthropocene · Climate change · Ecological impacts · Mangroves · Sea-level rise · Tropics

9.1 Introduction

Mangrove forests and their associated waterways, being at the interface between land and sea, may portend ecological changes in the coastal zone from climate change. The intertidal environment in low latitudes where mangroves live is physically and geologically dynamic. Wide variations in temperature, wave action, tides, salinity, anoxia, and rainfall are just some of the factors driving and shaping mangrove ecosystems over time and space (Duke et al. 1998; Alongi 2016).

Mangrove ecosystems are fairly robust and adaptable in the face of environmental uncertainty, surviving in normal circumstances by key adaptive features such as large below-ground storage and transformation rates of carbon, nitrogen, and other nutrients; simple architecture and self-design; highly efficient but complex biotic controls; species redundancy; and multiple feedbacks which serve to facilitate and augment recovery from, or resilience to, natural and human disturbances (Alongi 2008, 2015). Evidence for their high adaptability to disturbance comes from patterns of recovering stands being reminiscent of pioneer-phase forest characteristics, as forest composition and structure arise from a complex interplay of physiological tolerances and competitive interactions leading to a mosaic of arrested or interrupted succession sequences, in response to physiochemical gradients and changes in shoreline evolution (Lugo 1980; Fromard et al. 1998; Alongi 2008). Not all mangroves survive disturbances virtually unscathed, as the extent of impact depends on the extent, severity, and time scale of the disturbance event.

Climate change consists of a series of types of disturbance depending on whether the effects are direct or indirect. The global rise in sea-level, increases in atmospheric greenhouse gases and temperatures, and changing atmospheric moisture and precipitation patterns are direct effects of climate change. Indirect effects are changes in seasonal patterns; changes in extreme weather events such as extreme high water events, droughts, storms and cyclones; acidification; changes in coastal circulation affecting tidal cycles and tidal exchange of nutrients and dispersal of propagules; changing salinity gradients affecting tidal exchange and species distribution and composition; changes in freshwater inflow and allochthonous sediment input; and degradation of adjacent ecosystems closely linked to mangroves (Alongi 2015; Jennerjahn et al. 2017). These impacts must be considered against a plethora of other human impacts that affect many mangrove forests: deforestation; subsidence due to extraction of water and other natural resources; damming and diversion of freshwater inflows; changes in riverine sediment inflows by increased catchment erosion and damming; overharvesting of mangrove wood, fish, and shellfish; pollution; changes in hydrology and tidal flushing due to construction of roads, levees, and dredging of navigation channels.

This chapter assesses the responses of mangroves to climate change and the likely outcomes over long temporal scales against the background of other human-induced disturbances. Also, it is important to consider the evidence for how mangroves have responded to earlier environmental perturbations, such as historical changes in sea-level, in order to gain some insight into how they may respond to global change in the Anthropocene.

9.2 Past, Present, and Future Responses to Sea-Level Change

Since their first appearance along the shores of the Tethys Sea during the Late Cretaceous-Early Tertiary period, mangroves have had to adapt and endure numerous variations in sea-level and other climatic events up to the present. Mangroves in the Quaternary period experienced a sea-level that was 120–125 m lower than present at the Last Glacial Maximum, with two periods of very rapid rise (>20 m) at both 14 and 11 ky BP due to abrupt climatic shifts during the transition from the last glacial into the present interglacial (Endfield and Marks 2012).

The present distribution of mangroves is a legacy of previous responses to changes in sediment accretion or erosion, surface elevation, hydrology, weather, climate, and various disturbances, and interactions among and between populations and communities (Woodroffe et al. 2016). The existence of relic pollen and peat deposits, however, provides abundant evidence of dramatic changes in mangroves over geologic time, especially in relation to late Quaternary changes in sea-level (Kim et al. 2005; Yulianto et al. 2005; Li et al. 2012; Urrego et al. 2013; Berger et al. 2013), with an overall pattern of paleo-ecological mangrove succession. The geological record and present analyses of mangrove sediment accretion rates versus rates of sea-level rise (see Fig. 2 in Alongi 2015) indicate that mangroves have kept pace with rising sea-level (Woodroffe et al. 2016). Regionally, however, some mangroves have gone locally extinct in the face of very rapid sea-level rise; reconstructed historical patterns have shown significant change depending on rate of sea-level rise as well as rates of subsidence or uplift (McCloskey and Liu 2013; Punwong et al. 2013; Limaye et al. 2014).

Evidence from deep mangrove peat deposits indicates that over thousands of years mangroves have kept pace with sea-level rise, as found in the Caribbean (McKee et al. 2007) and the northern Indian Ocean (Rashid 2014). The fact that these peat deposits are buried beneath existing mangroves and Holocene sediments, however, indicates that eventually the rate of sea-level rise increased beyond a critical threshold at which mangroves were not able to keep pace (McIvor et al. 2013). Records of relic pollen reveal fluctuating responses of mangrove succession in synchrony with changing sea-level, as found on Borneo (Yulianto et al. 2005), the Caribbean coast of Colombia (Urrego et al. 2013), the Sunda Shelf of Southeast Asia

(Hanebuth et al. 2011), East Africa (Punwong et al. 2013), West Africa (Kim et al. 2005), and the Galápagos (Seddon et al. 2011).

Modern evidence from time series analysis of photos, remote sensing images, digital terrain models, and rates of sediment accretion implies that mangrove responses to sea-level rise correspond roughly to patterns of surface elevation changes. Along the Pacific coast of Mexico, for example, fringing mangroves drowned because of rising sea-level accompanied by warm waters of El Niño events but mangroves located in higher intertidal positions were drive inland for a net increase in mangrove area (Lopez-Medellin et al. 2011). Mangrove responses to rising sea-level are likely to be complex with local variations playing a key role in predicting whether mangroves of a given locality or region will survive. For instance, in Micronesia, sedimentation is sufficient to offset elevation losses on some islands but not on others as mangroves set low on the shoreline are more susceptible than forests at higher elevations (Krauss et al. 2010).

The future is in doubt for mangroves occupying tropical river deltas and low islands, the former currently at risk of changing drastically due to damming, changes in coastal ocean circulation, and sedimentation and the latter at risk due to their low profile relative to sea-level. On low islands in the Pacific, mangroves are migrating landwards with rising sea-level but are constrained by coastal developments such as roads and levees (Lovelock et al. 2015). In Micronesia and Melanesia local extinction of mangroves is occurring as sea-level rises and mangroves cannot migrate landwards (Lovelock et al. 2015). Similarly, many deltaic islands of the Sundarbans along the Indian and Bangladesh coast are undergoing subsidence and disappearance as a decline in sediment input from the Ganges and other rivers has caused subsidence and a concomitant decline in mangroves on the central and eastern islands of the Sundarbans (Rahu et al. 2012; Rashid 2014). In some river deltas sea-level rise, storms and cyclones coupled with diminishing sediment and freshwater supply have enhanced subsidence, causing a shift of mangroves to higher elevations. In deltas across Africa and Asia, about 2% of mangroves disappeared between 2000–2016, mainly through erosion and conversion to agriculture (Lagomasino et al. 2019). However, in some regions, there was rapid expansion of mangroves over this 16-yr period, resulting in new, taller, and more carbon-dense forests. In the Amazon Macrotidal Mangrove Coast, large, migrating mud banks cyclically accrete and erode in synchrony with oscillations in tides and sediment erosion and accretion, resulting in mangroves migrating frequently to adapt to changes in the position of the shoreline (Schettini et al. 2020).

Are mangroves keeping pace with current rates of sea-level rise? An updated analysis (Fig. 9.1) of mangrove accretion rates versus local mean sea-level rise suggests that mangroves (i.e. below the solid line in Fig. 9.1), particularly in Australia, New Zealand, the Caribbean, Central America and on low Pacific islands and in subsiding river deltas such as the Sundarbans and in Southeast Asia are not keeping pace currently. However, mangroves (i.e. data points located above the solid line in Fig. 9.1) located in other areas of Southeast Asia and the Pacific (e.g. Papua New Guinea), South America, Africa, the Middle East, South Asia, and South Asia (primarily China) are keeping pace with current sea-level rise. Many of these latter

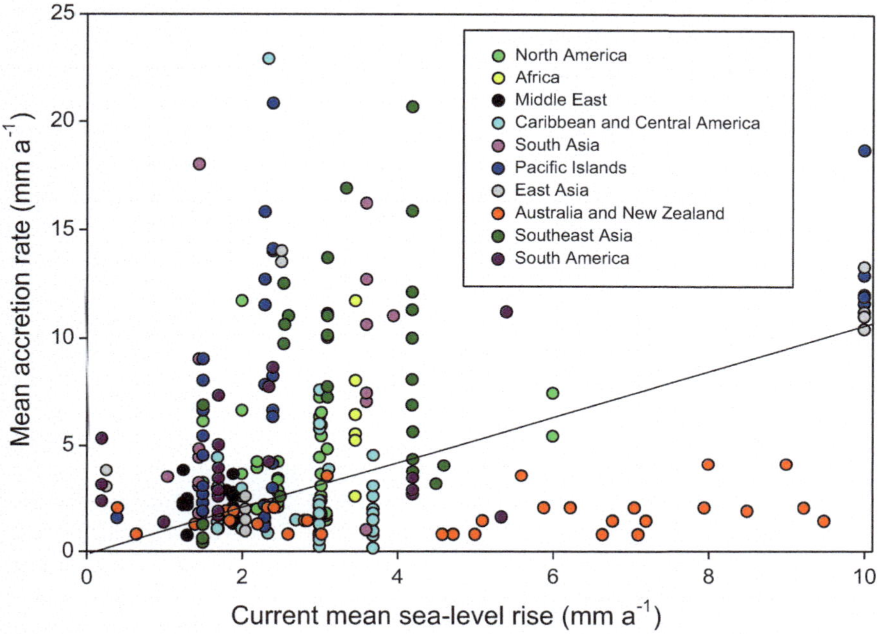

Fig. 9.1 The relationship between measured rates of mangrove soil accretion (mm a^{-1}) and current rates of mean sea-level rise (mm a^{-1}) worldwide. The sea-level rise data are from satellite altimetry or tide gauge data available on the website: http://www.nodc.noaa.gov/General/sealevel.html. Mangrove sedimentation data are from references in Alongi (2009, 2012, 2015, 2018) and Sasmito et al. (2016) and updated from data in Hayden and Granek (2015), Hoque et al. (2015), Carnero-Bravo et al. (2016, 2018), MacKenzie et al. (2016), Marchio et al. (2016), Almahasheer et al. (2017), Phillips et al. (2017), Ruiz-Fernández et al. (2018), Chappel (2018), Cusack et al. (2018), Murdiyarso et al. (2018), Pérez et al. (2018), Saderne et al. (2018), Fu et al. (2019), Hale et al. (2019), Kusumaningtyas et al. (2019), Soper et al. (2019) Swales et al. (2019), Bomer et al. (2020), Hapsari et al. (2020), Matos et al. (2020), Hatje et al. (2021), Passos et al. (2021). The solid line delimits a 1:1 relationship between accretion rate and rate of sea-level rise

forests occur in areas of rapid accretion due to highly impacted and populated catchments, especially in China, Brazil, and India. The wide scatter of data points reflects how mangroves in disparate coastal settings in different parts of world respond so differently to the same rate of sea-level rise. Most of the mangrove accretion rates were measured using radionuclides such as ^{137}Cs and ^{210}Pb and are not highly reliable indicators of accretion compared with the use of surface elevation tables; there is also considerable uncertainty in the rates of sea-level rise as there are large variations seasonally and with changes in atmospheric pressure and weather. The accretion rates also do not reflect the importance of changes in surface elevation gain; a mangrove forest may be rapidly accumulating soil on the surface but the local area may be subsiding, resulting in a net decrease relative to sea-level, as is occurring in the Sundarbans (Hanebuth et al. 2013). A detailed analysis of recent trends in mangrove surface elevation changes across the Indo-Pacific region using data from surface elevation table instruments shows that for 69% of mangroves the current rate of sea-level rise exceeded the soil surface elevation gain (Lovelock et al. 2015). Further, model analysis suggests that mangrove forests located at low tidal range and low sediment delivery could be submerged by 2070. As shown for macrotidal mangroves in northern Australia, mangroves have migrated consistently over geological time in synchrony with post-glacial sea-level rise; how mangroves in this region adjust in future depends not only on sediment availability but also on local topography (Woodroffe 2018). In contrast, created mangroves in Tampa Bay, Florida, store below-ground carbon and rates of surface elevation change enable them to adjust to sea-level rise (Krauss et al. 2017). A more recent analysis indicates that while mangroves expanded between 9800 and 7500 years ago at a rate driven mainly by the rate of relative sea-level rise, it was very likely (90% probability) that mangroves were unable to sustain accretion when relative sea-level rise exceeded 6.1 mm a^{-1} (Saintilan et al. 2020), in agreement with the data in Fig. 9.1 at rates of sea-level rise greater than 6 mm a^{-1}. Mangrove forests are likely losers with respect to rises in sea-level, especially in regions of substantial subsidence (e.g. the Sundarbans, the Solomon Islands; Albert et al. 2017), a low tidal range, changes in precipitation, and poor ecological conditions (Cinco-Castro and Herrera-Silveira 2020). The reality is that mangroves may respond in complex ways to sea-level rise. Some mangroves will probably survive if the rate of sea-level rise is slow enough in some locations, but there will likely be significant changes in community composition, forest structure, morphology, and anatomy, including changes in vascular vessel densities, fibre wall thickness, bark anatomy, formation of hypertrophied lenticels, adventitious roots, and increased aerenchyma development (Ellison and Farnsworth 1997; Wang et al. 2007; Yáñez-Espinosa and Flores 2011; Reef and Lovelock 2014). Experimental studies indicate species-specific responses to sea-level rise and waterlogging (Ye et al. 2003, 2004, 2010; Cardona-Olarte et al. 2006; Chen and Wang 2017). High tolerance is exhibited by the cosmopolitan species, *Avicennia marina*, to waterlogging, but responses are highly variable in relation to immersion depth and length of time, salinity, temperature, and other factors (Lu et al. 2013; Mangora et al. 2014). Another cosmopolitan species, *Rhizophora stylosa,* similarly has a high tolerance to waterlogging as experiments simulating growth of *R. stylosa* seedlings to sea-level rise found that the species was flood-tolerant with high stem growth rate and leaf assimilation rate as well as efficient utilization of carbohydrate reserves stored in hypocotyls of seedlings; both growth and physiology were affected by salinity and by an increase in flooding time (Chen and Wang 2017). *R. stylosa* exhibited competitive dominance at high salinity, a good adaptation of seedlings to future sea-level rise.

Future survival of mangrove forests ultimately depends on the rapidity of future increases in sea-level rise. The current prediction (Jevrejeva et al. 2019) is that sea-level will continue to rise at a median rate of 6.2 mm a^{-1} until 2100, which is faster than most current rates of sea-level rise (Fig. 9.1); most mangroves at or above the rate of 6.2 mm a^{-1} are at a high risk of drowning.

9.3 Responses to Rising Atmospheric CO_2

Fifty years ago, the mean atmospheric CO_2 concentration was 326.17 ppm, 21% lower than the current CO_2 concentration of 416.39 ppm and will continue to rise to 600–800 ppm by 2100 (see https://www.ipcc.ch/data/). Experimental evidence has repeatedly shown that elevated CO_2 concentrations enhance photosynthesis, growth, and leaf chlorophyll *a* concentration in mangroves, with responses being species-specific and variable, depending on salinity, temperature, nutrient availability, and water-use efficiency (Farnsworth et al. 1996; Ball et al. 1997; Snedaker and Araújo 1998; McKee and Rooth 2008; Cherry et al. 2009; McKee et al. 2012; Reef et al. 2016; Jacotot et al. 2018; Yin et al. 2018; Tamimia et al. 2019; Manea et al. 2020; Maurer et al. 2020). The early studies in the 1990s showed that growth of *Rhizophora stylosa, R. apiculata*, and *R. mangle* was enhanced by increasing CO_2 concentrations, but only a low salinity. Snedaker and Araújo (1998) showed that *R. mangle, Avicennia germinans, Conocarpus erectus,* and *Laguncularia racemosa* exhibited increases in transpiration efficiency but a decline in stomatal conductance and transpiration with increasing CO_2 concentrations. The response of most mangrove species to increasing CO_2 levels will be complex, with many species thriving, but some species declining or exhibiting no or little change. Interactive effects among CO_2 concentrations, salinity, temperature, and nutrient availability imply that intertidal position and regional location may be an important determinant in the response of most species. Species patterns within an estuary may change based on the ability of each species to respond to spatial and temporal variations in the above-mentioned drivers.

More recent experiments have supported the findings of the earlier studies. In a glasshouse study, Reef et al. (2016) examined the effects of elevated CO_2 and nutrient availability on seedlings of *Avicennia germinans* and found large gains in growth and photosynthesis when seedlings grown under elevated CO_2 were supplied with elevated nutrient concentrations compared to their responses under ambient conditions. Growth was greatly enhanced only under high nutrient conditions and elevated CO_2; root volume doubled under low nutrient and elevated CO_2 conditions relative to ambient nutrient and CO_2 levels. Biochemical pathways play a role in mangrove responses to climate change as *A. germinans* produces the osmolyte, glycine betaine, which increases tolerance to environmental stressors. Under exposure to increasing salinity, ambient and high CO_2 concentrations, and a dose of glycine betaine, *A. germinans* seedlings exhibited increased salt tolerance and higher photosynthetic rates (Maurer et al. 2020); under elevated CO_2, the temperature optimum for photosynthesis increased by 4 °C. In a similar experiment, Reef et al. (2015) found confounding effects of salinity and elevated CO_2 concentrations on the survivorship, growth, photosynthesis, root architecture and leaf nutrient and ion concentrations in seedlings of *A. germinans*. The optimum salinity for growth shifted higher with greater CO_2 concentrations, with carbon assimilation rates significantly higher under elevated CO_2. However, at higher salinity, growth declined even at elevated CO_2 levels, although there was great water-use efficiency. This outcome was likely due to non-stomatal limitations to growth at high salinities.

Under conditions of elevated CO_2 concentration (800 ppm) and increased tidal flooding to simulate sea-level rise, Jacotot et al. (2018) found that net photosynthetic rates and water-use efficiency of *A. marina* and *R. stylosa* were enhanced and more pronounced in the warm season, suggesting that predicted temperature increases would further enhance photosynthesis. However, these gains were minimized with longer flooding duration but only by 5%.

Elevated CO_2 also alters the rhizosphere microbiome, as evidenced by growth experiments with *Kandelia obovata* (Yin et al. 2018). Over a period of 20 weeks, elevated CO_2 increased leaf chlorophyll *a* levels and root microbial biomass, with some alteration of ammonia-oxidizing archaea. Further, there was a shift in carbon utilization from the preferred carbon sources of sugars, amino acids, and carboxylic acids under ambient conditions to the use in the order of amino acids > carbohydrates > polymers > carboxylic acids > amines > phenolic acids, indicating a shift in metabolic pathways under elevated CO_2. When subjected to ambient CO_2 and a temperature of 38 °C, *R. apiculata* seedlings responded positively to the higher temperature but elevated CO_2 (650 ppm) enhanced growth only at a lower temperature (21 °C). Under conditions of high CO_2 and temperature, the seedlings nearly perished (Tamimia et al. 2019), indicating complex outcomes to elevated CO_2 concentrations when mangroves are subjected to other drivers such as increasing temperatures and flooding.

These complex responses may, nevertheless, offer a competitive advantage to mangroves as the increasing encroach upon saltmarshes (see next section). Manea et al. (2020) grew the mangroves, *Aegiceras corniculatum* and *A. marina* individually and in a model saltmarsh community under increasing CO_2 and reduced salinity. The mangroves experienced stronger competition from saltmarsh species under elevated CO_2, *A. marina* seedlings produced 48% more biomass under elevated CO_2 when grown in competition with saltmarsh species. *A. corniculatum* was not affected by elevated CO_2 but had 36% greater growth under seawater salinity compared to hypersaline conditions. Rising atmospheric CO_2 concentrations and lower salinity associated with sea-level rise may thus enhance the establishment of mangrove seedlings in saltmarshes, facilitating mangrove encroachment.

9.4 Responses to Increasing Temperature

Physiologically, mangrove responses to increasing temperature, as with most plant and animal species, follow a sigmoid curve with an initial linear rise in respiration, growth and photosynthesis slowing, plateauing, and then declining as a critical lethal threshold is reached. The critical temperature threshold is likely to be species-specific, although rates of leaf photosynthesis for most mangrove species peak at temperatures at or below 30 °C (Alongi 2009). Leaf CO_2 assimilation rates of many species decline as temperature increases from 33 to 35 °C.

Temperature increases alone are likely to result in increased growth, reproduction, photosynthesis and respiration, changes in forest community structure, diversity and are already resulting in an expansion of latitudinal limits in the southern

United States and the Caribbean (Kennedy et al. 2016; Osland et al. 2013, 2020; Cavanaugh et al. 2019; Macy et al. 2019), New Zealand (Stokes et al. 2010; Suyadi et al. 2019), Australia (Rogers et al. 2005; Williamson et al. 2011), southern Africa (de Boer 2002; Peer et al. 2018), southeastern Brazil (Ximenes et al. 2018), and southern China (Durango-Cordero et al. 2013). Limits to the expansion of mangroves are due to the frequency and intensity of extreme cold events (Cavanaugh et al. 2019). For example, an examination of 38 sites spread across the mangrove range in the Gulf of Mexico and Atlantic coasts of North America and found that for *A. germinans* near their northern range limit, the temperature threshold for leaf damage is close to −4 °C with mortality thresholds closer to −7 °C (Osland et al. 2020).

Mangrove expansion may be coupled to changes in precipitation (Wang et al. 2014) as temperature alone does not always delimit the latitudinal range of *Rhizophora* and *Avicennia* because of large regional differences in monthly temperature change, as warmest monthly temperatures are higher at the latitudinal limits in the northern than in the southern hemisphere (Quisthoudt et al. 2012). Mangrove expansion and subsequent saltmarsh contraction are consistent with the poleward increase in temperatures and the frequency and intensity of extreme cold events, but other factors such as extreme weather events, genetic plasticity, impacts of changes in ocean currents on dispersal abilities, and changes in precipitation are likely co-factors (Saintilan et al. 2014; Kennedy et al. 2016; Ximenes et al. 2018).

Rising temperatures and expansion of mangroves further into higher latitudes may impact other mangrove flora and fauna. Rising temperatures also affect animal physiology in different ways. For example, the generalist fiddler crab, *Minuca rapax*, loses less water than the fiddler crab, *Leptuca thayeri* (Principe et al. 2018) due to differences in carapace permeability; survivability is higher for *M. rapax* under desiccated conditions but is more affected by temperature increase on its physiology. The fiddler crab, *Leptuca uruguayensis* is more sensitive to warming than *Leptuca leptodactyla* showing physiological and behavioural differences (da Silva Vianna et al. 2020). The latter species was able to adjust its metabolic rate to temperature increase and reducing ammonia excretion, whereas the former species showed adaptation limits. Fiddler crabs inhabiting vegetated areas are more vulnerable to higher temperatures and may change its geographic range, while fiddler crabs on mud and sandflats are more tolerant to temperature rise and may have a competitive advantage as global temperatures rise. A large-scale mesocosm experiment found that an increase of 1.2 °C leads to the homogenization and flattening of mangrove root epibiont communities, leading to 24% increase in the overall cover of algal epibionts on roots but 33% decline in diversity of epibiont species and a decrease in structural complexity (Walden et al. 2019).

Along the South African coast, fiddler crabs have spread farther south while other fauna such as the gastropod, *Terebralia palustris*, has disappeared, although there does not appear to be a decrease in diversity with an increase in latitude (Peer et al. 2018). The transition to higher latitudes and the expansion of mangroves at the expense of saltmarshes indicates the complex interactions are occurring during the transition, most notably competition. Several studies have indicated that mangrove species that invade saltmarsh areas are superior competitors (Macy et al. 2019; Manea et al. 2020). Migration may be mediated by biotic interactions and may be facilitated by increasing propagule abundance from greater reproductive rates and greater genetic variation due to outcrossing (Proffitt and Travis 2014). Surveying the Atlantic and Gulf coasts of Florida, Proffitt and Travis (2014) found that reproductive frequencies varied significantly but increased with latitude and strongly along the Gulf coast with a concomitant increase in outcrossing. Adaptation to a new environment is perpetuated and promoted by the self-enforcing nature of migration as more colonizers lead to more propagules and outcrossing leads to greater genetic variation.

9.5 Responses to Changes in Precipitation and Extreme Weather Events

Mangrove responses to increasing, decreasing, or more variable precipitation may be less complex, but any responses must be considered with co-occurring drivers such as increasing temperatures, atmospheric CO_2 concentrations, and rises in sea-level. Generally, mangrove forests attain their peak biomass and productivity in the wet tropics due to continually warm temperatures and high rainfall (Alongi 2009). Thus, increases in precipitation will likely result in greater biomass and productivity. Predictions of shifts in tropical rainfall patterns indicate two views: one predicts that wet regions will get wetter and the other view suggests increased rainfall where the rise in sea surface temperature exceeds the mean surface warming in the tropics, i.e. “warmer-get-wetter”. Computer simulations indicate that the pattern of ocean warming induces ascending atmospheric flow at the Equator and subsidence with distance from the Equator, anchoring a band of annual rainfall increase near the Equator that reflects the “warmer-get-wetter” view (Huang et al. 2013). However, the ascending motion oscillates back and forth across the Equator with the Sun, pumping moisture upwards and causing seasonal rainfall anomalies following a “wet-get-wetter” pattern. Thus, seasonal mean rainfall in the tropics combines the “wet-get-wetter” and “warmer-get-wetter” patterns. Recently, rainfall-based thresholds have been identified for mangrove range limits in western North America, western Gulf of Mexico, western South America, western Australia, the Middle East, northwest Africa, east central Africa, and west central Africa (Osland et al. 2017).

Not only will there be changes in precipitation rates and in regional patterns, but the frequency and intensity of extreme weather events, such as cyclonic storms and prolonged droughts, is predicted to increase. Temperature anomalies, defined as extreme temperature events more than three standard deviations from the long-term mean from 1951–1980, have shifted more than one standard deviation towards higher values, leading to more extreme warming events (Hansen et al. 2012). The increased occurrence of such events is having a dramatic impact on mangroves, such as mass mortality events (Lovelock et al. 2017). This shift towards extreme temperature events has also resulted in a concomitant decrease in extreme freeze events (Hayes et al. 2020) resulting in a shift favouring mangrove expansion within the mangrove-marsh ecotone and expanding their range poleward, as studied most extensively along the Gulf of Mexico and the Florida Atlantic coastline (Cavanaugh et al. 2014; Saintilan et al. 2014; Osland et al. 2020). The clearest example of mangrove mortality due to an extreme weather event is the massive dieback of mangroves in the Gulf of Carpentaria in northern Australia (Duke et al. 2017; Sippo et al. 2018, 2020). Mangrove forests in the gulf suffered severe dieback (6% of vegetation) during the summer of 2015–2016 along 1000 km of shoreline. The onset of the dieback was coincident with unprecedented high temperatures, low precipitation, and lack of the normal summer wet season (Duke et al. 2017). An unusually lengthy severe drought coupled with a temporary drop in sea-level contributed to mass mortality. The dieback had severe consequences for mangrove functioning as evidenced by a shift from a dominance of oceanic carbon outwelling to increased atmospheric CO_2 emissions and decreased alkalinity exports (Sippo et al. 2020); this shift was likely driven by reduced mangrove productivity and increased oxygen soil permeation. Dieback of mangroves in the Gulf of Carpentaria also resulted in a trophic shift in crabs with a loss of litter-feeding crabs but an increase in crabs that feed on microphytobenthos; infauna was unaffected (Harada et al. 2019). An examination of data since the 1960s showed about 36,000 ha of mangrove mortality worldwide, about 70% of which was attributed to low frequency, high intensity weather events (typhoons, cyclones, hurricanes) and climate extremes (Sippo et al. 2018). Tropical cyclonic storms account for 45% of the reported mangrove mortality area since the 1960s, being the largest cause of mortality. Recent large-scale mortality, however, associated with extreme climatic events in Australia accounts for 22% of all reported forest loss over the past six decades, suggesting the increasing importance of extreme climatic events (Sippo et al. 2018). In Mangrove Bay in north Western Australia, there have been two dieback events over a 16-year period, with the most recent one coincident with the dieback in the Gulf of Carpentaria (Lovelock et al. 2017). The diebacks in Mangrove Bay were coincident with periods of very low sea-level due to intensification of El Niño-Southern Oscillation (ENSO) leading to increased soil salinities and subsequent canopy loss and reduced recruitment. In June 2016, a hailstorm with winds over 100 km h^{-1} caused mangrove dieback in eastern Brazil (Servino et al. 2018). The losses were extensive, corresponding to 29% of total forest area; 15-mo later some areas of dieback showed initial recovery while others continued to degrade. Recovery was not helped by the El Niño creating mild drought conditions. The impact of any one cyclonic disturbance is often significant, with some stands taking decades to fully recover, as evidenced from mangrove forests in the Caribbean and Florida (Imbert 2018; Rivera-Monroy et al. 2019); such climatic disturbances prevent these ecosystems from ever reaching a steady-state. However, not all mangroves are significantly affected by extreme climate events as found for mangroves on a Colombian Caribbean Island that appear to be resilient to short drought events related to ENSO (Galeano et al. 2017). At the other extreme, mangroves such as *A. germinans* may be resilient to extreme freeze events due to genetically based freeze tolerances (Hayes et al. 2020). Nevertheless, increasing mortality events are likely in future in light of forecasts of increasing frequency, intensity, and destructiveness of cyclonic storms and climatic extremes, such as heat waves and low and high sea-level episodes.

9.6 Responses to Coastal Ocean Acidification

Ocean acidification occurs because the global ocean takes up about one third of the atmospheric carbon released from fossil fuel combustion, cement production, and land-use change, with the subsequent hydrolysis of increasing amounts of CO_2 in seawater increasing the hydrogen ion concentration thereby reducing the pH of ocean water and causing wholesale shifts in seawater carbonate chemistry (Doney et al. 2009). Surface ocean Ph has declined since preindustrial times by about 0.1 unit and is predicted to decline a further 0.3–0.4 units by later this century. Acidification in the coastal ocean is a more complex process as carbonate chemistry in estuarine and coastal waters is strongly regulated by changes in biological activity related to increases in anthropogenic delivery of nutrients by rivers, groundwater, and eutrophication (Borges and Gypens 2010). Eutrophication strongly amplifies acidification due to the accumulation of algal biomass and its subsequent decomposition, decreasing dissolved oxygen (DO) levels and contributing to hypoxia (Wallace et al. 2014). Hypoxia increases pCO_2 values and upwelling processes, if any, can bring CO_2^- enriched water to coastal waters, amplifying the effects of acidification.

Acidification has a significant impact on tropical marine life, with several marine organisms of some taxonomic groups showing decreased growth and physiological tolerance to lower pH. These groups include tropical hermatypic corals which show a significant decline in calcification, the process involving the formation of calcium carbonate skeletons. Other calcifying organisms such as photosynthetic calcareous algae, photosynthetic symbiont-bearing foraminifera, and some species of molluscs, jellyfishes, fishes, echinoderms, and pteropods similarly show a decline in calcification as well as shell dissolution; reduction in shell mass; growth reductions; reduced metabolism, fertility, and embryo development; increased mortality; reduced thermal tolerance; reduced food intake and increased ventilation (Hofmann et al. 2010).

Not all estuarine, coastal, and shelf organisms exhibit a negative response to acidification, as many organisms show either a positive or mixed response to acidification. Seagrasses, brown macroalgae, kelps, sea anemones, fishes, and most non-calcifying organisms exhibit positive responses, while some Caribbean corals, plankton, fleshy macroalgae, upright calcareous algae, and crustose coralline algae show contrasting responses (negative, positive, no effect), depending on species-specific tolerances and the degree of acidification (Alongi 2020).

Seagrasses and mangroves will be the most resilient ecosystems to the effects of coastal acidification. Several seagrass species usually respond positively, or not at all, to lower pH by their capacity to modify pH within canopies and within their habitat. Mangrove ecosystems may prove to be the most resistant and resilient in the face of coastal acidification. As detailed by Alongi (2020), the pH of mangrove soils is ordinarily low, within the range of 4–7, especially in the mangroves of South and

Southeast Asia and Africa. Mangrove soils have low pH due to high rates of bacterial respiration, high concentrations of polyphenolic acids, and the net effects of metabolic processes associated with the trees and their root systems.

Nearly all estuarine and nearshore waters in the tropics, including mangrove tidal creek waters and other waterways, naturally exhibit very wide variations in salinity (range: 0.1–48), pH (range: 4–9), pCO_2 (range: 4–32,763μatm), and $[CO_3^{2-}]$ and are a strong source of CO_2 emissions to the atmosphere due to pCO_2 and $[CO_3^{2-}]$ oversaturation (see Table 4 in Alongi 2020). Oversaturation and highly variable pH are the net result of high rates of bacterial respiration, eutrophication, and the influence of fluvial discharge, including export of alkalinity, organic matter and dissolved inorganic carbon (DIC), deposition of anthropogenic acids and bases, intense weathering, land-use change, acid sulphate soil discharge, and acidic groundwater. These chemical factors predispose mangroves and other coastal ecosystems to resilience to coastal acidification.

Mangroves are apparent buffers of acidification in the tropical coastal zone (Sippo et al. 2016). Carbon (DIC, dissolved CO_2) and alkalinity dynamics measured in six Australian mangrove tidal creeks showed a mean export of DIC, but alkalinity fluxes ranged from a small import and a large export; a net import of free CO_2 was measured, equivalent to one third of the estimated air–water CO_2 flux. Sippo et al. (2016) upscaled these results globally and showed that mangrove alkalinity exports are equivalent to about 14% of global river or continental shelf alkalinity fluxes. The effect of DIC and alkalinity exports creates a measurable increase in pH indicating that mangroves partly counteract acidification in adjacent tropical coastal waters. Mangroves may thus be one of the largest sources of alkalinity to the tropical coastal ocean, providing buffering against acidification.

9.7 Predictions and Conclusions

Considerable uncertainty exists in predicting mangrove responses to climate change, but reasonable prognostications can be made based on our considerable knowledge of current mangrove responses to temperature, current sea-level rise, salinity, rainfall, and past and present impacts of tropical storm systems and extreme drought. Likely regional responses to projected changes in climate are listed in Table 9.1. Acidification is not included as the predicted decline in coastal ocean pH is within the boundaries of current pH variability in mangrove waters and soils. The likely scenarios assume that mangrove responses will include complex changes in floral and faunal species composition, morphology, biodiversity, biomass, physiology, growth and productivity, and that functionality (growth, respiration, productivity, etc.) will increase with increases in temperature, precipitation, and atmospheric CO_2 concentrations up to the critical physiological thresholds of individual species, but only when other environmental conditions are favourable.

Regionally, mangrove forests will continue to expand latitudinally along the Gulf of Mexico, the east coast of Florida, the coasts of South Africa, eastern Australia, northern New Zealand, southeastern Brazil, and China. On low islands in the Pacific

Table 9.1 Predicted responses of mangrove ecosystems to expected climate change impacts by 2100

Region	Sea-level rise (SLR) (mm a^{-1})	Temperature increase	Precipitation change	Salinity change	Tropical cyclone/ extreme events increase	Responses
North America	3.0–5.1	Very likely	Increase very likely	No change	Likely	Continued latitudinal expansion along Gulf of Mexico coast as °C and [CO_2] increase. Increase in damage/ destruction from increased frequency of hurricanes.
Africa	1.1–3.8	Very likely	Increase in North and South, decrease in West	Decrease (0–2.0) in Central West and East Africa	No change	Mangroves continue poleward expansion in South Africa. Most mangroves highly fragmented so at higher risk of losses due to continued deforestation and degradation.
Middle East	2.2–3.3	Very likely (up to 7 °C)	Likely decrease	Increase (0–2.0)	No Change	Landward expansion/migration unlikely. Likely losses in Red Sea and the Levant due to high °C, less rain, higher salinity.
Caribbean and Central America	1.0–2.5	Very likely	Increase very likely	Increase (0.5–1.0)	Likely	Most low island mangroves not keeping pace with current SLR (Fig. 9.1) but predicted increase in rainfall during ENSO likely to increase sediment delivery along S Caribbean coast of Central America. Arid-zone mangroves likely to decline as salinities and °C increase, especially both Mexico coasts.
South Asia	1.0–1.4	Very likely	Increase very likely in India and Sri Lanka	Increase (0–0.5)	Likely	Arid-zone mangroves and those in Indus delta of Pakistan likely to decline due to low rainfall, subsidence, and

						low sediment delivery. E India vulnerable due to low tidal range, subsidence, and increased strength and frequency of cyclones, especially the Sundarbans.
Pacific Islands	1.4–2.0	Very likely	Increase very likely	No change	Likely	Low islands of Oceania highly vulnerable to SLR due to lack of upland space for landward migration.
East Asia	1.4–3.4	Very likely	Increase very likely	Increase (0–0.5)	Likely	S China mangroves vulnerable to predicted SLR, low sediment input, and increased strength and frequency of typhoons.
Australia and New Zealand	2.0–3.8	Very likely	Increase very likely	No change	Likely	Most mangroves in SE Queensland and SE Australian coast not keeping pace with current SLR (Fig. 9.1) but continued encroachment into salt marshes. Mangroves in NW Australia likely to decline due to increasing aridity and SLR. New Zealand mangroves likely to continue to expand on north island as °C warm.
Southeast Asia	2.3–4.5	Very likely	Increase very likely	Decrease (0–1.0)	Likely	E Philippine mangroves vulnerable to increased strength and frequency of typhoons. E Sumatra, Sulawesi, and S Vietnam vulnerable to predicted SLR because of low tidal range. River delta mangroves (e.g. Mekong, Ayeyarwaddy deltas) likely to decline with SLR, subsidence, and decreased sediment supply from damming and river modification.

(continued)

Table 9.1 (continued)

Region	Sea-level rise (SLR) (mm a^{-1})	Temperature increase	Precipitation change	Salinity change	Tropical cyclone/ extreme events increase	Responses
South America	2.0–3.5	Very likely	Very likely increase in west and north, decrease in SE	Increase (0–1.0)	Likely	South American mangroves along north coast unlikely to be impacted as increased rainfall will result in increased sediment supply (e.g. Amazon, Orinoco Rivers). SE Brazil mangroves likely to expand south due to increasing °C. Mangroves on Pacific coast of South America likely to continue increasing stand size, but no latitudinal extension due to cold Peruvian current and arid conditions.

Sources: IPCC (2013, 2014), Alongi (2015), Gilman et al. (2016), He et al. (2016), Ward et al. (2016), Wilson (2017), Amuzu et al. (2018), Mukhopadhyay et al. (2018), Schuerch et al. (2018), Mafi-Gholami et al. (2020)

and in the Caribbean, mangrove will likely decline as sea-level rise traps mangroves that have no or little upland space to colonize. In the arid tropical regions of northwest Australia, Pakistan, the Arabian Peninsula, and the coasts of Mexico, mangrove forest will likely decline due to increased aridity as salinities increase, freshwater becomes scarcer, and critical temperature thresholds are reached more frequently. Mangroves will likely decline in tropical river deltas subject to subsidence such as the Sundarbans, Mekong, Zaire, Ayeyarwaddy, and Fly Rivers, as sediment and freshwater delivery declines due to damming and alterations to river flow and as sea-level rises. In areas that experience tropical storms or other extreme events, such as in northern Australia, the Gulf of Mexico, the east India coast, Florida, the Philippines, south China coast, and the northern Caribbean, mangroves will likely suffer increased damage, destruction, and loss. Some areas of Southeast Asia, such as the east coast of the island of Sumatra, north coast of Java, Sulawesi, and southern Vietnam are likely to lose mangroves because of their low tidal range. In contrast, mangroves in other parts of Southeast Asia may response positively to increased rainfall and warmer temperatures (e.g. west coast of Peninsular Malaysia, southwest coast of Thailand). Mangroves will likely respond positively or exhibit no significant change along the tropical west and north coasts of South America and the Caribbean and Pacific coasts of Central America (e.g. Costa Rica).

Mangroves are likely to survive further into the Anthropocene, but most impacts will be negative rather than positive. However, the greatest threat to mangroves continues to be, and will likely remain, deforestation and degradation which must be considered with climate change impacts.

References

Albert S, Saunders MI, Roelfsema, Leon JX, Johnstone E, Mackenzie JR, Hoegh-Guldberg O, Grinham AR, Phinn SR, Duke NC, Mumby PJ, Kovacs E, Woodroffe CD (2017) Winners and losers as mangrove, coral and seagrass ecosystems respond to sea-level rise in Solomon Islands. Environ Res Lett 12: 094009, https://doi.org/10.1088/1748-9326/aa7e68.

Almahasheer H, Serrano O, Duarte CM, Arias-Ortiz A, Masque P, Irigoien X (2017) Low carbon sink capacity of Red Sea mangroves. Sci Rept 7:9700, https://doi.org/10.1038/s41598-017-10424-9.

Alongi DM (2008) Mangrove forests: resilience, protection from tsunamis, and responses to global climate change. Estuar Coast Shelf Sci 76:1–13.

Alongi DM (2009) The energetics of mangrove forests. Springer, Netherlands.

Alongi DM (2012) Carbon sequestration in mangrove forests. Carbon Manage 3:313–322.

Alongi DM (2015) The impact of climate change on mangrove forests. Curr Clim Change Rep 1: 30–39.

Alongi DM (2016) Mangroves. In: Kennish M (ed) Encyclopedia of estuaries, Springer-Verlag, Berlin, p 393–404.

Alongi DM (2018) Blue carbon: coastal sequestration for climate change mitigation. Springer Briefs in Climate Studies, Springer Nature, Cham Switzerland.

Alongi DM (2020) Vulnerability and resilience of tropical coastal ecosystems to ocean acidification. Examines Mar Biol Oceanogr 3, https://doi.org/10.313031/EIMBO.2020.03.000562.

Amuzu J, Jallow BP, Kabo-Bah AT, Yaffa S (2018) The climate change vulnerability and risk management matrix for the coastal zone of the Gambia. Hydrology 5: 14, 10.10.3390/hydrology5010014.

Ball MC, Cochrane MJ, Rawson HM (1997) Growth and water use of the mangroves *Rhizophora apiculata* and *R. stylosa* in response to salinity and humidity under ambient and elevated concentrations of atmospheric CO_2. Plant Cell Environ 20: 1158–1166.

Berger JF, Charpentier V, Crassard R, Martin C, Davtian G, Lopez-Sáez JA (2013) The dynamics of mangrove ecosystems, changes in sea level and the strategies of Neolithic settlements along the coast of Oman (6000-3000 cal. B.C.). J Archaeol Sci 40:3087–3104.

Bomer EJ, Wilson CA, Hale RP, Hossain AN, Rahman FA (2020) Surface elevation and sedimentation dynamics in the Ganges-Brahmaputra tidal delta plain, Bangladesh: evidence for mangrove adaptation to human-induced tidal amplification. Catena 187:104312, https://doi.org/10.1016/j.catena.2019.104312.

Borges AV, Gypens N (2010) Carbonate chemistry in the coastal zone responds more strongly to eutrophication than to ocean acidification. Limnol Oceanogr 55: 346–353.

Cardona-Olarte P, Twilley RR, Krauss KW, Rivera-Monroy V (2006) Responses of neotropical mangrove seedlings grown in monoculture and mixed culture under treatments of hydroperiod and salinity. Hydrobiol 569; 325–391,

Carnero-Bravo V, Sanchez-Cabeza JA, Ruiz-Fernández AC, Merino-Ibarra M, Hillaire-Marcel C, Corcho-Alvarado JA, Röllin S, Diaz-Asencio M, Cardoso-Mohedano JG, Zavala-Hidalgo J (2016) Sedimentary records of recent sea-level rise and acceleration in the Yucatan Peninsula. Sci Tot Environ 573:1063–1069.

Carnero-Bravo V, Sanchez-Cabeza JA, Ruiz-Fernández AC, Merino-Ibarra M, Corcho-Alvarado JA, Sahli H, Hélie JF, Preda M, Zavala-Hidalgo J, Díaz-Asencio M, Hillaire-Marcel C (2018) Sea-level rise sedimentary record and organic carbon fluxes in a low-lying tropical coastal ecosystem. Catena 162:421–430.

Cavanaugh KC, Kellner JR, Forde AJ, Gruner DS, Parker JD, Rodriguez W, Feller IC (2014) Poleward expansion of mangroves is a threshold response to decreased frequency of extreme cold events. Proc Nat Acad Sci 111:723–727.

Cavanaugh KC, Dangremond EM, Doughty CL, Williams AP, Parker JD, Hayes MA, Rodriguez W, Feller IC (2019) Climate-driven regime shifts in a mangrove-salt marsh ecotone over the past 250 years. Proc Nat Acad Sci 116:21602–21608.

Chappel AR (2018) Soil accumulation, accretion, and organic carbon burial rates in mangrove soils of the Lower Florida Keys: a temporal and spatial analysis. MS Thesis, Univ S Florida, St Petersburg, USA.

Cherry JA, McKee KL, Grace JB (2009) Elevated CO2 enhances biological contributions to elevation change in coastal wetlands by offsetting stressors associated with sea-level rise. J Ecol 7:67–77.

Cinco-Castro S, Herrera-Silveira J (2020) Vulnerability of mangrove ecosystem to climate change effects: the case of the Yucatan Peninsula. Ocean Coast Manage 192: 105196, https://doi.org/10.1016/j.ocecoaman.2020.105196.

Chen L, Wang W (2017) Ecophysiological responses of viviparous mangrove *Rhizophora stylosa* seedlings to simulated sea-level rise. J Coast Res 33: 1333–1340.

Cusack M, Saderne V, Arias-Ortiz A, Masque P, Krishnakumar PK, Rabaoui L, Qurban MA, Qasem AM, Prihartato P, Loughland RA, Elyas AA (2018) Organic carbon sequestration and storage in vegetated coastal habitats along the western coast of the Arabian Gulf. Environ Res Lett 13:074007, https://doi.org/10.1088/1748-9326/aac899.

de Boer WF (2002) The rise and fall of the mangrove forests in Maputo Bay, Mozambique. Wetlands Ecol Manage 10:313–322.

de Silva Vianna B, Miyai CA, Augusto A, Costa TM (2020) Effects of temperature increase on the physiology and behavior of fiddler crabs. Physiol Beh 215; 112765, https://doi.org/10.1016/j.physbeh.2019.112765.

Doney SC, Fabry VJ, Feely RA, Kleypas JA (2009) Ocean acidification: the other CO_2 problem. Annu Rev Mar Sci 1: 169–192.

Duke NC, Ball MC, Ellison JC (1998) Factors influencing biodiversity and distributional gradients in mangroves. Glob Ecol Biogeogr Lett 7: 27–47.

Duke NC, Kovacs JM, Griffiths AD, Preece L, Hill DJE, van Oosterzee P, Mackenzie J, Morning HS, Burrows D (2017) Large-scale dieback of mangroves in Australia's Gulf of Carpentaria: a severe ecosystem response, coincidental with an unusually extreme weather event. Mar Freshw Res 68: 1816–1829.

Durango-Cordero JS, Satyanarayana B, Zhang J, Wang J, Chen M, Fanghong X, Chan JC, Kangying L, Bogaert J, Koedam N, Dahdouh-Guebas F (2013) Vegetation structure at Zhangiang Mangrove National Nature Reserve (ZMMNR), PR China: a comparison between original and non-original trees using ground-truthing, remote sensing and GIS techniques. http://www.vliz.be/imisdocs/publications/232700.pdf.

Ellison AM, Farnsworth EJ (1997) Simulated sea level change alters anatomy, physiology, growth, and reproduction of red mangrove (*Rhizophora mangle* L.). Oecologia 112: 435–446.

Endfield GH, Marks RB (2012) Historical environmental change in the tropics. In: Metcalf SE, Nash DJ (eds) Quaternary environmental changes in the tropics, Wiley, Chichester, p 360–391.

Farnsworth EJ, Ellison AM, Gong WK (1996) Elevated CO_2 alters anatomy, physiology, growth, and reproduction of red mangrove (*Rhizophora mangle* L.). Oecologia 108: 599–609.

Fromard F, Puig H, Mougin E, Marty G, Betoulle A, Cadamuro L (1998) Structure, above-ground biomass, and dynamics of mangrove ecosystems: new data from French Guiana. Oecologia 115: 39–53.

Fu H, Zhang Y, Ao X, Wang W, Wang M (2019) High surface elevation gains and prediction of mangrove responses to sea-level rise based on dynamic surface elevation changes at Dongzhaigang Bay, China. Geomorphol 334:194–202.

Galeano, A., Urrego, L.E., Botero, V. and Bernal, G. (2017). Mangrove resilience to climate extreme events in a Colombian Caribbean Island. Wetlands Ecology and Management 25:743–760

Gilman E, Van Lavieren H, Ellison J, Jungblut V, Wilson L, Areki F, Brighouse G, Bungitak J, Dus E, Henry M, Sauni Jr I, Kilman M, Matthews E, Teariki-Ruatu N, Tukia S, Yuknavage K (2016) Pacific Island mangroves in a changing climate and rising sea. UNEP Regional Seas Reports and Studies No. 179. UNEP, Regional Seas Programme, Nairobi, Kenya.

Hale RP, Wilson CA, Bomer EJ (2019) Seasonal variability of forces controlling sedimentation in the Sundarbans National Forest, Bangladesh. Front Earth Sci 7:211, https://doi.org/10.3389/feart.2019.00211.

Hanebuth TJJ, Voris HK, Yokoyama Y, Saito Y, Okuno J (2011) Formation and fate of sedimentary depocentres on Southeast Asia's Sunda Shelf over the past sea-level cycle and biogeographic implications. Earth Sci Rev 104: 92–110.

Hanebuth TJ, Kudrass HR, Lindstader J, Islam B, Zander AM (2013) Rapid coastal subsidence in the central Ganges-Brahmaputra Delta (Bangadesh) since the 17th century deduced from submerged salt-producing kilns. Geol 41:987–990.

Hansen J, Sato M, Ruedy R (2012) Perception of climate change. Proc Nat Acad Sci 109: E2415–E2423.

Hapsari KA, Jennerjahn TC, Lukas MC, Karius V, Behling H (2020) Intertwined effects of climate and land use change on environmental dynamics and carbon accumulation in a mangrove-fringed coastal lagoon in Java, Indonesia. Glob Change Biol 26: 1414–1431.

Harada Y, Fry B, Lee SY, Maher DT, Sippo JZ Connolly RM (2019) Stable isotopes indicate ecosystem restructuring following climate-driven mangrove dieback. Limnol Oceanogr 65: 1251–1263.

Hatje V, Masque P, Patire VF, Dorea A, Barros F (2021) Blue carbon stocks, accumulation rates, and associated spatial variability in Brazilian mangroves. Limnol Oceanogr 66:321–334.

Hayden HL, Granek EF (2015) Coastal sediment elevation change following anthropogenic mangrove clearing. Estuar Coast Shelf Sci 165:70–74.

Hayes MA, Shor AC, Jesse A, Miller C, Kennedy JP, Feller I (2020) The role of glycine betaine in range expansions; protecting mangroves against extreme freeze events. J Ecol 108: 61–69.
He YH, Mok HY, Lai EST (2016) Projection of sea-level change in the vicinity of Hong Kong in the 21st century. Intl J Climat 36: 3237–3244.
Hofmann GE, Barry JP, Edmunds PJ, Gates RD, Hutchins DA, Klinger T, Sewell MA (2010) The effect of ocean acidification on calcifying organisms in marine ecosystems: an organism to ecosystem perspective. Annu Rev Mar Sci 41: 127–147
Hoque MM, Abu Hena MK, Ahmed OH, Idris MH, Rafiqul HA, Billah MM (2015) Can mangroves help combat sea-level rise through sediment accretion and accumulation? Malay J Sci 34:78–86.
Huang P, Xie S-P, Hu K, Huang G, Huang R (2013) Patterns of the seasonal response of tropical rainfall to global warming. Nature Geosci 6: 357–361.
Imbert D (2018) Hurricane disturbance and forest dynamics in east Caribbean mangroves. Ecosphere 9: e02231, https://doi.org/10.1002/ecs2.2231.
IPCC (2013) Climate change 2013: the physical science basis. Cambridge University Press, New York, USA.
IPCC (2014) Climate change 2014: impacts, adaptation, and vulnerability. Cambridge University Press, New York, USA.
Jacotot A, Gensous S, Marchand C, Allenbach M (2018) Effects of elevated atmospheric CO_2 and increased tidal flooding on leaf gas-exchange parameters of two common mangrove species: *Avicennia marina* and *Rhizophora stylosa*. Photosynth Res 138: 249–260.
Jennerjahn, TC, Gilman E, Krauss KW, Lacerda LD, Nordhaus I, Wolanski E (2017) Mangrove ecosystems under climate change. In: Rivera-Monroy VH, Lee SY, Kristensen E, Twilley RR (eds), Mangrove ecosystems: a global biogeographic perspective, Springer, Cham, Switzerland, p 211–244.
Jevrejeva S, Frederikse T, Kopp RE, Cozannet GL, Jackson LP, van de Wal RSW(2019) Probabilistic sea level projections at the coast by 2100. Survey Geophys 40: 1673–1696.
Kennedy JP, Pil MW, Proffitt CE, Boeger WA, Stanford AM, Devlin DJ (2016) Postglacial expansion pathways of red mangrove, *Rhizophora mangle,* in the Caribbean Basin and Florida. Am J Bot 103: 260–276.
Kim J-H, Dupont L, Behling H, Versteegh GJM (2005) Impacts of rapid sea-level rise on mangrove deposit erosion: application of teraxeroland *Rhizophora* records. J. Quatern Sci 20: 221–225.
Krauss KW, Cahoon DR, Allen JA, Ewel KC, Lynch JC, Cormier N (2010) Surface elevation change and susceptibility of different mangrove zones to sea-level rise on Pacific high islands of Micronesia. Ecosystems 13: 129–143.
Krauss KW, Cormier N, Osland MJ, Kirwan ML, Stagg CL, Nestlerode JA, Russell MJ, From AS, Spivak AC, Dantin DD, Harvey JE, Almario AE (2017) Created mangrove wetlands store belowground carbon and surface elevation change enables them to adjust to sea-level rise. Sci Rept 7: 1030, https://doi.org/10.1038/s41598-017-01224-2.
Kusumaningtyas MA, Hutahaean AA, Fischer HW, Pérez-Mayo M, Ransby D, Jennerjahn TC (2019) Variability in the organic carbon stocks, sources, and accumulation rates of Indonesian mangrove ecosystems. Estuar Coast Shelf Sci 218:310–323.
Lagomasino D, Fatoyinbo T, Lee S, Feliciano E, Trettin, Shapiro A, Mangora MM (2019) Measuring mangrove carbon loss and gain in deltas. Environ Res Lett 14:025002, https:doi.org/10.1088/1748-9326/aaf0de.
Li Z, Saito Y, Mao L, Tamura T, Li Z, Song B, Zhang Y, Lu A, Sieng S, Li J (2012) Mid-Holocene mangrove succession and its response to sea-level change in the upper Mekong River delta, Cambodia. Quatern Res 78:386–399.
Limaye RB, Kumaran KPN, Padmalal D (2014) Mangrove habitat dynamics in response to Holocene sea level and climate changes along southwest coast of India. Quatern Intl 325: 116–125.
Lopez-Medellin X, Ezcurra E, González-Abraham C, Hak J, Santiago LS, Sickman JO (2011) Oceanographic anomalies and sea level rise drive mangroves inland in the Pacific coast of Mexico. J Veg Sci 22: 143–151.

Lovelock CE, Cahoon DR, Friess DA, Guntenspergen GR, Krauss KW, Reef R, Rogers K, Saunders ML, Sidik F, Swales A, Saintilan N, Thuyen LX, Triet T (2015) The vulnerability of Indo-Pacific mangrove forests to sea-level rise. Nature 526:559–563.

Lovelock, CE, Feller IC, Reef R, Hickey S, Ball MC (2017) Mangrove dieback during fluctuating sea levels. Sci Rept 7: 1680, https://doi.org/10.1038/s41598-017-01927-6.

Lu WZ, Chen LZ, Wang WQ, Tam NFY, Lin GH (2013) Effects of sea level rise on mangrove *Avicennia* population growth, colonization and establishment: evidence from a field survey and greenhouse manipulation experiment. Acta Oecol 49: 83–91.

Lugo AE (1980) Mangrove ecosystems: successional or steady state? Biotropica 12: 65–72.

Matos CR, Berredo JF, Machado W, Sanders CJ, Metzger E, Cohen MC (2020) Carbon and nutrient accumulation in tropical mangrove creeks, Amazon region. Mar Geol 429:106317, https://doi.org/10.1016/j.margeo.2020.106317.

MacKenzie RA, Foulk PB, Klump JV, Weckerly K, Purbospito J, Murdiyarso D, Donato DC, Nam VN (2016) Sedimentation and belowground carbon accumulation rates in mangrove forests that differ in diversity and land use: a tale of two mangroves. Wetlands Ecol Manage 24:245–261.

Macy A, Sharma S, Sparks E, Goff J, Heck KL, Johnson MW, Harper P, Cebrian J (2019) Tropicalization of the barrier islands of the northern Gulf of Mexico: a comparison of herbivory and decomposition rates between smooth cordgrass (*Spartina alterniflora*) and black mangrove (*Avicennia germinans*). PLoS ONE 14: e0210144, https://doi.org/10.1371/journal.pone.0210144.

Mafi-Gholami D, Zenner, EK, Jaafari A (2020) Mangrove regional feedback to sea level rise and drought intensity at the end of the 21st century. Ecol Ind 110: 105972, https://doi.org/10.1016/j.ecolind.2019.105972.

Manea A, Geedicke I, Leishman MR (2020) Elevated carbon dioxide and reduced salinity enhance mangrove seedling establishment in an artificial saltmarsh community. Oecologia 192: 273–280.

Mangora MM, Mtolera MSP, Björk M (2014) Photosynthetic responses to submergence in mangrove seedlings. Mar Freshw Res 65: 497–504.

Marchio DA, Savarese M, Bovard B, Mitsch WJ (2016) Carbon sequestration and sedimentation in mangrove swamps influenced by hydrogeomorphic conditions and urbanization in Southwest Florida. Forests 7:116, 10.30116.390/f706.

Maurer R, Tapis ME, Shor AC (2020) Exogenous root uptake of glycine betaine mitigates improved tolerance to salinity stress in *Avicennia germinans* under ambient and elevated CO_2 conditions. FASEB J 34, https://doi.org/10.1096/fasebj.2020.34.s1.07221.

McCloskey TA, Liu K-B (2013) Sedimentary history of mangrove cays in Turneffe Islands, Belize: evidence for sudden environmental reversals. J Coast Res 29: 971–983.

McIvor A, Spencer T, Möller I, Spalding M (2013) The response of mangrove soil surface elevation to sea level rise. Natural Coastal Protection Series: Report 3. Cambridge Coastal Research Unit Working Paper 42. The Nature Conservancy and Wetlands International.

McKee KL, Cahoon DR, Feller IC (2007) Caribbean mangroves adjust to rising sea level through biotic controls on change in soil elevation. Glob Ecol Biogeogr 16: 545–556.

McKee KL, Rooth JE (2008) Where temperate meets tropical: multi-factorial effects of elevated CO2, nitrogen enrichment, and competition on a mangrove-salt marsh community. Glob Change Biol 14:971–984.

McKee K, Rogers K, Saintilan N (2012) Response of salt marsh and mangrove wetlands to changes in atmospheric CO 2, climate, and sea level. In Middleton BA (ed), Global change and the function and distribution of wetlands. Springer, Dordrecht, p 63–96.

Mukhopadhyay A, Payo A, Chanda A, Ghosh T, Chowdhury SM, Hazra S (2018) Dynamics of the Sundarbans mangroves in Bangladesh under climate change. In Nicholls RJ, Hutton CW, Adgar WN, Hanson SE, Munsur Rahman Md, Salekin M (eds), Ecosystem services for well-being in deltas. Palgrave Macmillan, Cham, Switzerland, p 489–503.

Murdiyarso D, Hanggara BB, Lubis AA (2018) Sedimentation and soil carbon accumulation in degraded mangrove forests of North Sumatra, Indonesia. BioRxiv:325191, https://doi.org/10.1101/325191.

Osland MJ, Enwright N, Day RH, Doyle TW (2013) Winter climate change and coastal wetland foundation species: salt marshes vs. mangrove forests in the southeastern United States. Glob Change Biol 19:1482–1494.

Osland M, Feher L, Griffith K, Cavanaugh K, Enwright N (2017) Climatic controls on the global distribution, abundance, and species richness of mangrove forests. Ecol Monogr 87: 341–359.

Osland M, Day RH, Hall CT, Feher LC, Armitage AR, Cebrian J, Dunton KH, Hughes AR, Kaplan DA, Langston AK, Macy A, Weaver CA, Anderson GH, Cummins K, Feller IC, Snyder CM (2020) Temperature thresholds for black mangrove (*Avicennia germinans*) freeze damage, mortality and recovery in North America: refining tipping points for range expansion in a warming climate. J Ecol 108: 654–665.

Passos T, Penny D, Sanders C, De Franca E, Oliveira T, Santos L, Barcellos R (2021) Mangrove carbon and nutrient accumulation shifts driven by rapid development in a tropical estuarine system, northeast Brazil. Mar Pollut Bull 166:112219, https://doi.org/https://doi.org/10.1016/j.marpolbul.2021.112219.

Peer N, Rajkaren A, Miranda NAF, Taylor RH, Newman B, Porri (2018) Latitudinal gradients and poleward expansion of mangrove ecosystems in South Africa; 50 years after Macnae's first assessment. African J Mar Sci 40: 101–120.

Pérez A, Libardoni BG, Sanders CJ (2018) Factors influencing organic carbon accumulation in mangrove ecosystems. Biol Lett 14:20180237, https://doi.org/10.1098/rsbl.2018.0237.

Phillips DH, Kumara MP, Jayatissa LP, Krauss KW, Huxham M (2017) Impacts of mangrove density on surface sediment accretion, belowground biomass and biogeochemistry in Puttalam Lagoon, Sri Lanka. Wetlands 37:471–483.

Principe SC, Augusto A, Costa TM (2018) Differential effects of water loss and temperature increase on the physiology of fiddler crabs from distinct habitats. J Therm Biol 73: 14–23.

Proffitt CE, Travis S (2014) Red mangrove life history variables along latitudinal and anthropogenic stress gradients. Ecol Evol 4:2352–2359.

Punwong P, Marchant R, Selby K (2013) Holocene mangrove dynamics in Makoba Bay, Zanzibar. Paleogeogr Paleoclim Paleoecol 379–380: 54–67.

Quisthoudt K, Schmitz N, Randin CF, Dahdouh-Guebas F, Robert EM, Koedam N (2012) Temperature variation among mangrove latitudinal range limits worldwide. Trees 26:1919–1931.

Rahu A, Das S, Banerjee K, Mitra A (2012) Climate change impacts on Indian Sundarbans: a time series analysis (1924–2008). Biodivers Conserv 21: 1289–1307.

Rashid T (2014) Holocene sea-level scenarios in Bangladesh. Springer Briefs in Oceanography. https://doi.org/10.1007/978-981-4560-99-3.

Reef R, Lovelock CE (2014) Historical analysis on mangrove leaf traits throughout the 19th and 20th centuries reveal differential responses to increases in atmospheric CO_2. Glob Ecol Biogeogr 23: 1209–1214.

Reef R, Winter K, Morales J, Adame MF, Reef DL, Lovelock CE (2015) The effect of atmospheric carbon dioxide concentrations on the performance of the mangrove *Avicennia germinans* over a range of salinities. Physiol Plant 154: 358–368.

Reef R, Slot M, Motro U, Motro M, Motro Y, Adame MF, Garcia M, Aranda J, Lovelock CE, Winter K (2016) The effects of CO_2 and nutrient fertilization on the growth and temperature response of the mangrove *Avicennia germinans*. Photosynth Res 129: 159–170.

Rivera-Monroy VH, Danielson TM, Castañeda-Moya E, Marx BD, Travieso R, Zhao X, Gaiser EE, Farfan LM (2019) Long-term demography and stem productivity of Everglades mangrove forests (Florida, USA): resistance to hurricane disturbance. Forest Ecol Manage 440: 79–91.

Rogers K, Saintilan N, Heijnis H (2005) Mangrove encroachment of salt marsh in Western Port Bay, Victoria: the role of sedimentation, subsidence, and sea level rise. Estuar 28:551–559.

Ruiz-Fernández AC, Agraz-Hernández CM, Sanchez-Cabeza JA, Díaz-Asencio M, Pérez-Bernal LH, Keb CC, López-Mendoza PG, y Correa JB, Ontiveros-Cuadras JF, Saenz JO, Castellanos JR (2018) Sediment geochemistry, accumulation rates and forest structure in a large tropical mangrove ecosystem. Wetlands 38:307–325.

Saderne V, Cusack M, Almahasheer H, Serrano O, Masqué P, Arias-Ortiz A, Krishnakumar PK, Rabaoui L, Qurban MA, Duarte CM (2018) Accumulation of carbonates contributes to coastal vegetated ecosystems keeping pace with sea-level rise in an arid region (Arabian Peninsula). J Geophys Res: Biogeosciences 123:1498–1510.

Saintilan N, Wilson N, Rogers K, Rajkaren A, Krauss KW (2014) Mangrove expansion and salt marsh decline at mangrove poleward limits. Glob Change Biol 20: 147–157.

Saintilan N, Khan NS, Ashe E, Kelleway JJ, Rogers K, Woodroffe CD, Horton BP (2020) Thresholds of mangrove survival under rapid sea level rise. Science 368:1118–1121.

Sasmito SD, Murdiyarso D, Friess DA, Kurnianto S (2016) Can mangroves keep pace with contemporary sea-level rise? A global data review. Wetlands Ecol Manage 24:263–278.

Schettini CA, Asp NE, Ogston AS, Gomes VJ, McLachlan RL, Fernandes ME, Nittrouer CA, Truccolo EC, Gardunho DC (2020) Circulation and fine-sediment dynamics in the Amazon Macrotidal Mangrove Coast. Earth Surf Process Landform 45:574–589.

Schuerch M, Spencer T, Temmerman S, Kirwan ML, Wolff C, Lincke D, McOwen CJ, Pickering MD, Reef R, Vafeidis AT, Hinkel J, Nicholls RJ, Brown S (2018) Future response of global coastal wetlands to sea level rise. Nature 561: 231–234.

Seddon AWR, Froyd CA, Leng MJ, Milne G, Willis KJ (2011) Ecosystem resilience and threshold response in the Galápagos coastal zone. PLoS ONE 6: e22376, https://doi.org/10.1371/journal.pone.0022376.

Servino RN, de Oliveira Gomes LE, Bernardino AF (2018) Extreme weather impacts on tropical mangrove forests in the Eastern Brazil Marine Ecoregion. Sci Tot Environ 628–629: 233–240.

Sippo JZ, Maher DT, Tait DR, Holloway C, Santos IR (2016) Are mangroves drivers or buffers of coastal acidification? Insights from alkalinity and dissolved inorganic carbon export estimates across a latitudinal transect. Glob Biogeochem Cycles 30: 753–766.

Sippo JZ, Lovelock CE, Santos IR, Sanders CJ, Maher DT (2018) Mangrove mortality in a changing climate: an overview. Estuar Coast Shelf Sci 215: 241–249.

Sippo JZ, Sanders CJ, Santos IR, Jeffrey LC, Call M, Harada Y, Maguire K, Brown D, Conrad SR, Maher DT (2020) Coastal carbon cycle changes following mangrove loss. Limnol Oceangr. https://doi.org/10.1002/lno.11476.

Snedaker SC, Araújo RJ (1998) Stomatal conductance and gas exchange in four species of Caribbean mangroves exposed to ambient and increased CO2. Mar Freshw Res 9:325–327.

Soper FM, MacKenzie RA, Sharma S, Cole TG, Litton CM, Sparks JP (2019) Non-native mangroves support carbon storage, sediment carbon burial, and accretion of coastal ecosystems. Glob Change Biol 25:4315–4326.

Stokes DJ, Healy TR, Cooke PJ (2010) Expansion dynamics of monospecific, temperate mangroves and sedimentation in two embayments of a barrier-enclosed lagoon, Tauranga Harbour, New Zealand. J Coast Res 26:113–122.

Suyadi S, Gao J, Lundquist CJ, Schwendenmann L (2019) Land-based and climatic stressors of mangrove cover change in the Auckland Region, New Zealand. Aqua Conserv Mar Freshw Ecosyst 29, https://doi.org/10.1002/aqc.3146.

Swales A, Reeve G, Cahoon DR, Lovelock CE (2019) Landscape evolution of a fluvial sediment-rich *Avicennia marina* mangrove forest: insights from seasonal and inter-annual surface-elevation dynamics. Ecosystems 22:1232–1255.

Tamimia B, Wan Juliana WA, Nizam MS, Zain CRCM (2019) Elevated CO_2 concentration and air temperature impacts on mangrove plants (*Rhizophora apiculata*) under controlled environment. Iraqi J Sci 60: 1658–1666.

Urrego LE, Correa-Metrio A, González C, Castaño AR, Yokoyama Y (2013) Contrasting responses of two Caribbean mangroves to sea-level rise in the Guajira Peninsula (Colombian Caribbean). Paleogeogr Paleoclim Paleoecol 370: 92–102.

Walden G, Noirot C, Nagelkerken I (2019) A future 1.2 °C increase in ocean temperature alters the quality of mangrove habitats for marine plants and animals. Sci Tot Environ 690: 596–603.

Wallace RB, Baumann H, Grear JS, Aller RC, Gobler CJ (2014) Coastal ocean acidification: the other eutrophication problem. Estuar Coast Shelf Sci 148: 1–13.

Wang W, Xiao Y, Chen L, Lin P (2007) Leaf anatomical responses to periodical waterlogging in simulated semidiurnal tides in mangrove *Bruguiera gymnorrhiza* seedlings. Aquat Bot 86: 223–228.

Wang X, Piao S, Ciais P, Friedlingstein P, Myneni RB, Cox P, Heimann M, Miller J, Peng S, Wang T, Yang H (2014) A two-fold increase of carbon cycle sensitivity to tropical temperature variations. Nature 506:212–215.

Ward RD, Friess DA, Day RH, Mackenzie RA (2016) Impacts of climate change on mangrove ecosystems: a region by region overview. Ecosys Health Sustain 2: e01211, https://doi.org/10.1002/ehs2.1211.

Williamson GJ, Boggs GS, Bowman DM (2011) Late 20th century mangrove encroachment in the coastal Australian monsoon tropics parallels the regional increase in woody biomass. Reg Environ Change 11:19–27.

Wilson R (2017) Impacts of climate change on mangrove ecosystems in the coastal and marine environments of Caribbean Small Island Developing States (SIDS). Caribbean Mar Clim Change Rept Card: Sci Rev 61–82.

Woodroffe CD (2018) Mangrove response to sea level rise: palaeoecological insights from macrotidal systems in northern Australia. Mar Freshw Res 69: 917–932.

Woodroffe CD, Rogers K, McKee KL, Lovelock CE, Mendelssohn IA, Saintilan N (2016) Mangrove sedimentation and response to relative sea-level rise. Annu Rev Mar Sci 8:243–266.

Ximenes AC, Ponsoni L, Lira CF, Koedam N, Dahdouh-Guebas F (2018) Does sea surface temperature contribute to determining range limits and expansion of mangrove in Eastern South America (Brazil)? Remote Sens 10: 1787, https://doi.org/10.3390/rs10111787.

Yáñez-Espinosa L, Flores J (2011) A review of sea-level rise effect on mangrove forest species: anatomical and morphological modifications. In Casalengo E (ed) Global warming impacts- case studies on the economy, human health, and on urban and natural environments. InTech, New York, p 253–276

Ye Y, Tam NFY, Wong YSLCY (2003) Growth and physiological responses of two mangrove species (*Bruguiera gymnorrhiza* and *Kandelia candel*) to waterlogging. Environ Exp Bot 49: 209–221.

Ye Y, Tam NFY, Wong YSLCY (2004) Does sea level rise influence propagule establishment, early growth and physiology of *Kandelia candel* and *Bruguiera gymnorrhiza*? J Exp Mar Biol Ecol 306: 197–215.

Ye Y, Gu YT, Gao HY, Lu CY (2010) Combined effects of simulated tidal sea-level rise and salinity on seedlings of a mangrove species. Hydrobiol 641: 287–300.

Yin P, Yin M, Cai Z, Wu G, Lin G, Zhou J (2018) Structural instability of the rhizosphere microbiome in mangrove plant *Kandelia candel* under elevated CO_2. Mar Environ Res 140: 422–432.

Yulianto E, Rahardjo AT, Noeradi D, Siregar DA, Hirakawa K (2005) A Holocene pollen record of vegetation and coastal environmental changes in the coastal swamp forest at Batulicin, South Kalimantan, Indonesia. J Asian Earth Sci 25: 1–8.

Roles of Mangroves in Combating the Climate Change

10

Anupam Kumari and Mangal S. Rathore

Abstract

Mangrove ecosystem possesses great potential to mitigate the adverse impacts of climate change and shows a higher range of ecological stability. Mangroves grow luxuriantly in the harsh conditions such as high salinity, temperature, extreme tides, strong winds and muddy and anaerobic soil. They evolved with well-developed morphological, ecological as well as physiological adaptations that make them more resistant and resilient to overcome the effect of adverse condition. Mangroves have unique properties in terms of structure and function such as viviparous germination, well-developed aerial roots, lack of growth rings, adaptable to environmental changes and more efficient nutrient retention capabilities. Various important ecosystem services such as nutrient cycling, carbon storage, soil formation and ecotourism were provided by mangroves to support the livelihoods of coastal societies in the tropical and subtropical area. Mangrove ecosystem is the most productive ecosystem and stores more amount of carbon in their above- and below-ground parts than that of terrestrial forest. This ecosystem acts as a functional unit which involved plants, animals, microorganism and their interaction with a surrounding environment. Mangroves perform several important ecological functions such as coastal protection, carbon sequestration, enriching biodiversity of coastal areas; promote land accretion and support fisheries. They also have more economical significant. This chapter aims to provide deep insight of mangroves in combating climate change.

A. Kumari · M. S. Rathore (✉)
Academy of Scientific and Innovative Research, (AcSIR), Ghaziabad, India

Division of Applied Phycology and Biotechnology, CSIR-Central Salt and Marine Chemicals Research Institute (CSIR-CSMCRI), Bhavnagar, Gujarat, India
e-mail: mangalrathore@csmcri.res.in

R. P. Rastogi et al. (eds.), *Mangroves: Ecology, Biodiversity and Management*,
https://doi.org/10.1007/978-981-16-2494-0_10

Keyword

Mangroves · Carbon sequestration · Coastal ecosystem · Biodiversity · Climate change

10.1 Introduction

Climate change is described as long-term identifiable variations in the state of climate like atmospheric CO_2, elevated temperature, precipitations, etc. (Korres et al. 2016). Apart from this, several natural events that happened in Earth's history were also responsible for climate change. Hence, the concept of 'climate change' implies the changes in weather patterns happening across the last century and triggered by anthropogenic activities that release greenhouse gasses such as CO_2, CH_3, NO_2, etc. (De Ollas et al. 2019). Virtually, global agriculture is frequently influenced by environmental/climate change and these changes have become serious global issues with their various, extensive and persistent effects. The problems of food security imposed by climate/environmental changes/abiotic stresses have become a major challenge for the researchers of plant science (Nikalje et al. 2018). The word 'abiotic stress' involves various stresses occurring due to multifaceted environmental conditions like salinity, drought, heat, cold, freezing, heavy metals, UV and hypoxia (Hirayama and Shinozaki 2010). The responses of plants to abiotic stress are essential and imperative for their existence as well as to combat with the changing environment (Fancy et al. 2017). These responses occur at various levels, such as morphological, physiological, biochemical, molecular and cellular (Nakashima et al. 2009). For the survival, plant deploys various adaptive mechanisms, including synthesis and accumulation of osmo-protective molecules, up- and down-regulation of multiple genes, producing stress-responsive proteins and hormones, etc. (Nakashima et al. 2009; Fancy et al. 2017).

Among the various abiotic stresses, salinity, water (drought and flood) and heat stresses are the most severe threat to crop production in different parts of the world (Qadir et al. 2008). Elevated salt concentration in the soil causes destruction of the plant as well as soil and finally leads to desertification. The adverse impacts of salinity on plants are associated with osmotic, ionic and oxidative stresses. In the soil, higher concentrations of salt decrease soil water potential that imposes water scarcity or osmotic strain. Salt stress causes more accumulation of Na^+ and Cl^- and reduction in the concentration of K^+ and Ca^{2+} which in turn causes ionic stress and nutrient imbalance.

Water stress arises due to either scarcity of water (drought) or excess of water (flood). The most common occurring water stress is water deficit condition, i.e. drought stress. Drought is a common word used to explain atmospheric or weather phenomena and describe as a period without rainfall. Most commonly, according to agricultural and physiological perspectives, drought stress happens when the accessible water for plants in the soil declines due to decrease in soil moisture content at a particular time and low rainfall (Keyvan 2010; Dai 2011). The impacts of drought stress vary based on several aspects, like strength and extent of the stress, genotype of the plant or growing stage, and moreover the imprint of the earlier stress periods present on the plant (Fleta-Soriano and Munné-Bosch 2016; Joshi et al. 2016). The most common symptoms of drought are loss of leaf turgor, wilting, etiolation, yellowing and premature downfall of leaves (Jaleel et al. 2009; Bhargava and Sawant 2013). Some uncommon symptoms are also observed like crack of bark and twig, branch dieback, necrosis, impeded growth and ultimately death to plants (Sapeta et al. 2013).

Flood stress reduces the supply of oxygen to the roots. Due to the lower supply of oxygen critical functions of roots became out of order such as limited uptake of nutrients and respiration. Moreover, in the mangrove environment higher concentrations of methane are also linked to anoxia that affects the growth and development of plants. Decreased survival, photosynthesis and productivity are the major consequences of flooding (Ellison 2000). Stress arises because of flooding causing stomata closure; consequently it causes negative regulation of the photosynthetic apparatus leading to ROS generation in the chloroplast (García-Sánchez et al. 2007). During the waterlogged situation, mangroves recorded to be influenced by limitation of gases diffusion such as CO_2, O_2, reduced light intensity under water and accumulation of ethylene, etc. (Tomlinson 1986). Decreased light intensity in the water during flood conditions is known to cause oxidative stress in plants.

With fluctuating climate scenario and raising the global temperature, heat stress became a major issue for nowadays agriculture. Among the forever changing components of the climate, the aberrant rise in the global mean temperature in current years is the major factor for vulnerability of current agriculture (Chakraborty et al. 2014). The deleterious effect of elevated temperature impacts growth, development, metabolism and crop yield resulting in severe financial losses. Elevated temperature curtails the lifespan of plants, enhances senescence and severely influences the productivity. Higher temperature also tends to cause a modification in membrane fluidity resulting in changes in membrane function (Barnabás et al. 2008).

Biotic stress occurs in the plants due to the interaction of living organisms like especially bacteria, viruses, insects, nematodes, fungi and weeds. These biotic stress-causing factors directly deprive the nutrients of their host, cause damage to plants, resulting in less productivity or even plant death (Das et al. 2016). Moreover, water deficiency, cell wall damage, mechanical wounding, osmotic stress and impairment of plant metabolism also occur during biotic stress (Abood and Lösel 2003). They affect photosynthesis processes like the rate of photosynthesis per leaf area is reduced due to virus infection and reduction in leaf area occur because of infection of chewing insects (Kathiresan and Bingham 2001). Due to the pre as well as postharvest losses of the crop, biotic stress can become a major challenge for nowadays.

10.2 Mangroves

Mangroves belong to a special plant group that grows luxuriantly at the interface between land and sea of the tropical as well as subtropical areas (Kathiresan and Bingham 2001). The word 'mangrove' refers to both specific types of plants as well as their special types of habitat. These special types of habitats are known as tidal forest/mangrove forest/mangal/swamp/wetland (Saenger 2002; Spaulding et al. 2010). Mangroves act as true ecotones due to possessing some characteristics of both marine and terrestrial biomes (Alongi 2009). They have developed a lot of distinct structural, functional and physiological adaptation like viviparous germination, aerial roots, lack of growth rings, etc. These adaptations provide the capability to tolerate and thrive in very harsh conditions like high salinity, temperature, extreme tides and strong winds as well as muddy and anaerobic soil (Alongi 2009). Mangrove is one of the most productive ecosystems across the world and provides unique ecological environments that accommodate a rich collection of species. They have great importance in several fields like ecology, economy as well as social and offer a significant array of ecosystem goods and services (Gopal and Chauhan 2006). Mangroves evolved with well-developed morphological, ecological as well as physiological adaptations that make them more resistant and resilient to overcome the impact of adverse circumstance.

10.2.1 Spatial Distribution

Mangroves cover <1% of tropical and 0.4% of global forest areas (FAO 2007; Van Lavieren et al. 2012). They mostly occur on all the continents found between the Tropic of Cancer and Capricorn. Globally, mangroves occupy around 75% of tropical coastal line and mainly confined between 30 °C north and 30 °C south latitudes (Saenger 2002; Singh et al. 2012). The latitudinal distribution is restricted by essential climate factors like the occurrence of extreme cold weather and aridity (Osland et al. 2013; Saintilan et al. 2014). Moreover, the distribution and structural development are further restricted by the availability of freshwater or rainfall even within the areas with appropriate temperature (Osland et al. 2014; Alongi 2015). Across the world, mangroves are recorded in 123 countries and presently total areas occupied by mangroves are around 152,000 km^2 (Spaulding et al. 2010). Almost 75% of mangroves are recorded in 15 countries, among them just 6.9% are protected (Thomas et al. 2017). Mangroves belong to the South as well as Southeast Asia constitutes the world's utmost diverse and extensive groups of mangroves and account for 41.1% of global mangroves. Indonesia serves as the major source of mangroves and contributes more than 20% of the total mangroves (Hamilton and Casey 2016). These plants establish a unique and rare type of ecosystem. However, they are threatened as devastated almost five times faster than forests. For instance, the most threatened regions of mangrove are known as in America (North as well as Central) due to several activities like aquaculture, coastal development, hurricanes, etc. (Van Lavieren et al. 2012). India contributes 3.1% mangroves of entire world cover. Gujarat covered the second largest area of mangroves after Sundarbans in India. At present, the term 'mangrove' involves eighty four species from twenty four genera and sixteen families. In spite of this, just seventy species are considered as real mangroves whilst others are mangrove associates (Thatoi et al. 2016). All mangroves not only differ in the species composition, but also exhibit different proportions of shrubs, trees, and so on. Different species of mangroves are not distributed uniformly. An unique zonation of mangrove species is recorded. The zonation is determined by several factors like height of the land, salinity and tide at a certain level. Moreover, the capability to compete with other species is also a major factor considered in zonation. *Rhizophora* and *Xylocarpus* are highly salt as well as flood-tolerant mangrove, hence reported near to the shoreline as harsh conditions are found there. *Bruguiera* and *Sonneratia* are reported less salt tolerant and observed mainly in the inland area.

10.2.2 History

The word 'mangrove' is originated from the Guarani (official language of Paraguay). During the initial phase of the 1610_S the term was written like 'mangrow' deriving from Spanish 'mangle' or Portuguese 'mangue', however, later phase of 1690_S the word 'mangrow' moved into an English word 'mangrove' by etymology. Several terms are associated with mangrove, viz. mangrove ecosystem/mangrove forest community/mangal. Many other terms are also used for mangrove forest like oceanic rain forest, tidal forest, tidal swamp forest, mangrove swamp and coastland wood land (Kathiresan and Bingham 2001; Van Lavieren et al. 2012; Hamilton and Casey 2016; Thatoi et al. 2016). For example, mangal or mangrove forest community and mangrove ecosystem are formed by association of microbes and animals with plants, respectively (MacNae 1969). It is advised that the term 'mangrove' must be signified to exclusive mangrove species whilst the term 'mangal' to the forest community in its place (Nabeelah Bibi et al. 2019). Mangroves have been discovered and studied since the ancient period. First of all, mangrove was discovered by Nearchus in the Persian Gulf during 325 B.C. In the late 325 B.C., Theophrastus (Greek philosopher) also found the existence of mangrove plants and documented in the book 'Historia Plantarum' (Kathiresan and Bingham 2001; Spaulding et al. 2010; Schneider 2011). Both of them explained about *Rhizophora* tree that held up by roots as a polyp, leaves, flowers, etc. (MacNae 1969; Spaulding et al. 2010). Subsequently, the ancient Greeks know about three areas of mangroves, viz. Arabian Sea, Persian Gulf and Red Sea during 323 B.C. (Schneider 2011). In the realm of information, mangroves originated from Indo-Malayan areas and most of the mangrove species are documented there across the world (Hamilton and Casey 2016).

Mangroves are extremely old, probably arising immediately after the first angiosperms about 114 million years back (Duke 1992). The first genera of mangroves evolves were most possibly *Avicennia* and *Rhizophora* that emerging almost at the end of the Cretaceous period (Chapman 1976). The propagules and seeds of mangroves possess a unique characteristic that supports them to float on the water surface. Owing to this property, it was very simple/effortless for spreading of mangroves species by water dispersion from India and East Africa to Central and South America around 23–66 million years back.

10.2.3 Features of Mangroves

Mangroves have developed with unique characteristics as they occurred in the transition zone, where low soil oxygen contents and daily rises and falls of the tide and salinity are recorded. Consequently, to cope with this changing environment, they have evolved with both halophytic and xeromorphic features. The most distinct morphological feature of mangroves is associated with their root system, for instance, the buttress roots of *Heritiera* and *Xylocarpus*, the pneumatophores of *Avicennia*, *Lumnitzera* and *Sonneratia*, the root knees of *Xylocarpus, Ceriops* and *Bruguiera* and the slit roots of *Rhizophora* (Kathiresan and Bingham 2001). These specialized roots work like respiratory system and assist gas exchange in mangroves growing in anaerobic as well as in saline soil. Numerous mangrove roots do not penetrate deeply in the anaerobic rock layer. As an alternative, they produce plentiful lateral roots to support them.

Another distinct morphological appearance was observed in the leaves of mangroves. They have thick, leathery, wax-like, dark green (except *Nypa fruticans*), elliptical shaped and moderate size leaves. Wax-like leaves help them to retain more water content and prevent water loss, or stomata positioned in such a manner that causes less evaporation (Kathiresan and Bingham 2001). Most commonly, the leaves show dorsiventral symmetry; however, isolateral leaves are also observed in the *Phoenix paludosa* and *Sonneratia apetala* (Das et al. 1995). The leaves are arranged in adjusted decussate (bijugate) form in which less than 180 °C angle kept between each pair and proceeding pair. This type of arrangement prevents self-shading of the leaves and makes branch system (Tomlinson 1986). Mangrove leaves contain six kinds of stomata that only differ in the arrangement of guard and subsidiary cells. In the majority of species, either external or external and internal both the surface of the stomatal pore is covered with horn or beak-like cuticular outgrowth that lower the stomatal transpiration (Das and Ghose 1993). Generally, bundle sheath fibres and bundle sheath extensions are absent in the leaves. On the other hand, they have extended tracheids lasting in the ends of the vein. Branched sclereids are well developed and most abundant in *Aegiceras*, *Aegialitis, Rhizophora* and *Sonneratia.* Perhaps, mechanical support of leaves is provided by sclereids. It is assumed that sclereids and tracheids also participate in the storage of water (Tomlinson 1986).

Mangrove woods also possess the unique anatomical characteristics that are explained by Tomlinson (1986). They lack growth rings or conspicuously anomalous; for instance, in *Avicennia.* Thus, getting old is difficult. The wood has exclusive features that make it possible to overcome the elevated osmotic potential of seawater as well as the transpiration triggered by high temperatures (Kathiresan and Bingham 2001). Several narrow vessels are going from end to end of the wood. The density of these vessels ranges from 33 to 270 nm such as in *Excoecaria* and

Aegiceras, respectively. These vessels facilitate to generate high tensions in the xylem because a little reduction in the diameter of vessels creates improperly more rise in flow resistance (Scholander et al. 1964). Simple perforation plates are found in the vessel elements that form the vessels. In spite of this, scalariform perforation plates are seen in the member of *Rhizophoraceae* (expect *Kandelia*). Size as well as the allocation of the vessels is the key factor that vigorously influencing the water conduction via wood (Kathiresan and Bingham 2001). Movement of water occurs most quickly by ring-porous woods in which the largest vessels are present in the outermost growth layer. However, slowest water conduction occurs in the diffuse-porous woods in which vessels are present more uniformly in size and allocations. Most of the mangrove woods are diffuse-porous but only *Aegialitis rotundifolia* is a ring-porous wood (Kathiresan and Bingham 2001).

Some special characteristics are also observed in the seeds and during the seedling stage. Mostly, mangrove fruits are cigar-shaped, greenish in colour and length ranges from 2 to 25 cm. Endosperm haustoria are made by *Avicennia marina* at the time of early embryonic histodifferentiation. As soon as the growth phase is started extra-ovular type of embryonic development is noted subsequently. Thus, the mature seed is enclosed via a pericarp initiating completely from the wall of the ovary. Starting to the last of histodifferentiation up to the maturation of seeds, cotyledon cells turn into enormously vacuolated and have more quantity of soluble sugars and serve as major nutrient reserves in the mature seeds. Embryos require abundant oxygen to grow since the waterlogged soil is oxygen deficient. Fertilized seeds develop directly into seedlings even though attached to the parent tree. The phenomenon is known as vivipary. Just after germination, the seedling grows to form a propagule. This propagule falls over the water surface and floats till finding a suitable place to grow well (Miththapala 2008).

10.3 Factor Affecting the Growth of Mangroves

Climate change imposed severe threats to mangrove ecosystems. The climate change factors involve changes in sea level, temperature, atmospheric CO_2 concentration, precipitation, storms, high water events, circulation patterns of ocean as well as the health of functionally associated adjoining ecosystems (Gilman et al. 2008). These factors most commonly influence the growth of mangroves. However, mangroves show resistance and resilience to mitigate the potential impact of climate change (Kandasamy 2017).

10.3.1 Temperature

Global warming is an important challenge for threatening the integrity of the mangrove ecosystem. Mean global temperatures are expected to increase very quickly (1.1–6.4 °C) in the twenty-first century (Solomon et al. 2007). The warmer environment will create several injuries and possibilities for mangroves as well as salt marshes plants. Due to temperature variations such as average temperature and extremely low-temperature like frosts in North America, mangrove exhibits complex dynamics at their latitudinal distribution (Cavanaugh et al. 2014). It is very tough to predict the exact impact of elevated temperature. However, the poleward movement of a few species is known in response to high temperatures (Saintilan et al. 2014). It is expected that elevated surface temperature influences the mangroves in various ways like changes in the composition of species, phenological patterns (for example, flowering and fruiting timing), photosynthesis, respiration, microbial decomposition and in the case when the temperature does not go beyond upper threshold enhances the productivity of mangroves. Moreover, higher temperature causes expanding of the ranges of mangroves in higher latitudes where other factors such as appropriate physiographic conditions and supply of propagules are abundant and temperature acts only by limiting factor (Field 1995; Ellison 2000). The majority of the mangrove show maximum density of the shoot when the average air temperature reaches 25 °C whilst the production of leaves stops when the temperature falls below 15 °C (Hutchings and Saenger 1987). The optimal leaf temperature for photosynthesis is achieved when leaf temperature is assumed to be in the range from 28 to 32 °C. The photosynthesis ceases when the temperature of the leaf is between 38 and 40 °C (Andrews et al. 1984).

10.3.2 Atmospheric CO_2 Concentration

The level of atmospheric CO_2 has enhanced by 35% from the pre-industrial period. The CO_2 concentration was 280 ppm (parts per million by volume) in 1880 and it has reached 379 ppmv in 2005 as well as the acidity of the ocean has also increased by 25% (Solomon et al. 2007). In spite of large difference, each model forecast a further enhancement in the level of CO_2 by the century's end. Someone has anticipated that the CO_2 level will be either double or even triple than of the present level (Van Lavieren et al. 2012). It will be tough to envisage the responses; however, the rate of photosynthesis and respiration, nutrient availability, water use efficiency and salinity will probably change (Ball et al. 1997). Inside the estuaries, the patterns of species are more probable to vary depending on the interactive impacts of change in CO_2 concentration, sea level, temperature as well as weather patterns to the species-specific responses. However, there will possibly be few or even no changes in the production of the canopy (Alongi 2002, 2008). Higher CO_2 conditions are likely to improve the mangrove growth when the gain of carbon is restricted by demand for evaporation of the leaves however, not at the time it is restricted by root salinity. There is no proof that higher CO_2 concentration will enhance the growth of mangrove species in a higher range of salt concentration. Although all species may not react in the same way and other environmental factors like salinity, temperature, nutrient quantities as well as the hydrological regime can affect the response of mangrove to the elevated level of atmospheric CO_2 (Field 1995). The impact of increased CO_2 on mangrove is not well known and scarcity of research is found in this area. Enhanced concentration of CO_2 could modify the ability of competition, hence changing the composition of the community along gradients of salinity—humidity (Ball et al. 1997).

10.3.3 Sea Level

Sea level rise is one of the major consequences of global warming. During the twentieth century, 12–22 cm rise of sea level has documented (Holgate and Woodworth 2004; Thomas et al. 2004). Many climate models have anticipated a hastened rate of rising sea levels in the upcoming decades (Church and White 2006; Solomon et al. 2007). The range of rising in global sea level projection is 0.18–0.59 m as of 1980–1999 to the last of the twenty-first century. Latest findings on the acceleration of changes in global sea levels signify that the upper level of projections is expected to happen (Church and White 2006). The mangroves will face serious threats in the future due to sea level rise (Field 1995; Gilman et al. 2008). Expectations of loss of mangroves range from 30% to even extinction (Duke et al. 2007; IPCC 2007). Many physical, as well as biological factors, determine the extent of mangrove loss. According to the rates of sea level rise (SLR) in addition to vertical accretion, mangroves can move towards landward. SLR and vertical accretion both are interconnected to slope and space available at the edge of landward. Adaptation to the changes in sea level, range of tide, sediment supply, etc. mainly depends on specific species (Alongi 2002, 2008). For establishing mangroves in the new regions several factors are needed like appropriate hydrology, the composition of sediment, available waterborne seedling as well as potential to compete with other species of plants (Krauss et al. 2003). If sea level rise happens slowly enough and sufficient space exists for expansion, then mangrove can adapt to changes in the sea level. Mangroves would be likely to migrate to landward as the sea level rises. Furthermore, with the rise in sea level, the width of the mangrove system can reduce. The capability of landward migration of mangrove is dictated by local environments like infrastructure (dikes, roads, sea walls, urbanization) as well as topography (steep slope). It has been assumed that the mangroves found in riverine regions with the closely packed forest of mangroves are less prone to sea level rise. Mangroves located in small islands, scarcity of rivers, coastal development, ground water extraction, underground mining, tectonic movement and steep topography are assumed to be most vulnerable to sea level rise. (Ellison and Stoddart 1991).

The four main factors describing below are responsible for the resistance and resilience of mangroves in response to changes in sea level: (1) the vulnerability of mangroves is determined by the rate of sea level rise in proportion to the sediment surface of mangroves (Cahoon et al. 2006; Cahoon and Hensel 2006; Gilman et al. 2008). Change in sediment elevation occurs at different rates in different zones of mangrove vegetation, thus the composition of mangrove species affects the response of mangroves (Krauss et al. 2003; Rogers et al. 2005; McKee et al. 2007). In response to changing sea levels, certain zone shows more resistance and resilience. Moreover, different species of mangroves also needed different time periods for colonizing in the new habitat, which happens with change in relative sea level. The fast colonizing species may compete rapidly with slower colonizing species and become more dominant (Lovelock and Ellison 2007). (2) The physiographic location involving differences in the slope of land previously and presently occupies by mangroves as well as obstacles found during their landward migration influencing the resistance and resilience of mangroves (Gilman et al. 2008). Lastly (3) the collective impacts of all stress factors influence the resistance and resilience ability of mangroves. It is not anticipated that mangrove responds according to the assumptions of Bruun rule (a model used for prediction of erosion of beaches) as sediment budget processes of mangrove are differ from beaches. In addition, on a small scale and site-specific estimations inaccurate outcomes are generated by predictive models of coastal erosion (Bruun 1988; List et al. 1997; Pilkey and Cooper 2004). The accelerated rise of sea level likely to stimulate the replacement of salt marsh plants by mangroves when mangroves and salt marsh plants co-exist. In the many estuaries of SE Australia, the replacement of salt marsh plants by mangroves was observed by Saintilan and Williams (1999).

10.3.4 Precipitation

Global rainfall is expected to enhance almost 25% by 2050 due to climate change. Though extent and direction of change in precipitation will not be uniform, they would differ according to season and spatial distribution (Houghton et al. 2001; Solomon et al. 2007). Precipitation is more likely to increase in higher latitudes, while in most of the subtropical areas (mainly at the poleward margins of subtropics) precipitation is likely to decrease (Solomon et al. 2007). According to the recent prediction of the IPCC, substantial enhancement in precipitation is likely to happen in northern and central Asia, northern Europe as well as North Eastern and South eastern America. They have also anticipated drought conditions in southern Asia, southern Africa, the Mediterranean and in the Sahel (Solomon et al. 2007). The trends for long-term precipitation were not observed in the other parts. Mangrove growth and spatial distribution are expected to affect by changes in precipitation patterns (Field 1995; Ellison 2000). Excess of precipitations especially overflow of freshwater can change patterns of salinity which in turn affects vegetation of the coastal ecosystem. Primarily based on the links found in the rainfall trends and habitat condition of mangrove, low rainfall and more evaporation will cause higher salinity, lower seedling growth and survival, less net primary productivity, decreasing biodiversity. Moreover, due to the formation of hypersaline flats from upper tidal zones, a significant reduction in mangrove area will also be observed (Field 1995; Duke et al. 1998). The area having less precipitation will contribute less freshwater to mangroves resulting in more salinity. For instance, *Heritiera fomes* ('Sundari') and *Nypa fruticans* species of mangroves are continuously disappearing from the Sundarbans regions of India. It happens because the supply of freshwater is completely shut off due to the siltation of Bidyadhari (Mitra et al. 2009). Furthermore, highly salt-tolerant species like Ceriops (*Rhizophoraceae* family) occupied the place of *Heritiera fomes* and *Nypa fruticans* (VYAS 2012). Recently, less salt-tolerant species of mangroves are continuously vanishing from the entire Indian mangrove ecosystem (except Andaman and Nicobar islands) and more salt-tolerant species such as *Avicennia marina* are becoming dominant. Species of *Rhizophoraceae* (True mangrove) were dominant in Muthupet around 150 years back. However, they are extinct spatially these days (Kandasamy 2017). In the 1950s, mangrove wetlands of Godavari had around 90% populations of tall and closely packed tress of *Avicennia officinalis*, *Lumnitzera racemosa* and *Excoecaria agallocha*. However, their population have reduced to 37% and are replaced by salt marsh bushes of *Suaeda nudiflora* and *S. maritima*. The reductions in the consistency and quantity of freshwater supply to the mangroves are the main cause for variations in the species composition of mangroves (Kandasamy 2017).

The salt concentration in the mangrove tissue increases with increasing soil salinity while the water content decreases concomitantly, which leads to low productivity (Field 1995). The sulphate content of seawater will also increase with elevated levels of salinity which would further stimulate peat decomposition and cause the vulnerability of mangroves in response to changes in sea level (Snedaker 1993, 1995). At the same time, more rainfall causes enhanced growth and biodiversity of mangroves and increases in the area of mangrove (Duke et al. 1998). More diverse, taller and more productive mangroves are located on the shorelines having more rainfall than the shorelines having less rainfall, recorded in the majority of the locations worldwide along with Australia (Duke et al. 1998; Ellison 2000). Mangroves inhabiting the area of more rainfall have higher productivity and diversity most likely because of less exposure to salinity and sulphate content as well as more supply of nutrients and fluvial sediment (McKee 1993; Ellison 2000). More intense rainfall is expected to affect many physical processes and erosion in the tidal wetlands and catchments. In the arid regions, less rainfall combined with more evaporation which possibly leads to reduction in the area of mangroves, less diversity and conversion of the landward area to the barren hypersaline surface (Snedaker 1995).

Due to the variations in local weather, floods and droughts occur which may possess a significant impact on vegetation. The vegetation of salt-tolerant species is promoted during the duration of low to moderate flooding whilst catastrophic flooding causes mortality of plant and subsequently, the establishment of the annual plants occur. It is observed in the Nueces Estuary of Texas (USA) in the study of semi-arid, subtropical salt marsh (Forbes and Dunton 2006). Variations in the pattern of rainfall may promote the shift of vegetation at the distribution boundaries. A positive relationship was observed between variables of rainfall and landward migration of mangroves during the study of more than 32 years on patterns of rainfall and geographical distribution in the Moreton Bay mangrove forest (Southeast Queensland, Australia) (Eslami-Andargoli et al. 2009). Many other factors like geomorphology, local weather and disturbance can alter the patterns and rates of mangrove expansion. There are some more explanations regarding landward migration of mangroves towards salt marsh such as modified patterns of tidal, enhanced sedimentation and nutrient contents and anthropogenic disturbance (Saintilan and Williams 1999).

10.3.5 Storms

According to IPCC (Intergovernmental Panel on Climate Change) projections on climate change, peak intensities of wind and precipitation of the tropical cyclone along with tropical cyclone mean are expected to increase in the twenty-first century due to climate change (Houghton et al. 2001; Solomon et al. 2007). Moreover, the warmer climate is anticipated to produce stronger tropical cyclones having highly intense precipitation and strong speeds of winds (Solomon et al. 2007). The intensity of the effect will be depending upon strength, extent, size and frequency of storms. There are several effects of storms, viz. defoliation, up-rooting, the mortality of trees as well as enhance stress, which arises due to a changing sediment elevation of mangrove. Sediment elevation occurs because of erosion, compression and deposition of soil as well as peat collapse (Cahoon and Hensel 2006; Piou et al. 2006). The recovery rate of damage caused by the storm may be very slow. The heights of storm surge are expected to rise if the occurrence of low pressure and strong wind enhances. This happens due to the occurrence of more frequent and severe storms because of the changing climate (Church et al. 2001; Solomon et al. 2007). The more frequent and intense storms have great potential to mangroves damaged by means of defoliation and mortality. The area having a bulk mortality of trees along with fewer surviving plantlets and trees would suffer from the permanent conversion of the ecosystem. It occurs since recovery via recruitment of seedling possibly will not happen as a result of changes in the sediment elevation and associated changes in hydrology (Cahoon et al. 2003). Other natural incidents like a tsunami (Indian Ocean tsunami on 26th December 2004) may also cause serious impairment to mangroves (Danielsen et al. 2005; Kathiresan and Rajendran 2005; Dahdouh-Guebas et al. 2006). Several models have projected poleward storm shifts and highly intense (but less) tropical storm paths through a little latitudinal degree in both hemispheres (Geng and Sugi 2003; Yin 2005; Bengtsson et al. 2006). Variations in the average precipitation are expected to have little impacts on vegetation as compared to variations in the occurrence of extreme events. The elevation of wetland may also change due to episodic incidents like cyclones, hurricanes, lightning strikes, storms, storm surges and flooding associated with it, strong wind as well as excess flow of fresh water (Whelan et al. 2005; Cahoon et al. 2006). As the frequencies and intensity of these episodic incidents are predicted to rise in linked with climate change and their effect can increase on wetland resilience in the future (Christensen et al. 2007; Kundzewicz et al. 2007). Besides the rise in sea level, the intensity of storm surges is projected to enhance 5–10% by 2050 which may also serve as a flood for mangroves. The species composition and health of mangrove may be affected by storm because of changes in sedimentation, inundation, recruitment and salinity. *Avicennia* and *Sonneratia* are highly susceptible as compared to *Rhizophora* species in response to storm surge. It happens because the species of *Rhizophora* have slit roots that remain standing in the above rise in sea level whilst pneumatophores roots are mainly submerged during the rise in sea level and found in the *Avicennia* and *Sonneratia* species. Additionally, in the area of mangroves trapping of sediment and accumulation of peat are facilitated by slit roots. Storms and tropical cyclones occur most frequently in the Bay of Bengal, and east coast of India is severely affected than the west coast. Though, *Rhizophora spp.* works as protecting force against the natural calamity in India. In general, species of *Avicennia* regenerate after cyclones in these regions. Thus, Indian mangroves exhibit resistance to cyclones.

10.4 Response of Mangroves to Environmental Stresses/ Climate Change

Mangrove ecosystems are the rarest and toughest ecosystems on the Earth. Their natural environment is comprised of abiotic stress (salt, flood, heavy metal, low oxygen) and biotic hindrance of different nature and extent (Das et al. 2016). Mangroves have developed the distinctive capability to combat with this extraordinary magnitude and type of stress due to their continuous exposure. However, the susceptibility and responses of mangrove against climate change will be extremely impacted by human interference (Das et al. 2016).

10.4.1 Soil Characteristics

The soil characteristics exhibit a significant effect on the growth and nutrition of mangroves.

Electrical conductivity, cation exchange capacity, pH and saltiness are some important characteristics of the soil (Kathiresan and Bingham 2001). However, the concentrations of nutrients seem to be the most important factor. Mangrove ecosystems are well-balanced and most efficient sink of nutrients along with the actual import of dissolved nutrients like nitrogen, silicon and phosphorus (Kathiresan and Bingham 2001). The availability of nutrients can limit growth, development and ultimately the productivity of several mangrove forests. Moreover, different concentration of nutrients may also affect the distribution of species and their competitive abilities (Chen and Twilley 1998). The limitation of nutrients can differ according to individual habitats of mangrove. For instance, seedlings of *Rhizophora apiculata* perform significantly well in the cultivated area enriched with potassium (Kathiresan and Bingham 2001). Most commonly, mangroves found in the soil with low nutrient carbonate having the limitation of phosphorous. Available phosphorous is likely to be bound with calcium and efficiently retaining it inside the sediments. Almost seven and three-fold enhancements in the rate of stem elongation and in the leaf area were observed, respectively, in the seedling of *Rhizophora mangle* enriched with phosphorous during mesocosm and field experiments. No such types of responses were detected in the case of nitrogen addition (Koch and Snedaker 1997). Similarly, the growth of dwarf *R. mangle* in the Belizean mangal is limited by low availability of phosphorous (Feller 1995). They also encourage the growth of tough and long-lasting leaves known as sclerophyll. The development of sclerophylls possibly will be an adaption for conserving nutrients in the habitats of oligotrophic. Other mechanisms may also be found in the mangroves to retain useful nutrients like most abundant concentration of nitrogen and potassium have recorded in the mature and photosynthetically active leaves than senescent leaves. It appears as an outcome of the translocation of nutrients out of aged leaves into other parts of plants prior to the leaves fall. Due to the damage of the mangrove ecosystem, its capability of nutrient retention may compromise. For instance, substantial loss of nitrogen and phosphorous have detected in the severely damaged area of mangroves in North Queensland (Kaly et al. 1997). This may happen because of a dramatic decline in the burrows density and a decrease in the population of crab. The disturbance impact will vary according to habitat and will be determined by the characteristics of the sediments and patterns of flow on each site. As no differences was recorded in the soil nutrient concentration of healthy and degraded areas of mangroves in Pedada Strait (Triwilaida Intari 1990). The distinctive features of mangrove sediments are sulphides that affect the distribution of mangroves. The concentration and distribution of the sulphides are controlled by the mixing of tidal, bioturbation and also via mangrove themselves. Most often less reduction of soils is observed close to the aerial roots of some species that directs low sulphide content. Moderately reducing soil with fewer sulphide contents has recorded in the area dominated by *R. mangle* in neotropical Florida mangrove forest. On the other hand, highly reducing soil with more sulphide contents area has dominated by *Avicennia germinans* (McKee 1993). Unexpectedly, the same pattern of soil characteristics has not been detected in a similar mangrove ecosystem residing in Brazil. As strongly reducing soil with high sulphide contents are found in the area dominated by *R. mangle*. At the same time, soil sulphide content of *A. germinans* is extremely variable due to the change of rhizosphere from oxygenated to anoxic conditions (Lacerda et al. 1995). The more exchangeable trace metals are also found in the soil of *Avicennia* (Lacerda et al. 1995). In young forests, sulphate reduces very slowly into sulphide that leads to high nutrient contents and low toxicity of sulphide (Alongi et al. 1998). Higher contents of sulphide may damage seedlings of mangrove, cause closure of stomata, decrease gas exchange, decline growth and ultimately mortality increases. The rate of sulphate reduction can increase by disturbance and heavy organic input. Clearance of mangrove ecosystem or gaps in the formation of the canopy may alter the physical as well as chemical properties of the underlying soil. This led to anaerobiosis condition and also increases sediment sulphide contents. Most often, soils of mangroves are deficient in oxygen or without oxygen. So mangroves have distinct features to combat with this condition. The presences of pneumatophores (breathing roots) are the most prominent feature. This is above-ground roots which are occupied with spongy tissue. Several little holes are found in the bark that allows oxygen to be transported to the root parts.

10.4.2 Salinity

Salinity of coastal areas is controlled by many factors like climate, topography, hydrology and tidal flooding. These factors strongly influence the growth and productivity of the mangrove ecosystem. They may also affect competitive abilities of the species. In general, mangroves are facultative halophytes that tolerate both fluctuating and higher salt concentrations. Some researchers have also classified mangrove as obligate halophytes. Many species of mangrove attain optimal growth at 5–25% salinities of seawater. The growth of mangroves occurs in more or less waterlogged soil and water with varied concentration of salinity and may be equal to salt concentration of open sea. Soil salinization is the most severe environmental stresses mainly in the areas of arid, semi-arid and mangrove ecosystems (Das et al. 2016). Higher salt accretions lower down the soil water potential that makes more difficult for plant to obtain nutrients and water from soil. Hence, water deficit condition arises in plants as a result of salinity stress that lead to physiological drought. The most deleterious impacts of salinity on plants include ionic stress, osmotic stress, oxidative stress, enhanced ROS production, reduced growth, photosynthesis and ultimately low productivity (Doganlar et al. 2010; Kosová et al. 2013). Salt-tolerant mechanisms of mangroves are alike to those in glycophytes. However, most likely mangrove may sequester or exclude salt ions more effectively (Jithesh et al. 2006).

Mangroves developed various adaptive mechanisms like ion sequestration, salt accumulation and salt excreting to cope up with these adverse conditions. The leaves of mangroves tend to be thicker and smaller in response to salt stress. Small leaf releases more heat by way of convection thus enhances cooling of the leaf. *Sonneratia alba* and *Sonneratia lanceolata* (closely related species) showed interspecific differences in salinity tolerance in association with differences in their distribution besides natural seasonal salinization (Ball and Pidsley 1995). *Sonneratia alba* may grow in 100% salinities of sea water while *Sonneratia lanceolata* grow in 50% salinities of sea water. Moreover, vivipary characteristics of mangroves also play a critical role in imparting enhanced salinity stress tolerance (Zheng et al. 1999). Four genera of mangrove (*Rhizophoraceae* family) are known to have vivipary characteristics, viz. *Bruguiera*, *Ceriops*, *Rhizophora* and *Kandelia* (Tomlinson 1986). It has believed that vivipary feature of mangroves helps in germination during saline conditions. Typical salt glands structures are present in the species of *Acanthus*, *Avicennia*, *Aegialitis* and *Aegiceras* that remove excess salt concentration and imparting more salt tolerant. Furthermore, the analogous structure of salt gland is found in the species of *Conocarpus* and *Laguncularia* (Tomlinson 1986). Structures similar to 'Pimples' or 'tiny bumps' resembling a certain extent with salt glands are present in the petioles of the leaves *L. racemosa* and *C. erectus* (Tomlinson 1986). Salt leaks from salt glands of *Avicennia* and *Aegiceras* have deposited as a crystal of salts in their leaves. The species of *Avicennia*, *Sonneratia*, *Rhizophora* and *Xylocarpus* also accumulates salt in the bark of roots and stem (Scholander 1968). One more distinct property of mangroves is the development of the succulent structure by leaves thickening to store more water content (Suárez and Sobrado 2000). For instance, the thickness of the leaves as well as water content may increase in *Laguncularia racemosa* during salinity stress. The enhanced water content diluted the absorbed salt, thus reduced the damage caused by salinity stress to some extent (Sobrado 2005). Due to the development of stress-induced succulence, the rate of photosynthesis may also enhance by an increment in the surface of the internal leaf for gas exchange. The wax present in some mangroves leaves also acts as protective features and contributes less transpiration in them than other species lacking this trait. Mangroves possess numerous anatomical and morphological features that are much alike to higher water use efficiency features of terrestrial xerophytes. These features allow mangroves to grow in high saline or physiological dry condition devoid of any visible adverse impacts of severe water stress (Parida and Jha 2010). Many genera of mangrove retain sunken stomata and thickly layered of epidermis having waxy cuticle. The species of *Rhizophora*, *Avicennia*, *Ceriops* and *Bruguiera* have more prominent sunken stomata beneath the epidermis (Miller et al. 1975). The waxy cuticles are covered with various shaped hairs such as stellate scales in the species of *Heritiera* and tricellular peltale stellate hairs in *Avicennia* (Miller et al. 1975). To combat with the adverse impacts of ROS production mangroves have developed strongly regulated mechanisms involving ROS scavenging (enzymatic and non-enzymatic) pathways (Das et al. 2016). During salt stress, the steep increment in the SOD level was documented in the *Bruguiera parviflora* and *B. gymnorrhiza* (Takemura et al. 2000; Parida et al. 2004). Around 8.1 times increment in the SOD activity was observed in *B. gymnorrhiza* when subjected to salinity stress (500 mM for 9 days) (Takemura et al. 2000). In *A. marina* higher SOD as well as POX activity was detected under saline conditions (Cherian et al. 1999). The activity of catalase enzyme was varied in the species of *Bruguiera* under salt stress. Reduce catalase activity was recorded in the *Bruguiera parviflora* when exposed to salt stress (400 mM NaCl) (Jithesh et al. 2006). During salt stress, increased activity of antioxidant enzymes (SOD and POD) was also detected in the *Aegialitis rotundifolia*, *Bruguiera gymnorrhiza*, *Xylocarpus mekongensis*, *Xylocarpus granatum*, *Heritiera fomes*, *Phoenix paludosa* (Dasgupta et al. 2012). Higher activities of antioxidant enzymes signifying their protecting role in response to oxidative stress (Atreya et al. 2009; Foyer and Shigeoka 2011). Apart from enzymatic antioxidants, non-enzymatic antioxidants (ascorbic acid, carotenoids, tocopherol and glutathione) also perform an essential function in ROS scavenging under stress conditions.

10.4.3 Temperature

Climatic factors like temperature strongly impact diversity and the global distribution of mangrove forests (Duke et al. 2017). Generally, in the tropical wet climate, mangrove ecosystems are most diverse, abundant and highly productive. Chilling and freezing temperatures may cause reduced biomass, less productivity and mangrove mortality. Increased temperature is also likely to change the expansion of latitudinal limit, diversity and composition of the community (Duke et al. 2017). The poleward expansion of mangroves occurs most probably in response to the temperature and sea level rise (Reid and Beaugrand 2012). Elevated temperatures are expected to give an advantage to the Pacific Islands. As mangrove diversity is anticipated to increase at higher latitudes (at present, occupied by the only species of *Avicennia*) due to warming (Burns 2001). Warming is estimated to facilitate the expansion of mangroves into salt marshes of the Pacific Islands (Burns 2001). Thermal tolerance is varied in the mangroves species of China. Based on the temperature tolerance, Li and Lee (1997) classified mangrove species of China into three groups: (1) thermophilic eurytopic species (e.g. *Rhizophora stylosa, Bruguiera gymnorrhiza, B. sexangula, Acrostichum aureum* and *Excoecaria agallocha*), (2) thermophilic stenotopic species (e.g. *R. apiculata, R. mucronata, Pemphis acidula, Nypa fruticans* and *Lumnitzera littorea*) and (3) cold resistant eurytopic species (e.g. *Aegiceras corniculatum, Avicennia marina* and *Kandelia candel*). Changes in the soil temperature are likely to increase at the same rate and magnitude as that of sea surface temperature. However, changes in temperature of soil are mostly very less than that of atmospheric temperature, because saturated soil has a higher ability to retain heat. So, mangroves are not expected to be much adversely affected even the temperature of soil increases. The rate of bacterial growth and multiplication also increases due to the elevated temperature of sediment and ultimately enhances the rate of regeneration and nutrient cycling. The leaf inclination is most essential for the regulation of temperature. For instance, 10 °C differences were found in the leaf air temperature on a clear day when the leaf held horizontally. Moreover, photosynthetic gas exchange is most sensitive to vapour pressure deficits and temperature of leaf air. Mangroves apparently respond to elevated temperature; however, the exact temperature is not known at which their functionality plateaus or they die. Most of the species show maximum photosynthesis at temperature 30 °C or below. In many species rate of CO_2 assimilation decreases either gradually or sharply when the temperature rises from 33 to 35 °C (Ball and Sobrado 2002). Photosynthesis is often decline in exposed leaves because of photo-inhibition. Only elevated temperature is expected to enhanced photosynthesis, respiration and growth.

10.4.4 Metal and Organic Pollutant

Heavy metals are major environmental pollutants, their toxic effects exert severe problem with the growth and development of plants (Zornoza et al. 2002; Yadav 2010). Some important metals like iron (Fe), nickel (Ni), manganese (Mn), zinc (Zn) and copper (Cu) are needed for proper growth, development and functioning of plants. However, the higher concentration of these metals exhibits the toxic impact on plants (Rai et al. 2004). Even though highly selective transporters are found in the plants, but some other metals like metalloid arsenic (As), aluminium (Al), lead (Pb) and cadmium (Cd) are also taken by plants that may cause toxicity even at low concentration (Sebastiani et al. 2004). The most adverse effect of heavy metals on plants is lipid peroxidation that causes deterioration of membrane. Moreover, due to the heavy metal stress, reduction in the net photosynthesis occurs through the damage in the photosynthetic electron transport system and photosynthetic metabolism (Vinit-Dunand et al. 2002). ROS production is the major response of plant during heavy metal stress. However, mangroves have developed great ability to mitigate the deleterious effects of toxicity of heavy metal stress. They also have well-developed anti-oxidative defence system to combat with the adverse impacts of heavy metals caused oxidative damage. Higher activities of both enzymatic (APX, GPX, GR, SOD) and non-enzymatic antioxidants in heavy metal stressed plants have suggested their role in adaptive mechanisms (Verma and Dubey 2003; Yadav 2010). Due to the closeness of mangrove habitats with industrialized areas and population centres, they have continuous exposure/inputs of heavy metals. The sediments of mangroves also exhibit significant contamination of heavy metals. Several studies showed that mangrove has great potential to keep heavy metals and tolerate their higher concentration (Zhang et al. 2007; Huang and Wang 2010; Huang et al. 2010). During heavy metal stress, antioxidant activities of mangroves have been evaluated by many researchers (Macfarlane and Burchett 2001; Zhang et al. 2007; Caregnato et al. 2008). For instance, significantly higher activity of peroxidase was recorded in *A. marina* when exposed to Zn, Pb and Cu metal stress (concentration less than toxic effect) (Macfarlane and Burchett 2001). Enhanced activity of catalase enzyme was scored in the leaves and roots of *K. candel* and *B. gymnorrhiza* during stress (Hg, Cd and Pb). Moreover, the leaves of *B. gymnorrhiza* showed higher lipid peroxidation than *K. candel* under stress. However, higher catalase activity and lower lipid peroxidantion in *K. candel* validated its better tolerance of heavy metal stress than *B. gymnorrhiza* (Zhang et al. 2007). A significant increment in the quantity of glutathione, proline and phytochelatins has been recorded in the leaves of *B. gymnorrhiza* and *K. candel* under various heavy metal stresses such as Hg, Pb and Cd (Huang and Wang 2010). Leaf of *A. marina* exhibited a noteworthy constructive relationship between the activities of guaiacol peroxidase and zinc concentration during the studies of antioxidant (glutathione) and lipid peroxidation. Likewise, the lipid hydroperoxides concentration was also increased in proportion to increasing zinc concentration when subjected to 2 and 8 weeks of metal stress (Caregnato et al. 2008). Higher mortality (around 47% in the course of the first 4 years) was observed in the seedlings of *Rhizophora apiculata* when planted in the area earlier used for tin mining. However, mortality was attributed to the varied distribution of soil particle and microtopography instead of metal contamination.

Due to three characteristics of mangrove habitats have made it preferred sites from long ago for the dumping of sewage: (1) flow by waste disperses of habitat from an area to other parts, (2) filtration of nutrients itself by vegetation from the water and (3) physical processes as well as the soil of mangroves, microbes and algae associated with mangroves absorb a huge quantity of the pollutants (Wong et al. 1997). Nitrogen and phosphorous (nutrients) are found majorly in the pollutions. The soil of mangroves retained both phosphorus and nitrogen when treated with synthetic wastewater (Tam and Wong 1996). This observation suggests that mangroves tolerate organic pollutants, results must be seen very carefully as they cannot hold in other habitation. The sewage dumping impacts will be determined by sewage quantity, dumping duration and the distinct properties of each mangrove ecosystem. The patterns of water flow through habitat are especially important as it will determine the rate of flushing and pollutant residence times. Elevated levels of organic pollution may cause infection, demise and variation in the composition of species within mangrove forest. Sewage discharge into the Red sea killed pneumatophores of *A. mariana* (Mandura 1997). The pneumatophores deficiency reduced uptake of nutrient, the surface area for respiration and retarded the mangrove growth. Away from nutrients, organic pollution of the mangrove ecosystem can contain other debris and anthropogenic chemicals. The sediments of mangroves in Chengue Bay and Cienaga Grand de Santa Marta have noteworthy residues of organochlorine pesticide. The levels of some of these pollutants change according to season. The mangrove habitat of Jamaica contains a large quantity of non-mangrove wood and plastic. The amount of these solid wastes shows a strong relationship with rainfall in an adjacent metropolitan region.

10.4.5 Carbon Dioxide

The concentrations of atmospheric CO_2 have been rising continuously and their levels are likely to be more elevated in the upcoming century. It is assumed that increasing concentration of CO_2 can regulate and elevate the impacts of local environmental stress factors like changing patterns of salinity and tidal inundation upon the mangrove encroachment into nearby ecosystems (Saintilan and Rogers 2015). Enhanced concentrations of CO_2 are likely to increase water use efficiency and productivity of mangroves under certain environmental stress conditions. This could also enhance biomass, change biotic interaction and increase extent and coverage of mangroves (McKee and Rooth 2008; Cherry et al. 2009; Langley et al. 2009; McKee et al. 2012; Saintilan and Rogers 2015; Lovelock et al. 2016; Reef et al. 2016). These influences are also depending upon additional factors like salinity regimes, availability of nutrients and interactions with biotic factors (Lovelock et al. 2016; Reef et al. 2016). Elevated CO_2 may increase the mangrove growth; however, responses vary according to species. In many species, these responses are confounded by changes in water use efficiency, availability of nutrients and soil salinity. The *Rhizophora mangle*, *Rhizophora apiculata* and *Rhizophora stylosa* showed enhanced growth at elevated CO_2 level and less salt concentration, but not in the case of higher salt concentration. Moreover, all three species maturated earlier as compared to control plants (Ball et al. 1997). Net primary productivity of *A. germinans*, *R. mangle* and *Conocarpus erectus* was not influenced by elevated CO_2 level; however, productivity of *L. racemosa* was declined. Furthermore, all four species showed enhanced transpiration efficiency; but with increasing CO_2 concentration decline in transpiration and stomatal conductance was recorded (Snedaker and Araújo 1998). These outcomes suggest that responses of mangroves to increasing CO_2 will be complex, as some species blooming whereas others show declines or little or even no change. The effects of the interaction of elevated CO_2 with temperature, humidity, nutrient availability and salinity imply that location of the coastal area might be a key determinant in the mangrove response. For example, patterns of species in the estuaries can vary depending upon the capability of species to respond to temporal as well as spatial variations in the availability of nutrients, salinity and other factors in association to elevated CO_2 levels. The effects of double CO_2 concentration were analysed on the seedlings of *Rhizophora mangle* by Farnsworth et al. (1996). Seedling grown in double CO_2 concentration exhibited significant increment in the leaf area, branching activity, total stem length and biomass in comparison to seedling grown in normal CO_2 levels. In this experiment, under a high concentration of CO_2, the reproduction of *R. mangle* was attained only after 1 year of growth, whilst it naturally takes complete 2 years to become able for reproduction during field conditions. Hence, enhanced CO_2 also emerged to hasten the maturation of mangroves along with growth. It is predicted that the increased atmospheric CO_2 concentration elevates mangrove growth but it will not be enough to compensate the harmful effects of sea level rise. Other effects of elevated CO_2 and temperature on mangrove are the coral reefs degradation triggered by impaired growth and mass bleaching. Impairment of coral reefs can negatively affect the mangrove ecosystems that depend upon the reefs for providing shelter from wave action (where mangrove—reef couple system occurs).

10.4.6 Biotic Factors like Pests

Some of the animals and plants residing in the mangrove forest/mangal are severe pests. These pests cause impairment of the mangroves, reduction in the growth and productivity and killing the trees in extreme cases (Kathiresan and Bingham 2001). Certain harmful species may not harm directly to the mangroves rather than cause injury by competing for limited resources. Interspecific competition is the most common process in the forest of mangrove which is confirmed by allelopathic interaction among the species of mangrove. For instance, leaf litter of certain mangroves (*R. apiculata, Ceriops decandra* and *Lumnitzera racemosa*) secrete toxic leachates that impede the seedling growth of *R. mucronata* and *R. apiculata* (Kathiresan 1993). Generally, the situation of osmotic stress causes suberization and lignification in mangroves that prohibit the development of undergrowth of dense herbaceous in the forests of mangroves. Thus, they reduce intense competition between non-mangrove plants and mangroves. The *Acrostichum* (mangrove fern) is a weedy pest that causing substantial damages to mangrove forests (Kathiresan and Bingham 2001). Some pest injures the mangroves just by residing on their surface such as *Chiracanthium* and *Tetragnatha nitens* living on *Rhizophora*. On the leaves, they lay the eggs, which cause chlorosis, rolling of leaf, and wilting. Severe invasions may kill the mangroves. The *Phthirusa maritime* (semi-parasitic mistletoe) possess more intense and direct impacts on the trees. Higher rate of transpiration, lower water use efficiency, and rate of CO_2 assimilation were detected in *Coccoloba uvifera* and *Conocarpus erectus* when infected with this parasite (Orozco et al. 1990). Due to the feeding action of herbivores, most severe and extensive damage has been reported in the mangroves. The tidal regime and current are the main factors that controlled the distribution of pests. Indeed, insects are the most destructive among the animals feed on the canopy of mangroves. Some insect herbivores merely utilize mangroves as a substitute host since they are serious crop pests. Some examples of pests have the apparent choice for mangroves, viz. *Antestiopsis* on *Avicennia*, *Calliphara* on *Excoecaria*, *Glaucias* on *Lumnitzera* and *Mictis* on *Sonneratia*. Mangroves stands can be fully defoliated by insect herbivores (Kathiresan and Bingham 2001). When *Aspidiotus destructor* (scale insects) attacked on the leaves of *Rhizophora*, leaves turn yellow at the site of infection, subsequently brown and finally necrosis happens. In case of severe infection, leaves dry up, fall down and even the death of complete seedlings occurs. Certain herbivores especially infect to the seeds and reproductive tissues of mangroves. The *Afrocypholaelaps africana* (mite) feeds on the pollen of mangroves. Likewise, crabs are the main predators of seed that damage the seeds severely. However, mangrove forests cover a smaller portion of tropical forest; latest research demonstrated that herbivory levels of these forests were similar to other forests of the tropics and temperate. Although the survivals of mangroves are very well in response to an invasion because they have distinct phytochemicals such as derivatives of polyphenol, phenols, terpenoids, alkaloids and occurrence of numerous antioxidant defence mechanism (Bandaranayake 2002). Mutual association between mangroves and endophytes Aspergillus flavus (e.g. *E. agallocha L.*, *R. mucronata Lam.*, *A. officinalis L.*, *K. candel Druce*) improves the defence responses of mangroves against different abiotic and biotic stresses (Ravindran et al. 2012; Thatoi et al. 2014). The defence system of plants in response to pathogen attack includes oxidative burst process resulting in ROS production. Plants activated multiple defence process against biotic stress that prompts ROS scavenging system of plants. Insect herbivores can exhibit priority amongst mangrove hosts. For instance, the *Oiketicus kirbyi* (bagworm) eliminated around 5% foliage of the *Laguncularia racemosa*, 10% of the *Conocarpus erectus* and 80% of the *Avicennia germinans* in an Ecuadorian mangal (Gara et al. 1990). The physico-chemical properties may correlate with the susceptibility of individual mangroves plants or species of mangroves. More toughness of leaf decreases digestibility and palatability. Tannins also protect from infection with herbivores. Species of *Rhizophora* have higher levels of tannin, thus suffer less damage from herbivore attacks, than that of *Avicennia* species (having less content of tannins). The health of mangroves may also influence the feeding choice of insects. Trees enriched with nutrients tend to more suffering with herbivore attacks. *Rhizophora mangle* when treated with P and NPK exhibits significantly higher infection with herbivores *Marmara* (mines stem) and *Ecdytolopha* (feed on apical buds). However, treatment with N only did not enhance the attack of herbivory, but their seeds can also be infected by insects. The propagules of *Avicennia marina* are likely to injure by insect borers, but this does not kill them (Robertson and Duke 1990).

10.5 Ecological Importance of Mangroves

Mangroves are one of the most productive and valuable ecosystems in the world. They provide a lot of benefits to the local communities in addition to the ecology and environment surrounding them. In response to environmental inconstancy, mangroves show a higher range of community persistence and ecological stability. They provide an extensive array of economic and ecosystem services like resources of fuel and food; nursery ground for aquatic and terrestrial fauna; cycling of nutrients; formation of soil and wood; carbon sequestration and ecotourism (Alongi 2012). Mangroves stabilize and protect the coastline, enrich the waters of coastal area, supported fisheries of coastal area and yield commercial forest products. They are biodiversity-rich and protect from coastal erosion that happens because of intense tropical storm and tsunamis. The rates of carbon production in the mangroves ecosystem are equal to the tropical humid forest. Mangroves exhibit higher ratios of carbon mass in below- to above-ground than terrestrial trees and allocate correspondingly higher carbon in the below-ground part (Alongi 2012). Most carbon-rich biomes are mangroves, having an average of 937 tC ha^{-1}. They accelerate the rate of sediment accumulation (~5 mm $year^{-1}$), the burial of carbon (174 gC m^{-2} $year^{-1}$) and facilitating more accretion of fine particles (Alongi 2012). Mangroves contribute just around 1% (13.5 Gt year $^{-1}$) carbon sequestration of the world's forests. However, being habitats of the coastal area they contribute 14% carbon sequestration through the ocean. If the stocks of carbon are disturbed, it may result a huge emissions of gas (Alongi 2012). Mangroves modify the environment of soil which in turn influences the growth and survival of each and every functional variety of anaerobic as well as aerobic bacteria. Energetically efficient and highly evolved interactions of plant–soil–microbes are key factor account for high productivity of mangroves during the very harsh situation. The mangroves that are distinctly adjusted themselves to face the continuous tidal movement are also able to endure stronger energy of wind and force of waves happen during extreme events of weather. Based on the ecological situation, mangroves absorb minimum 70–90% of the waves energy and act like physical buffers between the shore and elements (UNEP-WCMC 2006). Organic matter and roots of mangroves absorb floodwater by working as a sponge and also play a role in to trap sediment. Roots of mangroves also act as filters to sort out pollutants that come to sea from inland waters. So, they help in improving the water quality reaching the ecosystem of the sea. The plant removes CO_2 from the atmosphere and accumulates it as biomass through a process that is recognized as carbon sequestration. That is why ocean and plants are known as carbon sinks. It is projected that a large number of carbons are sequestered by mangroves, around 25.5 million tonnes of carbon in a year (Eong 1993). There is also the anticipation that mangroves deliver greater than 10% of vital dissolved organic carbon that is provided to the world ocean from land (Dittmar et al. 2006). Due to the tangled and extensive supportive root systems of mangroves, they trap sediments more efficiently and prevent them from wash into the sea. The timber of mangrove is utilized to make and build furniture, houses, boats, fences and rafters all over the world. Around 300,000m^3 wood of mangrove is extracted every year from

Sundarbans (Miththapala 2008). The wood of mangroves is also employed as fuelwood and yet supply 90% of fuel utilized in Vietnam. The screw pine (*pandanus*) and mangrove palm (*Nypa*) leaves are used in weaving and thatching. The *Cerbera manghas* wood is used to crave puppets and masks in Sri Lanka. Due to the presence of salt glands, mangroves serve as a source of sodium. The ash of certain species like *Avicennia* is utilized as soap. The aerenchyma tissues of the several *Sonneratia spp* with their breathing roots are applied in the production of fish floats and corks. Tannins and gums are produced from the bark of various species that are yet utilized in the Indian subcontinents to cure fish nets and leather (Miththapala 2008). In India and Bangladesh, an important local industry is the production of honey from mangroves. They produce almost 20 tonnes of honey yearly from 200,000 ha of mangroves. Shoots, roots, leaves and fruits of mangroves serve as edible fruits and vegetables in several parts of the world. Moreover, drinks and sugars are extracted from the *Sonneratia* species (Miththapala 2008). Approximately 70 distinct mangrove plants are recorded to possess traditional medicinal utilization in the treatment of several diseases and ailments (Bandaranayake 1998). For instance, *Lumnitzera*, *Rhizophora* and *Bruguiera* are applied in many ailments like angina, diarrhoea and blood pressure (Upadhyay et al. 2002).

10.6 Future Prospective and Conclusion

Mangroves are a unique group of plants, located at the interface between land and sea. They grow luxuriantly in extreme environmental conditions like high salinity, extreme temperature and tides, flood, strong winds and anaerobic soil. For the growth and survival during harsh conditions, they developed several adaptations such as morphological, physiological and ecological. These adaptations make them more resistant and resilient to the adverse impacts of the harsh conditions. Mangroves provide a lot of economic and environmental services like nutrient cycling, carbon sequestration, soil formation, enrichment of biodiversity of coastal areas, supporting fisheries and ecotourism. They also protect coastal ecosystems from extreme environmental conditions. To overcome the deleterious impacts of salinity, mangroves possess well-developed ion sequestration, salt accumulation and salt excreting properties. To eliminate the excess ROS produced during stress conditions, they also have a variable extent of antioxidant enzyme and non-enzyme activity. These antioxidants help in ROS detoxification, protect the cellular and photosynthetic functions from the ROS induced oxidative stress and maintain healthy growth and productivity in the harsh conditions. Since very little information about the molecular mechanism of mangroves against climate change are available. A deep understanding and knowledge of the key stress-related gene and cross talk among the various signalling components must remain a strong research area in the future. By getting the complete picture of the fundamental mechanism of mangroves to cope with several biotic and abiotic stresses may be utilized in a logical and systemic way that will provide great vision in regard to their potential application. Moreover, the development of climate-resilient crops using the major stress-responsive genes from mangroves will be a better choice to overcome the problem of food security for ever-growing populations.

Acknowledgements CSIR-CSMCRI PRIS 172/2020.

The authors are heartily thankful to Council of Scientific and Industrial Research (CSIR), New Delhi for financial support in the form of infrastructure and different plan projects. AK is thankful to the Department of Biotechnology (DBT), Government of India and Academy of Scientific and Innovation Research (AcSIR), Ghaziabad for fellowship in the form of JRF/SRF and registration in the Ph.D program.

References

Abood JK, Lösel DM (2003) Changes in carbohydrate composition of cucumber leaves during the development of powdery mildew infection. Plant Pathology, 52(2):256–265.

Alongi D (2009) The energetics of mangrove forests. Springer Science & Business Media. Springer Netherlands.

Alongi DM (2002) Present state and future of the world's mangrove forests. Environ Conserv 29 (3):331–349.

Alongi DM (2008) Mangrove forests: resilience, protection from tsunamis, and responses to global climate change. Estuar Coast Shelf Sci 76(1):1–13.

Alongi DM (2012) Carbon sequestration in mangrove forests. Carbon Manage 3(3):313–322.

Alongi DM (2015) The impact of climate change on mangrove forests. Curr Clim Change Rep 1 (1):30–39.

Alongi DM, Sasekumar A, Tirendi F, Dixon P (1998) The influence of stand age on benthic decomposition and recycling of organic matter in managed mangrove forests of Malaysia. J Exp Mar Biol Ecol 225(2):197–218.

Andrews TJ, Clough BF, Muller GJ (1984) Photosynthetic gas exchange properties and carbon isotope ratios of some mangroves in North Queensland. In Physiology and Management of Mangroves (pp. 15–23). Springer, Dordrecht.

Atreya A, Vartak V, Bhargava S (2009) Salt priming improves tolerance to desiccation stress and to extreme salt stress in Bruguiera cylindrica. Inter J Integr Biol 6(2):68–73.

Ball MC, Sobrado MA (2002) Ecophysiology of mangroves: challenges in linking physiological processes with patterns in forest structure. In Physiological Plant Ecology: 39th Symposium of the British Ecological Society (Vol. 39, pp. 331–346), Cambridge University Press.

Ball MC, Pidsley SM (1995) Growth responses to salinity in relation to distribution of two mangrove species, *Sonneratia alba* and *S. lanceolata*, in northern Australia. Func Ecol 9:77–85.

Ball MC, Cochrane MJ, Rawson HM (1997) Growth and water use of the mangroves *Rhizophora apiculata* and *R. stylosa* in response to salinity and humidity under ambient and elevated concentrations of atmospheric CO_2. Plant Cell Environ 20(9):1158–1166.

Bandaranayake W (1998) Traditional and medicinal uses of mangroves. Mangroves and Salt Marshes 2(3):133–148.

Bandaranayake WM (2002) Bioactivities, bioactive compounds and chemical constituents of mangrove plants. Wet Ecol Manage 10(6):421–452.

Barnabás B, Jäger K, Fehér A (2008) The effect of drought and heat stress on reproductive processes in cereals. Plant cell Environ 31(1):11–38.

Bengtsson L, Hodges KI, Roeckner E (2006) Storm tracks and climate change. J Clim 19 (15):3518–3543.

Bhargava S, Sawant K (2013) Drought stress adaptation: metabolic adjustment and regulation of gene expression. Plant Breeding 132(1):21–32.

Bruun P (1988) The Bruun rule of erosion by sea-level rise: a discussion on large-scale two-and three-dimensional usages. J Coast Res 4:627–648.

Burns WC (2001) The possible impacts of climate change on Pacific island state ecosystems. Inter J Global Environ Issues 1(1):56–72.

Cahoon DR, Hensel P (2006) High-resolution global assessment of mangrove responses to sea-level rise: a review. In Proceedings of the Symposium on Mangrove Responses to Relative Sea Level Rise and Other Climate Change Effects (Vol. 13, pp. 9–17).

Cahoon DR, Hensel P, Rybczyk J, McKee KL, Proffitt CE, Perez BC (2003) Mass tree mortality leads to mangrove peat collapse at Bay Islands, Honduras after Hurricane Mitch. J Ecol 91 (6):1093–1105.

Cahoon DR, Hensel PF, Spencer T, Reed DJ, McKee KL, Saintilan N (2006) Coastal wetland vulnerability to relative sea-level rise: wetland elevation trends and process controls. In Wetlands and Natural Resource Management (pp. 271–292). Springer, Berlin, Heidelberg.

Caregnato FF, Koller CE, MacFarlane GR, Moreira JC (2008) The glutathione antioxidant system as a biomarker suite for the assessment of heavy metal exposure and effect in the grey mangrove, *Avicennia marina* (Forsk.) Vierh. Mar Poll Bull 56(6):1119–1127.

Cavanaugh KC, Kellner JR, Forde AJ, Gruner DS, Parker JD, Rodriguez W, Feller IC (2014) Poleward expansion of mangroves is a threshold response to decreased frequency of extreme cold events. Proc Nat Acad Sci 111(2):723–727.

Chakraborty K, Bhaduri D, Uprety DC, Patra AK (2014) Differential response of plant and soil processes under climate change: a mini-review on recent understandings. Proc Nat Acad Sci, India Section B: Biol Sci 84(2):201–214.

Chapman VJ (1976) Mangrove vegetation. Vaduz, Liechtenstein: Cramer, 447.

Chen R, Twilley RR (1998) A gap dynamic model of mangrove forest development along gradients of soil salinity and nutrient resources. J Ecol 86(1):37–51.

Cherian S, Reddy MP, Pandya JB (1999) Studies on salt tolerance in *Avicennia marina* (Forstk.) Vierh.: effect of NaCl salinity on growth, ion accumulation and enzyme activity. Indian J Plant Physiol 4(4):266–270.

Cherry JA, McKee KL, Grace JB (2009) Elevated CO_2 enhances biological contributions to elevation change in coastal wetlands by offsetting stressors associated with sea-level rise. J Ecol 97(1):67–77.

Christensen JH, Hewitson B, Busuioc A, Chen A, Gao X, Held R, Jones R, Kolli RK, Kwon WK, Laprise R, Magaña Rueda V, Mearns L, Menéndez CG, Räisänen J, Rinke A, Sarr A, Whetton P (2007) Regional climate projections. In Climate Change, 2007: The Physical Science Basis. Contribution of Working group I to the Fourth Assessment Report of the Intergovernmental Panel on Climate Change, University Press, Cambridge, Chapter 11 (pp. 847–940). Cambridge/New York.

Church JA, White NJ (2006) A 20th century acceleration in global sea-level rise. Geophys Res Lett 33(1): https://doi.org/10.1029/2005GL024826.

Church JA, Gregory JM, Huybrechts P, Kuhn M, Lambeck K, Nhuan MT, Qin D, Woodworth PL (2001) Changes in sea level. In: JT Houghton, Y Ding, DJ Griggs, M Noguer, PJ Van der Linden, X. Dai, K. Maskell, and CA Johnson (eds.): Climate Change 2001: The Scientific Basis: Contribution of Working Group I to the Third Assessment Report of the Intergovernmental Panel (pp. 639–694). Cambridge University Press, Cambridge, United Kingdom, and New York, NY, USA.

Dahdouh-Guebas F, Koedam N, Danielsen F, Sørensen MK, Olwig MF, Selvam V, Parish F, Burgess ND, Topp-Jørgensen E, Hiraishi T, Karunagaran VM, Rasmussen MS, Hansen LB, Quarto A, Suryadiputra N (2006) Coastal vegetation and the Asian tsunami. Science 311:37–38.

Dai A (2011) Drought under global warming: a review. Wiley Interdisciplinary Reviews: Climate Change 2(1):45–65.

Danielsen F, Sørensen MK, Olwig MF, Selvam V, Parish F, Burgess ND, Hiraishi T, Karunagaran VM, Rasmussen MS, Hansen LB, Quarto A (2005) The Asian tsunami: a protective role for coastal vegetation. Science 310(5748):643–643.

Das PK, Chakravarti V, Dutta A, Maity S (1995) Leaf anatomy and chlorophyll estimates in some mangroves. Indian Forester 121(4):289–294.

Das S, Ghose M (1993) Morphology of stomata and leaf hairs of some halophytes from Sundarbans, West Bengal. Phytomorphol 43(1–2):59–70.

Das SK, Patra JK, Thatoi H (2016) Antioxidative response to abiotic and biotic stresses in mangrove plants: A review. Internat Rev Hydrobiol 101(1–2):3–19.

Dasgupta N, Nandy P, Sengupta C, Das S (2012) Protein and enzymes regulations towards salt tolerance of some Indian mangroves in relation to adaptation. Trees 26(2):377–391.

De Ollas C, Morillón R, Fotopoulos V, Puértolas J, Ollitrault P, Gómez-Cadenas A, Arbona V (2019) Facing climate change: biotechnology of iconic Mediterranean woody crops. Front Plant Sci 10:427.

Dittmar T, Hertkorn N, Kattner G, Lara RJ (2006) Mangroves, a major source of dissolved organic carbon to the oceans. Global Biogeochem Cyc 20(1):GB1012, doi: https://doi.org/10.1029/2005GB002570.

Doganlar ZB, Demir K, Basak H, Gul I (2010) Effects of salt stress on pigment and total soluble protein contents of three different tomato cultivars. Afr J Agri Res 5(15):2056–2065.

Duke N, Ball M, Ellison J (1998) Factors influencing biodiversity and distributional gradients in mangroves. Glob Ecol Biogeogr Lett 7(1):27–47.

Duke NC (1992) Mangrove floristics and biogeography. Tropical mangrove ecosystems, (pp. 63–100). American Geophysical Union, Washington, D.C., USA.

Duke NC, Kovacs JM, Griffiths AD, Preece L, Hill DJ, Van Oosterzee P, Mackenzie J, Morning HS, Burrows D (2017) Large-scale dieback of mangroves in Australia's Gulf of Carpentaria: a severe ecosystem response, coincidental with an unusually extreme weather event. Mar Freshwater Res 68(10):1816–1829.

Duke NC, Meynecke JO, Dittmann S, Ellison AM, Anger K, Berger U, Cannicci S, Diele K, Ewel KC, Field CD, Koedam N (2007) A world without mangroves?. Science 317(5834):41–42.

Ellison JC, Stoddart DR (1991) Mangrove ecosystem collapse during predicted sea-level rise: Holocene analogues and implications. J Coast Res 7:151–165.

Ellison JC (2000) How South Pacific mangroves may respond to predicted climate change and sea-level rise. In Climate Change in the South Pacific: Impacts and Responses in Australia, New Zealand, and Small Island States (pp. 289–300). Springer, Dordrecht.

Eong OJ (1993) Mangroves-a carbon source and sink. Chemosphere 27(6):1097–1107.

Eslami-Andargoli L, Dale PER, Sipe N, Chaseling J (2009) Mangrove expansion and rainfall patterns in Moreton Bay, southeast Queensland, Australia. Estuar Coast Shelf Sci 85 (2):292–298.

Fancy NN, Bahlmann AK, Loake GJ (2017) Nitric oxide function in plant abiotic stress. Plant Cell Environ 40(4):462–472.

Farnsworth EJ, Ellison AM, Gong WK (1996) Elevated CO_2 alters anatomy, physiology, growth, and reproduction of red mangrove (Rhizophora mangle L.). Oecologia 108(4):599–609.

Feller IC (1995) Effects of nutrient enrichment on growth and herbivory of dwarf red mangrove (Rhizophora mangle). Ecol Monogr 65(4):477–505.

Field CD (1995) Impact of expected climate change on mangroves. In Asia-Pacific Symposium on Mangrove Ecosystems (pp. 75–81). Springer, Dordrecht.

Fleta-Soriano E, Munné-Bosch S (2016) Stress memory and the inevitable effects of drought: a physiological perspective. Front Plant Sci 7:143.

Food and Agriculture Organization of the United Nations, (2007) The world's mangroves 1980–2005. FAO Forestry Paper 153.

Forbes MG, Dunton KH (2006) Response of a subtropical estuarine marsh to local climatic change in the southwestern Gulf of Mexico. Estuar Coast 29(6):1242–1254.

Foyer CH, Shigeoka S (2011) Understanding oxidative stress and antioxidant functions to enhance photosynthesis. Plant physiol 155(1):93–100.

Gara RI, Sarango A, Cannon PG (1990) Defoliation of an Ecuadorian mangrove forest by the bagworm, *Oiketicus kirbyi* Guilding (Lepidoptera: Psychidae). J Tropic For Sci 3(2):181–186.

García-Sánchez F, Syvertsen JP, Gimeno V, Botía P, Perez-Perez JG (2007) Responses to flooding and drought stress by two citrus rootstock seedlings with different water-use efficiency. Physiol Plant 130(4):532–542.
Geng Q, Sugi M (2003) Possible change of extratropical cyclone activity due to enhanced greenhouse gases and sulfate aerosols-Study with a high-resolution AGCM. J Clim 16 (13):2262–2274.
Gilman EL, Ellison J, Duke NC, Field C (2008) Threats to mangroves from climate change and adaptation options: a review. Aquat Bot 89(2):237–250.
Gopal B, Chauhan M (2006) Biodiversity and its conservation in the Sundarban mangrove ecosystem. Aquat Sci 68(3):338–354.
Hamilton SE, Casey D (2016) Creation of a high spatio-temporal resolution global database of continuous mangrove forest cover for the 21st century (CGMFC-21). Glob Ecol Biogeogr 25 (6):729–738.
Hirayama T, Shinozaki K (2010) Research on plant abiotic stress responses in the post-genome era: Past, present and future. The Plant J 61(6):1041–1052.
Holgate SJ, Woodworth PL (2004) Evidence for enhanced coastal sea level rise during the 1990s. Geophys Res Lett 31(7).
Houghton J, Ding Y, Griggs D, Noguer M, van der Linden P, Dai X, Maskell K, Johnson C (Eds.), 2001. Climate Change 2001: The Scientific Basis (Published for the Intergovernmental Panel on Climate Change). Cambridge University Press, Cambridge, United Kingdom, and New York, NY, USA.
Huang GY, Wang YS (2010) Expression and characterization analysis of type 2 metallothionein from grey mangrove species (Avicennia marina) in response to metal stress. Aquat Toxicol 99 (1):86–92.
Huang GY, Wang YS, Sun CC, Dong JD, Sun ZX (2010) The effect of multiple heavy metals on ascorbate, glutathione and related enzymes in two mangrove plant seedlings (*Kandelia candel* and *Bruguiera gymnorrhiza*). Oceanol Hydrobiol Stud 39(1):11–25.
Hutchings P, Saenger P (1987) Ecology of mangroves. Ecology of mangroves. University of Queensland Press, St. Lucia.
IPCC (2007) Climate change 2007: synthesis report. Contribution of Working Groups I, II and III to the Fourth Assessment Report of the Intergovernmental Panel on Climate Change. IPCC, Geneva, Switzerland, p. 104.
Jaleel CA, Manivannan P, Wahid A, Farooq M, Al-Juburi HJ, Somasundaram R, Panneerselvam R (2009) Drought stress in plants: a review on morphological characteristics and pigments composition. Inter J Agri Biol 11(1):100–105.
Jithesh MN, Prashanth SR, Sivaprakash KR, Parida AK (2006) Antioxidative response mechanisms in halophytes: their role in stress defence. J Genet 85(3):237.
Joshi R, Wani SH, Singh B, Bohra A, Dar ZA, Lone AA, Pareek A, Singla-Pareek SL (2016) Transcription factors and plants response to drought stress: current understanding and future directions. Front Plant Sci 7:1029.
Kaly UL, Eugelink G, Robertson AI (1997) Soil conditions in damaged North Queensland mangroves. Estuaries 20(2):291–300.
Kandasamy K (2017) Mangroves in India and climate change: an overview. In: Participatory Mangrove Management in a Changing Climate (pp. 31–57). Springer, Tokyo.
Kathiresan K, Bingham BL (2001) Biology of mangroves and mangrove ecosystems. Adv Mar Biol 40:84–254.
Kathiresan K, Rajendran N (2005) Coastal mangrove forests mitigated tsunami. Estuar Coast Shelf Sci 65(3):601–606.
Kathiresan K (1993) Dangerous pest on nursery seedlings of Rhizophora. Indian Forester 119(12).
Keyvan S (2010) The effects of drought stress on yield, relative water content, proline, soluble carbohydrates and chlorophyll of bread wheat cultivars. J Ani Plant Sci 8(3):1051–1060.
Koch MS, Snedaker SC (1997) Factors influencing Rhizophora mangle L. seedling development in Everglades carbonate soils. Aquat Bot 59(1–2):87–98.

Korres NE, Norsworthy JK, Tehranchian P, Gitsopoulos TK, Loka DA, Oosterhuis DM, Gealy DR, Moss SR, Burgos NR, Miller MR, Palhano M (2016) Cultivars to face climate change effects on crops and weeds: a review. Agro Sustain Develop 36(1):12.

Kosová K, Prášil IT, Vítámvás P (2013) Protein contribution to plant salinity response and tolerance acquisition. Internat J Mole Sci 14(4):6757–6789.

Krauss KW, Allen JA, Cahoon DR (2003) Differential rates of vertical accretion and elevation change among aerial root types in Micronesian mangrove forests. Estuar Coast Shelf Sci 56 (2):251–259.

Kundzewicz ZW, Mata LJ, Arnell NW, Doll P, Kabat P, Jimenez B, Miller K, Oki T, Zekai S, Shiklomanov I (2007) Freshwater resources and their management. In Climate change 2007: impacts, adaptation and vulnerability contribution of Working Group II to the Fourth Assessment Report of the Intergovernmental Panel on Climate Change. (pp. 173–210) Cambridge University Press, Cambridge.

Lacerda LD, Ittekkot V, Patchineelam SR (1995) Biogeochemistry of Mangrove Soil Organic Matter: a Comparison Between Rhizophora and Avicennia Soils in South-eastern Brazil. Estuar Coast Shelf Sci 40(6):713–720.

Langley JA, McKee KL, Cahoon DR, Cherry JA, Megonigal JP (2009) Elevated CO_2 stimulates marsh elevation gain, counterbalancing sea-level rise. Proce Nat Acad Sci 106(15):6182–6186.

Li MS, Lee SY (1997) Mangroves of China: a brief review. Forest Ecol Manage 96(3):241–259.

List JH, Sallenger Jr AH, Hansen ME, Jaffe BE (1997) Accelerated relative sea-level rise and rapid coastal erosion: testing a causal relationship for the Louisiana barrier islands. Mar Geol 140 (3–4):347–365.

Lovelock CE, Ellison JC, (2007) Vulnerability of mangroves and tidal wetlands of the Great Barrier Reef to climate change. In: Climate Change and the Great Barrier Reef: A Vulnerability Assessment. (pp. 237–269) Great Barrier Reef Marine Park Authority and Australian Greenhouse Office, Australia.

Lovelock CE, Krauss KW, Osland MJ, Reef R, Ball MC (2016) The physiology of mangrove trees with changing climate. In Tropical Tree Physiology: Adaptations and Responses in a Changing Environment, (pp. 149–179). Springer, New York, New York, USA.

MacFarlane GR, Burchett MD (2001) Photosynthetic pigments and peroxidase activity as indicators of heavy metal stress in the grey mangrove, *Avicennia marina* (Forsk.) Vierh. Mar Pollut Bull 42 (3):233–240.

MacNae W (1969) A general account of the fauna and flora of mangrove swamps and forests in the Indo-West-Pacific region. In: Advances in Marine Biology (Vol. 6, pp. 73–270). Academic Press.

Mandura AS (1997) A mangrove stand under sewage pollution stress: Red Sea. Mangroves and Salt Marshes, 1(4):255–262.

McKee K, Rogers K, Saintilan N (2012) Response of salt marsh and mangrove wetlands to changes in atmospheric CO_2, climate, and sea level. In Global Change and the Function and Distribution of Wetlands (pp. 63–96). Springer, Dordrecht.

McKee KL, Rooth JE (2008) Where temperate meets tropical: multi-factorial effects of elevated CO_2, nitrogen enrichment, and competition on a mangrove-salt marsh community. Glob Change Biol 14:971–984.

McKee KL (1993) Soil physicochemical patterns and mangrove species distribution—reciprocal effects?. J Ecol 81:477–487.

McKee KL, Cahoon DR, Feller IC (2007) Caribbean mangroves adjust to rising sea level through biotic controls on change in soil elevation. Glob Ecol Biogeogr 16(5):545–556.

Miller PC, Hom J, Poole DK (1975) Water relations of three mangrove species in south Florida. Oecol Plant Fr Da 10:355–367.

Miththapala S (2008) Mangroves. Coastal ecosystems series, Volume 2. Mangroves. Coastal ecosystems series, Volume 2. IUCN, Sri Lanka.

Mitra A, Gangopadhyay A, Dube A, Schmidt AC, Banerjee K (2009) Observed changes in water mass properties in the Indian Sundarbans (northwestern Bay of Bengal) during 1980–2007. Curr Sci 1445–1452.

Nabeelah Bibi S, Fawzi MM, Gokhan Z, Rajesh J, Nadeem N, Rengasamy Kannan RR, Albuquerque RDDG, Pandian SK (2019) Ethnopharmacology, phytochemistry, and global distribution of mangroves-A comprehensive review. Marine Drugs, 17(4):231.

Nakashima K, Ito Y, Yamaguchi-Shinozaki K (2009) Transcriptional regulatory networks in response to abiotic stresses in Arabidopsis and grasses. Plant Physiol 149(1):88–95.

Nikalje GC, Srivastava AK, Pandey GK, Suprasanna P (2018) Halophytes in biosaline agriculture: Mechanism, utilization, and value addition. Land Degr Develop 29(4):1081–1095.

Orozco A, Rada F, Azocar A, Goldstein G (1990) How does a mistletoe affect the water, nitrogen and carbon balance of two mangrove ecosystem species?. Plant Cell Environ 13(9):941–947.

Osland MJ, Enwright N, Stagg CL (2014) Freshwater availability and coastal wetland foundation species: ecological transitions along a rainfall gradient. Ecology 95(10):2789–2802.

Osland MJ, Enwright N, Day RH, Doyle TW (2013) Winter climate change and coastal wetland foundation species: salt marshes vs. mangrove forests in the southeastern United States. Glob Change Biol 19(5):1482–1494.

Parida AK, Jha B (2010) Salt tolerance mechanisms in mangroves: a review. Trees 24(2):199–217.

Parida AK, Das AB, Mohanty P (2004) Investigations on the antioxidative defence responses to NaCl stress in a mangrove, Bruguiera parviflora: differential regulations of isoforms of some antioxidative enzymes. Plant Growth Regulation 42(3):213–226.

Pilkey OH, Cooper JAG (2004) Society and sea level rise. Science 303(5665):1781–1782.

Piou C, Feller IC, Berger U, Chi F (2006) Zonation patterns of Belizean offshore mangrove forests 41 years after a catastrophic hurricane 1. Biotropica: J Biol Conserv 38(3):365–374.

Qadir M, Tubeileh A, Akhtar J, Larbi A, Minhas PS, Khan MA (2008) Productivity enhancement of salt-affected environments through crop diversification. Land Degr Develop 19(4):429–453.

Rai V, Vajpayee P, Singh SN, Mehrotra S (2004) Effect of chromium accumulation on photosynthetic pigments, oxidative stress defense system, nitrate reduction, proline level and eugenol content of Ocimum tenuiflorum L. Plant Sci 167(5):1159–1169.

Ravindran C, Naveenan T, Varatharajan GR, Rajasabapathy R, Meena RM (2012) Antioxidants in mangrove plants and endophytic fungal associations. Bot Mar 55(3):269–279.

Reef R, Slot M, Motro U, Motro M, Motro Y, Adame MF, Garcia M, Aranda J, Lovelock CE, Winter K (2016) The effects of CO_2 and nutrient fertilisation on the growth and temperature response of the mangrove Avicennia germinans. Photosynth Res 129(2):159–170.

Reid PC, Beaugrand G (2012) Global synchrony of an accelerating rise in sea surface temperature. Marine Biological Association of the United Kingdom, 92(7):1435.

Robertson AI, Duke NC (1990) Mangrove fish-communities in tropical Queensland, Australia: spatial and temporal patterns in densities, biomass and community structure. Mar Biol 104 (3):369–379.

Rogers K, Saintilan N, Cahoon D (2005) Surface elevation dynamics in a regenerating mangrove forest at Homebush Bay, Australia. Wet Ecol Manage 13(5):587–598.

Saenger P (2002) Mangrove ecology, silviculture and conservation. Springer Science & Business Media Berlin, Germany.

Saintilan N, Rogers K (2015) Woody plant encroachment of grasslands: a comparison of terrestrial and wetland settings. New Phytol 205(3):1062–1070.

Saintilan N, Williams RJ (1999) Mangrove transgression into saltmarsh environments in south-east Australia. Glob Ecol Biogeogr 8(2):117–124.

Saintilan N, Wilson NC, Rogers K, Rajkaran A, Krauss KW (2014) Mangrove expansion and salt marsh decline at mangrove poleward limits. Glob Change Biol 20(1):147–157.

Sapeta H, Costa JM, Lourenco T, Maroco J, Van der Linde P, Oliveira MM (2013) Drought stress response in Jatropha curcas: growth and physiology. Environ Exp Bot 85:76–84.

Schneider P (2011) The discovery of tropical mangroves in graeco-roman antiquity: Science and wonder. Hakluyt Society pp. 1–16.

Scholander PF (1968) How mangroves desalinate seawater. Physiol Plant 21(1):251–261.
Scholander PF, Hammel HT, Hemmingsen EA, Bradstreet ED (1964) Hydrostatic pressure and osmotic potential in leaves of mangroves and some other plants. Proce Nat Acad Sci, USA 52 (1):119.
Sebastiani L, Scebba F, Tognetti R (2004) Heavy metal accumulation and growth responses in poplar clones Eridano (Populus deltoides x maximowiczii) and I-214 (P. x euramericana) exposed to industrial waste. Environ Exp Bot 52(1):79–88.
Singh AK, Ansari A, Kumar D, Sarkar UK (2012) Status, biodiversity and distribution of mangroves in India: an overview. Uttar Pradesh State Biodiversity Board. Marine Biodiversity: One Ocean, Many Worlds of Life, 59–67.
Snedaker S (1993) Impact on mangroves. In: Climate Change in the Intra-American Seas: Implications of Future Climate Change on the Ecosystems and Socio-economic Structure of the Marine and Coastal Regimes of the Caribbean Sea, Gulf of Mexico, Bahamas and N. E. Coast of South America. (pp. 282–305) Edward Arnold, London.
Snedaker SC, Araújo RJ (1998) Stomatal conductance and gas exchange in four species of Caribbean mangroves exposed to ambient and increased CO_2. Mar Freshwat Res 49 (4):325–327.
Snedaker SC (1995) Mangroves and climate change in the Florida and Caribbean region: scenarios and hypotheses. Hydrobiologia 295(1–3):43–49.
Sobrado MA (2005) Leaf characteristics and gas exchange of the mangrove Laguncularia racemosa as affected by salinity. Photosynthetica 43(2):217–221.
Solomon S, Qin D, Manning M, Chen Z, Marquis M, Averyt KB, Tignor M, Miller HL (2007) Climate change 2007: The physical science basis. Intergovernmental Panel on Climate Change (IPCC), Cambridge University Press, Cambridge.
Spaulding M, Kainuma M, Collins L (2010) World Atlas of Mangroves. A collaborative project of ITTO, ISME, FAO, UNEP-WCMC, Earthscan: London, UK.
Suárez N, Sobrado MA (2000) Adjustments in leaf water relations of mangrove (Avicennia germinans) seedlings grown in a salinity gradient. Tree Physiol 20(4):277–282.
Takemura T, Hanagata N, Sugihara K, Baba S, Karube I, Dubinsky Z (2000) Physiological and biochemical responses to salt stress in the mangrove, *Bruguiera gymnorrhiza*. Aquat Bot 68 (1):15–28.
Tam NFY, Wong YS (1996) Retention of wastewater-borne nitrogen and phosphorus in mangrove soils. Environ Technol 17(8):851–859.
Thatoi H, Samantaray D, Das SK (2016) The genus Avicennia, a pioneer group of dominant mangrove plant species with potential medicinal values: a review. Front Life Sci 9(4):267–291.
Thatoi HN, Patra JK, Das SK (2014) Free radical scavenging and antioxidant potential of mangrove plants: a review. Acta Physiol Plant 36(3):561–579.
Thomas N, Lucas R, Bunting P, Hardy A, Rosenqvist A, Simard M (2017) Distribution and drivers of global mangrove forest change, 1996–2010. PloS one 12(6):e0179302.
Thomas R, Rignot E, Casassa G, Kanagaratnam P, Acuña C, Akins T, Brecher H, Frederick E, Gogineni P, Krabill W, Manizade S (2004) Accelerated sea-level rise from West Antarctica. Science 306(5694):255–258.
Tomlinson PB (1986) The Botany of Mangroves. Cambridge University Press, United Kingdom.
Triwilaida Intari SE (1990) Factors affecting the death of mangrove trees in the Pedada Strait, Indragiri Hilir, Riau, with reference to the site conditions. Buletin Penelitian Hutan (531):33–48.
UNEP-WCMC (2006) In the front line: shoreline protection and other ecosystem services from mangroves and coral reefs. UNEP-WCMC, Cambridge, UK. pp. 33.
Upadhyay VP, Ranjan R, Singh JS (2002) Human-mangrove conflicts: The way out. Curr Sci 83 (11):1328–1336.
Van Lavieren H, Spalding M, Alongi DM, Kainuma M, Clüsener-Godt M, Adeel Z (2012) Securing the future of mangroves. A policy brief. United Nations University, Institute for Water, Environment and Health, UNESCO-MAB with ISME, ITTO, FAO, UNEP-WCMC and TNC. p. 53, Canada.

Verma S, Dubey RS (2003) Lead toxicity induces lipid peroxidation and alters the activities of antioxidant enzymes in growing rice plants. Plant Sci 164(4):645–655.

Vinit-Dunand F, Epron D, Alaoui-Sossé B, Badot PM (2002) Effects of copper on growth and on photosynthesis of mature and expanding leaves in cucumber plants. Plant Sci 163(1):53–58.

Vyas P (2012) Biodiversity conservation in Indian Sundarban in the context of anthropogenic pressures and strategies for impact mitigation. Saurashtra University, Rajkot, India.

Whelan KR, Smith TJ, Cahoon DR, Lynch JC, Anderson GH (2005) Groundwater control of mangrove surface elevation: Shrink and swell varies with soil depth. Estuaries 28(6):833–843.

Wong YS, Tam NFY, Lan CY (1997) Mangrove wetlands as wastewater treatment facility: a field trial. In Asia-Pacific Conference on Science and Management of Coastal Environment (pp. 49–59). Springer, Dordrecht.

Yadav SK (2010) Heavy metals toxicity in plants: an overview on the role of glutathione and phytochelatins in heavy metal stress tolerance of plants. South Afr J Bot 76(2):167–179.

Yin JH (2005) A consistent poleward shift of the storm tracks in simulations of 21st century climate. Geophys Res Lett 32(18).

Zhang FQ, Wang YS, Lou ZP, Dong JD (2007) Effect of heavy metal stress on antioxidative enzymes and lipid peroxidation in leaves and roots of two mangrove plant seedlings (Kandelia candel and Bruguiera gymnorrhiza). Chemosphere 67(1):44–50.

Zheng WJ, Wang WQ, Lin, P., (1999) Dynamics of element contents during the development of hypocotyles and leaves of certain mangrove species. J Exp Mar Biol Ecol 233(2):247–257.

Zornoza P, Vázquez S, Esteban E, Fernández-Pascual M, Carpena R (2002) Cadmium-stress in nodulated white lupin: strategies to avoid toxicity. Plant Physiol Biochemi 40(12):1003–1009.

Role of Mangroves in Pollution Abatement 11

Arumugam Sundaramanickam, Ajith Nithin, and
Thangavel Balasubramanian

Abstract

Mangroves are one of the significant categories of coastal vegetation, distributed in the shorelines of tropical and subtropical regions and performing a dynamic role in coastal ecosystems. The recent developments of coastal cities, industrialization, unplanned recreational activities and expansion of aquaculture at mangrove areas in various regions of the world have generated a threat on such significant ecosystems. Nevertheless, the mangrove vegetation has some adaptive features to mitigate the pollution. These ecosystems act as physical, chemical and biological barriers for the transference of pollutants. They play a vital role in trapping sediments; assimilate the excess nutrients by phyto- or bioremediation of toxic substances. The earlier studies revealed that mangroves can act as possible phytoremediators. They absorb a considerable quantity of toxic metal ions and store them in various parts such as stems and roots, consequently evade the transmission of heavy metal ions. In addition, the mangrove associated microorganisms are responsible for remediation of numerous toxic contaminants present within the mangroves. In view of the above, the present chapter offers a comprehensive discussion on the role of mangrove vegetation on abatement of pollutants.

Keywords

Mangrove ecosystem · Heavy metals · Plastics · Hydrocarbons

A. Sundaramanickam (✉) · A. Nithin · T. Balasubramanian
Faculty of Marine Sciences, CAS in Marine Biology, Annamalai University, Parangipettai, India

R. P. Rastogi et al. (eds.), *Mangroves: Ecology, Biodiversity and Management*,
https://doi.org/10.1007/978-981-16-2494-0_11

11.1 Introduction

Mangroves forests are composed of 73 species of trees and shrubs that grow along the coasts of temperate regions and are known for their ecological and socioeconomic importance (FAO 2007; Sandilyan and Kathiresan 2012). These comprise of woody trees, shrubs and flowering trees which are well adapted to the estuarine conditions (Spalding et al. 2010). Mangroves have highly adapted morphological and physiological features that acclimatize to the saline and anoxic environmental conditions, which enable the increase of biodiversity and biomass productivity proving to be ecologically and economically beneficial (Mukherjee et al. 2014; Duke and Schmitt 2015). These benefits include fisheries, maintaining coastal water quality, shoreline protection, primary production, aquaculture, tourism and recreation, whereas the most important global service is its role in pollution abatement (Duke et al. 2007; Richards and Fries 2016).

Mangroves have several ecological functions, which contribute to economy and improve quality of human life (Rodríguez-Rodríguez et al. 2016). Mangrove ecologies are regarded as barriers prohibiting the release of terrestrial contaminants into the oceans (Li et al. 2019). Mangroves play an essential role in climate regulation, soil stabilization and controlling erosion, regulating nutrient cycles and protecting seagrasses as well as coral reefs (Alongi 2014).

Mangroves act as protective barriers against several natural disasters however this role is dependent on several factors, viz. tree height, density and species composition, dominant species make up (roots, stems, branches and foliage), diameter of roots and trunks, and habitats elevation, nature of channel, pools as well as status of ecological degradation of the forests (McIvor et al. 2012). These natural barriers are also effective against storms, cyclones and tidal waves in several tropical regions (as reviewed by Sandilyan and Kathiresan 2015).

Over the past half century several diseases have occurred globally where the chief causative agents have been heavy metals. Heavy metal pollution especially in the marine environments is a serious threat owing to their persistence, toxicity and non-degradable nature. Heavy metals originate from mining, aquaculture, smelting, agriculture, printing and from industrial sources, viz. petrochemical and electronic industry while discharge of untreated municipal wastes into the marine environment is also one of the sources (Paz-Alberto and Sigua 2013). All discharges from these sources enter the marine habitats where they get accumulated in the organisms and are biomagnified to higher organisms (Rainbow and Luoma 2011). The primary sources for pollution in the mangroves include trace metals, oil residues, pesticides and untreated wastewater. These pollutants affect photosynthesis and growth of plants ultimately leading to mortality (Lovelock et al. 2009).

In short, it can be stated that mangroves offer numerous benefits; however, these may become retarded due to anthropogenic activities that affect mangroves. The following section describes some of the services rendered by the mangroves.

11.2 Solutions from the Mangroves to Pollution Problems

Mangrove ecosystem is of vast ecological, traditional and socioeconomic values (Kathiresan and Bingham 2001). They are actively involved in controlling of marine toxicants, such as heavy metals, pesticides, oils & greases, nutrients and plastics (Wang Q et al. 2019b). They are also involved in carbon sequestration and coastal production. Furthermore, the mangrove residing microorganisms are actively involved in degradation of a wide array of pollutants (Fig. 11.1).

11.2.1 Reducing Nutrient Loads

The ocean waves and currents constantly change the shapes of coastlines as sediment accretion and removal occurs during each wave. This may cause soil erosion along the coasts and this is where mangroves prove to be essential as they help to prevent soil erosion and bind the soil. The mangroves decrease wave intensity thereby reducing sediment discharge into the oceans. Decreased wave flow enables the deposition of bulk mass of sediment along the shores. Mangroves provide organic matter such as leaves and twigs which bind the sediments together and prevent subsidence (Spalding et al. 2014).

The domestic and industrial effluents are discharged into the estuaries; however, the mangroves are capable of purifying and improving the water quality thereby minimizing the effects of these discharges. However, if the pollutant concentrations in these discharges increase, it may harm the mangroves and surrounding environments (Sri Dattatreya et al. 2018). Mangroves can maintain the quality of

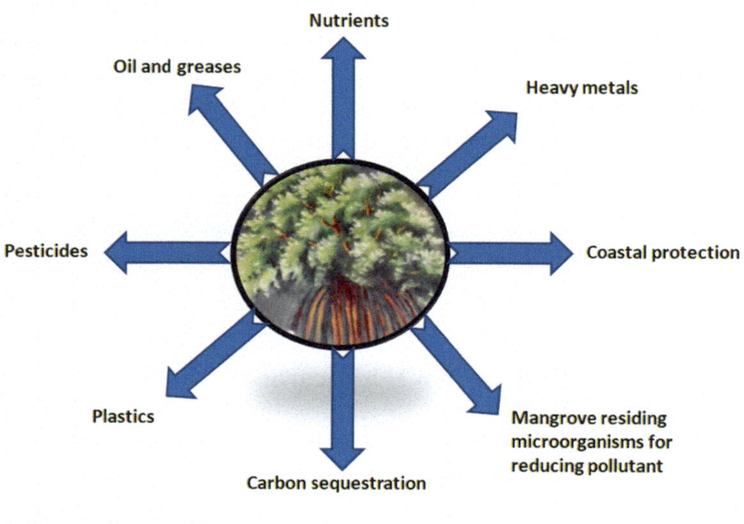

Fig. 11.1 Mangroves on pollution abatements

Table 11.1 Mangrove sources in carbon sequestration

Source	Common/Scientific name	Role in pollution abatement	References
Plant	*Ceriops decandra*	Carbon sequestration	Clough (1998), Alongi et al. (2000)
Plant	*Rhizophora stylosa* and *Avicennia marina*	Carbon sequestration	Alongi et al. (2003)
Plant	*Sonneratia caseolaris*	Carbon sequestration	Alongi et al. (2004b)
Plant	*Rhizophora apiculata,*	Carbon sequestration	Alongi et al. (2004a), Alongi (2009)
Plant	*Kandelia candel*	Carbon sequestration	Alongi et al. (2005)
Soil	–	Carbon sequestration	Ren et al. (2010)
Sediment	Mangrove peat	Carbon sequestration	Ezcurra et al. (2016)

estuaries by eliminating pollutants from effluents and transforming them into useful compounds (Wu et al. 2008).

The sedimentary characteristics such as soil grain, salinity and soil nutrients are also essential in regulating the benthic macro-organisms (Safahieh et al. 2012). Regions with fluctuating soil nutrients prevent growth of soil associated fauna. Hence, the soil nutrients are also detrimental in the mangrove productivity (Kathiresan and Bingham 2001). Dissanayake and Chandrasekara (2014) clearly explained the importance of physicochemical characteristics on the abundance of benthic organisms.

Nitrates and ammonia are two important compounds that have a role to play in regulating water quality. Nitrates are regularly maintained by the nitrogen cycle where nitrates are converted to nitrites, which are consumed by several organisms as nitrogen source. Excess concentrations of ammonia may prove toxic to the surrounding. During nitrification, ammonia is converted to nitrates by ammonia oxidizing bacteria such as *Nitrosomonas* sp., *Nitrococcus* sp. and *Nitrosospira* sp. (Purkhold et al. 2000; Ward and O'Mullan 2002; Purkhold et al. 2003). These species are abundantly found in all ecosystems including the mangroves. Moreover, *Nitrosopumilus maritimus* also plays a decisive role in eliminating ammonia from the mangrove ecosystems (Cao et al. 2011) (Table 11.1).

11.2.2 Reduction of Petroleum Hydrocarbons

Mangroves are known to attract tourists; therefore, the estuaries associated to the mangroves are subjected to constant boating activities apart from regular fishing. Such activities may yield oil spillage. Effluents from the paper and petroleum industries also contribute to oil pollution in mangroves (Lovelock et al. 2009).

Additionally, accidental spillage from oil containers also contributes to contamination of the mangrove system. These contaminations might cause several environmental effects on natural resources. The severity of oil spillages is dependent on volume of discharge, type of discharge and environmental conditions, which varies from season to season (Wang and Stout 2007). Total petroleum hydrocarbons (TPHs) are toxic to the aquatic organisms as they belong to one of the highly toxic group of persistent organic carbons. TPHs are released into various environments as a consequence of extraction, refining and consumption carried out by the petroleum industries (McNioll and Baweja 1995; USEPA 2000). Adequate and effective strategies are required to combat these hydrocarbons which can be achieved by attenuation, biostimulation and bioaugmentation within the soil which is possible in the mangroves. Mangrove areas are subjected to oil spills which affect the soils, plant surfaces and cause sub-lethal and lethal effects (Swan et al. 1994; Duke et al. 1999; Duke and Burns 2003; NOAA 2014). Such disruptions may disturb the socioeconomic benefits rendered by the mangroves.

The ubiquitous hydrocarbons are persistent in the environment due to anthropogenic activities which include oil and gas operations. Polycyclic aromatic hydrocarbons (PAHs) are the most common form of organic pollutant persistent in the environment (Tam and Wong 2008; Lu et al. 2011). Oil spillage may transpire during transportation of crude oil, discharges from petroleum products and during emissions from industries (Tam and Wong 2008). These petroleum hydrocarbons adhere to soil particles, hence degradation is tedious (Barathi and Vasudevan 2001).

Bioremediation of TPHs is a time consuming process since most of the TPH compounds are of high molecular weight (McNioll and Baweja 1995; Huang et al. 2005). Numerous environmental and physicochemical factors including nutrient availability, sunlight, temperature, soil grain size as well as microbial community influence the bioremediation of TPHs (Colombo et al. 2005; Lacerda 2006; Moreira et al. 2011). Some TPH removal by ex-situ methods such as Thermal Desorption, Chemical Oxidation and Incineration have been conducted with limited success. In situ technology involving air sparging, land farming, biosparging, bioventing, bioremediation, and phytoremediation has been carried out successfully (Seabra 2008). Among these, phytoremediation in combination with biotechnology has provided excellent results (Espinosa et al. 2005; Huang et al. 2005; Parrish et al. 2005; Doumett et al. 2008). Mangrove sediments enable to capture, transform and rhizodegrade TPHs during phytoremediation. Mangrove soils contain microbial communities suitable for rhizodegradation of TPHs and make it less bio-available to plants and other biota (Kamath et al. 2004). A former study (Moreira et al. 2011) has exhibited that phytoremediation using mangrove plants is more efficient than bioremediation.

Once an oil spill has occurred, the plants take some time to grow. However, plant recovery is faster than animal recovery as animals require the toxicity to be completely eliminated (Salmo and Duke 2010). Similarly, new seedlings can be planted only after the complete removal of toxicity (Duke 2001). Erosion and elevation are also important components which influence mangrove growth after an oil spill (Lewis 2005). Biotic and abiotic components are also essential to determine the mangrove recovery after each oil spillage. The biotic components include propagules and tolerances of the species towards oil spillage, whereas the chief abiotic components include erosion and elevation (Duke 2016).

11.2.3 Pesticides Degradation

Plants have been progressively used in bioremediation as a probable solution to pollution. Certain plants including mangroves are known to eliminate heavy metals from soil and water, while some grasses can provide a remedy for problems due to petroleum hydrocarbons (USEPA 2000). One such problem is pesticide spillage which may occur during its manufacture, distribution, formulation and application in the field. Excess levels of pesticide in soil may cause several problems such as leaching and running off into ground and surface water, respectively, while also affecting the soil organisms.

It is well known that bioremediation using plants is a time consuming process. During this extended time the pesticides may gradually be eliminated or become less bio-available to the biota. In certain cases, the pesticides may transform the soil structure which may cause drastic changes to such environments. A study (Belden et al. 2004) was conducted to upsurge the phytoremediation efficiency by employing prairie grasses along with mulberry trees. The study showed increased efficiency which suggested that similar techniques can be employed to the mangroves ecology to decrease the pollution due to pesticides.

Persistent Organic Pollutants (POPs), such as DDTs, chlordanes, endosulfans, etc. are ubiquitously found in waters or sediments in mangroves of the tropical areas. Even though these contaminants have been detected in various ecosystems globally, very little knowledge is available on the consequences of these compounds on the mangrove ecosystems (Bayen 2012).

The estuaries bring in these toxicants that are sprayed on to the agricultural fields. Since some estuaries inundate the mangroves, certain pesticides are deposited (Chilundo et al. 2008; Nagelkerken et al. 2008). However, the exact composition of these pollutants and their effects on various metabolic processes of the mangroves are largely unknown. This is where biosurfactants come into and play to combat the effects of several pollutants. Biosurfactants are bio-molecules that are retrieved from bacteria and fungi (Maier and Soberon-Chavez 2000; Mukherjee et al. 2006) and aid as natural choices to remove surface chemicals (Banat et al. 2010; Cameotra et al. 2010). Some studies have proved the ability of bio-surfactant as an effective surface agent (Kheiralla et al. 2013). However, their significance in complete degradation of pesticides is still under research.

11.2.4 Heavy Metals Remediation

Mangroves are essential in combating pollutants as they are known to sink anthropogenic pollutants and toxicants. In the mangrove soils, metal contaminants generally accumulate in the superficial water and pore water, and in solid phases like abiotic and biotic components (Lewis et al. 2011). A former study has highlighted the depletion of oxygen owing to the inundation of mangroves (Bayen 2012). The presence of sulphides and decreased oxygen concentration may contribute to co-precipitation of heavy metals in mangrove soils. In addition to sulphides, physicochemical concentrations are also associated with trace metals (Bayen 2012). Some mangrove plants like *Rhizophora* sp. can accumulate up to 95% more trace metals than other plants, thereby enabling the mangroves to retain and prevent heavy metals from infiltration into the environment (Silva et al. 1990). Salinity exchanges in the estuaries are also responsible for metal retention. The major cations such as Na, K, Ca and Mg intensify in the water during increased salinity and these ions bind with heavy metals (Laing et al. 2009).

Phytoremediation employing mangroves is used to remove/detoxify contamination due to heavy metals. The mangroves are employed to eliminate heavy metals and certain harmful organic compounds from soil while some volatile-organic compounds can be removed from the groundwater (Paz-Alberto and Sigua 2013).

An aggregate of 33 species of mangrove plants are adapted at consuming heavy metals. *Avicenna marina* has the utmost potential to degrade heavy metals than other mangrove plants. In general, roots are known to take up more heavy metals than the aerial parts, hence the upper plant tissues are not good indicators of metal pollution. However, in a study *A. marina* leaves had consumed 10% more metals than roots. The study had revealed that *A. marina* plants act as an indicator of Cu, Zn and Pb, as these metals were preferred (MacFarlane and Burchett 2002; MacFarlane et al. 2003).

The biodiversity of mangroves is also dependent on magnitude of pollution in the particular environment. Increased pollution causes the rise of pollution tolerant species while other species do not grow in these areas. *A. marina* plants are identified as superlative pollution tolerant species hence may be abundantly present in areas of high pollution. In simple words, it can be said that pollution hampers the mangrove biodiversity (Maiti and Chowdhury 2013).

The dispersion of metals varies with depth, coastal distance and typology of above ground mangrove vegetation (Kehrig et al. 2003; Marchand et al. 2006; Chatterjee et al. 2009). Trace metals may arise from natural and anthropogenic sources hence the exact source of the respective metals is difficult to differentiate. The anthropogenic sources of metals include mining (Marchand et al. 2006), gas (Antizar-Ladislao et al. 2011; Kruitwagen et al. 2008) and textile dye industries (Machado et al. 2002). The metals are brought to mangroves by rivers (Kehrig et al. 2003), or tidal exchange (Kruitwagen et al. 2008) or by atmospheric transfer (Rumbold et al. 2011). Mangroves efficiently retain the metals in plant tissues, hence reduce the metal bioavailability to the fauna.

11.2.5 Act as a Plastics Trap

Plastics are widely used in our day to day lives which have led to the improper disposal of plastics and the resultant mismanagement of wastes. The mismanaged plastic wastes accumulate in the beaches, mangroves and other aquatic ecosystems (Lebreton et al. 2017). Plastic debris has always garnered attention due to its low degradation rate, accumulation and bioavailability to the aquatic organisms (Iñiguez et al. 2016). Approximately 60–90% of marine debris constitutes plastics owing to its high production (Li et al. 2016). The plastic conundrum arises from mismanagement which according to a study ranged between 4.8 and 12.7 million metric tons in 2010 (Jambeck et al. 2015). In nature, compounds such as phthalates and persistent organic pollutants may adhere to plastic debris by virtue of which bioaccumulation and biomagnifications may occur (Moore and Phillip 2011; Jang et al. 2016; Clukey et al. 2018).

The larger plastic debris persists in these marine systems for a long period of time, thereby undergoing breakdown to minor particles referred to as microplastics which pose a bigger threat than larger plastics (Li et al. 2016). Microplastics are ingested by various organisms such as mollusks, crustaceans, fishes and others due to its smaller size and attractive colours (Wright et al. 2013; Antão-Barboza et al. 2018). Microplastics adsorb POPs and heavy metals leading to the bioaccumulations and biomagnifications, thereby causing their entry into the food chain (Ríos et al. 2007; Andrady 2011; Kühn et al. 2015; Bennecke et al. 2016; Wang et al. 2016; Massos and Turner 2017). Microplastics can release toxic materials which are consciously added during the manufacture of plastics, thereby causing environmental threat (Gallo et al. 2018). Moreover, microplastics are potential transporters of exotic species (Rech et al. 2016) and pathogenic microorganisms (Kovač Viršeka et al. 2017). However, information on the complete effects of microplastics on the marine environment is still insufficient (Auta et al. 2017).

Estuaries provide several ecological benefits to man (Barbier et al. 2011); however, they are constantly polluted due to domestic sewage discharges which act as an entry pathway for microplastics into the marine environment (Browne et al. 2011; Cesa et al. 2017). Cordeiro and Costa (2010) explained that the litter in the mangroves originated from terrestrial and freshwater sources. Mangroves are unique with regard to their retention and transformations capabilities of land-based litter and pollutants due to intense biomass and productivity (Fourqurean et al. 2012; Li et al. 2016; Booth and Sear 2018), hence are garnering much research attention (Martin et al. 2019). Some of the terrestrial origins of microplastics include personal care goods, for example, facial scrubs, cleansers, creams and toothpaste which reach the aquatic systems through municipal sewage and wastewater treatment plants (Auta et al. 2017; Murphy et al. 2016).

Mangroves have a complex root system consisting of pneumatophores and prop roots which enhance turbulence and wave energy (Horstman et al. 2014; Norris et al. 2017). These roots act as a filter system which traps materials transported by waves and currents. An earlier report exhibited the different retention capacities of various objects dumped in the mangrove forests (Ivar do Sul et al. 2014).

Plastics trapped by mangrove roots cause external damage which affects the tree and the dependent fauna. These plastics also prevent gas exchange, meanwhile release absorbed chemicals to the mangrove ecosystem (Cole et al. 2011). A number of natural factors such as estuarine hydrodynamics, mangrove characteristics, sediment grain size along with human activities such as tourism, mariculture and coastal dumping are some of the determining factors for plastic distribution and retention in the mangroves (Ivar do Sul et al. 2014; Lima et al. 2016). Ivar do Sul et al. (2014) explained that retention capacity of the mangroves is reliant on the hydrodynamics of the plastic materials while also mentioned that plastic bags are retained the most owing to its hydrodynamic property. Thus the hydrodynamics of semi-enclosed mangrove habitats are a regulating factor for the distribution of plastics (Boelens et al. 2018).

Polycyclic aromatic hydrocarbon (PAH) is a type of organic contaminant found in sediments which are transferred to organisms (Ramdine et al. 2012). The hydrophobic nature of plastics enables POPs such as dichlorodiphenyltrichloroethane (DDTs), polycyclic aromatic hydrocarbons (PAHs) and polychlorinated biphenyls (PCBs) to get adsorbed onto microplastic surfaces (Endo et al. 2005; Teuten et al. 2007; Frias et al. 2010; Rochman et al. 2013). These adsorbed compounds are consumed when the microplastics are ingested and released into the tissues of organisms (Engler 2012). Similarly, toxic monomers (Saido et al. 2009) and some plastic additives like bisphenol A and phthalates (Fries et al. 2013) leach from the microplastics and affect the marine organisms and environment.

Polyethylene (PE), Polypropylene (PP), Polyvinyl chloride (PVC), rayon and polyester (PES) are the some of the regular polymers recovered from the mangroves (Li et al. 2018; Zhu et al. 2019). However, Ajith et al. (2020) reviewed that apart from these Polystyrene (PS) and nylon are also commonly found in the marine environment and this could enter the mangroves. PP has a wide use in industrial and household applications which include textiles, stationery, packaging, automobile parts and laboratory equipment (Gewert et al. 2015). PE is used in packaging, plastic bottles, bags (Zhang et al. 2016) and fishing gears (Chen et al. 2018; Wang T et al. 2019a). The other polymers are also dominant, but their origins are not documented well (Ajith et al. 2020).

Plastic polymers are recalcitrant in nature which prohibits rapid microbial degradation enabling the long-term persistence of plastics in the environment (Longo et al. 2011). Additionally, the plastic surfaces act as substrates for microbial colonization (Dussud and Ghiglione 2014). During colonization, microbes initiate biodegradation by secreting extracellular enzymes (Eich et al. 2015; Sekhar et al. 2016). Certain microorganisms produce biofilms which also influence biodegradation of plastics (Sheik et al. 2015; Eich et al. 2015; Jeon and Kim 2016). Microbial enzymes are important in biodegradation of plastic substances, since some enzymes utilize plastics as carbon source (Auta et al. 2018).

11.2.6 Mangrove Residing Microorganisms

Mangroves occur at the interface of marine and terrestrial ecosystems which exemplify a prosperous biodiversity. Microorganisms are essential components of the mangrove ecosystem; they play crucial roles in the creation and upholding of this biosphere. They also supply biotechnologically potential and valuable products (Thatoi et al. 2013). This ecosystem offers a typical environment harbouring different assemblages of microbes, for instance, actinomycetes, bacteria, cyanobacteria, fungi, protozoa, algae, etc. (Sen and Naskar 2003). Among them, bacteria and fungi provide a noteworthy contribution towards reducing the pollutants (Fig. 11.2). These microbes contributing in different steps of decomposition and mineralization of mangrove leaves among other litters significantly contribute to the productivity of the mangrove environment. They play an important role in recycling nutrients, involve in carbon sequestration and devastate pollutants. The microorganisms degrade the pollutants by their enzymatic action while enzymes act as biocatalysts facilitating degradation. Among the mangrove residing microbes, the bacteria are the most dominated group as they are several fold higher than those of fungi (Kathiresan and Qasim 2005). Microbes utilize a broad range of substances to produce energy and nutrients from the surrounding environment for the growth and multiplications of cells. The mangroves are the extremely productive ecosystems hold rich nutrients which support the growth and multiplication of diverse groups of microorganisms.

Domestic wastes and industrial effluents are expelled into the estuaries and channels associated with the mangroves which may release excess nutrients in the surface water and soil. This may instigate eutrophication and algal blooms whilst excess nitrogenous wastes may also be harmful. Certain aquatic biota including bacteria and algae enable mangroves to eradicate the harmful nitrogenous wastes and unused nutrients present on the surface water and soil. This purified water can be reused in aquaculture (Spalding et al. 2014).

Mangroves are one of the main ‘hotspots’ for marine fungi (Shearer et al. 2007). Although the trunks and aerating roots are permanently or intermittently submerged the other parts remain free of salt water interference. The lichens and terrestrial fungi

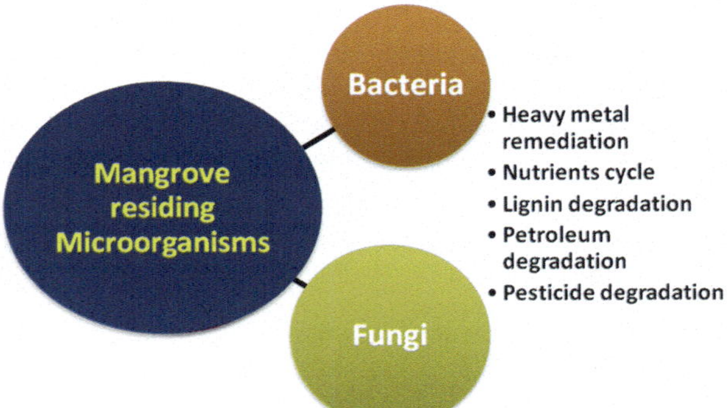

Fig. 11.2 Mangrove residing microorganisms and their dynamic role in pollutants reduction

inhabit the dry parts of plant while marine groups live in the bottom region; middle region overlies the marine fungi (Sarma and Hyde 2001).

Bacteria are well known to degrade petroleum hydrocarbons as they have been employed successfully during oil spills (Rahman et al. 2003; Brooijmans et al. 2009). Biodegradation of petroleum hydrocarbons relies on a number of factors, such as the type of compound and availability to microorganisms. Some bacteria consume hydrocarbons as their primary source of supplement (Yakimov et al. 2007). However, high molecular weight hydrocarbons can never be degraded (Atlas and Bragg 2009). Mangrove sediments host a variety of bacteria and fungi among which certain strains may have ability to degrade hydrocarbons. Some of these species include *Bacillus subtilis*, *Pseudomonas fluorescens* and *Rhodococcus erythropolis* (Duke 2016).

The *B. subtilis* is capable of degrading hydrocarbons and can survive in extreme conditions of heat. These spores are advantageous as they have the potential to produce a natural biosurfactant called surfactin which is synthesized in combination with crude oil (Queiroga et al. 2003). The strain *P. fluorescens* is also known to produce biosurfactants which have the potential to degrade hydrocarbons (Kaczorek and Olszanowski 2011). The bacterium *R. erythropolis* has a unique metabolism which consumes carbon from hydrocarbons, thereby degrading various compounds such as chlorinated phenols, steroids, lignin, coal and crude oil (Brandao et al. 2003).

The mangrove fungus has attracted attention of scientists due to their wealthy and diverse potentials. Wu (1993) screened 15 species of fungi from mangroves regions of Tanshui Estuary, Taiwan. He observed that they can secrete different enzymes that have potential to decompose mangrove litter. Xin et al. (2002) found that the marine white-rot fungus can degrade 50% of lignin incubated with entire sugarcanes.

Microorganisms adapt to all kinds of environment (Brooks et al. 2011; Aujoulat et al. 2012) and possess the potential to transform plastic polymers. The mangrove residing microbes can metabolize the pollutants which might enhance biotransformation and subsequent degradation (Luigi et al. 2007). Several studies have articulated their results on microbial degradation of plastic polymers. PET was degraded by *Ideonella sakaiensis* 201-F6 (Yoshida et al. 2016); degradation polystyrene by *Pseudomonas* sp. and *Bacillus* sp. (Mohan et al. 2016); the attenuation of molecular weight of PE under test conditions by a fungus *Zalerion maritimum* (Paco et al. 2017); degradation of PE by *Bacillus cereus* (Sowmya et al. 2014); *Kocuria palustris* M16, *Bacillus subtilis* H1584 and *Bacillus pumilus* M27 (Harshvardhan and Jha 2013). A number of other bacteria are also involved in polymer-degradation which includes *Pseudomonas stutzeri*, *Alcaligenes faecalis*, *Pseudomonas putida*, *Staphylococcus* sp., *Streptomyces* sp. and *Brevibacillus borstelensis* (Ghosh et al. 2013; Caruso 2015). Therefore, it may be affirmed that microbial degradation of plastics is an eco-friendly approach (Boelens et al. 2018). Mangroves host a wide variety of microbes (Kathiresan 2003; Thatoi et al. 2012) due to the suitable conditions of temperature, pH and salinity (Ghizelini et al. 2012).

11.2.7 Carbon Sequestration

Mangroves are the principal carbon-rich habitats of the tropic regions (Kristensen et al. 2008; Donato et al. 2011; Pendleton et al. 2012; Siikamäki et al. 2012b; Alongi 2014). Typically, mangroves garner quadruple amounts of carbon than other floral habitats (Donato et al. 2011). However, mangrove deforestation occurs at a significantly higher rate than the loss of other types of forests (FAO 2007). Mangrove deforestation causes carbon emission and reduces carbon sequestration (Alongi 2014). Carbon emissions coupled with other greenhouse gases (CH_4, N_2O) may have contributed to climate change (IPCC 2014). To combat climate change, it is necessary to prevent deforestation of mangroves and subsequent carbon emission (Kristensen et al. 2008; Donato et al. 2011; Houghton 2012; Alongi 2014).

Blue carbon emissions can be reduced significantly by the mangrove restoration programs (Pendleton et al. 2012; Siikamäki et al. 2012a). Through mangrove restoration, several environmental activities such as photosynthesis, respiration, water purification, decomposition and predation may also be restored (Cardinale et al. 2011; Parrotta et al. 2012). Blue carbon sequestration through mangrove restoration is gaining popularity due to its importance in climate change (Le 2008; Joffre and Schmitt 2010). Blue carbon sequestration enables to promote and safeguard mangroves as well as provide a solution to climate change (Robledo et al. 2004). However, mangrove restoration is possible only with an ecologically beneficial design and socioeconomic support for its construction (Mazda et al. 2006).

In the early days, carbon accumulation was conducted as short-term studies and the results expressed that a large amount of carbon is stocked in the soil. However, over time it was proved that short-term studies do not portray the actual picture of carbon sequestration. This is evidenced in a study where short-term assessment of 137Cs, 222Th and 210Pb radionuclides suggested a high carbon accumulation rate of 1.25 tC ha^{-1} $year^{-1}$ (Alongi et al. 2004). However, ground-truth studies and imaging studies have showed a lesser carbon accumulation rate of 0.5 tC ha^{-1} $year^{-1}$ (Alongi et al. 2004). This is due to simultaneous accretion and erosion of mangroves forests. The actual picture of carbon sequestration can be projected only by carrying out long-term studies (Alongi 2011).

Despite the numerous services provided by the mangroves, they are constantly subjected to human pressures as over 120 million people are known to inhabit areas surrounding the mangroves (UNEP 2014). A large area of mangroves is converted to aquaculture as this provides an income and means of food. The loss of mangroves may lead increased concentration of carbon in the atmosphere. Mangroves are essential sources of blue carbon sinks. Blue carbon is the organic carbon stored, sequestered, and released from mangroves, salt marshes, seagrasses and other marine environments (Nellemann et al. 2009; McLeod et al. 2011; Murray et al. 2011; Pendleton et al. 2012; Duarte et al. 2013; Alongi 2014). Annually, mangroves contribute to 30% of global carbon sequestration (Siikamäki et al. 2012a). Mangroves need to be restored by carrying out effective restoration programs such as Reduced Emissions during deforestation and degradation (REDD) (Ahmed and Glaser 2016) (Table 11.2).

Table 11.2 Mangrove sources in regulation of nutrient load/water quality

Source	Common/Scientific name	Role in pollution abatement	References
Soil	–	Treatment of water quality	Tam and Wong (1995)
Plant	*Kandelia candel, Bruguiera gymnorrhiza*	Salinity and nutrient removal	Ye et al. (2001)
Plant	*Kandelia candel*	Treatment of municipal wastewater	Wu et al. (2008)
Soil	Glomalin related protein	Improves water quality	Wang et al. (2018)

11.2.8 Coastal Protection

Mangroves act as natural barriers that prevent storms and cyclones or resist their intensity, thereby minimizing the damage on the inner areas. The constant exchange of sediments between oceans and mangroves strengthens the mangrove soil thereby preventing soil erosion. The dense mangroves growing along the coastal regions can act as a natural shield, help to reduce the impact of natural calamities like cyclone, tsunami and coastal storm by engrossing the waves' energy because of the flexibility of the mangrove stem. The study conducted by Kathiresan and Rajendran (2005) scientifically proved that the mangroves play a significant role in coastal protection. During excessive high tides or tsunamis, the coasts are protected by the mangroves, which reduce the wave intensity, hence the waves do not enter too far into the terrestrial areas (Spalding et al. 2014).

11.3 Conclusion

Though the mangroves provide solutions for several pollutants, the tolerance of mangroves towards toxicity necessitates testing to comprehend the toxicity threshold level of mangroves. This may be accomplished by testing the various stages of mangroves right from their juvenile stages. By such experiments, the information on tolerance of mangroves towards toxicity can be documented. Such information extends knowledge of the scientific community to apprehend effects of chemical and non-chemical stressors.

References

Ahmed N, Glaser M (2016) Coastal aquaculture, mangrove deforestation and blue carbon emissions: is REDD+ a solution?. Mar Policy 66:58–66.

Ajith N, Arumugam S, Parthasarathy S, Manupoori S, Janakiraman S (2020) Global distribution of microplastics and its impact on marine environment—a review. Environ Sci Pollut Res 27:25970–25986.

Alongi DM (2011) Carbon payments for mangrove conservation: ecosystem constraints and uncertainties of sequestration potential. Environ Sci Pol 14:462–470.

Alongi DM (2014) Carbon cycling and storage in mangrove forests. Annu Rev Mar Sci 6:195–219.

Alongi DM, Sasekumar A, Chong VC, Pfitzner J, Trott LA, Tirendi F, Dixon P, Brunskill GJ (2004) Sediment accumulation and organic material flux in a managed mangrove ecosystem: estimates of land–ocean–atmosphere exchange in peninsular Malaysia. Mar Geol 208:383–402.

Alongi DM, Tirendi F, Trott LA, Xuan TT (2000) Benthic decomposition rates and pathways in plantations of the mangrove *Rhizophora apiculata* in the Mekong delta, Vietnam. Mar Ecol Prog Ser 194:87–101.

Alongi DM (2009) The energetics of Mangrove forests. Springer, Amsterdam, The Netherlands.

Alongi DM, Clough BF, Dixon P, Tirendi F (2003) Nutrient partitioning and storage in arid-zone forests of the mangroves *Rhizophora stylosa* and *Avicennia marina*. Trees 17(1):51–60.

Alongi DM, Pfitzner J, Trott LA, Tirendi F, Dixon P, Klumpp DW (2005) Rapid sediment accumulation and microbial mineralization in forests of the mangrove *Kandelia candel* in the Jiulongjiang estuary, China. Est Coast Shelf Sci 63:605–618.

Alongi DM, Wattayakorn G, Tirendi F, Dixon P (2004a) Nutrient capital in different aged forests of the mangrove *Rhizophora apiculata*. Bot Mar 47(2):116–24.

Alongi DM, Sasekumar A, Chong VC, Pfitzner J, Trott LA, Tirendi F, Dixon P, Brunskill GJ (2004b) Sediment accumulation and organic material flux in a managed mangrove ecosystem: estimates of land–ocean–atmosphere exchange in peninsular Malaysia. Mar Geol 208:383–402.

Andrady A (2011) Microplastic in the marine environment. Mar Pollut Bull 62(8):1596–1605.

Antão-Barboza L, Vethaak AD, Lavorante B, Lundebye A, Guilhermino L (2018) Marine microplastic debris: an emerging issue for food security, food safety and human health. Mar Pollut Bull 133:336–348.

Antizar-Ladislao B, Sarkar SK, Anderson P, Peshkur T, Bhattacharya BD, Chatterjee M, Satpathy KK (2011) Baseline of butyltin contamination in sediments of Sundarban mangrove wetland and adjacent coastal regions, India. Ecotoxicol 20(8):1975–1983.

Atlas RM, Bragg J (2009) Bioremediation of marine oil spills: when and when not-the Exxon Valdez experience. Microb Biotechnol 2(2):213–221.

Aujoulat F, Roger F, Bourdier A, Lotte A, Lami B, Marchandin H, Jumas-Bilak E (2012) From environment to man: gene evolution and adaptation of human opportunistic bacterial pathogens. Genes 3:191–232.

Auta HS, Emenike CU, Fauziah SH (2017) Distribution and importance of microplastics in the marine environment: a review of the sources, fate, effects, and potential solutions. Environ Int 102:165–176.

Auta HS, Emenike CU, Jayanthi B, Fauziah SH (2018) Growth kinetics and biodeterioration of polypropylene microplastics by *Bacillus* sp. and *Rhodococcus* sp. isolated from mangrove sediment. Mar Pollut Bull 127:15–21.

Banat I, Franzetti A, Gandolfi I, Bestetti G, Martinotti MG, Fracchia L, Smyth TJ, Marchant R (2010) Microbial biosurfactants production, applications and future potential. Appl Microbiol Biotechnol 87:427–444.

Barathi S, Vasudevan N (2001) Utilization of petroleum hydrocarbons by *Pseudomonas fluorescens* isolated from a petroleum-contaminated soil. Environ Intern 26(5–6):413–416.

Barbier EB, Hacker SD, Kennedy C, Koch EW, Stier AC, Silliman BR (2011) The value of estuarine and coastal ecosystem services. Ecol Monogr 81:169–193.

Bayen S (2012) Occurrence, bioavailability and toxic effects of trace metals and organic contaminants in mangrove ecosystems: A review. Environ Intern 48:84–101.

Belden JB, Clark BW, Phillips TA, Henderson KL, Arthur EL (2004) IOWA State University, Digital Repository, Entomology Publicationshttp, pp. 156–157.

Bennecke D, Duarte B, Paiva F, Cacador I, Canning-Clode J (2016) Microplastic as vector of heavy metal contamination from the marine environment. Estuar Coast Shelf Sci 178:189–195.

Boelens T, Schuttelaars H, Schramkowski G, Mulder TD (2018). The effect of geometry and tidal forcing on hydrodynamics and net sediment transport in semi-enclosed tidal basins. Ocean Dynam. 68:1285–1309.

Booth DJ, Sear J (2018) Coral expansion in Sydney and associated coral-reef fishes. Coral Reefs 37:995.

Brandao PF, Clapp JP, Bull AT (2003) Diversity of nitrile hydratase and amidase enzyme genes in *Rhodococcus erythropolis* recovered from geographically distinct habitats. Appl Environ Microbiol 69:5754–5766.

Brooijmans RJW, Pastink MI, Siezen RJ (2009) Hydrocarbon-degrading bacteria: the oil-spill clean-up crew. Microb Biotechnol 2(6):587–594.

Brooks AN, Turkarslan S, Beer KD, Lo FY, Baliga NS (2011) Adaptation of new cells to new environments. Wiley Interdiscip. Rev Syst Biol Med 3(5):544–561.

Browne MA, Crump P, Niven SJ, Teuten E, Tonkin A, Galloway T, Thompson R (2011) Accumulation of microplastic on shorelines worldwide: sources and sinks. Environ Sci Technol 45:9175–9179.

Cameotra SS, Makkar RS, Kaur J, Mehta SK (2010) Synthesis of biosurfactants and their advantages to microorganisms and mankind. Adv Exp Med Biol 672:261–280.

Cao H, Li M, Hong Y, Ji-Dong Gu (2011) Diversity and abundance of ammonia-oxidizing archaea and bacteria in polluted mangrove sediment. Syst Appl Microbiol 34:513–523.

Cardinale BJ, Matulich KL, Hooper DU, Byrnes JE, Duffy E, Gamfeldt L, Balvanera P, O'connor MI, Gonzalez A (2011) The functional role of producer diversity in ecosystems. Am J Bot 98 (3):572–592.

Caruso G (2015) Plastic degrading microorganisms as a tool for the bioremediation of plastic contamination in aquatic environments. J Pollut Eff Control 3:112.

Cesa FS, Turra A, Baruque-Ramos J (2017) Synthetic fibers as microplastics in the marine environment: a review from textile perspective with a focus on domestic washings. Sci Total Environ 598:1116–1129.

Chatterjee M, Massolo S, Sarkar SK, Bhattacharya AK, Bhattacharya BD, Satpathy KK Saha S (2009) An assessment of trace element contamination in intertidal sediment cores of Sunderban mangrove wetland, India for evaluating sediment quality guidelines. Environ Monit Assess 150 (1–4):307.

Chen M, Jin M, Tao P, Wang Z, Xie W, Yu X, Wang K (2018) Assessment of microplastics derived from mariculture in Xiangshan Bay, China. Environ Pollut 242:1146–1156.

Chilundo M, Kelderman P, O'Keeffe JHO (2008) Design of a water quality monitoring network for the Limpopo river basin in Mozambique. Phys Chem Earth 33(8–13):655–665.

Clough BF (1998) Mangrove forest productivity and biomass accumulation in Hinchinbrook Channel, Australia. Mangroves Salt Marshes 2:191–198.

Clukey KE, Lepczyk CA, Balazs GH, Work TM, Li QX, Bachman MJ, Lynch JM (2018) Persistent organic pollutant in fat of three species of Pacific pelagic long line caught sea turtles: Accumulation in relation to ingested plastics marine debris. Sci Total Environ 610:402–411.

Cole M, Lindeque P, Halsband C, Galloway TS (2011) Microplastics as contaminants in the marine environment: a review. Mar Pollut Bull 62:2588–2597.

Colombo JC, Barreda A, Bilos C, Cappelletti N, Migoya MC, Skorupka C (2005) Oil spill in the Rıo de la Plata Estuary, Argentina: 2-hydrocarbon disappearance rates in sediments and soils, Environ Pollut 134:267–276.

Cordeiro CAMM, Costa TM (2010) Evaluation of solid residues removed from a mangrove swamp in the S~ao Vicente Estuary, SP, Brazil. Mar Pollut Bull 60(10):1762–1767.

Dissanayake N, Chandrasekara U (2014) Effects of Mangrove Zonation and the Physicochemical Parameters of Soil on the Distribution of Macrobenthic Fauna in Kadolkele Mangrove Forest, a Tropical Mangrove Forest in Sri Lanka. Adv Ecol. https://doi.org/10.1155/2014/564056

Donato DC, Kauffman JB, Murdiyarso D, Kurnianto S, Stidham M, Kanninen M (2011) Mangroves among the most carbon-rich forests in the tropics. Nat Geosci 4(5):293–297.

Doumett S, Lamperi L, Checchini L, Azzarello E, Mugnai S, Mancuso S, Petruzzelli G, Del Bubba M (2008) Heavy metal distribution between contaminated soil and *Paulownia tomentosa*, in a pilot-scale assisted phytoremediation study: Influence of different complexing agents. Chemosphere 72:1481–1490.

Duarte CM, Losada IJ, Hendriks IE, Mazarrasa I Marbà N (2013) The role of coastal plant communities for climate change mitigation and adaptation. Nat Clim Change *3*(11):961–968.

Duke NC (2001) Gap creation and regenerative processes driving diversity and structure of mangrove ecosystems. Wetl Ecol Manag 9:267–279.

Duke NC (2016) Oil spill impacts on mangroves: recommendations for operational planning and action based on a global review. Mar Pollut Bull 109(2):700–715.

Duke NC, Burns KA (2003) Fate and effects of oil and dispersed oil on mangrove ecosystems in Australia. Environmental Implications of Offshore Oil and Gas Development in Australia: Further Research. A Compilation of Three Scientific Marine Studies. Australian Petroleum Production and Exploration Association (APPEA), Canberra, pp. 232–363.

Duke NC, Burns KA, Swannell RPJ (1999) Research into the bioremediation of oil spills in tropical Australia: with particular emphasis on oiled mangrove and saltmarsh habitat. Report to the Australian Maritime Safety Authority and the Great Barrier Reef Marine Park Authority.

Duke NC, Meynecke JO, Dittmann S, Ellison AM, Anger K, Berger U, Cannicci S, Diele K, Ewel KC, Field CD, Koedam N, Lee SY, Marchand C, Nordhaus I, Dahdouh-Guebas F (2007) A world without mangroves. Science 317:41–42.

Duke NC, Schmitt K (2015) Mangroves: Unusual forests at the seas edge. In: Pancel, L., Köhl, M. (Eds.), Tropical forest handbook. Springer-Verlag, Berlin, Heidelberg.

Dussud C, Ghiglione JF (2014) Bacterial Degradation of Synthetic Plastics- CIESM Workshop Monograph, France.

Ezcurra P, Ezcurra E, Garcillán PP, Costa MT, Aburto-Oropeza O (2016) Coastal landforms and accumulation of mangrove peat increase carbon sequestration and storage. Proc Nat Acad Sci 113(16):4404–4409.

Eich A, Mildenberger T, Laforsch C, Weber M (2015) Biofilm and diatom succession on polyethylene (PE) and biodegradable plastic bags in two marine habitats: early signs of degradation in the pelagic and benthic zone? PloS one 10 (9):0137201.

Endo S, Takizawa R, Okuda K, Takada H, Chiba K, Kanehiro H, Date T (2005) Concentration of polychlorinated biphenyls (PCBs) in beached resin pellets: variability among individual particles and regional differences. Mar Pollut Bull 50(10):1103–1114.

Engler RE (2012) The complex interaction between marine debris and toxic chemicals in the ocean. Environ Sci Technol 46(22):12302–12315.

Espinosa E, Martinez ME, Torres EF, Rojas MG (2005) Improvement of the hydrocarbon phytoremediation rate by *Cyperus laxus* Lam. inoculated with a microbial consortium in a model system. Chemosph 59:405–413.

FAO (2007) The World's Mangroves 1980–2005. FAO, Rome.

Fourqurean JW, Duarte CM, Kennedy H, Marba N, Holmer M, Mateo MA, Apostolaki ET, Kendrick GA, Krause-Jensen D, McGlathery KJ, Serrano O (2012) Seagrass ecosystems as a globally significant carbon stock. Nat Geosci 5:505–509.

Frias JP, Sobral P, Ferreira AM (2010) Organic pollutants in microplastics from two beaches of the Portuguese coast. Mar Pollut Bull 60(11):1988–1992.

Fries E, Dekiff JH, Willmeyer J, Nuelle MT, Ebert M, Remy D (2013) Identification of polymer types and additives in marine microplastic particles using pyrolysis-GC/MS and scanning electron microscopy. Environm Sci Proc Impacts 15(10):1949–1956.

Gallo F, Fossi C, Weber R, Santillo D, Sousa J, Ingram I, Nadal A, Romano D (2018) Marine litter plastics and microplastics and their toxic chemicals components: the need for urgent preventive measures. Environ Sci Eur 30.

Gewert B, Plassmann MM, MacLeod M (2015) Pathways for degradation of plastic polymers floating in the marine environment. Environ Sci Process Impact 17:1513.

Ghizelini AM, Mendonça-Hagler LCS, Macrae A (2012) Microbial diversity in Brazilian mangrove sediments e a mini review. Brazilian J Microbiol 1242–1254.

Ghosh SK, Pal S, Ray S (2013) Study of microbes having potentiality for biodegradation of plastics. Environ Sci Pollut Res 20:4339–4355.

Harshvardhan K, Jha B (2013) Biodegradation of low-density polyethylene by marine bacteria from pelagic waters, Arabian Sea, India. Mar Pollut Bull 77(1):100–106.

Horstman EM, Dohmen-Janssen CM, Narra PMF, Van den Berg NJF, Siemerink M, Hulscher SJ (2014) Wave attenuation in mangroves: A quantitative approach to field observations. Coast Eng 94:47–62.

Houghton RA (2012) Carbon emissions and the drivers of deforestation and forest degradation in the tropics. Curr Opin Environ Sustain 4597–603.

Huang XD, El-Alawi Y, Gurska J, Glick BR, Greenberg BM (2005) A multi-process phytoremediation system for decontamination of persistent total petroleum hydrocarbons (TPHs) from soils. Microchem J 81:139–147.

Iñiguez ME, Conesa JA, Fullana A (2016) Marine debris occurrence and treatment: A review. Renew Sustain Energy Rev 64:394–402.

IPCC, Climate Change (2014) Synthesis Report–Summary for Policy makers, Intergovernmental Panel on Climate Change, Valencia.

Ivar do Sul JA, Costa MF, Silva-Cavalcanti JS, Araújo MCB (2014) Plastic debris retention and exportation by a mangrove forest patch. Mar Pollut Bull 78(1–2):252–257.

Jambeck JR, Geyer R, Wilcox C, Siegler TR, Perryman M, Andrady A, Narayan R, Law KL (2015) Plastic waste inputs from land into the ocean. Science 347:768–771.

Jang M, Shim WJ, Han GM, Rani M, Song YK, Hong SH (2016) Styrofoam debris as a source of hazardous additives for marine organisms. Environ Sci Technol 50(10):4951–4960.

Jeon HJ, Kim MN (2016) Isolation of mesophilic bacterium for biodegradation of polypropylene. Int Biodeterior Biodegrad 115:244–249.

Joffre OM, Schmitt K (2010) Community livelihood and patterns of natural resources uses in the shrimp-farm impacted Mekong Delta. Aquacult Res 41:1855–1866.

Kaczorek E, Olszanowski A (2011) Uptake of hydrocarbon by *Pseudomonas fluorescens* (P1) and *Pseudomonas putida* (K1) strains in the presence of surfactants: a cell surface modification. Water Air Soil Pollut 214:451–459.

Kamath R, Rentz JA, Schnoor JL, Alvare PJJ (2004) Phytoremediation of hydrocarbon contaminated soils: principles and applications, in: R. Vazquez-Duhalt, R. Quintero-Ramirez (Eds.), Studies in Surface Science and Catalysis, Elsevier.

Kathiresan K (2003) Polythene and plastic-degrading microbes from the mangrove soil. Rev Biol Trop 51(3):629–634.

Kathiresan K, Bingham L (2001) Biology of mangroves and mangrove ecosystems. Adv Mar Biol 40:82–251.

Kathiresan K Qasim SZ (2005) Biodiversity of mangrove ecosystems. Hindustan Publishing, New Delhi(India)

Kathiresan K, Rajendran N (2005) Coastal mangrove forests mitigated tsunami. Estuar Coast Shelf Sci 65:601–606.

Kehrig HA, Pinto FN, Moreira I, Malm O (2003) Heavy metals and methyl mercury in a tropical coastal estuary and a mangrove in Brazil. Org Geochem 34:661–669.

Kheiralla ZH, Ashour SM, Rushdy AA, Ahmed HA (2013) Characterization of Biosurfactants Produced by *Halobacillus Dabanensis* and *Pontibacillus Chungwhensi* Isolated from Oil_Contaminated Mangrove Ecosystem in Egypt. Appl Biochem Microbiol 49(3):263–269.

Kovač Viršeka M, Nika Lovšin M, Koren Š, Kržan A, Peterlin M (2017) Microplastics as a vector for the transport of the bacterial fish pathogen species *Aeromonas salmonicida*. Mar Pollut Bull 125(1–2):301–309.

Kristensen E, Bouillon S, Dittmar T, Marchand C (2008) Organic carbon dynamics in mangrove ecosystems: a review. Aquat Bot 89:201–219.

Kruitwagen G, Pratap HB, Covaci A, Bonga SEW (2008) Status of pollution in mangrove ecosystems along the coast of Tanzania. Mar Pollut Bull 56:1022–1031.

Kühn S, Bravo EL, Van Franeker JA (2015) Deleterious effects of litter on marine life. In: Bergmann, M., Gutow, L., Klages, M. (Eds.), Marine Anthropogenic Litter. Springer, Berlin, pp. 75–116.

Lacerda LD (2006) Manguezais do nordeste e mudanças ambientais, Ciência Hoje 39:24–28.
Laing GD, Rinklebe J, Vandecasteele B, Meers E, Tack FMG (2009) Trace Metal Behavior in Estuarine and Riverine Flood Plain Soils and Sediments: A Review. Sci Total Environ 407 (13):3972–3985.
Le H (2008) Economic reforms and mangrove forests in central Vietnam. Soc Nat Resour 21:106–119.
Lebreton L, van der Zwet J, Damsteeg J, Slat B, Andrady A, Reisser J (2017) River plastic emissions to the world's oceans. Nat Commun 8:15611.
Lewis III, RR (2005) Ecological engineering for successful management and restoration of mangrove forests. Ecol Eng 24:403–418.
Lewis M, Pryor R, Wilking L (2011) Fate and Effects of Anthropogenic Chemicals in Mangrove Ecosystems: A Review. Environ Pollut 159(10): 2328–2346.
Li J, Zhang H, Zhang K, Yang R, Li R, Li Y (2018) Characterization, source, and retention of microplastic in sandy beaches and mangrove wetlands of the Qinzhou Bay, China. Mar Pollut Bull 136:401–406.
Li R, Zhang L, Xue B, Wang Y (2019). Abundance and characteristics of microplastics in the mangrove sediment of the semi-enclosed Maowei Sea of the south China sea: New implications for location, rhizosphere, and sediment compositions. Environ Pollut 244: 685–692.
Li WC, Tse HF, Fok L (2016) Plastic waste in the marine environment: a review of source, occurrence and effects. Sci Total Environ 566–567:333–349.
Lima ARA, Barletta M, Costa MF, Ramos JAA, Dantas DV, Melo PAMC, Justio AKS, Ferreira GVB (2016) Changes in the composition of ichthyoplankton assemblage and plastic debris in mangrove creeks relative to moon phases. J Fish Biol 89(1):619–640.
Longo C, Savaris M, Zeni M, Brandalise RN, Grisa AMC (2011) Degradation study of polypropylene (PP) and bioriented polypropylene (BOPP) in the environment. Mater Res 14 (4):442–448.
Lovelock CE, Ball MC, Martin KC, Feller IC (2009) Nutrient enrichment increases mortality of mangroves. PLoS One 4(5):5600.
Lu H, Zhang Y, Liu B, Liu J, Ye J, Yan C (2011) Rhizodegradation gradients of phenanthrene and pyrene in sediment of mangrove (*Kandelia candel* (L.) Druce). J Hazard Mat 196:263–269.
Luigi M, Gaetano DM, Vivia B, Angelina LG (2007) Biodegradative potential and characterization of psychrotolerant polychlorinated biphenyl-degrading marine bacteria isolated from a coastal station in the Terra Nova Bay (Ross Sea, Antarctica). Mar Pollut Bull 54:1754–1761.
MacFarlane GR, Burchett MD (2002) Toxicity, growth and accumulation relationships of copper, lead and zinc in the grey mangrove Avicennia marina (Forsk.) Vierh. Mar Environ Res 54 (1):65–84.
MacFarlane GR, Pulkownik A, Burchett MD (2003) Accumulation and Distribution of Heavy Metals in the Grey Mangrove, *Avicennia marina* (Forsk.) Vierh: Biological Indication Potential. Environ Pollut 123(1):139–151.
Machado W, Moscatelli M, Rezende LG, Lacerda LD (2002) Mercury, zinc, and copper accumulation in mangrove sediments surrounding a large landfill in southeast Brazil. Environ Pollut 120:455–461.
Maier RM, Soberon-Chavez G (2000) *Pseudomonas aeruginosa* rhamnolipids: biosynthesis and potential applications. Appl Microbiol Biotechnol 54:625–633.
Maiti SK, Chowdhury A (2013) Effects of Anthropogenic Pollution on Mangrove Biodiversity: A Review. J Environ Prot 4:1428–1434.
Marchand C, Lallier-Verges E, Baltzer F, Alberic P, Cossa D, Baillif P (2006) Heavy metals distribution in mangrove sediments along the mobile coastline of French Guiana. Mar Chem 98:1–17.
Martin C, Almahasheer H, Duarte CM (2019) Mangrove forests as traps for marine litter. Environ Pollut 247:499–508.
Massos A, Turner A (2017) Cadmium, lead, and bromine in beached microplastics. Environ Pollut 227:139–145.

Mazda Y, Magi M, Ikeda Y Kurokawa T, Asano T (2006) Wave reduction in a mangrove forest dominated by *Sonneratia* sp. Wetl Ecol Manag 14:365–378.

McIvor AL, Spencer T, Möller I, Spalding M (2012) Storm surge reduction by mangroves. Natural Coastal Protection Series: Report 2. Cambridge Coastal Research Unit Working Paper 35. ISSN 2050–7941.

Mcleod E, Chmura GL, Bouillon S, Salm R, Björk M, Duarte CM, Lovelock CE, Schlesinger WH, Silliman BR (2011) A blueprint for blue carbon: toward an improved understanding of the role of vegetated coastal habitats in sequestering CO_2. Front Ecol Environ 9(10):552–560.

McNicoll DM, Baweja AS (1995) Bioremediation of Petroleum-contaminated Soils: An Innovative, Environmental Friendly Technology, Environment Canada.

Mohan AJ, Sekhar VC, Bhaskar T, Nampoothiri KM (2016) Microbial assisted high impact polystyrene (HIPS) degradation. Bioresour Technol 213:204–207.

Moore C, Phillip C (2011) Plastic Ocean: How a Sea Captain's Chance Discovery Launched a Determined Quest to Save the Oceans. First ed. Penguin Group (USA) Inc., New York.

Moreira ITA, Oliveira OMC, Triguis JA, dos Santos AMP, Queiroz AFS, Martins CMS, Silva CS, Jesus RS (2011) Phytoremediation using *Rizophora mangle* L. in mangrove sediments contaminated by persistent total petroleum hydrocarbons (TPH's). Microchem J 99:376–382.

Mukherjee N, Sutherland WJ, Dicks L, Hugé J, Koedam N, Dahdouh-Guebas F (2014) Ecosystem service valuations of mangrove ecosystems to inform decision making and future valuation exercises. PLoS One 9:1–9.

Mukherjee S, Das P, Sen R (2006) Towards commercial production of microbial surfactants. Trends Biotechnol 24:509–515.

Murphy F, Ewins C, Carbonnier F, Quinn B (2016) Wastewater treatment works (WwTW) as a source of microplastics in the aquatic environment. Environ Sci Technol 50(11):5800–5808.

Murray BC, Pendleton L, Jenkins WA, Sifleet S (2011) Green Payments for Blue Carbon: Economic Incentives for Protecting Threatened Coastal Habitats, Nicholas Institute for Environmental Policy Solutions, Duke University, North Carolina.

Nagelkerken I, Blaber SJM, Bouillon S, Green P, Haywood M, Kirton LG, Meynecke JO, Pawlik J, Penrose HM, Sasekumar A, Somerfield PJ (2008) The habitat function of mangroves for terrestrial and marine fauna: a review. Aquat Bot 89(2):155–185.

Nellemann C, Corcoran E, Duarte CM, Valdés L, De Young C, Fonseca L, Grimsditch G (2009) Blue Carbon: The Role of Healthy Oceans in Binding Carbon. Blue Carbon. A Rapid Response Assessment. United Nations Environment Programme, GRID-Arendal.

NOAA (2014) Oil Spills in Mangroves. Planning & Response Considerations. US Department of Commerce, National Oceanic and Atmospheric Administration (NOAA), National Ocean Service, Office of Response and Restoration, Seattle, Washington.

Norris BK, Mullarney JC, Bryan KR, Henderson SM (2017) The effect of pneumatophore density on turbulence: a field study in a Sonneratia-dominated mangrove forest, Vietnam. Cont Shelf Res 147:114–127.

Paco A, Duarte K, da Costa JP, Santos PSM, Pereira R, Pereira ME, Freitas AC, Duarte AC, Rocha-Santos, TAP (2017) Biodegradation of polyethylene microplastics by the marine fungus *Zalerion maritimum*. Sci Total Environ 586:10–15.

Parrish ZD, Banks MK, Schwab AP (2005) Assessment of contaminant liability during phytoremediation of polycyclic aromatic hydrocarbon impacted soil. Environ Pollut 137:187–197.

Parrotta JA, Wildburger C, Mansourian S (2012) Understanding Relationships between Biodiversity, Carbon, Forests and People: the Key to Achieving REDD+ Objectives. International Union of Forest Research Organizations, Vienna.

Paz-Alberto AM, Sigua GC (2013) Phytoremediation: A Green Technology to Remove Environmental Pollutants. Am J Clim Change 2:71–86.

Pendleton L, Donato DC, Murray BC, Crooks S, Jenkins WA, Sifleet S, Craft C, Fourqurean JW, Kauffman JB, Marbà N, Megonigal P (2012) Estimating global "blue carbon" emissions from conversion and degradation of vegetated coastal ecosystems. PloS One 7(9):43542.

Purkhold U, Pommerening-Roser A, Juretschko S, Schmid MC, Koops HP, Wagner M (2000) Phylogeny of all recognized species of ammonia oxidizers based on comparative 16S rRNA and amoA sequence analysis: implications for molecular diversity surveys. Appl Environ Microbiol 66:5368–5382.

Purkhold U, Wagner M, Timmermann G, Pommerening-Roser A, Koops HP (2003) 16S rRNA and amoA-based phylogeny of 12 novel betaproteobacterial ammonia-oxidizing isolates: extension of the dataset and proposal of a new lineage within the nitrosomonads. Int J Syst Evol Microbiol 53:1485–1494.

Queiroga CL, Nascimento LR, Serra GE (2003) Evaluation of paraffin biodegradation and biosurfactant production by *Bacillus subtilis* in the presence of crude oil. Brazilian J Microbiol 34(4):321–324.

Rahman KSM, Rahman TJ, Kourkoutas Y, Petsas I, Marchant R, Banat IM (2003) Enhanced bioremediation of n-alkane in petroleum sludge using bacterial consortium amended with rhamnolipid and micronutrients. Bioproc Technol 90(2):159–168.

Rainbow PS, Luoma SN (2011) Metal toxicity, uptake and bioaccumulation in aquatic invertebrates—modelling zinc in crustaceans. Aqua Toxicol 105(3–4):455–465.

Ramdine G, Fichet D, Louis M, Lemoine S (2012) Polycyclic aromatic hydrocarbons (PAHs) in surface sediment and oysters (*Crassostrea rhizophorae*) from mangrove of Guadeloupe: levels, bioavailability, and effects. Ecotoxicol Environ Saf 79:80–89.

Rech S, Borrell Y, García-Vázquez E (2016) Marine litter as vector of non-native species: what we need to know. Mar Pollut Bull 113(1–2):40–43.

Ren H, Chen H, Li Z, Han W (2010) Biomass accumulation and carbon storage of four different aged *Sonneratia apetala* plantations in Southern China. Plant Soil 327:279–291.

Richards DR, Fries DA (2016) Rates and drivers of mangrove deforestation in Southeast Asia, 2000–2012. PNAS 113 (2):344–349.

Ríos LM, Moore C, Jones PR (2007) Persistent organic pollutants carried by synthetic polymers in the ocean environment. Mar Pollut Bull 54(8):1230–1237.

Robledo C, Fischler M, Patiño A (2004) Increasing the resilience of hill side communities in Bolivia: has vulnerability to climate change been reduced as a result of previous sustainable development cooperation? Mt Res Dev 24:14–18.

Rochman CM, Hoh E, Hentschel BT, Kaye S (2013) Long-term field measurement of sorption of organic contaminants to five types of plastic pellets: implications for plastic marine debris. Environ Sci Technol 47(3):1646–1654.

Rodríguez-Rodríguez JA, Sierra-Correa PC, Gómez-Cubillos MC, Villanueva LVL (eds) (2016) Mangroves of Colombia. In: Finlayson, C., Milton, G., Prentice, R., Davidson, N. The Wetland Book. Springer, Dordrecht.

Rumbold DG, Evans DW, Niemczyk S, Fink LE, Laine KA, Howard N, Krabbenhoft DP, Zucker M (2011) Source identification of Florida Bay's methyl mercury problem: mainland runoff versus atmospheric deposition and in situ production. Est Coast 34(3):494–513.

Safahieh A, Nabavi MB, Vazirizadeh, A, Ronagh MT, Kamalifar R (2012). Horizontal zonation in macrofauna community of Bardestan mangrove Creek, Persian Gulf. W J Fish Mar Sci 4(2): 142–149.

Saido K, Itagaki T, Sato H, Kodera Y, Abe O, Ogawa N, Miyashita K (2009) New contamination derived from marine debris plastics. In: Paper presented at the 238th National Meeting of the American Chemical Society (ACS).

Salmo III, SG, Duke NC (2010) Establishing mollusk colonization and assemblage patterns in planted mangrove stands of different ages in Lingayen Gulf, Philippines. Wetl Ecol Manage 18: 745–754.

Sandilyan S, Kathiresan K (2012) Mangrove conservation: a global perspective. Biodivers Conserv 21(14):3523–3542.

Sandilyan S, Kathiresan K (2015) Mangroves as bioshield: An undisputable fact. Ocean Coast Manage 103:94–96.

Sarma VV, Hyde KD (2001) A review on frequently occurring fungi in mangroves. Fungal Divers 8:1–34.
Seabra PN (2008) Biorremediação de solos contaminados por petróleo e derivados, in: (Ed.), Microbiologia Ambiental.
Sekhar VC, Nampoothiria KM, Mohana AJ, Naira NR, Bhaskarb T, Pandey A (2016) Microbial degradation of high impact polystyrene (HIPS), an e-plastic with decabromodiphenyl oxide and antimony trioxide. J Hazard Mater 318:347–354.
Sen N, Naskar K (2003) Algal flora of Sundarbans Mangal. Daya, New Delhi.
Sheik, S, Chandrashekar KR, Swaroop K, Somashekarappa HM (2015) Biodegradation of gamma irradiated low density polyethylene and polypropylene by endophytic fungi. Int Biodeterior Biodegrad 105:21–29.
Shearer CA, Descals E, Kohlmeyer B, Kohlmeyer J, Marvanova L, Padgett DE, Porter D, Raja HA, Schmit JP, Thorton HA, Voglymayr H (2007) Fungal biodiversity in aquatic habitats. Biodivers Conserv 16:49–67.
Siikamäki J, Sanchirico JN, Jardine S, McLaughlin D, Morris DF (2012a) Blue Carbon: Global options for reducing emissions from the degradation and development of coastal ecosystems, resources for the future, Washington DC.
Siikamäki J, Sanchirico JN, Jardine SL (2012b) Global economic potential for reducing carbon dioxide emissions from mangrove loss. Proc Nat Acad Sci 109:14369–14374.
Silva CAR, Lacerda LD, Rezende CE (1990) Metals Reservoir in a Red Mangrove Forest. Biotropica 4:339–345.
Sowmya HV, Ramalingappa KM, Thippeswamy B (2014) Biodegradation of polyethylene by *Bacillus cereus*. Adv Polym Sci Technol 4(2):28–32.
Spalding M, McIvor A, Tonneijck FH, Tol S and van Eijk P (2014) Mangroves for coastal defence. Guidelines for coastal managers & policy makers. Published by Wetlands International and The Nature Conservancy. p42.
Spalding M, Kainuma M, Collins L (2010) World Atlas of Mangroves. Earthscan, London.
Sri Dattatreya P, Madhavi K, Satyanarayana B, Adnan Amin, Harini C (2018) Assessment of Physico-chemical Characteristics of Mangrove Region in the Krishnapatnam Coast, India. Int J Curr Microbiol App Sci 7(5):2326–2342.
Swan JM, Neff JM, Young PC (1994) Energy Research and Development Corporation (Australia), Australian Petroleum Production Exploration Association. Environmental Implications of Off-shore Oil and Gas Development in Australia: the Findings of an Independent Scientific Review, Sydney.
Tam NFY, Wong YS (2008) Effectiveness of bacterial inoculum and mangrove plants on remediation of sediment contaminated with polycyclic aromatic hydrocarbons. Mar Pollut Bull 57:716–726.
Tam NFY, Wong YS (1995) Mangrove soils as sinks for wastewater-borne pollutants Hydrobiol 295:231–241.
Teuten EL, Rowland SJ, Galloway TS, Thompson, RC (2007) Potential for plastics to transport hydrophobic contaminants. Environ Sci Technol 41(22):7759–7764.
Thatoi H, Behera BC, Mishra RR, Dutta SK (2013) Biodiversity and biotechnological potential of microorganisms from mangrove ecosystems: a review. Annal Microbiol *63*(1):1–19.
Thatoi HN, Behera BC, Dangar TK, Mishra RR (2012) Microbial biodiversity in mangrove soils of Bhitarkanika, Odisha, India.
UNEP (2014) The Importance of Mangroves to People: a Call to Action, UNEP World Conservation Monitoring Centre, Cambridge.
USEPA (2000) Introduction to Phytoremediation.Washington, DC. EPA/600/R-99/107.
Wang Z, Stout SA (2007) Oil Spill Environmental Forensics. Academic Press, California.
Wang M, Cao W, Jiang C, Yan Y, Guan Q (2018) Potential ecosystem service values of mangrove forests in southeastern China using high-resolution satellite data. Estuarine Coastal Shelf Sci., 209: 30-40.
Wang T Zou X Li B, Yao Y, Zang Z, Li Y, Yu W, Wang W (2019a) Preliminary study of the source apportionment and diversity of microplastics: taking floating microplastics in the South China Sea as an example. Environ Pollut 245:965–974.

Wang J, Tan Z, Peng J, Qiu Q, Li M (2016) The behaviors of microplastics in the marine environment. Mar Environ Res 113:7–17.

Wang Q, Mei D, Chen J, Lin Y, Liu J, Lu H, Yan C (2019b) Sequestration of heavy metal by glomalin-related soil protein: implication for water quality improvement in mangrove wetlands. Water Res 148:142–52.

Ward BB, O'Mullan GD (2002) Worldwide distribution of *Nitrosococcus oceani*, a marine ammonia-oxidizing gamma-proteobacterium, detected by PCR and sequencing of 16S rRNA and amoA genes. Appl Environ Microbiol 68:4153–4157.

Wright S, Thompson R, Galloway T (2013) The physical impacts of microplastic on marine organisms: a review. Environ Pollut 178:483–492.

Wu Y, Chung A, Tam NFY, Pi N, Wong MH (2008) Constructed mangrove wetland as secondary treatment system for municipal wastewater. Ecol Eng 34:137–146.

Wu RY (1993) Studies on the microbial ecology of the Tansui Estuary. Bot Bull Acad Sin 34:13–30.

Xin, Li, Ryuichiro, K and Kokki, S. 2002. Biodegradation of sugarcane bagasse with marine fungus *phlebia* sp. MG–60. J Wood Sci 48: 159–162.

Yakimov MM, Timmis KN, Golyshin PN (2007) Obligate oil-degrading marine bacteria. Curr Opin Biotechnol 18(3):257–266.

Ye Y, Tam NFY, Wong YS (2001) Livestock wastewater treatment by a mangrove pot- cultivation system and the effect of salinity on the nutrient removal efficiency. Mar Pollut Bull 42 (6):513–521.

Yoshida S, Hiraga K, Takehana T, Taniguchi I, Yamaji H, Maeda Y, Toyohara K, Miyamoto K, Kimura Y, Oda K (2016) A bacterium that degrades and assimilates poly (ethylene terephthalate). Science 351(6278):1196–1199.

Zhang K, Su J, Xiong X, Wu X, Wu C, Liu J (2016) Microplastic pollution of lakeshore sediments from remote lakes in Tibet plateau, China. Environ Pollut 219:450–455.

Zhu J, Zhang Q, Li Y, Tan S, Kang Z, Yu X, Lan W, Cai L, Wang J, Shi H (2019) Microplastic pollution in the Maowei Sea, a typical mariculture bay of China. Sci Total Environ 658:62–68.

Measurement and Modeling of Above-Ground Root Systems as Attributes of Flow and Wave Attenuation Function of Mangroves

12

Masaya Yoshikai, Takashi Nakamura, Rempei Suwa, Rene Rollon, and Kazuo Nadaoka

Abstract

Mangrove forests protect the coasts from natural disasters such as storm surges and tsunamis. These ecosystems also deposit sediments—a key process in carbon sequestration and adapting to rising sea level. These services provided by mangroves are related to their drag effect that significantly attenuates flow and waves, and enhances sedimentation. The drag force exerted by mangroves is due to the complex structures of their above-ground roots. A key parameter for the quantitative assessment of drag force is the projected area of vegetation. In this chapter, we focus on the above-ground root system (prop roots) of *Rhizophora*—the most dominant genus in the Asia-Pacific region and likely to exhibit the highest drag among mangrove species—in exploring the projected area of vegetation. We describe the methods of field measurement and an empirical model for the projected area of the *Rhizophora* prop root system. The results show the allometric relationships between the prop root projected area and tree size, and how the allometric relationships vary depending on the sites. The developed model shows its great ability as well as some limitations in accurately predicting the projected area of prop roots. We then discuss the environmental factors that may affect the allometric relationship, and prospects in developing the universal

M. Yoshikai (✉) · T. Nakamura · K. Nadaoka
School of Environment and Society, Tokyo Institute of Technology, Tokyo, Japan
e-mail: yoshikai.m.aa@m.titech.ac.jp; nakamura.t.av@m.titech.ac.jp; nadaoka.k.aa@m.titech.ac.jp

R. Suwa
Forestry Division, Japan International Research Center for Agricultural Sciences (JIRCAS), Tsukuba, Japan

R. Rollon
Institute of Environmental Science and Meteorology, College of Science, University of the Philippines, Quezon City, Philippines

R. P. Rastogi et al. (eds.), *Mangroves: Ecology, Biodiversity and Management*, https://doi.org/10.1007/978-981-16-2494-0_12

model, which can predict the prop root projected area in any type of environment. We also discuss perspectives in the measurement and modeling of other types of mangrove above-ground root system such as pneumatophores of *Avicennia* and *Sonneratia* species.

Keywords

Drag · *Rhizophora* · Root system · Allometry · Modeling · Morphological plasticity

12.1 Introduction

Mangroves are known to form complex structures in their above-ground roots. Examples are the prop roots of *Rhizophora* species (Fig. 12.1a), the pneumatophores of *Avicennia* and *Sonneratia* species that are usually denoted as pencil roots

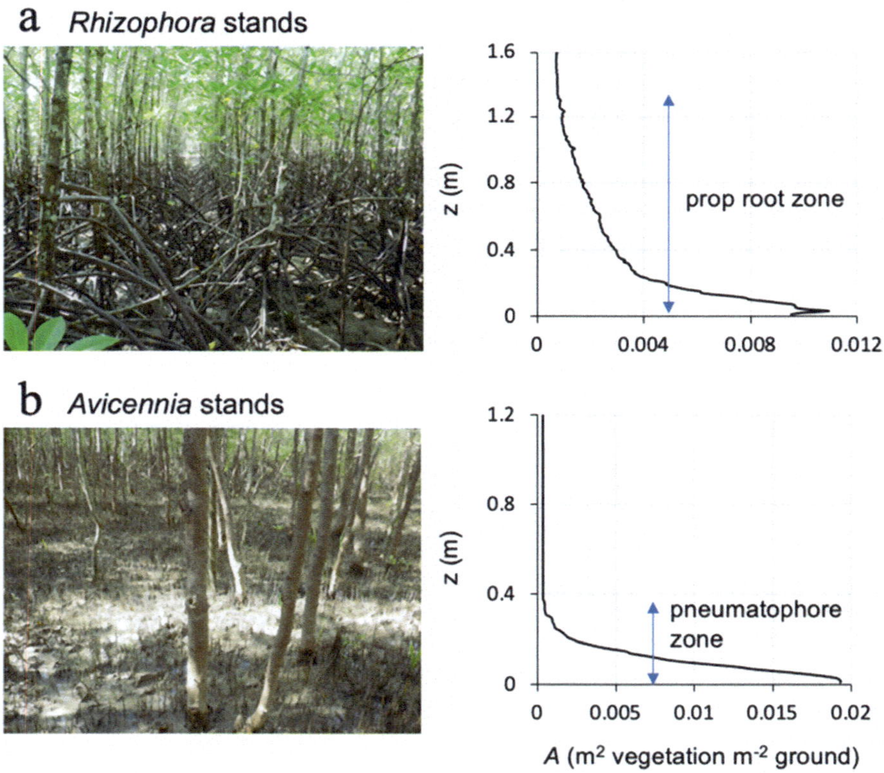

Fig. 12.1 (**a**) Prop root system of *Rhizophora* stands and (**b**) pneumatophores of *Avicennia* stands with vertical distribution of the projected area of vegetation (Yoshikai et al., unpublished data). z: height from the ground, *A*: projected area of vegetation per unit ground area per 1 cm vertical interval. The data on *A* was collected in a planted mangrove forest in Panay Island, the Philippines

(Fig. 12.1b), and the root knees and buttresses that are seen in *Bruguiera* species and some other species (Srikanth et al. 2016; Tomlinson 2016). These are considered to have evolutionally developed to deal with the anaerobic nature of substrates generated by the recurrent tidal water flooding by oxygenating the absorbing roots (fine roots) below-ground through the above-ground roots that are exposed in the atmosphere (Srikanth et al. 2016; Tomlinson 2016).

The complex structures of the mangrove above-ground root system exert drag force that substantially influences tidal flow (Furukawa et al. 1997; Mazda et al. 2005; Horstman et al. 2015; Chen et al. 2016). It has been recognized that drag effects of mangroves have an important role in protecting shorelines against damages by tsunami-induced waves (Dahdouh-Guebas et al. 2005; Danielsen et al. 2005; Yanagisawa et al. 2010) and storm surges (Das and Vincent 2009; Mclvor et al. 2012; Menéndez et al. 2018). While it is expected that the risk of coastal flooding is enhanced by increase in the occurrence of more intense tropical cyclones and sea-level rise in the future (Woodruff et al. 2013), ecosystem-based coastal protection has been proposed as sustainable and cost-effective option which could be alternative to gray structures such as seawalls (Temmerman et al. 2013). In this regard, the quantification of the effects of mangroves on wave attenuation and hydrodynamic load reduction is of great interest to properly assess the mangroves' function as coastal defense (Maza et al. 2017, 2019; Montgomery et al. 2019; Tomiczek et al. 2020).

The drag effects of mangroves also facilitate sedimentation by flow attenuation and sediment trapping by the surface of vegetation (Horstman et al. 2015; Willemsen et al. 2016; Chen et al. 2018). While mangroves have been threatened by sea-level rise, the vertical accretion of mangroves through sedimentation is considered as a key process in the survival of mangroves (Krauss et al. 2014; Lovelock et al. 2015; Woodroffe et al. 2016; Saintilan et al. 2020). Sedimentation also contributes to the mangrove carbon storage by trapping allochthonous carbon (Xiong et al. 2018), and thus may play an important role in carbon sequestration.

Several studies highlighted the importance of the quantification of mangrove drag effects. A key parameter is the projected area of vegetation as suggested in the following equation for the drag force exerted by vegetation:

$$F_D = \frac{1}{2} C_D A_f U^2 \tag{12.1}$$

where C_D is the drag coefficient, A_f is the total projected area of the submerged vegetation per unit volume, and U is the depth-averaged flow velocity (Chen et al. 2016; Maza et al. 2017). An example of the vertical profile of the projected area of vegetation of *Rhizophora* and *Avicennia* stands per unit ground is provided in Fig. 12.1, which was manually measured in Bakhawan Ecopark in Panay Island, the Philippines (Yoshikai et al., unpublished data), showing the significance of the above-ground root system on the projected area of vegetation. This kind of data is needed for estimating the forest-scale drag force considering the wide areal coverage of the target mangrove forest. On the other hand, the projected area of vegetation may vary significantly with the mangrove conditions such as tree density, age, and size of the individual trees. It may also be influenced by species composition as suggested by the different patterns in the vertical profile of the projected area between the prop root system and pneumatophores (Fig. 12.1). However, individual measurements of the above-ground root morphology for the whole mangrove forest are impractical. In this regard, data collection on the above-ground root morphological traits of different species in various sites, and development of a predictive model based on the collected data are effective ways to estimate the projected area of vegetation in a mangrove forest, thus moving forward our understanding on the forest-scale mangrove drag effects.

In this chapter, we describe the methods of field measurement and an empirical model used to estimate the projected area of the above-ground root system. We especially focus on the prop root system of *Rhizophora*, some knowledge and information on the morphological trait of which have been obtained by recent studies and the empirical models proposed (Ohira et al. 2013; Yoshikai et al. 2021). This species is also the most dominant species in the Asia-Pacific region (Ong et al. 2004) and likely to exhibit the highest drag among mangrove species (Horstman et al. 2014). We also discuss some perspectives on the field measurement and model development for the other above-ground root types such as pneumatophores of *Avicennia* and *Sonneratia* species, whose morphology have not been well investigated.

12.2 Field Measurement of Mangrove Above-Ground Root Morphology

In this section, we review the methods of field measurement of mangrove above-ground root morphology, including both manual and ground-based remote-sensing techniques. Although studies on the mangrove above-ground morphology are quite limited, there are several research works that investigated the *Rhizophora* prop root system and the pneumatophores of *Avicennia* and *Sonneratia* species. We therefore targeted these two root systems for the review presented here.

12.2.1 Manual Measurement

12.2.1.1 Prop Root System of *Rhizophora*

The projected area of *Rhizophora* stands varies vertically (Fig. 12.1a), and obtaining and predicting such vertical profile is the objective of the study. In this regard, we partitioned the prop root system into 0.1 m-thick vertical layers, and obtained the vertical profile of the projected area in each layer (A_i, where i is the layer number shown in Fig. 12.2) from field data.

Ohira et al. (2013) proposed a methodology of manual measurement of the prop root structure in the field to estimate A_i by approximating the individual prop root shapes as quadratic curves. For the approximation, three parameters are required,

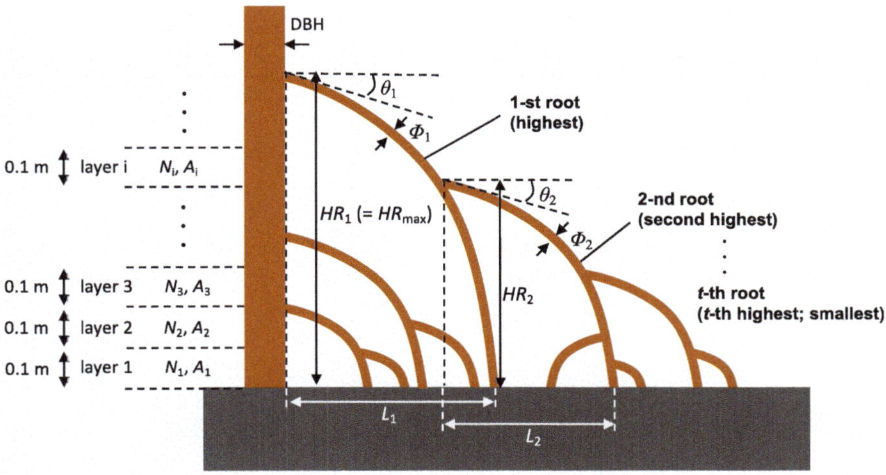

Fig. 12.2 Vertical layering of the prop root system with layer variables—number of prop roots (N_i) and projected area (A_i), numbering of individual prop roots, and geometric parameters of prop roots (HR_k: height, L_k: horizontal distance, θ_k: angle, Φ_k: diameter of k-th root), and stem (DBH: diameter at breast height). t-th root is the smallest prop root, thus t indicates the total number of prop roots. Modified from Yoshikai et al. (2021)

which are root height (HR), horizontal distance (L), and angle (θ) of individual prop roots (Fig. 12.2). The shape of each prop root was thus predicted as:

$$y = -\frac{HR + L\tan\theta}{L^2}x^2 + (\tan\theta)x + HR \tag{12.2}$$

where $y \geq 0$ and is the root height from the ground at position x, and x is the horizontal distance from the emergent point of the prop root. For example, for a primary (first-order) root that emerges from the stem, $x = 0$ refers to the position of the stem. The individual prop root projected area can be obtained by multiplying the prop root length obtained from Eq. (12.2) and the diameter (Φ). The whole-tree prop root projected area, which is a sum of the projected areas of individual prop roots, and its vertical distribution (A_i) can then be calculated. Note that Eq. (12.2) expresses the projected prop root area from the side view which does not consider the azimuth angle of prop roots to water flow direction. When the information on the azimuth angles of the individual prop roots are available, the projected areas of the individual prop roots from the flow direction can be calculated as:

$$y = -\frac{HR + L\tan\theta}{L^2}\left(\frac{x}{\cos\psi}\right)^2 + (\tan\theta)\left(\frac{x}{\cos\psi}\right) + HR \tag{12.3}$$

where ψ is the azimuth angle of the individual prop roots relative to the flow direction. The methodology proposed by Ohira et al. (2013) quantifies the individual root shapes, and therefore has the advantage of providing the three-dimensional

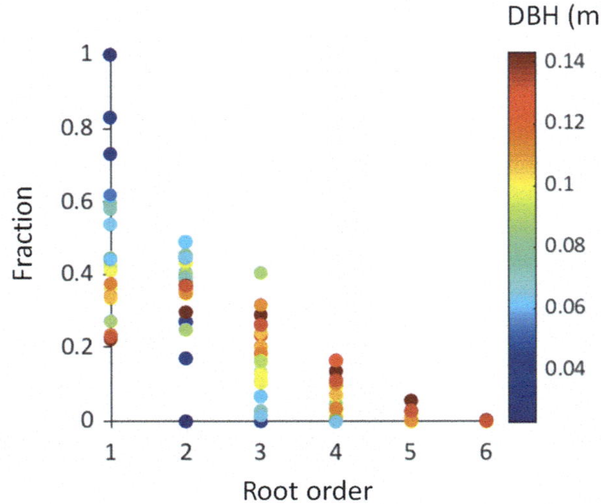

Fig. 12.3 The fractions of the projected areas of the first- to sixth-order prop roots to the whole-tree prop root projected area for 22 trees sampled from a natural mangrove forest in Ishigaki Island, Japan. The color indicates the DBHs of the individual trees. (Data source: Yoshikai et al. 2021)

(3D) information, which is useful when reconstructing 3D structures from the data. For example, the model *Rhizophora* stands used for the laboratory flume experiments done in Maza et al. (2017, 2019), Shan et al. (2019), and Tomiczek et al. (2020) were built by referring to the data collected by Ohira et al. (2013).

While the measurement by Ohira et al. (2013) is limited to the first-order roots only, Yoshikai et al. (2021) have measured the four geometric parameters (HR, L, θ, Φ; Fig. 12.2) for all prop roots, including the second-, third-, and even higher-order roots—where the root order indicates the level of branching from the stem—from one sampled tree. The whole-tree prop root projected area and its vertical distribution was shown for the first time, and the significance of the prop roots higher than the first-order in the whole-tree prop root projected area was revealed. Figure 12.3 shows the fractions of the projected areas of the first- to sixth-order prop roots to the whole-tree projected area. The data shows that branching rate of prop roots increases as DBH increases, and up to six-order prop roots were observed in trees with DBH $>$ 0.13 m. As a result, the presence of prop roots higher than the first-order becomes more significant in the whole-tree prop root projected area, with up to 80% significance in a tree with DBH = 0.14 m. This indicates that the non-inclusion of the prop roots higher than first-order will fall short in obtaining the actual projected area of the prop root system.

However, measuring the four prop root parameters (HR, L, θ, Φ) for all prop roots in a tree remains to be very laborious. For example, measurement of the prop root system of a tree with 30 prop roots by two persons would take 30–40 min, and it is not uncommon that some trees have more than 100 prop roots depending on tree size. Therefore, collecting a sufficient number of samples on the prop root system for developing a predictive model in a site would be labor-intensive and time-consuming. In this regard, the measurement of the prop root system was simplified from measuring the four prop root parameters to one parameter—HR in Yoshikai et al. (2021) for the purpose in estimating the vertical distribution of A_i without considering the individual prop root shapes. This simplification is possible using the strong relationship between the vertical distributions of the number of prop roots (N_i) and the prop root projected area (A_i), which can be expressed as a linear equation with intercept fixed at zero (coefficient of determination, $R^2 = 0.9632$):

$$A_i = 0.0043N_i \tag{12.4}$$

where the unit of A_i is m^2 (see Yoshikai et al. 2021 for the plot). This equation indicates that the averaged projected area of one prop root in vertical layers can be considered uniform with an area of 0.0043 m^2. The vertical profile of N_i can be obtained from the height data (*HR*) by counting the number of prop roots in each vertical layer. With this simplification, the time for measuring the prop root system morphology can be shortened by around one-third compared to measuring the four parameters. This was further simplified by measuring only the highest root height and the total number of prop roots in a tree to predict N_i, and thereby A_i (Yoshikai et al. 2021); this is described in Sect. 12.3.7. This further simplification greatly reduces the amount of field work needed to quantify the prop root system of a tree. It should be noted, however, that this method is valid only for statistically predicting N_i and A_i, which does not consider the detailed actual 3D structure of the prop root system.

12.2.1.2 Pneumatophores of *Avicennia* and *Sonneratia* Species

There are several studies that have measured the morphology of pneumatophores—aerial roots of *Avicennia* and *Sonneratia* species—in the field. The pneumatophore morphology such as spatial density (m^{-2}), height, and diameter are usually surveyed within quadrats of 0.5 m × 0.5 m or 1 m × 1 m with some replicates to characterize the structural complexity in this type of forest (Dahdouh-Guebas et al. 2007; Horstman et al. 2014; Lienard et al. 2016; Zhang et al. 2019). This methodology is used because the pneumatophores usually distribute relatively uniform in space in the *Avicennia*- and *Sonneratia*-dominated forests, unlike the prop roots of *Rhizophora* stands.

The individual shape of the pneumatophore is not very complicated. It has been approximated as a cylindrical shape (Horstman et al. 2014) or truncated cone shape (Zhang et al. 2015a; Lienard et al. 2016) when quantifying pneumatophore parameters such as volume and projected area. To validate this approximation, we have measured pneumatophore geometry (height, diameters at the base, one-third and two-third heights, and at the tip) in the field, and sampled the pneumatophore for laboratory analysis. The volume and the lateral surface area of the pneumatophore were geometrically estimated by assuming a truncated cone shape as in Zhang et al. (2015a) and Lienard et al. (2016). In the laboratory analysis, the volume of the sampled pneumatophore was measured using the water displacement method. The lateral surface area was measured using the aluminum foil method for surface area determination (Marsh 1970), which is commonly used for measuring the surface area of corals. The geometrically estimated volume (125 cm^3) and lateral surface area (152 cm^2) turned out to be very close to the measured values (120 cm^3 and 144 cm^2, respectively) with around 5% error, suggesting the validity of this approximation. Figure 12.1b shows the field-measured vertical distribution of the projected area of vegetation of *Avicennia* stands in Bakhawan Ecopark, Panay Island, the Philippines (Yoshikai et al. unpublished data). The near-ground area (around up to 0.3 m height) is densely vegetated with pneumatophores, and shows a high projected area of vegetation. Because the pneumatophore height is limited up to 0.4 m, the projected area of vegetation higher than that height is primarily due to the stem parts. However, the above-mentioned studies are limited to the quadrat-scale only, and quantification of the whole-tree pneumatophore morphological parameters such as total pneumatophore projected area, volume, and extent of individual trees remains unresolved. Therefore, it is still unclear how the projected area per unit ground shown in Fig. 12.1b will change with tree size and density.

The *Avicennia* and *Sonneratia* root system is developed below-ground through the cable roots. This makes it very difficult to identify which pneumatophore originates from which tree in the *Avicennia*- and *Sonneratia*-dominated forests, unlike the prop roots system of *Rhizophora* trees. Investigation of the pneumatophore root system requires a below-ground approach. Although some schematic descriptions on the above- and below-ground root system can be found in Gill and Tomlinson (1977), Ezcurra et al. (2016), and Tomlinson (2016), few studies have focused on the below-ground root system. Destructive measurement such as the full excavation of the root system is one traditional way of quantification (Danjon et al. 2005; Smith et al. 2014), but it is labor-intensive, time-consuming, and poses risks to the forest ecosystem. While some non-destructive methodologies have been developed for detecting the below-ground root system in terrestrial ecosystem such as ground penetrating radar (GPR, Zhu et al. 2014; Hardiman et al. 2017) and electric resistivity imaging (ERI, Amato et al. 2008; Rossi et al. 2011), there are no studies to our knowledge that have applied such techniques in mangrove forests. Unlike the substrate conditions for the terrestrial ecosystem, the substrate of mangrove forests is under the influence of saline water intrusion that may affect radar attenuation and electric resistivity below-ground. The effects of saline waters in soil should be examined for possible application of these non-destructive techniques. All these factors make the investigation of the pneumatophore root system challenging. Yet, some studies have added meaningful insights in this field. Yando (2018) measured the pneumatophore extent of *Avicennia germinans* at the mangrove-salt marsh boundary where the extent of pneumatophore of the individual trees can be visibly recognized, and found a correlation between the maximum pneumatophore extent and tree height. Vovides et al. (2016) developed a simple non-destructive methodology to trace the below-ground cable roots from tree trunk using an ultrasonic Doppler fetal monitor and steel rods. With this method, they measured the pneumatophore extent of *Avicennia germinans* stands, and obtained a significant relationship with tree size as with Yando (2018). These results suggest the scaling relations in the individual pneumatophore root system and the possibilities for the development of a predictive model. More comprehensive data such as the number of cable roots per tree, branching of the cable roots, and pneumatophore densities specific to the cable root surface area should be collected for realizing the model development.

12.2.2 Ground-Based Remote-Sensing Techniques

As described in Sect. 12.2.1.1, manual measurement of the detailed morphological structures of the above-ground root system (e.g., the four parameters *HR*, *L*, θ, and Φ of all the prop roots in a tree) could be labor-intensive and time-consuming. While the simplification of the measurement of the prop root system could reduce the amount of field works, it sacrifices some information on the morphological parameters such as the shape of the individual prop roots and the three-dimensional spatial extent of the prop root system. On the other hand, applications of novel ground-based remote-sensing techniques such as terrestrial laser scanning (TLS, also known as terrestrial LiDAR) and photogrammetry in the characterization of ecosystem structures have recently advanced. These techniques provide the 3D information on the morphological traits with low time and labor requirements, thus having the potential to overcome the limitations of manual measurement.

The TLS instrument emits small footprint laser pulses at a high rate and accurately measures the distance between the sensor and a target based on the elapsed time between the emission and the return of laser pulses (Yao et al. 2011). Hence, the 3D coordinates of the point cloud that represents object surfaces are obtained with great spatial resolution. The TLS has been used in numerous ecological studies, including forest biomass estimation (Yao et al. 2011; Hosoi et al. 2013), canopy structures (Greaves et al. 2015; Olsoy et al. 2016), and more complicated parameters like leaf angle distribution (Liu et al. 2019). There are some studies that have applied TLS for mangroves—Olagoke et al. (2016) used TLS for estimating the volume and biomass of large mangrove trees that are usually difficult to directly measure; Paynter et al. (2016) used TLS for capturing the 3D structure of a complex prop root system of *Rhizophora mangle* and reconstructed a 3D model. In Bakhawan Ecopark, Panay Island, the Philippines, a field survey using a portable-TLS instrument to scan the *Rhizophora apiculata* stands was conducted (Fig. 12.4a). Using the instrument, point cloud of the stands in an area wider than 5 m $\times$ 5 m was rapidly and easily acquired (Fig. 12.4b; Baloloy et al. personal communication). Figure 12.4c shows the scaled up point cloud of a tree, and a 3D model of the prop root system of this tree was reconstructed using a freeware MeshLab (Fig. 12.4d). The 3D model well represents the detailed morphological traits of the complex prop root system that cannot be described by the information taken by the manual measurement.

Photogrammetry is also often used to quantify the complex structure of mangrove ecosystems. This method produces a 3D model from a series of overlapping photographs taken from multiple perspectives using structure-from-motion algorithm (SfM) (Figueira et al. 2015). There are several free and open source software (e.g., VisualSfM, Regard3D) that allow the easy application of the algorithm for 3D model reconstruction. This method is also cost-effective as it technically requires

Fig. 12.4 (**a**) Scanning of *Rhizophora* stands using a portable-TLS instrument in the field, (**b**) obtained point cloud (Baloloy et al., personal communication), (**c**) point cloud upscaled to a tree, (**d**) 3D model of the prop root system reconstructed using the point cloud, (**e**) 3D-printed model of the prop root system, and (**f**) flume experiment using the 3D-printed models

only a digital camera and a reference scale for data collection in the field. In a large-scale, this method has been used to map the 3D canopy structures of mangroves from the photographs taken from UAVs (Navarro et al. 2019; Yaney-Keller et al. 2019). In a small-scale, Zhang et al. (2015b) has applied photogrammetry to capture the 3D information of a young *Rhizophora stylosa* tree with a relatively simple structure. Lienard et al. (2016) and Norris et al. (2017) applied photogrammetry to the pneumatophores of *Sonneratia* stands in deriving the projected area of vegetation and volume. Lienard et al. (2016) compared the photogrammetry-derived and manually measured pneumatophore morphological parameters, and showed a reasonable agreement. They also stressed the ability of photogrammetry in deriving the morphological parameters of pneumatophores having complex geometry due to the numerous barnacles attached to the pneumatophore surface, which cannot be resolved by manual measurement. However, the photogrammetry requires a number of photographs taken from various perspectives for one target object, thus making it difficult to collect 3D information of a relatively wide area in a short period of time compared to TLS (Fig. 12.4b).

Another 3D scanning technique is the use of RGB-D (Red, Green, Blue-Depth) sensor which is available at low cost (Kamal et al. 2014). Kamal et al. (2014) used this sensor to scan the 3D structure of a prop root system of *R. stylosa* and pneumatophores of *Avicennia marina*. From the reconstructed 3D models, they analyzed the fractal dimension in the root system. Yanagisawa and Miyagi (2020) also used the RGB-D sensor to scan the prop root systems of *Rhizophora apiculata* trees with different ages. Their results showed its successful application to both relatively simple and complex structures of prop root systems.

The studies that have utilized the ground-based remote-sensing techniques showed remarkable abilities in acquiring the detailed 3D information of the complex morphological structures of the above-ground root system. These are quite promising methods for the rapid data collection of the morphological parameters of above-ground roots with very high accuracy. Also, these have great potential for quantifying other types of mangrove above-ground root system, like root knees and buttresses of *Bruguiera gymnnorrhiza*, the morphological structure of which has not been investigated yet. Another advantage of these techniques is the possibility of creating 3D-printed models of the above-ground root system from the data (Fig. 12.4e). This makes it feasible to conduct flume experiments using more realistic physical mangrove models for investigating drag effects and hydrodynamics (Fig. 12.4f).

12.3 Modeling the Morphological Structure of the Prop Root System

In this section, we describe the models for the prop root system developed based on the field data. The models were designed to predict the vertical distribution of the projected area A_i using scaling relations. We review two models—a model for a prop root system with only primary roots developed by Ohira et al. (2013), and a model which can be applied to prop root system with multiple-order of prop roots developed by Yoshikai et al. (2021). We then discuss the effects of the environmental conditions on the complexity of the prop root system.

12.3.1 Model for the Prop Root System with Only Primary Roots

Ohira et al. (2013) proposed an empirical model of the prop root system with only the primary roots. The model predicts the four parameters of the individual prop roots (*HR*, *L*, θ, Φ) from DBH, which is needed to quantify the shapes and the projected areas of the individual prop roots. The vertical distribution of whole-tree prop root projected area (A_i; Fig. 12.2) is then obtained. Based on the field data for the prop root system of *Rhizophora mucronata* and *R. apiculata* collected from Thailand, they found the following relationships: DBH–HR_{max}, HR_{max}–t, HR_k–L_k, HR_k–θ_k, and (HR_{max}, DBH)–Φ_k relationships, where a uniform vertical interval of each root height in the tree was assumed. This indicates that the shapes of all the prop roots of a tree can be predicted only from DBH, thereby A_i. However, it should be noted that these relationships have been confirmed only for the primary prop roots; therefore, the model by Ohira et al. (2013) may not be valid for the prop root system with multiple-order prop roots which can be seen in any mangroves. When looking at the HR_k–L_k relationship for the prop root system in a mangrove forest in Ishigaki Island, Japan, significant correlation can be seen for the primary roots (Fig. 12.5a), but the relationship is no longer significant for the higher-order prop roots (Fig. 12.5b), highlighting the complicated structures of the prop root system with multiple-order prop roots. This fact limits the applicability of the model by Ohira et al. (2013) only to the prop root system with primary roots.

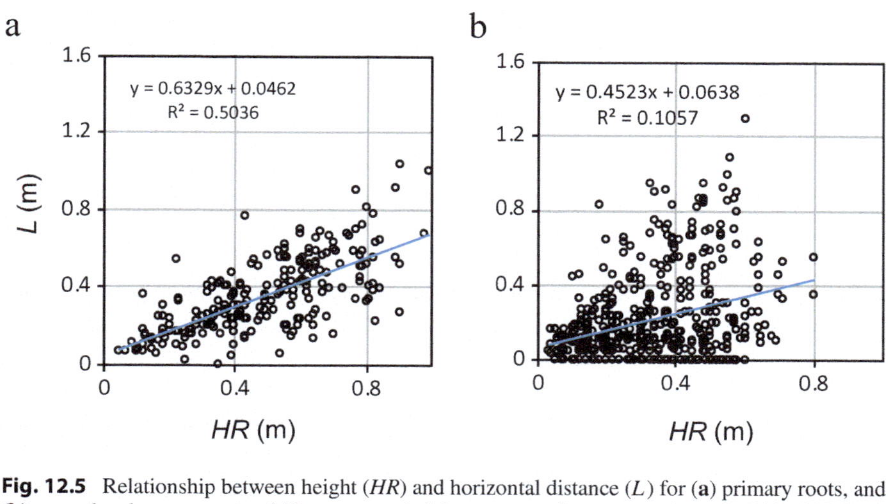

Fig. 12.5 Relationship between height (*HR*) and horizontal distance (*L*) for (**a**) primary roots, and (**b**) second-order prop roots of 22 trees sampled from a natural mangrove forest in Ishigaki Island, Japan. (Data source: Yoshikai et al. 2021)

12.3.2 Model for the Prop Root System with Multiple-Order Prop Roots

To characterize the complexity of the prop root system with multiple-order prop roots and predict A_i, a new model was proposed in Yoshikai et al. (2021). The model was designed to specifically predict the vertical distribution of the number of prop roots (N_i)—a good predictor of A_i as suggested by Eq. (12.4). A parameter—scaling factor (S), which determines the prop root structure—was introduced in the model. Here, the individual prop roots in a prop root system was numbered in the order of root height, from the highest to the lowest; the first root refers to the highest root, the second root refers to the second-highest, and k-th root refers to the k-th highest in the prop root system (Fig. 12.2). Then the height distribution of the prop roots is approximated as:

$$\frac{HR_{k+1}}{HR_k} = S \tag{12.5}$$

where HR_k and HR_{k+1} are the height of the k-th and (k + 1)-th root (m), and S is the scaling factor ranging from 0 to 1. Note that the height ratio of the k-th and (k + 1)-th root could be the ratio of roots coming from different branches. If the height of the first root (HR_1: hereafter denoted as HR_{max}), which is the maximum root height in a prop root system, is given, the height of the k-th root can be calculated as:

$$HR_k = HR_{max} S^{(k-1)} \tag{12.6}$$

The minimum root height is determined by a critical height (HR_{min}, m). If the t-th root is the one with the minimum height in a prop root system, t is the largest integer number that satisfies the following expression:

$$HR_t = HR_{max} S^{(t-1)} \geq HR_{min} \tag{12.7}$$

In this way, the height of prop roots in a tree can be modeled from the parameters HR_{max}, S, and HR_{min}. N_i is then predicted by counting the number of modeled prop roots in each vertical layer. Using fixed values of HR_{max} and HR_{min}, the number of prop roots increases as the value of S increases toward 1, which implies more primary (first-order) roots and/or more branching of prop roots (higher-order roots). Among the three model parameters, HR_{max} and HR_{min} can be directly measured in the field, however, HR_{min} was treated as a constant with a value of 0.05 m for simplicity. The variations in the values of HR_{min} are small with a mean value of 0.062 ± 0.039 m, based on the data collected from various mangroves in Indonesia, the Philippines, and Japan in Yoshikai et al. (2021), and these variations do not largely affect the model output. The values of S for the individual trees are unknown, but can be determined by searching the optimum value (optimized S) with which the modeled number of prop roots in the vertical layer fits best with the field-measured number of prop roots. Figure 12.6 shows some examples of the

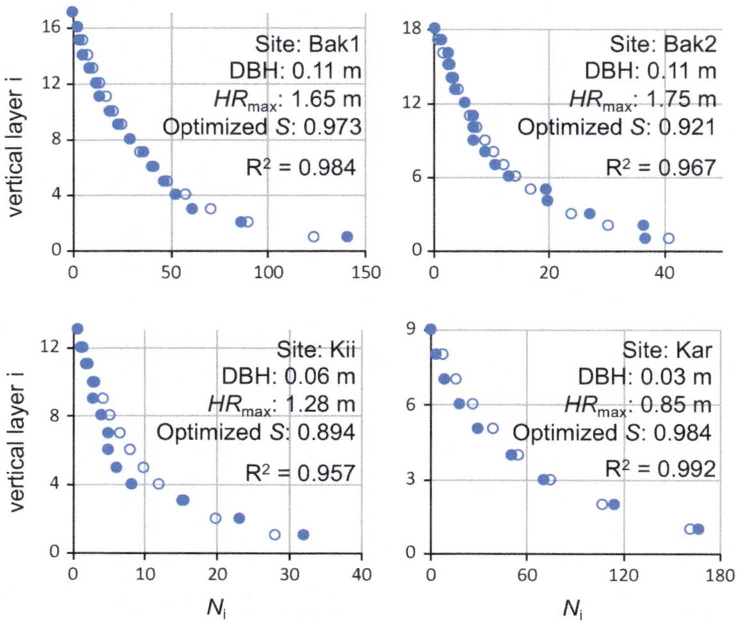

Fig. 12.6 Comparison of measured and modeled vertical distribution of the number of prop roots (N_i) for selected trees from different mangroves—Bak1 and Bak2: ~17 and ~ 30 years planted stands in Bakhawan Ecopark, Panay Island, the Philippines, respectively; Kii: planted stands in KII Ecopark in Panay Island, the Philippines; Kar: mix of natural and planted stands in Karimunjawa Island, central Java, Indonesia. The measured values for HR_{max}, a constant value for HR_{min} (0.05 m), and the optimized values for S are used to predict N_i for individual trees. (Data source: Yoshikai et al. 2021)

comparison of the vertical distribution of the field-measured and modeled number of prop roots for some selected trees sampled from different mangroves. The number of prop roots generally increases from the top of the prop root system to the bottom in the form of a power function. It was shown that the model can reproduce such patterns with high accuracy by adjusting the value of S. Among the 156 trees sampled in Yoshikai et al. (2021), the vertical distributions of N_i of 120 trees were reproduced by the model with R^2 values higher than 0.95. This suggests that the complexity of the prop root system with multiple-order prop roots can be well represented by the parameter S.

12.3.3 Analysis of the Scaling Relation in the Prop Root System

The optimized S values described in Sect. 12.3.2 have a strong relationship with tree size such as DBH, which can be expressed as:

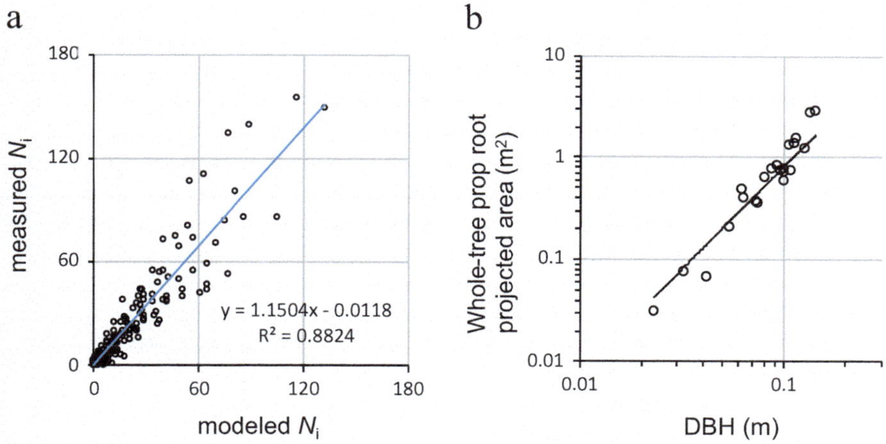

Fig. 12.7 (**a**) Comparison of measured and modeled number of prop roots (N_i) in vertical layers for 22 trees sampled from a natural mangrove forest in Ishigaki Island, Japan. All vertical layers with prop roots were used in the plot. (**b**) Comparison of measured and modeled whole-tree prop root projected area for the 22 trees. (Data source: Yoshikai et al. 2021)

$$S = 1 - \alpha \mathrm{DBH}^{\alpha_1} \tag{12.8}$$

where α is a constant, and α_1 is the scaling exponent; the unit of DBH is meter. The value "1" in this equation represents the asymptotic maximum value of S. Therefore, the S values of the individual trees, which cannot be directly measured in the field, can be predicted from DBH—the most fundamental and easy-to-measure parameter. It should be noted that the scaling relations represented by α and α_1 may significantly vary among sites and/or species; this is discussed in the following section.

HR_{max} of the individual prop root system also have a strong relationship with DBH as shown by Ohira et al. (2013), Mendez-Alonzo et al. (2015), and Yoshikai et al. (2021), which can be expressed as:

$$HR_{max} = \beta_1 \mathrm{DBH} + \beta \tag{12.9}$$

where β_1 and β are the slope and intercept terms, respectively. Like the case of optimized S, the relationship of HR_{max} with DBH may vary among sites and/or species.

These suggest that once the scaling relations for S and HR_{max} for a species in a site represented by Eqs. (12.8) and (12.9) are obtained, the values of S and HR_{max}, thereby N_i, of trees with various DBHs for the species in the site can be predicted from DBH. An example of the model prediction is provided in Fig. 12.7a which shows the comparison between measured and modeled N_i and the whole-tree prop root projected area for 22 trees sampled in site Fuk, where the scaling relations for S and HR_{max} were obtained as:

$S = 1 - 0.0006\text{DBH}^{-1.755}$ ($R^2 = 0.92$) and
$HR_{\max} = 2.71\text{DBH} + 0.50$ ($R^2 = 0.43$),

respectively (see Yoshikai et al. 2021 for the plots). The model predicted N_i with a reasonable accuracy ($R^2 = 0.8824$). Using the relationship between N_i and A_i (Eq. (12.4)), A_i can also be predicted, and the estimated whole-tree prop root projected area is plotted against DBH together with the field-measured values (Fig. 12.7b). The whole-tree prop root projected area shows the log-log relationship with DBH and the model result agrees well with this trend. The log-log relationship suggests the allometry in the whole-tree prop root projected area, and this is consistent with the other studies that showed the allometric relations in the prop root biomass and DBH (Ong et al. 2004; Comley and McGuinness 2005; Van Vinh et al. 2019).

12.3.4 Site and Species Differences in the Scaling Relation

The compilation of the S–DBH relationships of *R. apiculata* and *R. mucronata* stands obtained from the different sites is provided in Fig. 12.8. Note that the relationships in the form of the following equation are shown, which is a rearrangement of Eq. (12.8), after log-transformation:

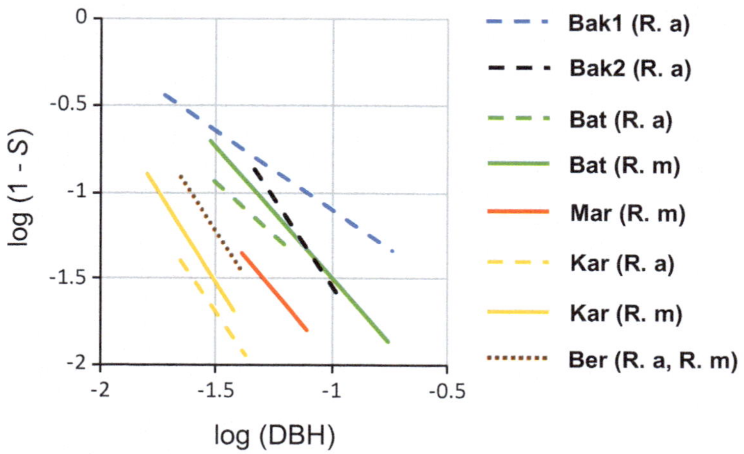

Fig. 12.8 Compilation of S–DBH relationships of *R. apiculata* and *R. mucronata* stands sampled from the different sites—Bak1 and Bak2: ~17 and ~ 30 years planted stands in Bakhawan Ecopark, Panay Island, the Philippines, respectively; Bat: ~20 years planted stands in an island in Batan Bay in Panay Island, Philippines; Mar: natural stands in Maratua Island, east Kalimantan, Indonesia; Kar: mix of natural and planted stands in Karimunjawa Island, central Java, Indonesia; Ber: natural stands at coast of Berau Continental Shelf in east Kalimantan. R. a: *R. apiculata*; R. m: *R. mucronata*. The unit of DBH is meter. (Data source: Yoshikai et al. 2021)

$$\log(1 - S) = \alpha_0 + \alpha_1 \log(\mathrm{DBH}) \tag{12.10}$$

where α_1 and α_0 are the slope and intercept term, respectively, and $\alpha_0 = \log(\alpha)$ of Eq. (12.8). Here and hereafter, log denotes the common logarithm. It clearly shows how the relationships of *S*–DBH could vary among sites even if the species is the same. The effects of the differences in the relationship for *S* can be seen variation of the total number of prop roots at the different sites. For example, at site Bak2, *R. apiculata* trees with DBH $= 0.03 \sim 0.04$ m (log DBH $= -1.52 \sim 1.40$) have around 10 prop roots ($\log(1-S) = -0.65 \sim 0.59$) while trees of the same species and same size at site Bat have around $15 \sim 30$ prop roots ($\log(1-S) = -1.07 \sim 0.77$), and $110 \sim 220$ prop roots ($\log(1-S) = -1.62 \sim 1.94$) in trees of the same species and size at site Kar. Therefore, the scaling relation obtained at a site should not be applied to other mangroves without validation; otherwise, it may lead to a large error in the prediction of A_i, and ultimately the drag force by the mangroves.

The phenomenon causing the site difference in scaling relation is considered as morphological plasticity of the prop root system, which may be the result of the mangroves' adaptation to the severe environment that they inhabit. When the *S*–DBH relationships of co-existing species in sites Kar and Bat—*Rhizophora apiculata* and *Rhizophora mucronata*—are compared, the values of *S* for *R. apiculata* tend to be slightly higher than *R. mucronata*, which indicates higher complexity in the prop root system of this species. This highlights the importance of environmental effects on the morphological plastic response and the interspecific differences in the prop root system that should be taken into consideration when predicting the prop root structures.

12.3.5 Representation of the Scaling Relations in Prop Root System

To assess the effects of environmental conditions and species variation, a representation of the site- and species-specific scaling relation is necessary. Equation (12.10), which represents the site- and species-specific scaling relation, has two parameters—the slope α_1 and the intercept α_0. In Yoshikai et al. (2021), it was found that α_1 and α_0 have a linear relationship that can be written as:

$\alpha_1 = 0.4192\alpha_0 - 0.2067$ ($R^2 = 0.82$).

Therefore, by substituting this equation into Eq. (12.10), the site- and species-specific scaling relation can be expressed as:

$$\log(1 - S) = \alpha_0 + (0.4192\alpha_0 - 0.2067)\log(\mathrm{DBH}) \tag{12.11}$$

In Yoshikai et al. (2021), the value of α_0, which minimizes the total error between the measured and modeled N_i in the vertical layers for all sampled trees, was derived for each species and each site, and was denoted as optimum α_0. By using the optimum α_0 in Eq. (12.11), it was shown that N_i was reproduced with a reasonable accuracy for all sampled trees. This suggests that the site and species differences in the scaling relations in the prop root system can be well represented by the single parameter α_0.

The values of α_0 optimized for each species at each site ranged from −5.38 to −2.23 (Yoshikai et al. 2021), where lower α_0 value stands for higher complexity of the prop roots system (higher *S* values) for trees having the same DBH. It is considered that species differences and environmental conditions create the variations in the optimized α_0 values. Compared to the range of the optimized α_0 for each species at each site, the differences of the optimized α_0 values between the two co-existing species—*R. apiculata* and *R. mucronata*—were small (0.20 ~ 0.29 in sites Bat and Kar), with lower values for *R. apiculata*. Therefore, the effects of environmental conditions on the prop root system are considered to be more significant than the species variation, which may be the result of the morphological plastic response in the prop root system.

12.3.6 Effects of Environmental Conditions on Prop Root System Complexity

Morphological plasticity is a common phenomenon in plants that is considered as plants' adaptive strategy to the environment (Bloom et al. 1985). One of the most typical morphological plastic responses is the different patterns in the biomass allocation between shoots and roots in response to nutrient availability (Bloom et al. 1985; Ågren and Franklin 2003; Mašková and Herben 2018). When nutrient availability decreases, it is likely that plants' growths are limited by nutrient rather than carbon availability through photosynthesis, thus plants allocate relatively more biomass to roots to compensate for the limited resources (Ågren and Franklin 2003). It has been shown that mangroves also respond to environments in a similar way—investing more biomass to below-ground parts under nutrient-poor conditions (Castañeda-Moya et al. 2011, 2013). Sherman et al. (2003) found the increase in the ratio of root to above-ground biomass with increased soil (pore-water) salinity, whose osmotic effects significantly regulate the resource uptake of mangroves in below-ground (Peters et al. 2014); this is consistent with the mangroves' response to nutrient-poor conditions (Castañeda-Moya et al. 2011, 2013). Vovides et al. (2014) showed that mangrove tree stems tend to be less slender with increasing soil salinity, which may be explained by the plant's adaptive strategy to enhance the water uptake ability by increasing the hydraulic conductivity in the stem.

Morphological responses to resource limitation such as those mentioned may be applied to the prop root system. In Yoshikai et al. (2021), it was implied that factors that affect the prop root system complexity are sediment thickness, sediment hardness, and species composition. Sediment thickness correlated well with the optimized α_0—trees on a shallow substrate (around 0.1 m) such as mangroves formed on a reef flat like sites Kar and Mar (Fig. 12.8) tend to have prop root systems with higher complexity than trees on thicker substrate. Although the mangrove below-ground root biomass is concentrated in the shallow part of the sediment (usually until 0.4 m) (Tamooh et al. 2008; Xiong et al. 2017), shallow substrates with 0.1 m thickness may constrain the development of the mangrove's below-ground root system in the vertical direction, thereby constraining the fine root biomass development that has a role in absorbing resources below-ground. This may lead to the limitation of the plant's growth due to the lowered potential for below-ground resource uptake. To compensate for the limited development of the below-ground root system in the vertical direction, plants may have needed to increase the number of prop roots anchoring to the ground to increase the fine root biomass in the limited space below-ground. This may be the mechanism for the formation of the prop root systems with higher complexity on shallow substrates.

Sediment hardness was not directly correlated with the optimized α_0 in Yoshikai et al. (2021), but the multivariate analysis which involved sediment thickness, sediment hardness, soil salinity, and species variation as independent variables, and the optimized α_0 as dependent variable revealed that the hard sediments such as sandy substrates possibly contribute to the higher complexity of the prop root system. Hence, it was hypothesized that the hydrodynamic force by tidal flow or waves that may have created the sandy substrate may also be the factor affecting the prop root system complexity. This should be further investigated in future studies to explain the site variations in the scaling relations of the prop root system.

Soil salinity, on the other hand, had no significant relationship with the optimized α_0 in Yoshikai et al. (2021), which seems contrary to previous studies that showed changes in the above- and below-ground biomass ratios with changes in soil salinity (Sherman et al. 2003; Peters et al. 2014). However, the regulation of the below-ground resources uptake due to high soil salinity may be alleviated by high nutrient concentrations; plants may need a smaller amount of water uptake if the pore-water is rich in nutrients. Similarly, even if the soil salinity is not high, limitation of below-ground resources could be severe if the nutrient concentration is low. The effects of soil salinity on the prop root system morphology should be therefore considered together with the pore-water nutrient concentrations. Thus, in future studies, nutrient concentrations, which were not measured in Yoshikai et al. (2021), should be measured to explain the effects of soil salinity and below-ground resource availability on the complexity of the prop root system.

The results in Yoshikai et al. (2021) implied that species composition may also affect the prop root system complexity. In a *Sonneratia alba*-dominated forest at site Ber, the prop root system of *Rhizophora apiculata* and *Rhizophora mucronata* showed complexity as high as the sites with the shallow substrates, e.g., sites Mar and Kar, even though the sediment thickness is more than 5 m. Here, the ground is intensively covered by pneumatophores of *Sonneratia alba* even inside the area of the prop root system of *Rhizophora* stands; thus, below-ground interspecific competition between *Sonneratia alba* and *Rhizophora* species for resources below-ground such as water and nutrients may be present (Pranchai et al. 2018; Peters et al. 2020). The below-ground interspecific competition may lead to below-ground resource limitation for growth of individual trees; this may have facilitated the production of prop roots anchoring to the ground to increase the fine root biomass and absorb more water and nutrients. The effects of species composition on the prop root morphology should also be investigated in future studies.

12.3.7 Prospective of the Development of the Universal Model for Prop Root System

Yoshikai et al. (2021) applied a multivariate model to predict the species- and site-specific relations from environmental variables and species variation, but the accuracy in predicting N_i was not sufficiently high. Therefore, the model has not yet been generalized for application to various *Rhizophora* stands with different environmental conditions and species variation. Further data collection on the scaling relations of the prop root system with environmental variables from varying environmental conditions and species composition is needed for development of a universal model. For rapid data collection of the scaling relations, it was suggested in Yoshikai et al. (2021) to use the following equation, which is a rearrangement of Eq. (12.7):

$$\log S = \frac{\log HR_{\min} - \log HR_{max}}{t - 1} \tag{12.12}$$

where t is the total number of whole-tree prop roots, and $HR_{\min}$ is a constant with a value of 0.05 m. Thus, only t and $HR_{\max}$ are required, which are relatively easy to measure in the field, to predict S values of individual trees.

For the environmental variables, it is desirable to measure the pore-water nutrient concentrations in addition to the sediment thickness and soil salinity, to properly assess the effects of below-ground resource limitation on the prop root system morphology. It is also advantageous to conduct sediment coring inside or near the area of the prop root system to measure the fine root biomass to confirm the hypothesis that the increased number of prop roots anchoring to the ground increases fine root biomass to compensate for the below-ground resource limitation. Also, to assess the effects of hydrodynamic loads on the prop root system morphology, parameters such as tidal amplitude, ground elevation, flow velocity, and wave height might be necessary. Prop root system morphology with and without interspecific competition in similar environments should also be investigated to quantify the effects of interspecific competition. These kinds of information may help in developing a universal model for the prop root system.

12.3.8 Perspectives of Model Development of the Other Types of the Above-Ground Root System

Due to the limited knowledge and data on the above-ground root system of other mangrove genus such as those with pneumatophores, it is still difficult to develop a predictive model for these types of above-ground root system. Development of methodologies to effectively measure the root system in the field is needed. Some studies have shown an allometric relationship of below-ground biomass with DBH in mangrove species including *Avicennia* and *Sonneratia* species (Comley and McGuinness 2005; Komiyama et al. 2005). Because the below-ground biomass is directly linked to pneumatophores (Gill and Tomlinson 1977; Ezcurra et al. 2016;

Tomlinson 2016), it is expected that the parameters of pneumatophores such as volume and projected area have also allometric relationships with DBH. The scaling relations in the pneumatophore root system were also implied by the significant relationship between the pneumatophore extent and tree size shown by Vovides et al. (2016) and Yando (2018). These results suggest the possibility of the development of the pneumatophore root system using the scaling relations.

On the other hand, Vovides et al. (2016) and Yando (2018) showed different patterns in the relationships between the pneumatophore extent and tree size—the extent of pneumatophore increased with tree size at first, but saturated at around 1.7 m in Vovides et al. (2016) while the saturation trend was not clearly seen in Yando (2018), and extension longer than 4 m was observed for some trees. This difference may be attributed to the different forest conditions—mixed forest of *A. germinans* and other species in Vovides et al. (2016) (where pneumatophore root development may be interfered by the root system of the neighbor trees), and monospecific stands of *A. germinans* at the mangrove-salt marsh boundary in Yando (2018) (where the interference of root system development may not be very severe). Dahdouh-Guebas et al. (2007) compiled literature data on pneumatophore density of *Avicennia marina* from the various mangroves and showed that the density could significantly vary among sites. They also revealed that depressions of the microtopography cause local increase in the pneumatophore density and length, which is probably a mangrove's adaptation to the places with longer inundation periods (Dahdouh-Guebas et al. 2007). These studies suggest that the pneumatophore root system also exhibits morphological plasticity to the environments. Therefore, as with the case of the *Rhizophora* prop root system, it is necessary to collect the data on the morphology, together with environmental variables from various mangroves with different environmental conditions and species composition to develop a universal model. There is a lack of studies quantifying the morphological structures not only for the pneumatophore root system, but also for other types of above-ground root system. We recommend that more research be focused on the mangroves' morphological trait as these attributes are significant in understanding the mangrove drag effects that influence coastal protection, adaptation to sea-level rise, and carbon dynamics of mangrove forests.

Acknowledgments We are grateful to the Japan International Cooperation Agency (JICA) and Japan Science and Technology Agency (JST) through the Science and Technology Research Partnership for Sustainable Development Program (SATREPS) for financially supporting the Project "Comprehensive Assessment and Conservation of Blue Carbon Ecosystems and their Services in the Coral Triangle (BlueCARES)." We also thank Alvin Baloloy and Dr. Ariel C. Blanco for provision of the TLS data, Reginald Argamosa for his help in the field survey, and Dr. Charissa Ferrera for her help in the laboratory analysis.

References

Ågren G, Franklin O (2003) Root : shoot ratios, optimization and nitrogen productivity. Anna Bot 92:795–800.

Amato M, Basso B, Celano G, Bitella G, Morelli G, Rossi R (2008) In situ detection of tree root distribution and biomass by multi-electrode resistivity imaging. Tree Physiol 28:1441–1448.

Bloom AJ, Chapin III FS, Mooney HA (1985) Resource limitation in plants—an economic analogy. Annu Rev Ecol Syst 16:363–392.

Castañeda-Moya E, Twilley RR, Rivera-Monroy VH, Marx BD, Coronado-Molina C, Ewe SM (2011) Patterns of root dynamics in mangrove forests along environmental gradients in the Florida Coastal Everglades, USA. Ecosystems 14:1178–1195.

Castañeda-Moya E, Twilley RR, Rivera-Monroy V (2013) Allocation of biomass and net primary productivity of mangrove forests along environmental gradients in the Florida Coastal Everglades, USA. For Ecol Manag 307:226–241.

Chen Y, Li Y, Cai T, Thompson C, Li Y (2016) A comparison of biohydrodynamic interaction within mangrove and saltmarsh boundaries. Earth Surf Process Landforms 41:1967–1979.

Chen Y, Li Y, Thompson C, Wang X, Cai T, Chang Y (2018) Differential sediment trapping abilities of mangrove and saltmarsh vegetation in a subtropical estuary. Geomorphology 318:270–282.

Comley BWT, McGuinness KA (2005) Above and below ground biomass, and allometry of four common northern Australian mangroves. Aust J Bot 53:431–436.

Dahdouh-Guebas F, Jayatissa LP, Di Nitto D, Bosire JO, Lo Seen D, Koedam N (2005) How effective were mangroves as a defence against the recent tsunami? Curr Biol 15:R443–R447.

Dahdouh-Guebas F, Kairo JG, De Bondt R, Koedam N (2007) Pneumatophore height and density in relation to micro-topography in the grey mangrove *Avicennia marina*. Belg J Bot 40 (2):213–221.

Danielsen F, Sorensen MK, Olwig MF, Selvam V, Parish F, et al (2005) The Asian tsunami: a protective role for coastal vegetation. Science 310:643.

Danjon F, Fourcaud T, Bert D (2005) Root architecture and wind-firmness of mature *Pinus pinaster*. New Phytol 168:387–400.

Das S, Vincent JR (2009) Mangroves protected villages and reduced death toll during Indian super cyclone. Proc Nat Acad Sci 106:7357–7360.

Ezcurra P, Ezcurra E, Garcillán PP, Costa MT, Aburto-Oropeza O (2016) Coastal landforms and accumulation of mangrove peat increase carbon sequestration and storage. Proc Nat Acad Sci 113:4404–4409.

Figueira W, Ferrari R, Weatherby E, Porter A, Hawes S, Byrne M (2015) Accuracy and precision of habitat structural complexity metrics derived from underwater photogrammetry. Remote Sens 7:16883–16900.

Furukawa K, Wolanski E, Mueller H (1997) Currents and sediment transport in mangrove forests. Estuar Coast Shelf Sci 44:301–310.

Gill A, Tomlinson P (1977) Studies on the growth of red mangrove (*Rhizophora mangle* L.) 4. The adult root system. Biotropica 9:145–155.

Greaves H, Vierling L, Eitel J, Boelman NT, Magney T, Prager C, Griffin K (2015) Estimating aboveground biomass and leaf area of low-stature Arctic shrubs with terrestrial LiDAR. Remote Sens Environ 164:26–35.

Hardiman BS, Gough CM, Butnor JR, Bohrer G, Detto M, Curtis PS (2017) Coupling fine-scale root and canopy structure using ground-based remote sensing. Remote Sens 9:182.

Horstman EM, Dohmen-Janssen CM, Narra PMF, Van den Berg NJF, Siemerink M, Hulscher SJMH (2014) Wave attenuation in mangroves: a quantitative approach to field observations. Coast Eng 94:47–62.

Horstman EM, Dohmen-Janssen CM, Bouma TJ, Hulscher SJMH (2015) Tidal-scale flow routing and sedimentation in mangrove forests: combining field data and numerical modelling. Geomorphology 228:244–262.

Hosoi F, Nakai Y, Omasa K (2013) 3-D voxel-based solid modeling of a broad-leaved tree for accurate volume estimation using portable scanning lidar. ISPRS J Photogram Remote Sens 82:41–48.

Kamal S, Lee SY, Warnken J (2014) Investigating three-dimensional mesoscale habitat complexity and its ecological implications using low-cost RGB-D sensor technology. Methods Ecol Evol 5:845–853.

Komiyama A, Poungparn S, Kato S (2005) Common allometric equations for estimating the tree weight of mangroves. J Tropic Ecol 21:471–477.

Krauss KW, McKee KL, Lovelock CE, Cahoon DR, Saintilan N, Reef R, Chen L (2014) How mangrove forests adjust to rising sea level. New Phytol 202:19–34.

Lienard J, Lynn K, Strigul N, Norris BK, Gatziolis D, Mullarney JC, Bryan KR, Henderson SM (2016) Efficient three-dimensional reconstruction of aquatic vegetation geometry: estimating morphological parameters influencing hydrodynamic drag. Estuar Coast Shelf Sci 178:77–85.

Liu J, Skidmore A, Wang T, Zhu X, Premier J, Heurich M, Beudert B, Jones S (2019) Variation of leaf angle distribution quantified by terrestrial LiDAR in natural European beech forest. ISPRS J Photogram Remote Sens 148:208–220.

Lovelock CE, Cahoon DR, Friess DA, Guntenspergen GR, Krauss KW, Reef R, Rogers K, Saunders ML, Sidik F, Swales A, Saintilan N, Thuyen LX, Triet T (2015) The vulnerability of Indo-Pacific mangrove forests to sea-level rise. Nature 526:559–563.

Marsh JA (1970) Primary productivity of reef-building calcareous red algae. Ecology 51 (2):255–263.

Mašková T, Herben T (2018) Root:shoot ratio in developing seedlings: how seedlings change their allocation in response to seed mass and ambient nutrient supply. Ecol Evol 8:7143–7150.

Maza M, Adler K, Ramos D, Garcia A, Nepf H (2017) Velocity and drag evolution from the leading edge of a model mangrove forest. J Geophys Res 122(11):9144–9159.

Maza M, Lara JL, Losada IJ (2019) Experimental analysis of wave attenuation and drag forces in a realistic fringe *Rhizophora* mangrove forest. Adv Water Resour 131:103376.

Mazda Y, Kobashi D, Okada S (2005) Tidal-scale hydrodynamics within mangrove swamps. Wetl Ecol Manag 13:647–655.

McIvor A, Spencer T, Möller I (2012) Storm surge reduction by mangroves. Nat Coast Prot Ser 35.

Mendez-Alonzo R, Moctezuma C, Ordonez VR, Angeles G, Martinez AJ, Lopez-Portillo J (2015) Root biomechanics in *Rhizophora mangle*: anatomy, morphology and ecology of mangrove's flying buttresses. Ann Bot 115:833–840.

Menéndez P, Losada IJ, Beck MW, Torres-Ortega S, Espejo A, Narayan S, Díaz-Simal P, Lange G-M (2018) Valuing the protection services of mangroves at national scale: the T Philippines. Ecosys Serv 34:24–36.

Montgomery JM, Bryan KR, Mullarney JC, Horstman EM (2019) Attenuation of storm surges by coastal mangroves. Geophys Res Lett 46:2680–2689.

Navarro JA, Algeet-Abarquero N, Fernández-Landa A, Esteban J, Rodríguez-Noriega P, Guillén-Climent ML (2019) Integration of UAV, Sentinel-1, and Sentinel-2 Data for mangrove plantation aboveground biomass monitoring in Senegal. Remote Sens 11:77.

Norris B, Mullarney J, Bryan K, Henderson SM (2017) The effect of pneumatophore density on turbulence: a field study in a Sonneratia-dominated mangrove forest, Vietnam. Continent Shelf Res 147:114–127.

Ohira W, Honda K, Nagai M, Ratanasuwan A (2013) Mangrove stilt root morphology modelling for estimating hydraulic drag in tsunami inundation simulation. Trees 27:141–148.

Olagoke A, Proisy C, Féret J, Blanchard E, Fromard F, Mehlig U, Menezes MM, Santos V, Berger U (2016) Extended biomass allometric equations for large mangrove trees from terrestrial LiDAR data. Trees 30:935–947.

Olsoy PJ, Mitchell J, Levia D, Clark P, Glenn N (2016) Estimation of big sagebrush leaf area index with terrestrial laser scanning. Ecol Indicat 61:815–821.

Ong JE, Gong WK, Wong CH (2004) Allometry and partitioning of the mangrove, *Rhizophora apiculata*. For Ecol Manag 188(1–3):395–408.

Paynter I, Saenz E, Genest D, Peri F, Erb A, Li Z, Wiggin K, Muir J, Raumonen P, Schaaf ES, Strahler A, Schaaf C (2016) Observing ecosystems with lightweight, rapid-scanning terrestrial lidar scanners. Remote Sens Ecol Conserv 2:174–189.

Peters R, Vovides AG, Luna S, Grueters U, Berger U (2014) Changes in allometric relations of mangrove trees due to resource availability—a new mechanistic modelling approach. Ecol Modell 283:53–61.

Peters R, Walther M, Lovelock C, Jiang J, Berger U (2020) The interplay between vegetation and water in mangroves: new perspectives for mangrove stand modelling and ecological research. Wet Ecol Manag 28:697–712.

Pranchai A, Jenke M, Vogt J, Grueters U, Yue L, de Menezes MM, Wagner S, Berger U (2018) Density-dependent shift from facilitation to competition in a dwarf *Avicennia germinans* forest. Wetlands Ecol Manage 26:139.

Rossi R, Amato M, Bitella G, Bochicchio R, Gomes J, Lovelli S, Martorella E, Favale P (2011) Electrical resistivity tomography as a non-destructive method for mapping root biomass in an orchard. Eur J Soil Sci 62:206–215.

Saintilan N, Khan NS, Ashe E, Kelleway JJ, Rogers K, Woodroffe CD, Horton BP (2020) Thresholds of mangrove survival under rapid sea level rise. Science 368:1118–1121.

Shan Y, Liu C, Nepf H (2019) Comparison of drag and velocity in model mangrove forests with random and in-line tree distributions. J Hydrol 568:735–746.

Sherman RE, Fahey T, Martinez P (2003) Spatial patterns of biomass and aboveground net primary productivity in a mangrove ecosystem in the Dominican Republic. Ecosystems 6:384–398.

Smith A, Astrup R, Raumonen P, Liski J, Krooks A, Kaasalainen S, Åkerblom M, Kaasalainen M (2014) Tree root system characterization and volume estimation by terrestrial laser scanning and quantitative structure modeling. Forests 5:3274–3294.

Srikanth S, Kaihekulani S, Lum Y, Chen Z (2016) Mangrove root: adaptations and ecological importance. Trees 30:451–465.

Tamooh F, Huxham M, Karachi M, Mencuccini M, Kairo JG, Kirui B (2008) Belowground root yield and distribution in natural and replanted mangrove forests at Gazi bay Kenya. For Ecol Manag 256(6):1290–1297.

Temmerman S, Meire P, Bouma TJ, Herman P, Ysebaert T, De Vriend HJ (2013) Ecosystem-based coastal defence in the face of global change. Nature 504:79.

Tomiczek T, Wargula A, Lomonaco P, Goodwin S, Cox D, Kennedy A, Lynett P (2020) Physical model investigation of mid-scale mangrove effects on flow hydrodynamics and pressures and loads in the built environment. Coast Eng 162:103791.

Tomlinson PB (2016) The botany of mangroves. 2nd ed. Cambridge University Press, Cambridge.

Van Vinh T, Marchand C, Linh TVK, Vinh DD, Allenbach M (2019) Allometric models to estimate above-ground biomass and carbon stocks in *Rhizophora apiculata* tropical managed mangrove forests (Southern Viet Nam). For Ecol Manag 434:131–141.

Vovides AG, Vogt J, Kollert A, Berger U, Grueters U, Peters R, Lara-Domínguez AL, López-Portillo J (2014) Morphological plasticity in mangrove trees: salinity-related changes in the allometry of *Avicennia germinans*. Trees 28:1413–1425.

Vovides AG, Marín-Castro B, Barradas G, Berger U, Lopez-Portillo J (2016) A simple and cost-effective method for cable root detection and extension measurement in estuary wetland forests. Estuar Coast Shelf Sci 183:117–122.

Willemsen PWJM, Horstman EM, Borsje BW, Friess DA, Dohmen-Janssen CM (2016) Sensitivity of the sediment trapping capacity of an estuarine mangrove forest. Geomorphology 273:189–201.

Woodroffe CD, Rogers K, McKee KL, Lovelock CE, Mendelssohn IA, Saintilan N (2016) Mangrove sedimentation and response to relative sea-level rise. Ann Rev Mar Sci 8:243–266.

Woodruff JD, Irish JL, Camargo SJ (2013) Coastal flooding by tropical cyclones and sea-level rise. Nature 504:44.

Xiong Y, Liu X, Guan W, Liao B, Chen Y, Li M, Zhong C (2017) Fine root functional group based estimates of fine root production and turnover rate in natural mangrove forests. Plant Soil 413:83–95.

Xiong YM, Liao BW, Wang FM (2018) Mangrove vegetation enhances soil carbon storage primarily through in situ inputs rather than increasing allochthonous sediments. Mar Pollut Bull 131:378–385.

Yanagisawa H, Miyagi T (2020) A study on the structural characteristics of prop root of mangrove trees using the 3-D laser scanner. Mangrove Sci 11:23–26. (in Japanese)

Yanagisawa H, Koshimura S, Miyagi T, Imamura F (2010) Tsunami damage reduction performance of a mangrove forest in Banda Aceh, Indonesia inferred from field data and a numerical model. J Geophys Res 115:C06032.

Yando ES (2018) Dispersal, establishment, and influence of black mangrove (*Avicennia germinans*) at the salt marsh-mangrove ecotone (Doctoral dissertation, University of Louisiana at Lafayette). Retrieved from ProQuest Dissertations and Theses database. (10814132)

Yaney-Keller A, Tomillo PS, Marshall J, Paladino F (2019) Using Unmanned Aerial Systems (UAS) to assay mangrove estuaries on the Pacific coast of Costa Rica. PLoS ONE 14(6): e0217310.

Yao T, Yang X, Zhao F, Wang Z, Zhang Q, Jupp D, Lovell J, Culvenor D, Newnham G, Ni-Meister W, Schaaf C, Woodcock C, Wang J, Li X, Strahler A (2011) Measuring forest structure and biomass in New England forest stands using Echidna ground-based lidar. Remote Sens Environ 115:2965–2974.

Yoshikai M, Nakamura T, Suwa R, Argamosa R, Okamoto T, Rollon R, Basina R, Primavera-Tirol Y, Blanco AC, Adi NS, Nadaoka K (2021) Scaling relations and substrate conditions controlling the complexity of *Rhizophora* prop root system. Estuar Coast Shelf Sci 248:107014.

Zhang X, Chua VP, Cheong H-F (2015a) Geometrical and material properties of *Sonneratia alba* mangrove roots. Trees 29:285–297.

Zhang X, Chua VP, Cheong H (2015b) Hydrodynamics in mangrove prop roots and their physical properties. J Hydro-environ Res 9:281–294.

Zhang Y, Ding Y, Wang W, Li Y, Wang M (2019) Distribution of fish among *Avicennia* and *Sonneratia* microhabitats in a tropical mangrove ecosystem in South China. Ecosphere 10(6): e02759.

Zhu S, Huang C, Su Y, Sato M (2014) 3D Ground penetrating radar to detect tree roots and estimate root biomass in the field. Remote Sens 6:5754–5773.

Mangrove as a Natural Barrier to Environmental Risks and Coastal Protection

13

Nazlin Asari, Mohd Nazip Suratman, Nurul Atiqah Mohd Ayob, and Nur Hasmiza Abdul Hamid

Abstract

Mangroves has been widely acknowledged for its role as a natural barrier to various environmental risks such as storms, tsunamis, waves, and coastal erosions by becoming the first defense in protecting the coastlines. In addition, this natural barrier also provides a wide range of ecosystem services including fisheries, timber productions, provision of foods, tourism, climate regulation, carbon storage that reduce the coastal communities' vulnerability to hazards. Unfortunately, in many parts of the world, mangroves have been lost due to the pressure of urban developments, rapid expansion of aquaculture and agriculture, mining, overexploitation of timbers, and an increasing environmental risk. Given the extensive damages caused by the impact of environmental risks, for instance, the recent tsunami occurred in 2018 at Sulawesi, mangroves have never been more important to be conserved and restored due their crucial roles in providing natural protection to the ecosystems. Therefore, mangroves need to be seen as valuable resources to be managed and sustainably. This chapter aimed at reviewing the roles of mangrove as coastal protection as well as a natural barrier to storm surges, tsunami, wind, waves, and erosion, and the benefits of protecting mangroves.

N. Asari · N. H. Abdul Hamid
Faculty of Applied Sciences, Universiti Teknologi MARA (UiTM), Shah Alam, Malaysia

Institute for Biodiversity and Sustainable Development, Universiti Teknologi MARA (UiTM), Shah Alam, Malaysia

M. N. Suratman (✉)
Faculty of Applied Sciences and Institute for Biodiversity and Sustainable Development, UiTM, Shah Alam, Malaysia
e-mail: nazip@uitm.edu.my

N. A. Mohd Ayob
Faculty of Applied Sciences, Universiti Teknologi MARA (UiTM), Shah Alam, Malaysia

R. P. Rastogi et al. (eds.), *Mangroves: Ecology, Biodiversity and Management*, https://doi.org/10.1007/978-981-16-2494-0_13

Keywords

Mangrove forest · Coastal protection · Environmental risks · Natural barrier · Mangrove ecosystem

13.1 Introduction

Mangrove forests are a unique ecosystem that exist between the land and the sea at low latitudes in the tropical or subtropical regions of the planet and able to withstand tidal differences. It has been widely known to be the first defenses against the environmental risks such as storms, tsunamis, waves, and coastal erosions that can save lives and property. Despite its coastal protection importance, this ecosystem also provides a variety of services to local community including as sources of income, food, fisheries, home to fish breeding, timber provision, tourism, climate regulation, and carbon storage.

However, it was estimated that around one quarter of the world's mangroves have been lost due to anthropogenic activities, mainly through conversion to aquaculture, agriculture and urban land uses (Barbier and Cox 2003; Duke et al. 2007; Spalding et al. 2010; Friess and Webb 2014; Barbier 2016). Asia has the largest mangrove extent which is about 42% from 1% of all tropical forests in the world (Menéndez et al. 2018); however, these mangrove forests continue to loss at an alarming rate of 2% a year (Spalding et al. 2010).

As the mangrove worldwide continues to disappear or degraded due to environmental risks, human population, and development pressures, it becomes essential for better understanding the roles of mangroves as a natural barrier to environmental risks and to assess the benefit of having mangrove forests as our natural coastal and livelihood protection. These existing protective values offered by mangrove forests will be sacrificed and coastal risk will be increased if these habitats continue to loss or degraded. Therefore, the fact that these mangrove forests are not just providing the protective values to the coastal, but also have economic importance to humans. The consideration for protection, restoration, and conservation of these habitats may be important in future coastal management decision. Mangrove forests need to be seen as a valuable resource to be managed and conserved on a sustainable basis. This chapter aimed at reviewing the roles of mangroves as coastal protection as well as a natural barrier to storm surges, tsunami, wind, waves, and erosion. Apart from that, this paper will also highlight the benefits of protecting mangroves.

13.2 An Overview of Mangrove Forests

Mangrove forests occur along sheltered coastlines in the tropics and sub-tropics. The first analysis of global mangrove area was conducted by the Food and Agriculture Organization (FAO) in 1980 with 14,653,000 ha (FAO 2007). However, the recent analysis performed by Hamilton and Casey (2016) indicated that the total mangrove areas globally are estimated to be 8,349,500 ha. The estimated figure by Hamilton and Casey (2016) was more accurate and reliable as compared to the compilation of statistical reports because they incorporated remotely-sensed technology. Remote sensing techniques are known to be essential for mangroves as it involves an extensive area which is difficult to access and larger topographic maps are not available (Suratman 2014). Various advancements in remote sensing technology for mangrove ecosystems have been discussed by Suratman (2014). The change detection analysis conducted by Hamilton and Casey (2016) has clearly proven that the mangrove area worldwide has significantly reduced about 57% in the last four decades from 1980.

Malaysia, which is one of the countries in South East Asia, has among the largest extents of mangroves (Hamdan 2012) despite consisting of only less than 2% of the total land area in Malaysia. According to a report by Hamdan et al. (2018) on the characterizing and monitoring of mangroves in Malaysia using Landsat, in 2017, there are a total of 627,567 ha of mangrove forest reserve in Malaysia, where mostly found in Sabah (60%), followed by Sarawak (22%) and Peninsular Malaysia (18%). This report showed the changes that occurred in mangrove areas over 27 years of monitoring from 1990 in which the rate of mangroves deforestation was about 0.1% per year between 1990 and 2017. Most of the changes occurred mainly outside the Permanent Forest Reserve and according to the states' structural planning. The major factors that contributed to these changes have been identified as direct conversion to other land uses, predominantly for aquaculture and agriculture, and coastal erosion (Hamdan et al. 2018).

Mangrove forests are able to withstand harsh environment. They have dense and massive root systems in which some of grow above the ground mainly to allow these plants to absorb nutrients and water, to breathe and to take root in the muddy soil that characterizes these regions. The roots and some parts of trunks of mangrove trees are covered by water during high tide and will rise above the water level during low tide. Some species of mangrove trees are resistant to saltwater or brackish water on tidal flats. The specific characteristics of the mangrove ecosystem make it a good habitat for a variety of terrestrial, aquatic, and amphibian animals (Rasmeemasmuang and Sasaki 2015). According to Spalding et al. (2010), such forests are fertile and create a balanced ecosystem. There are a total of 67 species of mangrove out of 27 genera from 16 families (Field 1995). Among the 67 species, 29 are from the families of Acanthaceae and Rhizophoraceae.

13.3 Mangroves as a Natural Barrier and Coastal Protection

In the past decades, various events occurred on the coastline that caused a huge impact to the coastal communities, properties, and lives, such as the Indian Ocean Tsunami in 2004 with a magnitude of 9.0 M_W, which caused severe damages to Sumatra, Nicobar and Andaman Islands, Bangladesh, Myanmar, Thailand, Malaysia, and Singapore. Hurricane Katrina occurred in the United States in 2005, Typhoon Haiyan had caused significant destruction in the Philippines and Southeast

Asia in 2013, and the recent tsunami that occurred at Sulawesi Indonesia in 2018. While there are growing concern on the possibility of more frequent and powerful tropical cyclones due to climate change (Webster 2005; Knutson et al. 2010; Marois and Mitsch 2015), the restoration, conservation, and creation of mangrove forests can provide the solution in reducing these environmental risks damages to the areas that are vulnerable to this natural disaster. Mangrove forests are highly tolerant in harsh environment, being daily subject to salt exposure, temperature, water, and varying degree of anoxia. Therefore, these forests exhibit a high degree of ecological stability as compared to other forests (Alongi 2008). There is growing evidence that mangrove served as a natural barrier to tsunami, wind, wave, storm, and coastal erosion, which will be discussed further in the next sections.

13.3.1 Mangroves as a Tsunami Barrier

According to Ahmadun et al. (2020), tsunami refers to as the series of waves with an extremely long wavelength that grow in height and velocity as they enter the port or near coastal area where the depth is relatively shallow. It usually occurs due to earthquakes, asteroid impact, or volcanic activities in the deep sea, which caused the sudden movements of the seabed (Egorov 2007; Ahmadun et al. 2020). The evidence from the Indian Ocean Tsunami 2004 indicated that mangrove can help in reducing tsunami damage, but it is depending on the critical factors such as wave depth (Cunningham 2019), width of forest, proportion of aboveground biomass vested to roots, tree height, soil texture, size and speed of the tsunami, distance from tectonic event, and angle of the tsunami incursion relative to the coastline (Alongi 2008). According to Cunningham's findings from Wood Environment & Infrastructure Solutions UK Limited, it is estimated that a 500 meter wide mangrove forest can significantly dissipate the force of tsunamis under 3 meters, but beyond this the mangroves are likely to become damaged. This was supported by the finding from modeling results of Yanagisawa et al. (2010) indicated that mangrove forests could potentially reduce the hydrodynamic force of tsunami by 70% when incoming waves remained under 3 meters.

Based on few initial post-impact studies in Southeastern India, Andaman Islands, Sri Lanka, and Malaysia indicated that mangroves offered a positive and significant defense against the full impact of the tsunami (Kathiresan and Rajendran 2005; Chang et al. 2006; Alongi 2008; Ahmadun et al. 2020). Cunningham (2019) stated that number of studies using hydraulic models and wave experiments that have been validated using the data collected in the aftermath of the Indian Ocean Tsunami 2004 and it has been widely reported (FAO n.d.) that the extensiveness of mangrove forests can help in reducing loss of life and property damage from tsunami by absorbing the first brunt of impact and dissipating wave energy as it approaches coastline. For instance, in the Penang State of Malaysia, large inland territories where the mangrove forests were intact, only slightly damaged as reported by Penang Inshore Fisherman Welfare Association. Similar examples for the positive role of mangrove to help in reducing the damages caused by tsunami and act as buffer zone during past natural hazards such as in Tamil Nadu, India, the well-established mangrove forest of Pichavaram acted as a protective belt slowing down the waves and protected around 1700 people living in hamlets build inland between 100 to 1000 m from the mangroves (FAO n.d.).

It is clear that the effectiveness of mangrove forests against tsunami depends on a few critical factors such as depth of forest (100–500 m), wave depth, and speed. However, the presence of mangrove forests could still help in protecting the properties and people even though it does not offer full protection against tsunami over certain wave depth.

13.3.2 Reduction of Wind and Swell Wave Damage

Wind and swell wave are formed by the action of the wind on the water surface in areas of open water. The wind waves are generated near the coast, while swell waves are generated away from the coast that have been agitated by the effects of wind (Pugh 1987; Woodroffe 2002; McIvor et al. 2012a). Wind and swell waves are commonly seen at the seashore, as they break on the beach or smash against rocks (Spalding et al. 2014). Number of reports indicated that the mangrove forests have the ability to reduce wave height and buffer wind speed (Mazda et al. 2006; Quartel et al. 2007; Bao 2011). While mangrove forests are receiving little incoming waves every day, it may receive larger waves during storms and periods of high winds which can cause flooding and damage to coastal infrastructure. The value of these mangrove forests in providing such protection for high speed and damaging winds and wave are often overlooked (Barbier 2016).

McIvor et al. (2012b) reviewed available information about the capacity of mangroves to reduce wind and swell waves, in order to inform decision makers, planners, and coastal engineers about the role mangroves can play in coastal defense against these hazards. According to their study, by reducing wave energy and height, mangroves can potentially reduce the height of wind and swell waves over relatively short distances in which wave height can be reduced by between 13 and 66% over 100 m of mangroves. The highest rate of wave height reduction per unit distance occurs near the mangrove edge, as the waves begin to pass through the mangroves.

There are a few critical factors that are affecting the wave attenuation in mangroves such as the depth of water in the mangrove forest, slope, topography, the size and age of trees, the height of trees, the types of mangrove in the area, density of mangrove trees as shown in Fig. 13.1 (McIvor et al. 2012b). Hence, the amount of protection that is offered by the mangrove forest depends greatly on the forest quality.

13.3.3 Protection against Storm Damage

Storm surges refer to as the abnormal rise of sea water level in coastal areas during a short-lived atmospheric disturbance such as storm or hurricane, measured as the

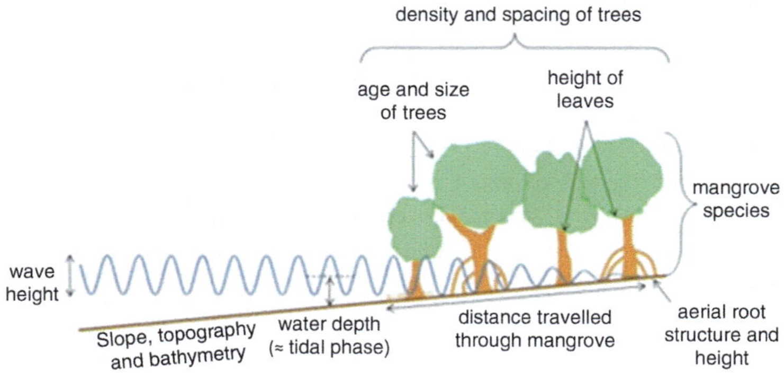

Fig. 13.1 Factors affecting the wave attenuation in mangroves. (Adapted from McIavor et al. 2012b)

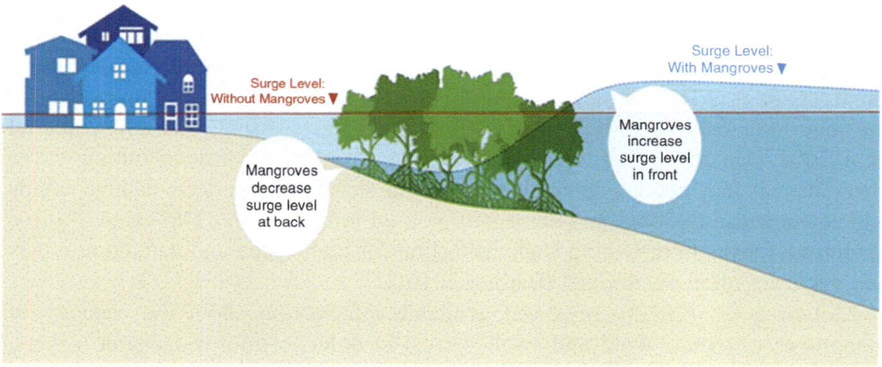

Fig. 13.2 General schematic of mangrove effects on surge level. (Source: Narayan et al. 2019)

height of the water over and above the normal predicted astronomical tide (NOAA 2020). It is a phenomenon mainly linked to wind, but also depending on other elements such as atmospheric pressure, swells, and tides as well as bathymetry and topography (Bertin 2016). Storm surge reduction rate can be measured as a certain number of centimeters of water level reduction per meter of inland distance (McIvor et al. 2012a).

Mangrove forests have been shown as particularly important ecosystem-based protection (first line defend) against storm surge (Krauss et al. 2009; Zhang et al. 2012; McIvor et al. 2015; Narayan et al. 2019). By acting as obstacles to storm waves and surges, they reduce flood extents and therefore protecting people and reduce property damages on coastlines (WAVES 2017; Menéndez et al. 2018; Dasgupta et al. 2019). Figure 13.2 shows the general schematic of mangrove effects on surge level. The reduction level of storm surge on the adjacent coastal to mangrove resulting from the mangrove "blocking" (increasing surge level in seaward and decreasing level on the back of mangrove) (Narayan et al. 2019).

The complex network of tree roots, trunks, and branches of the mangroves forest obstructs the flow of water and can serve to trap debris movement, even large moving objects which can reduce the damage to the adjacent coastal lands. Where mangroves are extensive, they are able to reduce storm surge water depths as the surge flows inland. Although the measured rates of storm surge reduction through mangrove range from 5–50 cm km^{-1} width of mangroves, a small reduction in water level can already greatly reduce the extent of flooding in adjacent coastal land (Spalding et al. 2014). In addition, surface wind waves are expected to be reduced by more than 75% over one kilometer of mangroves (McIvor et al. 2012b). The extent of mangrove protection depends on a number of factors, including mangrove width, density of mangrove forest, topography, storm characteristic such as the size and forward speed of the storm and spectral features of waves and tidal stage at which waves enter the forest (McIvor et al. 2012b; Dasgupta et al. 2019).

A study in Florida found that the surge height decreases at the rate of 23 cm km^{-1} through an area with a mixture of mangrove and open water, while in areas with less open water recorded surge height reduction rates ranged from 40 to 48 km. In contrast, the storm surges height at the front of the mangrove zone increases by about 10% to 30% because of the "blockage" of mangroves to surge water (Zhang et al. 2012). Studies have shown that Hurricane Wilma that hit South Western Florida in 2005 had flooded the area and extended to 70% further inland without the protection of the mangroves (Zhang et al. 2012; Liu et al. 2013).

Storm surge and associated flooding often lead to loss of human life, and destruction of property and infrastructure in populated, low coastal areas. Between 2005 and 2015, WAVES (2017) recorded 56% of property damage from natural hazards in the Philippines was due to typhoons and storms, and another 29% due to floods. Mangroves not only provide the most protection for frequent lower intensity storms, but for more catastrophic events, such as the 1-in-25-year storm, they provide more than US $1.6 billion in averting damages throughout the Philippines (WAVES 2017). The flood impact on Tanya Delta, Kenya experienced by villages unprotected by mangroves was 2.4 times greater than that of villages partially protected by mangroves and 14.7 times greater than that of villages completely protected by mangroves (Karanja and Saito 2018). Das and Vincent (2009) conducted a study following the cyclone with 9 m storm surge in 1999 in Orissa India. They found that villages with wider mangroves between them and the coast had significantly fewer deaths than villages with narrow mangroves or no mangroves at all.

13.3.4 Reduction of Coastal Erosion

Erosion refers to as the land ward displacement of the shoreline resulting in the loss of land and retreating shoreline. Erosion may be caused by storms, wave, tides, and winds, or in response to large-scale events such as glaciation or orogenic cycles that may significantly alter sea levels and tectonic activities that cause coastal land subsidence or emergence (Prasteya 2007). Wind, waves, and currents are natural forces that easily move the unconsolidated soils in the coastal area, resulting in rapid changes in the position of the shoreline. When the waves and current tap against the shore, it creates change where it can sometimes bring the sediment to the coast but sometimes causing erosion and the loss of land (Vo Quoc and Kuenzer 2012). Vo Quoc and Kuenzer (2012) study found that there is a clear occurrence of erosion-deposition processes but the erosion process is more dominant. Over the past decades, the erosion is the most important issue in the coastal areas all over the world which has threatened coastal community. Nearly 30% of the Malaysian coastline is undergoing erosion (Othman 1994). In Vietnam, the coastline has been eroded continuously at the rate of approximately 50 m per years since the early twentieth century (Mazda et al. 1997). The erosion can cause saline intrusion, affecting drinking water sources and agricultural production. They also can lead to massive flooding during storm surge, high tides, or periods of excessive rainfall (Winterwerp et al. 2014).

According to Danielsen et al. (2005), the establishment of mangroves serves as an important ecological function in enhancing the sedimentation and mitigating coastal erosion by reducing the energy of waves, tidal currents and storms that would otherwise erode the coastline. A number of factors affect the ability of mangrove forests to reduce erosion, which included forest width, degree of sediment compaction, and tree morphology (height, root structure, ratio of above- to below-ground biomass) (Alongi 2008). As discussed earlier, mangroves able to dissipate wind and ease the wave energy, thus slow the flow of current which reducing the water capacity to dislodge sediment and carry them out of the mangrove area. Slower water flow allowed already suspended sediment to settle out from the water, hence encourage sediment deposition (Spalding et al. 2014). Mangroves can trap more than 80% of incoming sediment brought by a wave due to the calmness of water (Furukawa and Wolanski 1996) and contributed at 1–8 mm sediment per year (Horstman et al. 2014).

Many species of mangrove trees have different types of aerial root such as stilt root, pneumatophores, root knees, and plank roots. The dense network of fine root systems helps consolidate the coastal soil (Mazda et al. 1997), binding the sediments in places (Saenger 2002; Alongi 2008), and preventing the sediment from being washed away by waves. Observation along the west coast of peninsular Malaysia shows mangrove has an important role to play in the sediment deposition and erosion cycle of the muddy coast (Othman 1994). *Rhizophora* and *Bruguiera* species are found to form a denser mat of roots compared to *Avicennia* species, therefore more consolidate soil and deep root system to reduce the erosion (Othman 1994). A similar study by Azlan and Othman (2009) stated species such as *Rhizophora apiculata* (Bakau Minyak), *Rhizophora mucronata* (Bakau Kurap), and *Sonneratia* (Perepat) found along the shoreline are able to serve as a natural buffer against the destruction of wave action and/or tide and *Rhizophora* sp. is one of the best attenuators among other species. Mangrove soils are rich in organic matter produced by the mangroves themselves, including living root, but also dead leaves and woody materials. The organic matter accumulates to form a peat that increase in thickness over time due to mangrove soils are often waterlogged and have very low oxygen content due to anaerobic condition (Spalding et al. 2014).

Thampanya's et al. (2006) study showed mangrove-covered coastlines are said to be less likely to erode or will erode more slowly than non-vegetated coastlines or former loss of mangroves contributed to the dramatically change in the sediment movement pattern where the land begins to erode so that land disappears into the sea. Conversion of mangrove area to agriculture or aquaculture purposes leads to the rapid breakdown of organic matter as oxygen became available in the soil which in turn causes subsidence. The conversion effect has been seen in coastal areas, such as along the coast in Central Java (Winterwerp et al. 2014), Gulf of Thailand, and Guyana where the coastal erosion and accretion occur irregularly along the coast, but an intensification of erosion has been noticed every year (Spalding et al. 2014).

13.4 Importance of Mangrove Forests

13.4.1 Act as Carbon Stores

Mangrove forests are present in 105 countries, including Indonesia, Brazil, Malaysia, and Papua New Guinea. It contains 50% of the global mangrove intensity and grouped as among the most carbon rich ecosystems in the tropics. Mangroves are important tropical carbon sinks, and their role in mitigating climate change is well documented across the globe. Different from tropical forests, mangrove forests are able to store more carbon per area as compared to other terrestrial ecosystem (Shaltout et al. 2020) which makes them an important natural carbon sink. One of the studies on mangrove carbon storage in North Sulawesi, Central Kalimantan, and Central Java, Indonesia conducted by Murdiyarso et al. (2009) indicated that that total carbon storage in mangrove ecosystems is exceptionally high compared with most forest types, with a mean of 968 Mg C ha^{-1} and a range of 863–1073 Mg C ha^{-1}. These carbon stocks were a result from a combination of well-structured forest with tree up to 2 m in diameter and organic rich peat soil to a depth of 5 m or more.

On other studies by Trumper et al. (2009) and Mcleod et al. (2011), it has been demonstrated that these coastal forests can store up to five times the carbon present in tropical forests per area, sequestering over 2 tons of carbon on average per ha/year. Moreover, the quantification of carbon stocks is necessary to evaluate their contribution to climate regulation, as it is one of the most important ecosystem services of mangroves (Palacios Peñaranda et al. 2019).

Several studies have reported that most of the carbon stocks in the mangrove forests are stored in below-ground biomass and sediment carbon reserves of over 80% from the total carbon stocks (Romañach et al. 2018; Palacios Peñaranda et al. 2019; Eid et al. 2020). Among the main sources of organic carbon for soil are litter production and dead wood debris from the plant. The decomposition and decay of organic material caused by bacteria increase the accumulation of organic matter in the soil sediment. The amount of carbon storage in mangrove ecosystems varies geographically, which is determined by the surroundings (Eid et al. 2020) such as structural forest properties (basal area and height), and water quality parameters such as salinity, and dissolved oxygen (Palacios Peñaranda et al. 2019).

It is also important to highlight that mangrove forests can be carbon sinks, but, if the ecosystem is disturbed, they can become a carbon source as well. The clearing of mangrove forests causes the drying up of mangrove sediments, which increase the microbial activity following the loss of anaerobic environment. This in turn causes an oxidation for the soil and leads to the release of stored carbon into the atmosphere which then contributes to the remission of atmospheric greenhouse gas emissions. A study conducted by Shaltout et al. (2020) found that carbon loss from deforestation of mangrove represents 0.6% of its global emission annually. It is also found that mangrove conversion could release to the atmosphere by 84 to 159 million tonnes of CO_2 with the values up to 90% of the carbon stored in the soils are lost after deforestation. For instance, by converting mangroves to pasture would release three times more CO_2 per hectare to the atmosphere than the conversion of Amazon forests. Therefore, higher rates of mangrove loss in recent decades put their carbon stocks at risk and reduce its ability to act as carbon stores. The impacts are obviously on the loss of the carbon sink function as well as the 94% of the potential carbon sequestration by mangrove forest worldwide (Friess 2016). Thus, it is essential to acknowledge the importance of the mangrove forests and to value their conservation. As mangrove forests can store sizeable volume of carbon, they need to be preserved and managed sustainably, to retain along with the increase in carbon storage.

13.4.2 Rural Livelihoods and Mangroves

Ecologically, mangrove ecosystem serves as a defense of ecological balance between life on the land and the sea. They form an important aspect of the livelihoods of coastal communities in developing countries. Mangrove forests have contributed significantly to the socio-economic lives of coastal resident and benefit various human populations from local and regional scales (Orchard et al. 2015; Jakovac et al. 2020). With the existing mangrove forest, its benefit returns to local people itself as one of main sources of income to the communities. It also expected to increase the added value for the people who live around the mangrove area, including fish farmers and fishermen (Saw and Kanzaki 2015). For example, the high price of crab in the world market has helped the locals in gaining extra income as reported by Saw and Kanzaki (2015) in the province of South Sulawesi. With the development of fish pond around the mangrove forest area, it increases the social income and at the same time creates jobs for locals.

Given that the mangrove forests influence marine fish production, this mangrove area produces a wide range of import and export commodities with greater number of fodders such as shrimp, crabs, shells, and fish (Jakovac et al. 2020). For instance, in 2013–2014, there were 3443 thousand tonnes of marine fish production at 36% of total fish production in the India. These marine fish species include cuttlefish, squid, lobster, shrimp, and certain types of finfish. As reported by Jakovac et al. (2020), marine fishes are mangrove-dependent and referring to as crustaceans such as prawns and crabs, mollusks, demersal finfish, snappers, catfishes, pomfrets, and croakers. According to Friess (2016), 1 km^2 increase in mangrove areas leads to a 185.84 tonnes increase in total marine fish production yearly. The annual per hectare contribution of mangroves to total marine fish production is therefore 1.86 tonnes.

Although mangrove forests become an energy source for many species of marine biota, they also act as shelter and habitat for some livelihoods (Umilia and Asbar 2016). An eco-farming aquaculture system in Guangxi, for example, allows the implementation of fishery systems without cutting mangrove trees. The system is based on a network of underground tubes and pipes in between mangrove roots that augments availability of habitat for fishes. The system generated between 27,000 and 45,000 US$ per hectare per year in fish production (Dat and Yoshino 2013). Besides habitat for fishes, mangrove forest also increases the yield of shrimp productivity as reported by Spalding and Parrett (2019). In reality, as in Dat and Yoshino (2013) study, shrimp productivity was great with the existence of mangrove forests as compared to without any mangrove forest (Jakovac et al. 2020). Hence, it is obvious that mangrove forest has provided sources of income and food to local communities as well as home for marine species breeding.

Apart from sources of income and food, mangrove ecosystem also serves as a supplier of products that bring economic benefits to humans. It supplies fuel (charcoal), construction materials, and also medicine (Anneboina and Kumar 2017). In addition to their contribution to human used, mangroves also provide raw materials such as timber and wood (Friess 2016).

13.4.3 Biodiversity of Mangroves

Mangrove forests are typically composed of shrubs and trees and form extensive forested wetlands along both muddy and carbonate coasts in tropical and subtropical climates (Eid et al. 2020). Mangrove has contributed to the present diversity in the mangrove ecosystems as old as 65 million years. According to Duncan et al. (2016), with only 0.12% of the world's total land area and 0.7% of the total global area of tropical forest, mangrove forest gives an effect on the biodiversity. Rich in sediments and nutrients in mangrove forest provides better nursery grounds and breeding sites for birds, reptiles, and mammals in this environment (Duncan et al. 2016). There are over 9000 ha of mangrove forests in Peninsular Malaysia which act as largest estuarine mangrove system. They provide special habitat for animals and also for marine life including seagrasses, algae, fungi, and fishes. The salinity changes with the tides and season influences the distribution of organism in this ecosystem (Pasquaud et al. 2015).

As a coastal intertidal wetland forest, mangroves composed of halophytic tree and shrub species. According to Pasquaud et al. (2015), there are approximately 70 vegetation species in 40 genera. Associated with this vegetation are a plethora of coastal and terrestrial fauna, including fish, crustaceans, snakes, and mammals. These flora and fauna diversity in mangrove forests provide scientific study and also offer tourism opportunities such as in Malaysia, there is a well-known habitat for twinkling fireflies (*Photuris lucicrescens*) that lives on mangrove trees along the river bank located at Kampung Kuantan (Suratman 2008). Table 13.1 shows the list of flora species found in mangrove forests in Malaysia.

At higher latitudes, mangroves intergrade into temperate intertidal habitats such as salt marsh ecology. In this ecology, it is composed of fine silts and clays, mud flats harbor burrowing creatures, including clams, mussels, oysters, fiddler crabs, sand shrimp, and bloodworms (Adi and Sari 2016). As reported by Nusantara et al. (2015) with the presence of mangrove, the shrimp productivity is at greatest point. It tends to improve the production of shrimp in the mangrove forest compared to other area. The increase density of fauna includes tiger prawn (*Penaeus monodon*), sea crab (*Scylla paramamosain*), and brackish fishes including *Latescal carifer* and *Oreochromis niloticus* as well as seagrasses (*Gracilaria* spp.).

Apart from that, mangroves also provide shelter for many types of birds (Romañach et al. 2018). They are acting as critical habitat for native and migratory bird species. It includes a wide variety of waterbirds and terrestrial birds. There are 150–250 bird species in each bio-geographical as reported by Hattam et al. (2021). The most common species includes waders, herons, egrets, storks, and birds or prey such as sea eagles, brahminy kites, ospreys, and fish eagles (Hattam et al. 2021).

13.5 Conclusions

The main objective of this chapter was to review and summarize the role of mangrove in coastal protection as well as a natural barrier to storm surges, tsunami, wind, waves, and erosion which continue to be debated globally. Apart from that, this paper also reviews the benefits of protecting mangroves. Evidences from few studies have shown that the mangrove forests provide the solution in reducing the environmental risks of damages to the areas that are vulnerable to this natural disaster. Mangrove forests act as a natural barrier to the tsunamis which help in reducing the damages impact of the tsunami. The presence of mangrove forests could still help in protecting the properties and people even though it does not offer full protection against tsunami over certain wave depth. Mangrove forests can potentially reduce the height of wind and swell waves over relatively short distances; however, the amount of protection depends greatly on the forest structure, depth, and composition. Mangrove forests not only provide protection from frequent lower intensity storms, but from more catastrophic events, which depends on mangrove width, density of mangrove forest, topography, size, and forward speed of the storm. Apart from that, mangrove forests also offer the protection against coastal erosion by reducing the wave energy and slow the flow of current which reducing the water capacity to dislodge sediment out of mangrove areas. Therefore, the conservation of healthy mangrove forests can be viewed as an adaptive measure that is important in reducing the risk as a form of sustainable coastal protection. It is important to note that forest quality influences the amount of protection that is offered by the mangrove forests. Hence, the mangrove forest with good structure, composition, depth,

Table 13.1 A list of plant species found in the mangrove forests in Malaysia (Saenger et al. 1983; Ashton and Macintosh 2002; MOSTI 2003 as cited by Suratman 2008)

No.	Family	Species	Category[a]	Life-form	Common name (in Malay)
1	Acanthaceae	*Acanthus ilicifolius*	MA	Shrub	Jeruju puteh
2	Arecaceae	*Nypa fruticans*	MA	Palm	Nipah
3	Asteraceae	*Pluchea indica*	MA	Shrub	Beluntas
4	Avicenniaceae	*Avicennia alba*	M	Tree	Api-api puteh
5	Avicenniaceae	*A. lanata*	M	Tree	Api-api bulu
6	Avicenniaceae	*A. marina*	M	Tree	Api-api jambu
7	Avicenniaceae	*A. officinalis*	M	Tree	Api-api ludat
8	Combretaceae	*Lumnitzera littorea*	M	Shrub/tree	Teruntum merah
9	Combretaceae	*L. racemosa*	M	Shrub/tree	Teruntum putih
10	Euphorbiaceae	*Excoecaria agallocha*	M	Tree	Buta-buta
11	Meliaceae	*Xylocarpus granatum*	M	Tree	Nyireh bunga
12	Meliaceae	*X. meluccensis*	M	Tree	Nyireh batu
13	Myrsinaceae	*Aegiceras corniculatum*	M	Shrub	Kachang-kachang
14	Myrsinaceae	*A. floridum*	M	Shrub	Kachang-kachang
15	Pteridaceae	*Acrostichum aureum*	M	Fern	Piai raya
16	Pteridaceae	*A. speciosum*	M	Fern	Piai lasa
17	Rhizophoraceae	*Bruguiera cylindrica*	M	Tree	Berus
18	Rhizophoraceae	*B. gymnorhiza*	M	Tree	Tumu merah
19	Rhizophoraceae	*B. parviflora*	M	Tree	Lenggadai
20	Rhizophoraceae	*B. sexangula*	M	Tree	Tumu putih
21	Rhizophoraceae	*Ceriops decandra*	M	Tree	Tengar
22	Rhizophoraceae	*C. tagal*	M	Tree	Tengar
23	Rhizophoraceae	*Rhizophora apiculata*	M	Tree	Bakau minyak
24	Rhizophoraceae	*R. mucronata*	M	Tree	Bakau kurap
25	Rubiaceae	*Scyphiphora hydrophyllacea*	M	Shrub	Chigam
26	Sapotaceae	*Planchonella obovata*	MA	Tree	Menasi
27	Sonneratiaceae	*Sonneratia alba*	M	Tree	Perepat
28	Sonneratiaceae	*S. caseolaris*	M	Tree	Berembang
29	Sonneratiaceae	*S. ovata*	M	Tree	Gedabu
30	Sterculiaceae	*Heritiera littoralis*	MA	Tree	Dungun
31	Leguminosae	*Caesalpinia crista*	MA	Tree	Unak
32	Leguminosae	*Derris trifoliata*	MA	Tree	Tuba laut
33	Leguminosae	*D. uliginosa*	MA	Tree	Setui
34	Malvaceae	*Thespesia populnea*	MA	Tree	Bebaru

(continued)

Table 13.1 (continued)

No.	Family	Species	Category[a]	Life-form	Common name (in Malay)
35	Pandanaceae	*Pandanus odoratissimus*	MA	Palm	Pandan
36	Tiliaceae	*Brownlowia argentata*	MA	Shrub/tree	Kiei

MA: mangrove associate
[a]M: true mangrove

width, and high density could potentially become the protective layers in any catastrophic events associated with environmental risks damages as compared to small and scattered mangrove forests. In assessing the overall value of these ecosystems, it is important not to focus just on their protective value, but at the exclusion of the wide range of benefits and synergistic relationships that these vital habitats provide which supports thousands of marine life species breeding, rich in biodiversity and sources of livelihood for coastal communities. This habitat is also important in carbon storing as compared to the other tropical forests in which contribute to the reduction of CO_2 and their conservation can be look upon as investing in climate mitigation. Improving the value of the protective service of mangroves, as well as the other benefits provided by these critical habitats, may prove important in the future coastal management decisions.

Given the enormous benefits of mangrove forests, proper management and conservation is therefore necessary to ensure the continued existence of mangrove forests. The Conservation of mangroves can be enhanced by few initiatives such as gazetting all remaining mangrove forests within forest reserves or protected areas. Some mangrove forests are already gazetted but many other mangroves have yet to be protected. Other measures are by retaining protective mangrove buffers along coastlines and rivers to prevent erosion. Managing mangrove forests as fishery reserves to encourage the environmentally sensitive commercial aquaculture activities could help in conservation of mangrove. Public awareness and educating the community by enlightenment campaigns should be conducted to sensitize the local population on the effect of unsustainable removal of the mangrove vegetation as well as on the sustainable management and utilization of the forest and forest resources. The loss of mangrove forest will not only contribute to the rapid loss of biodiversity and shoreline, but also negatively impact human livelihoods and ecosystem function. Mangroves conservation should not be overlooked, especially as they are important for speciation and can be significant drivers of diversification over time.

References

Adi W, Sari SP (2016). Detection of mangrove distribution in Pongok Island. Procedia Environ Sci 33: 253–257.

Ahmadun F, Wong RMM, Said MA (2020) Consequences of the 2004 Indian Ocean Tsunami in Malaysia. Saf Sci 121: 619–631.

Alongi DM (2008) Mangrove forests: resilience, protection from tsunamis, and responses to global climate change. Estuar Coast Shelf Sci 76: 1–13.

Anneboina LR, Kumar KV (2017) Economic analysis of mangrove and marine fishery linkages in India. Ecosyst Serv 24: 114–123.

Azlan NI, Othman R (2009) Monitoring of mangroves area using remote sensing towards shoreline protection. Proceedings of 16th International Symposium GIS, Ostrava, 85–95.

Bao TQ (2011) Effect of mangrove forest structures on wave attenuation in coastal Vietnam. Oceanologia 53: 807–818.

Barbier EB (2016) The protective service of mangrove ecosystems: a review of valuation methods analysis. Mar Pollut Bullet 109 (2): 676–681.

Barbier EB, Cox M (2003) Does economic development lead to mangrove loss: a cross country. Contemp Econ Policy 21: 418–432.

Bertin X (2016) Storm surges and coastal flooding: status and challenges. La Houille Blanche 2: 64–70.

Chang SE, Adams BJ, Alder J, Berke PR, Chuenpagdee R, Ghosh S, Wabnitz C (2006) Coastal ecosystems and tsunami protection after the December 2004 Indian Ocean tsunami. Earthquake Spectra 22: 863–887.

Cunningham E (2019) The role of mangrove in coastal protection. Published by Wood Environment & Infrastructure Solutions UK Limited.

Danielsen F, Sorensen MK, Olwig MF, Selvam V, Parish F, Burgess ND, Hiralshi T, Karunagaran VM, Rasmussen MS, Hansen LB, Quarto A, Suryadiputra N (2005) The Asian tsunami: a protective role for coastal vegetation. Science 310: 643.

Das S, Vincent JR (2009) Mangroves protected villages and reduced death toll during Indian super cyclone. Proc Natl Acad Sci USA 106: 7357–7360.

Dasgupta S, Islam MS, Huq M, Khan ZH, Hasib MR (2019) Quantifying the protective capacity of mangroves from storm surges in coastal Bangladesh. PLoS One 14: e0214079.

Dat PT, Yoshino K (2013) Comparing mangrove forest management in Hai Phong City, Vietnam towards sustainable aquaculture. Procedia Environ Sci 17: 109–118.

Duke NC, Meynecke JO, Dittmann S, Ellison AM, Anger K, Berger U, Cannicci S, Diele K, Ewel, KC, Field CD, Koedam N, Lee SY, Marchand C, Nordhaus I, Dahdouh-Guebas F (2007) A world without mangroves? Science 317: 41–42.

Duncan C, Jurgenne HP, Nathalie P, Julian RT, Rona J, Loma A, Heather JK (2016) Rehabilitating mangrove ecosystem services: a case study on the relative benefits of abandoned pond reversion from Panay Island, Philippines. Mar Pollut Bullet 109: 772–782.

Egorov Y (2007) Tsunami wave generation by the eruption of underwater volcano. Nat Hazards Earth Syst Sci 7 (1): 65–69.

Eid ME, Khaled MK, Hamdi A, Muhammad A, Adel M, Abir Mc (2020) Evaluation of carbon stock in the sediment of two mangrove species, *Avicennia marina* and *Rhizophora mucronata*, growing in the Farasan Islands, Saudi Arabia. Oceanologia 62: 200–213.

Field C (1995) Journey amongst mangroves. International Society for Mangrove Ecosystem, Okinawa, Japan.

Food and Agriculture Organization (FAO) (2007) The world's mangroves 1980–2005, FAO Forestry Paper 153: 77.

Food and Agriculture Organization (FAO) (n.d.) Tsunami mitigation by mangroves and coastal forests. http://www.fao.org/forestry/tsunami/27285@69434/en/

Friess DA (2016) Mangrove forests. Magazine, National University of Singapore. Curr Biol 26: 748.

Friess DA, Webb EL (2014) Variability in mangrove change estimates and implications dominated by *Sonneratia* sp. Wet Ecol Manag 14 (4): 365–378.

Furukawa K, Wolanski E (1996) Sedimentation in mangrove forest. Mangroves Salt Marshes 1: 3–10.

Hamdan O, Misman MA, Linggok V (2018) Characterizing and monitoring of mangroves in Malaysia using Landsat-based spatial-spectral variability. IOP Conf Ser: Earth Environ Sci 169: 10.
Hamdan O (2012) Status of Mangroves in Peninsular Malaysia. Hamdan, O (Ed.), Forest Research Institute Malaysia, Gemilang Press Sdn Bhd.
Hamilton SE, Casey D (2016) Creation of a high spatio-temporal resolution global database of continuous mangrove forest cover for the 21st century (CGMFC-21). Global Ecology and Biogeography 25(6) 729–738. doi:10.1111/geb.12449.
Hattam C, Goh HC, Then AY, Edwards-Jones A, Ruslan NF, Yap JS, Moh HH (2021) Using nexus thinking to identify opportunities for mangrove management in the Klang Islands, Malaysia. Estuar Coast Shelf Sci 248: 107157.
Horstman EM, Dohmen-Janssen CM, Narra PMF, van den Berg NJF, Siemerink M, Hulscher SJMH (2014) Wave attenuation in mangroves: a quantitative approach to field observations. Coast Eng 94: 47–62.
Jakovac CC, Latawiec AE, Lacerda E, Lucas IL, Korys K , Iribarrem A, Malaguti GA, Turner RK, Luisetti T, Strassburg BBN (2020) Costs and carbon benefits of mangrove conservation and restoration: A global analysis. Ecol Econ 176 (2020) p 106758 10.1016/j.ecolecon.2020.106758
Karanja JM, Saito O (2018) Cost–benefit analysis of mangrove ecosystems in flood risk reduction: a case study of the Tana Delta, Kenya. Sustain Sci 13: 503–516.
Kathiresan K, Rajendran N (2005) Coastal mangrove forests mitigated tsunami. Estuar Coast Shelf Sci 65: 601–606.
Knutson TR, McBride JL, Chan J, Emanuel K, Holland G, Landsea C, Held I, Kossin JP, Srivastava AK, Sugi M (2010) Tropical cyclones and climate change. Nat Geosci 3: 157–163.
Krauss KW, Doyle TW, Doyle TJ, Swarzenski CM, From AS, Day RH, Conner WH (2009) Water level observations in mangrove swamps during two hurricanes in Florida. Wetlands 29 (1): 142–149.
Liu H, Zhang K, Li Y, Xie L (2013) Numerical study of the sensitivity of mangroves in reducing storm surge and flooding to hurricane characteristics in southern Florida. Cont Shelf Res 64: 51–65.
Marois DE, Mitsch W (2015) Coastal Protection from tsunamis and cyclone provided by mangrove wetlands-A review. Int J Biodivers Sci Ecosyst Serv Manag 11 (1): 71–83.
Mazda Y, Magi M, Ikeda Y, Kurokawa T, Asano T (2006) Wave reduction in a mangrove forest for the assessment of ecosystem service provision. Glob Ecol Biogeogr 23: 715–725.
Mazda Y, Magi M, Kogo M, Hong PN (1997) Mangroves as a coastal protection from waves in the Tong King Delta, Vietnam. Mang. Salt Marsh 1(2): 127–135.
McIvor A, Spencer T, Möller I, Spalding M (2012a) Storm surge reduction by mangroves. Natural Coastal Protection Series: Report 2. Cambridge Coastal Research Unit Working Paper 41. The Nature Conservancy and Wetlands International.
McIvor AL, Möller I, Spencer T, Spalding M (2012b) Reduction of wind and swell waves by mangroves. Natural Coastal Protection Series: Report 1. Cambridge Coastal Research Unit Working Paper 40. Published by The Nature Conservancy and Wetlands International. p 27.
McIvor A, Spencer T, Spalding M, Lacambra C, Moller I (2015) Mangroves, tropical cyclones, and coastal hazard risk reduction. In: Coastal and Marine Hazards, Risks, and Disasters, J.F. Shroder, J.T. Ellis, and D.J. Sherman, eds. Elsevier. pp 403–429.
Mcleod E, Chmura GL, Bouillon S, Salm R, Björk M, Duarte CM, Lovelock CE, Schlesinger WH, Silliman BR (2011) A blueprint for blue carbon: toward an improved understanding of the role of vegetated coastal habitats in sequestering CO_2. Front Ecol Environ 9: 552–560.
Menéndez P, Losada IJ, Beckb MW, Torres-Ortegaa S, Espejoa A, Narayanc S, Díaz-Simala P, Langed G (2018) Valuing the protection services of mangroves at national scale: The Philippines. Ecosys Serv 34: 24–36.
Murdiyarso D, Donato D, Kauffman JB, Kurnianto S, Stidham M, Kanninen M (2009) Carbon storage in mangrove and peatland ecosystems. A preliminary account from plots in Indonesia. In working paper 48 Center for International Forest Research (CIFOR): p. 39.

Narayan S, Thomas C, Matthewman J, Shepard CC, Geselbracht L, Nzerem K, Beck MW (2019) Valuing the Flood Risk Reduction Benefits of Florida's Mangroves (The Nature Conservancy).
National Oceanic and Atmospheric Administration, NOAA (2020) From https://oceanservice.noaa.gov/facts/stormsurge-stormtide.html
Nusantara MA, Malikusworo H, Helmi P (2015) Evaluation and planning of mangrove restoration programs in Sedari Village of Kerawang District, West Java: Coastal Development Programs. Procedia Environ Sci 23: 207–214.
Orchard SE, Stringer CL, Quinn CH (2015) Mangrove system dynamics in Southeast Asia: Linking livelihoods and ecosystem services in Vietnam. Reg.Environ. Change Open access (Available online from: http://link.springer.com/article/10.1007%2Fs10113-015-0802-5 http://dx.doi.org/10.1007/s10113-015-0802-5
Othman MA (1994) Value of mangroves in coastal protection. In: Sasekumar A., Marshall N., Macintosh D.J. (eds) Ecology and conservation of Southeast Asian Marine and freshwater environments including Wetlands. Developments in Hydrobiology, vol 98. Springer, Dordrecht.
Palacios Peñaranda, ML, Kintzb JRC, Salamancac NJP (2019) Carbon stocks in mangrove forests of the Colombian Pacific. Estuar Coast Shelf Sci 227: 106299.
Pasquaud S, Vasconcelos RP, França S, Henriques S, Costa MJ, Cabral H 2015 Worldwide patterns of fish biodiversity in estuaries: Effect of global vs. local factors Estuarine. Coastal and Shelf Science 154: 122–128.
Prasteya G (2007) Protection from coastal erosion. In: Braatz, S., Fortuna, S., Broadhead, J., & Leslie, R. (Eds.), Coastal protection in the aftermath of the Indian Ocean tsunami: What role for forests and trees? Bangkok: Food and Agriculture Organization.
Pugh D (1987) Tides, surges and mean sea level. Published by John Wiley and Sons Ltd.
Quartel S, Kroon A, Augustinus P, Van Santen P, Tri NH (2007) Wave attenuation in coastal mangroves in the Red River Delta, Vietnam. J Asian Earth Sci 29 (4): 576–584.
Rasmeemasmuang T, Sasaki J (2015) Wave reduction in mangrove forests: general information and case study in Thailand. In: Chapter 24, Handbook of Coastal Disaster Mitigation for Engineers and Planners. Elsevier: p 511–535.
Romañach SS, Donald L De A, Hock LH, Yuhong L, Su YT, Raja SRB, Lu Z (2018) Conservation and restoration of mangroves: Global status, perspectives, and prognosis. Ocean Coast Manag 154: 72–82.
Saenger P (2002) Mangrove ecology, silviculture and conservation. Kluwer Academic Publishers, Netherlands: p 360.
Saw AA, Kanzakia M (2015) Local livelihoods and encroachment into a mangrove forest reserve: a case study of the Wunbaik reserved mangrove forest, Myanmar. Procedia Environ Sci 28: 483–492.
Shaltout KH, Ahmed MY, Alrumman Sulaiman A, Ahmed Dalia A, Eid Ebrahem M (2020) Evaluation of the carbon sequestration capacity of arid mangroves along nutrient availability and salinity gradients along the Red Sea coastline of Saudi Arabia. Oceanologia 62 (1): 56–69.
Spalding M, McIvor A, Tonneijck FH, Tol S, van Eijk P (2014) Mangroves for coastal defence. Guidelines for coastal managers & policy makers. Published by Wetlands International and The Nature Conservancy: p 42.
Spalding M, Kainuma M, Collins L (2010) World Atlas of mangroves. A collaborative project of ITTO, ISME, FAO, UNEP-WCMC, UNESCO-MAB, UNU-INWEH and TNC. Earthscan, London, UK: p 319.
Spalding M, Parrett CL (2019) Global patterns in mangrove recreation and tourism. Mar Pol 110: 103540.
Suratman MN (2008) Carbon sequestration potential of mangroves in southeast Asia. In: Bravo F, LeMay V, Jandl R, von Gadow K (eds) Managing forest ecosystems. The challenge of climate change. Springer Science, The Netherlands: p 297–315.
Suratman MN (2014) Remote sensing technology: recent advancements for mangrove ecosystem. In Faridah Hanum et al. (Eds.), Mangrove ecosystems of Asia, Springer Science and Business: p 295–317.

Thampanya U, Vermaat JE, Sinsakul S, Panapitukkul N (2006) Coastal erosion and mangrove progradation of Southern Thailand. Estuar Coast Shelf Sci 68: 75–85.
Trumper K, Bertzky M, Dickson B, van der Heijden G, Jenkins M, Manning P (2009) The natural fix? The role of ecosystems in climate mitigation. A UNEP Rapid Response Assessment. United Nations Environment Programme, UNEPWCMC, Cambridge, UK: p 65.
Umilia E, Asbar (2016) Formulation of mangrove ecosystem management model based on eco-minawisata in the Coastal Sinjai, South Sulawesi. Social Behav Sci 227: 704–711.
Vo Quoc T, Kuenzer C (2012) Can Gio mangrove biosphere reserve evaluation 2012: current status, dynamics and ecosystem services. IUCN, Ha Noi, Vietnam: p 100.
Wealth Accounting and the Valuation of Ecosystem Services, WAVES (2017) Valuing the protection services of mangroves in the Philippines. WAVES Technical Report 2017.
Webster PJ (2005) Changes in tropical cyclone number, duration, and intensity in a warming environment. Science 309: 1844–1846.
Winterwerp H, Wesenbeeck BV, Dalfsen JV, Tonneijck F, Astra A, Verschure S, Eijk PV (2014) A sustainable solution for massive coastal erosion in Central Java. Towards Regional Scale Application of Hybrid Engineering ~Discussion paper.
Woodroffe CD (2002) Coasts: form, process and evolution. Cambridge University Press, Cambridge: p 623.
Yanagisawa H, Koshimura S, Miyag T, Imamura F (2010) Tsunami damage reduction performance of a mangrove forest in Banda Aceh, Indonesia inferred from field data and a numerical model. J Geophys Res-Oceans 115.
Zhang KQ, Liu H, Li Y, Hongzhou X, Jian S, Rhome J, Smith III TJ (2012) The role of mangroves in attenuating storm surges. Estuar Coast Shelf Sci 102: 11–23.

Diversity and Community Structure of Polychaetes in Mangroves of Indian Coast

14

P. Murugesan and T. Balasubramanian

Abstract

This review article provides a comprehensive account of the diversity of polychaetes associated to Indian mangrove ecosystems. Polychaetes constitute an important component in marine benthic communities and play a major ecological role in mangrove ecosystem. Information on polychaete diversity is available only from 12 mangroves out of 19 on the east coast and only from 11 out of 21 those of west coast of India. Altogether 385 species of polychaetes belonging to 176 genera under 49 families, representing 35.22% of the total polychaetes in India, have been listed from mangroves skirting east and west coasts of India. Of this total, as many as 231 species of polychaetes in east coast and 230 species in mangroves of west coast of India were reported to occur. That way, species of the following families such as Nereididae, Spionidae, Terebellidae Sabellidae, Phyllodocidae, Eunicidae, Serpulidae, Lumbrinereidae, Onuphidae, Syllidae, Maldanidae, Capitellidae, and Glyceridae were found to be dominant in the mangrove environment. Among these, Genera *Nereis*, *Glycera*, *Lumbriconereis*, *Prionospio* were found to be dominant followed by *Phyllodoce*, *Nephtys*, *Onuphis*, *Polydora*, and *Syllis* in the next level of abundance. Regrettably, despite its immense role, studies related to polychaete diversity have been made only in 23 mangroves out of 40 mangroves in India. Compared to the studies carried out in other parts of the world, especially Western hemisphere, continuous survey is to be made along the mangroves of both the coasts of India focusing polychaetes, which attach quite a lot of ecological significance and economic importance.

P. Murugesan (✉) · T. Balasubramanian
Faculty of Marine Sciences, Centre of Advanced Study in Marine Biology, Annamalai University, Parangipettai, India

R. P. Rastogi et al. (eds.), *Mangroves: Ecology, Biodiversity and Management*,
https://doi.org/10.1007/978-981-16-2494-0_14

Keywords

Polychaetes · Mangroves · West coast · East coast · Diversity

14.1 Introduction

Mangroves form a habitat for a wide variety of species, some occurring in high densities and form food and shelter for a plethora of commercially important fin and shell fishes. According to Forest Survey of India, mangroves occupy around 4921 km^2 of the Indian coast. Of this total, 71.5% of the mangrove cover is found in east coast and the remaining 28.5% in the west coast (FSI 2017). India has three types of mangrove habitats, namely deltaic, backwater-estuarine, and insular. Comparing both the coasts of India, the east coast is shallow and the shore is quietly shelving with several back water region and thus the mangroves development is quite high in the East coast compared to West coast.

14.1.1 Importance of Mangroves and its Associated Faunal Groups

Mangroves are the distinctive source of ecology and are extremely important coastal resources, which are vital for socio-economic development of the region (Murugesan et al. 2018). They are productive habitats and support coastal fisheries. As a detritus-based ecosystem, leaf litter from the mangroves provides the base for adjacent aquatic and terrestrial food webs. Mangroves afford provisions and shelter for a large number of commercially important fin and shellfishes. True to its sense, the mangrove ecosystem serves as breeding, feeding, and nursery grounds for most of the commercially important fin and shellfishes, on which thousands of coastal folk depend for their livelihood (Manson et al. 2005). It is considered to have physical, chemical, and biological processes which promote the adaptation of inhabiting organisms to tolerate both the extremes of environmental variables both morphologically and physiologically.

Krom and Berner (1980) have reported that the decomposition of organic matter consists of nutrients such as nitrogen and phosphorus, which play a vital role in the establishment of healthy mangroves. However, sediment where the animals inhabit often acts as buffer either as a source or sink of nutrients especially phosphorus by adsorption–desorption reactions. Hence, the sediment plays a crucial role on benthic faunal diversity in the mangrove ecosystem. In recent years, the mangrove ecosystems have been the cynosure of Benthic Ecologists world over as many researches is being focused towards carbon sequestration, one of the vogue words of the day. Understandably, this ecosystem is known to sequester around 22.8 million metric tons of carbon annually, covering 0.1% of the earth's forests, which is equal to 11% of terrestrial carbon into the ocean (Jennerjahn and Ittekkot 2002).

Under this circumstance, a detailed and complete knowledge of the bottom fauna is not only important for the determination of productivity (Raveenthiranath Nehru

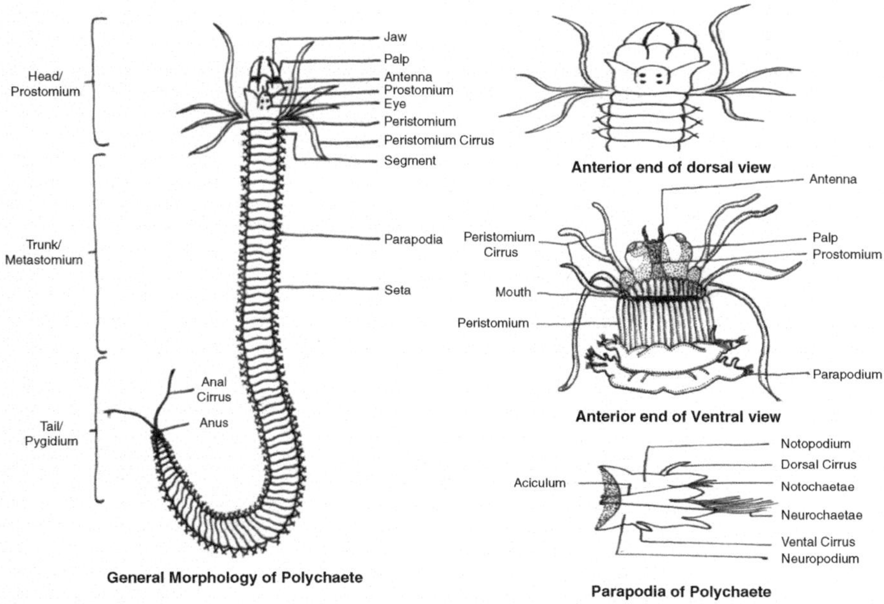

Fig. 14.1 Diagnostic characters of a typical polychaete worm

1990) but also helpful in understanding the diversity of the habitat. That way, the macrofauna are the most widely studied group which is retained on 500 micron sieve. They reside beneath the sediment surface in burrows and tubes. Thus, seemingly, the bottom of the mangrove substratum habitats forms for an array of macro-benthic organisms of various size and taxonomic categories. Unsurprisingly, Indian mangrove ecosystems are known to have a total of 3985 biological species that include 919 (23%) floral species and 3066 (77%) faunal species. Thus, the faunal species component is about three times greater than the floral component of the mangrove ecosystem (Kathiresan 2000; Kathiresan and Quasim 2005). Mangroves are inhabited by a variety of macrobenthic invertebrates, such as polychaetes, brachyuran crabs, gastropods, bivalves, hermit crabs, barnacles, sponges, tunicates, and sipunculids. Mangrove invertebrates often exhibit marked zonation patterns and colonize a variety of specific micro-environments. While some species dwell on the sediment surface or reside in burrows, others live on pneumatophores and lower tree trunks or prop-roots, burrow in decaying wood, or can even be found in the tree canopies (Sasekumar 1974; Smith et al. 1991; Ashton 1999). Of these, polychaetes are found to be the major macro-benthic organisms in mangrove environment. They are segmented worms belonging to the phylum Annelida and Class Polychaeta. The characteristic features of typical polychaete worm are shown in Fig. 14.1. They are habitually found to be the predominant taxa of macro-benthic communities, which they typically constitute a major proportion of the total macrofaunal diversity owing to their high adaptability (Fauchald 1977;

Ward and Hutchings 1996; Brunel 2005; Prabakaran et al. 2019) and their diet consist of microbial, meiobial, and organic substances, in mangrove soil in which they have certain special adaptations for survival, such as mucus secreting devices which is used to protect themselves in unfavorable conditions in the estuarine environment (Bandekar et al. 2017).

14.1.2 Role of Polychaetes in Mangrove Environment

Polychaetes form an important links between the primary detritus at the base of the food web and consumers of higher trophic levels (Macintosh 1984). In view of their greater abundance and biomass (secondary production), the energy assimilated by the polychaetes play a significant role in nutrient recycling in the mangrove ecosystem. They also serve as the food for a variety of demersal organisms of higher trophic level. Most of the macrobenthos assist in the breakdown of particulate organic material by exposing them to microbes and their waste materials contain rich nutrients forming the food for other consumers. Thus, the macrobenthos in general plays a major ecological role in the mangrove ecosystem (Warren and Underwood 1986). Further, they are also secondary producers of mangroves subsoil habitat production, which is crucial for tracing the biotic stability of the area from the fisheries point of view as has been described by Murugesan et al. (2018). Justifiably, they play a fundamental role in recycling of organic materials formed during the decomposition of mangrove litter and subsequent detritus formation and nutrient release, thereby improving the soil structure and its productivity through nutrient cycling, which eventually exposed to microbes and their wastes forming food for other consumers (Kumar 2003). Added to this, polychaetes are also used as veritable pollution indicator groups for the detection of disturbance in an environment. Based on their adaptability, certain groups of polychaetes, namely Spionids, Capitellids, Maldanids are reported to occur exceedingly large in number in polluted environments (Pearson and Rosenberg 1978; Prabakaran et al. 2019) and those of Hesionids, Terebellids, Syllids reported to occur more in relatively pristine environments (Khan and Murugesan 2005). Rightly, Samidurai et al. (2012) reported that progression in macro-benthic communities is highly related to the organic fortification and pollution in any given ecosystem. In addition to ecological role, commercial prospects also exist for wealth creation as these worms are used widely as feed for fin and shell fish brooders (Wang et al. 2018). True to this, their lipid content may provide a source of essential PUFA (poly unsaturated fatty acids), which are fundamental for the production of high quality seedlings of shell fishes. On the whole, these worms have a pivotal role in the mangrove environments (Muniasamy et al. 2013; Parvez et al. 2018).

Taking into account the facts stated supra, a ken on benthic faunal diversity in general and polychaete diversity, in particular occurring in Indian mangroves will be an immediate concern to understand the status of mangrove ecosystems. Accordingly, an extensive literature survey on polychaete diversity in Indian mangroves has been made and based on the available information and intensive Google search, this review account pertinent to polychaetes diversity along the Indian mangroves has been given.

14.2 Method of Collection of Polychaetes

There have been various kinds of broad relative studies regarding sampling strategy and their capabilities (Ankar 1977; Eleftheriou and Holme 1984; Gage 1975; Rosenberg 1978; Rumohr 1990). For professionally collecting the marine benthic macro-faunal samples, the sampling instrument must be capable of penetrating to an enough depth to collect the organisms present in the desired area of sampling. Most of the earlier reports state that the greater part of species and individuals are to occur in the upper 5–10 cm, even though large burrowing molluscs and crustaceans may be found deeper and the pelagic polychaetes even collected from the plankton samples. Accordingly the protocol is formulated and samples are collected in intertidal, sub-tidal and shelf and slope regions. In the intertidal region, 625 cm^2 wooden quadrate is kept and the surface is dug out for 10 cm. Sub-tidal polychaetes can be quantitatively collected using Peterson grab covering an area of about 1 m^2. The Peterson grab is found to be an efficient gear for sampling in shallow water environments, both the mangrove and estuarine environments. In addition to sediment Grabs, Dredges, trawls and traps are also used for collecting benthos for qualitative purpose. Pelagic forms will be collected by filtering around 200 l of surface water and the sample will be calculated under a microscope using Sedgwick rafter and the number of animals per cubic meter of water will be calculated. In addition to these, a few groups of polychaetes can also be collected by hand-picking method from intertidal region, floating wood, bottom mud, gastropod and bivalve shells and various forms of hard substrates such as stones, cement boulders, wooden piers, and boat hulls in the coastal zone.

14.2.1 Sieving, Treatment, Sorting, and Identification of Polychaetes

After collection of sediment samples, the faunal components are to be segregated from the sediment using set of sieves. The cost-effective and accurate size of sieve has to be used. Once the mesh size has been set, the sediment samples collected will be sieved through 0.5 mm aperture size mesh and the number of polychaetes obtained in the sieve is counted. If there is any damage in fauna, the head alone will be taken into account. After sieving, the sieve retains are to be collected in a marked container and preserved in 5–7% (neutral) formaldehyde. Prior to extraction of polychaetes, selective staining of the sample is to be done with a few drops of Rose Bengal dye for the improved visibility and recognition of the species at the time of identification (Pfannkuche and Thiel 1988). Although the use of rose Bengal is not recommended by veteran benthic ecologists, it is undoubtedly helpful in sorting and identification of the faunal communities. After adding the stain, the container is to be turned upside down gently for complete mixing. As it is acidic, leaving the fauna in it for too long may decay the faunal samples. To avoid this, the animals are washed carefully with fresh water and then are preserved in 75% alcohol. The concentration of alcohol should be gradually increased, i.e. 30%, 50%, 75%. Magnesium chloride or propylene phenoxetol or menthol is also used as narcotics prior to alcohol so as to prevent the distortion of the body parts.

After a day or two, the preserved faunal samples are smoothly but thoroughly washed in fresh water to remove the formalin and salt concentration, preventing the former from dissolving the shells of weak molluscs. For initial sorting, white background trays with bench lights, needles, and pen brushes will be enough. Dissection microscope (40 x) can also be used for sorting purposes. The key factor for the achievement of sorting is not to add more samples in the sorting tray at a given point of time as more the materials, the greater is the chance of overlooking specimens. Therefore, adequate care is to be taken in every step since there is every possibility of missing of many species.

For the sake of convenience, the Indian mangroves have been categorized into (i) East coast and (ii) West coast mangroves and accordingly based on the available published reports and intensive Google search, information on polychaete diversity, state-wise and coast-wise is provided here.

14.3 Diversity of Mangrove Associated Polychaetes along the Indian Coast

Globally, ~ 11,500 species of polychaetes belonging to 1417 Genera and 85 families have been reported to occur, which account for >60% of the total macro-benthic community. Of these species, 6033 species belong to the group Errantia, whereas 5085 and 158 species belong to the groups Sedentaria and Echiura, respectively. Additionally, 180 species were from families currently outside of or as yet unassigned to, groupings in WoRMS, and referred to as Polychaeta *incertae sedis*, newly created group (Pamungkas et al. 2019). In them, as many as 1093 species, representing 8.66% of the total number of polychaete species, are known in India (Achary et al. 2005). Among the twelve different mangroves habituated States or Union Territories, which comprises of ~40 different mangrove ecosystems, information on polychaete diversity is available only from 12 mangroves out of 19 on the east coast and only from 11 out of 21 those of the west coast of India. Among the mangroves of both the coasts of India, the maximum number of polychaetes was recorded from the mangroves of Tamil Nadu (213) followed by those of Maharashtra (189), West Bengal (85), Karnataka (64), Odisha (50), Andhra Pradesh (46), Kerala (34), Gujarat (8), Pondicherry (5) in that order (Fig. 14.2). Regrettably, there is no detailed study on the polychaete taxonomy from mangrove habituated states such as Gujarat, Goa, Andaman and Nicobar, and Daman and Diu. Altogether 385 polychaete species were recorded from the mangroves of east and west coasts of India. Of these, 213 species belonged to group Errantia and the remaining 172 species to group Sedentaria as the species are categorized into groups Errantia and Sedentaria

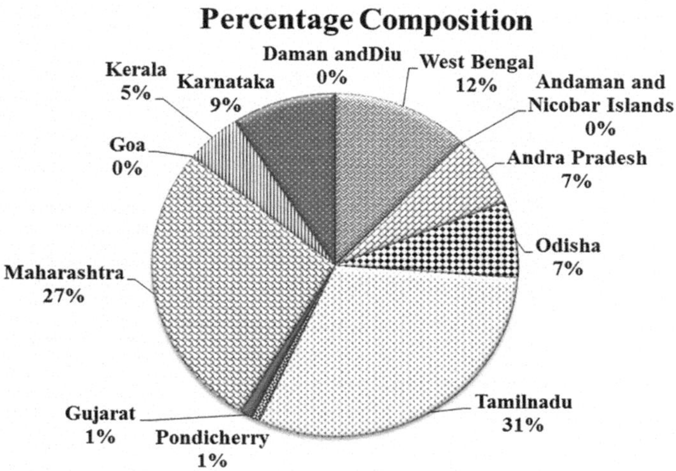

Fig. 14.2 State-wise contribution of polychaetes reported in Indian mangroves

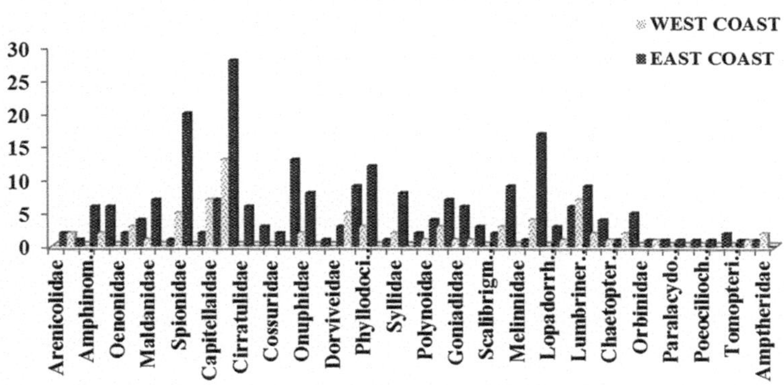

Fig. 14.3 Family-wise contribution of polychaetes recorded in Indian mangroves

based on their mobility. Going through the available published information, it is revealed that representatives of the following families', viz., Phyllodocidae, Syllidae, Nereidae, Eunicidae, Onuphidae, Glyceridae, Spionidae, Cirratulidae, Terebellidae, Sabellidae, and Serpulidae are reported to occur very commonly in the mangrove ecosystems (Fig. 14.3). Similarly, the following species were found to be common in mangrove which includes *Lepidonotus kumara, Dendronereis aestuarina, Dendronereis arborifera, Diopatra neapolitana, Nereis chingringhattensis, Malacoceros indicus, Marphysa mossambica, Heteromastus similis, Paraheteromastus tenuis, Prionospio cirrifera, Prionospio pinnata, and Euclymene annandalei.*

14.4 Diversity of Polychaetes in Mangroves of East Coast of India

In respect of mangroves of the east coast, only a few studies are available which include the monumental work of Nandi and Choudhury (1983) who have made pioneering attempt of quantitative assessment of benthic macrofauna of Sagar Island skirting Sundarbans mangroves. They have reported 17 species of macrofauna of which the polychaetes were one of the best represented taxa with the following species, namely *Lumbrinereis* sp., *L. polydesma*, *L. notocirrata*, *Diopatra neapolitana*, and *Talehsapia annandalei.* After them, Misra and Choudhury (1985) studied the benthic fauna in selected locations of Sundarbans mangroves and reported 30 species of polychaetes belonging to 24 genera under 13 families. Kasinathan and Shanmugam (1986) made preliminary studies on the benthic macrofaunal communities of Pichavaram mangroves, Tamil Nadu. Subsequently Kathiresan (2000) recorded four different dominant polychaete species such as *Heteromastus similis, Euclymene annandale, Perinereis* sp., *Mercierella enigmatica* from Pichavaram mangroves, Tamil Nadu. Kumar (2001) reported 25 species of polychaetes belonging to 20 genera from the selected locations of Sundarbans mangroves, West Bengal. Samidurai et al. (2012) studied on the spatial and temporal distribution of macrobenthos in different mangrove ecosystems, namely (i) artificially developed mangroves along the stretch of Vellar estuary, (ii) riverine (true) mangroves, and (iii) island mangrove ecosystems of Tamil Nadu, India. Among the three ecosystems, a total of 46 macro-benthic species comprising four groups of which polychaetes were found to be the most dominant group with 27 species followed by gastropods, bivalves, and crustaceans. Kumar and Khan (2013) reported the following 6 species of polychaetes, namely *Capitella capitata*, *Marphysa* sp., *M. macintoshi*, *Namalycastis indica*, *Nereis* sp., and *Pseudonereis variegata* from artificially developed mangroves in Pondicherry. In their study, Thilagavathi et al. (2013) listed as many as 188 species of polychaetes from three different kinds of mangrove ecosystems of Tamil Nadu stretching ~650 km long coast, namely (i) developing mangrove ecosystems (from Chennai to Pondicherry), (ii) Riverine mangrove ecosystems (Parangipettai to Pichavaram and partly in Muthupettai), and (iii) Island mangrove ecosystems (Palk bay and Gulf of Mannar region). Contrastingly, Sekar et al. (2013) reported the following polychaete species such as *Chaetopterus* sp. *Pista* sp. *Nephtys* sp. exceedingly large in number in selected places along the southeast coast of India. Pravinkumar et al. (2013) reported 16 species of polychaetes from three different mangrove zones of Pichavaram, Tamil Nadu. Among the polychaetes, *Polydora* sp. *Exogone clavator*, *Pygospio elegans*, and *Eyclymene* sp., were found to be common throughout the study. Rao et al. (2015) studied the macrofaunal diversity in the mangroves of the Port Blair Bay, South Andaman Islands and of the various faunal taxa reported; polychaetes were one of the best represented taxa with 13 species. Murugesan et al. (2016) made a comparative study on benthic biodiversity of natural vis-à-vis artificially developed mangroves of south east coast of India. The study revealed as many as 23 species in natural (Pichavaram mangroves) mangroves and 19 species in artificially developed mangroves (developed in Vellar estuary). Of both the regions, polychaetes emerged as the dominant group with 14 species in Pichavaram followed by other taxa. Very recently, Murugesan et al. (2018) studied the polychaete taxonomy along the mangroves of the east coast of India and reported as many as 58 species of polychaetes, in which Sundarbans mangroves stood at top with 58 species followed by Pichavaram with 51, Bhitarkanika with 42 and Muthupettai mangroves came last in the list with 34 species of polychaetes.

14.5 Diversity of Polychaetes in Mangroves of West Coast of India

With respect to the mangroves skirting west coast, as early as during 1970's, Parulekar (1971) carried out pioneering studies on marine benthos of mangroves of Maharashtra and reported as many as 46 species of polychaetes. After him, Padmakumar (1984) studied on the ecology of Mangrove swamp near Juhu beach of Bombay and reported 49 species of macrofauna and of which the annelids were the dominant group with the following polychaetes as *Nereis* sp., *Lumbrinereis polydesma, Polydora* sp., *Goniodopsis incerta*, *Glycera* sp., and *Scolelepis squamata*. Subsequently, Kumar and Antony (1994) and Kumar (1995) recorded 33 polychaete species belonging to 20 genera under 10 families from mangrove Swamps of Cochin, west coast of India. The errant polychaetes were found to be more common than sedentaria group. Of these 33 species, *Marphysa gravely*, *Paraheteromastus tenuis*, *Dendronereis aestuarina*, and *D. heteropoda* were the most dominant species that occurred throughout the study. Further, Kumar (1999) reported five different species of polychaetes from the Cochin mangrove ecosystem. Subsequently, Kumar (2003) recorded as many as 43 species of polychaetes along the major mangrove ecosystems such as Cochin—Bombay with 33 species in Cochin mangroves and 10 in Bombay. Among the 15 families identified, the species of the following families, viz., Nereidae (24 species), Eunicidae (20 species), Glyeridae (8 species), Capitellidae (7 species), and Spionidae (7 species) were found to be dominant. Saravanakumar et al. (2007) studied the benthic macrofaunal assemblages of mangroves of Gulf of Kachchh and reported 62 species comprising 5 taxa. Of which, the polychaetes consisted of the following nine species, namely *Diopatra neapolitana*, *Eunice* sp., *Glycera alba, Lumbriconereis latreilli, Marphysa stragulum, Nereis* sp. *Perinereis* sp., *Pulliella armata, Thalehasapia tenuis*. Pati et al. (2015) recorded a total of 180 species of polychaetes belonging to 113 genera under 41 families and six orders from Maharashtra coast covering 6 coastal districts (Thane, Mumbai-suburban, Mumbai, Raigad, Ratnagiri, and Sindhudurg) stretching over 653 km long coastline. There are about 18 prominent creeks/back waters, many of which having strong presence of mangrove cover and thus the compilation of Patil et al. revealed a surge in the species richness to considerable species diversity from 46 species of Parulekar's study to 180 species by Patil et al. Moving a little towards north, in Karnataka, Bandekar et al. (2017) studied on the macro-benthic polychaetes in mangroves regions along Kali estuary and reported as many as 61 species of

Fig. 14.4 Common polychaete species recorded in mangroves of east and west coast of India

polychaetes under 13 families. Of this, members of Nereidae, Spionidae, Eunicidae, Glyceridae, Sabellaridae, and Terebellidae were found to be higher in number. Similarly, species of genera *Nereis*, *Lumbrinereis*, and *Glycera* outnumbered the other genera. Similarly, Satish et al. (2018) studied temporal variations in polychaetes of the mangrove region of Shirgaon, Ratnagiri and reported a total of six polychaetes species such as *Nereis* sp. *Perinereis* sp. *Lumbrinereis* sp. *Marphysa* sp. - I, *Marphysa* sp. - II and a few un-identified polychaete sp. from the intertidal mangrove area of Shirgaon, Ratnagiri, Maharashtra. Of this species, *Nereis* sp. and *Marphysa* sp. showed their consistency in their occurrence throughout the study.

Looking at the commonness of the species occurrence, among the families, the representative of the following families, viz., Nereididae (50 species), Spionidae

Fig. 14.5 Common polychaete species reported from east and west coast mangroves of India

(28 species), Lumbrineridae (21 species), Terebellidae (20 species), Sabellidae (19 species), Syllidae (17 species), Capitellidae (17 species), Eunicidae (16 species), Serpulidae (16 species), and Cirratulidae (12 species) were found to be dominant in that order (Figs. 14.4 and 14.5).

Based on the foregoing account and available published reports, a total of 385 species of polychaetes belonging to 176 genera under 49 families are known to occur in mangroves of east and west coasts of India (Table 14.1). This total covers 35.22% of the total polychaete species recorded in India. Comparing mangroves, as many as 231 species of polychaetes were reported in various mangrove of east coast and 230 in west coast mangroves.

Table 14.1 List of polychaetes reported in mangroves of east and west coasts of India

S. No	Polychaete species	Family	Group	East coast						West coast					
				West Bengal	Andaman and Nicobar Islands	Andra Pradesh	Odisha	Tamil Nadu	Pondicherry	Gujarat	Maharashtra	Goa	Kerala	Karnataka	Daman and Diu
1	*Abarenicola gilchristi*	Arenicolidae	Sedentaria	–	–	–	–	+	–	–	–	–	–	–	–
2	*Acoetes melanonota*	Acoetidae	Errantia	–	–	–	–	–	–	–	+	–	–	–	–
3	*Aglaophamus dibranchis*	Nephtyidae	Errantia	–	–	–	–	–	–	–	+	–	–	–	–
4	*Ammotrypane aulogaster*	Opheliidae	Sedentaria	–	–	–	–	–	–	–	–	–	–	–	–
5	*Ampharete capensis*	Ampharetidae	Sedentaria	–	–	–	–	–	–	–	+	–	–	–	–
6	*Amphicteis gunneri*	Ampharetidae	Sedentaria	–	–	–	–	+	–	–	–	–	+	+	–
7	*Amphicteis* sp.	Ampharetidae	Sedentaria	–	–	–	–	–	–	–	–	–	–	+	–
8	*Amphinome rostrata*	Amphinomidae	Errantia	–	–	–	–	+	–	–	–	–	–	–	–
9	*Amphinome* sp.	Amphinomidae	Errantia	+	–	–	+	+	–	–	–	–	–	–	–
10	*Ancistrosyllis constricta*	Pilargidae	Errantia	+	–	–	–	+	–	–	–	–	–	–	–
11	*Ancistrosyllis groenlandica*	Pilargidae	Errantia	–	–	–	–	+	–	–	–	–	–	–	–
12	*Ancistrosyllis parva*	Pilargidae	Errantia	–	–	–	–	+	–	–	–	–	–	–	–
13	*Ancistrosyllis* sp.	Pilargidae	Errantia	+	–	+	+	+	–	–	–	–	–	–	–
14	*Ancistrosyllis constricta*	Pilargidae	Errantia	–	–	–	–	–	–	–	–	–	–	–	–
15	*Aphrodita alta*	Aphroditidae	Errantia	–	–	–	–	–	–	–	+	–	–	+	–
16	*Arabella iricolor*	Oenonidae	Errantia	–	–	–	–	+	–	–	+	–	–	–	–

17	*Arenicola bombayensis*	Arenicolidae	Sedentaria	–	–	–	–	–	–	–	+	–	–	–	–
18	*Arenicola loveni*	Arenicolidae	Sedentaria	–	–	–	–	+	–	–	–	–	–	–	–
19	*Armandia lanceolata*	Opheliidae	Sedentaria	–	–	–	–	+	–	–	+	–	–	–	–
20	*Armandia longicaudata*	Opheliidae	Sedentaria	–	–	–	–	+	–	–	–	–	–	–	–
21	*Axiothella obockenisis*	Maldanidae	Sedentaria	–	–	–	–	+	–	–	+	–	–	–	–
22	*Bhawania cryptocephala*	Chrysopetalidae	Errantia	–	–	–	–	–	–	–	+	–	–	+	–
23	*Bhawania goodei*	Chrysopetalidae	Errantia	–	–	–	–	+	–	–	–	–	–	–	–
24	*Boccardia polybranchia*	Spionidae	Sedentaria	+	–	+	–	+	–	–	+	–	–	–	–
25	*Brada villosa*	Flabelligeridae	Sedentaria	+	–	+	+	+	–	–	–	–	–	–	–
26	*Branchiocapitella singularis*	Capitellidae	Sedentaria	–	–	–	–	+	–	–	+	–	+	–	–
27	*Capitella capitata*	Capitellidae	Sedentaria	+	–	+	+	+	+	–	+	–	–	+	–
28	*Capitella* sp.	Capitellidae	Sedentaria	–	–	–	–	–	–	–	–	–	+	+	–
29	*Ceratonereis burmensis*	Nereididae	Errantia	–	–	–	–	–	–	–	+	–	–	–	–
30	*Ceratonereis costae*	Nereididae	Errantia	–	–	–	–	+	–	–	+	–	+	+	–
31	*Ceratonereis keiskama*	Nereididae	Errantia	–	–	–	–	+	–	–	–	–	–	–	–
32	*Ceratonereis mirabilis*	Nereididae	Errantia	–	–	–	–	+	–	–	+	–	–	–	–
33	*Chaetopterus variopedatus*	Chaetopteridae	Sedentaria	–	–	–	–	–	–	–	+	–	–	–	–
34	*Chaetozone setosa*	Cirratulidae	Sedentaria	–	–	–	–	+	–	–	–	–	–	–	–
35	*Chloeia flava*	Amphinomidae	Errantia	–	–	–	–	+	–	–	–	–	–	–	–

(continued)

Table 14.1 (continued)

S. No	Polychaete species	Family	Group	East coast						West coast					
				West Bengal	Andaman and Nicobar Islands	Andra Pradesh	Odisha	Tamil Nadu	Pondicherry	Gujarat	Maharashtra	Goa	Kerala	Karnataka	Daman and Diu
36	*Chloeia parva*	Amphinomidae	Errantia	–	–	–	–	+	–	–	–	–	–	–	–
37	*Chloeia rosea*	Amphinomidae	Errantia	–	–	–	–	–	–	–	+	–	–	–	–
38	*Chone collaris*	Sabellidae	Sedentaria	+	–	–	+	+	–	–	–	–	–	–	–
39	*Chone filicaudata*	Sabellidae	Sedentaria	–	–	–	–	+	–	–	–	–	–	–	–
40	*Chone letterstedti*	Sabellidae	Sedentaria	+	–	+	+	+	–	–	–	–	–	–	–
41	*Cirratulus chrysoderma*	Cirratulidae	Sedentaria	–	–	–	–	+	–	–	–	–	–	–	–
42	*Cirratulus cirratus*	Cirratulidae	Sedentaria	–	–	–	–	–	–	–	+	–	–	–	–
43	*Cirratulus concinnus*	Cirratulidae	Sedentaria	–	–	–	–	+	–	–	+	–	–	–	–
44	*Cirratulus gilchristi*	Cirratulidae	Sedentaria	–	–	–	–	+	–	–	–	–	–	–	–
45	*Cirratulus* sp.	Cirratulidae	Sedentaria	+	–	–	–	+	–	–	–	–	–	–	–
46	*Cirriformia afer*	Cirratulidae	Sedentaria	–	–	–	–	–	–	–	+	–	–	–	–
47	*Cirriformia chrysoderma*	Cirratulidae	Sedentaria	–	–	–	–	–	–	–	+	–	–	–	–
48	*Cirriformia filigera*	Cirratulidae	Sedentaria	–	–	–	–	–	–	–	+	–	–	–	–
49	*Cirriformia limnoricola*	Cirratulidae	Sedentaria	–	–	–	–	–	–	–	+	–	–	–	–
50	*Cirriformia tentaculata*	Cirratulidae	Sedentaria	–	–	–	–	+	–	–	+	–	–	–	–
51	*Cirrophorus branchiatus*	Paraonidae	Sedentaria	+	–	+	+	+	–	–	–	–	–	–	–
52	*Cossura coasta*	Cossuridae	Sedentaria	+	–	+	+	+	–	–	+	–	–	–	–
53	*Cossura delta*	Cossuridae	Sedentaria	–	–	–	–	+	–	–	–	–	–	–	–

54	*Cossura longocirrata*	Cossuridae	Sedentaria	–	–	–	–	–	–	–	+	–	–	–	–
55	*Dasychone cingulata*	Sabellidae	Sedentaria	–	–	–	–	+	–	–	+	–	–	–	–
56	*Dasychone serratibranchis*	Sabellidae	Sedentaria	–	–	–	–	–	–	–	+	–	–	+	–
57	*Dendronereides heteropoda*	Nereididae	Errantia	+	–	–	–	–	–	–	+	–	+	+	–
58	*Dendronereis aestuarina*	Nereididae	Errantia	+	–	–	–	+	–	–	+	–	+	+	–
59	*Dendronereis arborifera*	Nereididae	Errantia	+	–	+	+	+	–	–	+	–	+	+	–
60	*Diopatra claparedii*	Onuphidae	Errantia	–	–	–	–	–	–	–	+	–	–	+	–
61	*Diopatra cuprea*	Onuphidae	Errantia	+	–	–	–	+	–	–	+	–	–	–	–
62	*Diopatra neapolitana*	Onuphidae	Errantia	+	–	–	–	+	–	+	+	–	+	+	–
63	*Diopatra* sp.	Onuphidae	Errantia	–	–	–	–	–	–	–	–	–	–	+	–
64	*Dipolydora coeca*	Spionidae	Sedentaria	–	–	–	–	–	–	–	+	–	–	+	–
65	*Dipolydora giardi*	Spionidae	Sedentaria	–	–	–	–	–	–	–	+	–	–	+	–
66	*Disoma orissae*	Trochochaetidae	Sedentaria	–	–	–	–	+	–	–	–	–	–	–	–
67	*Dorvillea incertus*	Dorvilleidae	Errantia	–	–	–	–	+	–	–	–	–	–	–	–
68	*Dorvillea neglecta*	Dorvilleidae	Errantia	–	–	–	–	+	–	–	–	–	–	–	–
69	*Enice pennata*	Eunicidae	Errantia	–	–	–	–	+	–	–	–	–	–	–	–
70	*Enice tentaculata*	Eunicidae	Errantia	–	–	–	–	+	–	–	–	–	–	–	–
71	*Eteone barantollae*	Phyllodocidae	Errantia	+	–	–	–	–	–	–	–	–	–	–	–
72	*Eteone ornata*	Phyllodocidae	Errantia	–	–	–	–	+	–	–	+	–	–	–	–
73	*Eteone siphodonta*	Phyllodocidae	Errantia	–	–	–	–	+	–	–	–	–	–	–	–
74	*Eteone* sp.	Phyllodocidae	Errantia	–	–	–	–	–	–	–	–	–	–	+	–
75	*Eteone syphodonta*	Phyllodocidae	Errantia	–	–	–	–	–	–	–	+	–	–	–	–

(continued)

Table 14.1 (continued)

S. No	Polychaete species	Family	Group	East coast						West coast					
				West Bengal	Andaman and Nicobar Islands	Andra Pradesh	Odisha	Tamil Nadu	Pondicherry	Gujarat	Maharashtra	Goa	Kerala	Karnataka	Daman and Diu
76	*Euchone rosea*	Sabellidae	Sedentaria	–	–	–	–	+	–	–	–	–	–	–	–
77	*Euclymene annandalei*	Maldanidae	Sedentaria	–	–	–	–	+	–	–	+	–	–	–	–
78	*Euclymene oerstedii*	Maldanidae	Sedentaria	–	–	–	–	–	–	–	–	–	–	–	–
79	*Euclymene sp.*	Maldanidae	Sedentaria	+	–	–	+	+	–	–	–	–	–	–	–
80	*Eulalia macroceros*	Phyllodocidae	Errantia	–	–	–	–	+	–	–	–	–	–	–	–
81	*Eulalia sanguinea*	Phyllodocidae	Errantia	–	–	–	–	+	–	–	–	–	–	–	–
82	*Eunice antennata*	Eunicidae	Errantia	–	–	–	–	–	–	–	+	–	–	–	–
83	*Eunice australis*	Eunicidae	Errantia	–	–	–	–	+	–	–	–	–	–	–	–
84	*Eunice guttata*	Eunicidae	Errantia	–	–	–	–	–	–	–	+	–	–	–	–
85	*Eunice indica*	Eunicidae	Errantia	–	–	–	–	+	–	–	–	–	–	–	–
86	*Eunice laticeps*	Eunicidae	Errantia	–	–	–	–	–	–	–	+	–	–	–	–
87	*Eunice savignyi*	Eunicidae	Errantia	–	–	–	–	–	–	–	+	–	–	–	–
88	*Eunice* sp.	Eunicidae	Errantia	+	–	+	+	+	–	+	–	–	+	+	–
89	*Eunice tubifex*	Eunicidae	Errantia	–	–	–	–	+	–	–	–	–	+	+	–
90	*Euphrosine capensis*	Euphrosinidae	Errantia	–	–	–	–	+	–	–	+	–	–	–	–
91	*Eurythoe complanata*	Amphinomidae	Errantia	+	–	+	+	+	–	–	+	–	–	–	–
92	*Eurythoe parvecarunculata*	Amphinomidae	Errantia	+	–	–	–	+	–	–	+	–	–	–	–
93	*Exoene* sp.	Syllidae	Errantia	–	–	–	–	–	–	–	–	–	–	+	–
94	*Exogone clavator*	Syllidae	Errantia	+	–	+	+	+	–	–	–	–	–	–	–

95	*Exogone verugera*	Syllidae	Errantia	–	–	–	–	+	–	–	–	–	–	–	–
96	*Fabrica filamentosa*	Fabriciidae	Sedentaria	+	–	–	+	+	–	–	–	–	–	–	–
97	*Fabricia bansei*	Fabriciidae	Sedentaria	–	–	–	–	–	–	–	+	–	–	–	–
98	*Fabriciola mossambica*	Fabriciidae	Sedentaria	–	–	–	–	+	–	–	–	–	–	–	–
99	*Ficopomatus uschakovi*	Serpulidae	Sedentaria	–	–	–	–	–	–	–	+	–	–	–	–
100	*Galathowenia oculata*	Oweniidae	Sedentaria	–	–	–	–	–	–	–	+	–	–	–	–
101	*Gattyana deludens*	Polynoidae	Errantia	–	–	–	–	–	–	–	+	–	–	–	–
102	*Gaudichaudius cimex*	Polynoidae	Errantia	–	–	–	–	–	–	–	+	–	–	–	–
103	*Glycera africana*	Glyceridae	Errantia	–	–	–	–	+	–	–	–	–	–	–	–
104	*Glycera alba*	Glyceridae	Errantia	+	–	–	–	+	–	+	+	–	+	+	–
105	*Glycera benguellana*	Glyceridae	Errantia	+	–	+	+	+	–	–	–	–	–	–	–
106	*Glycera convoluta*	Glyceridae	Errantia	–	–	–	–	–	–	–	+	–	–	–	–
107	*Glycera emerita*	Glyceridae	Errantia	–	–	–	–	–	–	–	–	–	–	–	–
108	*Glycera incerta*	Glyceridae	Errantia	–	–	–	–	–	–	–	+	–	–	–	–
109	*Glycera longipinnis*	Glyceridae	Errantia	+	–	–	+	+	–	–	+	–	+	+	–
110	*Glycera onicornis*	Glyceridae	Errantia	–	–	–	–	+	–	–	–	–	–	–	–
111	*Glycera rouxii*	Glyceridae	Errantia	+	–	–	–	–	–	–	–	–	–	–	–
112	*Glycera* sp.	Glyceridae	Errantia	+	–	–	–	+	–	–	–	–	–	+	–
113	*Glycera unicornis*	Glyceridae	Errantia	+	–	–	+	+	–	–	+	–	–	–	–
114	*Glycinde capensis*	Goniadidae	Errantia	–	–	–	–	+	–	–	–	–	–	–	–
115	*Glycinde multidens*	Goniadidae	Errantia	–	–	–	–	–	–	–	+	–	–	–	–
116	*Glycinde oligodon*	Goniadidae	Errantia	–	–	–	–	+	–	–	+	–	–	–	–
117	*Goniada emerita*	Goniadidae	Errantia	+	–	+	+	+	–	–	+	–	–	–	–
118	*Goniada goniada*	Goniadidae	Errantia	–	–	–	–	+	–	–	–	–	–	–	–

(continued)

Table 14.1 (continued)

				East coast						West coast					
S. No	Polychaete species	Family	Group	West Bengal	Andaman and Nicobar Islands	Andra Pradesh	Odisha	Tamil Nadu	Pondicherry	Gujarat	Maharashtra	Goa	Kerala	Karnataka	Daman and Diu
119	*Goniada* sp.	Goniadidae	Errantia	–	–	–	–	–	–	–	–	–	+	+	–
120	*Goniadella gracilis*	Goniadidae	Errantia	–	–	–	–	+	–	–	+	–	–	–	–
121	*Goniadopsis longicirrata*	Goniadidae	Errantia	–	–	–	–	–	–	–	+	–	–	–	–
122	*Goniadopsis maskallensis*	Goniadidae	Errantia	–	–	–	–	+	–	–	–	–	–	–	–
123	*Haploscoloplos kerguelensis*	Orbiniidae	Sedentaria	–	–	–	–	–	–	–	+	–	–	–	–
124	*Haplosyllis spongicola*	Syllidae	Errantia	–	–	–	–	–	–	–	+	–	–	–	–
125	*Harmothoe africana*	Polynoidae	Errantia	–	–	–	–	+	–	–	–	–	–	–	–
126	*Hermundura annandalei*	Pilargidae	Errantia	–	–	–	–	–	–	–	+	–	–	–	–
127	*Hesione intertexta*	Hesionidae	Errantia	–	–	–	–	+	–	–	–	–	–	–	–
128	*Hesione pantherina*	Hesionidae	Errantia	–	–	–	–	–	–	–	+	–	–	–	–
129	*Heteromastides bifidus*	Capitellidae	Sedentaria	–	–	–	–	–	–	–	+	–	–	–	–
130	*Heteromastus filiformis*	Capitellidae	Sedentaria	–	–	–	–	–	–	–	+	–	–	–	–
131	*Heteromastus similis*	Capitellidae	Sedentaria	–	–	–	–	+	–	–	+	–	–	–	–
132	*Heteromastus* sp.	Capitellidae	Sedentaria	–	–	–	–	–	–	–	–	–	–	+	–
133	*Hololepidella maculata*	Polynoidae	Errantia	–	–	–	–	+	–	–	–	–	–	–	–

134	*Hyalinoecia tubicola*	Onuphidae	Errantia	+	–	+	+	–	–	–	–	–	–	–	–
135	*Hyboscolex longiseta*	Scalibregmatidae	Sedentaria	+	–	+	–	+	–	–	–	–	–	–	–
136	*Hydroides albiceps*	Serpulidae	Sedentaria	–	–	–	–	+	–	–	+	–	–	–	–
137	*Hydroides diramphus*	Serpulidae	Sedentaria	–	–	–	–	–	–	–	+	–	–	–	–
138	*Hydroides heteroceros*	Serpulidae	Sedentaria	–	–	–	–	+	–	–	+	–	–	–	–
139	*Hydroides homoceros*	Serpulidae	Sedentaria	–	–	–	–	+	–	–	–	–	–	–	–
140	*Hydroides norvegicus*	Serpulidae	Sedentaria	–	–	–	–	–	–	–	+	–	–	–	–
141	*Hydroides operculatus*	Serpulidae	Sedentaria	–	–	–	–	–	–	–	+	–	–	–	–
142	*Inermonephtys inermis*	Nephtyidae	Errantia	–	–	–	–	–	–	–	+	–	–	–	–
143	*Irmula spissipes*	Syllidae	Errantia	–	–	–	–	–	–	–	+	–	–	–	–
144	*Isolda pulchella*	Melinnidae	Sedentaria	–	–	–	–	+	–	–	+	–	–	–	–
145	*Lanice socialis*	Terebellidae	Sedentaria	–	–	–	–	+	–	–	–	–	–	–	–
146	*Laonice cirrata*	Spionidae	Sedentaria	+	–	+	+	+	–	–	–	–	–	–	–
147	*Laonome indica*	Sabellidae	Sedentaria	+	–	–	–	+	–	–	–	–	–	–	–
148	*Lavensia* sp.	Nereididae	Errantia	–	–	–	–	–	–	–	–	–	–	+	–
149	*Leanira hystricis*	Sigalionidae	Errantia	–	–	–	–	+	–	–	–	–	–	–	–
150	*Leocrates claparedii*	Hesionidae	Errantia	–	–	–	–	+	–	–	+	–	–	–	–
151	*Leocratides ehlersi*	Hesionidae	Errantia	–	–	–	–	–	–	–	+	–	–	–	–
152	*Lepidonotus carinulatus*	Polynoidae	Errantia	–	–	–	–	–	–	–	+	–	–	–	–

(continued)

Table 14.1 (continued)

				East coast						West coast					
S. No	Polychaete species	Family	Group	West Bengal	Andaman and Nicobar Islands	Andra Pradesh	Odisha	Tamil Nadu	Pondicherry	Gujarat	Maharashtra	Goa	Kerala	Karnataka	Daman and Diu
153	*Lepidonotus tenuisetosus*	Polynoidae	Errantia	+	–	–	–	+	–	–	+	–	–	+	–
154	*Loimia medusa*	Terebellidae	Sedentaria	–	–	–	–	+	–	–	–	–	–	–	–
155	*Lopadorhynchus henseni*	Lopadorrhynchidae	Errantia	–	–	–	–	+	–	–	–	–	–	–	–
156	*Lopadorhynchus nationalis*	Lopadorrhynchidae	Errantia	–	–	–	–	+	–	–	–	–	–	–	–
157	*Lumbriconereis heteropoda*	Lumbrineridae	Errantia	–	–	–	–	+	–	–	–	–	–	+	–
158	*Lumbriconereis impatiens*	Lumbrineridae	Errantia	–	–	–	–	+	–	–	–	–	–	–	–
159	*Lumbriconereis latreilli*	Lumbrineridae	Errantia	–	–	–	–	+	–	+	–	–	+	+	–
160	*Lumbriconereis polydesma*	Lumbrineridae	Errantia	+	–	–	–	+	–	–	–	–	–	+	–
161	*Lumbriconereis pseudobifilaris*	Lumbrineridae	Errantia	–	–	–	–	+	–	–	–	–	+	+	–
162	*Lumbriconereis simplex*	Lumbrineridae	Errantia	–	–	–	–	+	–	–	–	–	+	+	–
163	*Lumbriconereis* sp.	Lumbrineridae	Errantia	–	–	–	–	–	–	–	+	–	+	+	–
164	*Lumbriconeries aberrans*	Lumbrineridae	Errantia	–	–	–	–	–	–	–	–	–	–	+	–
165	*Lumbrinereis brevicirra*	Lumbrineridae	Errantia	–	–	–	–	+	–	–	–	–	–	–	–

166	*Lumbrinereis magalhaensis*	Lumbrineridae	Errantia	–	–	–	–	+	–	–	–	–	–	–	–
167	*Lumbrineris albidentata*	Lumbrineridae	Errantia	+	–	+	+	–	–	–	–	–	–	–	–
168	*Lumbrineris bifilaris*	Lumbrineridae	Errantia	–	–	–	–	–	–	–	+	–	–	–	–
169	*Lumbrineris brevicirra*	Lumbrineridae	Errantia	–	–	–	–	–	–	–	+	–	–	–	–
170	*Lumbrineris hartmani*	Lumbrineridae	Errantia	–	–	–	–	–	–	–	+	–	–	–	–
171	*Lumbrineris heteropoda*	Lumbrineridae	Errantia	–	–	–	–	–	–	–	+	–	–	–	–
172	*Lumbrineris japonica*	Lumbrineridae	Errantia	–	–	–	–	–	–	–	+	–	–	–	–
173	*Lumbrineris polydesma*	Lumbrineridae	Errantia	–	–	–	–	–	–	–	+	–	–	–	–
174	*Lumbrineris pseudobifilaris*	Lumbrineridae	Errantia	–	–	–	–	–	–	–	+	–	–	–	–
175	*Lumbrineris simplex*	Lumbrineridae	Errantia	–	–	–	–	–	–	–	+	–	–	–	–
176	*Lycastis indica*	Nereididae	Errantia	–	–	–	–	–	–	–	–	–	–	+	–
177	*Lycastonereis indica*	Nereididae	Errantia	+	–	–	–	–	–	–	–	–	–	–	–
178	*Magelona cincta*	Magelonidae	Sedentaria	+	–	–	–	+	–	–	+	–	–	+	–
179	*Magelona papillicornis*	Magelonidae	Sedentaria	–	–	–	–	+	–	–	–	–	–	+	–
180	*Magelona rosea*	Magelonidae	Sedentaria	–	–	–	–	–	–	–	+	–	–	–	–
181	*Malacoceros indicus*	Spionidae	Sedentaria	+	–	+	+	+	–	–	+	–	–	+	–
182	*Maldane sarsi*	Maldanidae	Sedentaria	+	–	–	+	+	–	–	+	–	–	–	–

(continued)

Table 14.1 (continued)

S. No	Polychaete species	Family	Group	East coast						West coast					
				West Bengal	Andaman and Nicobar Islands	Andra Pradesh	Odisha	Tamil Nadu	Pondicherry	Gujarat	Maharashtra	Goa	Kerala	Karnataka	Daman and Diu
183	*Maldanella capensis*	Maldanidae	Sedentaria	–	–	–	–	–	–	–	–	–	–	+	–
184	*Maldanella grossa*	Maldanidae	Sedentaria	–	–	–	–	–	–	–	+	–	–	–	–
185	*Marphysa gravelyi*	Eunicidae	Errantia	–	–	–	–	+	–	–	–	–	+	–	–
186	*Marphysa macintoshi*	Eunicidae	Errantia	–	–	–	–	–	+	–	–	–	–	–	–
187	*Marphysa mossambica*	Eunicidae	Errantia	–	–	–	–	–	–	–	+	–	–	–	–
188	*Marphysa sanguinea*	Eunicidae	Errantia	–	–	–	–	–	–	–	+	–	–	–	–
189	*Marphysa* sp.	Eunicidae	Errantia	–	–	–	–	–	+	–	+	–	–	+	–
190	*Marphysa stragulum*	Eunicidae	Errantia	–	–	–	–	–	–	+	–	–	+	–	–
191	*Mediomastus capensis*	Capitellidae	Sedentaria	–	–	–	–	–	–	–	+	–	–	–	–
192	*Mediomastus* sp.	Capitellidae	Sedentaria	–	–	–	–	–	–	–	–	–	–	+	–
193	*Megaloma* sp.	Magelonidae	Sedentaria	–	–	–	–	–	–	–	–	–	–	–	–
194	*Megalomma quadrioculatum*	Magelonidae	Sedentaria	+	–	+	+	–	–	–	–	–	–	–	–
195	*Mercierella enigmatica*	Serpulidae	Sedentaria	–	–	–	–	+	–	–	–	–	+	–	–
196	*Mesochaetopterus*	Chaetopteridae	Sedentaria	–	–	–	–	+	–	–	–	–	–	–	–
197	*Micronephthys sphaerocirrata*	Nephtyidae	Errantia	–	–	–	–	–	–	–	+	–	–	–	–

198	*Micronereides capensis*	Nereididae	Errantia	–	–	–	–	–	–	–	+	–	–	–	–
199	*Minuspio cirrifera*	Spionidae	Sedentaria	+	–	–	–	+	–	–	+	–	–	–	–
200	*Namalycastis indica*	Nereididae	Errantia	–	–	–	–	–	–	–	+	–	–	–	–
201	*Neanthes chilkaensis*	Nereididae	Errantia	–	–	–	–	–	–	–	+	–	–	–	–
202	*Neanthes chingrighattensis*	Nereididae	Errantia	–	–	–	–	–	–	–	+	–	–	–	–
203	*Neanthes cricognatha*	Nereididae	Errantia	–	–	–	–	–	–	–	+	–	–	–	–
204	*Neanthes* sp.	Nereididae	Errantia	+	–	+	+	+	–	–	–	–	–	–	–
205	*Neanthes unifasciata*	Nereididae	Errantia	–	–	–	–	–	–	–	+	–	–	–	–
206	*Neosabellaria cementarium*	Sabellariidae	Sedentaria	–	–	–	–	–	–	–	+	–	–	–	–
207	*Nephtys bucera*	Nephtyidae	Errantia	–	–	–	–	+	–	–	–	–	–	–	–
208	*Nephtys capensis*	Nephtyidae	Errantia	–	–	–	–	+	–	–	–	–	–	–	–
209	*Nephtys dibranchis*	Nephtyidae	Errantia	+	–	+	+	–	–	–	–	–	–	–	–
210	*Nephtys lyrochaeta*	Nephtyidae	Errantia	–	–	–	–	–	–	–	+	–	–	–	–
211	*Nephtys oligobranchia*	Nephtyidae	Errantia	–	–	–	–	–	–	–	+	–	–	–	–
212	*Nephtys polybranchia*	Nephtyidae	Errantia	–	–	–	–	+	–	–	+	–	–	+	–
213	*Nephtys* sp.	Nephtyidae	Errantia	–	–	–	–	–	–	–	–	–	–	+	–
214	*Nephtys sphaerocirrata*	Nephtyidae	Errantia	–	–	–	–	+	–	–	–	–	–	–	–
215	*Nereis abbreviata*	Nereididae	Errantia	–	–	–	–	+	–	–	–	–	–	–	–
216	*Nereis aibuhitensis*	Nereididae	Errantia	–	–	–	–	–	–	–	+	–	–	–	–
217	*Nereis chilkaensis*	Nereididae	Errantia	–	–	–	–	–	–	–	–	–	+	–	–

(continued)

Table 14.1 (continued)

S. No	Polychaete species	Family	Group	East coast						West coast					
				West Bengal	Andaman and Nicobar Islands	Andra Pradesh	Odisha	Tamil Nadu	Pondicherry	Gujarat	Maharashtra	Goa	Kerala	Karnataka	Daman and Diu
218	*Nereis chingrighattensis*	Nereididae	Errantia	+	–	–	–	–	–	–	–	–	–	–	–
219	*Nereis cricognatha*	Nereididae	Errantia	+	–	–	–	–	–	–	–	–	–	–	–
220	*Nereis diversicolor*	Nereididae	Errantia	+	–	–	+	+	–	–	–	–	–	–	–
221	*Nereis falcaria*	Nereididae	Errantia	–	–	–	–	–	–	–	+	–	–	–	–
222	*Nereis glandicincta*	Nereididae	Errantia	–	–	–	–	–	–	–	–	–	–	–	–
223	*Nereis granulata*	Nereididae	Errantia	–	–	–	–	+	–	–	–	–	–	–	–
224	*Nereis jacksoni*	Nereididae	Errantia	–	–	–	–	+	–	–	–	–	–	–	–
225	*Nereis kauderni*	Nereididae	Errantia	–	–	–	–	–	–	–	–	–	+	–	–
226	*Nereis* sp.	Nereididae	Errantia	+	–	+	+	+	+	+	+	–	+	+	–
227	*Nereis talehsapensis*	Nereididae	Errantia	–	–	–	–	–	–	–	+	–	–	–	–
228	*Nereis virens*	Nereididae	Errantia	–	–	–	–	+	–	–	–	–	–	–	–
229	*Neries oligobranchia*	Nereididae	Errantia	–	–	–	–	–	–	–	–	–	+	+	–
230	*Ninoe lagosiana*	Lumbrineridae	Errantia	–	–	–	–	–	–	–	+	–	–	–	–
231	*Ninoe notocirrata*	Lumbrineridae	Errantia	–	–	–	–	–	–	–	+	–	–	–	–
232	*Notocirrus brevicirrus*	Oenonidae	Errantia	–	–	–	–	+	–	–	–	–	–	–	–
233	*Notomastus aberans*	Capitellidae	Sedentaria	+	–	+	+	+	–	–	+	–	–	–	–
234	*Notomastus fauveli*	Capitellidae	Sedentaria	–	–	–	–	+	–	–	+	–	–	–	–
235	*Notomastus giganteus*	Capitellidae	Sedentaria	–	–	–	–	+	–	–	–	–	–	–	–

236	*Notomastus latericeus*	Capitellidae	Sedentaria	+	–	–	+	–	–	–	+	–	–	–	–
237	*Notoproctus pacificus*	Maldanidae	Sedentaria	+	–	+	+	+	–	–	–	–	–	–	–
238	*Onuphis eremita*	Onuphidae	Errantia	–	–	–	–	+	–	–	+	–	–	–	–
239	*Onuphis eremite*	Onuphidae	Errantia	–	–	–	–	+	–	–	–	–	–	–	–
240	*Onuphis geophiliformis*	Onuphidae	Errantia	–	–	–	–	+	–	–	+	–	–	–	–
241	*Onuphis holobranchiata*	Onuphidae	Errantia	–	–	–	–	+	–	–	+	–	–	–	–
242	*Onuphis* sp.	Onuphidae	Errantia	+	–	–	–	+	–	–	–	–	–	+	–
243	*Ophelia africana*	Opheliidae	Sedentaria	–	–	–	–	+	–	–	–	–	–	–	–
244	*Ophelia capensis*	Opheliidae	Sedentaria	–	–	–	–	+	–	–	–	–	–	–	–
245	*Ophelid* sp.	Opheliidae	Sedentaria	–	–	–	–	–	–	–	–	–	–	+	–
246	*Ophelina acuminata*	Opheliidae	Sedentaria	–	–	–	–	–	–	–	+	–	–	–	–
247	*Ophiodromous* sp.	Hesionidae	Errantia	–	–	–	–	–	–	–	+	–	–	–	–
248	*Opisthosyllis* sp.	Syllidae	Errantia	–	–	–	–	–	–	–	–	–	–	+	–
249	*Orbinia angrapequensis*	Orbiniidae	Sedentaria	+	–	+	–	+	–	–	+	–	–	–	–
250	*Oriopsis eimeri*	Sabellidae	Sedentaria	–	–	–	–	+	–	–	–	–	–	–	–
251	*Oriopsis neglecta*	Sabellidae	Sedentaria	–	–	–	–	+	–	–	–	–	–	–	–
252	*Owenia fusiformis*	Oweniidae	Sedentaria	–	–	–	–	+	–	–	+	–	–	–	–
253	*Owenia* sp.	Oweniidae	Sedentaria	–	–	–	–	–	–	–	–	–	–	+	–
254	*Panthalis oerstedi*	Acoetidae	Errantia	–	–	–	–	–	–	–	+	–	–	–	–
255	*Paraheteromasus tenuis*	Capitellidae	Sedentaria	–	–	–	–	–	–	–	–	–	+	–	–
256	*Paralacydonia weberi*	Paralacydoniidae	Sedentaria	–	–	–	–	+	–	–	–	–	–	–	–

(continued)

Table 14.1 (continued)

				East coast						West coast					
S. No	Polychaete species	Family	Group	West Bengal	Andaman and Nicobar Islands	Andra Pradesh	Odisha	Tamil Nadu	Pondicherry	Gujarat	Maharashtra	Goa	Kerala	Karnataka	Daman and Diu
257	*Paralepidonotus ampulliferus*	Polynoidae	Errantia	–	–	–	–	–	–	–	+	–	–	–	–
258	*Paraonidea* sp.	Paraonidae	Sedentaria	+	–	+	–	+	–	–	–	–	–	–	–
259	*Paraonis* sp.	Paraonidae	Sedentaria	+	–	+	–	+	–	–	–	–	–	–	–
260	*Paraprionospio cristata*	Spionidae	Sedentaria	–	–	–	–	–	–	–	+	–	–	–	–
261	*Paraprionospio pinnata*	Spionidae	Sedentaria	–	–	–	–	–	–	–	+	–	–	–	–
262	*Parheteromastus tenuis*	Capitellidae	Sedentaria	–	–	–	–	–	–	–	+	–	–	–	–
263	*Pectinaria crassa*	Pectinariidae	Sedentaria	–	–	–	–	+	–	–	–	–	–	–	–
264	*Pelagobia longicirrata*	Lopadorrhynchidae	Errantia	–	–	–	–	+	–	–	–	–	–	–	–
265	*Perinereis cavifrons*	Nereididae	Errantia	–	–	–	–	–	–	+	+	–	+	+	–
266	*Perinereis cultrifera*	Nereididae	Errantia	–	–	–	–	+	–	–	+	–	–	–	–
267	*Perinereis falsovariegata*	Nereididae	Errantia	+	–	+	–	+	–	–	–	–	–	–	–
268	*Perinereis helleri*	Nereididae	Errantia	–	–	–	–	–	–	–	+	–	–	–	–
269	*Perinereis nigropunctata*	Nereididae	Errantia	+	–	–	–	–	–	–	+	–	–	–	–
270	*Perinereis nuntia*	Nereididae	Errantia	–	–	–	–	–	–	–	+	–	–	–	–
271	*Perinereis nuntia bombayensis*	Nereididae	Errantia	–	–	–	–	–	–	–	+	–	–	–	–

272	*Perinereis nuntia brevicirris*	Nereididae	Errantia	–	–	–	–	–	–	–	+	–	–	–	–
273	*Perinereis nuntia typica*	Nereididae	Errantia	–	–	–	–	–	–	–	+	–	–	–	–
274	*Perinereis nuntia vallata*	Nereididae	Errantia	–	–	–	–	–	–	–	+	–	–	–	–
275	*Perinereis* sp.	Nereididae	Errantia	+	–	+	+	+	–	–	+	–	+	+	–
276	*Perinereis vancaurica*	Nereididae	Errantia	–	–	–	–	–	–	–	+	–	–	–	–
277	*Petaloproctus terricolus*	Maldanidae	Sedentaria	–	–	–	–	–	–	–	+	–	–	–	–
278	*Pherusa monroi*	Flabelligeridae	Sedentaria	–	–	–	–	+	–	–	–	–	–	–	–
279	*Phyllochaetopterus socialis*	Chaetopteridae	Sedentaria	–	–	–	–	–	–	–	+	–	–	–	–
280	*Phyllochaetopterus* sp	Chaetopteridae	Sedentaria	–	–	–	–	–	–	–	–	–	–	–	–
281	*Phyllodoce longipes*	Phyllodocidae	Errantia	–	–	–	–	+	–	–	–	–	–	–	–
282	*Phyllodoce madeirensis*	Phyllodocidae	Errantia	+	–	–	–	+	–	–	+	–	–	–	–
283	*Phyllodoce malmgreni*	Phyllodocidae	Errantia	–	–	–	–	+	–	–	–	–	–	–	–
284	*Phyllodoce* sp.	Phyllodocidae	Errantia	–	–	–	–	–	–	–	–	–	+	+	–
285	*Phyllodoce tenuiss*	Phyllodocidae	Errantia	–	–	–	–	+	–	–	–	–	–	–	–
286	*Phyllodoce tubicola*	Phyllodocidae	Errantia	–	–	–	–	+	–	–	–	–	–	–	–
287	*Pisione africana*	Sigalionidae	Errantia	–	–	–	–	+	–	–	–	–	–	–	–
288	*Pisionidens indica*	Sigalionidae	Errantia	–	–	–	–	+	–	–	+	–	–	–	–
289	*Pista cristata*	Terebellidae	Sedentaria	+	–	+	+	+	–	–	–	–	–	–	–
290	*Pista herpini*	Terebellidae	Sedentaria	–	–	–	–	–	–	–	+	–	–	–	–
291	*Pista indica*	Terebellidae	Sedentaria	–	–	–	–	–	–	–	+	–	+	+	–

(continued)

Table 14.1 (continued)

S. No	Polychaete species	Family	Group	East coast						West coast					
				West Bengal	Andaman and Nicobar Islands	Andra Pradesh	Odisha	Tamil Nadu	Pondicherry	Gujarat	Maharashtra	Goa	Kerala	Karnataka	Daman and Diu
292	*Pista pachybranchiata*	Terebellidae	Sedentaria	–	–	–	–	–	–	–	+	–	–	–	–
293	*Pista quadrilobata*	Terebellidae	Sedentaria	–	–	–	–	+	–	–	–	–	–	–	–
294	*Pista typha*	Terebellidae	Sedentaria	–	–	–	–	+	–	–	–	–	–	–	–
295	*Platynereis calodonta*	Nereididae	Errantia	–	–	–	–	+	–	–	–	–	–	–	–
296	*Platynereis dumerilii*	Nereididae	Errantia	+	–	+	+	+	–	–	–	–	–	–	–
297	*Podarke angustifrons*	Hesionidae	Errantia	–	–	–	–	–	–	–	+	–	–	–	–
298	*Poecilochaetus serpens*	Poecilochaetidae	Sedentaria	–	–	–	–	+	–	–	+	–	–	–	–
299	*Polycirrus plumosus*	Terebellidae	Sedentaria	–	–	–	–	+	–	–	–	–	–	–	–
300	*Polycirrus tribullata*	Terebellidae	Sedentaria	–	–	–	–	+	–	–	–	–	–	–	–
301	*Polydora capensis*	Spionidae	Sedentaria	–	–	–	–	–	–	–	+	–	–	–	–
302	*Polydora ciliata*	Terebellidae	Sedentaria	–	–	–	–	+	–	–	+	–	–	–	–
303	*Polydora hophura*	Terebellidae	Sedentaria	–	–	–	–	+	–	–	–	–	–	–	–
304	*Polydora kempi*	Terebellidae	Sedentaria	+	–	–	–	–	–	–	–	–	–	–	–
305	*Polydora* sp.	Terebellidae	Sedentaria	+	–	+	+	+	–	–	–	–	+	+	–
306	*Polydra ciliata*	Terebellidae	Sedentaria	–	–	–	–	+	–	–	–	–	–	–	–
307	*Polyphysia crassa*	Scalibregmatidae	Sedentaria	+	–	–	+	+	–	–	–	–	–	–	–

308	*Pomatoceros triqueter*	Serpulidae	Sedentaria	–	–	–	–	+	–	–	–	–	–	–	–
309	*Pomatoleios kraussii*	Serpulidae	Sedentaria	+	–	+	+	+	–	–	–	–	–	–	–
310	*Potamilla leptochaeta*	Sabellidae	Sedentaria	–	–	–	–	–	–	–	+	–	–	–	–
311	*Praxillella affinis pacifica*	Maldanidae	Sedentaria	–	–	–	–	–	–	–	+	–	–	–	–
312	*Prionospio aucklandica*	Spionidae	Sedentaria	–	–	–	–	–	–	–	+	–	–	–	–
313	*Prionospio cirrifera*	Spionidae	Sedentaria	+	–	–	–	+	–	–	+	–	+	+	–
314	*Prionospio cirrobranchiata*	Spionidae	Sedentaria	+	–	+	+	+	–	–	–	–	–	–	–
315	*Prionospio malmgreni*	Spionidae	Sedentaria	–	–	–	–	+	–	–	–	–	–	–	–
316	*Prionospio pinnata*	Spionidae	Sedentaria	+	–	+	+	+	–	–	–	–	+	+	–
317	*Prionospio polybranchiata*	Spionidae	Sedentaria	–	–	–	–	+	–	–	+	–	–	+	–
318	*Prionospio saldanha*	Spionidae	Sedentaria	+	–	+	+	+	–	–	–	–	–	–	–
319	*Prionospio sexoculata*	Spionidae	Sedentaria	+	–	+	+	+	–	–	–	–	–	–	–
320	*Prionospio* sp.	Spionidae	Sedentaria	+	–	–	–	+	–	–	–	–	–	–	–
321	*Protodorvillea biarticulata*	Dorvilleidae	Errantia	–	–	–	–	–	–	–	+	–	–	–	–
322	*Protodorvillea egena*	Dorvilleidae	Errantia	–	–	–	–	+	–	–	–	–	–	–	–
323	*Protula tubularia*	Serpulidae	Sedentaria	–	–	–	–	–	–	–	+	–	–	–	–
324	*Pseudonereis variegata*	Nereididae	Errantia	–	–	–	–	+	+	–	–	–	–	–	–

(continued)

Table 14.1 (continued)

S. No	Polychaete species	Family	Group	East coast						West coast					
				West Bengal	Andaman and Nicobar Islands	Andra Pradesh	Odisha	Tamil Nadu	Pondicherry	Gujarat	Maharashtra	Goa	Kerala	Karnataka	Daman and Diu
325	*Pseudopolydora kempi*	Nereididae	Errantia	–	–	–	–	+	–	–	–	–	–	–	–
326	*Pulliella armata*	Capitellidae	Sedentaria	–	–	–	–	–	–	+	+	–	+	–	–
327	*Rhodine gracilior*	Maldanidae	Sedentaria	–	–	–	–	+	–	–	–	–	–	–	–
328	*Sabella* sp.	Sabellidae	Sedentaria	+	–	+	+	+	–	–	–	–	–	–	–
329	*Sabella spallanzanii*	Sabellidae	Sedentaria	–	–	–	–	–	–	–	+	–	–	–	–
330	*Sabellaria cementarium*	Sabellidae	Sedentaria	–	–	–	–	+	–	–	–	–	–	–	–
331	*Sabellaria intoshi*	Sabellidae	Sedentaria	+	–	+	+	+	–	–	–	–	–	–	–
332	*Sabellaria pectinata*	Sabellidae	Sedentaria	+	–	–	–	–	–	–	–	–	–	–	–
333	*Sabellaria* sp.	Sabellidae	Sedentaria	–	–	–	–	–	–	–	–	–	–	+	–
334	*Sabellaria spimnulosa*	Sabellidae	Sedentaria	–	–	–	–	+	–	–	–	–	–	–	–
335	*Sabellastarte longa*	Sabellidae	Sedentaria	–	–	–	–	–	–	–	+	–	–	–	–
336	*Salmacina dysteri*	Serpulidae	Sedentaria	–	–	–	–	–	–	–	+	–	–	–	–
337	*Scolelepis squamata*	Spionidae	Sedentaria	+	–	+	–	+	–	–	+	–	–	–	–
338	*Scoloplella capensis*	Spionidae	Sedentaria	–	–	–	–	+	–	–	+	–	–	–	–
339	*Scoloplos armiger*	Orbiniidae	Sedentaria	–	–	–	–	–	–	–	+	–	–	–	–
340	*Scoloplos marsupialis*	Spionidae	Sedentaria	–	–	–	–	+	–	–	+	–	–	–	–

341	*Scoloplos* sp.	Spionidae	Sedentaria	–	–	–	–	–	–	–	–	–	–	+	–
342	*Scyphoproctus djiboutiensis*	Capitellidae	Sedentaria	–	–	–	–	–	–	–	+	–	–	–	–
343	*Serpula vermicularis*	Serpulidae	Sedentaria	–	–	–	–	+	–	–	+	–	–	+	–
344	*Sigalion squamatum*	Sigalionidae	Errantia	–	–	–	–	+	–	–	–	–	–	–	–
345	*Sigambra constricta*	Pilargidae	Errantia	–	–	–	–	–	–	–	+	–	–	–	–
346	*Sigambra parva*	Pilargidae	Errantia	–	–	–	–	–	–	–	+	–	–	–	–
347	*Sphaerosyllis erinaceu*	Syllidae	Errantia	–	–	–	–	+	–	–	–	–	–	–	–
348	*Sphaerosyllis sublaevis*	Syllidae	Errantia	–	–	–	–	–	–	–	+	–	–	–	–
349	*Spio bengalensis*	Spionidae	Sedentaria	+	–	–	–	–	–	–	–	–	–	–	–
350	*Spio filicornis*	Spionidae	Sedentaria	+	–	+	+	+	–	–	–	–	–	–	–
351	*Spiochaetopterus costarum*	Spionidae	Sedentaria	–	–	–	–	+	–	–	–	–	–	–	–
352	*Spiophanes bombyx*	Spionidae	Sedentaria	–	–	–	–	–	–	–	+	–	–	–	–
353	*Spiophomianes soderstromi*	Spionidae	Sedentaria	–	–	–	–	+	–	–	–	–	–	–	–
354	*Spirobrachus teraceros*	Serpulidae	Sedentaria	–	–	–	–	–	–	–	–	–	–	–	–
355	*Spirobranchus kraussii*	Serpulidae	Sedentaria	–	–	–	–	–	–	–	+	–	–	–	–
356	*Spirorbis foraminosus*	Serpulidae	Sedentaria	–	–	–	–	+	–	–	+	–	–	–	–
357	*Sternaspis scutata*	Sternaspidae	Sedentaria	+	–	+	+	+	–	–	+	–	–	+	–
358	*Sthenelais boa*	Sigalionidae	Errantia	–	–	–	–	+	–	–	+	–	–	+	–
359	*Sthenelais japonica*	Sigalionidae	Errantia	–	–	–	–	+	–	–	+	–	–	–	–

(continued)

Table 14.1 (continued)

S. No	Polychaete species	Family	Group	East coast						West coast					
				West Bengal	Andaman and Nicobar Islands	Andra Pradesh	Odisha	Tamil Nadu	Pondicherry	Gujarat	Maharashtra	Goa	Kerala	Karnataka	Daman and Diu
360	*Streblosoma Persia*	Terebellidae	Sedentaria	+	–	–	+	+	–	–	–	–	–	–	–
361	*Streblospio benedicti*	Spionidae	Sedentaria	–	–	–	–	–	–	–	+	–	–	–	–
362	*Stylarioides stylarioides*	Terebellidae	Sedentaria	–	–	–	–	+	–	–	–	–	–	–	–
363	*Syllidia armata*	Hesionidae	Errantia	–	–	–	–	+	–	–	–	–	–	–	–
364	*Syllis armillaris*	Syllidae	Errantia	–	–	–	–	–	–	–	+	–	–	–	–
365	*Syllis benguellana*	Syllidae	Errantia	+	–	+	+	+	–	–	–	–	–	–	–
366	*Syllis cornuta*	Syllidae	Errantia	–	–	–	–	–	–	–	+	–	–	–	–
367	*Syllis gracilis*	Syllidae	Errantia	+	–	+	+	+	–	–	+	–	–	–	–
368	*Syllis hyalina*	Syllidae	Errantia	–	–	–	–	–	–	–	+	–	–	–	–
369	*Syllis longocirrata*	Syllidae	Errantia	–	–	–	–	+	–	–	–	–	–	–	–
370	*Syllis* sp.	Syllidae	Errantia	+	–	+	+	+	–	–	–	–	–	–	–
371	*Syllis trifalcata*	Syllidae	Errantia	–	–	–	–	+	–	–	–	–	–	–	–
372	*Syllis variegata*	Syllidae	Errantia	–	–	–	–	–	–	–	+	–	–	–	–
373	*Talehsapia annandalei*	Pilargidae	Errantia	+	–	–	–	–	–	–	–	–	+	–	–
374	*Tambalagamia orientalis*	Nereididae	Errantia	–	–	–	–	–	–	–	+	–	–	–	–
375	*Terebella pterochaeta*	Terebellidae	Sedentaria	–	–	–	–	+	–	–	–	–	–	–	–
376	*Terebellid* sp.	Terebellidae	Sedentaria	–	–	–	–	–	–	–	–	–	–	+	–
377	*Terebellides stroemi*	Terebellidae	Sedentaria	+	–	+	+	+	–	–	+	–	–	+	–

378	*Tharyx filibranchia*	Cirratulidae	Sedentaria	–	–	–	–	–	–	–	+	–	–	–	–
379	*Thelepus setosus*	Terebellidae	Sedentaria	–	–	–	–	+	–	–	–	–	–	–	–
380	*Tylonereis fauveli*	Nereididae	Errantia	–	–	–	–	+	–	–	–	–	–	–	–
381	*Tomopteris helgolandica*	Tomopteridae	Errantia	–	–	–	–	+	–	–	–	–	–	–	–
382	*Travisiopsis lobifera*	Typhloscolecidae	Errantia	–	–	–	–	+	–	–	–	–	–	–	–
383	*Typhloscolex muelleri*	Typhloscolecidae	Errantia	–	–	–	–	+	–	–	–	–	–	–	–
384	*Vanadis Formosa*	Phyllodocidae	Errantia	–	–	–	–	+	–	–	–	–	–	–	–
385	*Vermiliopsis glandigerus*	Sabellidae	Sedentaria	–	–	–	–	–	–	–	+	–	–	–	–

14.6 Unknown/Way Forward

Going through the information stated above, it is quite apparent that detailed studies related to the macrobenthos in general and polychaete diversity, in particular, have been made only less than 60% of the mangroves skirting both the coasts of India. Added to this woe, there is a great shortage of updated taxonomical information and recent literature barring the classical works of Fauvel (Fauvel, 1953) and Day (Day, 1967) to carry out benthic investigations. Advantageously, this group occurs almost in all the mangrove environments. Nevertheless, the only sketchy information is available in many mangroves. To tide over this problem, the following aspects on biodiversity of polychaetes of mangroves need to be given focus:

1. Continuous survey on cataloguing and spatial variability of polychaetes along the mangroves of Indian coast.
2. The extent of their utility in assessing habitat deterioration, especially in a few mangroves like Pulicat stretch, Pichavaram, and Muthupettai mangroves (shrimp farm discharges), Godavari belt in Andhra, etc., is to be studied.
3. Comparison of polychaete diversity in relation to various mangrove zones like *Avicennia*, *Rhizophora*, etc., so as to understand the harboring nature of the mangroves.
4. Information on pollution indicator species (r-selected species/opportunistic species and t-selected species from intensively exploited mangrove environs).
5. Genetic diversity of polychaetes among the various mangroves.
6. Co-variations in polychaete diversity with that of other taxa like nematodes and amphipods and their utility in ecological health assessment.

14.7 Conclusion

Based on the forgoing account, it is concluded that the present review yielded a fairly good amount of information on the benthic biodiversity in general and polychaete diversity in particular in the mangroves of the Indian coast. Further, as there is no comprehensive account of the diversity of polychaetes from the mangroves of the Indian coast, a comparison was done only based on the existing sporadic information and therefore a definite conclusion could not be drawn. On the other hand, reports related to the taxonomy of benthic faunal community are limited as the researchers worldwide do not evince interest in this line besides the enrolment of a young generation of benthic taxonomists has also been poor in the recent past. There are quite a lot of reasons for this: (i) unconcerned perspectives, both in society and educational systems, and (ii) organisms that are "unseen" from the outlook of instant, cost-effective and medical interest to man and more significantly poor financial support from the Government. To accomplish this, there is a dire need to establish a strong collaboration of benthic researchers among the Asian countries is the need of the hour. Once this is done and studies are initiated, certainly this will throw an important beam of light on the Polychaete taxonomy in the mangroves with a scope to develop management strategies as well as to attain momentous conclusions for the policy makers. The conservation troubles pertaining to this community can also be addressed when biosystematics studies are well recognized.

Acknowledgments The authors are thankful to the Director and Dean, Centre of Advanced Study in Marine Biology, Annamalai University, for the encouragement and facilities provided. We gratefully acknowledge the Ministry of Environment, Forest and Climate Change, Govt. of India, New Delhi, for the financial support (MoEn&F No:22-11/2009-CS.I).

References

Achary GP, Chakravarty G, Chakraborty SK, Jayasurya PK, Sarala Devi K (2005) Benthos-polychaetes. CMFRI Special Publication Mangrove ecosystems: a manual for the assessment of biodiversity. 83: 25–145.

Ankar S (1977) Digging profile and penetration of the Van Veen grab in different sediment types. University of Stockholm, Department of Zoology and Askö Laboratory. 16: 3–22.

Ashton EC (1999) Biodiversity and community ecology of mangrove plants, molluscs and crustaceans in two mangrove forests in Peninsular Malaysia in relation to management practices. Ph.D. Thesis. University of York, UK.

Bandekar PD, Naik UG, Haragi SB (2017). Diversity status of benthic macro polychaetes species in estuarine region of Karwar, West Coast of India. Int J Fish Aqua Stud 5(1): 216–219.

Brunel P (2005). Visages de la Biodiversite marine. Vertigo 6: 1–13.

Eleftheriou A, Holme NA (1984) Macro fauna techniques. Methods for the study of marine benthos. p 140–216.

Fauchald K (1977) The polychaete worms. Definitions and keys to the orders, families and genera. Nat Hist Mus LA County Mus Sci Ser 28: 1–190.

FSI (2017) India state of forest report. Forest Survey of India, Dehradun, p 55–61.

Gage JD (1975) A comparison of the deep-sea epibenthic sledge and anchor-box dredge samplers with the van Veen grab and hand coring by diver. Deep-Sea Res Oceanogr Abstr 22(10): 693–702.

Jennerjahn TC, Ittekkot V (2002) Relevance of mangroves for the production and deposition of organic matter along tropical continental margins. Naturwissenschaften 89(1): 23–30.

Kathiresan K, Quasim SZ (2005) Biodiversity of mangrove ecosystems (p. 251). New Delhi: Hindustan Publishing Corporation.

Kasinathan R, Shanmugam A (1986) Benthic macrofauna of the Pichavaram mangroves, South India. J Annamalai University Sci 34: 109–119.

Kathiresan K (2000) A review of studies on Pichavaram mangrove, southeast coast of India. Hydrobiol 430(1–3): 185–205.

Khan SA, Murugesan P (2005) Polychaete diversity in Indian estuaries. Indian J Geo-Mar Sci 34 (1): 114–119.

Krom MD, Berner RA (1980) Adsorption of phosphate in anoxic marine sediments. Limnol Oceanogr 25(5): 797–806.

Kumar, RS (1995) Macrobenthos in the mangrove ecosystem of Cochin backwaters, Kerala (southwest coast of India). Indian J Geo-Mar Sci 24(2): 56–61.

Kumar R (1999) New record of five annelids (Class: Polychaeta) from the mangrove habitat of the south west coast of India. J Mar Biol Assoc India 41(1&2): 116–118.

Kumar R (2001) Biodiversity and affinity of polychaetous annelids within the mangrove ecosystem of Indo-pacific region. J Mar Biol Ass India 43(1&2): 206–213.

Kumar RS (2003) A checklist of polychaete species some mangroves of Asia. Zoos' Print J 18: 1017–1020.

Kumar RS, Antony A (1994) Impact of environmental parameters on polychaetous annelids in the mangrove swamps of Cochin, South West coast of India. Indian J Geo-Mar Sci 23(3): 137–142

Kumar PS, Khan AB (2013). The distribution and diversity of benthic macro invertebrate fauna in Pondicherry mangroves, India. Aqua Biosyst 9(1): 15.

Macintosh DJ (1984) Ecology and productivity of Malaysian mangrove crab populations (Decapoda: Brachyura). Proceedings of the Asian Symposium on Mangrove Environmental Research and Management 354–377.

Manson FJ, Loneragan NR, Skilleter GA, Phinn SR (2005) An evaluation of the evidence for linkages between mangroves and fisheries: a synthesis of the literature and identification of research directions. Oceanogr Mar Biol 43: 483.

Misra A, Choudhury A (1985) Polychaetous annelids from the mangrove swamps of Sunderbans, India. In Pro Nat Symp Biol Util Cons Mangroves (Ed. LJ Bhoslae), p 448–452.

Muniasamy M, Muthuvelu S, Balachandar K, Murugesan P (2013) Diversity of benthic fauna in Coleroon estuary, south east coast of India. Inter J Recent Sci Res 4(10): 1617–1621.

Murugesan P, Pravinkumar M, Muthuvelu S, Ravichandran S, Vijayalakshmi S, Balasubramanian T (2016) Benthic biodiversity in natural vis-a-vis artificially developed mangroves of south east coast of India. Indian J Geo-Mar Sci 45(8): 1049–1058.

Murugesan P, Sarathy PP, Muthuvelu S, Mahadevan G (2018) Diversity and distribution of polychaetes in mangroves of east coast of India. In: Sahadev Sharma (ed) Mangrove ecosystem ecology and function, IntechOpen, UK, p 107–130.

Nandi S, Choudhury A (1983) Quantitative studies on the benthic macrofauna of Sagar Island, intertidal zones, Sunderbans, India. Mahasagar 16(3): 409–414.

Padmakumar KG (1984) Ecology of a mangrove swamp near Juhu Beach, Bombay with reference to sewage pollution. Unpublished PhD thesis submitted to the University of Bombay.

Pamungkas J, Glasby CJ, Read GB, Wilson SP, Costello MJ (2019) Progress and perspectives in the discovery of polychaete worms (Annelida) of the World. Helgol Mar Res 73(1): 4.

Parulekar AH (1971) Polychaetes from Maharashtra and Goa. J Bombay Nat History Soc 68: 726–749.

Parvez S, Long MJ, Poganik JR, Aye Y (2018) Redox signaling by reactive electrophiles and oxidants. Chem Rev 118(18): 8798–8888.

Pati SK, Swain D, Sahu KC, Sharma RM (2015) Diversity and distribution of polychaetes (Annelida: Polychaeta) along Maharashtra Coast, India. In: Rawat M, Dookia S, Sivaperuman C (eds) Aquatic ecosystem: biodiversity, ecology and conservation, Springer, New Delhi, p 53–65.

Pearson TH, Rosenberg R (1978) Macrobenthic succession in relation to organic enrichment and pollution of the marine environment. Oceanogr Mar Biol 16: 229–311.

Pfannkuche O, Thiel H (1988) Sample processing. Introduction to the study of meiofauna, 9: 134–145.

Prabakaran JR, Vibin A, Deivakumari M, Muruganantham M, Ramasubburayan R, Palavesam A, Immanuel G (2019) Comparison of polychaete diversity and distribution along the south Tamil Nadu coast (Lat. 8.08∘ to 10.79∘ N), India. Regional Stud Mar Sci 28: 100564.

Pravinkumar M, Murugesan P, Prakash RK, Elumalai V, Viswanathan C, Raffi SM (2013) Benthic biodiversity in the Pichavaram mangroves, Southeast Coast of India. J Oceanog Mar Sci 4(1): 1–11.

Rao VM, Kumar AT, Ghosh S (2015) Studies on the mangrove macro faunal diversity and assessment among different sites in port Blair Bay, South Andaman Islands. Fisher Aqua J 6 (2): 1.

Raveenthiranath N (1990) Ecology of macrobenthos in and around Mahandrapalli region of Coleroon estuary, Southeast coast of India. Ph.D. thesis, Annamalai University, India: 231.

Rosenberg DM (1978) Practical sampling of freshwater macro zoobenthos: a bibliography of useful texts, reviews, and recent papers. Canadian Fisheries and Marine Service Technical Report No. 790, p 15.

Rumohr H (1990) Soft bottom macrofauna: Collection and treatment of samples. ICES Tech Mar Environ Sci p 19.
Samidurai K, Saravanakumar A, Kathiresan K (2012) Spatial and temporal distribution of macrobenthos in different mangrove ecosystems of Tamil Nadu Coast, India. Environ Monit Assess 184(7): 4079–4096.
Saravanakumar A, Serebiah JS, Thivakaran GA, Rajkumar M (2007) Benthic macrofaunal assemblage in the arid zone mangroves of Gulf of Kachchh-Gujarat. J Ocean U China 6(3): 303–309.
Satish KM, Adsul AD, Indulkar ST, Pai R (2018) Seasonal variation in polychaetes of intertidal mangrove area of Shirgaon, Ratnagiri, Maharashtra, India. Int J Life Sci 6(1): 156–160.
Sasekumar A (1974) Distribution of macrofauna on a Malayan mangrove shore. J Anim Ecol 43: 51–69.
Sekar V, Prithiviraj N, Savarimuthu A, Rajasekaran R (2013) Macrofaunal assemblage on two mangrove ecosystems, southeast coast of India. Inter J Recent Sci Res 4(5): 530–535.
Smith TJ, Boto KG, Frusher SD, Giddins R (1991) Keystone species and mangrove forest dynamics: the influence of burrowing by crabs on soil nutrient status and forest productivity. Estuar Coast Shelf Sci 33: 19–32.
Thilagavathi B, Varadharajan D, Babu A, Manoharan J, Vijayalakshmi S, Balasubramanian T (2013) Distribution and diversity of macrobenthos in different mangrove ecosystems of Tamil Nadu Coast, India. J Aqua Res Develo 4(6): 1–12.
Wang CA, Hu G, Sun P, Gu W, Wang B, Xu Q, Liu H (2018) Effects of dietary protein at two lipid levels on growth, gonadal development, body composition and liver metabolic enzymes of brown trout (*Salmo trutta fario*) broodstock. Aqua Nutr 24(5): 1587–1598.
Ward TJ, Hutchings PA (1996) Effects of trace metals on infaunal species composition in polluted intertidal and sub-tidal marine sediments near a lead smelter, Spencer Gulf, South Australia. Mar Ecol Prog Ser 135: 123–135.
Warren JH, Underwood AJ (1986) Effects of burrowing crabs on the topography of mangroves swamps in New South Wales. J Exp Mar Biol Eco 102: 223–235.

Structure and Diversity of Plants in Mangrove Ecosystems

15

Nurun Nadhirah Md Isa and Mohd Nazip Suratman

Abstract

Mangroves are among the most productive, prominent and complex ecosystems which comprised of salt-tolerant trees and shrubs. These complex ecosystems are found between the latitudes of 32° north and 38° south, along the tropical coast of Africa, Australia, Asia and Americas. Mangrove forests are rich with diversity of flora and fauna. However, these unique ecosystems also challenged with destruction as a result from a variety of human activities such as aquaculture and effects from global climate change. This review paper highlights the distribution of world mangroves, including their structure and diversity as well as the threats facing by them. It is recommended that the mangrove ecosystems should always be protected and restored in order to sustain these ecosystems for biodiversity conservation and other ecosystem services.

Keywords

Conservation, Diversity · Mangrove · Structure · Threat

N. N. Md Isa
Faculty of Applied Sciences, Universiti Teknologi MARA (UiTM) Pahang Branch, Jengka, Pahang, Malaysia

M. N. Suratman (✉)
Faculty of Applied Sciences and Institute for Biodiversity and Sustainable Development, UiTM, Shah Alam, Malaysia
e-mail: nazip@uitm.edu.my

R. P. Rastogi et al. (eds.), *Mangroves: Ecology, Biodiversity and Management*,
https://doi.org/10.1007/978-981-16-2494-0_15

15.1 Introduction

Mangrove forest is one of the primary features of coastal ecosystems throughout the tropical and subtropical regions of the world. It has been previously described as "Coastal Woodland" and "Intertidal forest" (Aksornkoae 1995). Joshi and Ghose (2014) referred mangrove as a group of taxonomically heterogeneous woody shrubs and trees growing in the intertidal zone of tropical and subtropical coasts. Usually, mangrove vegetations are dicotyledonous woody shrubs or trees that are virtually confined to the tropics. They often form dense forests that dominate intertidal muddy shores, frequently consisting of virtually monospecific patches (Hogarth 2007).

The vascular plants in mangrove forests have special morphological, physiological and other non-visible adaptations to live in a saline intertidal environment dominated by low dissolved oxygen or sometimes anoxic fine sediments. These plants, composed with their complement of microorganisms and animals, form the mangrove ecosystem. The term mangrove thus refers both to the plants themselves as well as to the ecosystem. Regularly, plants which occur in the non-mangrove ecosystem and with none or only a few of these morphological adaptations are also found in the mangrove forests (Ong and Gong 2013). This chapter reviews the structure and composition of mangrove ecosystem along with the species diversity and distribution in different locations in the world.

15.2 Distribution of World Mangroves

Mangroves are located in a highly productive intertidal zone in more than 120 countries in the tropics and subtropics (Gandhi and Jones 2019). Worldwide, there are a total of 114 species of true mangroves belonging to 66 genera with species richness being greatest in the Indo-Pacific region (Tomlinson 1986). Figure 15.1 shows mangrove distribution zones and the number of mangrove species within each region. The most diverse mangrove species and largest mangrove areas in the world is in Southeast Asia with 6.8 million hectares. The largest areas of mangrove are found in Indonesia, Malaysia, Myanmar, Papua New Guinea and

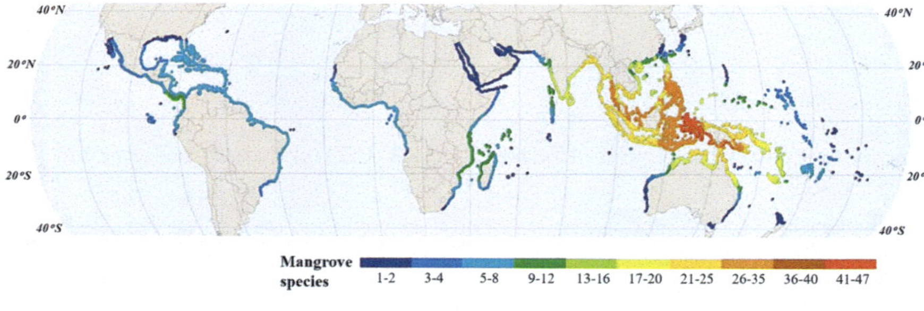

Fig. 15.1 World map of the mangrove distribution zones and the number of mangrove species along each region (Deltares 2014)

Thailand (Faridah-Hanum et al. 2012). Most of the largest mangrove areas, particularly in Asia, are in the deltas of major rivers. Examples include the Indus delta of Pakistan, the Sundarbans where the Ganges and Brahmaputra Rivers flow into the Bay of Bengal, the Merbok and Matang deltas of Peninsular Malaysia, the Fly River of Papua New Guinea and, in South America, the mangroves of the Amazon delta.

Thirty-five percent of the total 18 million ha of global mangrove forests are found in the Southeast Asian countries of Brunei Darussalam, Cambodia, Indonesia, Malaysia, Myanmar, the Philippines, Thailand and Vietnam. Indonesia alone has 4.5 million ha of mangroves (Primavera et al. 2000). Characteristically, they have a low tidal range and strong freshwater flow carrying a considerable load of sediment, much of which is deposited in the mangroves (Hogarth 2007).

15.3 Structure of Mangrove Ecosystems

Generally, mangroves can be divided into two groups which are exclusive and non-exclusive. Exclusive mangroves are restricted to intertidal areas, while non-exclusive mangroves are distributed in terrestrial or aquatic but also occur in the typical mangrove's environment. True or exclusive mangrove only can be found in certain habitats of mangroves (Hogarth 2007). While, non-exclusive mangrove also recognized as semi-mangrove, back mangrove or mangrove associates (Wang et al. 2011). Mangrove forests literally live in two worlds at once, acting as the interface between land and sea.

Adaptation to salt tolerance is of three types which are salt excluders, salt secretors and salt accumulators. The salt excluders are from the members of the genera like *Rhizophora*, *Bruguiera* and *Ceriops* from the family of Rhizophoraceae. The species in the genera *Acanthus*, *Aegialitis*, *Aegiceras* and *Avicennia* have salt-secreting glands on the leaf surface. The species like *Sonneratia*, *Xylocarpus* and *Excoecaria* are the salt accumulators. Various types of root adaptations of mangroves in the habitat are lenticels (*Bruguiera* spp.), pneumatophores (*Sonneratia* spp. and *Avicennia* spp.), knee roots (*Bruguiera* spp.), cable roots (*Avicennia* spp.), and stilt roots (*Rhizophora* spp.) (Ong and Gong 2013).

Zonation in mangrove forests often has been attributed to the responses of individual species to variation in degree of tidal inundation, salinity or other measurable edaphic gradients that vary across the intertidal (Ellison et al. 2000). Mangrove species zonation can be considered at different scales. On a tide-dominated shore, a clear vertical sequence of species often appears. An example of a general mangrove profile is shown in Fig. 15.2. One of mangrove main species, *Avicennia* often has a bimodal distribution, being abundant near the seaward margin and some way up to the shore areas. *Rhizophora* is distributed next to the seaward zone of the profile, followed by *Bruguiera* and *Ceriops*.

Mangrove vegetation in Southeast Asia may range from 1 to 2 m tall. *Avicennia alba* and *A. marina* stands on the seaward side of accreting shores can achieve up to 30 – 40 m tall stands along with the mixed species of *Bruguiera-Rhizophora* stands. On more exposed areas, but not eroding coastlines, one may find *Sonneratia alba*

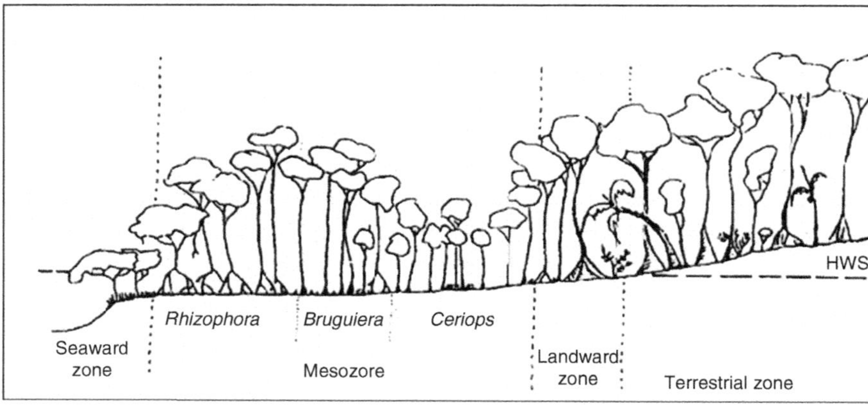

Fig. 15.2 Profile of a shore in north-eastern Australia (an area of high rainfall) that shows mangrove zonation (Tomlinson 1986)

and *A. alba*. While, along waters of lower salinity *Nypa fruticans*, *Cerbera odollam* and *S. caseolaris* are common. Apart from saplings, undergrowth has been often scarce, but certainly not absent, and species such as sea holly *Acanthus ilicifolius* and mangrove fern, *Acrostichum aureum* may be common along the banks of streams and in disturbed area. However, mangrove vegetation structure may vary depending on the location of the mangrove ecosystem, management and disturbance (Sarno et al. 2015). Moreover, climatic factors are the key in controlling the worldwide distribution of mangroves including topography, soil properties and tide fluctuations (Abou Seedo et al. 2017).

15.4 Diversity of Mangrove Plants

Mangroves have a diverse group of trees and shrubs that flourish in flooded and saline habitats (Hogarth 2007). Diversity is a community that has different characteristics with other communities. The characteristics of the community in an environment are biodiversity, the more diverse the biotic component (biodiversity), the higher the diversity. The greatest diversity of mangrove plant species exists in Southeast Asia. According to Giesen et al. (2006), the largest plant families recorded in Southeast Asia mangroves are Rhizophoraceae and followed by Orchidaceae and Asclepiadaceae. The smallest number of species was from Sonneratiaceae family with five species. According to Tomlinson (1986), Rhizophoraceae is the most dominant family because it has widespread distribution globally. Moreover, the family has also the adaptability to both extreme as well as non-extreme conditions. Therefore, in terms of floristic composition, species in this family is common and easily found in mangrove habitats.

From the total of 52 mangrove species recorded globally, 48 of the species occur in Indonesia, which is considered to be the most diverse of the Southeast Asia

Table 15.1 The true mangroves and mangrove associates endemic species to Southeast Asia

No.	Botanical Name	Family	Groups
1.	*Aegiceras floridum*	Primulaceae	True mangroves
2.	*Avicennia eucaltyptifolia*	Acanthaceae	True mangroves
3.	*Avicennia lanata*	Acanthaceae	True mangroves
4.	*Azima sarmentosa*	Salvadoraceae	Mangrove associates
5.	*Barringtonia conoidea*	Lecythidaceae	Mangrove associates
6.	*Blumeodendron tokbrae*	Euphorbiaceae	Mangrove associates
7.	*Camptostemon philippinense*	Malvaceae	Mangrove associates
8.	*Croton heterocarpus*	Euphorbiaceae	Mangrove associates
9.	*Fagraea crenulata*	Gentianaceae	Mangrove associates
10.	*Gluta velutina*	Anacardiaceae	Mangrove associates
11.	*Heritiera globosa*	Malvaceae	Mangrove associates
12.	*Ilex cymosa*	Aquifoliaceae	Mangrove associates
13.	*Ilex maingayi*	Aquifoliaceae	Mangrove associates
14.	*Ixora timoriensis*	Rubiaceae	Mangrove associates
15.	*Podocarpus polystachyus*	Podocarpaceae	Mangrove associates
16.	*Ochthocharis borneensis*	Melastomataceae	Mangrove associates
17.	*Quassia harmandiana*	Simaroubaceae	Mangrove associates
18.	*Rapanea porteriana*	Myrsinaceae	Mangrove associates
19.	*Scolopia macrophylla*	Flacourtiaceae	Mangrove associates
20.	*Serianthes grandiflora*	Leguminosae	Mangrove associates
21.	*Sindora siamensis*	Leguminosae	Mangrove associates
22.	*Symplocos celastrifolia*	Symplocaceae	Mangrove associates

Source: Suratman (2008)

countries, followed by Malaysia, with 42 species. The 35 mangrove species occurring in Southeast Asia are reported as uncommon or rare. Around 18 percent (51 species) of mangrove flora of Southeast Asia are endemic to the region, and includes 22 trees and shrubs, 13 epiphytes, eight ferns, four palms and climbers, respectively. Table 15.1 shows the list of true mangroves and mangrove associates species that are endemic in Southeast Asia. According to Tomlinson (1986), true mangroves refer to as plants species that occur only in mangrove forests, play a major role of the mangrove community, are not found in the terrestrial communities and have morphological specializations to the mangrove environment (i.e. aerial roots, vivipary) and some mechanism for salt exclusion. In contrast, the group of species that belong to the mangrove associates do not the criteria of true mangroves specified by Tomlinson above. These plants are mostly representing non-herbaceous, sub-woody and climbers that are found mostly in the regions adjoining the tidal periphery of mangrove forests. Most of them are either naturally or accidentally dispersed from next the adjacent forest types such as beach forests and lowland forests.

Malaysia has approximately 645,852 ha of mangroves which is the third largest in the Asia Pacific region. Peninsular Malaysia consists of 38 exclusive and

57 non-exclusive and associate mangrove species (Rozainah and Mohamad 2006). Mangrove species from *Rhizophora*'s genus is abundantly found throughout the Peninsula Malaysia and the Borneo island of Malaysia (Mojiol et al. 2019).

Rapid development in Singapore has affected the adjacent wetland area of the island. The areas have been destroyed and degraded due to mangrove area conversion into land development. The mangroves have long been reported to be undervalued for their utilitarian and intrinsic values. For example, only about 1% (6 km^2 out of 647.8 km^2) mangrove area remains from the original area of 13% (Hsiang 2000). Findings from the various studies on the present extent of mangrove forests show inconsistencies. For example, Hsiang (2000) reported that seven mangrove species were extinct in Singapore in the last half-century. In another study, Hsiang (2000) through his personal communication with Hugh Tan of National University of Singapore reported that only four mangrove species, i.e., *Barringtonia conoidea*, *Ochthocharis borneensis*, *Bruguiera sexangular* and *Brownlowia argentata*, were extinct. However, Shufen et al. (2012) reported only one species extinct, which is *B. argentata* whereas a total 35 true mangrove species can still be found in Singapore.

In Brunei, plant species composition in the mangroves was monitored using the Advanced Land Observation Satellite (ALOS). From the observation, it was reported that the total extent of mangrove cover was found to be 35,183.74 ha, of which Weston and Menumbok areas occupied more than two-folds (58%), followed by Sundar (27%) and Limbang (15%). The medium resolution of ALOS data was efficient for mapping dominant mangrove species such as *Nypa fruticans*, *R. apiculata*, *S. caseolaris*, *S. alba* and *Xylocarpus granatum* in the vicinity with classification accuracy of 80% (Satyanarayana et al. 2018).

In Thailand, mangrove forests occur on the muddy tidal flats at the river mouths and along the coast of southern and eastern parts of the country, both on the extensive coasts along the Gulf of Thailand and on the Andaman Sea one, where they are most heavily concentrated. Large mangrove stands are found along the Chao Phraya delta. Mangroves form a two-storeyed forest, with an upper layer generally growing up to 20 m in height, dominated by *Rhizophora apiculata* and, to a lesser extent, *R. mucronata*, *Heritiera littoralis* and *Xylocarpus mekongensis*. Common species of the lower layer are *Bruguiera cylindrica*, *B. parviflora*, *B. sexangula*, *Ceriops decandra* and *C. tagal*. *B. gymnorrhiza* is a common emergent of up to 40 m in height and 2 m in girth (FAO 2005). Landwards, where mud has accumulated, dryer soils are overgrown with ferns and herbs and can give way to evergreen forest. On the edge of stream *N. fruticans* is common.

Indonesia is one of the countries with an extensive mangrove forests in the world. In addition, Indonesia has a very high level of mangrove diversity. In Indonesia, *N. fruticans* also displays high density structure in Sembilang and Bungin Rivers in Sumatra. Twelve species of mangroves also found in both areas (Sarno et al. 2015). Indonesia comprised largest mangrove areas with approximately 31,890 km^2 in Southeast Asia. In Kumbewaha, Buton Island in Indonesia, 20 species and 17 mangrove tribes are commonly found. In terms of plant diversity, *R. mucronata* recorded the highest diversity value while the lowest is the *S. ovata* (Iksan et al. 2019).

Panda et al. (2017) reported that in Bhitarkanika National Parks, India a total 29 true mangrove species and 72 associate species were found. The recorded true mangroves belong to 11 families and 15 genera, and the associates were recorded from 39 families and 56 genera. Among the true mangrove families, Rhizophoraceae showed maximum richness both at species and generic levels with 10 true mangrove species.

15.5 Threat to Mangrove Forests

Despite the important socio-economic and environmental roles offered by mangrove forests, they are facing the same threats as terrestrial forests. Recently, there have been an increased demand of land for development, settlement and aquaculture activities by the surrounding communities that have resulted in overharvesting mangrove poles for piling that has made mangroves becoming plant in peril and degrading at an alarming rate. The conversion of mangroves has led to the destruction which is believed to be exceeding the terrestrial forest such as lowland and hill forests. The destruction of mangrove forests could be a catastrophic disaster especially to the local community livelihoods that rely on mangrove resources. If not handled wisely, the mangrove destruction can create a new source of carbon in the atmospheres.

From the past decade, mangroves received constant threats from the global climate change factors. The rapid anthropogenic activities coupled with sea level rises, storm and tsunami seem to be the new menace that could be the catalyst for the destruction of mangroves worldwide. The ad-hoc changes due to anthropogenic disturbance have altered the mangrove habitats entirely. The relevant stakeholders and whole communities need to make sure that these phenomena need to address and given serious attention to ensure these majestic ecosystems survive.

Changes in mangroves have been reported in many studies throughout the world. For instance, in Malaysia, Khuzaimah et al. (2013) in their study at Merbok mangrove forests indicated that during the period of 2000–2010 found that mangroves have experienced loss about 1246 ha due to an intense aquaculture farming, land development and agriculture sector. Another mangroves study conducted by Suratman and Ahmad (2012) in Pulau Indah, Selangor using Landsat images from 1995–1999 and 1999–2005 found that the net reduction during the two intervals were 2.73% and 0.24%, per annum, respectively. The study concluded that the main reasons for the decreased of were caused by rapid land development (port and ship terminals) and land settlement. Meanwhile, a study that was conducted in Johor, Malaysia using the remote sensing approach found out that between 1989 and 2014, the estimated loss of mangroves in the area was about 6030 ha (Kanniah et al. 2015). The underlying causes reported were uncontrollable land development, aquaculture activities, intensified erosion and urbanization.

Mangrove is the host with a wide variety of biodiversity, providing habitats for flora and fauna, including aquatic and terrestrial insects, fish, crustacean, mammalian, amphibian, reptilian and avian species (Thomas et al. 2017). Although mangroves contribute many returns to other living creatures, this ecosystem, the ecosystems also face their own threats and degradation that affect the socio-economic livelihood fulfillment. Therefore, understanding and awareness on the mangrove structure and diversity may provide insights for this precious ecosystem to be protected.

15.6 Conclusion

The rich mangrove ecosystems provide support to the people for their livelihoods and planet against global warming in the unique ways by preserving habitats for flora and fauna and providing protection against storm and flooding. Protecting mangrove forests not only help local communities that are depending on their resources, but also helps in providing the breeding ground for marine life, carbon storage and preserving mangrove plant biodiversity. Understanding about structure and species diversity in mangroves is crucial for management and conservation purposes. The mangrove plant structure has a significant effect on the ecosystem functions. In addition, understanding structural attributes of mangrove vegetation is important with respect to their productivity. Moreover, planning any species conservation initiatives and scientific management of mangroves requires an understanding of their community structure, species, diversity and composition. Therefore, the need to emphasize the documentation of this information is fundamental for long-term management and sustainability of these magnificent yet degraded ecosystems.

Considering mangroves to be one of the rich habitats with an endemic species, while offering so many benefits, both ecological and socio-economic, there is an urgent need to implement conservation efforts to control their losses. Uncontrolled anthropogenic activities in the mangrove forests will make them as a threatened ecosystem in the world.

References

Abou Seedo K, Abido MS, Salih A, Abahussain A (2017) Structure and Composition of Mangrove Associations in Tubli Bay of Bahrain as Affected by Municipal Wastewater Discharge and Anthropogenic Sedimentation. Int J Biodivers Conserv 2017:1–9.

Aksornkoae S (1995) Ecology and management of mangrove restoration and regeneration in East and Southeast Asia. Proceedings of the Ecotone IV, 18, 22.

Deltares (2014) Habitat requirements for mangroves. Available at: https://publicwiki.deltares.nl/display/BWN/Building+Block+Habitat+requirements+for+mangroves

Ellison AM, Mukherjee BB, Karim A (2000) Testing patterns of zonation in mangroves: Scale dependence and environmental correlates in the Sundarbans of Bangladesh. J Ecol 5: 813–824.

FAO (2005) Global Forest Resources Assessment 2005: Thematic Study on Mangroves. Forest Resources Development Service Forest Resources Division Forestry, Forestry Department FAO, Rome (Italy), p 10.

Faridah-Hanum I, Kudus KA, Saari NS (2012) Plant diversity and biomass of marudu bay mangroves in Malaysia. Pak J Bot 44: 151–156.

Gandhi S, Jones TG (2019) Identifying mangrove deforestation hotspots in South Asia, Southeast Asia and Asia-Pacific. Remote Sens 728: 1–27.

Giesen W, Wulffraat S, Zieren M, Scholten L (2006) Mangrove Guidebook for South East Asia. FAO and Wetlands International.

Hogarth PJ (2007) The Biology of Mangroves and Seagrasses. Oxford University Press, London.

Hsiang LL (2000) Mangrove conservation in Singapore: A physical or a psychological impossibility? Biodivers Conserv 3: 309–332.

Iksan M, Aba L, Taharu FI, Alfian A, Ardyati DPI, Jumiati, Alzarliani WOD, Hardin, Larekeng SH (2019) The diversity of mangrove forests in Kumbewaha, Buton Island, Indonesia. IOP Conference Series: Earth and Environmental Science 1:1–6.

Joshi HG, Ghose M (2014) Community structure, species diversity, and aboveground biomass of the Sundarbans mangrove swamps. Tropic Ecol 3: 283–303.

Kanniah KD, Sheikhi A, Cracknell AP, Goh HC, Tan KP, Ho CS, Rasli FN (2015) Satellite images for monitoring mangrove cover changes in a fast growing economic region in southern Peninsular Malaysia. Remote Sens 11: 14360–14385.

Khuzaimah Z, Ismail MH, Mansor S (2013) Mangrove changes analysis by remote sensing and evaluation of ecosystem service value in Sungai Merbok's mangrove forest reserve, Peninsular Malaysia. In Lecture Notes in Computer Science, Proceedings of the 13th International Conference in Computational Science and Its Applications–ICCSA. Springer, Berlin, Germany, p 611–622.

Mojiol AR, Wan Maisyahirah MMN, Muhammad Ali SH (2019). Tree Species Diversity of Mangrove at Tunku Abdul Rahman Park, Sabah Malaysia. Acta sci Agric 9: 181–187.

Ong JE, Gong WK (eds) (2013) Structure, Function and Management of Mangrove Ecosystems. International Society for Mangrove Ecosystems (ISME) Book Series No. 2, Okinawa, Japan, and International Tropical Timber Organization (ITTO), Yokohama, Japan.

Panda M, Murthy T, Samal R, Lele N, Patnaik A, Chand P (2017) Diversity of True and Mangrove Associates of Bhitarkanika National Park (Odisha), India. Int J Adv Res 1: 1784–1798.

Primavera JH, Garcia LMB, Castanos MT, Surtida MB (eds) (2000) Mangroves of Southeast Asia. Mangrove-Friendly Aquaculture: Proceedings of the Workshop on Mangrove-Friendly Aquaculture. 12–14 October 1999, Iliolo City, Philippines, p 231–239.

Rozainah MZ, Mohamad MR (2006) Mangrove Forest Species Composition and Density in Balok River, Pahang, Malaysia. Ecoprint: An International Journal of Ecology 13:23–28.

Sarno, Suwignyo RA, Zulkifli D, Munandar, Ridho MR (2015) Primary Mangrove Forest Structure and Biodiversity. Int J Agri Sys 2:135–141.

Satyanarayana B, Muslim AM, Horsali NAI, Zauki NAM, Otero V, Nadzri MI, Ibrahim S, Husain ML, Dahdouh-Guebas F (2018) Status of the undisturbed mangroves at Brunei Bay, East Malaysia: A preliminary assessment based on remote sensing and ground-truth observations. PeerJ 6:e4397 https://doi.org/10.7717/peerj.4397.

Shufen Y, Lim RLF, Chiou-rong S, Yong JWH (2012) The Current Status of Mangrove Forests in Singapore. Proceedings of Nature Society, Singapore's Conference on 'Nature Conservation for a Sustainable Singapore,' October 2011, p 99–120.

Suratman MN, Ahmad S (2012) Multi temporal Landsat TM for monitoring mangrove changes in Pulau Indah, Malaysia. In Proceedings of the 2012 IEEE Business, Engineering and Industrial Applications (ISBEIA) Symposium, Bandung, Indonesia, 23–26 September 2012, p 163–168.

Suratman MN (2008) Carbon sequestration potential of mangroves in southeast Asia Managing forest ecosystems: The challenge of climate change. Springer, p 297–315.

Thomas N, Lucas R, Bunting P, Hardy A, Rosenqvist A, Simard M (2017) Distribution and drivers of global mangrove forest change, 1996–2010. PLoS ONE 6: 1–14.

Tomlinson PB (1986) The botany of mangroves. J Tropic Ecol 2: 188–189.

Wang L, Mu M, Li X, Lin P, Wang W (2011) Differentiation between true mangroves and mangrove associates based on leaf traits and salt contents. J Plant Ecol 4: 292–301.

16 Livelihood of Forest Dependent Dwellers in Relation to the Exploitation of Resources at the Fringe of Indian Sundarban

Chandan Surabhi Das

Abstract

Sundarban, a single largest mangroves block in the world, shared by India and Bangladesh, has a rich biodiversity that provides staggering ecosystem services to local inhabitants for livelihood. Poor agriculture in one way and rich bio-resources in other have compelled local inhabitants to depend upon forest resources from time immortal. Mangroves are diverse and highly productive ecological communities, but the pristine Sundarban forest had been dwindled during the last two centuries because of committed destruction. The present study assesses the status of coastal communities who exclusively depend on the Sundarban forest for subsistence and livelihoods. For identifying the livelihood pattern of these communities a household survey ($n = 1079$) and in-depth interviewing ($n = 157$) were done in three villages of the Sundarban, such as Rajatjubille in Gosaba, Samsernagar in Hingalganj, and Kisorimohanpur in Kultali block. A Relative Livelihood Index (RLI) was developed among major occupant groups ($n = 14$), for assessing the degree of strength among primary occupations. The strength of fishing and crab collection as a major primary occupation is observed in Samsernagar and Rajatjubille, whereas in Kisorimohanpur agricultural labour dominates others. According to RLI, the primary occupation like honey collection, fishing, and carpentry scores negatives which justify these occupations were alone robust to support the occupants continuing livelihood. So the category of forest users and the type of resource collected from the forest should be considered as the main criteria for designing any conservation plan as well as a sustainable livelihood programme in Sundarban.

C. S. Das (✉)
Department of Geography, Barasat Government College, Kolkata, India

R. P. Rastogi et al. (eds.), *Mangroves: Ecology, Biodiversity and Management*,
https://doi.org/10.1007/978-981-16-2494-0_16

Keywords

Mangroves forest · Coastal communities · Primary occupation · Relative livelihood index · Sustainable livelihood

16.1 Introduction

According to the British Department for International Development (DFID 2000) livelihood comprises the capabilities, assets (including both material and social resources), and activities required for a means of living. A livelihood is sustainable when it can cope with and recover from stresses and shocks, and maintains or enhances its capacities and assets both now and in the future, while not undermining the natural resource base (Ashley and Carney 1999). A sustainable and vibrant livelihood system helps to build up 'layers of resilience' to overcome 'waves of adversity', and enables people to transform multiple adversities into opportunities (Glavovic et al. 2003; Chambers and Conway 1992). In the mangrove forest, the term 'livelihood' refers to the capabilities and assets that the households have at their disposal to cope with the change caused by the conservation policy and strategy within the framework (Darwin 2014). Mangrove is a shrub or small tree that grows in tidal coastal or brackish saline water of the tropical world. Chapman (1976) used the term 'mangrove' as intertidal forests or plant communities or 'mangal'. The term 'mangal' was also commonly used in French and in Portuguese to refer to both forest communities and individual plants. They, sometimes, have been described as 'coastal woodlands', '*mangals*', 'tidal forests', or 'mangroves' or 'mangrove forests' (Saenger 2002). So, in general, mangroves are salt-tolerant plant species that struggle every moment against tides as they straddle at the interface zone of the land and the sea. Mangroves are uniquely adapted to the intertidal zone in the tropics and sub-tropics ranging from mean sea level to high tide level, or in zones where moderate salinity, average temperature, and high rainfall are experienced (Alongi 2002). Under this environment, mangroves extend in more than 118 countries in the tropics, sub-tropics, and temperate regions covering, respectively, 24 million hectares (Twilley et al. 1992), 10 million hectares (Bunt et al. 1992), and 14–15 million hectares (Schwamborn and Saint-Paul 1996). However, the mangroves are confined to tropical or sub-tropical climates where the average monthly minimum temperature is 20 °C (Chapman 1976).

Sundarban is the largest patch of mangrove forest found anywhere in the world. The vegetation of the Sundarban mangrove forest appeared during 31,750 ± 2030 years before the past (BP) (Chaudhuri and Choudhury 1994). But the origin of Bengal basin wherein present-day Sundarban mangrove forests exist started as early as 126 million years BP (Naskar and Mandal 1999). Sundarban mangrove forests have been painstaking as one of the seven most significant wetlands globally, based on biological diversity (Junk et al. 2006). Importantly, Indian Sundarban alone represents 30 true mangroves out of 40 true mangroves reported in the Old World Tropics. In National perspective, Indian Sundarban mangrove covers 62% area (about 2400 km^2), which maintains 90% species diversity, followed by Andaman and Nicobar Island with 76.5% and Bhitarkanika, Orissa with 72.3% in respect of existing major mangroves diversity (Mandal and Naskar 2008). This densely populated reclaimed region, on the other hand, is one of the immature low lands of the world with a multitude of socio-economic problems (Das and Bandyopadhyay 2012). About 95% of the population in Sundarban is directly or indirectly dependent on water bodies for their daily activities (agriculture, aquaculture, and fishery) (Chowdhury et al. 2017). As a result, the virgin mangrove forest of Sundarban was destructed for meeting the needs of the human population for the last two hundred years (Mandal et al. 2010). Claude Russel, the British administrator in the Bengal province of India, initiated to reduce mangrove forest areas for the purpose of reclamation for human inhabitation since 1770 (Chaudhuri and Choudhury 1994). Most of the labours who were engaged in the reclamation activities were not sent back to their earlier homeland and rather rehabilitated in the forest cleared areas. The reclaimed forest areas were utilized for agriculture, fishing, and allied activities pertaining to livelihood purposes. Until 1971, the reclamation of the forest continued (Chaudhuri and Choudhury 1994). Since then the need for conservation of mangroves awoke, albeit lately. Nonetheless, degradation of the forest was not stopped, rather perpetuated slowly by illegal encroachment because of the dense population (Mandal et al. 2010).

People in the fringe zone of Sundarban are basically forest dwellers who traditionally could spend their livelihood through a collection of resources emanating from the forest. Since the Sundarban mangrove forest is full with varieties of resources which are available easily from land and water through minimal efforts spent. These resources comprise three items: (i) food, fodder, honey, leafy vegetable (ii) Tannin, wax, wood, thatching materials, timber for construction of the house, boat, fence, etc., and (iii) fuel wood. Most of the habitants in this region belong to SC (Scheduled Caste), ST (Scheduled Tribe), and OBC (Other Backward Class) communities accounting about 89% of the total population of Sundarban (Das and Mandal 2016). They are, by and large, poor, down trodden, diffident, and have a relatively least amount of inherent properties and wealth of economic benefit. Most of them are landless daily wage labours or marginal labours. In this perspective, the present study was intended to assess the patterns of livelihood of dwellers in relation with exploitation of the forest's resources as well as to analyse the present scenario of the collection of NTFPs for sustainable livelihood.

16.2 Study Area

The Sundarban, the world's largest continuous patch of mangrove forest, is formed at the delta of the Ganges, Brahmaputra, and Meghna rivers on the Bay of Bengal. The total area of the entire Sundarban is about 10,000 sq. km, 62% of which is found in Bangladesh and the rest, 38%, lies in the southeast of West Bengal, India (Spalding et al. 2010; Das 2012). The UNESCO Man and the Biosphere programme in 1989 declared 9630 km^2 of the western part of the Sundarban region as the

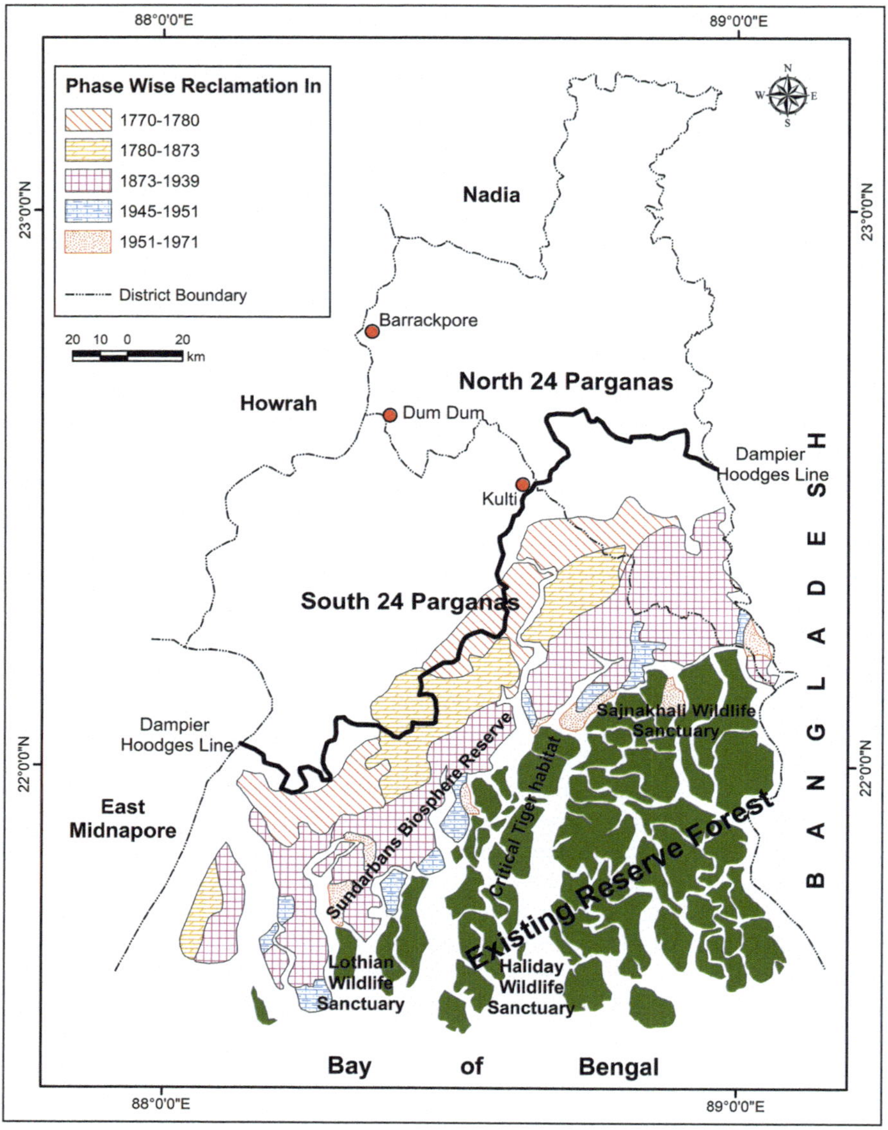

Fig. 16.1 Phase wise reclamation and loss of forest in Sundarban

Sundarban Biosphere Reserve (SBR), out of which about 5364 km^2 (c.56 percent) of the northwestern part have been cleared or reclaimed for human habitation, development of agriculture fields, and brackish water fisheries (Naskar and Guha Bakshi 1987) (Fig. 16.1). The northern boundary of the Sundarban is demarcated by the Dampier Hodges line of 1829–1830 (Danda et al. 2011; World Bank 2014). The

Table 16.1 Legal and Institutional overview of Indian Sundarban

Legal designation	Year	Area (sq. km)	Activities prohibited/regulated
Sundarban Tiger Reserve (STR)	1973	2584.89	Divided into two parts—Core (1699.62 km^2) and buffer area (885.27 km^2). The core area is prohibited from human interference, fishing and tourism activities allowed in buffer areas
Sajnakhali wildlife sanctuary	1976	362.40	Part of the buffer area of the STR but fishing and hunting prohibited
Lothian wildlife sanctuary	1976	38	Part of the buffer area of SBR. Fishing and hunting prohibited
Haliday wildlife sanctuary	1976	6	Part of the buffer area of SBR. Fishing and hunting prohibited
Sundarban National Park	1984	1330	Part of the buffer area of STR. Fishing and hunting activities prohibited
Sundarban Biosphere Reserve (SBR)	1989	9630	Total Sundarban (STR + SBR) divided into 3 parts—Core area (prohibited: 1692 km^2), buffer area (restricted: 2233 km^2), and transition zone (inhabitated: 5705 km^2).
Critical tiger habitat	2007	1699.62	Rename of the STR Core area where all anthropogenic activities prohibited

Source: Ghosh (2014), Sahana and Sajjad (2019), STR (2014)

reclaimed Sundarban now supports a fast rising population of 4.42 million with an average density of 1074 person km^{-2} and a growth rate of 18.23 persons yr^{-1} (Census 2011).

4226 km^2 of the southeastern part of the Sundarban was included under the Sundarban Reserve Forests in 1911. Out of these, 2585.89 km^2 of forests and tidal waterways of the extreme southeastern part of the Sundarban were declared as the Sundarban Tiger Reserve (STR) in December 1973 to promote a major national project for tiger conservation, the Project Tiger. The core area of the STR, measuring 1330 km^2 was declared as National Park on 4 May 1984. This core area (1700 km^2) has been renamed to Critical Tiger habitat in 2007 (Table 16.1). The STR was included in the list of UNESCO World Heritage Site in 1987.

16.3 Productivity of the Mangroves

Mangroves are diverse and highly productive ecological communities (Mandal et al. 2009; McDowell 1995) having important ecosystem services in the tropical coastal land of the world (Day et al. 1987). Mangroves are also a significant supplier of nutrients in the wetland ecosystems (Kamruzzaman et al. 2017). Biomass and net primary productivity of mangrove forest have been studied previously in different mangrove forests across the world (Putz and Chan 1986; Day et al. 1987, 1996; Saintilan 1997; Komiyama et al. 2000; Kamruzzaman et al. 2017). Measuring of these components indicates the ecosystem regulation and estimate of carbon stock

Table 16.2 Net primary production and related characteristics of major ecosystems

Ecosystem type	Area (10^6 km^2)	Net primary production: Normal range ($g\ m^{-2}\ yr^{-1}$)	Mean ($g\ m^{-2}\ yr^{-1}$)	Total (10^9 $t\ yr^{-1}$)
Tropical rainforest	17.0	1000–3500	2200	37.4
Tropical deciduous forest	7.5	1000–2500	1600	12.0
Temperate evergreen forest	5.0	600–2500	1300	6.5
Temperate deciduous forest	7.0	600–2500	1200	8.4
Boreal forest	12.0	400–2000	800	9.6
Woodland and scrubland	8.5	250–1200	700	6.0
Savanna	15.0	200–2000	900	13.5
Temperate grassland	9.0	200–1500	600	5.4
Tundra and alpine	8.0	10–400	140	1.1
Desert and semi-desert scrub	18.0	10–250	90	1.6
Extreme desert: rock, sand, and ice	24.0	0–10	3	0.07
Cultivated land	14.0	100–4000	650	9.1
Swamp and marsh	2.0	800–6000	3000	6.0
Lake and stream	2.0	100–1500	400	0.8
Open ocean	332.0	2–400	125	41.5
Upwelling zones	0.4	400–1000	500	0.2
Continental shelf	26.6	200–600	360	9.6
Algal beds and reefs	0.6	500–4000	2500	1.6
Estuaries (excluding marsh)	1.4	200–4000	1500	2.1

Source: Whittaker and Likens (1975)

(Tamai et al. 1986; Komiyama et al. 1987, 2000) is significant in mangroves compared to other varieties. Naskar and Ghosh (1989) opined that the mangrove ecosystem is the most productive ecosystem of the world. The biological productivity of an ecosystem may be defined as the rate of appearance of matter as living tissue (Simmons 1981). As shown by Whittaker and Likens (1975), swamps and marshes occupy the top position among the world's major ecosystem types in terms of mean net primary production of dry matter. This amounts to 3000 g m^{-2} yr^{-1} in swamp and marsh ecosystems compared to 2200 g m^{-2} yr^{-1} of tropical rain forests, 1600 g m^{-2} yr^{-1} of tropical seasonal forests, and 1300 g m^{-2} yr^{-1} of temperate evergreen forests (Table 16.2). It was estimated in the mangroves of Southeast Asia that rates of biomass accumulation range between 6.3 and 45.4 t ha^{-1} yr^{-1} (Hogarth 1999). The available literature on the productivity of mangroves also indicate a wide range of values between 2 and 16 m^3 ha^{-1} yr^{-1} of mean wood increments (Tomlinson 1986). The mean biomass increment and mean litterfall were 7.1 and 10.1 Mg ha^{-1} yr^{-1}, respectively, for the mangrove community in Bangladesh (Kamruzzaman et al. 2017).

Litter fall from mangrove vegetation, decomposition of litter and thus nutrient release into soil and water, all are cumulative effects of ecosystem services, which steadily reduce with the gradual loss of vegetation. The mangrove ecosystem alone is rather productive as measured to yielding as 350–500 g C/m2/yr (Mann 1982) that provide a considerable contribution to the food chain that leads to keep the sustainability of coastal fisheries (Upadhyay et al. 2002). The *Avicennia* spp. and *Sonneratia* spp. dominated forest area of Indian Sundarban has been assessed to produce about 212 tons/ha of biomass (Chakraborty 1985). It is estimated that a full-grown mangrove stand of 10 years old may add to soil the following amounts of nutrients through litter decomposition: N, 46.6 kg/ha/yr; K, 25.6 kg/ha/yr; Ca, 99.3 kg/ha/yr; Mg, 34.1 kg/ha/yr; and Na, 31.8 kg/ha/yr (Gong et al. 1984).

16.4 Reclamation and Livelihood in Sundarban

The pristine Sundarban forest had been dwindled during the last two centuries because of committed destruction. Degradation of Sundarban mangrove forests started since 1770 during British India. Claude Russel and later Tilmen Henckell started to reclaim Sundarban forests, which were then partly exploited for rehabilitation of the human population and remaining for rice cultivation and brackish water fisheries (Mandal et al. 2010). Settlement in the Sundarban included migrant populations of adjoining Midnapore district and from the district of Jharkhand (then Bihar) and Chhattisgarh (then Madhya Pradesh), who came in search of work and land (Danda 2007). Reclamation as well as settlements occurred in five phases (Fig. 16.1). At the end of the eighteenth century, the mangrove forest extended up to Kolkata (Ghosh et al. 2015); and at the time of independence in 1947 the forest was only 50% of its pre-colonial size (Giri et al. 2007). In 1947 India witnessed the destiny of her partition. As a result, an enormous incursion of refugees was settled in reclaimed Sundarban (Naskar and Guha Bakshi 1987) until 1971. Since then the need for conservation of mangrove awoke, albeit late. Nonetheless, degradation of the forest was not stopped, rather perpetuated slowly by illegal encroachment because of the dense population (Mandal et al. 2010).

The human population in the Sundarban increased rapidly in the post-colonial era, especially following the partition of India and Bangladesh. Since the 1970s, the Indian Sundarban mangroves have been protected under various legal measures which were established primarily to protect and help increase the threatened tiger population. After independence, the population of Indian Sundarban grew from 1.15 million in 1951 to 4.42 million in 2011, which led to an increase of 343% in last 60 years. The population was increased almost 15 times from 1872 to 2011. The decadal population growth rate of Sundarban is 18.23% (2001 to 2011). From 1901 to 1951 the population was increased almost double. After independence from 1951 to 2011, the population increased almost 4 times due to political phenomena. A mass of the population has migrated from Bangladesh to the Sundarban. As a result, mangrove forests had been decreased by 20.58% on account of the conversion of forest to agricultural land and settlements in the last 70 years (1968 to 2014) (Ghosh et al. 2015).

Salinity of the Sundarban is another concern. Embankment construction together with forest clearing made extensive human habitation possible in the Sundarban (Sánchez-Triana et al. 2018). The embankment construction process started in the late nineteenth century and continued through to the twentieth century. It made reclamation possible by preventing saline water from inundating land that was otherwise suitable for cultivation (Sánchez-Triana et al. 2014).

16.5 Poor Land Utilization

Sundarban, having 56% landless family, is considered to be one of the most backward regions in India (Singh et al. 2010). Historically, the region was dominated by a single crop farming system with the local Amon rice which was cropped in the cropping pattern: 'fallow-rice-fallow' (BARC 1998). But in the early 1980s brackish water shrimp farming came out as a vital land use leading to the emergence of a 'fallow-shrimp-rice' (Miah et al. 2003) farming. But all the rivers, creeks, and canals of the region have saline water, and therefore unsuitable for agriculture. Developing tube well irrigation proved difficult as the underground water up to 20–25 ft. from the surface is saline (German Agro Action Plan 2009). Besides, regular occurrence of *high tides twice* a day induces salinity in the water of the canals. In the Indian Sundarban, around 56% of land mass lies within the coastal low-lying ecosystems with an elevation of <5 m above mean sea level (Mandal et al. 2019b). Some parts are even below the mean sea level. All these factors force the region into a monoculture as well as poor agricultural productivity. As such, every year more and more agricultural lands were converted into more profitable shrimp cultivation resulting into gradual declining rice cultivation. This situation urges the landless and marginal people of the Sundarban to move frequently into the forests in search of livelihood.

16.5.1 Agriculture

The main economic activity in the Sundarban, rain-fed paddy agriculture, is made possible by the construction of earthen embankments to keep brackish tidal water at bay. Limited irrigation facilities lead to mono-cropping agriculture in Sundarban which are also constrained by the salinity of both ground water and surface water. Historically, in the Sundarban a variety of salt-tolerant paddy were cultivated on raised sections of the islands without embankments. However, according to the National Bureau of Plant Genetic Resources (NBPGR) only two varieties of them—Matla and Hamilton are cultivated now; others are believed to have been lost under the onslaught of green revolution (Ghosh 2010).

Agriculture in the reclaimed areas is also constrained by excess water during the rainy monsoon period (*kharif*) accompanied by two other phenomena like high humidity and less sunshine hours. High humidity lures pests and diseases, whereas low sunshine hours limit plant growth. The main *kharif* crop is the rice of a traditional variety, which is able to survive deep water-logging. Basically, farmers have no choice for alternate crops other than tall traditional rice varieties in kharif (Sarangi et al. 2015). During the dry winter months acute shortage of irrigation water along with an increase in soil and water salinity limits the agricultural production (Mandal et al. 2019a). The dominant soluble salts identified in the study region were NaCl and MgSO4 (Mandal et al. 2019b). As a result, during *Rabi* season, most of the land is left fallow and is used to graze cattle, but the little amount of areas are irrigated from village ponds stored with monsoon water. Paucity of sweet water in the region poses a major problem for irrigation (Pakrashi 2016). Only a limited net cropped area is irrigated in Sundarban. Sources of irrigation in the Sundarban are primarily tanks (71%) and shallow tube-wells (13%), with the former method being the preferred choice (World Bank 2014). The cropping intensity was low (114%) and more than 80 per cent of the farm lands remained fallow during the *Rabi* season (Mandal et al. 2017).

Because of non-remunerative agriculture and steady growth in the fishery sector, many agricultural lands are getting converted to high output-intensive brackish water farming (Mandal et al. 2019b) during last three or four decades, which poses severe environmental threats to the region. According to Hazra and Samanta (2016) because of increasing demands of shrimp cultivation, agricultural land of Indian Sundarban has been reduced from 2149 km^2 in 2001 to 1691 km^2 in 2008.

16.5.2 Non-Timber Forest Products/Non-Wood Forest Products

Because of the extensive and diverse resources available in Sundarban, the forest generates large-scale livelihood opportunities. Forest resources are christened as NTFP (non-timber forest product) and NWFP (Non-wood forest products). A biological product collected from forested area, including fish, crab, prawn seed, tiger prawn, firewood, construction wood, thatching leaves, all these constitute a major share of NTFP, with honey and bees wax being part of NWFP (Shackleton and Shackleton 2004). In general, the term Non-Timber Forest Products (NTFPs) covers all tangible products of forest origin, except wood (Ros-Tonen et al. 1995). Sometimes, these resources include parts of plants, fungi, and other biological materials or natural resources collected from forests apart from sawn timber. NTFPs may provide local job opportunities and contribute significantly to rural people as a source of income and facilitate the subsistence living in the fringe areas of Sundarban (Peters et al. 1989; Hegde et al. 1996).

16.5.2.1 Collection of Plant Resources

Utilization and exploitation of mangroves can take many forms (Tomlinson 1986) and most are present in the Indian Sundarban. The mangrove trees are used as swan timber, poles, fuel, and pulp wood. Tannins and dyes are also extracted from them. Some species are used as thatching material. Leaves of *Nypa fruticans* ('Golpata') is a major source of thatching material, which is extensively used by the poorer section of the rural population in Sundarban. Resources like tannin, wood, timber, etc., will amount to a staggering figure in terms of pecuniary benefit, but local people use these resources as their integral part of daily life without economic botheration. Mangrove flora also provides important medicinal benefits: leaves of *Bruguiera gymnorrhiza* are used for remedy of diarrhoea and blood pressure, *Rhizophora mucronata* for angina, *Acanthus ilicifolius* for asthma and rheumatism, *Lumnitzera racemosa* for herpes and itches, *Cynometra ramiflora* and *Excoecaria agallocha* for leprosy (Kothari and Rao 1995; Untawale 2006; Rodrigues 2006; Mandal and Bar 2018). *Nypa fruticans*, commonly known as the nipa palm (or simply nipa) or mangrove palm (Nishat 2019). The fruits of *Sonneratia apetala* are widely used as food and in treating various diseases like asthma, febrifuge, ulcers, swellings, sprains, bleeding, haemorrhages, and piles (Bandaranayake 1998; Hossain et al. 2016). The fruits of *Sonneratia apetala* are marketed now and also preferable food items to the Rhesus monkey (Sanyal 1992). Frond of *Acrostichum aureum* is used as leafy vegetables (Pillai and Ong 1999; Badhsheeba and Vadivel 2020). Most of the mangrove leaves are suitable fodder for the domestic livestock, apart from wild animals like deer, monkey, and wild boar. *Heritiera fomes* is the principal timber species in the Sundarban (Dasgupta et al. 2017; Khan et al. 2020). However, collection of plant resources are now restricted in Sundarban. Earlier, Golpata (*Nypa* sp.) and Hental (*Phoenix* sp.) which were collected by the fringe villagers were discontinued in 1978 and 1991, respectively. The coupe operation has been discontinued since the year 2001 (STR 2018).

16.5.2.2 Fuelwood Collection

After food, fuel occupied an important place for daily livelihood. The two major fuel wood species in the Sundarban are *Heritiera fomes* and *Ceriops decandra*. However, there are a number of other species, having good quality fuelwood include *Amoora cucullata*, *Aegiceras majus*, *Rhizophora mucronata*, *Hibiscus tiliaceus*, *Ceriops candolleana*, and *Cynometra ramiflora* (Islam et al. 2019a, 2019b). 75% of the total households of Sundarban were dependent on the fuel wood exclusively for energy source or in combination with other forms of biomass (Das and Mandal 2016). As a result, a large number of people are engaged in collecting firewood from the forest almost on a daily basis. Fuel wood collection is done by both solitary and group activity. The groups usually invade the forest with small traditional boats, but in solitary activity, people explore the riverbank of SRF and collect fuel wood from the vicinity (Chowdhury et al. 2020). Nearly 70% households' treat fuel wood, cow dung, and rice straw as the major fuel sources for cooking and related activities (Situmorang et al. 2020). Fuel wood was an energy source covering 41% of households, accompanied by 25% of households, depending upon cow dung and rice straw. Rural women of Sundarban generally collect shrimp fry in the river or creeks, and at the same time they also collect fuel wood from the forest edge (Kibria et al. 2018).

16.5.2.3 Apiculture

Apiculture is the science and practice of bee keeping which plays a valuable part in rural livelihoods of Sundarban. Forests with a variety of flowering plants provide a centre for beekeeping and honeybees (Hill and Webster 1995). Beekeeping generates much more than just honey. The maintenance of biodiversity and pollination of crops are the most valuable services provided by bees. Honey is just one of several different products that can be harvested: others are beeswax, pollen and propolis, royal jelly and venom, and the use of bees in apitherapy, which are medicine using bee products (Bradbear 2003, 2005). During the flower season of mangroves the swarms of honeybee colonies with honey are developed. Honey and the pollen in it are used as medicines, high energy food, and as a source of vitamins and minerals.

Sundarban mangrove areas have high potential for honey production, but honey collectors often use destructive methods of harvesting (Burgett 2000). Sundarban harbours numerous hives of rock bees (*Apis dorsata*). *The* floristic composition of the Sundarban is very favourable for honey production. The best quality unifloral honey of the Sundarban is kulshi-type followed by amur, goran, and keora (Table 16.3). Honey and Bee-wax in the Sundarban Reserve Forest are collected by a traditional occupational group called mawallis *or moulis* during the months of April–May every year. Honey collection from the Sundarban forest is a seasonal activity occurring usually in the period of 15th April to 31st May every year. In an average year, nearly 1000 honey collectors with a population of more than 8000 enter the forests with permits from forest department (Vyas 2012; Das and Mandal 2016; Sen and Pattanaik 2017). Generally a group comprising 10 members in one venture may collect about 10 to 12 quintals of honey (Mandal et al. 2010) costing Rs. 120,000 @ Rs.100.00/kg as per Govt. rate; so that one can earn Rs. 12,000 per trip; whereas in the local market one kg of honey cost ranging between Rs. 500 and 600.

Collecting honey from the deep forest is always a risky job from man-eaters (Kothari 2015) which occasionally result in fatalities (Table 16.4). The West Bengal Government has taken a unique initiative of providing beekeeping with apiary boxes

Table 16.3 Important Sundarban plants, flowering times, and quality of honey

Local name	Scientific name	Peak season (flowering time)	Quality of honey
Kulshi	*Aegiceras majus*	Feb to Mar	Very high
Amur	*Amoora cucullata*	Feb to Mar	High
Goran	*Ceriops decandra*	Mar to Apr	Very high
Keora	*Sonneratia apetala*	Mar to Apr	High
Passur	*Xylocarpus mekongensis*	Mar to Apr	Very high
Sundri	*Heritiera fomes*	Apr to May	Low
Gewa	*Excoecaria agallocha*	Apr to May	Low
Kakra	*Bruguiera gymnorrhiza*	Apr to May	Low
Baen	*Avicennia officinalis*	May to June	Low

Source: Field Study and Zohora (2011)

Table 16.4 Honey collection from STR areas of Sundarban

Honey Collection (STR)	2010–2011	2011–2012	2012–2013	2013–2014	2014–2015	2015–2016	2016–2017	2017–2018
Crude honey (kg)	14,300	18,025	24,750	20,950	47,412	33,515	19,050	15,000
Persons involved with permits	765	879	715	735	1155	979	604	486
Injury due to tiger attack	5	1	5	4	0	1	5	3
Death due to tiger attack	3	2	5	2	4	3	2	2

Source: STR (2018), GoWB (2018)

to the local people inside the protected forest camps in several places of Sundarban (Thakur et al. 2016). It is not only safer for the villagers, but also the quality of the collected honey is better. While collecting honey from the forest, the local squeeze the entire beehives and in process several bees and maggots also get squished in the honey resulting into low-quality impure honey (Ray 2000). This initiative is highly acceptable by the villages as the pure honey collected by this new method is sold to the forest department at a rate of Rs 600/per kg which is roughly five times higher than the earlier attempt. It is 100% pure natural honey, often recommended by the medical practitioners on account of its high therapeutic value (Kumar et al. 2010; Ajibola et al. 2012).

16.5.3 Fishery and Aquaculture

Fishing has been perhaps the most common and staple livelihood options for the people of Sundarban since their habitation of the area (Ghosh et al. 2018). Next to agriculture, fisheries provide a distinct source of employment and income for small and marginal farmers of Sundarban. Sundarban is considered as a gold mine for the fishery (Ray 2000). The Sundarban water supports 208 species of fish and crustaceans belonging to 84 families (IUCN 1994), a higher total than that of other tropical mangroves (Robertson et al. 1992). According to a survey, 312 bony fishes restricted to 214 Genera, 71 Families, and 18 Orders especially are found in marine and estuarine waters of the Sundarban mangroves (Kar et al. 2017). During the agricultural lean season, people resort to fishing and the collection of prawn seeds, even risking their lives from man-eating tigers and crocodiles. The Sundarban is a rich fishing ground. The fishing nets are still often knotted by hand, the weirs plaited manually. Fishing is still handwork, yet overfishing increasingly becomes a challenge. A large section of people in and around Sundarban earn their livelihood from fishing and pisciculture. But fish cultivation is not carried on the scientific method.

The shallow water, creeks, small and big rivers crossing mangrove forests support many species of fish. Over 120 species of fish are caught routinely by commercial fishermen (Hasan and Naser 2016; Chatterjee 2017). Some species such as *Hilsa ilisha* are exclusively marine in nature, but travel through estuaries to the upstream areas for breeding and then return to the sea. Shrimps and prawns constitute the most important fishery of the zone. The most important crustaceans' species are *Penaeus monodon* and *Macrobrachium rosenbergii.* Mud crab (*Scylla serrata*) is the largest edible crab found in the forest area and has a high economic value because of its very tasty meat and very high nutrient content.

At present, there are 5550 registered boats to fish within the forest area of Sundarban (GoWB 2019). Each of these boats, which are quite small in size, has a capacity of carrying about four persons and make several trips to the forests. Most of the fishing takes place from November to January; a lesser amount of catch is generally made between March and June. Due to the rough condition of the weather, fishing activity does not take place during other times of the year.

Sundarban is the top producer of fish and prawn, producing roughly 31% of the total inland fish/prawn production of West Bengal (Das et al. 2016). Brackish water prawn aquaculture started along the creeks of reclaimed Sundarban from the late 1980s and developed quite rapidly. Prawn farming is basically a capital-intensive mechanized undertaking that does not require large pools of labours. The people of Sundarban, however, got involved in the activity in a different way, by collecting seeds of tiger prawns from the coastal water during high tides. These then are sold to the prawn farms through brokers for nourishing. Mostly carried out by the women and children, the prawn seed collection became an economic sidekick for many families. In the fringe areas of the forest, close to 6000 people sieve the shallow intertidal water for prawn seeds regularly. In a study of Bangladesh, Angell (1994) estimates that one small-scale hatchery with three cycles of production per year can produce about 2 million post larvae, with fixed costs of about US$23,000 and operating costs of about US$9000. West Bengal being the largest producer of tiger shrimp among Indian maritime states and the majority of this was produced in Indian Sundarban covering 0.21 million ha potential brackish water areas in the state (Ghoshal et al. 2019). The areas where fishing is permitted within the buffer zones (522.85 km^2) are congested and overfished. The fishermen have to enter the core areas secretly, risking a fine of Rs 500 (US$7.76) for the first offence, Rs 1000 (US$15.52) for the second, and Rs 1100 (US$17.08) for the third if caught by a forest guard during patrolling (Sen and Pattanaik 2017).

Fishing activities in Sundarban have increased from 19% of the total workforce in 2000–2001 to 33.0% in 2010–2011 (Mistri and Das 2020). After Aila in 2009, fishing increases due to the loss of huge agricultural land by sudden increasing salinity. Approximately 11% of households in the Sundarban listed 'fishing' as one of the family occupations (Sanchez-Triana et al. 2014). This percentage goes up to 60–70% in areas with easy access to rivers (Sen 2019). However, according to the government report, more than one lakh families are engaged in inland fishing (GoWB 2005).

Steady declination of mangrove fisheries resources either due to overfishing or to habitat degradation or both is observed in Sundarban with other parts of Asia (Hasan and Naser 2016). Many residents of the Sundarban are dependent on income from gathering of natural *Penaeus monodon* post larvae wild-caught brood stock and seed for aquaculture, despite the formal ban on natural fry collection. This is one of the main sources of earning for the small and landless fisherman and women of this area. 7–99 mm post larvae and juveniles were available throughout the year with peaks in June, July, and December (De et al. 1978). The prawn seed collection is a highly destructive practice with a high by-catch rate that results in the capture and discard of non-target species and exerts a heavy toll on the sustainability of marine, estuarine, and freshwater fish species. Various studies in Sundarban suggest that an alarming declining rate in various fish species was observed in Sundarban (Hoq 2007). For every tiger prawn seed collection, about 400 of others species (fishes, crabs, other prawn, mollusc, etc.) are destroyed (Santhakumar et al. 2005; Mahapatra et al. 2014). Tiger shrimp seeds are fast dwindling away from the natural waters of Sundarban due to overfishing at various stages of its life cycle. As its post larval stage in estuaries, it is trapped by fine push and drag nets and fine meshed bag nets (*meen jal*); the juveniles are trapped by bag nets (*behundi jal*) in estuaries; the juveniles and pre adults are caught in marine waters by large bag nets; the pre adults and adults by trammel nets. Even the spawns are not spared and are caught from the open seas by trawl nets. (Mahapatra et al. 1999). Overexploitation and unscientific practices of aquatic resources were found to be the key factors for the declination of fishery resources in Sundarban.

16.6 Conservation and Livelihood Conflicts

Human survival and economic well-being are fully dependent upon biological diversity that encompasses all life forms, ecosystems, and ecological processes, acknowledging the hierarchy at genetic, taxa, and ecosystem levels (Wackernagel et al. 1997). The more is the biodiversity the greater is to access available resources, along with increased net primary production and decreased nutrient loss (Singh 2002). Globally top-down approaches suffer from the ailment of failing to factor in the participation of local people as part of the conservation process (Hazarika and Kalita 2019).

In the year 1973, the Government of India created the Sundarban Tiger Reserve, and subsequently the creation of buffer and core area (Nishat 2019) which restrict the forest users leading to deprived livelihood. It gave rise to conflict between forest managing authorities and local habitants. Excluding the livelihood issues of fringe dwellers from the conservation, the process makes it difficult to enforce any conservation strategies successfully. However, fringe dwellers of Sundarban had invariably participated in the conservation process long before it had started. Any type of disruption of traditional ways of living can trigger unwanted adverse impacts on local communities (García-Frapolli et al. 2009) which are tantamount to the poaching of wildlife. Fishing in Sundarban has a long history and is a generational

Table 16.5 Forest entrants in STR area in a typical year between 2013 and 2017

Category	Legal entrants (permit holders[a]: authorized for multiple entries)	Illegal entrants (approximate estimation by Forest Department staff/local survey)	Total (approximate)
Fishermen	3840	c. 7000	10,840
Honey collectors	750	250	1000
Nypa palm collector	1500	c. 600	2100
Tiger prawn collectors	None	c. 12,000 (during high tide)	12,000
Crab collectors	None	c.18000	18,000
Forest Department staff	237	None	237
Poachers	–	c. 100	100
Total	6267	c. 37,900	45,167

Source: Field Study
[a]No fresh permits are being issued by the Forest Department since 1992. It at present only *renews* the permits granted before 1992

occupation. The fishers have engaged in forest-based economic activities since civilization began in Sundarban. During the formation of the STR, its 'Management Plan' mentioned that only non-motorized boats would be allowed in the buffer areas for fishing. The people, especially communities, who depend on the natural environment of their locality to meet most of their material needs are known as 'ecosystem people' (Kothari and Patel 2006). Any type of human interference or boatings is strictly prohibited in the core areas. In the buffer that surrounds the core, access is allowed for livelihood purposes with a proper license and permit. But the major issue is that most of the forest entrants in Sundarban are illegal as they have no permits of resource collection issued by the forest department because of livelihood demands. According to a field survey estimate in 2018 near about 45 thousand forest users enter into the forest (both in restricted and non-restricted areas) for livelihood issues and surprisingly, only 14 percent are legal entrants (Table 16.5) As a result, success of different conservation attempts is very limited and there is an imbalance between resource collection and regeneration as well.

16.7 Livelihood vs Animal Attacks

Animal attacks on humans are common in Sundarban. Among the attacks from animals of the protected areas, tigers top the list followed by attacks by crocodiles and sharks. Attacks from crocodiles and tigers often prove fatal. The straying of tigers from reserve forests into the human habitations also poses a major problem for the residents living along the forest boundary. Conflicts between wildlife and humans in Sundarban are evidently owing to an increase in human population, extensive loss of natural habitats, and increase in dependency of forest resources. Conflicts are most acute when a species involved is critically imperiled while its presence in an area poses a significant threat to human welfare (Saberwal et al. 1994). Human-wildlife conflict is potentially any situation where: (1) the behaviour of people negatively impacts wildlife (this includes human impacts on habitat); (2) the behaviour of wildlife creates a negative impact for some stakeholders or is perceived by some stakeholders to impact themselves or others adversely, or (3) the wildlife focused behaviour of some people creates a negative interaction with other people, often in the form of a values clash. Thus, a people-wildlife problem can involve a people–wildlife interaction or a people–people interaction (i.e. a controversy) or both (Decker and Chase 1997).

Between 1998 and 2017, 546 forest entrants were attacked by tigers out of which 485 succumbed to their injuries with an average of 27.2 events per year (Das 2018). Some 14 percent of the victims were honey collectors, 5 percent were woodcutters, and as much as 80 percent were fishermen including crab collectors. About 1 percent of the victims were forest staff (Das 2018).

Crocodile victims are generally of two types—fishermen and tiger prawn seed collectors. In Sundarban, hundreds of people, mostly women, and young children get engaged in the prawn seed collection every day. Wading through waist-deep or even neck-deep water, they filter out of spawn of shrimps using fine nylon nets. In an area where the scope of alternative employment is limited, this activity has become popular in Sundarban during the last two decades as it yields very high returns (Ghosh et al. 2018). It is also done on a commercial scale, with nets spread across almost the entire width of the river with help of boats and buoys. According to a survey, some 132 people were attacked by crocodiles during 2000 to 2019; out of these, 61.16 percent succumbed to death on an average 7.9 persons every year. Almost 80 per cent of victims were prawn seed collectors and belonged to the age group of 11 to 50. They were mostly children and women. Male victims are slightly lower (46.60%) than the females (53.40%). Most of the cases were recorded from Gosaba (34%), followed by Patharpratima (25.24%) and Namkhana (18.45%) (Das 2018). Apart from the crocodiles, the persons being exposed to the creeks of Sundarban are also vulnerable to attack from sharks locally called *kāmots*. Shark bite is a relatively recent phenomenon in Sundarban that started to arise about 15 to 20 years ago (Das 2017). Increasing population pressure and dire poverty urge the people to take the risk of facing natural hazards as well as attack from wild animals as they venture into the jungle. Trespassers take undue advantage of this human presence in the zone for pilferage of forest produces and poaching of wild animals. It is also not uncommon for the animals to stray into human habitations and to cause depredations. All these lead to conflict between humans and animals—the root cause of which is socio-economical.

On an average, one forest user spent the time as 3–4 h/day in the minimum and 10–12 h/day in the maximum range, which accounted for about 15 days/month. In the purpose of crab collection, the duration used to vary from 3–4 h/day to 10–12 h/day, accounted for about 20–25 days/month. The collection of tiger prawn seeds (prawn: *Penaeus monodon*) was considered to be a hectic job for which one used to spend 6–7 h/day in the water, utilizing 21 days/month. In this activity, individuals put their maximum effort with high concentration to segregate small seeds of tiger prawn from heterogeneous species of juvenile fishes. Tiger prawn seeds had cost around Rs. 4/piece. Comparatively, the wood collection was recorded to spend about 5 h/day and treated as a concurrent activity done simultaneously with a collection of fishing and crab. Analyses of activities that led to a maximum exposure clearly showed that entering into creeks in forested areas for fishing, collection of crab, prawn seed, and fuelwood compelled the people to become most vulnerable. Analyses of seasonal incidences showed that maximum exposure normally happened during winter (59.7%), followed by monsoon period (18.6%) and then the summer months being remarkably low (1.7%) (Das 2017, 2018).

16.8 Case Study: Livelihood of Three Fringe Villages of Sundarban

For identifying the livelihood pattern of the people of the fringe areas of Sundarban, a household survey was done in three villages of the Sundarban under three blocks such as Gosaba, Hingalganj, and Kultali (Fig. 16.2) with a total area of 22.95 km^2 and a population of 17,728 (Census 2011). These *mouzas* are, here, known as fringe villages of Sundarban as these are bordered with the Sundarban Reserve Forest (Das 2015). Second, these villages were selected because a large section of dwellers depends on the forest resources of Sundarban for their daily livelihood.

16.8.1 Materials and Methods

A comprehensive database of the forest users was made through a door-to-door household survey with questionnaires interviewed with earning member(s) of all households during a period of May 2017 to April 2018. Collection of data on land use and other socio-economic aspects like land holding (Table 16.7), occupation (primary, secondary, and tertiary), caste, literacy, availability of drinking water, use of electricity, and 100 days works initiated by Govt. of India for provision of jobs to all. Total number of respondents was 1079 of 4161 households of three selected villages in the fringe areas in Sundarban (Table 16.6). Of the total respondents, only 505 (46.8%) were female. The eldest respondent was 95 years old when the mean age was 46.31 years.

Simultaneously, another in-depth interviewing was conducted on marginal workers (livelihood for less than 6 months) with a separate questionnaire to 157 respondents [Samsernagar (52), Rajatjubille (68) and Kisorimohanpur (47)] with a criteria of at least one member directly involved in activities like fishing, honey collection, crab collection as the source of primary occupation. Data was collected on dependency upon forest and forest resources, frequency of entering forest, type of resources collected, rate of daily utilization of resources, amount of

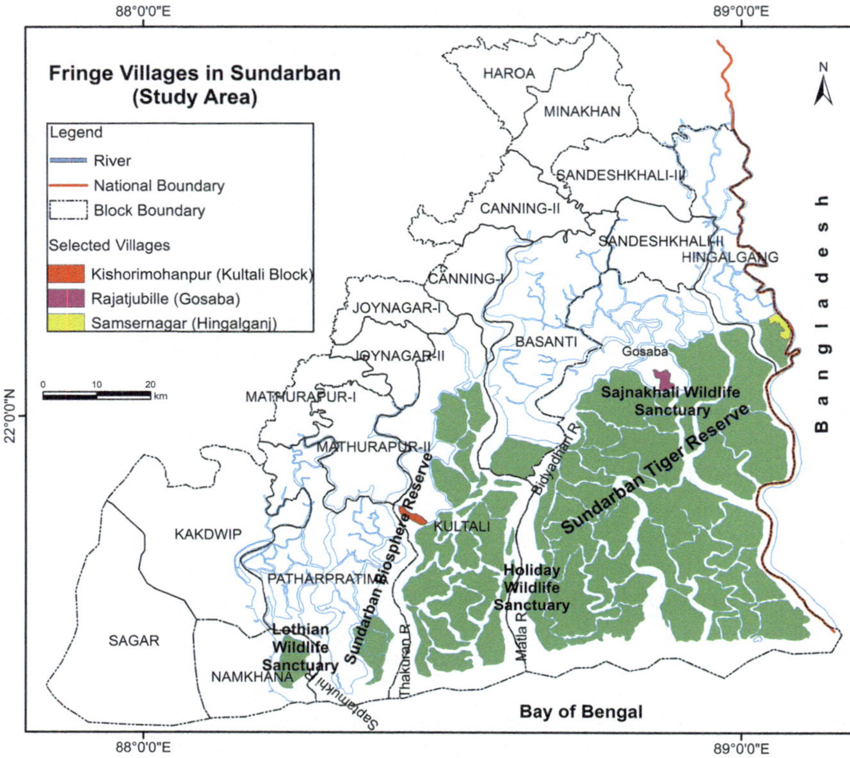

Fig. 16.2 Location of Fringe villages in Sundarban

Table 16.6 Sample study related population data

Village	No. of households	Population (2011)	Area (km^2)	No of samples	Mean age	Marginal workers
Samsernagar	1145	4394	6.83	345	48	558
Rajatjubillee	1745	6851	8.51	422	56	1937
Kisorimohanpur	1271	6483	7.65	372	52	1373
Total	4161	17,728	22.95	1079	50.3	3868

resources sold in the markets, income from sale of resources, type of fuel used, average use of fuel on daily basis, financial source for procurement of essential items, exposure to human-wildlife and its conflict, consequence of exposure to wild life, and role of eco-development committee (EDC) or forest protection committee (FPC) (Table 16.7).

Table 16.7 Land status of households of selected locations

Direct forest users category	Percentage of landless people			Average
	Samsernagar (n = 167)	Rajatjubille (n = 160)	Kisorimohanpur (n = 155)	
Fishermen	26.19	32.23	29.35	39.25
Crab catcher	30.91	19.20	31.44	27.18
PS collection	22.91	28.43	17.86	23.06
Honey collector	5.71	14.88	17.90	12.83
Nypa palm collector	14.28	5.26	3.45	7.66
Total	100.00	100.00	100.00	21.96

Source: Field Survey

16.8.2 Category of Occupation and Resources

The study listed 14 primary occupations (A), which constitute further 8 occupational types (Table 16.8). Each primary occupation was distinct from another based on its activity and tools used, though resources might be similar as biological perspective. The term 'primary occupation' referred to the activities which supported the major livelihood of forest dwellers and maximum time spent. The 'secondary occupation' (B) referred to those activities which were usually performed during the lean phase of primary occupation and were treated as the supplementary revenue generation. The 'tertiary occupation' (C) happened very rarely and was not much support for livelihood. However, the concept of primary, secondary, and tertiary occupations was relative to respective dwellers. The study considered the collection of aquatic resources (collection of finfish, shellfish, prawn seed, and crab) as a primary occupation to those having major earning for livelihood and maximum time spent. When the same occupants took agriculture during the lean phase, the activity was considered as a secondary occupation. All these occupants were made four groups based on their characteristic of dependency like resources dependent (RD) includes five activities, wage earner (WE) to four activities, self-resilient (SR) to four activities and Government servant (GS).

The major occupation of the villagers was based on a collection of aquatic bio-resources available in surrounding creeks, rivers, and estuaries. Resource dependency activities were the major occupations in all the regions of Sundarban with nearly 60% population being engaged followed by wage earners and self-resilience. Collection of aquatic bio-resources (fishing, crab, and prawn collection) was the principal occupation of the study areas followed by important secondary occupations like agriculture, honey collection, fuel wood collection working as labours and engaging in business and trading.

Table 16.8 Occupation categories and types in three villages of Sundarban

Occupational category	Occupation type	Resources	Category of occupation (%)		
			Samsernagar	Rajatjubile	Kisorimohanpur
Resource dependent	Collection of aquatic resource	Fin fish, shell fish crab, prawn seed	62.96	59.94	56.94
	Collection of forest resource	Honey, wax, wood, timber, leaf			
	Agriculture	Paddy and vegetables			
Wage labour	Wage earning	Wood processing daily useable, goods domestic works	26.2	31.60	31.6
Self-resilience	Business	Domestic items, chemical fertilizer	8.67	6.29	9.46
	Self-earning	Wood and timber, fermented palm juice			
	Tuition	School children			
Government service	Permanent salaried job	Government department	2.17	2.17	2.0
Total			100	100	100

Source: Field Survey

16.8.3 Relative Livelihood Index

Relative Livelihood Index (RLI) was developed to have a comparative assessment of the degree of strength among occupations (Fig. 16.3). The objective of this index was to assess whether the strength of primary occupation was alone robust to support the occupants continuing livelihood. If the sum of the total strength of secondary and tertiary occupations were greater than that of the primary one, then primary occupation was considered to be weak and dwellers belonging to it manifested a poor state of livelihood.

Our of 8 major occupations (Fishing and crab collection, Agricultural labour, Crab collection, Prawn seed collection, Casual labour, Honey collection, Fishing, Carpentry) four occupations scored positively, whereas the other four showed a negative trend (Table 16.9 and Fig. 16.4). Practically, occupants engaged in primary occupations measured with negative scores depended on secondary and tertiary

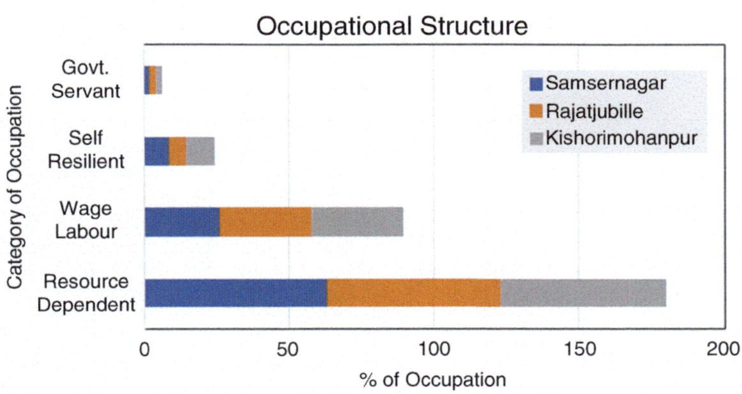

Fig. 16.3 Occupational structure of selected villages of Sundarban

Table 16.9 Calculation of relative livelihood index (RLI)

Sl. No.	Primary occupation	Score (number)		
		Samsernagar	Rajatjubille	Kisorimohanpur
1	Fishing and crab collection	60.97	23.35	19.80
2	Agricultural labour	21.38	4.82	24.75
3	Crab collection	17.6	10.85	−4.95
4	Prawn seed collection	14.59	8.88	4.95
5	Casual labour	−4.82	12.62	−9.90
6	Honey collection	−23.68	−27.78	−43.56
7	Fishing	−50.08	−31.4	−19.8
8	Carpentry	−34.99	−43.32	−52.78

RLI = [A × 3 − {(B × 2) + (C × 1)}], where A = Primary occupation, B = Secondary occupation and C = Tertiary occupation
Source: Field Survey

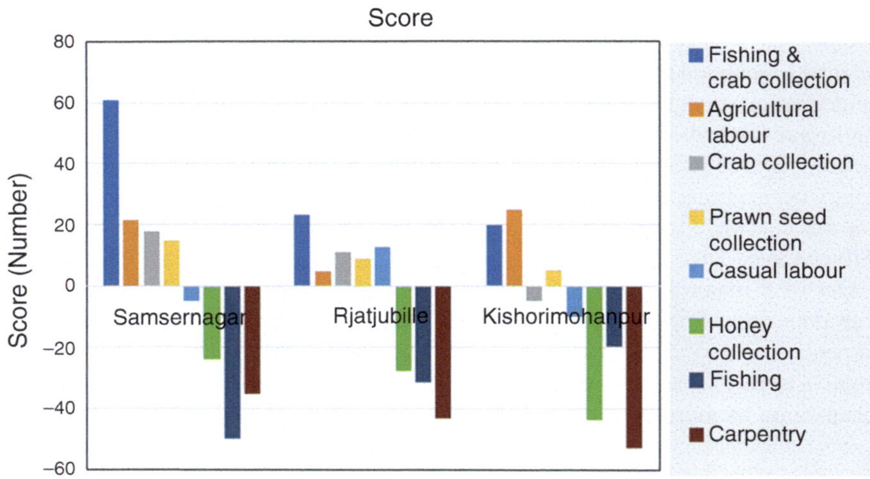

Fig. 16.4 Livelihood index of selected villages of Sundarban

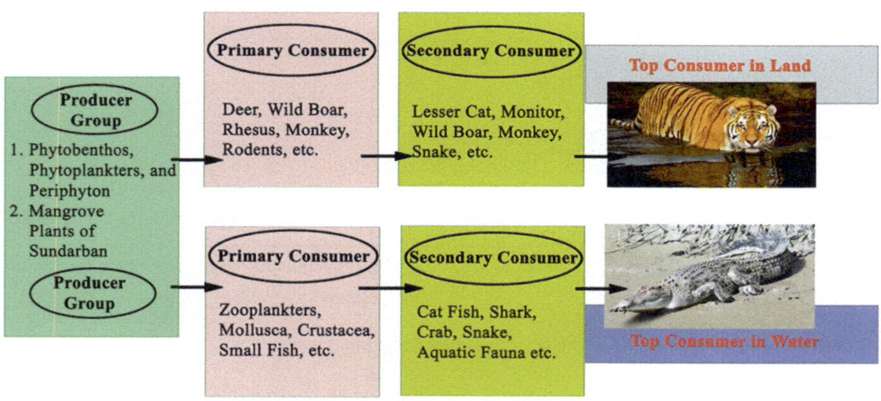

Fig. 16.5 Food chain of mangrove ecosystem of Sundarban

occupations for livelihood, since primary occupations of that particular livelihood were weak. Three occupations such as 'Fishing & crab collection', 'Crab collection', and 'Prawn seed collection' which had positive scores among others were exclusively resourced dependence. Occupants were inherently skillful in these occupations in one way and found many earnings/catches in collecting natural resources in others. On the other hand, occupations like casual labour, honey collection, fishing and carpentry scored negatively meaning that these belonged to a poor state of livelihood and for sustenance occupants had to depend on other occupations as secondary or tertiary in spite of their lack of skillfulness (Figs. 16.4 and 16.5).

Strength of fishing and crab collection as a major primary occupation is observed in Samsernagar and Rajatjubille, whereas in Kisorimohanpur agricultural labour dominates others. However, in the entire region, the primary occupation like honey collection, fishing, and carpentry scores negatives which justify that this occupation was alone robust to support the occupants continuing livelihood.

16.9 Conclusion

Sundarban mangrove ecosystem warrants maintaining the food chain of terrestrial and aquatic systems simultaneously apart from other ecosystem services necessary for the livelihood of the rural people living in the fringe areas (Fig. 16.5). Their major collection includes honey, fuelwood, fishes, and crabs both legally and illegally. The dwellers of the Sundarban have to face both visible and not-so-visible uncertainties in their traditional livelihoods like agriculture and fishing. On the other hand, huge population pressure forces the dwellers to make use of these natural resources terribly and indiscriminately, which gave rise to conflict between forest managing authorities and local habitants. Over exploiting of resources cannot be controlled without the execution of appropriate alternate livelihood arrangements including satisfactory marketing transactions. As such, there is a need for appropriate water and land care system for the entire Sundarban to make this fragile ecosystem into climatically more resilient and sustainable. Cooperative basis modern aquaculture practice and crab rearing, excavation of ponds for rainwater harvesting along pisciculture are some of the alternatives to provide independency of villagers for livelihood. Plantation with fast-growing trees of high calorific value can be initiated as bio-resource which may be common property to be available largely on the roadside, canal side, and river embankments, under the control of *Panchayat*. These attempts will make the villagers become independent as well as self-resilient in the collection of fuelwood in one way and discourage them to enter into the forest in others and will thereby minimize the risk of exposure to man-eaters both in terrestrial and water bodies. More research will promote more widespread efforts to develop conservation and sustainable development policies that integrate steady increasing salinity of the region, changes in mangrove dynamics, and the welfare impacts on poor and marginal communities. The socio-economic characteristics of these communities should be considered as the main criteria for designing any conservation plan as well as a sustainable livelihood programme which would reproduce both benefit-sharing and incentive measures applied to all stakeholders.

Acknowledgement I am grateful to Swapan Paul, a Research scholar of West Bengal State University for helping to prepare Figures no.1 and 2. Thanks are also due to Semonti Das, Assistant professor of Chandrakona Vidyasagar Mahavidyalaya for checking citation as well as references of the manuscripts.

References

Ajibola A, Chamunorwa JP, Erlwanger KH (2012) Nutraceutical values of natural honey and its contribution to human health and wealth. Nutr Metab 9(1): 61.

Alongi DM (2002) Present state and future of the world's mangrove forests. Environ Conserv 29(3): 331–349.

Angell CL (1994) Promotion of Small-scale Shrimp and Prawn Hatcheries in India and Bangladesh. Bay of Bengal Program (BOBP), Madras, India.

Ashley C, Carney D (1999) Sustainable livelihoods: Lessons from early experience (Vol. 7, No. 1). Department for International Development. London.

Badhsheeba MA, Vadivel V (2020) Physicochemical and phytochemical contents of the leaves of *Acrostichum aureum* L. J Glob Biosci 9(4): 7003–7018.

Bandaranayake WM (1998) Traditional and medicinal uses of mangroves. Mangroves Salt Marshes 2: 133–148.

BARC (1998) Changes in cropping pattern in Sundarban over last two decades. Bangladesh Agriculture Research Council, Dhaka, Bangladesh, p 10.

Bradbear N (2003) Bees and rural livelihoods. 16pp booklet in English, Portuguese and Spanish editions. Bees for Development, Monmouth, UK.

Bradbear N (2005) Bees for Development Journal 74. Bees for Development, UK.

Bunt JS, Robertson AI, Alongi DM (1992) Introduction. In: Tropical mangrove ecosystem. American Geophysical Union, Washington, DC, p 1–6.

Burgett M (2000) Honey hunters of the Sundarban, Bees Dev Journal 56: 6–7.

Census (2011) Primary census abstracts, Registrar General of India, Ministry of Home Affairs, Government of India, Available at: http://www.censusindia.gov.in/2011census/PCA/pca_highlights/pe_data.

Chakraborty K (1985) In: Abstract, Sundarbans Mangrove Biomass Productivity and Resources of Utilization of mangrove. Proceedings of National Symposium of Biology, Utilization and Conservation of mangrove.

Chambers R, Conway G (1992) Sustainable rural livelihoods: Practical concepts for the 21st century. IDS Discussion Paper 296, IDS, Brighton.

Chapman VJ (1976) Mangrove vegetation. In: Cramer J, p 447.

Chatterjee T (2017) Climate change and its collision on fisheries resource in sundarban region of south 24 pargana, West Bengal, Suresh Gyan Vihar University. Int J Environ Sci Technol 3(2): 10–13.

Chaudhuri AB, Choudhury A (1994) Mangroves of the Sundarbans. Volume 1, The IUCN Wetland Program. India.

Chowdhury A, Naz A, Maiti SK (2017) Health risk assessment of 'tiger prawn seed' collectors exposed to heavy metal pollution in the conserved mangrove forest of Indian Sundarbans: A socio-environmental perspective. Hum Ecol Risk Assess 23(2): 203–224.

Chowdhury AN, Mondal R, Brahma A, Biswas MK (2020) Eco-psychiatry and environmental conservation: Study from Sundarban Delta, India. Environ Health Insights 2(1): 61.

Danda AA (2007) Surviving in the Sundarbans: Threats and responses. Unpublished PhD, University of Twente. p 203.

Danda AA, Sriskanthan G, Ghosh A, Bandyopadhyay J, Hazra S (2011) Indian Sundarbans Delta: A vision. World Wide Fund for Nature-India, New Delhi, p 40.

Darwin NT (2014) Livelihood strategies of people surrounding the Sundarban mangrove forest (Doctoral dissertation, Charles Darwin University).

Das CS (2012) Tiger straying incidents in Indian Sundarban: Statistical analysis of case studies as well as depredation caused by conflict. Eur J Wildl Res 58(1): 205–214.

Das CS (2015) Causes, consequences and cost-benefit analysis of the conflicts caused by tiger straying incidents in Sundarban, India. Proceedings Zoological Soc 68(2): 120–130.

Das CS (2017) Human-wildlife conflicts in the Sundarbans, In: Danda AA, Joshi AK, Ghosh A, Saha, R (eds) State of art report on biodiversity in Indian Sundarbans. World Wide Fund for Nature-India, New Delhi. p 325–340.

Das CS (2018) Pattern and characterisation of human casualties in Sundarban by Tiger attacks, India. J Sustain For 1(2): 1–10.

Das CS, Bandyopadhyay S (2012) Sharing space—human animal conflicts in Indian Sundarban.

Das P, Das A, Roy S (2016) Shrimp fry (meen) farmers of Sundarban Mangrove Forest (India): A tale of ecological damage and economic hardship. Int J Agricult Food Res 5(2): 28–41.

Das CS & Mandal RN (2016) Coastal people and mangroves ecosystem resources vis-à-vis management strategies in Indian Sundarban. Ocean & Coastal Management, 134, 1–10.

Dasgupta S, Sobhan I, Wheeler D (2017) The impact of climate change and aquatic salinization on mangrove species in the Bangladesh Sundarbans. Ambio 46(6): 680–694.

Day Jr JW, Conner WH, Ley-Lou F, Day RH, Navarro AM (1987) The productivity and composition of mangrove forests, Laguna de Terminos, Mexico. Aquat Bot 27(3): 267–284.

Day Jr JW, Coronado-Molina C, Vera-Herrera FR, Twilley R, Rivera-Monroy VH, Alvarez-Guillen H, Day R, Conner W (1996) A 7 year record of above-ground net primary production in a southeastern Mexican mangrove forest. Aquat Bot 55(1): 39–60.

De DK, Chakaraborty RK, Subramanian M (1978) Preliminary survey on the availability of culturable brackishwater prawn seed in the Haldia estuary around Baughata, West Bengal. J Inland Fish Soc India 10: 142–143.

Decker DJ, Chase LC (1997) Human dimensions of living with wildlife: A management challenge for the 21st century. Wildlife Soc Bullet (1973–2006) 25(4): 788–795.

DFID GS (2000) Sustainable livelihoods guidance sheets, Section 2. Framework.

García-Frapolli E, Ramos-Fernández G, Galicia E, Serrano A (2009) The complex reality of biodiversity conservation through Natural Protected Area policy: Three cases from the Yucatan Peninsula, Mexico. Land Use Pol 26(3): 715–22.

German Agro Action Rural India (2009) Land shaping for crop irrigation and to prevent water stagnation in Sunderbans, West Bengal https://www.indiawaterportal.org/articles/ramakrishna-mission-lokasiksha-parishad-rkmlsp-promotes-land-shaping-crop-irrigation-and-0

Ghosh A (2010) The shadow lines. Penguin Books India.

Ghosh A, Schmidt S, Fickert T, Nüsser M (2015) The Indian Sundarban mangrove forests: History, utilization, conservation strategies and local perception. Diversity 7(2): 149–169.

Ghosh P (2014) Subsistence and biodiversity conservation in the Sundarban Biosphere Reserve, West Bengal, University of Kentucky, Theses and Dissertations-Geography. p. 26. https://uknowledge.uky.edu/geography_etds/26

Ghosh U, Bose S, Bramhachari R (2018) Living on the edge: Climate change and uncertainty in the Indian Sundarbans. STEPS Working Paper 101, Brighton: https://opendocs.ids.ac.uk/opendocs/handle/20.500.12413/13597

Ghoshal TK, De D, Biswas G, Kumar P, Vijayan KK (2019) Brackishwater aquaculture: Opportunities and challenges for meeting livelihood demand in Indian sundarbans. In The Sundarbans: A Disaster-Prone Eco-Region. Springer, pp. 321–349.

Giri C, Pengra B, Zhu Z, Singh A, Tieszen LL (2007) Monitoring mangrove forest dynamics of the Sundarbans in Bangladesh and India using multi-temporal satellite data from 1973 to 2000. Estuar Coast Shelf Sci 73(1–2): 91–100.

Glavovic B, Scheyvens R, Overton J (2003) Waves of adversity, layers of resilience. Exploring the sustainable livelihoods approach. In: Overton DJ, Nowak B (eds) Contesting development: Pathways to better practice. Proceedings of the Third Biennial Conference of the Aotearoa New Zealand International Development Studies Network (Dev Net), Massey University, Dec 5–7, 2002. Institute of Development Studies, p 289–293.

Gong WK, Ong JE, Wong CH, Dhanarajan G (1984) Productivity of mangrove trees and its significance in a managed mangrove ecosystem in Malaysia. In: Proceedings of the Asian Symposium on Mangrove Ecosystem in Research and Management (Eds. Soepadmo E., Rao A.N. and Macintosh D.J.) Kuala Lumpur: University of Malaya. p 216–225.

GoWB (2005) Administrative Report: 2004–2005: Sundarbans Development Board, Sundarbans Affairs Department, Government of West Bengal.
GoWB (2019) Annual report 2017–2018. Wildlife wing, Directorate of Forest, Govt. of West Bengal: p 227.
Hasan R, Naser N (2016) Fishermen livelihood and fishery resources of the Sundarbans reserved forest along the Mongla port area Bangladesh. Int J Fish Aquat Stud 4(3): 468–475.
Hazarika AK, Kalita U (2019) Conservation and Livelihood Conflict of Kaziranga National Park: A World Heritage Site of Assam, India. Space Culture, India 7(3): 224–232. https://doi.org/10.20896/saci.v7i3.656
Hazra S, Samanta K (2016) Temporal Change Detection (2001–2008): Study of Sundarban. Working Papers. ID: 10526, eSocialSciences.
Hegde R, Suryaprakash S, Achoth L, Bawa KS (1996) Extraction of NTFPs in the forests of Biligiri Rangan Hills, India: Contribution to rural income. Econ Bot 50(3): 243–251.
Hill DB, Webster TC (1995) Apiculture and forestry (bees and trees). Agroforest Syst 29 (3): 313–320.
Hogarth PJ (1999) The biology of mangroves. Oxford University Press (OUP).
Hoq ME (2007) An analysis of fisheries exploitation and management practices in Sundarbans mangrove ecosystem, Bangladesh. Ocean Coast Manage 50(5–6): 411–427.
Hossain SJ, Iftekharuzzaman M, Haque MA, Saha B, Moniruzzaman M, Rahman MM, Hossain H (2016) Nutrient compositions, antioxidant activity, and common phenolics of Sonneratia apetala (Buch.-Ham.) fruit. Int J Food Propert 19(5): 1080–1092.
Islam MM, Borgqvist H, Kumar L (2019a) Monitoring mangrove forest landcover changes in the coastline of Bangladesh from 1976 to 2015. Geocarto Int 34(13): 1458–1476.
Islam MN, Ratul SB, Sharmin A, Rahman KS, Ashaduzzaman M, Uddin GM (2019b) Comparison of calorific values and ash content for different woody biomass components of six mangrove species of Bangladesh Sundarbans. J Indian Acad Wood Sci 16(2): 110–117.
IUCN (1994) The World Conservation Union. In: Hossain Z, Acharya G (eds) Mangroves of Sundarbans vol. 2: Bangladesh. Bangkok, Thailand: IUCN-The World Conservation Union, p 230.
Junk WJ, Brown M, Campbell IC, Finlayson M, Gopal B, Ramberg L, Warner BG (2006) The comparative biodiversity of seven globally important wetlands: A synthesis. Aquat Sci 68(3): 400–414.
Kamruzzaman M, Ahmed S, Osawa A (2017) Biomass and net primary productivity of mangrove communities along the Oligohaline zone of Sundarbans, Bangladesh. Forest Ecosys 4(1): 1–9.
Kar A, Raut SK, Bhattacharya M, Patra S, Das BK, Patra BC (2017) Marine fishes of West Bengal coast, India: Diversity and conservation preclusion. Reg Stud Mar Sci 16: 56–66.
Khan MN, Khatun S, Azad MS, Mollick AS (2020) Leaf morphological and anatomical plasticity in Sundri (Heritiera fomes Buch.-Ham.) along different canopy light and salinity zones in the Sundarbans mangrove forest, Bangladesh. Glob Ecol Conserv 23: e01127.
Kibria AS, Costanza R, Groves C, Behie AM (2018) The interactions between livelihood capitals and access of local communities to the forest provisioning services of the Sundarbans Mangrove Forest, Bangladesh. Ecos Serv 32: 41–49.
Komiyama A, Havanond S, Srisawatt W, Mochida Y, Fujimoto K, Ohnishi T, Ishihara S, Miyagi T (2000) Top/root biomass ratio of a secondary mangrove (Ceriops tagal (Perr.) CB Rob.) forest. Forest Ecol Manage 139(1–3): 127–134.
Komiyama A, Ogino K, Aksornkoae S, Sabhasri S (1987) Root biomass of a mangrove forest in southern Thailand. 1. Estimation by the trench method and the zonal structure of root biomass. J Tropic Ecol 1: 97–108.
Kothari A, Patel A (2006) Environment and human rights: An introductory essay and essential readings. National Human Rights Commission.
Kothari MJ, Rao KM (1995) Dolichandrone spathacea (Lf) K. Schum.(Bignoniaceae)-a little known plant for Goa. J Econ Taxon Bot 19(2): 323–324.

Kothari N (2015) The plight of the honey collectors in Sundarban: Challenges, uncertainties strategies and survivals. J Human Soc Sci 20(2): 56–59.

Kumar KS, Bhowmik D, Biswajit C, Chandira MR (2010) Medicinal uses and health benefits of honey: An overview. J Chem Pharm Res 2(1): 385–395.

Mahapatra BK, Chatterjee P, Saha D, Datta NC (1999) Declining trend in the abundance of seeds of tiger shrimps, *Penaeus monodon* (fabricus) in the Sundarbans with suggestion for restorations. Sundarbans Mangal 532–538.

Mahapatra BK, Sarkar UK, Lakra WS (2014) A review on status, potentials, threats and challenges of the fish biodiversity of West Bengal. J Biodivers Biopros Dev 2: 140. https://doi.org/10.4172/2376-0214.1000140

Mandal RN, Das CS, Naskar KR (2010) Dwindling Indian Sundarban mangrove: The way out. Sci Cult 76(7–8): 275–282.

Mandal RN & Bar R (2018) Mangroves for Building Resilience to Climate Change. CRC Press.

Mandal RN, Meher PK, Naskar KR (2009) Effect of salinity on germination and seedling development of Heritiera fomes Buch. Ham. Proc Nat Acad Sci India, B-Biol Sci 79: 153–160.

Mandal RN, Naskar KR (2008) Diversity and classification of Indian mangroves—a review. Trop Ecol 49(2): 131–146.

Mandal RN, Saenger P, Das CS, Aziz A (2019a) Current status of mangrove forests in the trans-boundary Sundarbans. In: Sen HS (Ed.) The Sundarbans: A disaster-prone eco-region. Springer Nature Switzerland, p 93–131.

Mandal S, Burman D, Mandal UK, Lama TD, Maji B, Sharma PC (2017) Challenges, options and strategies for doubling farmers' income in West Bengal–reflections from coastal region. Agricul Econ Res Rev 30: 89–100.

Mandal UK, Burman D, Bhardwaj AK, Nayak DB, Samui A, Mullick S, Mahanta KK, Lama TD, Maji B, Mandal S, Raut S (2019b) Water logging and coastal salinity management through land shaping and cropping intensification in climatically vulnerable Indian Sundarbans. Agric Water Manage 216: 12–26.

Mann KH (1982) Ecology of coastal waters: A systems approach. University of California Press.

McDowell MC (1995) The development of the river Hoogli for navigation Commemoration volume of Port Calcutta, p 69.

Miah G, Bari N, Rahman A (2003) Agricultural activities and their impacts on the ecology and biodiversity of the Sundarban area of Bangladesh. J Natl Sci Foundation 31: 175–199.

Mistri A, Das B (2020) Environmental change, livelihood issues and migration. Springer.

Naskar K, Ghosh A (1989) Mangrove forest of the Sundarbans: Its impact on estuarine fisheries. Proceeding in Coast Zone Management of West Bengal, Sea Explores Institute, Calcutta, p A.47–59.

Naskar KR, Guha Bakshi DN (1987) Mangrove swamps of the Sundarbans–an ecological perspective. Naya Prokash, Calcutta.

Naskar KR, Mandal RN (1999) Ecology and biodiversity of Indian mangroves. Daya Books.

Nishat B (2019) Landscape narrative of the Sundarban: Towards collaborative management by Bangladesh and India. The World Bank.

Pakrashi H. (2016) Is there a tomorrow?: The story of survival of Sunderbans against climate change. In: Nautiyal S., Schaldach R., Raju K., Kaechele H., Pritchard B., Rao K. (eds) Climate change challenge (3C) and social-economic-ecological interface-building. Environmental science and engineering. Springer, Cham. https://doi.org/10.1007/978-3-319-31014-5_21

Peters CM, Gentry AH, Mendelsohn RO (1989) Valuation of an Amazonian rainforest. Nature 339: 655–656.

Pillai RS, Ong BL (1999) Effects of inorganic nitrogen availability on the sporophytes of *Acrostichum aureum L.* Photosynthetica 36(1–2): 259.

Putz FE, Chan HT (1986) Tree growth, dynamics, and productivity in a mature mangrove forest in Malaysia. Forest Ecol Manage 17(2–3): 211–230.

Ray P (2000) Sundarban–A goldmine for prawn fisheries. Paper presented at the Celebration of Fourth Sundarbans Day, Kolkata Wildlife Society. Kolkata, p 59–64.

Robertson AI, Alongi DM Boto KG (1992) Food chains and carbon fluxes. In: Robertson AI, Alongi DM (eds), Tropical mangrove ecosystems. American Geophysical Union, Washington. p. 193–326.
Rodrigues BF (2006) Biodiversity loss and its impacts on rural health/alternate systems of medicine. In: Sonak S (ed) Multiple dimensions of global environmental change, p. 170–181.
Ros-Tonen M, Dijkman W, Lammerts van Bueren E (1995) Commercial and sustainable extraction of non-timber forest products. Towards a policy and management oriented research strategy. The Tropenbos Foundation, Wageningen.
Saberwal VK, Gibbs JP, Chellam R & Johnsingh AJT (1994) Lion-human conflict in the Gir Forest, India. Conservation Biology, 8(2), 501–507.
Saenger P (2002) Mangrove ecology, silviculture and conservation. Springer Science & Business Media.
Sahana M, Sajjad H (2019) Vulnerability to storm surge flood using remote sensing and GIS techniques: A study on Sundarban biosphere reserve, India. Remote Sensing Appl: Soc Environ 13: 106–120.
Saintilan N (1997) Above-and below-ground biomasses of two species of mangrove on the Hawkesbury River estuary, New South Wales. Mar Freshwat Res 48(2): 147–152.
Sánchez-Triana E, Enriquez S, Afzal J, Nakagawa A, Khan AS (2014) Cleaning Pakistan's air: Policy options to address the cost of outdoor air pollution. The World Bank.
Sánchez-Triana E, Ortolano L, Paul T (2018) Managing water-related risks in the West Bengal Sundarbans: Policy alternatives and institutions. Int J Wat Resour Develop 34(1): 78–96.
Santhakumar V, Enamul Haque AK, Bhattacharya R (2005) An economic analysis of mangroves in South Asia. In: Khan M (ed) Economic development in South Asia. Tata McGraw Hill, New Delhi, p 368–437.
Sanyal P (1992) Tropical Ecosystem: ecology and Management, p. 309–313, (K. P. Singh and J. S. Singh eds, Wiley Eastern Ltd. New Delhi).
Sarangi SK, Maji B, Singh S, Burman D, Mandal S, Sharma DK, Singh US, Ismail AM, Haefele SM (2015) Improved nursery management further enhances the productivity of stress-tolerant rice varieties in coastal Rainfed lowlands. Field Crops Res 174: 61–70.
Schwamborn R, Saint-Paul U (1996) Mangroves-forgotten forests? Nat Resour Develop 43(44): 13–36.
Sen A, Pattanaik S (2017) How can traditional livelihoods find a place in contemporary conservation politics debates in India? Understanding community perspectives in Sundarban, West Bengal. J Political Ecol 24(1): 861–880.
Sen HS (2019) Climate-risk Sundarbans needs multi-pronged and unified approach for ecological sustenance a necessity for improved livelihood: Summary and concluding remarks. In: The Sundarbans: A disaster-prone eco-region, Springer, pp. 611–625.
Shackleton C, Shackleton (2004). The importance of non-timber forest products in rural livelihood security and as safety nets: A review of evidence from South Africa. South Afr J Sci 100(11): 658–664.
Simmons IG (1981) The ecology of natural resources, 2nd edition. Edward Arnold (Pub) Ltd, 16–19: p 436.
Singh OP (2002) Interannual variability and predictability of sea level along the Indian coast. Theor Appl Climatol 72(1–2): 11–28.
Singh A, Bhattacharya P, Vyas P, Roy S (2010) Contribution of NTFPs in the livelihood of mangrove forest dwellers of Sundarban. J Hum Ecol 29(3): 191–200.
Situmorang YA, Zhao Z, Yoshida A, Abudula A, Guan G (2020) Small-scale biomass gasification systems for power generation (<200 kW class): A review. Renewable Sustainable Energy Rev 117: 109486.
Spalding M, Kainuma M, Collins L (2010) World Atlas of mangroves. Earthscan: London, UK, p. 319.
STR (2018) Annual report 2017–2018. Conservator of forest & field director. Sundarban Tiger Reserve. p 236.

STR (2014) Annual report 2013–2014. Sundarban Tiger Reserve. Retrieved from www.sundarbantigerreserve.org/news/Annual%20Report%202013-14.pdf.

Tamai S, Nakasuga T, Tabuchi R, Ogino K (1986) Standing biomass of mangrove forests in southern Thailand. J Jpn Fore Soc 68(9): 384–388.

Thakur S, Ghosh S, Ghosh T (2016) Scope and constraint of beekeeping in Satjalia Island within Indian sundarban delta. Int J Agric Environ Res 2(1): 109–114.

Tomlinson PB (1986) The botany of mangroves. Cambridge University Press, London.

Twilley RR, Chen RH, Hargis T (1992) Carbon sink in mangrove and their implications to Carbon budget of tropical coastal ecosystem. Wat Air Soil Pollut 64: 265–288.

Untawale AG (2006) Change of coastal land use, its impact and management options. Multiple dimensions of global environmental change, TERI Press, New Delhi, India. p 23–43.

Upadhyay VP, Ranjan R, Singh JS (2002) Human-mangrove conflicts: The way out. Curr Sci 83 (11): 1328–1336.

Vyas P (2012) Biodiversity conservation in Indian Sundarban in the context of anthropogenic pressures and strategies for impact mitigation (Doctoral dissertation). Saurashtra University.

Wackernagel M, Larry O, Linares AC, Falfan ISL, Garcia JM, Geurrero AIS, Geurrero MGS (1997) Ecological footprints of nations: How much nature do they use? How much nature do they have? (Commissioned by the Earth Council, New Society Publishers).

Whittaker RH, Likens GE (1975) The biosphere and man. In: Lieth H, Whittaker RH (eds) Primary productivity of the biosphere, Springer-Verlag Ecological Series 14: 305–328.

World Bank (2014) Building resilience for sustainable development of the Sundarban. Strategy Report. Report. No. 88061, Sustainable Development Department, Environment and Water Resources Management Unit, South Asia Region.

Zohora FT (2011) Non-timber forest products and livelihoods in the Sundarbans. Rural livelihoods and protected landscapes: Co-management in the Wetlands and Forests of Bangladesh. Nishorgo Network, Dhaka, p. 99–117.

The Roles of Mangroves in Sustainable Tourism Development

17

Yarina Ahmad and Mohd Nazip Suratman

Abstract

Mangrove and its ecosystem have contributed significantly to the environment, community, and economy. Mangrove forests became a tourism attraction for decades, and the demand for nature-based tourism of mangrove forests is increasing throughout the world. Unsustainable mangrove tourism efforts may lead to many drawbacks, which include forest loss, depletion of natural resources, and also increase of pollution in mangrove forests and their surroundings. While aiming to maximise the economic earnings from mangrove tourism, it should be balanced with mangrove ecosystem conservation's efforts to ensure sustainable tourism development. This situation can be realised through implementing effective strategies, including continuous conservation and restoration efforts of mangroves; enhancing the policies and legislation on the use and management of mangroves; enhancing infrastructure and facilities, and efficient use of resources through sound technologies; and encouraging community participation and engagement. Balancing the economic activity of mangrove tourism while maintaining mangrove forests through preservation, conservation, and restoration are the key success to achieve sustainable tourism development.

Y. Ahmad
Faculty of Administrative Science and Policy Studies, Universiti Teknologi MARA (UiTM), Shah Alam, Malaysia

Institute for Biodiversity and Sustainable Development, Universiti Teknologi MARA (UiTM), Shah Alam, Malaysia

M. N. Suratman (✉)
Faculty of Applied Sciences and Institute for Biodiversity and Sustainable Development, UiTM, Shah Alam, Malaysia
e-mail: nazip@uitm.edu.my

R. P. Rastogi et al. (eds.), *Mangroves: Ecology, Biodiversity and Management*,
https://doi.org/10.1007/978-981-16-2494-0_17

Keywords

Mangrove · Tourism · Sustainable tourism development · Conservation

17.1 Introduction

Mangrove forests offer myriad benefits to the environment, community, and economy; however, conversely, they have received major threats from imbalanced and destructive industrial development initiatives such as excessive aquaculture and agriculture activities as well as high-scale urban expansion. While noting that mangrove forests have contributed to the local socio-economy sector for centuries, the demand for larger economic contributions through the tourism industry, for one, has placed the mangrove ecosystem in jeopardy. Balancing between economic earnings and mangrove ecosystem conservation efforts needs to be scrutinised to ensure sustainable tourism development.

This chapter presents the roles of mangrove in ensuring the sustainability of the coastal ecosystem; and the tourism industry needs to be enhanced by emphasising precisely four critical strategies. Firstly, through the continuous conservation and restoration efforts of mangroves to help achieve a sustainable tourism development and further promote the 2030 Sustainable Development Goals (SDGs). Secondly, there should be a focus on enhancing the policies and legislations of the use and management of mangroves, particularly in highlighting the importance of regulating mangrove forests. Thirdly, by enhancing infrastructure and facilities, efficient use of resources through sound technologies can be ensured to promote environmental cleanliness. Finally, community engagement and participation in utilising mangrove forests as a tourism attraction should be encouraged in maintaining and conserving the area. Thus, all parties must play their roles to help establish an environmentally-responsible community. This can be done by educating and creating awareness through various initiatives.

Mangrove conservation and restoration efforts alongside tourism development initiatives can be initiated by optimising the potential and values of mangrove forests and highlighting their economic and social benefits while minimising environmental impact. This effort will ensure a genuine sustainable tourism development through utilising mangroves as economic earning, as well as conserving the mangrove ecosystem for future generations.

17.2 Mangroves

The academic discourse surrounding mangrove and its ecosystem has been discussed globally for centuries (Bacon 1987; Suratman 2008; Webber et al. 2016; Spalding and Parrett 2019). According to Webber et al. (2016), the word "mangrove" refers to both a specific vegetation type and the unique habitat which is also called tidal forest, swamp, wetland, or mangal. Mangroves are a group of trees and shrubs which are formed in swampy areas with low oxygen conditions prevailing below the first few centimetres (Suratman 2008). Further, mangrove trees have a special mechanism which allows them to survive water logged soil conditions and to grow in saline and brackish water (Pons and Fiselier 1991).

Mangrove forests consist of mangrove trees which grow in a muddy coastal wetland distributed worldwide in tropical and warm temperate coastal areas (Hakim et al. 2017; Spalding and Parrett 2019). Notably, mangrove forests represent transitional ecosystems where the ocean, land, and freshwater meet (Suratman 2008), which collect nutrients to create their own growing sides as well as forming breeding grounds for shellfish and fish (Pons and Fiselier 1991). Mangrove forests are unique and considered as a complex major component of coastal zones in the tropical and sub-tropical regions (Suratman 2008) which play a major role in environmental services, economy, and social benefits (Hakim et al. 2017).

Mangrove and its ecosystem have been contributed to the socio-economy development of coastal community for many years. Traditionally, mangrove forests have been utilised by the indigenous and local people for a variety of purposes. The coastal community has been dependent on mangrove waters for centuries for fishes, shrimps, crabs, mollusks (Suratman 2008). In recent times, the community continues to rely on these products from the mangrove habitats not only for their daily living, but also trading for economic development.

While human has gained many benefits from mangrove ecosystem for centuries; they also have exploited these natural resources from various aspects. Mangrove forests continue to be degraded rapidly through different types of human activities (Saenger et al. 1983; Hamilton and Snedaker 1984). Large areas of mangrove had been lost in recent decades to aquaculture, agriculture, and urban expansion (Spalding et al. 2010; Richards and Friess 2016; Castellanos-Galindo et al. 2015). Undoubtedly, mangrove degradation affects the ecological stability of coastal zones (Suratman 2008); gives a negative impact on the coastal economy (marine resources and fisheries industry) (Hakim et al. 2017); and depletion of the natural resources (Amir 2019). These situations open debates among environmentalists and scholars on the needs to protect and conserve mangrove forests through balancing the economic development and ensuring the sustainability of mangrove forests and its ecosystem.

17.3 Mangrove Tourism

Mangrove is now recognised as being among the most important ecosystem in the world that provide various benefits to environment, community, and the economy (Spalding and Parrett 2019). Recently, mangrove forests are among the popular tourism attractions, particularly for nature-based tourism. This is not something new, as mangrove ecosystem has been a tourism attraction for decades. Since the 1970s, mangrove tourism was already found in the Caribbean (Bacon 1987). Since then, mangrove tourism and attractions have been associated with the general trend of ecotourism (Balmford et al. 2009; Cisneros-Montemayor and Sumaila 2010).

In one of the recent studies undertaken by Spalding and Parrett (2019), mangrove forests are found as tourist's attractions in many countries across the globe. The article explored mangrove recreation and tourism worldwide through online search of popular travel website TripAdvisor. In their study, which involved 3,985 mangrove "attractions" located at 93 countries and territories, revealed that mangrove tourism attracts tens to hundreds of millions of visitors annually; and thus, considered as a multi-billion industry.

Mangrove tourism is associated with other attractions including facilities, activities, and wildlife (Spalding and Parrett 2019). Among the facilities provided at mangrove tourism and attraction include a boardwalk, viewing tower, information centre, and information boards. The activities offered in the mangrove tourism areas are mostly related to boating, such as airboat, canoe and kayak, stand up paddle boarding; and other activities such as fishing and hiking. Wildlife is also another attraction associated with mangrove tourism; among them are birdlife, bioluminescence, fireflies, monkeys, proboscis monkey, manatee/dugong, crocodile/alligator, and many other wildlife attractions (Spalding and Parrett 2019). These are among the strategies to utilise mangrove forests as a tourism attraction undertaken by the tourism industry to benefit the community and boost the economy. However, rapid development of mangrove tourism may lead to drawbacks which are not only for short-term, but also will give long-term impact to mangrove habitat and its environment.

There are always two sides of a coin. Despite the widespread discussion on the values and benefits of mangrove forests and its ecosystem; there are debates about the loss of mangrove areas, and this happens in many countries across the globe (Webber et al. 2016). Unsustainable use for economic development purposes on mangroves has led to an alarming loss of global mangrove forests (WWF International 2017). Notably, the loss of mangrove forests has been reported since the 1980s (FAO 2007; Polidoro et al. 2010). Nearly half of all mangrove forests have disappeared since the mid-twentieth century (WWF International 2017). Further, the loss of mangrove forests globally was reported at a mean rate of 1-2 per cent per year (Duke et al. 2007; FAO 2007). In addition, the global loss rate of mangroves is 3 to 5 times higher than terrestrial forests (WWF International 2017). However, the loss of mangrove forests has been declining over the last three decades (Spalding et al. 2010). This could be an indication of increasing resilience of the remaining mangroves or the result of effective conservation and restoration/rehabilitation efforts (Webber et al. 2016).

Mangrove ecosystems provide a leading exemplar of the potentially high value to be obtained from multiple ecosystem services (Salem and Mercer 2012); and thus, provide a critical argument for both protection and restoration (Spalding and Parrett 2019). Utilising mangrove as tourism attraction and other species attraction might in turn generate powerful arguments for conservation and management (The Economist 2018); in particular for preserving the environment for future generations. Hence, four critical strategies should be undertaken to balance mangrove conservation efforts alongside tourism development in achieving a genuine mangrove sustainable tourism development.

17.4 Sustainable Development of Mangrove Tourism

Countries around the world are responsible in improving the planet and life of citizens to achieve a better and more sustainable future. The Sustainable Development Goals (SDGs) address the global challenges with the 17 Goals, which are all interconnected, with the aim of no one left behind (United Nation 2020). In promoting sustainable tourism development of mangrove, Goal 14 of the sustainable development agenda is highly related. The aim of Goal 14 is to "conserve and sustainably use the oceans, sea and marine resources for sustainable development". In this context, the world's ocean—their temperature, chemistry, and life—drive the global systems that make the earth habitable for humankind (UNDP 2020). The world's ocean covers mangrove forests and its ecosystem; and thus, through this goal, there is a high requirement for mangrove forests to be protected, conserved, and managed effectively, to benefit the community, society, as well as to preserve the environment for future generations. Other than that, promoting sustainable development of mangrove tourism also helps to achieve other SDGs including Goal 1: "No poverty"; Goal 2: "Zero hunger"; Goal 8: "Decent work and economic growth"; Goal 13: "Climate action"; and Goal 15: "Life on land" (Blum and Herr 2017).

Hence, in achieving a genuine sustainable development of mangrove tourism through balancing the economic needs and protecting the mangrove forests and its ecosystem; four critical strategies are emphasised. These four strategies are interrelated to each other—conservation and restoration of mangroves require a clear policy and legislation to ensure effective management of mangrove tourism; subsequently, infrastructure, facilities, community participation, and engagement must be enhanced—these strategies will support the agenda to achieve sustainable development of mangrove tourism holistically. Figure 17.1 presents the four critical strategies to achieve sustainable development of mangrove tourism.

Fig. 17.1 Promoting sustainable tourism development through four critical strategies

17.4.1 Continuous Conservation and Restoration Efforts of Mangroves

In debating this issue, the first choice in hand is to preserve the mangrove forests. In other words, no human activity is conducted at and/or nearby the mangrove surroundings. In some remote areas, there are mangrove forests which are not being utilised and exploited. However, once the mangrove forests are discovered, their benefits will be taken advantage by human, community, and industry for a source of food, income, and also increasing profit. Are utilising mangrove forests for a living and boosting the economy considered wrong? There is a degree of continuum in debating about this issue. Generating income activities by community and industry can be considered acceptable if the mangrove forests are protected, and conservation and restoration efforts of mangroves are done appropriately and continuously.

In recent trends, mangrove tourism is one of the earning strategies utilised by many countries as well as to promote their culture (Hakim et al. 2017; Spalding and Parrett 2019). While mangrove ecosystem is identified as among the most threatened habitats on the Earth, and with limited research on mangrove and conservation efforts; this resulted in a limited understanding of mangrove ecosystem among many parties (Suratman 2008). In Indonesia, for instance, there are abundant of mangrove forests; and their government highly encourages for tourism development in natural areas, including mangrove ecosystem (Hakim et al. 2017). While acknowledging that the mangrove forests in Indonesia are threatened, the country and its community have begun to take efforts to restore mangrove ecosystem. Among the activities conducted including corporate social responsibilities (CSR) performed by companies focusing on rehabilitation of mangrove forests and its ecosystem, and continuous meeting with various government officers to take actions to protect the environment (Hakim et al. 2017).

In Malaysia, for instance, actions taken by the government to gazette all remaining mangrove forests within forest reserves and protected areas are considered as an effective initiative to preserve mangrove forests. Other initiative to conserve mangrove forests is through devising well-balanced coastal land-use plans, such as maintaining sustainable limits in logging and other harvesting activities of its resources. The Malaysian government has acted in retaining protective mangrove buffers along coastlines and rivers to prevent erosion; managing mangrove forests as fishery reserves to encourage environmentally-sensitive commercial aquaculture activities; and finally introducing the social forestry schemes where damaged forest areas can be planted and managed for small-scale village timber enterprises (WWF-Malaysia 2020).

In terms of restoration, mangrove planting is considered as a popular action; however, it was not always effective (Wetlands International 2016). Mangrove planting has been widely recognised following the Indian Ocean Tsunami in 2004. Since then, mangrove planting has become very popular, and many parties, including the government, non-governmental organisations (NGOs), private sectors, and communities have planted mangrove trees, and many fund-raising activities were undertaken to support the idea of mangrove planting. Although thousands of hectares of mangrove have been actively planted across the globe, most planting efforts fail to effectively restore functional mangrove forests. Among the factors leading to the failure of mangrove planting include lack of support from local communities, mono-species planting leading to non-functional mangroves, planting at the area where the recovered mangrove would block sediment and water flow, planting at the area of original cause of loss, at the area where mangroves are settling naturally, and at the area where not previously covered by mangroves (Wetlands International 2016).

While taking experiences on the failure of mangrove planting across the globe, Wetlands International (2016) suggested considering two principles to ensure successful mangrove restoration. These include ensuring biography conditions are appropriate for mangrove recovery and the socio-economic conditions allow for mangrove recovery. These two principles are the cornerstone of the Ecological Mangrove Approach developed by Lewis III (2005). This approach has been successfully applied in mangrove planting in Indonesia, which was called Community-Based Ecological Mangrove Restoration; and is considered as a best practice to be adopted by other countries (Wetlands International 2016).

The conservation and restoration of mangrove forests is thus an important contribution to the achievement of the Agenda 2030 of the United Nations and therein defined SDGs (WWF International 2017). Further, restoring and protecting mangroves help fulfil multiple SDGs, including SDG14 which focuses on sustainably governing of the oceans and coasts and recognises the immense value of mangroves to local communities; SDG1 and SDG2 on eliminating poverty and hunger; SDG8 in ensuring livelihoods and economic growth; SDG13 taking actions against climate change impacts; and SDG15 on halting biodiversity loss (Blum and Herr 2017).

17.4.2 Enhancing the Policies and Legislations of the Use and Management of Mangroves

The loss of mangrove forests across the globe and its impact to the mangrove ecosystem and environment has led to the need for policy and legislation on the use and management of mangrove to be enhanced. In some countries, mangroves are protected through legislation that limits or prohibits mangrove clearing (Webber et al. 2016). For instance, in Brazil, there is legislation on Brazil's Federal Forestry Code (Brazil 2012) which prohibits the use of any components of mangrove trees or plants. In the United States of America, there is an act to regulate trimming, disturbance, or removal of mangroves in the state, which is referred to as the Mangrove Trimming and Preservation Act enacted in 1996 in the 1 United Nations, Treaty Series¸ vol. 996, No. 14583 © 2016 United Nations 7 state of Florida (Webber et al. 2016).

Notably, not all countries, states, or regions have a specific legislation to protect mangroves. In Tonga, there is no policy and legislation specifically designated to regulate the use, management, and conservation of mangroves (The International

Union for Conservation of Nature n.d.); hence, the country proposed for law reform on mangrove conservation and management. Similarly, in Malaysia, there is no specific policy and act for mangrove protection and management. The National Forestry Act 1984 provides for the administration, management, and conservation of forests and forest developments within the states in Peninsular Malaysia. Meanwhile, the Environmental Quality Act 1974 (Amended 1985) protects the forest environment and biodiversity, in particular the logging of natural forests. In terms of policy, there are two policies that cover environmental issues generally, which are National Policy on the Environment, and National Coastal Resource Management Policy (Abd Shukor 2004). Other countries that have no specific policy and regulation on mangrove protection and the environment are Pakistan, Thailand, and Vietnam (Beresnev et al. 2016).

The absence of a policy and legislation on mangroves can be considered as a pertinent issue that hinders mangrove protection, conservation, and thus lead to weak management of mangrove forests. This situation will further provide unclear guidance to relevant parties who deal with mangrove forests, not only for tourism purposes. At present, the countries that have no specific policy and legislation rely on various relevant policies and legislation that cover mangrove generally under the name of "environment", "forests", "land", and many others. These will create confusion and further may lead to a weak understanding of mangrove, its ecosystem, and how to protect them. Hence, there is an urgent need for the respected countries to take action to develop a specific policy and legislation as undertaken in Brazil and USA. This will support a clear agenda to protect, conserve, and manage mangroves effectively.

While Van Lavieren et al. (2012) has emphasised on the need to establish framework policy and legislation for mangroves at the national level; the authors also have suggested that laws and regulations related to mangrove protection, conservation, and management must be enacted and enforced accordingly. Further, there should be a framework of mangrove management, which emphasised on the right of ownership, access and use of mangrove forest; and enhance human, technical, legal, and financial capacity for mangrove management at different level. Among the management measures and tools to maximise the benefits and help secure the long-term future of mangroves recommended by Van Lavieren et al. (2012) include:

- Increase mangrove restoration;
- Increase community involvement in mangrove management;
- Implement sustainable mangrove forestry practices;
- Encourage sustainable aquaculture practices;
- Establish protected areas in ensuring the protection of mangrove biodiversity;
- Develop cohesive management plans;
- Promote managed realignment to (re-) establish landward expansion of mangrove habitat;
- Encourage and support mangrove ecotourism to generate income and employment for local communities and to improve outreach and education;

- Enhance existing carbon stocks and reverse CO2 emissions by increasing protection and restoration of mangrove ecosystems, and build mangroves into emissions trading and climate change mitigation planning;
- Utilise multilateral environmental agreements, together with the establishment of national legal protection measures, to support mangrove management.

Once the policy, legislation, and management framework of mangrove are in place to ensure effective mangrove management and conservation, the way forward agenda is to coordinate action towards the protection and restoration of mangroves needs to be embedded within the international policy arena, notably under biodiversity, wetlands, sustainable development, and climate change agreements. Aligning the national agenda of mangrove with the international standards will ensure a holistic achievement of SDGs, in particular to promote sustainable development of mangrove tourism in the world.

17.4.3 Enhancing Infrastructure and Facilities, and Efficient Use of Resources through Sound Technologies to Environmental Cleanliness

The healthy environment of mangrove forests attracts local and international tourists to visit and vacation. The forests are able to offer "nature therapy" for people who are always busy with lives in cities and urban areas. Wildlife, panorama views, and stress-free environment encourage people to experience life at mangrove forests. This situation urges the local authority to open the door for tourists. Various efforts and initiatives need to be undertaken to ensure there is an efficient use of mangrove-related resources in promoting environmental cleanliness. As demands for visit and vacation at mangrove forests are increasing, tourism development is also growing. Due to this situation, infrastructure and facilities provided at mangrove forests must be eco-friendly and suitable to prevent mangroves from destruction. Otherwise, mangrove forests remain in extreme risks due to degradation, erosion, and the extinction of wildlife.

In today's globalised world, the tourism development is integrated with sound technologies and development plans, especially in the complex area likes mangrove forest. One of the rapid tourism developments that have taken place in mangrove forests is resort construction. In maintaining the environmental cleanliness in the mangrove forests, site planning of resort development is extremely important (Said and Rahman 2000). Site planning must consider the estuarine or intertidal wetland to conserve the relationship between physical and biotic factors. The conservation of the relationship is vital to safeguard the food chains from being broken as well as to protect the microclimate of the mangrove ecosystem. This huge responsibility depends mostly on the ability and capability of an architect who is accountable for the resort development. Further, the architecture planning designed at mangrove forests must take into account the infrastructure and facilities related to intensive human activity and building development. For example, vehicle parking, registration

Fig. 17.2 'Building with Nature' activity at Timbul Sloko village, Indonesia (Source: Spalding et al. 2014b)

area, outdoor games, and dining should be concentrated at the core building area to protect mangrove-related resources. This planning concept can optimise the opportunity for guests to experience the serenity of the forest's natural setting (Said and Rahman 2000). In ensuring tourists are able to experience memorable events at mangrove forests, comprehensive tourist guideline must be developed and designed alongside the architecture planning to regulate tourists' behaviours while at the forests. This initiative helps the resort management in promoting environmental cleanliness and further maintains an eco-friendly infrastructure and facilities for the long-term usage as well as to conserve and protect the mangrove ecosystem.

The "Building with Nature" concept is another approach that can be considered in promoting environmental cleanliness at mangrove forests (Spalding et al. 2014a, 2017). This approach has shown positive results by combining green (nature-based) with grey (engineered) solutions. This initiative represents a remarkable shift in environmental paradigm—from fighting against nature, to collaboratively cooperate with natural processes. For instance, Northern Java's coasts are suffering from severe erosion whereby the areas are retreated by three kilometres. Improper development of infrastructure and aquaculture disrupts the protection of mangrove forests and the sediment flows towards the coast and floodplains. Due to this situation, a group of researchers used "Building with Nature" approach to encounter the destructive erosion of the coast, particularly in Timbul Sloko village in Central Java. The approach entails the placement of permeable dams that break the waves and trap sediment, thus reclaiming land. Once the land is back, mangroves can recolonise the area and help protect the coastline against erosion. The waves are clearly much lower inside the grid of permeable dams than outside. In some cases, pioneering mangrove trees are already becoming established. In a direct response to these initial tests, the village signed a decree in 2014, establishing 100 hectares of the recently lost land as protected area, ensuring that, upon recovery, it will not be damaged or destroyed again. Figure 17.2 shows the activity of "Building with Nature" at Timbul Sloko village (Spalding et al. 2014b).

In addition, mangrove forests and its restoration sites can also be strategically placed to contribute to upgrading infrastructure with greater adoption of environmentally sound technologies through applying "green-grey infrastructures" for coastal protection (Blum and Herr 2017). This is in line with the SDGs target to upgrade infrastructure and retrofit industries to make them sustainable, with increased resource-use efficiency and greater adoption of clean and environmentally sound technologies and industrial processes, with all countries taking action in accordance with their respective capabilities by 2030. These actions further build the resilience of vulnerable coastal communities by reducing their exposure to climate-related extreme events and environmental shocks and disasters (Blum and Herr 2017). Similarly, the SDGs targeted to build the resilience of the poor and those in vulnerable situations as well as to reduce their exposure and vulnerability to climate-related extreme events and other economic, social, and environmental shocks and disasters by 2030.

17.4.4 Encouraging Community Participation and Engagement

The benefits of mangrove forests bring immeasurable value to the local community such as providing business opportunities, source of protein, and wild attraction among tourists. For this reason, it is important to discuss the roles of local communities in protecting and conserving mangroves forests to retain the long-standing benefits of mangroves. The local community has been recognised as the cornerstone in managing and protecting mangrove forests from extinction (Valenzuela et al. 2020). The continuous protection and conservation of mangrove forests does not only require the participation of local community; nonetheless, trust, norms, and networks are also important in coordinating the actions which are associated as "social capital". This concept incorporated with two important aspects, which are (1) it is created by building social relationships, and (2) it is a mechanism to acquire more resources that can further offer more opportunities (Valenzuela et al. 2020). The combination of human capital and social capital helps in upgrading the local community participation to the next level which is community engagement. This situation further creates strong responsibility of local community in protecting and conserving mangrove forests from harm and destruction.

The role of local communities in protecting and conserving mangrove forests is closely discussed with Community-Based Conservation (CBC), which has been developed as a viable alternative for sustainably managing the forests (Datta et al. 2012). This concept is aimed at empowering local community in protecting and conserving mangrove forests (Walters 2004). Notably, there are no absolute strategies and approaches outlined by the CBC. This is because different approaches and strategies to achieve sustainability will be implemented by considering various aspects including ecological, economic, and social sustainability in particular places (Boyer et al. 2016).

However, there are various barriers that prevent the effectiveness of CBC in promoting alternatives to manage mangrove forests, such as socio-cultural transformation of the local community (Datta et al. 2012). In today's challenging world, globalisation and modernisation have provided significant impact towards mangrove forest management. Examples of poor forest management may lead to excessive fishing activity in tropical coastlines among local community and others in the destruction of the mangrove forests. The rapid tourism development also creates countless problems towards the growth of mangroves. Again, this situation has two sides of the coin, which are (1) devalue the role of the local community, and (2) open for more opportunity of research and innovation in properly managing the forests. Ironically, ineffective management of mangrove forests does not only affect the sustainability of mangrove tourism; but it has been argued to bring the local community back into poverty (Wesenbeeck et al. 2015).

In order to encourage participation and engagement among the local community, several alternatives can be undertaken. One of the highly preferred alternatives among the local community is by receiving incentives. However, debates related to this aspect revealed that generating participation of the local community with incentives is considered as challenging due to the difficulty in identifying their real needs and preferences (Pons and Fiselier 1991). If the actions of local communities in protecting and conserving mangrove forests are determined by incentives, questions arise in the situation if and when the incentives are withdrawn. This situation does not guarantee the continuity and successfulness of mangrove restoration. While agreeing that incentives can act as "carrots" to motivate such behaviour; however, incentives cannot create self-volunteerism behaviours among local communities in protecting and conserving mangrove forests.

Another alternative that can be conducted to encourage local community participation and engagement is through implementing relevant regulations (Astuti et al. 2017; Majesty and Fadmastuti 2018). Mangrove is regarded as a threatened species on the Earth, which requires legal enforcement (Official Website of Global Mangrove Alliance n.d.). This initiative requires a clear direction from the government to state and local authorities. Since mangrove forests are the meeting point of land, sea water, and fresh water, the legal responsibility lies on various entities and agencies. This situation creates conflict in the inter-agency cooperation between relevant agencies which further reduces the effectiveness of policies and regulations implemented (Beresnev et al. 2016).

In Thailand, for instance, mangroves are regulated and managed by various agencies, including the Ministry of Natural Resources and Environment (MoNRE), Ministry of Interior, Department of Fisheries, and Ministry of Agriculture and Cooperatives (Beresnev et al. 2016). Historically, policies and regulations pursued by the different entities have come into conflict, and the national decentralisation process has also led to disruptions. Conflict has been related primarily to the simultaneous pursuit of mangrove conservation and shrimp farm expansion, although in recent years, the decline in the profitability of shrimp production has reduced demands on land in mangrove areas (Memon and Chandio 2011). Further, this situation weakens the community participation and engagement in protecting and conserving mangrove forests.

The next alternative that can be undertaken has received less attention and preference; yet, it is a powerful strategy in encouraging the community participation and engagement in protecting and conserving mangrove forests. It is education and knowledge (Quarto 1999). There are a number of education programmes undertaken by the NGOs across the globe about conservation and restoration of mangroves. Among the organisations with projects around the world include the Mangrove Action Project, Western Indian Ocean (WIO) Mangrove Network, the Mangrove Alliance, and Mangrove Watch, as well as domestic organisations, including Honko, mangrove conservation and education organisation in Madagascar, and the Mangrove Forest Conservation Society of Nigeria, and many others (Webber et al. 2016). These actions can be examples for other countries to continuously educate and create awareness among relevant parties with direct or indirect responsibility to protect and conserve mangrove forests particularly for tourism attractions.

17.5 Conclusions and Recommendations

Mangroves and its ecosystem undoubtedly have contributed numerous benefits to environment, community, and the economy. Mangrove forests nowadays are among popular tourism attractions across the globe. Their natural environment and wildlife attractions offer economic opportunities for community, businesses, and developers to take advantage for income generation and profit making. Ideally, the aim is to achieve sustainable tourism development through balancing the economic activity of mangrove tourism while maintaining mangrove forests through preservation, conservation, and restoration (Refer to Fig. 17.3).

While mangrove tourism help boost the country's economy and promote their culture; it also has its drawbacks, which include forest loss, depletion of natural resources, and also an increase of pollution in mangrove forests and its surroundings. Overemphasising on the economic aspect, as well as for the development and urbanisation purposes will lead to an imbalance in sustainable tourism development. This is particularly when mangrove preservation, conservation, and restoration are not done appropriately and effectively. The situation continues to worsen when there is no specific or clear policy, regulation, and ineffective management of mangrove undertaken to utilise mangrove forests (Fig. 17.4). Additionally, many mangrove

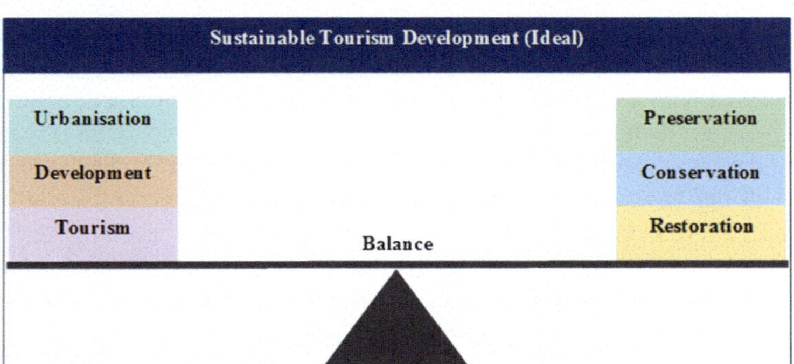

Fig. 17.3 Sustainable tourism development (Ideal)

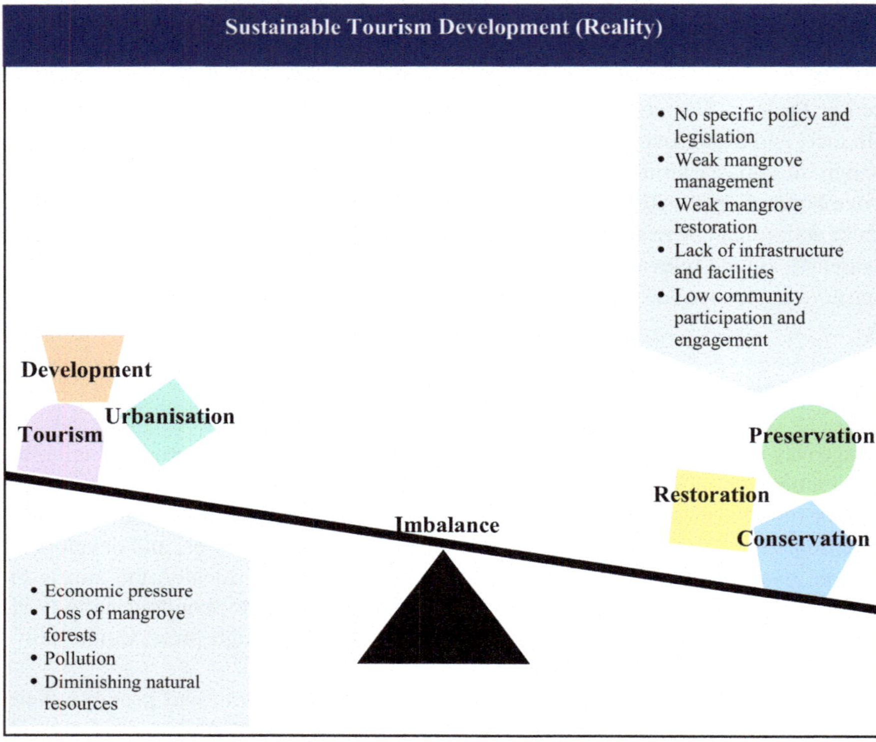

Fig. 17.4 Sustainable tourism development (Reality)

restoration or planting are also found ineffective due to inappropriate strategies and actions. Further, when mangrove forests are utilised as tourism attractions, suitable infrastructure and facilities are not provided. This contributes to other problems such as pollution. Finally, low community participation and engagement in the management of mangrove, as well as in promoting mangrove tourism hinders the effectiveness of sustainable tourism development. Undermining these aspects creates gaps for an imbalance of sustainable development of mangrove tourism.

Hypothetically, the ideal model of sustainable tourism development is the aim of any mangrove tourism attractions. With this ideal model in mind, there is a need for balancing the two aspects of economic pressure (tourism, development, and urbanization) with the environmental aspect (mangrove preservation, conservation, and restoration). This can be realised through implementing effective strategies: (1) Continuous conservation and restoration efforts of mangroves; (2) Enhancing the policies and legislations on the use and management of mangroves; (3) Enhancing infrastructure and facilities, and efficient use of resources through sound technologies; and (4) Encouraging community participation and engagement.

References

Abd. Shukor AH (2004) The use of mangroves in Malaysia. Promotion of mangrove-friendly shrimp aquaculture in Southeast Asia . Aquaculture Department, Southeast Asian Fisheries Development Center 136-144.

Amir AA (2019, May 17) Preserving mangroves and sustainable development. New Straits Times. https://www.nst.com.my/opinion/columnists/2019/05/489425/preservingmangrooves-and-sustainable-development.

Astuti S, Muryani C, Rindarjono MG (2017) The community participation on mangrove conservation in Sayung, Demak from 2004-2016. IOP Conference Series: Earth and Environmental Science 145.

Bacon PR (1987) Use of wetlands for tourism in the insular Caribbean. Ann Tour Res 14(1): 104-117.

Balmford A, Beresford J, Green J, Naidoo R, Walpole M, Manica A (2009). A global perspective on trends in nature-based tourism. PLoS Biol 7(6): e1000144.

Beresnev N, Phung T, Broadhead J (2016) Mangrove-related policy and institutional frameworks in Pakistan, Thailand and Viet Nam. Food and Agriculture Organizations of the United Nations Regional Office for Asia and the Pacific International Union for Conservation of Nature.

Blum J, Herr D (2017) Can restoring mangroves help achieve the Sustainable Development Goals? International Union for Conservation of Nature. Retrieved August 2, 2020. https://www.iucn.org/news/forests/201703/can-restoring-mangroves-help-achievesustainable-development-goals.

Boyer R, Peterson ND, Arora P, Caldwell K (2016) Five approaches to social sustainability and an integrated way forward. Sustainability 8.

Castellanos-Galindo GA, Cantera J, Saint-Paul U, Ferrol-Schulte D (2015). Threats to mangrove social-ecological systems in the most luxuriant coastal forests of the Neotropics. Biodivers Conserv 24: 701-704.

Cisneros-Montemayor AM, Sumaila UR (2010) A global estimate of benefits from ecosystem-based marine recreation: potential impacts and implications for management. J Bioecon 12: 245-268.

Datta D, Chattopadhyay RN, Guha P (2012) Community based mangrove management: A review on status and sustainability. J Environ Manag 107: 84-95.

Duke NC, Meynecke JO, Dittmann S, Ellison AM, Anger K, Berger U, Cannicci S, Diele K, Ewel KC, Field CD, Koedam N, Lee SY, Marchand C, Nordhaus I, Dahdouh-Guebas F (2007) A world without mangroves? Science 317(5834): 41–42.

FAO (2007) The World's Mangroves 1980-2005. FAO Forestry Paper No. 153. Rome, Forest Resources Division, FAO 77.

Hakim L, Siswanto D, Makagoshi N (2017) Mangrove conservation in East Java: The ecotourism development perspectives. J Tropic Life Sci 7(3): 277-285.

Hamilton LS, Snedaker SC (1984) Handbook for mangrove area management, IUCN/ UNESCO/ UNEP, Honolulu: East-West Center.

Lewis III R (2005). Ecological engineering for successful management and restoration of mangrove forests. Ecol Eng 24(2005): 403-418.

Majesty KI, Fadmastuti M (2018) Degree of community participation in mangrove resources management as livelihood support in West Java, Indonesia. International Conference Series on Life Cycle Assessment: Life Cycle Assessment as a Metric to Achieve Sustainable Development Goals (ICSoLCA 2018).

Memon J, Chandio A (2011) Critical appreciation of restoration and conservation of degraded mangroves in Thailand. Int J Environ Rural Dev 2: 108-113.

Official Website of Global Mangrove Alliance (n.d.) The Global Mangrove Alliance. http://www.mangrovealliance.org/wp-content/uploads/2017/08/global-mangrove-alliance_strategy.pdf.

Polidoro BA, Carpenter KE, Collins L, Duke NC, Ellison AM, Ellison JC, Joana C, Ellison JC, Farnsworth EJ, Fernando ES, Kathiresan K, Koedam NE, Livingstone SR, Miyagi T, Moore

GE, Nam VN, Ong JE, Primavera JH, Salmo III SG, Sanciangco JC, Sukardjo S., Wang Y, Yong JWH (2010) The Loss of Species: Mangrove Extinction Risk and Geographic Areas of Global Concern. PLoS ONE 5(4): e10095.

Pons LJ, Fiselier JL (1991) Sustainable Development of Mangroves. Landsc Urban Plan 20(1-3): 103-109.

Quarto A (1999) Local community involvement in mangrove rehabilitation: Thailand's Yadfon. In Streever, WJ (ed.), An international perspective on wetland rehabilitation. Springer, Dordrecht 139-142.

Richards DR, Friess DA (2016) Rates and drivers of mangrove deforestation in Southeast Asia, 2000-2012. Proc Natl Acad Sci, USA 113(2): 344–349.

Saenger P, Hegerl EJ, Davie JDS (1983) Global status of mangrove ecosystems. The Environmentalist 3 (supplement 3).

Said IA, Rahman MA (2000) Site planning and architectural design considerations of resorts in mangroves forest. Proceedings of the CASLE 2000.

Salem ME, Mercer DE (2012) The economic value of mangroves: A meta-analysis. Sustainability.

Spalding M, Kainuma M, Collins L (2010) World atlas of mangroves. Earthscan.

Spalding MD, Ruffo S, Lacambra C, Meliane I, Hale LZ, Shepard CC, Beck MW (2014a) The role of ecosystems in coastal protection: Adapting to climate change and coastal hazards. Ocean Coast Manag 90: 50-57.

Spalding M, McIvor A, Tonneijck FH, Tol S, Eijk VP (2014b) Mangroves for coastal defence: Guidelines for coastal managers & policy makers. Wetlands International and the nature conservancy 42.

Spalding MD, Burke L, Wood S, Ashpole J, Hutchison J, Ermgassen PZ (2017) Mapping the global value and distribution of coral reef tourism. Mar Policy 82: 104—113.

Spalding M, Parrett CL (2019). Global patterns in mangrove recreation and tourism. Mar Pol 110 (2019).

Suratman MN (2008). Carbon sequestration potential of mangroves in Southeast Asia. In Bravo F, Jandl R, LeMay V, Von Gadow K. (eds.), Managing forest ecosystems: The challenge of climate change. managing forest ecosystems. Springer, Dordrecht 17: 297-315.

The Economist (2018). Money swinging from trees. The Economist Magazine. London.

The International Union for Conservation of Nature (n.d.). Review of policy and legislation relating to mangroves, the use and management of mangroves in Tonga. The International Union for Conservation of Nature. https://www.iucn.org/sites/dev/files/content/documents/Tonga_policy_and_legislative_review_report.pdf.

United Nations (2020). About the sustainable development goals. Sustainable Development Goals. https://www.un.org/sustainabledevelopment/sustainable-development-goals/

UNDP (2020). Goal 14: Life below water. United Nations Development Programme. https://www.undp.org/content/undp/en/home/sustainable-development-goals/goal-14-life-below-water.html

Valenzuela RB, Youn Y-C, Mi SP, Chun J-N. (2020). Local people's participation in mangrove restoration projects and impacts on social capital and livelihood: A case study in the Philippines. Forests 11(580).

Van Lavieren H, Spalding M, Alongi D, Kainuma M, Clüsener-Godt M, Adeel Z (2012). Securing the future of mangroves: A policy brief, UNU-INWEH, UNESCO-MAB with ISME, ITTO, FAO, UNEP-WCMC and TNC. London: Earthscan.

Walters BB (2004). Local management of mangrove forests in the Philippines: Successful conservation or efficient resource exploitation? Hum Ecol 32(2): 177-195.

Webber M, Calumpong H, Ferreira B, Granek E, Green S, Ruwa R, Soares M (2016). Mangroves. United Nations. https://www.un.org/Depts/los/global_reporting/WOA_RPROC/Chapter_48.pdf.

Wesenbeeck BV, Balke T, Eijk PV, Tonneijck FH, Siry HY (2015). Aquaculture induced erosion of tropical coastlines throws coastal communities back into poverty. Ocean Coast Manag 116: 466-469.

Wetlands International (2016, November 7). Mangrove restoration: To plant or not to plant? Wetlands.org. https://www.wetlands.org/publications/mangrove-restoration-to-plant-or-not-toplant/.

WWF International (2017). 'Mangroves – A Life-saving Coastal Ecosystem: Scaling up protection and restoration for achieving SDGs', https://www.mangrovealliance.org/wp-content/uploads/2017/10/more-information-on-Save-Out-Mangroves-Now-.pdf.

WWF-Malaysia (2020). 'Mangrove Forests', https://www.wwf.org.my/about_wwf/what_we_do/forests_main/the_malaysian_rainforest/types_of_forests/mangrove_forests/.

Aquaculture in Mangroves

18

Tengku Mohd Zarawie Tengku Hashim,
Engku Azlin Rahayu Engku Ariff, and Mohd Nazip Suratman

Abstract

Ever since man realized the benefits of mangroves, the habitat has been impacted, but peaked in the twentieth century. Approximately 35% of the world's mangrove area was lost between 1980s and 1990s with deforestation rates ranging from 1% to 8%. The major drivers of mangrove deforestation in recent times include aquaculture, agriculture, urban expansion, forest product extraction, salt pond conversion, and the oil and gas industry. The boom in the aquaculture industry from 1970s onwards resulted in almost 28% of the habitat being lost in Asia (Bangladesh, India, China, Thailand, Vietnam, and Indonesia) and South America (Ecuador, Brazil, Peru) but by country, wise losses ranged from 7% to 63%. In South East Asia alone mangrove loss to aquaculture amounted to approximately 30% (1.66 million hectares). The total global economic value of mangrove loss to aquaculture is amounted at US$3.78–17.01billion/year. Three types of organisms are generally cultured in mangroves, namely fish, shrimp/prawns, and crabs. Mangrove conversion to aquaculture is a response to food security which is mainly to an increase in demand for protein and a decrease in marine capture fisheries. This is however is not without costs, such as habitat destruction, loss of ecosystem services, water quality reduction, exotic species introduction, and disease.

T. M. Z. Tengku Hashim · M. N. Suratman (✉)
Faculty of Applied Sciences and Institute for Biodiversity and Sustainable Development, UiTM, Shah Alam, Malaysia
e-mail: nazip@uitm.edu.my

E. A. R. Engku Ariff
Faculty of Applied Sciences, Universiti Teknologi MARA (UiTM) Pahang Campus, Bandar Tun Razak Jengka, Malaysia

R. P. Rastogi et al. (eds.), *Mangroves: Ecology, Biodiversity and Management*,
https://doi.org/10.1007/978-981-16-2494-0_18

Keywords

Mangrove · Aquaculture · Ecosystem services · Food security · Livelihoods

18.1 Introduction

The growth of the world population in the past century does not receive much attention from researchers, as it does not consider as a major threat to the human survivors until the end of the twentieth century. However, since the exponential growth of the human population exceeded 1 billion in the nineteenth century and world population growth projection is expected to reach 9.3 billion in the year 2060, uncertainties and the impacts of the growing population are started to emerge and visible (Goujon 2019). The significant increase in human population will put irreversible pressure on natural ecosystems as demand for the increase of natural resources in order to satisfy the consumption needs for the human population. One of the problematic scenarios that need to be considered by human throughout this era as consequences from the increase in human population is to address the enormous gap on how the world solve the arising issues in maintaining food security.

Since the great migration and cultural changes of the human race that took place in the mid-Holocene, the coastal area has always been the focal place for this purpose. With an increase in population density, the coastal area holds countless benefits that always generate interest from the human perspective to live and exploit the availability of resources in the area especially the mangrove ecosystems. Mangroves are recognized as a group of trees, shrubs, and palms that can endure and survive in harsh coastal conditions and can easily be distinguished by their unique characteristics in the coastal line of subtropical and tropical areas across the globe (Giri et al. 2011). The adaptability of mangroves to survive in a harsh condition is tremendously admired as it can be considered as the dominant ecosystem in the coastal area as compared to the other types of vegetation.

The impacts of saline seawater, tidal inundation period, and elevation play a major role in the distribution, zonation pattern, and growth rate of mangroves (Win et al. 2019). To endure the rigorousness in surviving in an unfriendly environment such as in the coastal climate, mangroves develop numerous adaptation strategies. One of the important morphological adaptations of mangroves is by possessing a unique root structure that functions as a stabilizing mechanism that spreads and acts like cable roots that anchor the mangroves to the soft muddy soils. Perhaps the most remarkable root adaptations of the mangroves are the stilt roots of *Rhizophora*, the pneumatophores of *Avicennia*, *Sonneratia,* and *Lumnitzera*, the root knees of *Bruguiera*, *Ceriops,* and *Xylocarpus,* and the buttress roots of *Xylocarpus* and *Heritiera* (Kathiresan and Bingham 2001). The root system of mangroves does not only act as a stabilizer, but also function as a gas exchange and breathing tool for mangroves to survive in the muddy and anaerobic place (Suratman 2008; Naidoo 2016). To cope with the saline environment, mangroves develop certain mechanisms to excrete the excessive salt caused by the seawater. For example, *Avicennia marina* secretes the excessive salt by a special salt gland that exists on both of its leaf surface (Krauss and Ball 2013) while other species such as *Rhizophora, Bruguiera*, and *Ceriops* possess ultra-filters in their root systems that filter the excessive salt while maintaining water uptake from the soil (Kathiresan and Bingham 2001).

Despite existing in the extreme coastal climate, mangroves are considered as natural jewel to the coastal ecosystem in terms of providing crucial ecological functions and pivotal roles in socioeconomic needs. Mangroves play a remarkable role in sheltering and reducing the damage from the impacts of the tsunami and strong wave-current in the coastal areas while offering protection to the coastal communities (Ahmadun et al. 2020). Mangroves also provide essential benefits in stabilizing coastal banks from erosion that is caused by strong tidal wave by trapping and binding sediment through its unique root systems (Sánchez-Núñez et al. 2019). Another crucial ecological function provided by the mangroves which could be a key component in mitigating the ever-challenging global climate change is their ability to sequester carbon and store it as part of their biomass for a long time period (Suratman 2008; Hashim et al. 2015, 2020). In terms of socioeconomic values, mangroves are considered as an important ecosystem to the coastal community by providing woods for timber, charcoal, and other commercial products (Abdul Aziz et al. 2015). Moreover, mangroves are proven to be an important ecosystem for commercial fishing and coastal aquaculture industries as they home many high grade marine habitats (Hutchison et al. 2014). Nevertheless, mangroves also provide job opportunities like generating household income to the local communities through coastal aquaculture industries and fishing activities in the mangroves.

Aquaculture activity is of the fastest growing industries on the planet as it contributes to the world food production system. As human population trend is exponentially increased in the upcoming decade, aquaculture seems to be one of the important solutions in maintaining, reducing the gap, and fulfilling the increasing demand for food supply. According to FAO (2020), it was estimated that in 2018, the global fish production had reached 179 million tonnes where aquaculture was among the highest contributors for this production with 82 million tonnes which was valued at USD 250 billion. As things stand right now, aquaculture is a more favourable technique to be used to produce mass food products as compared to the traditional method which is becoming less effective for supplying fishery resources to the market in the next decade (FAO 2018, 2020). With aquaculture industry gaining pace to expand even further, the coastal area especially mangroves always attract the attention of the governments, entrepreneurs, and local community throughout the globe as an area that has humongous potential to be transformed into an aquaculture hub. Furthermore, the condition of mangroves clicks all the necessary criteria such as brackish water, water quality suitability, and protection against natural elements such as storms and tsunamis (Hutchison et al. 2014; Ahmadun et al. 2020), adding some additional values why this area can be a successful aquaculture farming hub.

Despite the importance of mangroves to the coastal communities in providing numerous ecological and socioeconomic supports, this ecosystem has degraded at an alarming rate mainly due to global climate change and anthropogenic activities.

Although previous studies indicated that the loss of mangroves worldwide are caused by unprecedented climate change such as rising of sea level, storms, and tsunamis as well as the increase of sea currents (Richards and Friess 2016; Das and Mandal 2016) throughout the decade, evidence suggested that rapid and uncontrollable anthropogenic activities especially the aquaculture activities and land clearing are the main culprits for the reduction of mangroves worldwide (Truong and Do 2018). The unsustainable management and high conversion rate of mangroves to aquaculture pond throughout the decade have left the mangrove ecosystem in complete chaos. According to Valiela et al. (2001), in the past decade, mangroves had reduced to half of their size in Southeast Asia, South and Central America to make way for fish and shrimp farming.

Given the massive impacts of aquaculture activities to the mangroves and adjacent communities, it is crucial to find a solution to address this problem in order to ensure the sustainability and survival of mangrove ecosystems. This chapter aims at reviewing the roles of mangroves in supporting the aquaculture industries. The benefits and impacts of aquaculture industries to the mangrove ecosystem and the surrounding community are also highlighted.

18.2 The Role of Mangroves in Fisheries Habitat

Mangroves provide unique ecosystem functions that exist between the land and the sea as they consist of a rich assemblage of biodiversity. The unique ecosystems are not only crucial in providing a sustainable environment for many marine species, but also contribute to important commercial fisheries resources such as fishes, crustaceans (crabs and shrimps), and molluscs, which are sources of protein and income for the local community (Hutchison et al. 2014; Barbier et al. 2011). Mangroves play an essential ecological function to a large proportion of aquatic species and marine habitat by serving as a nursery ground, good breeding ground and act as shelter for juvenile and small fishes from predation (Hutchison et al. 2014). The effective and complex structure of mangrove habitats provides shelter for the aquatic organism from predation thus creating more microhabitat availability and increasing the amount of food (Mason et al. 2005).

Mangroves play a crucial role as the basis of the food chains that support a wide range of marine habitats in the coastal area. The high level of primary productivity from the mangrove vegetation such as litter, branches, and trunks as well as other primary producers is important in establishing a complex food web (Hutchison et al. 2014). With mangroves being considered as one of the highly productive ecosystems (Alongi 2012; Hutchison et al. 2014; Kamruzzaman et al. 2017), the main source of the primary productivity that acts as the base of the coastal ecosystem food chain comes from the mangrove trees, periphyton (algal that grows on tree roots), and phytoplankton (Hutchison et al. 2014). Meanwhile, other additional sources of food that enrich the mangroves come from adjacent ecosystems such as rivers, coral reefs, and sea grass by depositing material from upstream rivers and carrying material from the ocean by wave action (Igulu et al. 2013). In a study to determine the usage of mangroves by juvenile fish, Verweij et al. (2006) found that species richness of herbivore juvenile fish such as *Acanthurus chirurgus and A. Bahianus* was more abundant in the mangroves due to the presence of algae that grow on top of the mangrove roots. Another study that was conducted in Ryukyu Island, Japan in an effort to study the dependence of fish on mangroves also reported that the abundance of fish that can be found in the mangroves is due to the presence of food and shelter (Pantallano et al. 2018). Both of these studies are consistent with the hypothesis highlighted by Nanjo et al. (2014a), where the greater availability of food that accumulates in mangroves attracts fish and other marine habitats to the area.

Another important role provided by the mangroves in supporting fish habitat is by providing a suitable and safe area for juvenile fish and small fish against predators. The unique structures of mangroves such as the complex prop root system and canopy shades that exist from the combination of branches and leaves form a strategic hideout spot from predators (Nanjo et al. 2014b; Hutchison et al. 2014; Verweij et al. 2006). This comes to an agreement with a study conducted in South East Queensland, Australia using an artificial mangroves structure in a tank which concluded that abundance of marine fishes, especially the juvenile tends to be higher in a place that have more complex structure such as the mangroves in order to limit the risk of predation (Laegdsgaard and Johnson 2001). Another study also suggested that mangroves are used by small and juvenile fish as a refuge strategy by staying in mangrove's border for feeding purposes and flee into the mangroves when a large predator approaches them (Sheaves et al. 2016). Pantallano et al. (2018) in their study also revealed that fish species such as *Apogon amboinensis, Neopomacentrus taeniurus,* and *Pomacentrus taeniometopan* are the highest species of fish that can be found in the study area where they depend largely on the complexity of mangroves' roots and structures in order to seek protection against bigger predators. More recent studies conducted by Bradley et al. (2019) in Hinchinbrook, Northern Australia found that complex mangrove ecosystem in the coastal water provides a refuge area and a predominant habitat for coastal and reef juvenile fish species such as the *Acanthopagrus pacificus, A. australis, Lutjanus argentimaculatus, L. russellii, Epinephelus coioides,* and *E. malabaricus*. This is consistent with the main findings in a study that was conducted in Nahoon Estuary, South Africa, which compared the fish abundance in both salt marsh and mangroves indicated that fish species richness were higher in the mangroves as compared to salt marshes due to the complex structural and ecological benefit of the mangroves (Keur et al. 2019). Shrimps also utilize the mangroves as a place for seeking protection against predation. According to Rönnbäck et al. (1999) shrimps such as *Penaeus merguiensis* use the complex structure of mangrove roots as a place for seeking shelter from predators. Furthermore, the study found out that the abundance of mangroves in the coastal area could be the key in increasing the density of shrimp as it reduced the rate of predation while the murky water in the mangroves during tide helps in reducing the visibility of predator to prey.

Changes of water level in the mangrove forests which are directly influenced by the tide activity have created a variation of the water level that helps the aquatic species to navigate freely from an open water into parts of the forest thus enabling them to receive the benefits from the mangroves (Truong and Do 2018; Hutchison et al. 2014). The complex structure of the mangrove trees that directly come in contact with the water generates friction that slows the water speed in the mangroves and creates an ideal habitat for small juveniles fishes and prawns (Hutchison et al. 2014). These kinds of conditions create an ideal environment for breeding and nursery grounding for many commercial corals and freshwater fish where mangroves' structure traps fine sediment and creates a soft muddy floor that is crucial for larvae and egg to settle down in the mangrove environment and eventually will increase fish and invertebrate's population. According to Jamizan and Chong (2017), a study conducted in five different mangrove areas across West Peninsular Malaysia, Malaysia revealed that the geomorphological condition of mangroves which is connected by creeks has a strong relationship with abundance and distribution of fish (families of Lutjanidae, Haemulidae and Serranidae) and shrimp (penaeid shrimps). Furthermore, the study also stated that a large mangrove area that had many upstream rivers could be an ideal place for nursery grounding for many high commercial fishes and shrimps. This comes to an agreement with a study conducted by Tanaka et al. (2011) in Matang Mangrove Reserve, Malaysia where they found out that large mangrove areas play an important role in life cycle stages for many juvenile *Lutjanus johnii*. The study also revealed that the young juvenile of *L. johnii* could migrate up to 13 km upstream from the coastal area towards the mangroves using rich connective rives and were able to shift their diet from coastal food web to mangroves' dependence food web. El-Regal and Ibrahim (2014) also concluded that mangroves are the essential ground for nursery area where they found a large number of *Gerres oyena* and *Sardinella maderensis* juvenile during their study in three coastal mangroves area along the Egyptian Red Sea.

18.3 Aquaculture Activities

Aquaculture is believed to be the fastest growing and profitable industry around the globe that is responsible for supplying food resources to satisfy the growing needs of human food consumption (FAO 2018, 2020; Ahmed and Thompson 2019). The intense growth of cultivation of aquatic animals (fishes, shrimps, mussels, and oysters) sometimes plants (seaweed) in marine, brackish, and fresh water worldwide are considered as a blue revolution (Bavington and Banoub 2016; Ahmed and Thompson 2019). The blooming of the aquaculture industry throughout this decade is corresponding with the meteoric increase of world population and shifting pattern in human consumption behaviour towards fishery products. According to FAO (2020), the pattern of global fish consumption by human had outpaced the rate of annual world population growth (1.6%) which saw an increased rate of 3.1% from 1961 to 2017. In terms of per capita fish food consumption, the rate of increase was about 1.5% per year, which grew from 9 kg in 1961 to 20.5 kg in 2018 (FAO 2020).

The world of the aquaculture industry has contributed significantly in minimizing the gap between food supply and demand for human consumption. As mentioned previously, the aquaculture industry contributed about 82 million tonnes in 2018 where the industry was dominated by finfish (54.3 million tonnes), molluscs, mainly bivalve (17.7 million tonnes) and crustacean (9.4 million tonnes) (FAO 2020). With aquaculture activity seems to be one of the major tools to address the food security scenario, rapid expansion of aquaculture farm is visible throughout the globe. Mangroves are one of known areas which are synonym with aquaculture activities that are rapidly explored worldwide for the purpose of coastal aquaculture activity (Hutchison et al. 2014; Ahmed and Glaser 2016; Truong and Do 2018). Even though coastal aquaculture is significantly practiced all over the globe, this type of activity is more concentrated in South, Southeast and East Asia, and Latin America (FAO 2018, 2020). The condition of mangrove environment and suitable climate in those regions can be an ideal place to cultivate and practice the aquaculture activity.

18.3.1 Mangrove Ecosystem Services and Aquaculture Activities

Mangrove ecosystems play a tremendous role by providing numerous ecological supports that give an impact towards a successful coastal aquaculture activity. The role of mangroves in anticipating and reducing the impacts of tsunami and storms in the coastal area can reduce the damage and ensure the safety of aquaculture products and structures. Tsunamis and storms could have a devastating impact on aquaculture activities as it increases the water level in the area which could cause over flow to the aquaculture pond that might alter the salinity level and eventually will affect the growth and production of the aquaculture products (Nguyen et al. 2018). The catastrophic cyclone Sidr occurred in 2007 that hit Bangladesh had resulted in fatal impact to shrimp farming in the local area where the loss was about USD 36 million (IRIN 2008). According to Sarker and Azam (2007), the catastrophe of cyclone Sidr had resulted more than 90% loss of aquaculture farm in the regions of Morelgonj and Saronkhola Upazilas. Even though mangroves do not have the capability to captivate all the impacts of storms and tsunamis, the complex structure of mangroves can help in reducing the turbulence impact from storms and tsunamis thus benefitting the aquaculture industry in the area (Truong and Do 2018). The complex root structure of mangroves along the aquaculture site could be important in protecting the aquaculture pond by shielding and binding the soil from wave action thus reducing the erosion problem.

Mangroves are considered as an ecosystem that has high biodiversity that consists of many organisms that inhabit the area. As mentioned previously, mangroves are strategic place for many aquatic habitats for nursery ground, feeding area and seeking refuge from predators (Hutchison et al. 2014). The abundance of organisms in the mangroves will benefit the aquaculture product (fish or shrimp) by providing additional nutrients for feeding. According to Nagelkerken et al. (2008), after the process of spawning finish, the eggs dispersed all around mangroves and after a certain period of time, they would turn to planktonic larvae. The planktonic larvae would move or carried by currents into other parts of mangroves or in the aquaculture cage in the mangrove area that could be an additional source of food for other aquatic habitats as well as for fish and shrimp in the aquaculture cage.

The complex biodiversity of mangroves does not only provide additional food to fish or shrimp in aquaculture ponds, but also provide natural production of larvae and juveniles. Even though juveniles and larvae nowadays are spawned using breeding technique in the captivity facilities, the dependent on mangroves in providing wild seeds for the aquaculture industry is largely practiced in many countries. According to Beveridge et al. (1997), the aquaculture industry, especially shrimp farming was still largely depending on wild caught seeds or gravid females for the aquaculture industry. Rönnbäck et al. (1999) stated that the natural production of larvae and juveniles that scattered abundantly in mangroves is an important source of seeds for aquaculture industry which they entered naturally by flowing current or artificially caught. In addition, the ecological functions of mangroves in providing suitable climates such as sustaining water quality mitigate variation of salinity and turbidity also influence the growth rate and survival of aquaculture product.

According to Venkatachalam et al. (2018), a study they conducted in Tamil Nadu, India to determine the survival and growth of *Lates calcarifer* found out that the integration between mangroves and aquaculture produced a higher rate of survival (11%) and had a higher growth rate (12.5%) as compared to an aquaculture area that did not have mangroves. In Vietnam, mangroves are regarded as an important ecosystem for integration with aquaculture activity due to their ecological functions provided. According to Vietnamese and international experts, the integration between mangroves and aquaculture activity does not only improve the resilience of aquaculture product, but also reduce disease outbreak because mangroves act as a bio-filtering entity that improves water quality and buffers the impact of temperature fluctuation (Joffre et al. 2015).

18.3.2 Importance of Aquaculture Activities

18.3.2.1 Job Opportunities

The coastal areas always attract the attention of human beings due to the attractive socioeconomic function they offered. It is believed that the landscape of coastal area is immensely altered from time to time due to vigorous land use/cover changes that is caused by human growth and pressure, especially in the mangrove areas. According to Thomas (2017), a total of 120 million people worldwide occupied the mangroves and highly depending on mangroves' resources for survival. However, the communities that live in a coastal area are always regarded as communities surrounded by poverty. The blooming of the aquaculture industry in recent decade benefits the local community by creating a positive impact on the employment sector where it generates business opportunities and job creations hence boosting the socioeconomic livelihoods for the surrounding coastal community (Mialhe et al. 2013; Hussain et al. 2018). It was estimated that in 2018, a total of 59.51 million of people was directly involved in the fishery sector where from that number, 20.58 million of people engaged in aquaculture sector (FAO 2020). The rapid growth of the fishery sector, especially the aquaculture industry in the Asia region has resulted in an increase in job opportunity where 85% of the jobs in the fishery sector are concentrated in this region, followed by Africa (9%), America (%), Europe (1%), and Oceania (1%) (FAO 2020).

According to a study conducted by Mialhe et al. (2013), in their effort to determine the impact of shrimp farming to land use, employment and migration in Peru found out that there was a significant increase in job employment due to the extensive shrimp farming activities in the area. The study revealed that during the year 2001, only 439 people was directly involved in aquaculture activity. However, with the economic revenue that was brought by this industry in terms of household income, a total of 2660 people opted to join this industry as a full-time job in 2006. In Bangladesh, the blue economy phenomenon, especially aquaculture industry holds enormous potential in generating job opportunities for local coastal community and this could hold the essential key in reducing the shackle of poverty within this community (Hussain et al. 2018). According to Moni et al. (2018), in Bangladesh alone a total of 600,000 people were working in the shrimp aquaculture sector and benefited almost 3.5 million dependents. A study conducted in Bangladesh concluded that the aquaculture sector through shrimp farming production in the county has benefited almost 4.8 billion households by providing jobs to 1.2 million of people (Islam 2008). A similar trend of interim job opportunity was observed in Vietnam after the introduction of shrimp aquaculture activities. It was reported that the rapid production of aquaculture products in the CA Mau province had resulted an exponential increase in aquaculture workers from 85,000 in 1997 to 312,000 in 2003 (Moni et al. 2018).

In Malaysia, the aquaculture sector has received much attention from the government as it holds a potential economic gain as well as one of the alternative measures to improve the economic gap for rural society especially the coastal community. One of the alternatives that is implemented under the government policy (National Agro Food Policy NAP (2011–2020)) in conjunction with the 10th Malaysian Plan is to increase the production of aquaculture activities and products thus create and increase job opportunities in this sector (Yusoff 2014). According to Yusoff (2014), in 2012 the involvement of people in the aquaculture sector in Malaysia was 29,494 and from this total about 22.7% or 6715 were directly involved in aquaculture which was concentrated in mangrove areas.

18.3.2.2 Food Security

One of the main important criteria of the aquaculture industry towards human sustainability is minimizing the gap for food consumption (FAO 2018, 2020). The importance of aquaculture production in providing a source of food cannot be ignored as it contributes to almost 50% of world fish consumption alongside deep sea and inland fish capture (Barua and Rahman 2020). Recent advancement and evolution of aquaculture technology and vast global market trade in recent years have resulted in an increase in the production of aquaculture products and have a significant impact to the consumption rate of human worldwide. According to FAO (2020), in 2018 the per capita of fish consumption worldwide was about 20.5 kg, and this figure had outpaced other types of animal protein consumption products such as milk, meat, egg, etc. This scenario is significantly impacted by the recent technological development in processing, cold chain transport as well as diverse shipping and distribution hub that are efficiently used to exploit the high market demand for aquaculture products in the United States of America (USA), China, and the European Union (EU) (FAO 2020). Moreover, with fast changing and improvement in aquaculture distribution chain, it can have a significant increase to human fishery consumption of 36 kg per capita by the year of 2020 (FAO 2016).

According to Pradeepkiran (2019), it was estimated that a total of 1 billion people worldwide are largely depending on fish products. The high quality source of proteins, essential amino acids, vitamins, and minerals that exist in fishery products provide 16% of animal protein to human diet globally due to the abundance of resources and cheaper as compared to other types of animal protein in the market (Pradeepkiran 2019). According to Kawarazuka and Béné (2011) in 2010, 22 countries that could be categorized into low income and food deficient countries were highly dependable on fish product as the source of animal protein in their diet. In Malaysia, it was estimated that the fish consumption index among Malaysians is expected to increase from 53.1 kg in 2011 to 61.1 kg in 2020 due to the shifting patterns of diet for more healthy food based on fish protein (Yusoff 2014). The increasing pattern of consumption that is based on fish protein among Malaysians is also resulted from its growing population where the demand for fish products is expected to increase from 1.7 million tonnes in 2011 to 1.93 million tonnes by 2020 (Yusoff 2014).

18.3.2.3 Aquaculture Monetary Benefits

The aquaculture sector is considered as one of the highest profitable business in this decade that has a high market demand throughout the globe. According to the new statistic reported by FAO (2020), the aquaculture sector has recorded another milestone where in 2018 the aquaculture industry production produced an astonishing record of 114.5 million tonnes in live weight which was equal to the sale value of USD 263.6 billion. From the total aquaculture production, finfish was the highest producer with 53.3 million tonnes or USD 139.7 billion where it harvested from inland aquaculture (47 million tonnes, 104.3 billion) and marine and coastal aquaculture (7.3 million tonnes, USD 35.4 billion) (FAO 2020). In Southeast Asia, the aquaculture sector is considered as an important economic activity and source of income to the nation in this region as it contributed to almost USD 10.87 billion in 2017 as compared to 0.31 billion in 1984 (FAO 2018).

In Bangladesh, aquaculture activity, especially shrimp farming is the second most important export of goods that significantly contributes to the economy of the nation. The shrimp farming industry in Bangladesh is highly influenced by the high market demand from developed countries such as the USA, United Kingdom (UK), Belgium, Germany, and Japan due to the high quality grade of shrimp species such as *Penaeus monodon* that cultivate in Bangladesh mangroves (Islam and Bhuiyan 2016). According to Uddin et al. (2013), it was estimated that the export of shrimp in Bangladesh for the years of 2011–2012 was about USD 449.56 million as compared to export for years of 2001–2002 which was about USD 276.11 million.

In Malaysia during the year of 2012, brackish water aquaculture activity, especially shrimp farming activity (*P. monodon* and *P. vannamei*) contributed to 139129.51 tonnes of product that worth USD 0.39 million to the economy (Yusoff 2014). Being blessed with an abundance of resources, Malaysia considers aquaculture as one of the important sectors that can contribute to the development of the nation. According to Queiroz et al. (2013), with the financial benefit gained from the aquaculture industry, it has transformed Brazil's approach on aquaculture activity from the experimental production of shrimp farming to an extensive industry that was based on framework of rapid expansion and highly profitable. During this booming era of shrimp aquaculture, this approach has resulted Brazil to produce approximately 61,000 tonnes of shrimp production that valued USD 244.5 million in 2013.

With coastal aquaculture activity being regarded as an important source of income to the nation while producing economic activity for the coastal community, it is expected that this sector will give a huge impact to the coastal ecosystem especially the mangroves. By acknowledging the strategic and suitability of mangroves in coastal area as a hub for aquaculture activity, it can be estimated that more mangrove areas need to be cleared and converted to make way for these industries. Many studies reported that this activity contributed to the largest mangrove losses elsewhere in the world (Romañach et al. 2018; van Wesenbeeck et al. 2015; Richards and Friess 2016; Ottinger et al. 2016). The next section in this chapter will highlight the impacts of aquaculture activity to mangrove ecosystems.

18.4 Impact of Aquaculture in Mangroves

During the last few decades, aquaculture has benefited from scientific progress and the growth has been exponential. This increasing trend will need to remain in order to support the demand of the food supply of more than 9 billion people by the year of 2050. Unfortunately, upon the successful aquaculture, there have also been undesirable social and environmental issues, including unsustainable water or natural feed use, mangrove destruction, and biodiversity loss. According to Vo et al. (2013), the destruction of mangrove forest is due to the development of settlements, aquaculture and agriculture.

18.4.1 Eutrophication

Nutrient enrichment is one of the most serious threats to the coastal ecosystem (Downing et al. 1999; Cloern 2001). Previous studies have shown that growth of intertidal mangrove forests is stimulated with enhanced nutrient availability; however, the growth of plants is favoured in the aboveground parts compared to the root parts (Tilman 1991; Lambers and Poorter 1992). Consequently, plants exposed to high levels of nutrient availability must face greater vulnerability to environmental stressors that require large investment in roots for tolerance such as during the drought (Chapin 1991). Previous study was conducted by Lovelock et al. (2009) on 12 study sites where mangrove trees have been fertilized by using urea (nitrogen, N) or triple superphosphate (Phosphorus, P) for 3 years. The study found that N fertilization elevated the tree mortality in the region of high salinity and aridity and trees that fertilized with P tended to have higher probability of survival than those fertilized with N. From this increasing pattern across species and biogeographic regions, it solidly determines the influence of eutrophication in climatic interactions with the intertidal landscape. In addition, results showed that mangrove ecosystems that are exposed to the excess nutrient availability enhance the mortality rate, especially during drought, and that nutrient-induced mortality is higher in sites which has low rainfall, low humidity, and high sediment salinity.

The analysis of sediment cores collected over the Chinese island Hainan revealed that the land conversion had a significant impact on the biogeochemistry of the estuarine bay (Herbeck et al. 2020). The study found that the increase in C_{org} content up to 2.7% in the mangrove core (1) which was being collected directly adjacent to a riverine mangrove forest, and <0.7% in the lagoon and channel cores (2) which was taken in the estuarine lagoon approximately 500 m from the now aquaculture-dominated shore and core (3) which was collected from the middle of the estuarine channel approximately 5 km from the outlet. Furthermore, in the lagoon and channel cores (2 and 3), the increase in $\delta^{15}N$ values was found due to a simultaneous increase in land cover change to aquaculture. The higher value of $\delta^{15}N$ was a result of strong relation with the uptake and assimilation of dissolved inorganic nitrogen (DIN) by phytoplankton in the aquaculture ponds as well as in the estuary (Herbeck et al. 2020). The increment of $\delta^{15}N$ was 7.8% at the top sediment layer of the lagoon core (2) are comparable to those suspended matters that were measured inside aquaculture ponds and suspended matters at estuarine (Herbeck et al. 2011; Herbeck and Unger 2013; Bao et al. 2013).

Water discharge from aquaculture shrimp farm and sewage from the developing urban areas near mangroves which contain high level of nutrients are ordinary sources that contribute to pollution in the Northeast of Brazil (Queiroz et al. 2013; Suárez-Abelenda et al. 2014). A study related to eutrophication from aquaculture was done at three selected mangroves in Ceará State, Brazil, at the estuaries of Jaguaribe, Cocó, and Pacoti River by analysing water and soil physiochemical and biogeochemical parameters, and soil solid-phase P fractionation (Barcellos et al. 2019). The mangrove at Jaguaribe River was chosen because this river has been intensively impacted by shrimp farming ponds (Lacerda et al. 2008; Kauffman et al. 2018). However, the mangrove at the Cocó River was the region that received diffuse contamination from urban sewage (Molisani et al. 2007; Nóbrega et al. 2015); and the last region was mangrove at the Pacoti River that had some degree of disturbance with ponds for salt production, but has been reclaimed and protected by Brazilian environmental agencies since 1990 (Lacerda et al. 2007). Results from discriminant analysis showed that the most labile P forms increased gradually and significantly from control (0.073 mg L^{-1}) to sewage (0.042 mg L^{-1}) to shrimp farm (0.167 mg L^{-1}) impacted mangroves as observed by increasingly dissolved orthophosphate (PO_4^{3-}) content in water and the exchangeable/soluble P (Exch-P) extracted from soils (Barcellos et al. 2019). Hence, these aquaculture activities would increase the P forms soluble and readily available to algae and plant growth, with direct impacts for eutrophication and pollution in mangrove ecosystems (Bai et al. 2015).

A large number of nutrients such as N and P discharged into the coastal water can cause many ecological problems, such as severe eutrophication with frequent occurrence of harmful algal blooms (Wang et al. 2011). A study on the impact of nutrient release from fish cage aquaculture has been done in Daya Bay, located in Guangdong Province in southern China. The main components of DIN released by fish were NH_4–N and NO_3–N. Those chemical compounds that contained nitrogen were beneficial for the growth of phytoplankton and macroalgae (Troell et al. 2003, 2009). From the findings, the annual amount of N released from fish cage culture was 205.6 metric tons (hereafter tons) including 142.7 tons of dissolved inorganic nitrogen (DIN) and 39.2 tons of P which included 15.1 tons of dissolved inorganic phosphorus (DIP). Hence, the contribution of DIN and DIP from fish culture was about 7.0% and 2.7%, respectively (Qi et al. 2019). The ratio of cage-derived N and P was 21.1, higher than the ratio of coastal seawater which was 27.1, thus the cage culture may enhance the growth of phytoplankton by changing the nutrient structure (with regard to the N:P ratio) of seawater around farming regions (Qi et al. 2019).

18.4.2 Microplastic Pollution

Recent studies have identified high abundance of microplastics in mangrove and salt marsh habitats and suggested that the vegetation of wetlands is an effective way to retaining media of microplastics (Nor and Obbard 2014; Sutton et al. 2016; Weinstein et al. 2016). Qinzhou Bay in Guangxi Province, southwest China is well known as location for natural largest breeding area of *Magallana rivularis*. Wide range of mollusc farming in this area can contribute to microplastics pollutions. Styrofoam was one of the common microplastics found which was widely used in hanging-culture farms for mollusc. Published study by Li et al. (2018) in Quizhou Bay found that high concentrations of microplastics were observed near mollusc farms and polystyrene (PS) was the major microplastics polymer found in the area which was >98%. The result was expected since PS was broadly used to set up mollusc rafts in the Qinzhou Bay. Besides PS, polypropylene (PP) and polyethylene (PE) were also observed in the area since this polymer was used in woven bag and fish net. Study concluded that microplastic pollution in sediments of Qinzhou Bay was mainly a result of intensive mollusc aquaculture (Li et al. 2018).

18.4.3 Destruction of Mangrove Ecosystems

The issues on impact of coastal aquaculture on the environment, biodiversity, and society are argued and debated around the world (Primavera 2006; Hall 2011).

Globally, mangrove conversion to shrimp farming has been fully condemned because of its environmental and socioeconomic impacts (Primavera 1997; Deb 1998; Naylor et al. 1998; Páez-Osuna 2001; Lebel et al. 2002; Bush et al. 2010). In addition, high demands and economic return in the international market, as well as unplanned and unregulated shrimp farming have caused major destruction of mangroves in many countries, including Malaysia, Philippines, Bangladesh, China, Indonesia, Sri Lanka, Myanmar, India, Thailand, Brazil Mexico, and Vietnam (FAO 2007). In Asia, governments of many countries have been encouraging the conversion of mangrove forest to aquaculture pond in order to deal with food security and livelihoods since the 1960s and 1970s (Hishamunda et al. 2009). Consequently, a study by Hamilton (2013) found out that over 54% of mangrove deforestation (28% of the former mangrove areas) were due to conversion of aquaculture ponds. Southeast Asia has 35% of the world's 18 million ha of mangrove forests (Spalding et al. 1997), however, between 2000 and 2012, aquaculture was accounted for 30% of the mangrove forest loss. Besides aquaculture, the relevance of other land uses has also increased, such as oil palm plantations in Malaysia, Indonesia, rice agriculture in Myanmar, and urban development in Vietnam (Richards and Friess 2016)

18.4.4 Introduction of Non-Native Species

Introduction of new species for aquaculture has led to an enormous invasion of species diversity and trophic food chain in mangrove ecosystem, for e.g., introduction of the Indo-Pacific lionfish to the Atlantic (Whitfield et al. 2002) and the killer algae, *Caulerpa taxifolia* to the Mediterranean (Jousson et al. 1998). In addition, tropical white shrimp (*Penaeus vannamei*) native to the eastern Pacific coast of Central and South America has been strongly established around the world since the 1970s and in 2000. Presently, it has become the principle cultured shrimp species in Asia. However, diseases from shrimp aquaculture are much concern since the release of untreated shrimp farm discharge water can increase the nutrient load in the open water thus may lead to a disease outbreak in shrimp farms. This condition will provide mangrove unhealthy water bodies (Wolanski et al. 2000; Senarath and Visvanathan 2001). Tilapias are native to Africa and were introduced from 1946 onwards to Asia for breeding to support aquaculture industry. Unfortunately, they are now widely colonizing brackish water areas with mangroves, even though they are a freshwater group of fishes (Bagarinao and Primavera 2005).

18.5 Conclusion

Mangrove forests are found in sheltered saline coastal environments in tropical and subtropical latitudes and are among the most productive ecosystems in the world as they fulfil a range of ecologically and socioeconomically important functions including shoreline stabilization, storm and wave protection, reduction of coastal erosion as well as feeding and nursery habitat for numerous commercially important invertebrate and vertebrate species such as fishes and prawns. Besides, mangroves play a vital role in nutrient dynamic and coastal sediment. Mangroves also act as a carbon sink for atmospheric CO_2 and crucial source of organic carbon to the coastal sea. Since 1960s and 1970s, the conversion of mangrove forests to aquaculture ponds was encouraged by the governments of many countries and aquaculture production is the fastest growing industry on the planet which also responsible for the provision of high quality animal protein worldwide. As the demand for food supply increases due to the increasing of human population, aquaculture seems to be one of the important solutions in maintaining, reducing the gap and fulfilling the need. Nevertheless, the increase in aquaculture on the other hand tends to negatively impact the environment through the spread of diseases, destruction of wetlands and mangroves, declining in biodiversity of natural fish populations by the escape of non-native fish, and the pollution of surface and groundwater by effluent discharge. Hence, sustainable management can confirm a sustainable progress and the advantage of aquaculture.

References

Abdul Aziz A, Dargusch P, Phinn S, Ward A (2015) Using REDD+ to balance timber production with conservation objectives in a mangrove forest in Malaysia. Ecol Econom 120: 108–116.

Ahmadun FL-R, Wong MMR, Mat Said A (2020) Consequences of the 2004 Indian Ocean Tsunami in Malaysia. Saf Sci 121: 619–631.

Ahmed N, Glaser M (2016) Coastal aquaculture, mangrove deforestation and blue carbon emissions: is REDD+ a solution? Mar Pol 66: 58–66.

Ahmed N, Thompson S (2019) The blue dimensions of aquaculture: A global synthesis. Sci Total Environ 652: 851–861.

Alongi DM (2012) Carbon sequestration in mangrove forests. Carbon Manag 3: 313–322.

Bagarinao TU, Primavera JH (2005) Code of practice for sustainable use of mangrove ecosystems for aquaculture in Southeast Asia, aquaculture department, Southeast Asian Fisheries Development Center.

Bai XL, Zhou YK, Sun JH, Ma JH, Zhao HY, Liu XF (2015) Classes of dissolved and particulate phosphorus compounds and their spatial distributions in the water of a eutrophic lake: A 31 P NMR study. Biogeochemistry 126: 227–240.

Bao H, Wu Y, Unger D, Du J, Herbeck LS, Zhang J (2013) Impact of the conversion of mangroves into aquaculture ponds on the sedimentary organic matter composition in a tidal flat estuary (Hainan Island, China). Cont Shelf Res 57: 82–91.

Barbier EB, Hacker SD, Kennedy C, Koch EW, Stier AC, Silliman BR (2011) The value of estuarine and coastal ecosystem services. Ecol Monogr 81: 169–193.

Barcellos D, Queiroz HM, Nóbrega GN, De Oliveira Filho RL, Santaella ST, Otero XL, Ferreira TO (2019) Phosphorus enriched effluents increase eutrophication risks for mangrove systems in northeastern Brazil. Mar Pollut Bull 142: 58–63.

Barua P, Rahman SH (2020) aquatic health index of coastal aquaculture activities at South-Eastern Coast Of Bangladesh. Water Conserv Manag 4: 53–69.

Bavington D, Banoub D (2016) Marine fish farming and the blue revolution: Culturing cod fisheries. London Journal of Canadian Studies 31: 35–44.

Beveridge M, Phillips M, Macintosh D (1997) Aquaculture and the environment: the supply of and demand for environmental goods and services by Asian aquaculture and the implications for sustainability. Aquac Res 28: 797–807.

Bradley M, Baker R, Nagelkerken I, Sheaves M (2019) Context is more important than habitat type in determining use by juvenile fish. Landsc Ecol 34: 427–442.
Bush SR, Van Zwieten PA, Visser L, Van Dijk H, Bosma R, De Boer WF, Verdegem M (2010) Scenarios for resilient shrimp aquaculture in tropical coastal areas. Ecol Soc 15:15.
Chapin FS (1991) Integrated responses of plants to stress. BioScience 41: 29–36.
Cloern JE (2001) Our evolving conceptual model of the coastal eutrophication problem. Mar Ecol Prog Ser 210: 223–253.
Das CS, Mandal RN (2016) Coastal people and mangroves ecosystem resources vis-à-vis management strategies in Indian Sundarban. Ocean Coast Manag 134: 1–10.
Deb AK (1998) Fake blue revolution: environmental and socio-economic impacts of shrimp culture in the coastal areas of Bangladesh. Ocean Coast Manag 41: 63–88.
Downing J, Mcclain M, Twilley R, Melack J, Elser J, Rabalais N, Lewis W, Turner R, Corredor J, Soto D (1999) The impact of accelerating land-use change on the N-cycle of tropical aquatic ecosystems: current conditions and projected changes. Biogeochemistry 46: 109–148.
El-Regal MAA, Ibrahim NK (2014) Role of mangroves as a nursery ground for juvenile reef fishes in the southern Egyptian Red Sea. Egypt J Aquat Res 40: 71–78.
FAO (2007) The world's mangroves 1980–2005. FAO Rome, Italy.
FAO (2016) Aquaculture Department (2010) The state of world fisheries and aquaculture. Food and Agriculture Organization of the United Nations, Rome.
FAO (2018) The State of World Fisheries and Aquaculture 2018. Meeting the sustainable development goals.Rome.
FAO (2020) The State of World Fisheries and Aquaculture 2020. Sustainability in Action. FAO Rome, Italy.
Giri C, Ochieng E, Tieszen LL, Zhu Z, Singh A, Loveland TR, Masek J, Duke NC (2011) Status and distribution of mangrove forests of the world using earth observation satellite data. Glob Ecol Biogeogr 20: 154–159.
Goujon A (2019) Human Population Growth. In: Fath, B. (ed.) Encyclopedia of Ecology (Second Edition). Oxford: Elsevier.
Hall SJ (2011) Blue frontiers: managing the environmental costs of aquaculture, WorldFish.
Hamilton S (2013) Assessing the role of commercial aquaculture in displacing mangrove forest. Bull Mar Sci 89: 585–601.
Hashim T, Suratman M, Jaafar J, Hasmadi IM, Abu F (2015) Field assessment of above ground biomass (AGB) of mangrove stand in Merbok, Malaysia. Malays Appl Biol 44: 81–86.
Hashim T, Suratman M, Singh H, Jaafar J, Bakar A. (2020) Predictive model of mangroves carbon stocks in Kedah, Malaysia using remote sensing. IOP Conference Series: Earth and Environmental Science, 2020. IOP Publishing, 012033.
Herbeck LS, Krumme U, Andersen TJ, Jennerjahn TC (2020) Decadal trends in mangrove and pond aquaculture cover on Hainan (China) since 1966: mangrove loss, fragmentation and associated biogeochemical changes. Estuar Coast Shelf Sci 233: 106531.
Herbeck LS, Unger D (2013) Pond aquaculture effluents traced along back-reef waters by standard water quality parameters, δ15N in suspended matter and phytoplankton bioassays. Mar Ecol Prog Ser 478: 71–86.
Herbeck LS, Unger D, Krumme U, Liu SM, Jennerjahn TC (2011) Typhoon-induced precipitation impact on nutrient and suspended matter dynamics of a tropical estuary affected by human activities in Hainan, China. Estuar Coast Shelf Sci 93: 375–388.
Hishamunda N, Ridler NB, Bueno P, Yap WG (2009) Commercial aquaculture in Southeast Asia: Some policy lessons. Food Pol 34: 102–107.
Hussain MG, Failler P, Karim AA, Alam MK (2018) Major opportunities of blue economy development in Bangladesh. J Indian Ocean Reg 14: 88–99.
Hutchison J, Spalding M, Ermgassen P (2014) The role of mangroves in fisheries enhancement. The Nature Conservancy and Wetlands International: 54.

Igulu M, Nagelkerken I, Van Der Velde G, Mgaya Y (2013) Mangrove Fish Production is Largely Fuelled by External Food Sources: A Stable Isotope Analysis of Fishes at the Individual, Species, and Community Levels from Across the Globe. Ecosystems 16, 1336–1352.

Irin (2008) Bangladesh: Cyclone-hit Shrimp Farmers Face Uncertain Future, Integrated Regional Information Network, Nairobi.

Islam MS (2008) From pond to plate: towards a twin-driven commodity chain in Bangladesh shrimp aquaculture. Food Pol 33: 209–223.

Islam SDU, Bhuiyan MAH (2016) Impact scenarios of shrimp farming in coastal region of Bangladesh: an approach of an ecological model for sustainable management. Aquac Int 24: 1163–1190.

Jamizan AR, Chong V (2017) Demersal Fish and Shrimp Abundance in Relation to Mangrove Hydrogeomorphological Metrics. Sains Malaysiana 46: 9–19.

Joffre OM, Bosma RH, Bregt AK, Van Zwieten PA, Bush SR, Verreth JA (2015) What drives the adoption of integrated shrimp mangrove aquaculture in Vietnam? Ocean Coast Manag 114: 53–63.

Jousson O, Pawlowski J, Zaninetti L, Meinesz A, Boudouresque CF (1998) Molecular evidence for the aquarium origin of the green alga Caulerpa taxifolia introduced to the Mediterranean Sea. Mar Ecol Prog Ser 172: 275–280.

Kamruzzaman M, Ahmed S, Osawa A (2017) Biomass and net primary productivity of mangrove communities along the Oligohaline zone of Sundarbans, Bangladesh. For Ecosys 4: 1–9.

Kathiresan K, Bingham BL (2001) Biology of mangroves and mangrove ecosystems. Adv Mar Biol 40, 81–251.

Kauffman JB, Bernardino AF, Ferreira TO, Bolton NW, Gomes LEDO, Nobrega GN (2018) Shrimp ponds lead to massive loss of soil carbon and greenhouse gas emissions in northeastern Brazilian mangroves. Ecol Evol 8: 5530–5540.

Kawarazuka N, Béné C (2011) The potential role of small fish species in improving micronutrient deficiencies in developing countries: building evidence. Public Health Nutr 14: 1927–1938.

Keur R, James N, Rajkaran A (2019) A tale of two habitats: preliminary comparison of fish abundance and diversity between saltmarsh- and mangrove-dominated creeks in the Nahoon Estuary, South Africa. Afr J Mar Sci 41: 449–454.

Krauss KW, Ball MC (2013) On the halophytic nature of mangroves. Trees 27: 7–11.

Lacerda LD, Molisani MM, Sena D, Maia LP (2008) Estimating the importance of natural and anthropogenic sources on N and P emission to estuaries along the Ceará State Coast NE Brazil. Environ Monit Assess 141: 149–164.

Lacerda LDD, Menezes MOTD, Molisani MM (2007) Changes in mangrove extension at the Pacoti River estuary, CE, NE Brazil due to regional environmental changes between 1958 and 2004. Biota Neotropica 7: 67–72.

Laegdsgaard P, Johnson C (2001) Why do juvenile fish utilise mangrove habitats? J Exp Mar Biol Ecol 257: 229–253.

Lambers H, Poorter H (1992) Inherent variation in growth rate between higher plants: a search for physiological causes and ecological consequences. Advances in ecological research. Elsevier.

Lebel L, Tri NH, Saengnoree A, Pasong S, Buatama U (2002) Industrial transformation and shrimp aquaculture in Thailand and Vietnam: pathways to ecological, social, and economic sustainability? AMBIO: A Journal of the Human Environment 31: 311–323.

Li HX, Ma LS, Lin L, Ni ZX, Xu XR, Shi HH, Yan Y, Zheng GM, Rittschof D (2018) Microplastics in oysters Saccostrea cucullata along the Pearl River estuary, China. Environ Pollut 236: 619–625.

Lovelock CE, Ball MC, Martin KC, Feller IC (2009) Nutrient enrichment increases mortality of mangroves. PLoS One 4: e5600.

Mason NW, Mouillot D, Lee WG, Wilson JB (2005) Functional richness, functional evenness and functional divergence: the primary components of functional diversity. Oikos 111: 112–118.

Mialhe F, Gunnell Y, Mering C (2013) The impacts of shrimp farming on land use, employment and migration in Tumbes, northern Peru. Ocean Coast Manag 73: 1–12.

Molisani M, Lisieux R, Cavalcante M, Maia L (2007) Effects of water management on hydrology and water quality of a semi-arid watershed in the Northeast of Brazil. Braz J Aquat Sci Tech 11: 43–49.

Moni NN, Al Masud MM, Hossen SS (2018) Trade liberalization and socio-economic vulnerability: a cross country analysis on the shrimp sector. Khulna University Studies Volume 15 (1): 71—90

Nagelkerken I, Blaber SJM, Bouillon S, Green P, Haywood M, Kirton LG, Meynecke JO, Pawlik J, Penrose HM, Sasekumar A, Somerfield PJ (2008) The habitat function of mangroves for terrestrial and marine fauna: A review. Aquat Bot 89: 155–185.

Naidoo G (2016) The mangroves of South Africa: An ecophysiological review. South Afr J Bot 107: 101–113.

Nanjo K, Kohno H, Nakamura Y, Horinouchi M, Sano M (2014a) Differences in fish assemblage structure between vegetated and unvegetated microhabitats in relation to food abundance patterns in a mangrove creek. Fish Sci 80: 21–41.

Nanjo K, Kohno H, Nakamura Y, Horinouchi M, Sano M (2014b) Effects of mangrove structure on fish distribution patterns and predation risks. J Exp Mar Biol Ecol 461: 216–225.

Naylor RL, Goldburg RJ, Mooney H, Beveridge M, Clay J, Folke C, Kautsky N, Lubchenco J, Primavera J, Williams M (1998) Nature's subsidies to shrimp and salmon farming. American Association for the Advancement of Science.

Nguyen XH, Nguyen XT, Nguyen HH, Tran TT, Le DQ (2018) Assessment of storm surge risk in aquaculture in the Northern coastal area of Vietnam. Vietnam J Sci Technol Engineer 60: 89–94.

Nóbrega GN, Ferreira TO, Artur AG, De Mendonça ES, Raimundo ADO, Teixeira AS, Otero XL (2015) Evaluation of methods for quantifying organic carbon in mangrove soils from semi-arid region. J Soils Sediments 15: 282–291.

Nor NHM, Obbard JP (2014) Microplastics in Singapore's coastal mangrove ecosystems. Mar Pollut Bull 79 (1–2): 278–283.

Ottinger M, Clauss K, Kuenzer C (2016) Aquaculture: Relevance, distribution, impacts and spatial assessments–A review. Ocean Coast Manag 119: 244–266.

Páez-Osuna F (2001) The environmental impact of shrimp aquaculture: causes, effects, and mitigating alternatives. Environ Manag 28: 131–140.

Pantallano ADS, Bobiles RU, Nakamura Y (2018) Dependence of fish on subtropical riverine mangroves as habitat in the Ryukyu Islands, Japan. Fishe Sci 84: 613–625.

Pradeepkiran JA (2019) Aquaculture role in global food security with nutritional value: a review. Transl Anim Sci 3: 903–910.

Primavera JH (1997) Socio-economic impacts of shrimp culture. Aquac Res 28: 815–827.

Primavera JH (2006) Overcoming the impacts of aquaculture on the coastal zone. Ocean Coast Manag 49: 531–545.

Qi Z, Shi R, Yu Z, Han T, Li C, Xu S, Xu S, Liang Q, Yu W, Lin H (2019) Nutrient release from fish cage aquaculture and mitigation strategies in Daya Bay, southern China. Mar Poll Bull 146: 399–407.

Queiroz L, Rossi S, Meireles J, Coelho C (2013) Shrimp aquaculture in the federal state of Ceará, 1970–2012: Trends after mangrove forest privatization in Brazil. Ocean Coast Manag 73: 54–62.

Richards DR, Friess DA (2016) Rates and drivers of mangrove deforestation in Southeast Asia, 2000–2012. Proc Natl Acad Sci 113: 344–349.

Romañach SS, Deangelis DL, Koh HL, Li Y, Teh SY, Raja Barizan RS, Zhai L (2018) Conservation and restoration of mangroves: Global status, perspectives, and prognosis. Ocean Coast Manag 154: 72–82.

Rönnbäck P, Troell M, Kautsky N, Primavera JH (1999) Distribution pattern of shrimps and fish among Avicennia and Rhizophora micro habitats in the Pagbilao Mangroves, Philippines. Estuar Coast Shelf Sci 48: 223–234.

Sánchez-Núñez DA, Bernal G, Pineda JEM (2019) The Relative Role of Mangroves on Wave Erosion Mitigation and Sediment Properties. Estuar Coast 42: 2124–2138.

Sarker T, Azam M (2007) Super Cyclone SIDR 2007: climate change adaptation mechanisms for coastal communities in Bangladesh. Sarker, T. and Azam, M.(2012). Super cyclone SIDR: 85–105.

Senarath U, Visvanathan C (2001) Environmental issues in brackish water shrimp aquaculture in Sri Lanka. Environ Manag 27: 335–348.

Sheaves M, Johnston R, Baker R (2016) Use of mangroves by fish: new insights from in-forest videos. Mar Ecol Prog Ser 549: 167–182.

Spalding MD, Blasco F, Field CD eds (1997) World mangrove atlas. The international Society for Mangrove Ecosystems, Okinawa, Japan. 178 pp.

Suárez-Abelenda M, Ferreira T, Camps-Arbestain M, Rivera-Monroy V, Macías F, Nóbrega GN, Otero X (2014) The effect of nutrient-rich effluents from shrimp farming on mangrove soil carbon storage and geochemistry under semi-arid climate conditions in northern Brazil. Geoderma 213: 551–559.

Suratman MN (2008) Carbon sequestration potential of mangroves in southeast Asia. Managing forest ecosystems: The challenge of climate change. Springer.

Sutton PC, Anderson SJ, Costanza R, Kubiszewskic I (2016) The ecological economics of land degradation: Impacts on ecosystem service values. Ecological Economics Volume 129: 182–192.

Tanaka K, Hanamura Y, Chong VC, Watanabe S, Man A, Kassim FM, Kodama M, Ichikawa T (2011) Stable isotope analysis reveals ontogenetic migration and the importance of a large mangrove estuary as a feeding ground for juvenile John's snapper Lutjanus johnii. Fish Sci 77: 809.

Thomas N (2017) The importance of mangroves to people: a call to action.

Tilman D (1991) Relative growth rates and plant allocation patterns. Am Nat 138: 1269–1275.

Troell M, Halling C, Neori A, Chopin T, Buschmann A, Kautsky N, Yarish C (2003) Integrated mariculture: asking the right questions. Aquaculture 226: 69–90.

Troell M, Joyce A, Chopin T, Neori A, Buschmann AH, Fang JG (2009) Ecological engineering in aquaculture—potential for integrated multi-trophic aquaculture (IMTA) in marine offshore systems. Aquaculture 297: 1–9.

Truong TD, Do LH (2018) Mangrove forests and aquaculture in the Mekong river delta. Land Use Policy 73: 20–28.

Uddin MB, Sultana T, Rahman MM, Uddin MB, Sultana T, Rahman MM (2013) Shrimp export marketing of Bangladesh. Bangladesh Res Pub J 8: 146–151.

Valiela I, Bowen JL, York JK (2001) Mangrove forests: one of the world's threatened major tropical environments. Bioscience 51: 807–815.

Van Wesenbeeck BK, Balke T, Van Eijk P, Tonneijck F, Siry HY, Rudianto ME, Winterwerp JC (2015) Aquaculture induced erosion of tropical coastlines throws coastal communities back into poverty. Ocean Coast Manag 116: 466–469.

Venkatachalam S, Kathiresan K, Krishnamoorthy I, Narayanasamy R (2018) Survival and growth of fish (Lates calcarifer) under integrated mangrove-aquaculture and open-aquaculture systems. Aquac Rep 9: 18–24.

Verweij M, Nagelkerken I, De Graaff D, Peeters M, Bakker E, Van Der Velde G (2006) Structure, food and shade attract juvenile coral reef fish to mangrove and seagrass habitats: a field experiment. Mar Ecol Prog Ser 306: 257–268.

Vo QT, Oppelt N, Leinenkugel P, Kuenzer C (2013) Remote sensing in mapping mangrove ecosystems—An object-based approach. Remote Sens 5: 183–201.

Wang X, Hao F, Cheng H, Yang S, Zhang X, Bu Q (2011) Estimating non-point source pollutant loads for the large-scale basin of the Yangtze River in China. Environ Earth Sci 63: 1079–1092.

Weinstein JE, Crocker BK, Gray AD (2016) From macroplastic to microplastic: Degradation of high-density polyethylene, polypropylene, and polystyrene in a salt marsh habitat. Environ Toxicol Chem 35: 1632–1640.

Whitfield PE, Gardner T, Vives SP, Gilligan MR, Courtenay Jr WR, Ray GC, Hare JA (2002) Biological invasion of the Indo-Pacific lionfish Pterois volitans along the Atlantic coast of North America. Mar Ecol Prog Ser 235: 289–297.

Win S, Towprayoon S, Chidthaisong A (2019) Adaptation of mangrove trees to different salinity areas in the Ayeyarwaddy Delta Coastal Zone, Myanmar. Estuar Coast Shelf Sci 228: 106389.

Wolanski E, Spagnol S, Thomas S, Moore K, Alongi D, Trott L, Davidson A (2000) Modelling and visualizing the fate of shrimp pond effluent in a mangrove-fringed tidal creek. Estuar Coast Shelf Sci 50: 85–97.

Yusoff A (2014) Published. Status of resource management and aquaculture in Malaysia. Resource Enhancement and Sustainable Aquaculture Practices in Southeast Asia: Challenges in Responsible Production of Aquatic Species: Proceedings of the International Workshop on Resource Enhancement and Sustainable Aquaculture Practices in Southeast Asia 2014 (RESA), 2015. Aquaculture Department, Southeast Asian Fisheries Development Center, 53–65.

19 Ecological Valuation and Ecosystem Services of Mangroves

Hong Tinh Pham, Thi Hong Hanh Nguyen, and Sy Tuan Mai

Abstract

Knowledge of the economic value of mangrove ecosystem plays an important role in assisting managers and policy-makers in determining priorities, policies, and actions for conservation of the mangrove ecosystem. Methods to evaluate natural and environmental resources have been widely applied to raise awareness as to the value of mangrove ecosystems. In this chapter, we describe the economic and non-economic contribution of mangrove standings to both communities and society. The services which are related to the various components and ecological functions of a mangrove ecosystem are summarized. The chapter also describes how mangrove ecosystem services are valuated and what their estimated values are towards the provision of these services.

Keywords

Mangrove ecosystem · Ecological functions of mangroves · Ecosystem service · Ecological value · Valuation Method

H. T. Pham (✉) · T. H. H. Nguyen
Faculty of Environment, Hanoi University of Natural Resources and Environment, Hanoi, Vietnam
e-mail: phtinh@hunre.edu.vn

S. T. Mai
Mangrove Ecosystem Research Center, Hanoi National University of Education, Hanoi, Vietnam

R. P. Rastogi et al. (eds.), *Mangroves: Ecology, Biodiversity and Management*,
https://doi.org/10.1007/978-981-16-2494-0_19

19.1 Introduction

Globally mangroves are distributed in the inter-tidal zone of many tropical and subtropical coastal wetlands, forming some of the most productive and biologically diverse ecosystems. Mangroves cover an estimated area of 137,600 km^2 (Fig. 19.1) with Asia accounting for 38.7% (53,278 km^2), followed by Latin America and the Caribbean with 27,939 km^2 (20.3%), Africa with 27,465 km^2 (20.0%), Oceania with 16,329 km^2 (11.9%), North America with 11,563 km^2 (8.4%), and Europe with 1026 km^2 (0.7%) (Bunting et al. 2018). Mangroves exist at the terrestrial and marine interface and are considered to be one of the most productive and biologically diverse of all ecosystems. Mangrove leaf litter makes up the majority of net primary production, acting as a key source of organic matter within directly-associated food chains. Detritus-based food is also exported by water currents, providing nutrients for marine/aquatic organisms within a large coastal area (Naylor et al. 2000; Sheridan and Hays 2003; Aburto-Oropeza et al. 2008). Mangroves reduce the effects of larger waves and elevated water levels as a result of storm surges and tsunamis (Mazda et al. 2006; Zhang et al. 2012) providing coastal protection and helping local communities become more resilient. Mangroves also serve as a protected nursery for many marine species (Rönnbäck 1999; Walters et al. 2008).

The ecological importance of mangroves is widely recognized and the services that they provide are recognized by academics. However, there is often a lack of understanding or underestimation of the value of mangrove ecosystems and the benefits they provide by policy makers. There is also an increasing pressure to exploit resources by the local communities with mangroves sacrificed to make way for shrimp ponds, aquaculture, and other short-term land use purposes, and the impacts of anthropogenic pollution. This has led to a serious decline in the total area and quality of mangrove forests and resulted in the development and refinement of tools and models to assess and highlight the economic value of this precious ecosystem.

It is acknowledged that ecosystems in general and mangrove ecosystems, in particular, provide people with non-economic benefits (Tri 2006) and placing a monetary value on a natural ecosystem do not always recognize the tangible and non-use benefits of the system. However, economic valuation is a useful tool for

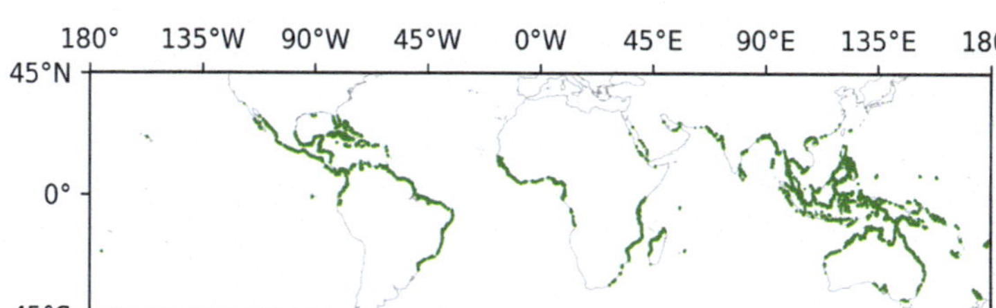

Fig. 19.1 Mangrove extent per country (Adapted from Bunting et al. 2018)

policy makers to make rational decisions against the pressures of economic development and contributing to the sustainable management of coastal resources.

Providing an economic valuation of the mangrove ecosystem requires interdisciplinary expertise and knowledge of both mangrove ecology and a theoretical understanding of valuation science. Experience and practical knowledge of both fields is also essential for application in specific cases. This chapter presents the current knowledge and a theoretical basis for developing methods for valuing mangrove ecosystems and application of a systematised approach to utilising research findings.

19.2 Components and Functions of the Mangrove Ecosystems

Mangroves are mainly woody trees and shrubs, characterized by the aerial roots and seedlings that germinate on the plant. They grow at the land–sea interface in the tropical and subtropical regions of the world (Latitude 30°N and 30°). Mangroves, along with associated organisms (e.g. microorganisms, fungi, and animals), interact with abiotic factors, regulating a distinct coastal tide-inundated ecosystem (Kathiresan and Bingham 2001) (Figs. 19.2 and 19.3, Table 19.1) that undergoes spatial-temporal movement (Tri et al. 2002).

The mangrove ecosystem is characterized by the coastal climate, alluvial soil, and tidal mud flats that are inundated by salt water from diurnal or semi-diurnal tides and mixing with fresh water from rivers resulting in brackish water and the physico-chemical properties of the water (e.g. salinity, pH, and temperature) is constantly changing with space and time. The biotic components of the mangrove ecosystem include marine and terrestrial organisms, and organisms adapted to the mangrove forest (Table 19.1).

Complex, functional, and dynamic interactions take place within the system which are capable of self-regulating and sustaining the mangrove ecosystem in a relatively steady state. However, negative anthropogenic activities can affect

Fig. 19.2 Mangrove ecosystem at Ba Lat estuary, Vietnam (Image source: Pham Hong Tinh)

Fig. 19.3 A woman catching crabs under mangrove canopy (Image source: Pham Hong Tinh)

ecosystem stability. The basic functions of the ecosystem are determined by the following interactive relationships:

Interaction between abiotic factors: This is clearly shown in such relationships as: salinity decreasing during periods of extended rainfall; turbidity increases when the amount of sediment input from the river is high; the substrate may become highly anaerobic during extended periods of flooding caused by sea water inundation.

Interaction between organisms: This is shown in food chains and webs, trophic levels, and moving energy flow in the ecosystem. These are the relationships between producers and consumers (mangrove plants - insects; mangrove leaves fallen on the forest floor - crabs, algae - zooplankton), predators and preys (seabass eat shrimp, shrimp eat plankton), symbiotic organisms, saprophytes (pathogenic fungi), parasites (worms, pathogenic bacteria), commensalism.

Interaction between biotic and abiotic factors: For example, abiotic factors (soil, water, air environment) affect the distribution, zoning, growth and development of mangrove species and other organisms in the forest. While mangroves can influence the microclimate, sequester carbon and the creation of burrows by crabs improves soil conditions. The relationship between biotic and abiotic factors is also reflected in the biogeochemical cycle, such as cycles of water, carbon, nitrogen, phosphorus, and sulfur. For example, the nutrient cycle in the mangrove ecosystem shows that mangroves and algae synthesize organic matter via photosynthesis using sunlight, CO_2 in the air, water and minerals obtained from the soil. These organic matters are stored in plant body tissues that are leaves, stems, and roots. Plant organic matters are directly eaten by other organisms (worms, bugs, and crabs) or after being decomposed, (decomposed food) help build up body tissues and energy for the movement of animals. When these animals die, they are broken down by microorganisms and fungi into minerals, returned to the soil, and then sucked up by plants as nutrients. This cycle is called the nutrient cycle or biogeochemical cycle because of the involvement of biological processes (food chains and food webs),

Table 19.1 Basic characteristics of components in a mangrove ecosystem

Component in the ecosystem		Sea factor	Land factor	Mangrove factor
Abiotic components	Atmosphere	Sea climate, wind, storms, high humidity, . . .	The inland climate depends on the distribution of landscape and vegetation factors	A place where there is a mixture of marine and inland climate.
	Soil	Anaerobic soils dependent on frequency and level of tidal inundation	Mud flats are gradually elevated	Mud flats in estuarine and coastal areas: anaerobic, salty, acid sulfate.
	Water	Saline seawater	Fresh water from the river	Brackish water (a mixture of seawater and fresh water)
Biotic components	Micro-organisms, fungi, moss, algae	Marine micro-organisms, fungi, moss, algae in sea water	Inland micro-organisms, fungi, moss, in mangrove plants	Micro-organisms, fungi, algae in brackish water, on mangrove roots
	Mangrove trees		Associate mangrove species	True mangrove species
	Invertebrates	Shrimps, oysters, clams, crabs, worms, ... living in the marine environment.	Insects and bugs (ants, butterflies, bees) living in mangrove plants	Some species of insects, crabs, ... can live both in mangrove plants and in seawater
	Vertebrates	Fish, reptiles, birds, animals living in the marine environment	Reptiles, birds, animals living in mangrove trees and those from the inland area coming here for food, nesting	Typical species of mangrove forests and seasonal migratory species (fish, sea turtles, snakes, birds, dugongs)

chemical reactions in metabolic, decomposition, mineralization, processes, physical, geographic, geomorphologic processes of the environment.

19.3 Goods and Services of Mangrove Ecosystems

In society, the relationships between people are economically reflected in the form of goods and services supplied by one person to the other and vice versa. In an ecosystem, goods and services are products of nature. These products have been developed during a long-term evolution process. While many people mistakenly believe that these products are readily available in nature for supply and continued exploitation by humans (Lebel et al. 2002).

Mangrove ecosystems provide goods to meet local people's demand for building materials, energy for cooking and food. These types of goods are traded and exchanged at the market, including plant products such as wood for house construction, furniture and firewood, nipa leaves for roofing. In addition, food resources, including honey, shrimp, crabs, fish, and oysters are an important source of nutrition for local people, and a valuable source of export income (Adger et al. 1995).

However, the value of the mangrove ecosystem is not only shown in goods, but also in its ability to provide essential and important services (Barbier et al. 1991) closely related to ecosystem function. The services ecosystems provide are divided into groups, including: provisioning services (food, raw materials), regulating services, supporting services, and cultural services (de Groot et al. 2012; MEA 2005). The services provided by mangroves are summarized in Table 19.2.

19.4 Valuation Methods of Mangrove Ecosystem Services

The valuation of an ecosystem in general and a mangrove ecosystem in particular is difficult, with authorities involved in managing and formulating policies from the central to the local level facing difficulties in making decisions on forest conservation or economic development. Mangrove ecosystem valuation is very useful if it is used as a tool to contribute to identifying priorities, policies, and actions for conservation. Economists have developed theoretical methods for evaluating natural resources and the environment. Those methods are divided into four groups (Table 19.3).

Detailed descriptions of some of the most widely adopted methods are given below:

Market price method: uses the market price to estimate the value of an ecosystem good or service, such as firewood or aquatic products sold at the market. Other ecosystem services such as absorbing wastes that help clean water sources, increasing fisheries productivity, etc. are determined through the benefits of end products. Intangible services as the aesthetics of the landscape, inspiration for the creation of art and maintenance of moral fortitude cannot be sold at market, but the prices which people want to pay or the degree of willingness to pay for these services can be valued. Pricing, quantity, preference, and cost data can all be easily collected and used to provide an economic evaluation. However, this method can only be applied to a limited number of goods and services brought about by ecosystems. It does not reflect the value of all ecosystem functions, nor it is fully reflective of the economic value of goods and services, as market prices are dependent on the policies, seasons, and other factors.

Benefit transfer method: estimates economic values for ecosystem services by applying assessment findings of previous research from other locations provided that these locations have the same conditions to reduce finance, time, and human resources. When this method is applied, it is necessary to carefully assess the transferability of findings; especially the features of residents, tradition, educational level (knowledge and responsibilities) as well as their interests. This requires the

Table 19.2 Ecosystem services provided by mangroves

Ecosystem components	Provisioning services	Regulating services	Supporting services	Cultural services
Mangrove trees (*Avicennia, Rhizophora, Sonneratia, Ceriops, . . .*)	– Provide: • Building materials • Fuel (firewood, charcoal) • Herbal medicine – Mangrove resin is used in the processing technology of varnish, ink	– Absorb carbon and pollutants, ameliorate the effects of climate change – Filter water – Reduce the impacts of winds, storms, and cyclones – Regulate temperature, humidity – Limit inland saline intrusion – Maintain Ecological balance	– Nutrient cycle – Seed dispersal – Trap alluvial sediment, expand the area of mud flats – Reduce Erosion – Provide a spawning ground for aquatic species, and nesting sites for birds; – Provide a refuge for vertebrates	– Scientific research – Maintain cultural heritage – Historical value, Ecotourism and landscape tourism – Create jobs for local people
Aquatic and marine resources (*fish, shrimp, crabs, oysters, snails, . . .*)	– Provide food for people – Animal feed to serve husbandry – Herbal medicine – Genetic resources for aquaculture	– Filter water, reduce water pollution – Maintain Ecological balance – Maintain Soil structure	– Food sources for other species – Nutrient cycle – Reserve source – Organic decomposition	– Scientific research – Create jobs for people
Other species (*Pythons, snakes, turtles, otters, birds, monkeys, honeybees, ...*)	– Food for people – Herbal medicine – Genetic resource	– Maintain Ecological balance – Maintain Soil structure	– Food sources for other species – Gene reserves – Help flowers pollinate – Seed dispersal	– Scientific research – Maintain cultural heritage – Create jobs for people
Soil and water in the coastal mangrove area	– Shelter – Farmland – Supply fresh water from rain, groundwater	– Regulate climate – Conserve water and land resources – Ecological balance	– Soil improvement – Prevention against forest fire – Shelter for wild animals	– Historical value – Maintain cultural heritage – Create jobs for people

consultation/comments of experts, scientists, officials, and local people. Researchers have to understand fully the study that is being utilized, including raw data, actual survey data, number of people interviewed, etc. However, this method should only

Table 19.3 Methods of economic valuation of mangrove ecosystem services

Category	Method	Application
Real market methods	Market price	Economic values of a mangrove product or service are estimated through its prices in commercial markets.
	Productivity change	Economic values of a mangrove product or service are estimated through its contribution to the production of the mangrove ecosystem products or services that are bought/sold in commercial markets.
	Replacement cost	Economic values of a mangrove product or service are estimated by using the costs of replacing ecosystem products or services.
	Avoided cost	Economic values of a mangrove product or service are estimated by using the costs of avoided damages resulting from lost ecosystem services.
Surrogate market methods	Travel cost	Economic values of mangrove products or services are estimated by asking people about their willingness to pay to travel to visit the mangrove ecosystem.
	Hedonic price	Economic values of a mangrove product are decomposed its total value into the value of its several attributes.
	Production function	Economic values of a mangrove product or service are estimated by modelling the physical contribution of the that product or service to economic output.
Hypothetical market methods	Contingent valuation	Economic values of a mangrove product or service are estimated by asking people about their willingness to pay for it in a hypothetical scenario.
	Choice modeling	Economic values of a mangrove product or service are estimated by asking people to make tradeoffs among them; willingness to pay is inferred from tradeoffs that include cost.
Other methods	Benefit transfer	Economic values of a mangrove product or service are estimated by using existing estimates from similar studies in other sites
	Ecosystem value coefficient	Economic values of a mangrove service are estimated by multiplying the mangrove area by global averaged value.

be applied in special cases because collecting and evaluating primary data for a particular area is more reliable. Costanza et al. (1997) and de Groot et al. (2012) used the benefit transfer method to estimate the global value of ecosystem services. These authors used values from many other studies to calculate average values which were then multiplied by biome area estimated from a global land use map.

Productivity change method: estimates economic values of ecosystem products or services that contribute to the production of commercially marketed goods (input). For example, the economic benefits of improving water quality can be measured by increases in crop yields, increased aquatic species due to the availability of organic humus as a suitable source of food and the habitat mangrove forests provide. This method, while not complicated, requires coordination between ecologists and economists; the data collected should reflect changes in the quality and quantity of raw materials from natural resources, including costs of production for the final goods; supply and demand for end products; supply and demand for other factors of the production process. This information is used in relation to the impact of quantitative and qualitative changes leading to amendments to producer or consumer surpluses, based on which economic benefits can be evaluated.

Travel cost method: is used to value ecosystem services related to tourism and entertainment, in which the cost–benefit analysis is based on: entrance fee and the area and quality of the environment in the tourist area. This approach is based on the time and costs tourists spend on travel to a given location. Therefore, the degree of willingness of tourists to pay for their visit to a given location can be determined by the number of times they visit and associated travel costs. This method is quite simple, inexpensive, and easily applied, however, one must assume that tourists visit a site for only one purpose. For travellers who visit a location with multiple purposes, valuing services for each purpose is more complex. On the other hand, travel time can be used in many ways and the time spent visiting a tourist site can be viewed as an "opportunity cost" (Tri 2006).

Methods of avoided cost, replacement cost: are economic valuation methods based on the cost of avoided damages due to preventing the loss of ecosystem services, the cost of ecosystem services replacement or the cost of providing additional ecosystem services. These methods do not rigorously measure the economic value, but are based on willingness to pay for specific goods or services. These methods can be applied in cases such as: valuing erosion prevention services of a forest or wetland by assessing the cost of dredging downstream river beds; valuing the water cleaning service that wetlands provide by calculating the cost of filtering and chemically treating water; valuing sea dyke protection services of mangrove ecosystems by calculating the costs of constructing permanent dyke sections; valuing the service of providing nurturing and breeding grounds for coastal marine/aquatic species through cost assessment of spawning ground rehabilitation programs and companies providing nurseries for and supplying aquatic seed source.

Contingent valuation method: is also known as the "stated preference" method or "revealed preference" because it is directly related to interviewing people; to establish, for example, willingness to pay a certain amount of money for specific environmental services (Tri et al. 1998). These values are not related to the purchasing and selling at the market and are therefore also known as passive values. They are composed of everything from basic life support associated with ecosystem health or biodiversity, to the enjoyment of natural beauty and responsibilities for leaving a legacy to future generations... For example, voluntarily contributing to mangrove conservation today will help improve knowledge for future generations such as the salt secreting mechanism of *Avicennia* mangroves and the existence of the otters in the coastal area (Tri 2006). When awareness is raised, people are willing to pay for environmental benefits regardless of the passive use value or non-use value. How much money each person is willing to pay depends on behavioral, psychological, cognitive, and cultural factors which limits the reliability of this method of assessment.

Hedonic price method: is used to estimate the economic value of environmental or ecosystem services that are directly related to and affect market prices. The basis of this approach is the determination of a price for a marketed good associated with

Table 19.4 Number of studies of mangrove ecosystem service valuation during 2007–2016 (Adapted from Himes-Cornell et al. 2018)

	Number of studies of evaluation of mangrove ecosystem services				
Evaluation methods	Provisioning services	Regulating services	Supporting services	Cultural services	Total
Avoided cost		3	1		4
Damage cost		2			2
Market price	35	10	2	11	58
Net price method	1				1
Replacement cost		10			10
Production function	8	3	4	1	16
Travel cost				2	2
Choice modeling	2			1	3
Contingent valuation	1	1		2	4
Benefit transfer	29	49	15	18	111
Others	11	20	6	9	46
Total	87	98	28	44	257

the characteristics and services it provides. Therefore, it is possible to evaluate each characteristic or determine the price a customer is willing to pay when each characteristic is improved in accordance with their preferences.

The application of this method is flexible and relevant to one or a group of mangrove ecosystem services. As discussed in a previous review study (Himes-Cornell et al. 2018) on economic evaluation methods for mangrove ecosystem services, market price is predominantly used to value directly paid mangrove ecosystem services (e.g., food, raw materials, carbon accumulation, leisure travel/tourism), while the benefit transfer method is widely used to estimate the value of all other ecosystem services. The other remaining valuation methods are not very frequently used.

The number of studies on the evaluation of mangrove ecosystem services by each research method is shown in Table 19.4.

19.5 Economic Value of Mangrove Ecosystem Services

Over the past two decades, many studies assessing the economic value of mangrove ecosystems have been carried out. These studies have mainly focused on a specific ecosystem or geographic area (Camacho-Valdez et al. 2014; Amarnath and Mouna 2016; Atkinson et al. 2016; Jerath et al. 2016; Mashayekhi et al. 2016; Mojiol et al. 2016; Sopheak and Hoeurn 2016; Susilo et al. 2016). However, some studies have

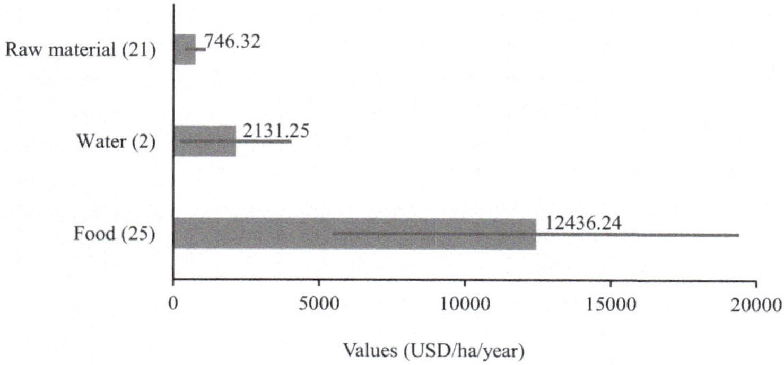

Fig. 19.4 Values of providing services of mangrove ecosystem (Based on data compiled by Himes-Cornell et al. 2018)

Fig. 19.5 Seadykes protected by mangroves (**a**) and seadyke eroded due to absence of protection mangroves (**b**) in Northern Vietnam (Image source: Phan Hong Anh)

attempted to calculate the global value of mangrove ecosystem services (Costanza et al. 1997; de Groot et al. 2012). The value of providing food, raw materials, moderating extreme events, preventing erosion, and maintaining life cycles of migratory species has been widely documented. However, other ecosystem services such as pollination, ornamental resources, and cultural benefits have received less attention (Himes-Cornell et al. 2018).

Mangroves provide a wide range of products (as detailed previously) are directly related to net primary production (Saenger 2002). This provisioning service has been evaluated by many scientists in different geographic regions around the world.

Figure 19.4 shows estimates of the economic value of ecosystem services provided by mangroves, ranging from 746.32 USD/ha /year (raw material provisioning service) or 2131.25 USD/ha/year (water related services) and up to 12,436.24 USD/ha/year (aquatic product provisioning service). Estimates shown in Figs. 19.4 and 19.5 are compiled and calculated from 48 different studies, concerning the value of services providing food, raw materials, and water.

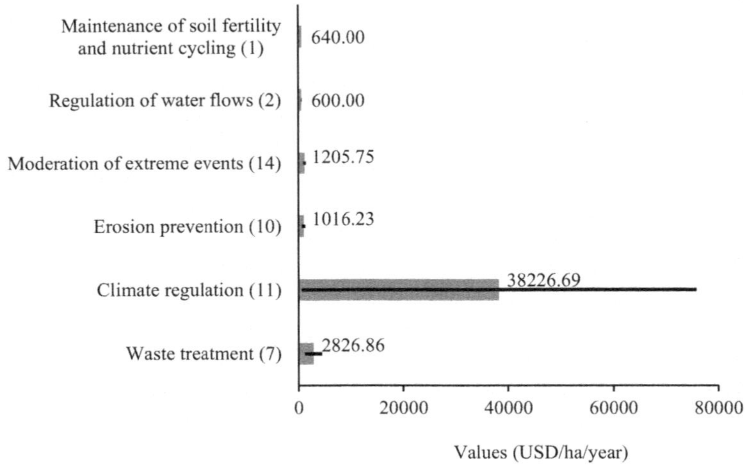

Fig. 19.6 Values of regulating services provided by mangrove ecosystems (Based on data compiled by Himes-Cornell et al. 2018)

Regulating services of a mangrove ecosystem, namely, quality maintenance, storm, flood and erosion control and climate regulation have been also widely studied and evaluated (Saenger 2002; Walters et al. 2008). In contrast limited attention (in terms of quality and quantity analysis) has been given to other regulating services including air quality, water flow (salt water intrusion), and biological control (Ilman et al. 2011).

Figure 19.6 lists the values of regulating services and highlights significant variation, not only between services but also within the same service. For example, the value of climate regulation services ranges from 2.2 USD/ha/year to 414,411 USD/ha/year, and the value of water treatment service varies from 31 USD/ha/year to 11,000 USD/ ha/year. Figure 19.6 also shows that climate regulation has the largest economic value of 38,226.69 USD/ha/year, the remaining services are much less such as water treatment 2826.86 USD/ha/year, erosion control 1016.23 USD/ha/year or flow regulation 600 USD/ha/year.

Mangrove ecosystems are considered to play a very important role in supporting coastal and offshore fishing as well as aquaculture by providing a breeding and nursery ground and habitat to various fish and crustaceans (Rönnbäck 1999; Walters et al. 2008). Therefore, supporting services are valued through the amount of fish or other species caught or available per area of mangrove (Baran 1999; Pauly and Ingles 1999) or the relative contribution of an area to a given harvest (Pauly and Ingles 1999; Rönnbäck 1999).

Figures 19.7 and 19.8 presents estimates of supporting service values of mangrove ecosystems based on 14 reference materials. Mangroves play a very significant role in maintaining the life cycle of migratory species with an estimated average value of 1725 USD/ha/year. Meanwhile, the value of maintaining biodiversity of mangrove ecosystems is estimated at 82.41 USD/ha/year.

Fig. 19.7 A crab shielded in a mangrove tree hollow (Image source: Vo Van Thanh)

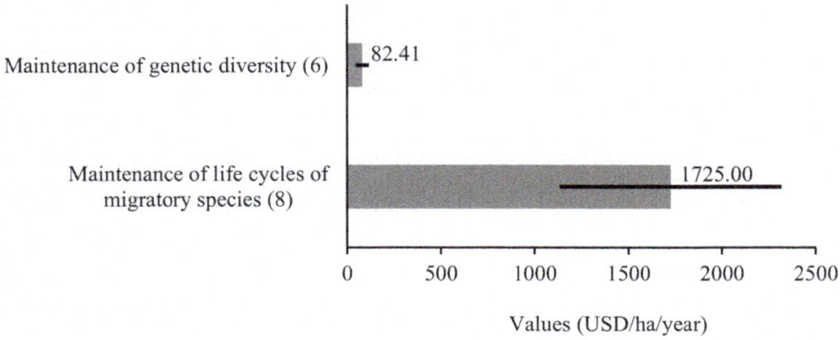

Fig. 19.8 Values of supporting services of mangrove ecosystem (Based on data compiled by Himes-Cornell et al. 2018)

A mangrove ecosystem provides local communities with many opportunities for aesthetic and recreational experiences, cultural and artistic inspiration, and spiritual and religious enrichment (Mastaller 1997; Rönnbäck et al. 2007). These cultural services are often intimately tied to local culture, heritage, and traditional knowledge

Fig. 19.9 Mangrove ecotourism in Morondova, Madagascar (Image source: Pham Hong Tinh)

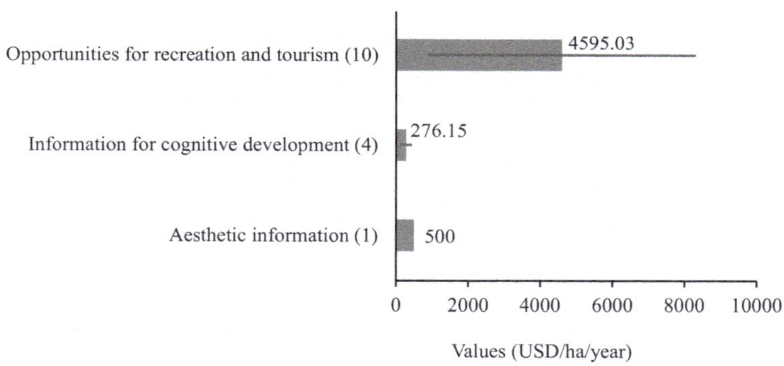

Fig. 19.10 Values of cultural services of mangrove ecosystems (Based on data compiled by Himes-Cornell et al. 2018)

(Walters et al. 2008). However, cultural services are generally difficult to quantify or map, as they are determined by the appreciation and sentiments of local people (Van Oudenhoven et al. 2014).

A number of studies have been conducted to value cultural services provided by mangrove ecosystems. These values vary greatly among different cultural services or among different studies of the same service. Figures 19.9 and 19.10 shows estimated cultural values of a mangrove ecosystem, in which opportunities for recreation and tourism see the highest estimated average value of 4595.03 USD/ha/year, followed by aesthetic information at 500 USD/ha/year and information for cognitive development at 276.15 USD/ha/year.

19.6 Perspective and Conclusion

Mangrove ecosystems play important roles in the sustainable development of coastal zones through their ecosystem services, such as provisioning services, regulating services, supporting services, and cultural services. The economic values of those mangrove ecosystem services have been estimated in many parts of the world. In which, some services such as providing foods and raw materials, moderation of extreme events, erosion prevention, and climate regulation are frequently evaluated, while others, such as pollination and ornamental resources and the cultural category of ecosystem services have not yet been given adequate attention. Various evaluation methods have been applied to estimate economic values of mangrove ecosystem services with market price and benefit transfer methods most frequently used. However, other methods like avoided/damage, choice modelling, travel cost, contingent valuation, etc. are rarely used to evaluate one or several mangrove ecosystem services. The estimated values also vary greatly across different ecosystem services or within an ecosystem service, but with different locations and evaluation methods.

References

Aburto-Oropeza O, Ezcurra E, Danemann G, Valdez V, Murray J, Sala E (2008) Mangroves in the Gulf of California increase fishery yields. Proc Natl Acad Sci 105: 10456–10459.

Adger N, Brown K, Cervigini R, Moran D (1995) Total economic value of forests in Mexico. Ambio 24: 286–296.

Amarnath JS and Mouna A (2016) Environmental impact assessment of coastal ecosystem in Tamil Nadu, India with hedonic and travel cost models. Int. J. Mar. Sci 66: 1–8.

Atkinson SC, Jupiter SD, Adams VM, Ingram JC, Narayan S, Klein CC, Possingham HP (2016) Prioritising mangrove ecosystem services results in spatially variable management priorities. PLoS One. doi: 10.1371/journal.pone.0151992.

Baran E (1999) A review of quantified relationships between mangroves and coastal resources. Phuket Marine Biological Center Res Bull 62: 57–64.

Barbier EB, Costanza E, Twilley RR (1991) Guidelines for tropical wetland evaluation. CATIE: Turrialba, Costa Rica, p 1-57.

Bunting P, Rosenqvist A, Lucas RM, Rebelo L-M, Hilarides L, Thomas N, Hardy A, Itoh T, Shimada M, Finlayson CM (2018) The Global Mangrove Watch—A New 2010 Global Baseline of Mangrove Extent. Remote Sens 10(10):1669, doi:https://doi.org/10.3390/rs10101669.

Camacho-Valdez V, Ruiz-Luna A, Ghermandi A, Berlanga-Robles CA, Nunes PALD (2014) Effects of land use changes on the ecosystem service values of coastal wetlands. Environ Manage 54: 852–864.

Costanza R, D'Arge R, de Groot R, Farber S, Grasso M, Hannon B, Limburg K, Naeem SV, O'Neill R, Paruelo J, Raskin RG, Sutton P, van den Belt M (1997) The value of the world's ecosystem services and natural capital. Nature 387: 253–260.

De Groot R, Brander L, Van Der Ploeg S, Costanza R, Bernard F, Braat L, et al (2012) Global estimates of the value of ecosystems and their services in monetary units. Ecosyst Serv 1: 50–61

Himes-Cornell A, Grose SO and Pendleton L (2018) Mangrove ecosystem service values and methodological approaches to valuation: Where Do We Stand? Front Mar Sci 5: 376.

Ilman M, Wibisono ITC, Suryadiputra INN (2011) State of the art information on mangrove ecosystems in Indonesia. Wetlands International - Indonesia Programme, Bogor.

Jerath M, Bhat M, Rivera-Monroy VH, Castaneda-Moya E, Simard M, Twilley RR (2016) The role of economic, policy, and ecological factors in estimating the value of carbon stocks in Everglades mangrove forests, South Florida, USA. Environ. Sci Policy 66: 160–169.

Kathiresan K, Bingham BL (2001) Biology of mangroves and mangrove ecosystems. Adv Mar Biol 40: 81–251.

Lebel L, Tri NH, Saengnoree A, Pasong S, Buatama U, Thoa LK (2002) Industrial transformation and shrimp aquaculture in Thailand and Vietnam: pathways to ecological, social and economic sustainablity? AMBIO: A Journal of the Human Environment 31(4): 311-323.

Mashayekhi Z, Danehkar A, Sharzehi GA, Majed V (2016) Coastal communities WTA compensation for conservation of mangrove forests: a choice experiment approach. Knowl Manag Aquat Ecosyst 417: 20, doi: 10. 1051/kmae/2016007.

Mastaller M (1997) Mangroves, the Forgotten Forest between Land and Sea. Tropical Press Sdn. Bhd, Kuala Lumpur, Malaysia.

Mazda Y, Magi M, Ikeda Y, Kurokawa T, Asano T (2006) Wave reduction in a mangrove forest dominated by Sonneratia sp. Wet Ecol Manag 14: 365–378.

MEA (2005) Millennium Ecosystem Assessment. Washington, DC: World Resources Institute.

Mojiol AR, Guntabid J, Lintangah W, Ismenyah M, Kodoh J, Chiang LK, Sompud J (2016) Contribution of mangrove forest and socio-economic development of local communities in Kudat District, Sabah Malaysia. Int J Agric For Plant 2: 122–129.

Naylor RL, Goldburg RJ, Primavera JH, Kautsky N, Beveridge MCM, Clay J, Folke C, Lubchenco J, Mooney H, Troell M (2000) Effect of aquaculture on world fish supplies. Nature 405: 1017–1024.

Pauly D, Ingles J (1999) The relationship between shrimp yields and intertidal vegetation (mangrove) areas: a reassessment. In Yáñez-Arancibia A, Lara-Domínguez AL (eds) Ecosistemas de Manglar en América Tropical. Instituto de Ecología A.C., México, UICN/ORMA, Costa Rica, NOAA/NMFS Silver Spring (MD), p 311-318.

Rönnbäck P (1999) The ecological basis for economic value of seafood production supported by mangrove ecosystems. Ecol Econ 29: 235–252.

Rönnbäck P, Crona B, Ingwall L (2007) The return of ecosystem goods and services in replanted mangrove forests: perspectives from local communities in Kenya. Environmental Conservation 34: 313-324.

Saenger P (2002) Mangrove ecology, silviculture and conservation. Kluwer, Dordrecht.

Sheridan P, Hays C (2003) Are mangroves nursery habitat for transient fishes and decapods? Wetlands 23: 449–458.

Sopheak K, Hoeurn C (2016) An estimation of the production function of fisheries in Peam Krasaob Wildlife Sanctuary in Koh Kong Province, Cambodia. EEPSEA Research Report No 2016022. Laguna: Economy and Environment Program for Southeast Asia (EEPSEA).

Susilo E, Purwanti P, Lestariadi RA (2016) Mangrove management in Dumas beach: economic and institutional analysis. Int J Manag Adm Sci 3: 11–24.

Tri NH (2006) Mangrove ecosystem service valuation – principle and application. Pubishing House of Vietnam National Economic University, Hanoi, Vietnam (in Vietnamese).

Tri NH, Adger WN, Kelly PM (1998) Natural resource management in mitigating climate impacts: the example of mangrove restoration in Vietnam. Glob Environ Change 8(1): 49-61.

Tri NH, Hong PN, Adger WN, Kelly PM (2002) Mangrove conservation and restoration for enhanced resilience. In Rapport DJ, Lasley WL, Rolston DE, Nielsen NO, Qualset CO, Damania AB (eds) Managing for healthy ecosystems. Lewis Publishers, Boca Raton, Florida, USA, p 389–402.

Van Oudenhoven APE, Siahainenia AJ, Sualia I, Tonneijck FH, Van der Ploeg S, de Groot RS (2014) Effects of different management regimes on mangrove ecosystem services in Java, Indonesia. Wageningen University (Wageningen) and Wetlands International (Ede & Bogor).

Walters BB, Rönnbäck P, Kovacs JM, Crona B, Hussain SA, Badola R, Primavera JH, Barbier E, Dahdouh-Guebas F (2008) Ethnobiology, socio-economics and management of mangrove forests: A review. Aquat Bot 89: 220–236.

Zhang K, Liu H, Li Y, Xu H, Shen J, Rhome J, Smith TJ III (2012) The role of mangroves in attenuating storm surges. Estuar Coast Shelf Sci 102-103: 11–23.

Management Action Plans for Development of Mangrove Forest Reserves

20

Mohamad Danial Md Sabri, Mohd Nazip Suratman, and
Nur Hajar Zamah Shari

Abstract

Mangroves are recognized as a productive and pivotal ecosystem by providing habitats for many types of biodiversity, sources of food and medicines, heighten socio-economic of local communities, contribute towards climate regulations and coastal protection from sporadic windstorm events. These ecosystems are widely distributed throughout tropical and subtropical regions of the world. Mangroves are basically generated from sediment complexes which has evolved over the last several thousand years. This resulted in a combination of geomorphic processes operating at a sea level close to its current position and ultimately acting as a bridge connecting the land and sea. Among tree communities found in this productive area are from some species such as *Rhizophora* spp., *Avicennia* spp., and *Bruguiera* spp. Overexploitation of mangroves through human activities such as illegal logging and encroachment, wildlife poaching, expansion of agricultural as well as aquaculture practices, apparently becomes among rapid drivers for degradation to this majestic ecosystem. Without proper control over these activities will serve the extinction on mangrove forests. Therefore, proper management action plans towards establishment of forest reserves in mangrove areas are needed to overcome such threats. Moreover, the management action plan is critical to guide policy and decision makers, local authorities, stakeholders, and communities to play their roles to ensure the sustainability of mangrove forests

M. D. Md Sabri · N. H. Z. Shari
Forestry and Environment Division, Forest Research Institute Malaysia, Kepong, Selangor, Malaysia

M. N. Suratman (✉)
Faculty of Applied Sciences and Institute for Biodiversity and Sustainable Development, UiTM, Shah Alam, Malaysia
e-mail: nazip@uitm.edu.my

R. P. Rastogi et al. (eds.), *Mangroves: Ecology, Biodiversity and Management*,
https://doi.org/10.1007/978-981-16-2494-0_20

and their resources. This chapter reviews related available policies, process, and stages that encapsulated within recent scientific studies on mangrove forests globally towards the development of effective management action plans to promote the sustainability of mangrove forest reserves.

Keywords

Mangroves forest · Management action plan · Productive ecosystem · Forest reserves

20.1 Introduction

Mangrove ecosystems ecologically inhabiting the interface between land and sea at low latitudes which occupy a harsh environment, regularly subject to tidal changes in temperature, water, and salt exposure, and varying degrees of conditions. Mangrove forests and their inhabitants are fairly robust and highly adaptable to life in waterlogged saline soils within warm, subtropical, and tropical seascapes. The forests occupy approximately 4.79 million ha of global coastline where the largest areas are found in Asia (5,547,000 ha) followed by Africa (3,240,000 ha) and North and Central America (2,571,000 ha) regions (FAO 2020). These forests provide various benefits and services, from the provision of materials (such as timber, medicinal plants, and fisheries) to being a regulating influence (such as protection from sporadic windstorm events, and carbon sequestration). Moreover, these forests offer cultural and spiritual benefits to the surrounding communities.

Albeit crucial for supporting local livelihoods, mangrove ecosystems are becoming increasingly threatened all over the world as they are undergoing rapid degradation. These mainly due to anthropogenic activities such as urban construction, infrastructure development for tourism, conversion for aquaculture and agriculture, overharvesting of mangrove resources, pollution, and human-induced climate change (Barbier 2016; Arumugam et al. 2020). According to Food and Agriculture Organization (FAO), the area of global mangroves declined by 1.04 million ha between 1990 and 2020 where the rate of loss more than halved over three decades. The global disappearance of mangroves poses a major impact on the vulnerability of coastal populations and property in developing countries, especially with respect to damaging and life-threatening storms and floods (Alongi 2008; Spalding et al. 2014).

As mangroves are declining over decades, understanding the consents of a diverse range of stakeholders in mangrove forest managements is crucial (Arumugam et al. 2020). The impact of forest management within mangroves areas is depending on the way of local policies and legislation that are implemented. Without proper management planning and actions, these important forest resources will deplete over time. To ensure the sustainable services provide by the mangroves, proactive actions are needed to compensate mangrove loss by gazetting more mangroves areas to become permanent forest reserves. Through this initiative, mangrove resources will be managed properly according to current legislation. Therefore, this chapter reviews previous studies on mangrove management and policies in order to develop an action plan for sustainability of mangrove forest reserves.

20.2 Current Distribution and Status of Global Mangroves

Mangrove ecosystems are widely recognized for their pivotal role in ecological sustainability by providing habitat for fish and crustacean of commercial value as well as for effective sediment trapping, nutrient recycling, and protection of shorelines from erosion (Datta et al. 2012). Besides that, freshwater floods occur during the monsoon seasons, while brackish water or seawater floods in other periods occasionally happened at these productive forest areas (Koh et al. 2018; Yahya et al. 2020). This situation will be affecting the trend of plant communities, diversity, and conditions in mangrove forests. These forests are mostly dominated by halophytic plants and predominantly found along the tropical and subtropical coastlines which eventually offer important ecosystem functions and services for local and regional communities (Biswas et al. 2018).

Globally, mangrove forests are distributed within 113 countries and territories (FAO 2020). In 2010, the mangrove areas were estimated to be 137,600 km^2 of coastline area globally (Bunting et al. 2018) which contributed to the economic value on the order of 200,000–900,000 USD ha^{-1} (UNEP-WCMC 2006). Asia region has recognized to have the largest extent and the highest species diversity of mangrove ecosystems in the world (ITTO 2002) with the five largest mangrove-holding nations include Indonesia, Brazil, Malaysia, Papua New Guinea, and Australia which represent nearly 61.6% of global mangrove cover (Blasco et al. 2001; Hamilton and Casey 2016; Martínez-Espinosa et al. 2020). In addition, other major mangrove areas are also found in Guinea Bissau, Mozambique, Madagascar, the Philippines, Thailand, and Vietnam (Giri et al. 2011). Mangrove areas recorded absent in Europe region as shown in Table 20.1. Based on previous information, Oceania has the smallest areal extent and Africa has the least species diversity (Ricklefs and Latham 1993).

Table 20.1 Global mangrove areas by region (FAO 2020)

Region	Mangrove area (ha)	Percentage (%)
Africa	3,240,000	21.91
Asia	5,547,000	37.52
Europe	0	0.00
North and Central America	2,571,000	17.39
Oceania	1,298,000	8.78
South America	2,130,000	14.41
World	*14,786,000*	*100.00*

Many developing countries nowadays are concerned with the importance of conserving the mangrove forests and coastal populations for the benefit towards the local livelihood and ecological-economic values (Datta et al. 2012). These countries are taking efforts to maintain the extension of their mangrove forests. The largest extension of mangrove forest recorded in neo-tropical latitudes which found in Brazil (9627 km^2), Mexico (741,917 km^2), and Cuba (421,538 km^2) (Suárez-Abelenda et al. 2014). While in Asia, Bangladesh is having a huge area of mangrove forest around 601,700 ha, which equal to 38.12% of the entire forest land and 4.13% of the land mass in Sundarbans (Khan et al. 2020). In Pakistan, 1050 km long coastline encompassing mangrove forests that provide many valuable ecosystem services. These mangroves are found along the entire coastline of Sindh but occur occasionally on the coast of Balochistan (Farooq and Siddiqui 2020). Meanwhile, in India, vast coastline of 8118 km has a rich mangrove ecosystem that offers a livelihood to several thousand people across the nation. These mangrove forests cover 4740 km^2 area along the Indian coastline of which 617 sq. km occur in the Andaman and Nicobar Islands (Kiruba-Sankar et al. 2018). These are among developing countries which thrive to maintain the existence of mangrove forest within their political territories.

Mangroves are mainly divided into three major categories which include true mangroves, mangroves, and mangrove associates (Wan Juliana et al. 2010). The total number of true mangrove plant species in the world is about 70 and which belong to 17 families. Apart from that, 51 species are present in the South Asian mangrove forests (Wan Ismail et al. 2018). Interestingly, Southeast Asia, which can be perceived as home for world largest mangrove where constitute area approximately over 6.8 million ha or equal to 34%–42% out of total mangroves worldwide (Giesen et al. 2007). A study by Giesen et al. (2007) also described that the greatest areas were found in Indonesia, which has the most extensive coverage of mangroves (60%), followed by Malaysia (11.7%), Myanmar (8.8%), and Thailand (5%). Besides that, 268 flora species were recorded, including 129 trees and shrubs, 50 terrestrial herbs consisting of 27 types of grass and grass-like plants, 28 climbers, 28 epiphytes, 24 ferns, 7 palms, 1 screw pine (pandan), and 1 cycad. However, from all of these, only 52 species were known as true mangrove species where the habitat is exclusively in mangrove ecosystems.

Ecologically, mangroves which consist of salt tolerant plants comprised of tree, palm, shrub, and fern communities were found in the intertidal zone, at the interface of land and sea, transitional zone of the coasts, estuaries, and along rivers draining into the sea (Hutchison et al. 2014; Shah et al. 2016). Among countries with a high diversity of mangroves are Indonesia with 45 species followed by Malaysia (40 species), Thailand (34 species), and Singapore (31 species) (Kathiresan and Rajendran 2005). In the case of Malaysia, mangrove forests are mainly found in the marine alluvium along sheltered coasts and estuaries both in Sabah, Sarawak, and Peninsular Malaysia. Among tree genus distribute along the coast are *Rhizophora*, *Avicennia*, *Bruguiera*, *Sonneratia*, and *Xylocarpus* spp. in species-specific belts depending on soil and inundation patterns (Wilkie and Fortuna 2003). The presence of mangrove species and diversity are depending on the soil types, predations, disturbance, weather stability, intertidal patterns, management regimes, and more based on geographical restricted, the region of their origin. Malaysia has shown the effort to conserve mangrove forests through the establishment of the Matang Mangrove Forest Reserve where this reserve is considered as the best managed mangrove forest in the world (Walters et al. 2008).

20.3 Drivers of Mangrove Loss

Mangrove forests, like other ecosystems, are subject to various disturbances that vary in their nature (e.g., geological, physical, chemical, biological) in time and space (Alongi 2008). The drivers of these disturbances may come from natural and anthropogenic sources. However, the great loss of mangrove forests is likely threatened by a range of anthropogenic activities. Loss of mangrove cover has increased significantly since 1970 (Giri et al. 2011; Richards and Friess 2016), and the remaining forest patches are under pressure from different sources of disturbance. According to previous records, mangroves are disappearing at a global loss rate of 1–2% per year (Spalding et al. 2010), and the loss rate reached 35% during the last 20 years (Polidoro et al. 2010; Carugati et al. 2018). These rates may be as high as or higher than rates of losses of upland tropical wet forests, and current loss rates are expected to continue unless mangrove forests are protected as a valuable resource (Polidoro et al. 2010).

The drivers in discussing this chapter are perceived as major threats for mangrove loss in many places around the world. Previous studies showed that one quarter of the world's mangroves have been lost due to human action, mainly through forest conversion to aquaculture, agriculture, and urban land uses (Barbier 2016). Other studies also showed that over 35% of mangroves worldwide have disappeared since the 1980s, mainly due to clear-cutting and conversion to aquaculture ponds (Valiela et al. 2001; Richards and Friess 2016). In this context, aquaculture was identified as the main source of degradation of mangrove forests. Tropical coastlines in Southern Thailand where the mangroves can be found have been converted into aquaculture ponds, consequently lost the important ecosystem services (Thampanya et al. 2006). The depreciation services such as the loss of sediment stabilization is currently causing problems along multiple aquaculture coasts worldwide in the form of coastal erosion. Additionally, erosion of aquaculture areas can be as severe as several kilometres per year where finally putting coastal communities in danger (Van Wesenbeeck et al. 2015). Furthermore, many large companies abandoned the ponds once their profits decrease, turning to other pristine mangrove areas for development of new and more productive ponds (Huitric et al. 2002).

Indonesia had lost more than 200,000 ha of its mangroves by the 1960s due to various threats where the largest losses were reported from Java and Sumatra (Koh et al. 2018). Besides that, Indonesia was reported among ten largest fish producers in the world. In the year of 1999 to 2003, fish production in Indonesia increased by an average of 8.5% per year, from 4952 thousand tonnes in 1999 to an estimated 5961 thousand tonnes in 2003, with one-fifth of the production coming from aquaculture.

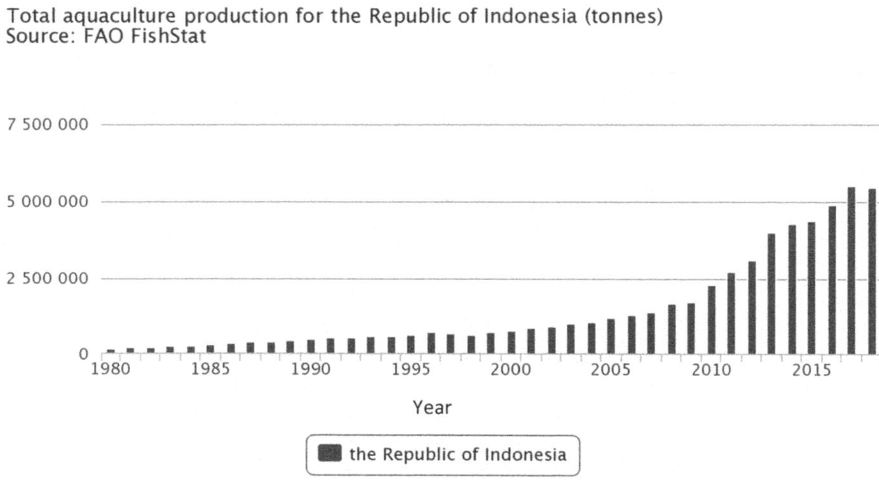

Fig. 20.1 Aquaculture trend in Indonesia from year 1980–2018 (FAO undated)

Aquaculture is deemed an important source of employment, providing livelihoods to an estimated around 2.4 million households in this country (http://www.fao.org/fishery/countrysector/naso_indonesia/en). Figure 20.1 shows that the trend of aquaculture in Indonesia from the year 1980 to 2018 as reported by FAO FishStat. Halting the aquaculture activity will affect some part of local economics, thus, proper management and action are believed to be crucial to ensure the availability of these mangrove forests to upkeep the ecosystem services.

Another major driver in mangrove loss is shrimp farming. Shrimp farming is among popular aquaculture activity which leads to land use changes that occurred in mangrove areas. Eastern African mangroves are currently facing additional threat from shrimp aquaculture, where the practices were originating from Southeast Asia. This shrimp aquaculture practices are now reaching to West Africa (Ajonina et al. 2008). While in Brazil, shrimp farming production in tropical coastal regions has increased significantly during the last 30 years due to food demand and high economic value (Queiroz et al. 2013). Despite it is now being recognized that the conversion of mangrove forests to shrimp farming becomes unsustainable in the long term due to the environmental impacts (e.g., eutrophication) and strong disruption of local economies (Suárez-Abelenda et al. 2014), some countries are politically extending the area to provide values for locals. Conversely, recent studies came out with shrimp farming model development for Vietnamese Mekong Delta to reduce the ecological impact from this activity. This model is beneficial to the local livelihoods and together with enhancement of knowledge in coastal protection (Nguyen et al. 2020). In this context, shrimp farming also plays greater role in mangrove forests with some pros and cons on the impacts to the ecosystem or socio-economic well-being.

Forest degradation is also believed to be a serious issue in management of mangrove ecosystems. Although the rate of global mangrove loss has declined, the future of global mangrove still remains uncertain as new territories of deforestation are opening up, mainly in Southeast Asia and West Africa (Arumugam et al. 2020). Southeast Asia, the region of largest and most biodiverse mangrove ecosystems on Earth (Spalding et al. 2010; Giri et al. 2011; Appeltans et al. 2012), is currently facing a highest rate of deforestation since 2000 (Hamilton and Casey 2016; Richards and Friess 2016). While in Africa, mangroves also have been subjected to enormous pressures and threats within the last past decades with great losses, for example, over 20–30% of the mangroves notably in the west and central Africa have lost since past 25 years. The loss happened through many factors, especially deforestation for fish-smoking (Ajonina et al. 2008). Another study in Bangladesh, Sundarbans mangrove forests were reported as being continuously degraded where total tree cover has been reduced by 50 per cent over the past 20 years (Khan et al. 2020). High demand, poor management, and overexploitation within mangrove forests without replacement plan were associated with severe degradation to the ecosystems.

20.4 Importance of Mangrove Conservation

Mangroves are forested wetlands that are uniquely adapted to the intertidal zone. These forests need to be conserved for the benefits of the local population and environment. Among salient services offered by these majestic ecosystems are carbon sequestration, supporting fisheries, providing timber and coastline protection against erosion and flood protection from storm surges and tsunamis (Alongi 2008; Van Bijsterveldt et al. 2020). Besides that, salt marshes and mangroves contribute to coastal protection by reducing wave energy, increasing sedimentation, and/or reducing erosion and movement of sediments (as reviews in Gedan et al. 2011; Spalding et al. 2014). Furthermore, thick vegetation cover reduces water flow velocities, turbulent flows, and shear stress over the sea bed, promoting sediment deposition, which can create accretion. In some cases, deposition stimulates below ground root production (McKee and Cherry 2009), and these roots scientifically found to improve soil cohesion and tensile strength, slowing rates of erosion at marsh edges. Hence, the roots themselves can also present a physical barrier between the water and soil, particularly in places where root systems extend below low tide levels (Gedan et al. 2011; Giri et al. 2011). These ecosystems also play a role by creating habitats for a variety of terrestrial fauna and providing various supplies for local communities (Shrestha et al. 2019). By focussing to fauna, as birds migrate to shallow water and muddy areas such as mangrove forests for food, resting areas, and protection as stated in a review by Azimah and Tarmiji (2018). Thus, fauna also needs mangroves for shelter and resting to sustain their species. Meanwhile, high level of fauna biodiversity as over 80% of commercial fisheries and other aquatic species spent most or part of their life cycle in the mangroves; ecologically, they play a crucial role in fertilization, stabilization, filtration, regulation of microclimate and acting as food chain support and as nurseries for many fish and invertebrate species economically, they provide a wide range of timber and non-timber forest products that support rural economies and having high ecotourism potentials (Ajonina et al. 2008).

From global climate regulation point of view, mangrove forests contribute their roles as a part of coastal carbon sequestration. Mangroves have been estimated to have higher amounts of carbon than the other types of forest, with a storage capacity of between 990 and 1074 tonne C ha^{-1} (Donato et al. 2011). Another impressive fact that although plant biomass in the ocean and coastal areas comprises only 0.05% of the total plant biomass on land, it cycles a comparable amount of carbon each year (Bouillon et al. 2008). Surprisingly, a typical hectare of mangrove has the potential to release carbon similar up to 3–5 ha of terrestrial tropical forest (Eong 1993). This shows that mangrove forests are having the capability to store large carbon that useful for climate regulation and climate change mitigation. In summary, important services provided by mangrove forests that benefit coastal communities is shown in Fig. 20.2.

20.5 Management Action Plans for Mangrove Forest Reserves Development

Mangroves like other natural resources require supportive and adaptive policies, which are adequately enforced. These policies have to be accompanied by initiatives such as plans to facilitate conflict resolution between stakeholders and encourage consultations with administrative officials (Campredon and Cuq 2001; Feka and Ajonina 2011). As discussed previously, mangroves provide many services to coastline communities where the extension of the area is deemed requisite for all mangrove countries. Meanwhile, there were so many drivers in mangrove loss and ultimately reduce the area of these majestic ecosystems. Hence, the proposed draft action plan is a strategic guide with a sequence of steps that forest managers could follow in order to develop future mangrove forest reserves. However, any action plan to be used should refer to the implemented legislation which taken locally. Different countries may have different approach to establish or maintain their mangrove areas. The next section proposes an action plan to develop mangrove forests reserve that may provide some insights for decision and policy makers and forest managers.

20.5.1 Action Plans

The action plan for development of mangrove forest reserves template follows six main steps, viz; (i) Site selection, (ii) Scientific expedition, (iii) Stakeholder consultation, (iv) Determining boundaries, (v) Gazettement, and (vi) Enforcement. Each of the steps then leads to a number of sections which help to define the context, operationalize the development of the plan, and implement the measures

Fig. 20.2 Actual and potential services are offered by mangroves (Datta et al. 2012)

Ecological/protective

- Sediment trapping
- Source of nutrients and organic matter
- Carbon sequestration
- Wastewater treatment
- Protection of shorelines
- Wildlife habitat

Commercial/rawmaterial/subsistence

- Plant parts as food
- Fodder
- Fuel (Charcoal)
- Construction material
- Medicine
- Honey, wax and alcohol
- Tannin and dyes
- Rayon and paper
- Industrial chemicals
- Ornamental goods
- Fishery
- Shrimp culture
- Tourism
- Butterfly farming

Aesthetic/miscellaneous

- Landscaping
- Shelter or fishing boats during storms
- Cultural/religious importance

(Fig. 20.3). This preceding section describes the steps and actions in detail. The proposed template is defined with the circular approach to foster the continuous improvement of the developed action plan.

20.5.2 Site Selection

The first step focuses on the context of identifying suitable sites which is fundamental to understand the main features and characteristic of the forest, in terms of its situation and function. They are described in the preceeding subsections.

Fig. 20.3 Proposed action plan overview

20.5.2.1 Windstorm Barrier and Coastal Protection

The aim is to analyse the frequency of the windstorms and cyclone events on particular coastal areas and their damaging effects to the local communities. In fact, more than 1.4 billion people live in coastal areas at a risk for tropical cyclone (Dilley et al. 2005). Cyclone and windstorm events are expected to increase as a result of more frequent high-intensity storms created by climate change and increased exposure created by the ongoing movement of people and assets to high-risk coastal areas (Del Valle et al. 2020). As an alternative coastal defence, conservation and restoration of natural habitats that can provide protection and act as a barrier against windstorm, mangroves forest may be considered as their plant's aerial root and canopy structure makes them capable of reducing wave action, wind velocity, and storm surge (Massel et al. 1999; Mazda et al. 2006; Barbier et al. 2008, 2011; Krauss et al. 2009; Zhang et al. 2012; Das and Crépin 2013; Liu et al. 2013; Horstman et al. 2014). Thus, the function of this forest as a protector of the coastline should be considered on priority to conserve the areas.

20.5.2.2 Act as Carbon Storage for Climate Regulation

Mangroves forests play an important role as a substantial carbon sink and there are categorized among the most carbon-rich forests in the tropics (Laffoley and Grimsditch 2009; Hamdan et al. 2013). The degradation of this forest will be possibly intensifying decomposition of stored carbon and finally, emitting large amounts to serve greenhouse gases in the atmosphere. In terms of ecological view, carbon from the atmosphere is absorbed by trees and plants through photosynthesis process and stored the carbon in the form of food within their stems and other organs. By this process, mangrove trees and plants also play role in climate regulation and climate change mitigation. Conservation of the mangrove forests would be vital in any efforts to address climate change. Responsible authorities may identify large mangrove areas in their region through modern equipment such as remote sensing. The targeted large area will probably store large amount of carbon.

20.5.2.3 Coastal Demographic Challenges

The coastal demographic context in which the mangrove lies is crucial both with reference to present and future scenarios. This information would contribute significantly to the understanding of future suburban and territorial development. Mangrove ecosystems provide numerous benefits to local communities and adjacent environments. They represent great direct economic importance to the livelihoods of millions of coastal residents throughout the world who harvest marine resources; extract timber for construction, firewood, and charcoal production; and cultivate mangrove honey (Suman 2019). Mangrove ecosystems must compete with powerful economic interests supported by national and local elites, such as urban residential developments, airports, power plants, construction and expansion of ports, aquaculture activities, and coastal tourism projects (Dale et al. 2014). This action aims to provide demographic challenge resulted from local communities within proximity to the mangrove areas in which can be useful in decision makings in development of mangrove forest reserves. Mangrove areas which are close to these challenging areas may not be worth considered as mangrove forest reserves. However, this idea is totally depending on how those natural habitats are being used.

20.5.2.4 Ecotourism and Recreation Prospects

The selection of future mangrove forest reserves may consist of ecotourism and recreation values. A previous study by Friess (2017) and numerous local case studies show that tourist activities can have several important benefits for mangrove education, protection, and conservation. However, the benefits of ecotourism must be balanced against its direct and indirect negative impacts on the mangrove ecosystem and the local communities that use them. However, strategies are needed to minimize the impacts from both activities, for example, mitigating the direct environmental impacts of increased numbers of tourist and boat traffic, and the indirect impacts that a sudden and large source of tourism income can have on local communities and economic enterprises. Hence, this action aimed to the authorities to identify areas which may allow public to enjoy the recreation activities within the forest areas and ultimately create awareness on the importance of mangrove conservation among them and provide attractive economic returns to the local communities.

20.5.3 Scientific Expedition

This section describes the next step for development of the mangrove forest reserve where the scientific information may be useful to provide scientific figures, supports and advise for all stakeholders.

20.5.3.1 Speeding up Scientific Research Activities and Collaborative Research

This action aimed to record all biological diversity, scientific findings, and social studies within the selected mangrove areas. Scientific research by one agency or collaboration with other agencies such as universities and research institutes may bring compilation of observations, knowledge, invent solutions, and create problem solving within those mangrove areas. It is important for scientific research to occur at a local level because research from one area may not be applicable to the context and needs of another region. For example, not all flora and fauna exist in all areas, sometime, they are endemic to some areas. An endemic species is important because they are in the habitats restricted to a particular area due to climate change, urban development or other occurrences.

20.5.3.2 Strengthen International Cooperation in Research, Protection, and Restoration

Related ministerial and government authorities which own the natural resources may encourage international involvement and cooperation to achieve conservation efforts. Engagement of international bodies, for example, the United Nations Forum on Forests (UNFF), Convention of International Trade in Endangered Species of Wild Fauna and Flora (CITES), Convention on Biological Diversity (CBD), United Nations Framework Convention on Climate Change (UNFCCC), and other international authorities could gain result in the protection of large areas of mangrove forests locally and globally. They may also encourage global private sectors to support through monetary values for research, conservations, forest restoration, and climate change mitigations efforts through a number of useful mechanisms.

20.5.4 Stakeholders' Consultation

This step is fundamental for the recognition of the importance of seeking feedback and understanding the views of those who affected from any alteration of policies, enforcement, and way of life. Public services have embraced the approach, seeking the involvement of the public in the development and shaping future services to particular communities. Therefore, stakeholder consultation becomes a requirement in the successful development of public policy and services. Following are the actions under the third step of proposed action plan in a way to develop mangrove forest reserves.

20.5.4.1 Reinforce Knowledge the Role, Value of Mangroves Management and Sustainability

This action will promote sustainable management by developing knowledge on effective protection and rehabilitation of mangrove forests to all stakeholders especially to local communities. Authorities should increase public awareness and education on the benefits of mangroves forests and the importance to sustain these areas. A study in Guyana has described if community members do not have a very good knowledge of the roles of mangroves, it could severely impact the way that the community values the mangrove resources and hence affect their willingness to be involved in conservation efforts (Da Silva 2015). Participation from local communities together with authorities will probably provide a chance for mangrove areas being protected from severe anthropogenic activities.

20.5.4.2 Transform Perception of Managers in Localities on the Role and Values of Mangrove

The objective of this action is to train and facilitate all managers from government or private agencies in localities on the importance of mangrove forest protection. In addressing government policies to the local communities, they are responsible to have a comprehensive view of the services from these forest areas provided to humankind, especially on income generation, poverty alleviation, and the provision of rural food security. For this action, ministerial level should provide them with necessary trainings to establish the administrative capacity for the management of mangroves and ability to balance between human needs and the limit of nature.

20.5.4.3 Socialize Forestry and Upgrade Living Standard for Local People

Local people basically are using mangroves as sources of economic well-being such as establishment of aquaculture ponds, timber products, source of medicine, infrastructure developments, and more. However, lack of basic knowledge on the importance of conservation can lead them to think that conservation efforts may be curbing their daily activities. In addition, conservation actions are underfunded in many areas, especially in Southeast Asia, particularly for the marginalized ecosystems such as mangrove forests (Friess 2016). This situation may drive away the initiative of local communities to conserve mangrove areas. Action from relevant authorities to socialize forestry and efforts to upgrade living standards for local communities is required. Effort of preserving mangrove areas may provide many benefits through future collaboration approaches between government and private sectors such as corporate social responsibility. This effort will boost economic values of these mangrove areas to be an education centres and ecotourism hotspots and ultimately increase the income of the local communities.

20.5.4.4 Consolidation of Mangrove Management Systems at all Levels

In this action, all relevant authorities from different agencies, which are responsible in management of mangrove areas, should be gathered with local communities to discuss and review current policies in aspect of forestry production, ecological, and

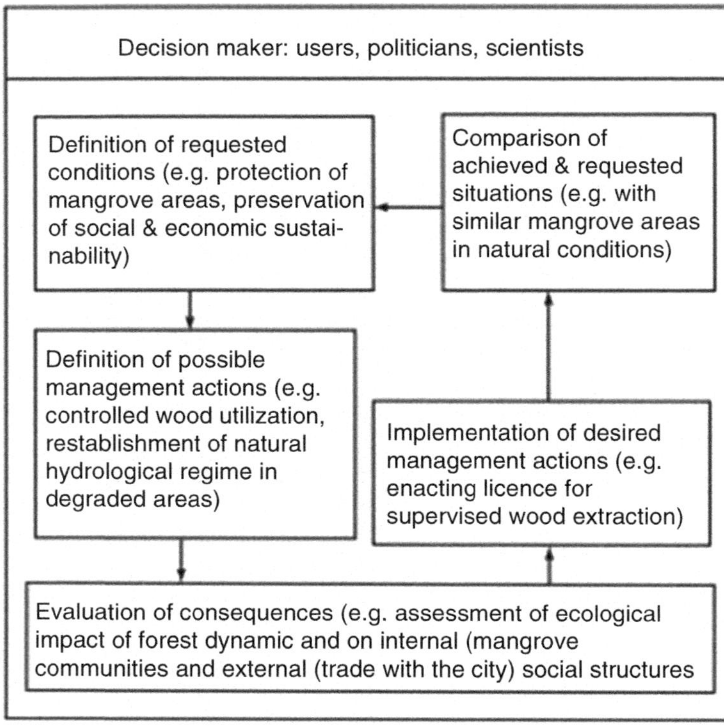

Fig. 20.4 Example of a decision support scheme, applied to a specific situation combining mangrove protection and use (Lara et al. 2010)

economic sustainability. Feedback from all these intersectoral entities should be consolidated prior the establishment of mangroves forest reserves. For example, any decision making have to include of many various parties, for example, resource users, politicians, and scientists who should jointly define the requested systems conditions, such as the protection of mangrove areas and the preservation of the social and economic sustainability of a low-income population (Lara et al. 2010) (Fig. 20.4).

20.5.4.5 Brief about the Proposed Action Plan to all Stakeholders

This section gives some insight to all stakeholders about measures and implementation of the action plan to their areas. These management action plans for development of mangrove forest reserves should be discussed through a participative process and through agreements among all stakeholders in the particular mangrove's areas. This is to ensure successful implementation of the action plan and lessen for future public resistance.

20.5.5 Determining Boundaries

The next step is determining the boundaries of selected mangrove areas. Boundaries are important components of spatially heterogeneous areas. Boundaries are the zones of contact that arise whenever areas are partitioned into patches (Cadenasso et al. 2003). Natural geographic features should be selected to define the boundaries of a forest management unit, including of rivers, streams, shorelines, ridges, and spurs. Besides that, permanent and clearly defined roads, railways, and tracks may also be used to mark boundaries (Armitage 1998).

20.5.5.1 Marking Selected Mangrove Areas on the Map

This section is to mark on paper using geographic information system (GIS) and remote sensing about the proposed mangrove areas that will be submitted for gazetting process. Maps fulfil a wide range of function in forest management depending on the type, the amount of detail features represented and their scale. A map can be used to portray all of the information consist of a "picture" which can be drawn into the mind of a planner or manager about the shapes and slopes of the ground, the pattern of streams and rivers, the vegetation cover, and the location and nature of man-made features of a targeted area (Armitage 1998). The map is important to be used as supporting documents in the decision makings.

20.5.5.2 Mangrove Areas Demarcation

This action is to determine the internal boundaries for selected mangrove areas that proposed as forest reserves and the boundaries should be demarcated clearly at the field. For demarcation process, roads, cut lines, pillars, painted standing trees, and poles can be used to define the internal boundaries. All trees within internal boundaries are suggested to be inventoried and mapped. Notices should be erected showing boundaries of mangrove forest reserves to avoid encroachment by local communities.

20.5.6 Gazettement

This section describes the next step to develop mangrove forest reserves through gazetting the selected areas. This gazettement is crucial to ensure all activities within proposed mangrove forest reserves will be protected by local legislation. The process gazettement an area needs to be undergoing the step according to available policies, legislations, and jurisdiction of the countries. A previous study explained that gazetting is a legal act that integrates the forest in the Permanent Forests Estate as a part of pre-requisite for a state's land title. Thus, the gazetted forestland can be allocated depending to the consent of private domain of other entities, such as a local municipality (commune) for public interest, social and economic activities or a private company for commercial logging (Ongolo and Karsenty 2015).

20.5.6.1 Preparing Authorized Documents for Gazettement of Mangrove Areas as Forest Reserves

This action aimed to prepare the authorized documents for proposing the mangrove areas to the national or state forest councils' consideration as the decision makers. At this stage, all information gathered through research expedition, study on feedback of local communities and map is required as supportive documents to be submitted prior the council meeting.

20.5.6.2 Approval of Gazettement of Mangroves Areas

After getting approval from the council, the next action involves relevant authorities where in this case, forest manager is to set up the management plan to be implemented development of the mangroves forest reserve. Land use zoning in the forest reserves may sound crucial for monitoring purposes. The zoning approach provides important information for potential stakeholders to identify suitable zones for the optimal allocation of resources and minimization of conflicts among users. Besides that, forest managers need to inform the stakeholders about the gazettement areas through advertisement, workshops, and convention. At this stage, forest manager and other authorities need to plan activities that may be useful for local communities.

20.5.7 Enforcements

This section is the final step which is related to the enforcement to take place at the forest reserves to control illegal usage of the areas. Such a management measure would aim to avoid a system shift towards an ecologically unsustainable situation. The national or state might confront elites (e.g., tourist entrepreneurs, land speculators, and logging companies) and take action against legislative abuses by these powerful groups. The motivation for strict law enforcement should be needed, perhaps to warn other mangrove users to respect the law. This may support the ecological sustainability of the system. Forest managers need to set up enforcement unit to monitor the forest reserve areas.

20.6 Conclusion

There are many ways to protect mangrove forests from being degraded by anthropogenic activities. However, despite the enforcement with strict law is implemented to protect these forests, many previous studies shown above described that the number of mangrove areas is declining every year globally. The loss of these mangroves should be replaced with available mangroves forest. However, no standardized plans are found to be referred by forest managers to develop mangroves forest reserves. This important gap in the literature has been identified for this purpose. The proposed action plan is considered to be a comprehensive guideline for the development of mangrove forest reserves as a part of the management plan.

The step-by-step action plans will facilitate forest managers and all stakeholders to address the technical, economic, social, and environmental sustainability that are useful for sustainable mangrove forest management.

References

Ajonina G, Diamé A, Kairo J (2008) Current status and conservation of mangroves in Africa: An overview. World Rainforest Movement Bulletin No 133.

Alongi DM (2008) Mangrove forests: Resilience, protection from tsunamis, and responses to global climate change. Estuar Coast Shelf Sci 76:1–13. https://doi.org/10.1016/j.ecss.2007.08.024

Appeltans W, Ahyong ST, Anderson G, et al (2012) The magnitude of global marine species diversity. Curr Biol 22:2189–2202. https://doi.org/10.1016/j.cub.2012.09.036

Armitage I (1998) Guidelines for the management of tropical forests 1. The production of wood. FAO, Rome

Arumugam M, Niyomugabo R, Dahdouh-Guebas F, Hugé J (2020) The perceptions of stakeholders on current management of mangroves in the Sine-Saloum Delta, Senegal. Estuar Coast Shelf Sci. https://doi.org/10.1016/j.ecss.2020.106751

Azimah AR, Tarmiji M (2018) Habitat requirements of migratory birds in the Matang Mangrove Forest Reserve, Perak. J Trop For Sci 30:304–311. https://doi.org/10.26525/jtfs2018.30.3.304311

Barbier EB (2016) The protective service of mangrove ecosystems: A review of valuation methods. Mar Pollut Bull 109:676–681. https://doi.org/10.1016/j.marpolbul.2016.01.033

Barbier EB, Hacker SD, Kennedy C, et al (2011) The value of estuarine and coastal ecosystem services. Ecol Monogr 81(2):169–193

Barbier EB, Koch EW, Silliman BR, et al (2008) Coastal ecosystem-based management with nonlinear ecological functions and values. Science (80–) 319:321–323. https://doi.org/10.1126/science.1150349

Biswas SR, Biswas PL, Limon SH, et al (2018) Plant invasion in mangrove forests worldwide. For Ecol Manage 429:480–492. https://doi.org/10.1016/j.foreco.2018.07.046

Blasco F, Aizpuru M, Gers C (2001) Depletion of the mangroves of Continental Asia. Wetl Ecol Manag 9:245–256. https://doi.org/10.1023/A:1011169025815

Bouillon S, Borges A V., Castañeda-Moya E, et al (2008) Mangrove production and carbon sinks: A revision of global budget estimates. Global Biogeochem Cycles 22:1–12. https://doi.org/10.1029/2007GB003052

Bunting P, Rosenqvist A, Lucas RM, et al (2018) The global mangrove watch—A new 2010 global baseline of mangrove extent. Remote Sens 10. https://doi.org/10.3390/rs10101669

Cadenasso ML, Pickett STA, Weathers KC, Jones CG (2003) A framework for a theory of ecological boundaries. Bioscience 53:750–758. https://doi.org/10.1641/0006-3568(2003)053[0750:AFFATO]2.0.CO;2

Campredon P, Cuq F (2001) Artisanal fishing and coastal conservation in West Africa. J Coast Conserv 7:91–100. https://doi.org/10.1007/BF02742471

Carugati L, Gatto B, Rastelli E, et al (2018) Impact of mangrove forests degradation on biodiversity and ecosystem functioning. Sci Rep 8:1–11. https://doi.org/10.1038/s41598-018-31683-0

Da Silva P (2015) Exploring a community's knowledge and use of a coastal mangrove resource: The case of Wellington Park, Guyana. Int J Sci Environ Technol 4:759–769

Dale PER, Knight JM, Dwyer PG (2014) Mangrove rehabilitation: a review focusing on ecological and institutional issues. Wetl Ecol Manag 22:587–604. https://doi.org/10.1007/s11273-014-9383-1

Das S, Crépin AS (2013) Mangroves can provide protection against wind damage during storms. Estuar Coast Shelf Sci 134:98–107. https://doi.org/10.1016/j.ecss.2013.09.021

Datta D, Chattopadhyay RN, Guha P (2012) Community based mangrove management: A review on status and sustainability. J Environ Manage 107:84–95. https://doi.org/10.1016/j.jenvman.2012.04.013

Del Valle A, Eriksson M, Ishizawa OA, Miranda JJ (2020) Mangroves protect coastal economic activity from hurricanes. Proc Natl Acad Sci U S A 117:265–270. https://doi.org/10.1073/pnas.1911617116

Dilley M, Chen RS, Deichmann U, et al (2005) Natural Disaster Hotspots A Global Risk Analysis. The International Bank for Reconstructio and Development/The World Bank and Columbia University, Washington, DC

Donato DC, Kauffman JB, Murdiyarso D, et al (2011) Mangroves among the most carbon-rich forests in the tropics. Nat Geosci 4:293–297. https://doi.org/10.1038/ngeo1123

Eong OJ (1993) Mangroves—A carbon source and sink. Chemosphere 27:1097–1107. https://doi.org/10.1016/0045-6535(93)90070-L

FAO (2020) Global Forest Assessment Resources 2020: Main report. Rome

Farooq S, Siddiqui PJA (2020) Assessment of three mangrove forest systems for future management through benthic community structure receiving anthropogenic influences. Ocean Coast Manag 190:105162. https://doi.org/10.1016/j.ocecoaman.2020.105162

Feka NZ, Ajonina GN (2011) Drivers causing decline of mangrove in West-Central Africa: A review. Int J Biodivers Sci Ecosyst Serv Manag 7:217–230. https://doi.org/10.1080/21513732.2011.634436

Friess DA (2016) Mangrove forests. Curr. Biol. Mag. R746–R748

Friess DA (2017) Ecotourism as a Tool for Mangrove Conservation. Sumatra J Disaster, Geogr Geogr Educ 1:24–35

Gedan KB, Kirwan ML, Wolanski E, et al (2011) The present and future role of coastal wetland vegetation in protecting shorelines: Answering recent challenges to the paradigm. Clim Change 106:7–29. https://doi.org/10.1007/s10584-010-0003-7

Giesen W, Wulffraat S, Zieren M, Scholten L (2007) Mangrove guidebook for Southeast Asia. FAO Regional Office for Asia and the Pacific, Bangkok

Giri C, Ochieng E, Tieszen LL, et al (2011) Status and distribution of mangrove forests of the world using earth observation satellite data. Glob Ecol Biogeogr 20:154–159. https://doi.org/10.1111/j.1466-8238.2010.00584.x

Hamdan O, Hasmadi IM, Aziz HK (2013) Mangrove carbon stock assessment by optical satellite imagery. J Trop For Sci 25:554–565

Hamilton SE, Casey D (2016) Creation of a high spatio-temporal resolution global database of continuous mangrove forest cover for the 21st century (CGMFC-21). Glob Ecol Biogeogr 25:729–738. https://doi.org/10.1111/geb.12449

Horstman EM, Dohmen-Janssen CM, Narra PMF, et al (2014) Wave attenuation in mangroves: A quantitative approach to field observations. Coast Eng 94:47–62. https://doi.org/10.1016/j.coastaleng.2014.08.005

Huitric M, Folke C, Kautsky N (2002) Development and government policies of the shrimp farming industry in Thailand in relation to mangrove ecosystems. Ecol Econ 40:441–455. https://doi.org/10.1016/S0921-8009(02)00011-3

Hutchison J, Spalding M, zu Ermgassen P (2014) The role of mangroves in fisheries enhancement. The Nature Conservancy, Bethesda, Maryland and Wetlands International, Wageningen, Netherlands. 54 pp.

ITTO (2002) ITTO Mangrove Workplan 2002–2006. Yokohama, Japan

Kathiresan K, Rajendran N (2005) Coastal mangrove forests mitigated tsunami. Estuar Coast Shelf Sci 65:601–606. https://doi.org/10.1016/j.ecss.2005.06.022

Khan MFA, Rahman MS, Giessen L (2020) Mangrove forest policy and management: Prevailing policy issues, actors' public claims and informal interests in the Sundarbans of Bangladesh. Ocean Coast Manag 186:105090. https://doi.org/10.1016/j.ocecoaman.2019.105090

Kiruba-Sankar R, Krishnan P, Dam Roy S, et al (2018) Structural complexity and tree species composition of mangrove forests of the Andaman Islands, India. J Coast Conserv 22:217–234. https://doi.org/10.1007/s11852-017-0588-3
Koh H., Teh S., Kh'ng X, Raja Barizan R (2018) Mangrove Forests : Protection Against and Reselience to Coastal Disturbance. J Trop For Sci 30:446–460
Krauss KW, Doyle TW, Doyle TJ, et al (2009) Water level observations in mangrove swamps during two hurricanes in Florida. Wetlands 29:142–149. https://doi.org/10.1672/07-232.1
Laffoley D, Grimsditch G (2009) The Management of Natural Coastal Carbon Sinks. Gland, Switzerland, Switzerland
Lara RJ, Cohen M, Szlafsztein C (2010) Drivers of Temporal Changes in Mangrove Vegetation Boundaries abd Consequences of Land Use. In: Caldwell MM, Heldmaier G, Jackson RB, et al. (eds) Mangrove Dynamic and Management in North Brazil. Springer-Verlag, London, pp. 127–141
Liu H, Zhang K, Li Y, Xie L (2013) Numerical study of the sensitivity of mangroves in reducing storm surge and flooding to hurricane characteristics in southern Florida. Cont Shelf Res 64:51–65. https://doi.org/10.1016/j.csr.2013.05.015
Martínez-Espinosa C, Wolfs P, Vande Velde K, et al (2020) Call for a collaborative management at Matang Mangrove Forest Reserve, Malaysia: An assessment from local stakeholders' view point. For Ecol Manage 458:117741. https://doi.org/10.1016/j.foreco.2019.117741
Massel SR, Furukawa K, Brinkman RM (1999) Surface wave propagation in mangrove forests. Fluid Dyn Res 24:219–249. https://doi.org/10.1016/S0169-5983(98)00024-0
Mazda Y, Magi M, Ikeda Y, et al (2006) Wave reduction in a mangrove forest dominated by Sonneratia sp. Wetl Ecol Manag 14:365–378. https://doi.org/10.1007/s11273-005-5388-0
McKee KL, Cherry JA (2009) Hurricane katrina sediment slowed elevation loss in subsiding brackish marshes of the Mississippi river delta. Wetlands 29:2–15
Nguyen HQ, Tran DD, Luan PDMH, et al (2020) Socio-ecological resilience of mangrove-shrimp models under various threats exacerbated from salinity intrusion in coastal area of the Vietnamese Mekong Delta. Int J Sustain Dev World Ecol 4509:1–14. https://doi.org/10.1080/13504509.2020.1731859
Ongolo S, Karsenty A (2015) The Politics of Forestland Use in a Cunning Government: Lessons for Contemporary Forest Governance Reforms. Int For Rev 17:195–209. https://doi.org/10.1505/146554815815500561
Polidoro BA, Carpenter KE, Collins L, et al (2010) The loss of species: Mangrove extinction risk and geographic areas of global concern. PLoS One 5. https://doi.org/10.1371/journal.pone.0010095
Queiroz L, Rossi S, Meireles J, Coelho C (2013) Shrimp aquaculture in the federal state of Ceará, 1970–2012: Trends after mangrove forest privatization in Brazil. Ocean Coast Manag 73:54–62. https://doi.org/10.1016/j.ocecoaman.2012.11.009
Richards DR, Friess DA (2016) Rates and drivers of mangrove deforestation in Southeast Asia, 2000–2012. Proc Natl Acad Sci U S A 113:344–349. https://doi.org/10.1073/pnas.1510272113
Ricklefs RE, Latham RE (1993) Global patterns of diversity in mangrove floras. In: Species Diversity in Ecological Communities: Historical and Geographical Perspectives. University of Chicago Press, Chicago, pp. 215–229
Shah K, Mustafa Kamal AH, Rosli Z, et al (2016) Composition and diversity of plants in Sibuti mangrove forest, Sarawak, Malaysia. Forest Sci Technol 12:70–76. https://doi.org/10.1080/21580103.2015.1057619
Shrestha S, Miranda I, Kumar A, et al (2019) Identifying and forecasting potential biophysical risk areas within a tropical mangrove ecosystem using multi-sensor data. Int J Appl Earth Obs Geoinf 74:281–294. https://doi.org/10.1016/j.jag.2018.09.017
Spalding M, Kainuma M, Collins L (2010) World Atlas of Mangroves. Erathscan Ltd, London
Spalding MD, Ruffo S, Lacambra C, et al (2014) The role of ecosystems in coastal protection: Adapting to climate change and coastal hazards. Ocean Coast Manag 90:50–57. https://doi.org/10.1016/j.ocecoaman.2013.09.007

Suárez-Abelenda M, Ferreira TO, Camps-Arbestain M, et al (2014) The effect of nutrient-rich effluents from shrimp farming on mangrove soil carbon storage and geochemistry under semi-arid climate conditions in northern brazil. Geoderma 213:551–559. https://doi.org/10.1016/j.geoderma.2013.08.007

Suman DO (2019) Mangrove Management: Challenges and Guidelines. In: Perillo G, Wolansk E, Cahoon D, Hopkinson C (eds) Coastal Wetland An Integrated Ecosystem Approach, Second edi. Elsevier, pp. 1055–1079

Thampanya U, Vermaat JE, Sinsakul S, Panapitukkul N (2006) Coastal erosion and mangrove progradation of Southern Thailand. Estuar Coast Shelf Sci 68:75–85. https://doi.org/10.1016/j.ecss.2006.01.011

UNEP-WCMC (2006) In the Front Line: Shoreline Protection and Other Ecosystem Services from Mangroves and Coral Reefs. UNEP-WCMC, Cambridge

Valiela I, Bowen JL, York JK (2001) Mangrove forests: One of the world's threatened major tropical environments. Bioscience 51:807–815. https://doi.org/10.1641/0006-3568(2001)051[0807:MFOOTW]2.0.CO;2

Van Bijsterveldt CEJ, Van Wesenbeeck BK, Van der Wal D, et al (2020) How to restore mangroves for greenbelt creation along eroding coasts with abandoned aquaculture ponds. Estuar Coast Shelf Sci 235:106576. https://doi.org/10.1016/j.ecss.2019.106576

Van Wesenbeeck BK, Balke T, van Eijk P, et al (2015) Aquaculture induced erosion of tropical coastlines throws coastal communities back into poverty. Ocean Coast Manag 116:466–469. https://doi.org/10.1016/j.ocecoaman.2015.09.004

Walters BB, Rönnbäck P, Kovacs JM, et al (2008) Ethnobiology, socio-economics and management of mangrove forests: A review. Aquat Bot 89:220–236. https://doi.org/10.1016/j.aquabot.2008.02.009

Wan Ismail WN, Wan Ahmad WJ, Salam MR, Latiff A (2018) Structural and floristic pattern in a disturbed mangrove tropical swamp forest: A case study from the langkawi UNESCO global geopark forest, peninsular Malaysia. Sains Malaysiana 47:861–869. https://doi.org/10.17576/jsm-2018-4705-01

Wan Juliana WA, Damanhuri A, Razali MS, et al (2010) Mangrove flora of Langkawi. In: Langkawi Research Centre, Institute for Environment and Development (LESTARI)

Wilkie ML, Fortuna S (2003) Status and trends in mangrove area extent worldwide. Rome

Yahya N, Idris I, Rosli NS, Bachok Z (2020) Mangrove-associated bivalves in Southeast Asia: A review. Reg Stud Mar Sci 38:101382. https://doi.org/10.1016/j.rsma.2020.101382

Zhang K, Liu H, Li Y, et al. (2012) The role of mangroves in attenuating storm surges. Estuar Coast Shelf Sci 102–103:11–23. https://doi.org/10.1016/j.ecss.2012.02.021

Geospatial Tools for Mapping and Monitoring Coastal Mangroves

21

L. Gnanappazham, Kumar Arun Prasad, and V. K. Dadhwal

Abstract

Mangroves are biologically important and productive ecosystems endowed with diverse flora and fauna. Despite their economic and ecological services, mangroves face threats from anthropogenic as well as coastal hazards. Assessment and monitoring of this critical and vulnerable habitat of the coastal ecosystem could play a major role in implementing conservation and management plan compatible with Sustainable Development Goals (SDGs). Mangroves by their geographic confinement in coastal/marshy areas are well suited to study with RS & GIS tools given the difficulty to conduct extensive field trials. Developments in the field of remote sensing and Geographic Information System (GIS) in the last three decades have substantially facilitated smart and efficient use of field surveys for assessing different parameters of mangrove ecosystem such as mapping, monitoring the health of mangrove cover, assessing their diversity, characterization of their biophysical and biochemical properties, as well as monitoring the conservation and restoration activities. Recent advancements in sensor technologies, providing very high spatial resolution multispectral, hyperspectral, microwave, and LiDAR data have substantially improved characterization and monitoring of mangroves. Contemporary *Data Science* methods on storage, geospatial data analytics, and advanced automated algorithms in handling the BIG Data available (archive and real-time data) facilitate a better understanding and assessment of spatio-temporal behaviour of the mangrove ecosystem. This

L. Gnanappazham (✉) · V. K. Dadhwal
Indian Institute of Space Science and Technology (IIST), Trivandrum, India
e-mail: gnanam@iist.ac.in; vkdadhwal@iist.ac.in

K. A. Prasad
Department of Geography, Central University of Tamil Nadu, Thiruvarur, India
e-mail: arunprasad@cutn.ac.in

R. P. Rastogi et al. (eds.), *Mangroves: Ecology, Biodiversity and Management*,
https://doi.org/10.1007/978-981-16-2494-0_21

manuscript presents an overview of how the RS and GIS technologies being evolved in the context of their use for scientific and quantitative studies on mangroves.

Keywords

Geospatial tools · Remote sensing · Image processing · Classification · Geographic information system · Spatial modelling · Floristic mapping · Biophysical and biochemical · Shoreline changes

21.1 Introduction

Mangroves are salt-tolerant woody halophytes generally thrive in marshy inter-tidal zones of tropical and sub-tropical coastlines inundated with diurnal (inundated once a day) or semi-diurnal (inundated twice a day) tides. The mixing of fresh and saline water as well as muddy and anaerobic soil substratum in the estuarine regions favour its growth (Tomlinson 1994).

21.1.1 Distribution of Mangrove Ecosystem

Mangroves are distributed along the low energy, sedimentary tropical coastlines largely between 30°N and 30°S latitudes of over 123 nations covering an area of around 0.15 million sq.km (FAO 2007; FSI 2019), with most of its proportion seen in South and South East Asian countries. Mangroves could survive in extreme habitat conditions such as high temperature, strong winds, turbulent inundation, high salinity, muddy anaerobic substratum. This is possible because they develop some special physiological and morphological adaptations such as saline water regulating root cell membranes, prop roots, aerial roots, and viviparous seedlings (Lugo and Snedaker 1974). The mangrove forest structure is generally characterized by zones of tree species depending on the inundation frequency, tidal range, resultant salinity levels, microtopography, and coastal morphology (Woodroffe 1992; Dalrymple et al. 1992; Ellison 2018). For instance, the world's largest mangrove ecosystem, Sundarbans is located in the extensive deltaic region shared by India and Bangladesh. The region is characterized by flat terrain formed from the sedimentation of an intricate system of rivers such as the Ganges, Brahmaputra, Meghna, and many other tidal channels (Nazrul-Islam 1993).

Recently, Global Mangrove Watch (GMW) Initiative estimated the mangrove extent for the year 2010 as 137,600 sq. km., of which Asia holds the major share with 38.7% succeeded by Latin America and the Caribbean (20.3%), Africa (20%), Oceania (11.9%), North America (8.4%), and the European Overseas Territories (0.7%) (Bunting et al. 2018). Of the global total, three-fourth of the mangrove cover is seen along the coastlines of just 15 nations, mostly in Southeast and South Asia. The eastern countries were found to have more species diversity with 58 species

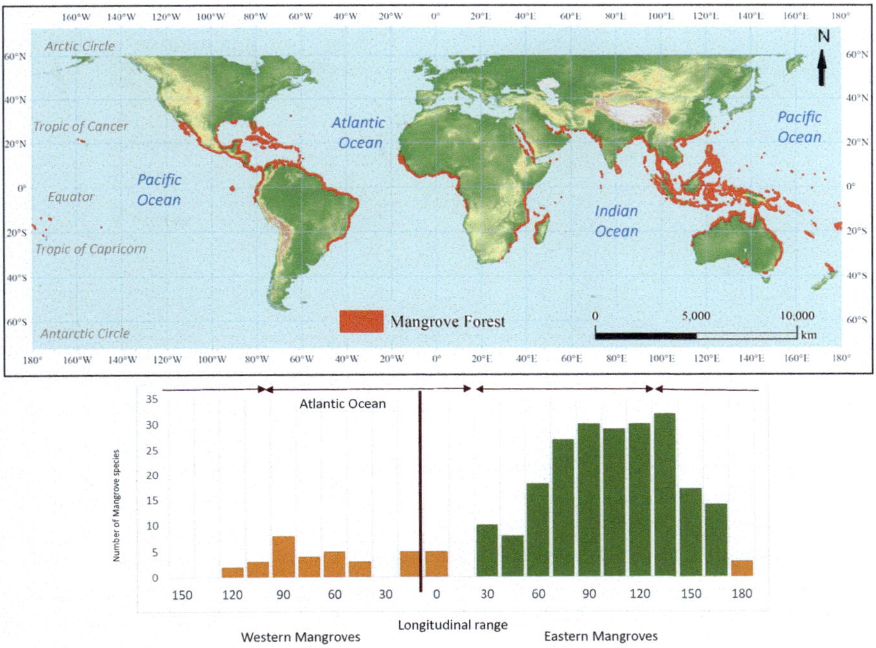

Fig. 21.1 Global distribution and diversity of mangroves. (Map Source: Tang et al. 2018 and Chart data from Tomlinson 1994)

under 23 genera when compared to the western countries with only 12 species under 7 genera (Spadling et al. 2010) (Fig. 21.1).

As per the India State of Forest Report 2019, the total mangrove forest cover of the country is 4,975 sq. km. which accounts for 0.70% of the nation's forest cover. This includes a very dense forest of 1,476 sq. km. (29.66%), moderately dense forest of 1,479 sq. km. (29.73%), and open forest of 2,020 sq. km. (40.61%). Among states and UT's, West Bengal holds the major share (42.45%) of mangrove area followed by Gujarat (23.66%), Andaman and Nicobar Islands (12.39%), Andhra Pradesh (8.12%), Maharashtra (6.44%), Odisha (5.04%), and other states (Fig. 21.2). Based on the geomorphologic characters, mangrove habitat in India can be classified into three categories: East coast, West coast, and Island mangroves. Major mangroves occur in the east coast with 57% of the total area and 88% of Indian species count when compared to the west coast. This is mainly because the east flowing rivers form extensive deltas with a dense network of creeks and muddy levees favouring the growth of mangroves (Kathiresan 2018; Mitra 2020).

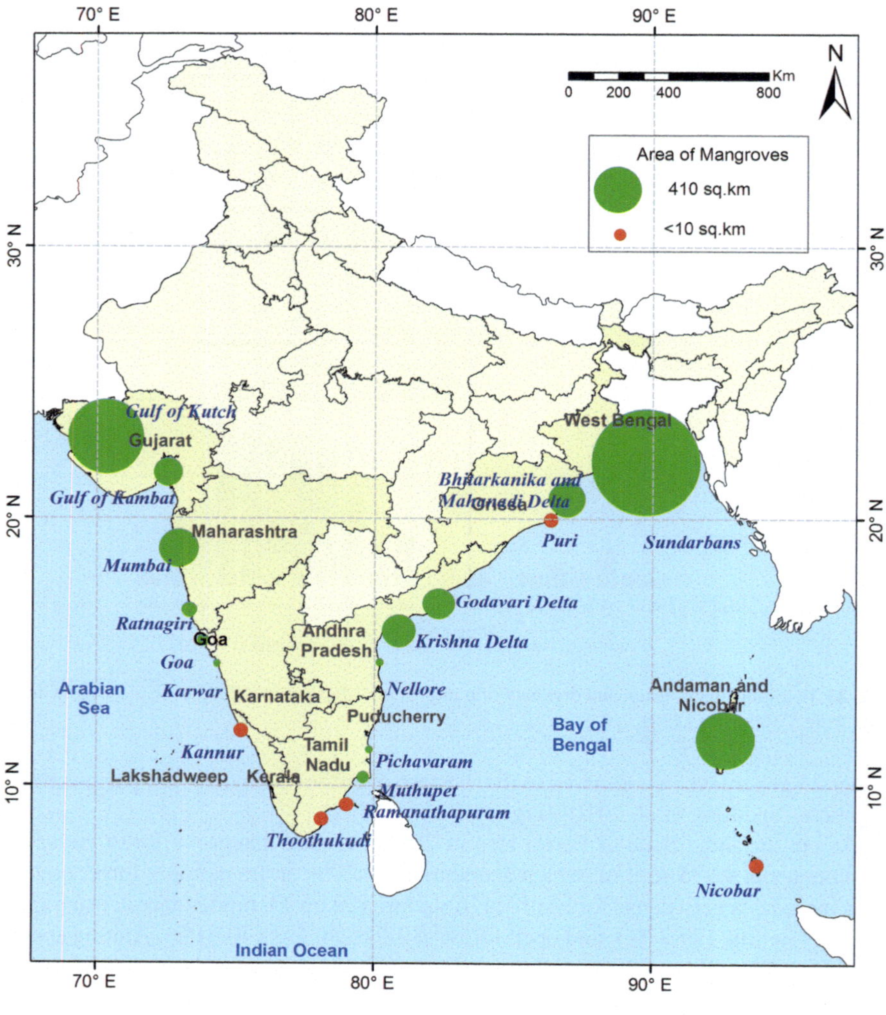

Fig. 21.2 Distribution of mangroves along Indian coastline. (Map prepared from data source: FSI 2019)

21.1.2 Ecological Function

Mangrove is complex yet one of the most productive ecosystems in the world due to its survival capability through self-adaptation, habitat hosting, and diverse range of flora present in the community. (i) The ecosystem acts as a breeding and feeding ground for a large variety of fishes, mollusks, crustaceans, and other related fauna. Therefore, mangroves are credited for its autotrophic nature which helps to maintain the *coastal food chain* (Alongi 2002). (ii) Most recently, the efficient carbon sequestration capability of mangroves attracts global attention and often regarded as "Carbon Sink" with the annual carbon sequestration potential of two to four times higher than that of the terrestrial tropical forests, which is highly significant in the perspective towards climate change mitigation and adaptation measures (Murdiyarso et al. 2015). (iii) Another significant characteristic of mangroves is that they act as a natural defence against coastal hazards like cyclones, storms, tsunami, and coastal erosion. (iv) Mangroves can withstand the sea level rise through vertical accretion under suitable conditions (Menéndez et al. 2020). Besides these, mangroves are considered to have medicinal value for acute diseases and be a resource of household material, fuel and fodder for the coastal community (Mitra 2020).

21.1.3 Threats on Mangroves

Despite its ecological, biological, and economic relevance, mangroves face threats from natural and anthropogenic factors. During the period between 1980 and 2005, approximately 35,600 sq. km of mangrove area had been lost which is much higher than the global forest cover loss (FAO 2007). Between the years 2000 and 2012, the global annual average mangrove deforestation rate was ranging between 0.16% and 0.39%, while it was much higher (between 3.58% and 8.08%) in Southeast Asian and South Asian countries. The loss was mainly due to the conversion of mangrove area into aquaculture ponds, palm plantations, rice cultivation, timber logging, and urban development (Sahu et al. 2015; Hamilton and Casey 2016; Jayanthi et al. 2018; Friess et al. 2019) which also reduced the species richness (Duke et al. 2007). The intense deforestation also causes severe impacts on carbon stored in mangrove soil and sizable below ground pools of dead roots. It is estimated that the loss of carbon storage due to mangrove deforestation would be ranging between 90 and 970 Million Ton/year, which is much higher than their annual carbon sequestration rate (Alongi 2014).

There exist indirect threats in the form of reduction in upland fresh water available to the mangroves by the construction of water storage reservoirs and associated with polluted water carried away by the discharge from upland agriculture practices with the usage of fertilizers and insecticides.

21.1.4 Conservation and Management

Several international conservation policies and global rehabilitation targets framed for the conservation of mangroves became successful to a certain extent. For example, the intergovernmental treaty "Ramsar Convention of Wetlands" intends to conserve global wetlands, including mangroves and their resources through strict regulatory policies, participatory management, and international cooperations (Kuenzer et al. 2011). Furthermore, "Blue Carbon Initiative" project funded by the United Nations Environmental Programme (UNEP) aims to restore and protect the vegetated coastal habitats to improve carbon sequestration through wetlands (Nellemann et al. 2009; Ellenbogen 2012).

Conservation and management of mangrove are highly relevant with several of the United Nations Sustainable Development Goals (SDGs) including Goal 13: Take urgent action to combat climate change and its impact; Goal 14: Conserve and sustainably use the oceans, seas, and marine resources for sustainable development; and Goal 15: Sustainably manage forests, combat desertification, halt and reverse land degradation, halt biodiversity loss (Chow 2018). However, it is known that the no-net-loss restoration of mangroves at global level requires mass restoration strategies such as site selection, species selection, artificial regeneration, etc., through the involvement of local stakeholders and institutions (Bosire et al. 2008). Mapping the spatial distribution and species composition of mangrove ecosystem is a prerequisite to generate baseline information for planning effective management strategies.

Remote sensing has replaced traditional field survey methods by providing cost-effective, continuous data for assessing different parameters of wetlands such as their spatial distribution, species identification, health status, invasive species identification, water quality, biomass characterization, anthropogenic impacts as well as monitoring and management of conservation and restoration activities (Kuenzer et al. 2014). Moreover, the availability of archived and real-time acquisition data facilitates the evaluation of seasonal and long-term changes in the coastal wetland ecosystem and its causative factors.

21.1.5 Information Needs for Monitoring Mangrove Ecosystems

The information that plays a crucial role in the decision making process and devising efficient conservation and management practices of mangrove ecosystem are listed below. And eventually, most of them could be derived through remote sensing and GIS tools.

a. Spatial Extent and the dynamics of mangroves and associated land cover features.
b. Interrelations with changing coastal landscapes.
c. Threats on mangroves in terms of either degradation or conversion due to anthropogenic and natural causes.
d. Species Distribution/dominance and its temporal variation.
e. Mangrove phenology (associations and species variation).
f. Mangrove biophysical characterization (height, biomass, leaf areas, etc.).
g. Mangrove Functional Attributes (Net Primary Productivity, Net Carbon Exchange, Evapotranspiration, etc.).
h. Mangrove Ecosystem Services (e.g., protection from the tsunami).

This chapter intends to give a brief overview covering the fundamentals of remote sensing, Geographic Information System (GIS) and their potential and demonstrated applications in assessing and monitoring the mangroves, as well as for devising efficient conservation and management strategies.

21.2 Concepts of Geoinformatics

The current field of Geoinformatics or Geospatial Science and technology is a synthesis of concepts from traditional and/or individually developed fields such as Geography, Cartography, aerial photogrammetry, remote sensing, database, computer science, global positioning system, navigation, etc. (Fig. 21.3).

Practically there are a few basic types of information products expected from the application of geoinformatics:

- Map of a particular theme or set of themes. For example, thematic maps like Mangrove distribution, vegetation map with mangrove types/association/species.
- Resultant map(s) by the combination of more than one theme of our interests. For example, mangrove density classes falling under different age groups or the range of soil salinity in which *Avicennia* species are dominant.
- Resultant statistics/tabulations of the attributes and associated attributes of the above two.
- With the advantage of developments in the contemporary computing and hardware techniques, visualization of above outputs with best possible 2D and 3D capabilities.
- Visualization on mobile devices leading to Virtual Reality maps.

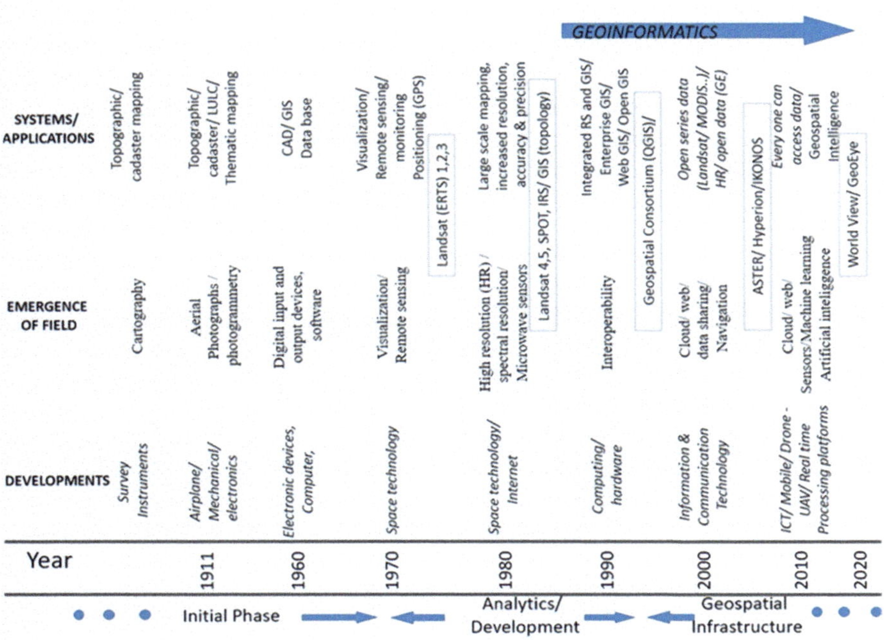

Fig. 21.3 Evolution of the field of geoinformatics or geospatial technology or simply remote sensing and GIS

21.2.1 Basics of Maps

A map is defined as the abstract of a real-world scenario and the degree of abstraction depends on the

- Reduction: Intended application and area of interest (mainly the type and the size of the region) will decide the scale of the map (we will discuss the scale later in this section).
- Simplification: Level of position and their attributes to be included in the map with the use of technologies and instruments.

However, one should find the optimum reduction and simplification requirements of the application.

Cartography, a branch of geography, is defined as the art and science of making and study of maps in all aspects. All maps can be characterized by two basic elements: location of the feature/object in two-/three-dimensional space (x, y, z) and its attributes, the qualities or magnitudes of mapped feature, for example, type of land or man-made structures like road, buildings, etc., in mapped location (Fig. 21.4). Depending on the objective, only the features to be represented are (i) filtered out for data collection and further processing, and (ii) transformed/reduced to the scale.

21.2.1.1 Scale

While representing the earth's feature on the map plane, the dimension of the feature on map is necessarily smaller than the mapped area on the ground. Hence, the amount of reduction from the measurements on real world to the map needs to be established using a linear metric and this could be achieved by *map scale*. Theoretically, scale is the ratio between the distance represented in map to the actual distance on the reality/ground and the scale is unit less. Map scale can be represented in three ways namely,

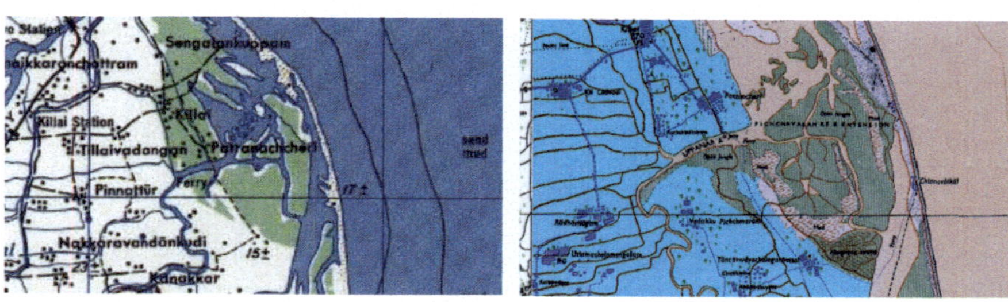

Fig. 21.4 Survey of India topographic maps of Pichavaram mangroves: Left: 1:250,000 scale map surveyed in 1935 showing over view of the wetland and Right: 1:50,000 scale map surveyed in 1970 showing intricacies of wetland including drainage details

a. *Representative Fraction* (RF) is represented as the ratio of the map distance to the ground distance in which both distances are represented in the same unit. For example, if the scale is 1:5000, then one unit in the map is equal to 5000 units in ground, i.e., 1 cm length in the map is 5000 cm (50 m) of length in reality or 1 mm in the map is equal to 5000 mm (5 m) in reality.
b. *Statement Scale* or verbal notation of expressing the scale as a simple statement for representing the number of map units as a fraction corresponding to earth units. The scale 1:50000 may be defined 1 cm to 500 m.
c. *Graphical Scale* is the method of expressing the scale by bar graph calibrated to express RF or statement scale visually. Representing scale as Bar is ideal to avoid any error occurring due to the enlargement or reduction of the map. However, just enlarging a small scale map to a bigger size does not mean that the resultant map gives all the positional details of an equivalent large scale map. Figure 21.4 (Left) shows a small scale map enlarged to match with the large scale map given in the right of the same area and extent. However, the details are more on the right-side map than the left-side map.

With this one can understand that the resizing of the map does not affect the physical measurements, but the details given in the map are specific to a representative fraction or statement scale.

In general, the map scale indicates what details a reader could expect from the map and to estimate distance. The selection of map scale, primarily depends on the size of the area to be mapped, the extent of details to be shown and the size of the paper. Based on the cartographic scale, maps can be categorized into:

a. Large scale map—where specific features of the landscape in a small area will be shown on a map and it provides detailed information. For example, cadastral maps, town plans, etc. try to show every detail of the small area selected.
b. Small scale maps—where more earth area is covered in each map unit, and hence the details are generally generalized or limited in these maps. For example, wall maps and maps in school atlas are small scale maps.

Currently with the use of GIS technology, maps can be displayed at various scale (zoom) levels and thus scale and maps have become digital which also facilitates their viewing on web.

21.2.1.2 Coordinate System and Map Projections

A place or location is identified on the earth surface by either two- or three-dimensional coordinates. It may be either (i) geographical or spherical coordinates in terms of latitude, longitude, and altitude or (ii) Cartesian or planar coordinate system using any of projection methods (http://geokov.com/education/map-projection.aspx).

Projection is a mathematical concept of transferring the locations of the features on the 3-dimensional global surface to a 2-dimensional map plane surface. The transformation from the surface of the earth (near ellipsoid) to the flat surface always involves distortion and no map projection is perfect (Orange peel cannot be flattened perfectly even after tearing). Every map projection tries to preserve one of the spatial properties of earth features and the neighbourhood relation such as *true direction, true distance, true area, and true shape*. A map projection is called *conformal* when the shape of the features is preserved. *Equal-area* (Equivalent) projection tries to represent the true area of the earth features on the map. *Equidistant* projections preserve the true distance between points in a particular direction, while *True-directional* or Azimuthal projection represents the true direction from the centre or particular point of the mapped area.

There is a long history behind the evolution of projections (Snyder and Maling 1993). As far as India is concerned, till 2000 the Survey of India and other agencies were using Polyconic projection. The reason could have been that the widely used reference map prepared by the Survey of India till 2005 used Polyconic projection aiming equivalent area upon projection (Later SOI starts releasing digital Open Series Maps in UTM projection). Similar norms were prevailing in other countries also with varied types of projection, which are suitable to represent the respective region. However, Universal Transverse Mercator (UTM) projection is being universally accepted by remote sensing and GIS community because of less error upon projection.

21.2.1.3 Surveying

Technically, surveying is the measurement of the location of a feature or object in an area or on the earth surface in terms of Cartesian coordinates or projection coordinates of a location having known coordinate (*Bench Mark or Control point*). Traditional survey methods include measuring using *chain, compass or Theodoloite* while modern methods include the use of instruments such as *Total Station, Global Positioning System (GPS),* and advanced GPS instruments that help the surveyors to make measurements at sub-cm accuracy. As such remote sensing technology is also called as one of the methods of surveying.

21.2.1.4 Global Positioning System and Geo-location

In recent decades, Global Positioning System has substantially reduced the laborious process of conventional survey methods and analytics. The user of a GPS receiver/instrument can get his three-dimensional position on the earth's coordinate system (latitude, longitude, and altitude) and the time in a precise manner with the help of signals received from a constellation of satellites. Hence, it is been useful for navigational purpose in the land, air as well as on sea. The positional accuracy of such GPS varies from a few centimetres to metres depending on the specifications of instruments used and the availability of open space to receive the satellite signal.

However, Assisted GPS (AGPS) which is installed in smart mobile phones help us to precisely get our location using GPS first and increase its accuracy, using the positions of connected cell phone towers thus enabling us to get the location even inside our home or under the trees (as long as the connectivity exists). Such AGPS gives us the traffic information, finding the nearest facility, etc. through wireless network.

21.3 Remote Sensing: An Introduction

In earlier times, conventional field survey methods were adopted for mapping and recording forest characteristics which is time consuming, labour intensive, expensive, and practically difficult in the case of temporal data collection in the inhospitable environment of the mangrove ecosystem. Remote sensing can be defined as "the science and art of obtaining information about an object, area, or phenomenon through the analysis of data acquired by a device that is not in contact with the object, area, or phenomenon under investigation" (Lillesand et al. 2004) (Fig. 21.5). Evelyn L. Pruitt of the U.S. Office of Naval Research has first introduced the term "remote sensing" in the 1950s, to denote the technology which is observing the earth's landscape from the distance without any physical contact.

21.3.1 Electromagnetic Radiation

The understanding of the electromagnetic radiation (EMR) characteristics and its interaction with the physical environment is much needed before looking into the details as EMR is the principal source of remote sensing. Light travels at a speed of 2.99×10^8 m/sec and attains the form of electromagnetic radiation having an electric and a magnetic field that are right angle to each other and perpendicular to the direction of travel (Rees 2013).

EMR is specified in terms of either wavelength (*μm* or *nm)* or frequency *hertz* (*Hz*). The relationship between wavelength and frequency of electromagnetic radiation could be derived as $c = \nu \lambda$, where c is the speed of light. Based on the wavelength, different regions of the EMR spectrum, including Gamma rays, X-rays, Ultraviolet, Visible, Infrared, microwaves, and radio waves have been recognized (Fig. 21.6).

- Optical spectrum includes
  i. *Visible* spectrum (0.30–0.7μm) comprising of UV (0.3–0.4μm) and visible light (0.4–0.7μm) which is a narrow waveband yet very important in photogrammetry and remote sensing and
  ii. *Reflective Infrared* region (0.7–3.0μm) including Near-Infrared (0.7–1.3μm) and Mid-Infrared (1.3–3.0μm) are used along with the visible region in optical remote sensing.
- The *thermal Infrared* region (3–5μm and 8–14μm) involves the measurement of emitted long wave radiation from the earth's surface. Understanding the black body radiation concept and associated laws will give more insight on this (Lillesand et al. 2004; Jensen 2014).
- The *Microwave* region (1 mm–1 m) of the electromagnetic spectrum is used in passive and active modes. The longer wavelength of microwave can penetrate through clouds and thick atmosphere hence acquire images in all weather conditions (Awange and Kyalo Kiema 2013).

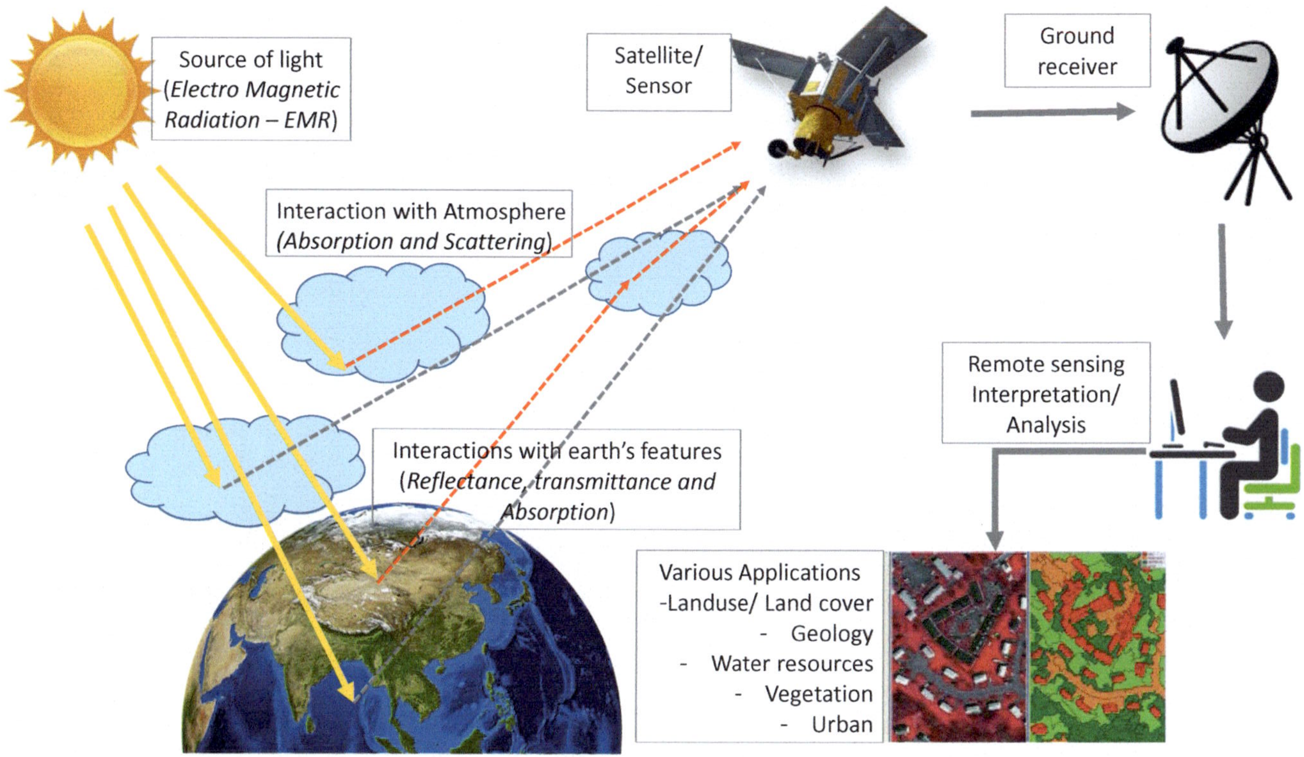

Fig. 21.5 Components of remote sensing process

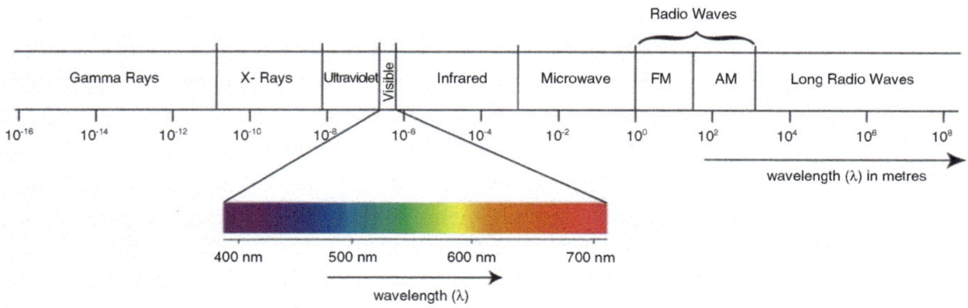

Fig. 21.6 Regions in the electromagnetic spectrum categorized based on their wavelength. The region between 400 nm and 700 nm is the only region visible to human eye. (Source: Jensen 2014)

21.3.2 Interaction of EMR

While EMR enters from the vacuum space to the atmosphere it interacts with the gas molecules and particles in the atmosphere including thin clouds and rain drops it gets redirected from its original path called *Scattering*. *Absorption* occurs when certain gases in the atmosphere absorbs certain wavelengths of EMR blocking them to reach the earth surface. The wavelength region that could pass through the atmosphere without much attenuation is called *atmospheric windows* and these regions could be effectively used in remote sensing. When the incoming radiation hits a target on the earth surface, it either gets *reflected, transmitted* or *absorbed* by the target based on the properties of the target and the incident wavelength (Jensen 2014; Lillesand et al. 2004).

21.3.3 Spectral Signatures

The remote sensing sensors measure the reflected energy of the incident solar radiation from various objects of the earth's surface depending on the texture, physical and chemical properties of the incident target, and the prevailing atmosphere. Different objects reflect and absorb differently at different wavelengths, and when we plot this we get a continuum curve with crests and troughs implying the reflectance peaks and absorption troughs. This is called a *Spectral Signature*, which is unique to an object like a fingerprint to every individual (Fig. 21.7). This premise provides the basis for multispectral remote sensing (Lillesand et al. 2004).

21.3.4 Evolution of Remote Sensing

In 1863, Maxwell proposed the electromagnetic theory followed by some of the significant findings in understanding the electromagnetic radiation beyond the visible spectrum. The first aerial photograph was reportedly made by Gaspard-Felix Tournachan (also called Nadar) from a balloon tethered over the Bievre Valley,

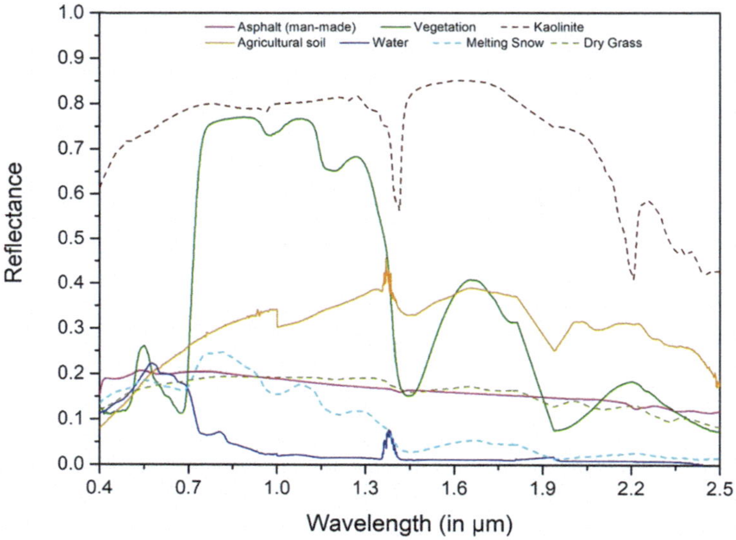

Fig. 21.7 Spectral signatures of selected natural and man-made features for the wavelength range of 0.4μm–2.5μm. (Data Source: USGS Spectral Library Version 7)

France in 1820. By the time of the Second World War in the 1940's, significant developments in the field of photographic reconnaissance were observed including the development of infrared and radar sensors (Rees 2013).

Zhou (2001) characterized the development of remote sensing satellites broadly into four stages: The *first generation* space borne (1960–1972) satellites acquired images primarily for reconnaissance survey (CORONA, ARGON, and LANYARD). *Second generation* satellites are successful for various applications using multispectral images (series of Landsat, SPOT, and IRS). *Third generation* satellites have been launched with high end sensor technology to acquire very high resolution images (SPOT HRV, IRS L4) and all weather capability with active sensors like ERS-1, JERS-1, and RISAT-2. The development in very high spatial (sub-metre) and spectral resolution (hundreds of bands) sensors denotes the fourth generation with the additional possibility of multiangle and three-dimensional observations with in-built GPS. The next generation technology expects to focus on an integrated system for a real-time earth observation requiring sophisticated imaging sensors, storage, communication hardware, and automated algorithms (Fu et al. 2020).

In the context of ecological and environmental mapping and monitoring the following categorization of sensors may be useful for a beginner to start working with the satellite data.

- Coarse resolution optical (VIS-NIR-SWIR) sensors having high repetivity (more frequent acquisition) are more suitable for environmental monitoring of large patches (e.g., NOAA-AVHRR, MODIS, MERIS, NPP having > 250 m resolution)

- Moderate resolution optical multispectral sensors are generally used for mangrove separation from other vegetation as well as discriminating associated vegetation and land cover features (~25 to 250 m). Most of such data is free and open, thus very popular among users.
- High (~5–25) and very high resolution (0.25 to 5 m) used for discrimination, mapping inventory, and change detection. High resolution sensors can provide very accurate information, but have a low swath which increases the gap between revisits to the same region. However, a new class of small satellites in large constellation (e.g., Planet Labs) is capable of providing daily visits.
- Active sensors (Sect. 21.3.4.1.1) SAR in C, X, and L band obtain texture, dielectric and overcome cloud and thick atmosphere which makes it operable in all weather conditions.
- Hyperspectral sensors with >50 bands and 10 nm spectral resolution are available for large area mapping as archive though not available for monitoring as they are not operational. Future missions like EnMAP (DLR, Germany) may overcome this limitation.

The main properties of the sensors widely used for earth observation are summarized in Annexure 1.

21.3.4.1 Types of Remote Sensing

Remote Sensing can be categorized into different types based on different criteria such as the source of EMR, platforms on which the sensor is being mounted, and its orbit.[1] Low Earth Orbit, Medium Earth, Geostationary, Sun Synchronous are the types of RS based on orbit.

Based on the Source of EMR

Based on the source of energy used by the sensor, remote sensing can be of two types. Sensors recording the reflected or emitted EMR from the object which is irradiated from artificially generated energy sources are *Active sensors* and from a natural source (Sun) are *Passive sensors* similar to the camera used with and without flash light. Most of the earth observation multispectral sensors are of Passive type and LiDAR, RADAR fall in Active type

Based on the Platform

Platforms are structures or vehicles on which remote sensing instruments are being mounted providing varied scale range.

a. Ground based platforms include portable hydraulic platforms and masts, and non-portable towers and weather surveillance stations.

[1]Orbit is the circular or elliptical path of the satellite above the earth for imaging.

b. Airborne platforms include hot air balloons, Unmanned Air Vehicles (UAV), drones, and aircrafts.
c. Spaceborne platforms included rockets, space shuttles, and satellites on which the sensors are mounted. These sensors provide global and periodical coverage of the earth's surface.

21.3.5 Resolutions in Remote Sensing Data

The data collected by the sensor can be characterized in terms of its resolution and that can be of four types: spatial, radiometric, spectral, and temporal resolution.

21.3.5.1 Spatial Resolution

The spatial resolution, analogous to map scale, referring to the size of the smallest possible object that can be identified in the image (Picture Element or Pixels) and it depends on the Instantaneous Field of View[2] (IFOV) of the sensor/camera and height of the satellite from the ground surface. Usually, geostationary weather satellites provide data with very coarse spatial resolution in a kilometre scale while polar orbiting earth observation satellites often provide finer resolution in metres. Examples for such satellites include Landsat 7 ETM+ (30 m), IRS LISS-III (23.5 m), Resourcesat (5.6 m), etc. Due to the recent advancements in sensor and imaging technology, remote sensing data with very high spatial resolution less than 5 m are also available but at a higher cost.

21.3.5.2 Radiometric Resolution

It refers to the sensitivity of a sensor to the differences in signal strength while recording the radiant flux reflected, emitted, or back-scattered off the target. Radiometric resolution is measured in bits to refer to the number of brightness levels (black–grey–white) used to record the reflected energy or capture the image. The higher the bit value, the more details can be obtained from the recorded image. For example, Landsat 7 ETM+ image has the radiometric resolution of 8 bits which means objects within the image can be differentiated using $2^8 = 256$ brightness/grey levels (0 to 255) depending on the reflectance of the objects at the particular wavelength.

21.3.5.3 Spectral Resolution

The spectral resolution refers to the ability of the sensor to acquire the image using more number of bands at finer bandwidth. Based on the number of spectral bands used and the spectral interval (band width) for which a sensor is sensitive, remote sensing images are categorized into,

[2]IFOV is the angular cone of sensor's/ camera's visibility to earth features (encompassing pixel size).

a. Panchromatic image is the acquisition of an image in a wider spectral range generally encompassing visible and or NIR spectrum and usually displayed in grey scale image.
b. Multispectral image is the most commonly used multi-band image with a minimum number of bands (usually less than 10), comparatively broader bandwidth, with higher spatial resolution.
c. Hyperspectral image is also a multi-band image having more than hundreds of spectral bands with relatively smaller bandwidth to record continuous spectrum. Analysing hyperspectral data requires a solid understanding of the spectral signatures of the objects under study.

21.3.5.4 Temporal Resolution

The temporal resolution of a remote sensing system refers to the frequency of image acquisition of the same region, i.e., repetivity of the satellite which is dependent on the orbital[3] parameters of the satellite, swath[4] width, and latitude of the region. High altitude orbit, wider swath, and higher latitudes allow more frequency than low altitude, narrow swath, and lower latitudes. The analysis of multi-date data can be used to study the temporal behaviour of the object or phenomenon. Geostationary satellites have a continuous view of the target (weather satellites), while in case of polar orbiting satellites for earth observation, satellites are required to have high temporal resolution for natural resource monitoring for example, Landsat 7 ETM+ sensor provides data with the temporal resolution of 16 days.

Some satellites are designed to acquire multiple images of the same area in a shorter interval of time either in the same orbit (Cartosat-2 series) or on the subsequent orbits (SPOT, IRS series) using a steering mechanism which is known as the revisit capability of the satellite. Current days with the entry of private sectors in space technology, a constellation of small satellites launched can accommodate the revisit capability by acquiring images on required time (Planet Labs series of satellites).

21.3.6 Multispectral Remote Sensing

Advancement in space technology and frequent availability of medium resolution data from satellites like Landsat, SPOT, IRS, ASTER, etc. (Annexure 1) and developments in computer algorithms for image preprocessing and classification have made the regional mapping of earth resources including mangroves an easy job and helped in monitoring and management activities in the last three decades.

---

[3]Orbital parameters include the altitude, inclination, and duration of satellite orbit around the earth.

[4]Swath is the width of the ground imaged by the satellite during the orbital pass.

21.3.6.1 Thematic Mapping

Mapping is the process of converting the image to information readable by a common user. The image collected by a remote sensor would be used either in analog or digital formats depending on the user's preference to map using visual interpretation or digital mapping.

Visual Interpretation

Image interpretation could be defined as the examination of images to identify the objects and judging them with the inherent skill of the human eye-brain system (McGlone 2004) by applying multi-concepts like multispectral, multi-temporal, multi-scale, multi-source, etc. Usually, visual interpretation is performed on the digitally pre-processed satellite data to map or derive the best results out of that. In general, satellite data is viewed either as a black and white image if a single band or True Colour Composite (TCC) or False Colour Composite (FCC) for multispectral bands while the latter is very common among ecologists and RS experts. Visual interpretation constitutes the fundamental principles of the *object identification or discrimination* (Jensen 2014; Konecny 2014) and the interpretation elements include (Please refer Fig. 21.8).

a. Location: It gives the precise location information of the image through x, y coordinates, and this could be obtained through traditional survey methods or GPS receivers.
b. Colour: The amount of energy reflected at different wavelength regions varies for different materials on the earth's surface. In Fig. 21.8, the mangrove vegetation appears red in colour because the vegetation reflects more NIR radiation, while sand and other concrete features (buildings) appear as white as it reflects almost equally in all bands. The water reflects comparatively more in blue and green bands than red and NIR bands and so appears blue.
c. Tone: the variation in the same colour depending on the amount of energy reflected in a particular wavelength. That is within vegetation itself, depending on its types the tone of red colour varies from bright red to dull red similarly from bright blue to dull blue for water bodies.
d. Size: The size of an object in the image is scale dependent. The determination of relative size of the target to that of nearby objects aids in the identification of the target.
e. Shape: It represents the structure or outline of the features and is an important element while determining any distinctive feature. Shape may be regular, for example, in the case of man-made features like aquaculture ponds and canals, and irregular in case of natural rivers and mangrove vegetation.
f. Texture: It indicates the characteristic arrangement or repetitions of colour in an image, and is scale dependent. Figure 21.8 represents heterogeneous vegetation cover that could be identified by the difference in its texture.
g. Pattern: It is the spatial arrangement of objects in the area, either randomly (sparse vegetation) or systematically arranged (aquaculture ponds).

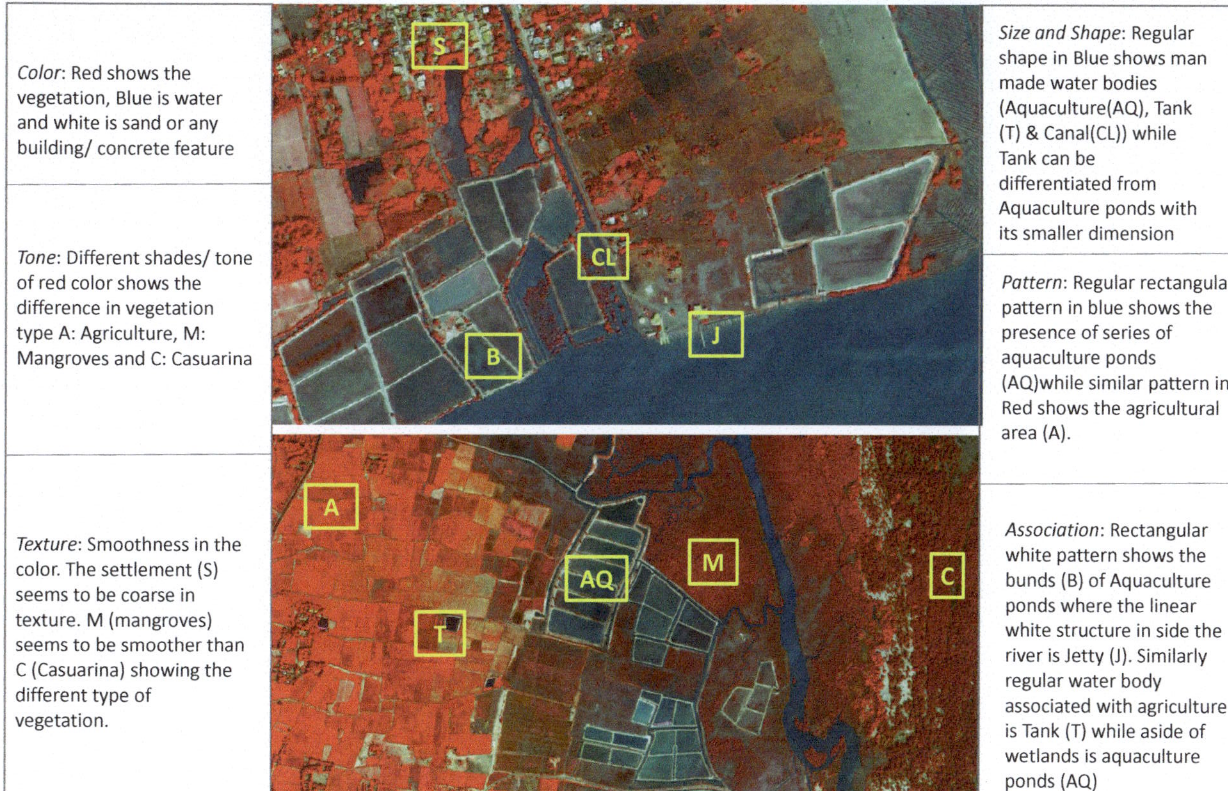

Fig. 21.8 Elements of visual image interpretation as explained through IKONOS (MS + PAN) 1 m resolution image in FCC (Red: Band 4; Green: Band 3; Blue: Band 2) of Pichavaram, Tamil Nadu India

h. Shadow: Remote sensing images are acquired mostly in time when the sun is in near nadir position (between 10 AM and 2 PM local time) to avoid shadows. Shadows may hinder the interpretation, but at specific instances, they can give distinctive clues on the target object.
i. Association: Association refers to the circumstance when a certain feature or object along with other related features or activities.

Minimum Mapping Unit

When satellite remote sensing data is used either by visual interpretation method of digital classification discussed below, a minimum size of object or feature on the earth which can be mapped depends on the spatial resolution of the sensor. For a map being prepared by visual interpretation, an object of 3×3 pixel size can be mapped as a polygon and for a digital classification group of 3×3 or 2×2 pixels is the minimum size of a feature to be classified which is called Minimum Mapping Unit (MMU). MMU is at least twice the size of resolution of the data, i.e., if the resolution of satellite data is 30 m, then MMU will be at least 3x3 times $30 \times 30 = 90 \times 90$ m and thus minimum size of an object of 90×90 m can be mapped and pixels below this size are not mapped due to inherent error in the data.

21.3.6.2 Digital Image Analysis

While using visual interpretation technique, spectral characteristics beyond visible region of EMR spectrum could not be fully utilized because of the limited ability of the eye to distinguish the tonal difference and associated spectral reflectance. Therefore, over the years, scientists developed advanced algorithms and methods to process the satellite remote sensing data in digital image format for information extraction with little to no user intervention make it time and cost efficient. Generally visual and digital image interpretation techniques could complement each other and the combination approach is widely accepted as the best for remote sensing application. A detailed description of many of the digital image analysis methods could be found in literature in recent times (Lillesand et al. 2004; Reddy 2008; Jensen 2016).

During satellite data acquisition, the radiated (reflected) light energy from the earth surface features received by the sensor is stored as Digital Number (DN) of pixels (picture element) of the imaged area which is a function of spatial scale, spectral band used and radiometric resolution (grey scale). Digital analysis involves the computation of DN value of the satellite image by considering it as a matrix (rows $\times$ column) of pixels.

Image Preprocessing

Images collected through space borne sensors are often attributed with radiometric and geometric errors and hence corrected/pre-processed to convert them to a usable format before visual interpretation or further digital analysis. In fact, it can be compared with the correction techniques available in any handheld or mobile cameras which are similar but substantially at a lower level. Preprocessing could be categorized into the following three broad operations.

i. Radiometric Correction: Radiometric correction involves the removal or reduction of noise introduced into the data due to the electronic noise of the sensor system and attenuation due to atmospheric conditions. Radiometric corrections could be done through simple image normalization or advanced radiometric calibration to radiance or scaled surface reflectance.
ii. Geometric Correction: The geometric errors may arise due to the platform instability resulting in skewing or distortions. Geometric correction of an image is important to precisely locate the point of interest in the geographic or projected coordinate system (Sect. 21.2.1.2), to perform overlay analysis with other thematic and temporal maps of the same area and to match with the adjacent images.
iii. Atmospheric Correction: Topographic attenuation and atmospheric conditions in the target area highly influence the image quality. However, the correction of atmospheric error is not mandatory for all the applications unless the remotely sensed is used for comparing the output maps/parameters with that of a different area at different time scales.

Though the above preprocessing techniques are to be applied on raw satellite data, when it is received by the user, these corrections are already carried out by the respective space agency unless the raw data is needed by a researcher to work with. Hence, the space agency provides the data from raw level to pre-processed level depending upon the user's requirement. It is advisable to use pre-processed data for any ecological applications (Sect. 21.3.10).

Image Classification

When the image is corrected for errors, the image is directly used for thematic mapping using classification methods or improved by enhancement, filtering, and transformation algorithms based on the theme of mapping (Please refer to Lillesand et al. 2004; Jensen 2016 for more details). Image classification involves the identification of spatial and spectral patterns in the remote sensing data corresponding to different land use/land cover and presents the results as thematic information through the analytical process called *pattern recognition*. Several image classification algorithms have been developed based on parametric, non-parametric, and non-metric methods. Parametric methods are used to analyse ratio and interval type of data (Sect. 21.4.4) while Non-Parametric methods could also be used to analyse the nominal data type. The classification algorithms can be broadly classified into unsupervised and supervised classification methods based on the underlying principle of user intervention in training sample selection.

In *unsupervised classification*, the image pixels are grouped into a specified number of classes based on the spectral reflectance of resultant classes. In this case, the analysts need only to specify the number of classes and the algorithm itself clusters the pixels based on a statistically determined measure on the pixels' DN/radiance values. The analyst could then reiterate the number of classes if the result is found unsatisfactory and repeat the classification until it clusters into informative classes. ISODATA and K-means are two popularly used unsupervised classification methods.

In Supervised classification, the analyst would be able to select training sites for each of the class within the image to train the classification algorithm. The analyst could utilize elements of visual interpretation and field knowledge to select such *training samples* representing known classes of interest. The analyst should collect at least N + 1 training pixels for each of the class when N numbers of bands are used in multispectral or hyperspectral image analysis to derive reliable output of supervised classification. This is called *Hughes Phenomenon* (Richards and Jia 2005). So to obtain satisfactory results it is advisable to identify the maximum number of training samples throughout the image on which the RS analyst is confident. Classification results can be represented as thematic maps, tables, and digital data files. Following are some prominent supervised classification algorithms prevalently used in digital image analysis.

i. *Minimum Distance Classifier (MDC)* calculates the mean of each training class in spectral space[5] and then it measures the spectral distance between each of the pixels in the image to that of the mean of each training class. Euclidean distance is the most common distance measure used in this method. Then, the input pixel will be assigned to the respective spectral class for which the measured distance is minimum (Fig. 21.9a).
ii. *Maximum Likelihood Classification (MLC)* works based on the assumption that the statistics for each class in each band are normally distributed. The classifier calculates the probability that a given pixel belongs to a specific class and each pixel is assigned to the class that has the highest probability, i.e. maximum likelihood (Fig. 21.9b).
iii. *Spectral Angle Mapper (SAM)* calculates the similarity of the spectrum of unknown pixel to the spectrum of the reference pixel in the spectral space of dimensions corresponding to the number of spectral bands used. It calculates the angular difference between the reference and target pixels in vector space and is usually smaller the angle closer to the classes (Fig. 21.9c).
iv. *Support Vector Machine (SVM)* is a non-parametric supervised classification method derived from statistical learning theory suitable for complex and noisy data. It employs optimization algorithms to locate optimal boundaries with the least errors among all possible boundaries separating classes (Fig. 21.9d linear SVM). When the training samples are not linearly separable, the samples are mapped using "non-linear SVM" (Schölkopf and Smola 2001; Tso and Mather 2009).

Accuracy Assessment

Regardless of visual or digital mapping, the classification output needs to be assessed for its accuracy to allow a degree of confidence to be attached to the result.

[5]Spectral/feature space is *n* dimensional space where the DN/spectral values of *n* bands are plotted.

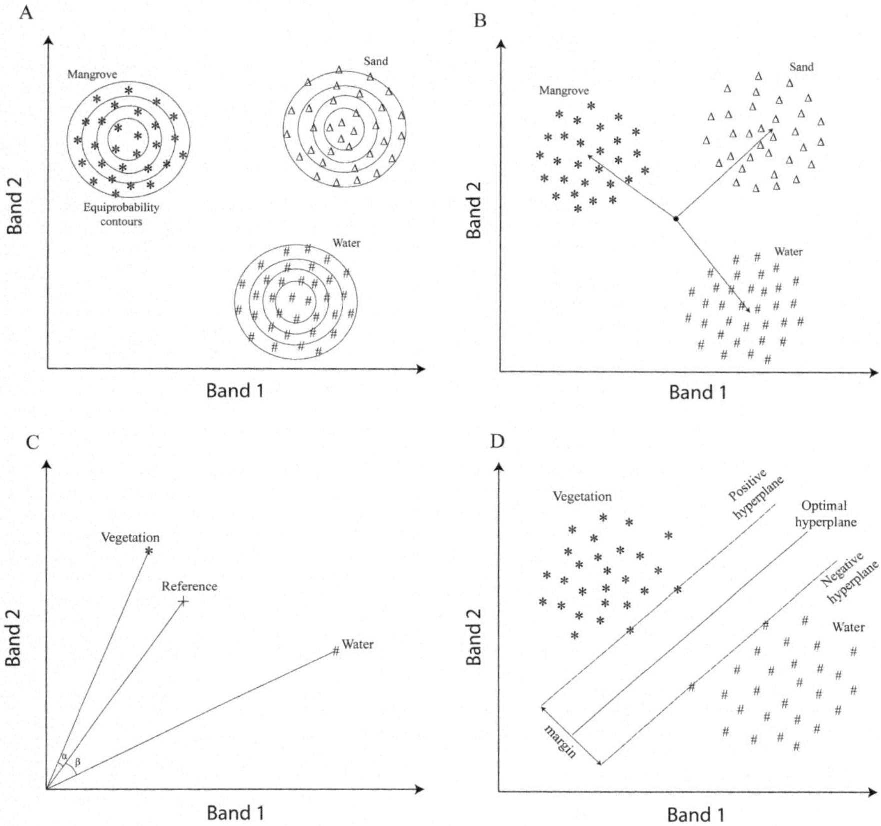

Fig. 21.9 Graphical representation of (**a**) Maximum Likelihood Classifier, (**b**) Minimum Distance Classifier, (**c**) Spectral Angle Mapper, and (**d**) Support Vector Machine supervised classification algorithms. (Image Source: Lillesand et al. 2004; Richards and Jia 2005)

This could be done through the preparation of *classification error matrix* or *confusion matrix*. Confusion matrix (Table 21.1) compares the relationship between known reference data (ground truth) and corresponding classification results on class-by-class basis. *Commission or inclusion error* indicates the incorrect assignment of class (A) to pixels that belong to classes other than A and the corresponding accuracy is called User's accuracy (in user's perspective how many pixels of other classes are included in a particular class) while *omission or exclusion error* occurs when the pixels belonging to a particular class (A) are missed to get assigned to that class (A) and the corresponding accuracy is Producer's accuracy (in map maker's perspective how many pixels are omitted from a class).

Table 21.1 Confusion matrix to assess the accuracy of classification for a total of 95 pixels/ samples of 3 classes: Mangrove, agriculture, and water

Reference	Mangrove	Agriculture	Water	Classified Total	User accuracy
Classified					
Mangrove	21	6	0	27	21/27 = 0.78
Agriculture	5	31	1	37	31/37 = 0.84
Water	7	2	22	31	22/31 = 0.71
Reference total	33	39	23	95	
Producer Accuracy	21/ 33 = 0.64	31/ 39 = 0.8	22/ 23 = 0.96		OA = 74/ 95 = 0.78 Kappa = 0.67

$$\text{Kappa} = \frac{N \times \Sigma\text{Cii} - \Sigma(\text{CTi} \times \text{RTi})}{1 - \Sigma(\text{CTi} \times \text{RTi})}; \text{Overall accuracy (OA)} = \frac{\Sigma Cii}{N}$$

Where N—total number of pixels, C_{ii}—number of correctly classified pixels/ samples, CT_i—total number of the classified pixels for ith class, and RT_i—total number of reference pixels for the ith class.

The *overall accuracy* is computed by dividing the total number of pixels correctly classified by the total number of reference pixels. Kappa coefficient calculation, a discrete multivariate technique used in accuracy assessment, is a measure of the difference between the actual agreement between reference data and an automated classifier and the chance agreement between the reference data and a random classifier (Lillesand et al. 2004).

21.3.6.3 Vegetation Indices (VI)

Regardless of classification or mapping, the information about the vegetation traits could be studied using *Vegetation indices* (VI) which are used by application scientist from the basic understanding of vegetation to multiple analyses, which uses the spectral reflectance characteristics of different vegetation groups in different spectral bands. VI indicates the vegetation phenology associated with the health condition, stress and biophysical characteristics and is an important parameter widely used in studying agricultural and vegetation monitoring from local to global scale. Scientists have developed generally applicable, vegetation specific, and spectral band specific vegetation indices for effective monitoring of global vegetation (Richards and Jia 2005; Jensen 2016). The most widely used VI is Normalized Difference Vegetation Index (NDVI) which is derived using Red and Near-Infrared bands of multispectral data (Townshend and Justice 1986; Tucker and Sellers 1986). The values of NDVI varied between −1 and +1. Higher NDVI values represent the higher vegetation vigour and health (Fig. 21.10). Specific to wetlands, Normalized Difference Wetland Vegetation Index (NDWVI) was derived (Kumar et al. 2019) by making use of the sensitivity of wetlands to shortwave infrared which helps in mapping the mangroves very accurately.

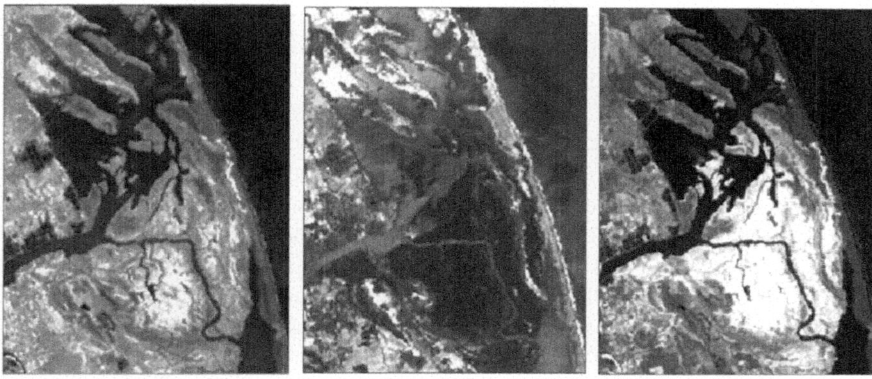

Fig. 21.10 NDVI image derived from NIR and Red band of Pichavaram mangroves

21.3.7 High Resolution Imaging and analysis

Recent advances like *high resolution multispectral data* such as IKONOS, Quick Bird, and Worldview and recent classification algorithms such as *Object-Based Image Analysis* (OBIA) have refined the classification accuracy of mangroves further. OBIA, unlike pixel-based classification, the images are first segmented into "objects" representing groups of pixels belonging to ground patches, entities or their elements (primitives) which can be then classified by unsupervised, supervised, or rule-based algorithms. Some of the important properties of OBIA method over pixel-based classification methods include (1) ability to include object-level shape, texture, and relevant contextual properties of an image in addition to its spectral reflectance values into the classification framework, (2) ability to reduce the salt-and-pepper noise and enhancement of classification accuracy through smoothening of local variation within objects, and (3) ability to analyse with multiple object layers nested within each other at different spatial scales to identify landscape patch, cover type, and ecosystem structure at multiple hierarchies. This estimation of ground objects through image objects finds them more ecologically relevant and potentially more resilient to minor geospatial positioning subject to image registration error (Dronova 2015).

21.3.8 Hyperspectral Remote Sensing

Hyperspectral remote sensing is otherwise called as imaging spectrometry as it combines two sensing modalities, imaging and spectrometry. Normally the spectral wavelength range of 400–2500 nm is used in imaging spectrometry technique which typically acquires images in more spectral bands in a narrow bandwidth approximately ranging from 0.5 to 25 nm. This difference in reflectance properties among materials has made hyperspectral remote sensing, a potential technique for identifying the intricacies in micro-level in earth observation.

The reflectance acquired in narrow contiguous bands of hyperspectral data (which is lacking in multispectral) provide the analyst with very peculiar information in relation to the leaf biochemical and biophysical characters which could be utilized to characterize the mangrove ecosystem such as species discrimination, plant health monitoring, nutrient intake characterization, invasive species monitoring, etc. (Fig. 21.11). Differences in spectral reflectance among different plant species provided by the contiguous bands can be used for (a) species level classification, (b) biochemical (chlorophyll and carotenoid content estimation), and (c) biophysical characterization (Leaf Area Index, Biomass, etc.). For instance, we take the example of spectral signatures of eight mangrove species from the same family *Rhizophoraceae* at multispectral bandwidth of Landsat 8 OLI sensor (left) and hyperspectral bandwidth of ASD Fieldspec 3 radiometer (right) (Fig. 21.11).

Though hyperspectral data has an enormous amount of spectral information, its analysis has some technical difficulties in deriving useful information out of such voluminous data. Apart from this, hyperspectral data also possesses multi-collinearity problem due to the redundant information from more number of bands. Appropriate *dimensionality reduction methods* and *advanced classification algorithms* must be chosen for developing an efficient framework for information extraction (More details on HS image processing methods can be referred in Richards and Jia 2005).

21.3.9 Microwave Remote Sensing

Mangroves and flooded vegetation usually have distinct microwave signatures when compared to associate land use land cover. In *microwave remote sensing*, the signal received is measured for its intensity and is termed as backscatter coefficient in decibels (dB) unit. Differences in wavelength and polarization[6] of transmitted and received signals, and incidence angle on the vegetated surface can exhibit varying backscatter coefficients based on the different transmission capability of microwaves under various configurations. Also, internal properties of plant such as moisture content, cell structure, biochemical content, etc. and biophysical properties such as size, geometry, leaf, and branch orientation results in unique backscatter signal value) (Kuenzer et al. 2011). The most important property of microwave data is that it is not prone to cloud cover, haze, and other atmospheric disturbances and this property makes it suitable for mapping mangroves as they locate in tropical and sub-tropical areas. Microwave data were used for mangrove cover mapping, bio-physical parameters retrieval, health status monitoring, and biomass estimation

[6]Electromagnetic waves consist of an electric and a magnetic field vibrating at right angles to each other and it is necessary to adopt a convention to determine the polarisation of the signal. For this purpose, the plane of the electric field is used as plane of Polarization.

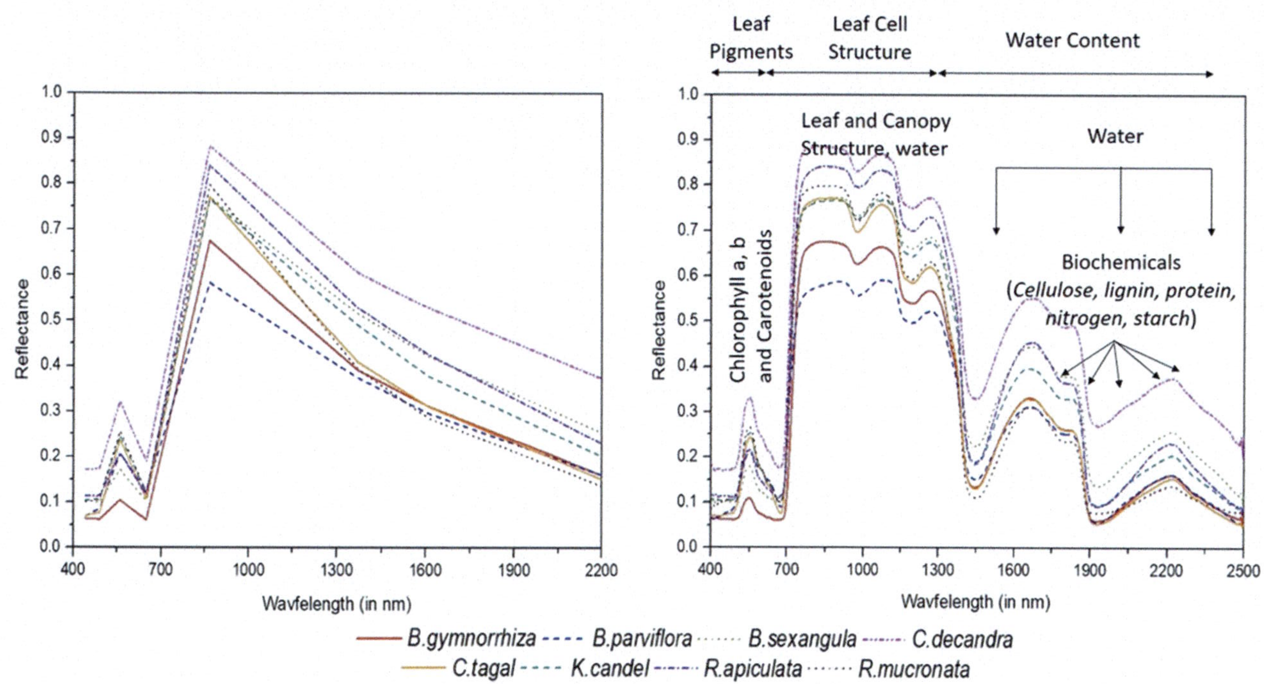

Fig. 21.11 Spectral reflectance of eight mangrove species from Rhizophoraceae family at the spectral resolution of Landsat 8 OLI multispectral sensor (left) and ASD Fieldspec 3 hyperspectral sensor (right). (Source: Prasad and Gnanappazham 2016)

21.3.10 Evolution of RS Data Access

Remote sensing satellite data has been used for wide spread applications in India, after the launch of first ever EO satellite IRS 1A launched in 1988 and National Remote Sensing Centre (NRSC), Indian Space Research Organization (ISRO) was identified as the authorized distribution agency of Indian and foreign satellite data and the same continues. Since then the satellite data are accessed in the form of either hard copy or in digital mode. Images in hard copy form were supplied either to match with the Survey of India topographic maps (1; 250,000 or 1:50000) or in terms of scenes[7] wise images. Such images are mostly used in False Colour Composite (FCC—Please refer Fig. 21.8) format. Digital mode of data was supplied in digital media such as tapes/floppies/CD/DVDs. In both cases, the user was able to get either pre-processed data (corrected for geometric and radiometric errors) or raw data.

Later with the development of technology and the huge amount of archive on earth observation data, Analysis Ready Data (ARD) was made available by the data providers including USGS, BHUVAN that pave the way to the access of Data Cube (spatio-temporal data with space on 2 dimensions and time as the third dimension (Fig. 21.12). Such data set are not limited to just remote sensing data but also, derived maps such as landuse/land cover, water resources, infrastructure, hazard maps, etc. Development of data cube involves data preprocessing, integration and optimization of data and its storage and make it ready for sharing and analysis (Kopp et al. 2019) (See also 5 Open Geospatial data).

The Open Geospatial Consortium (OGC), which was established in 1994, brought together different academic, private and public sector parties and soon focused on the standardization of interfaces for accessing geospatial data resources and analytical procedures (Reichardt 2017). It enabled web based services for data search, access, analysis, and output generation with or without downloading the original data (Guo and Onstein 2020)

21.4 Geographic Information System

Theoretically, GIS could be defined as the organized system of hardware, software, data, and people to enter the data, edit it, analyse or manipulate and model the data to derive useful output in terms of maps, statistics or charts. Initially, GIS had been widely used along with remote sensing technology mainly for the earth resource observation, assessment, and management. However, the application of GIS has diversified growth in different sectors from administrative management to stock market and business sector. In this section, we will see the basics of different components of GIS.

[7] Scenes are regular division of continuous image acquired for the specified swath along the track of the satellite's orbit.

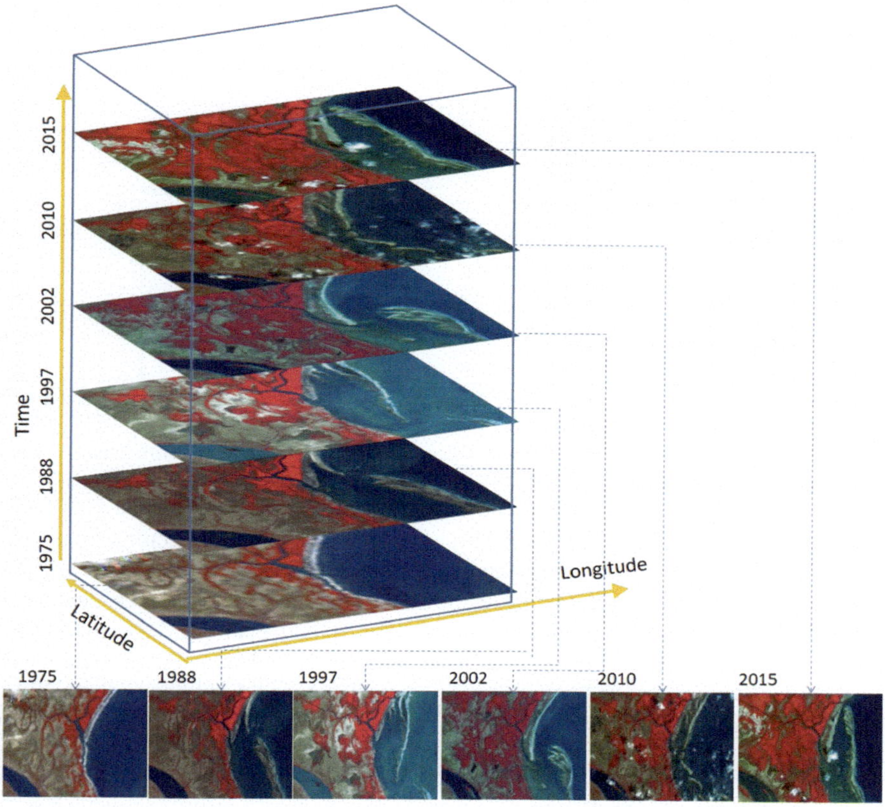

Fig. 21.12 Data cube: Representation of spatio—temporal data showing the dynamics in mangroves and coastal geomorphology of part of Krishna delta

21.4.1 GIS/Spatial Data Model and Data

There are two types of spatial data models, namely raster and vector depending on the type of data handled. *Raster* is the data in image format, i.e., regular arrangement of pixels in continuous rows and columns which are usually used to represent *continuous data* or gradually varying data (mostly numeric data) and *vector* data possess the data in the form of points and collection of points as lines or polygons which are called *discrete data* (Fig. 21.13) which are of categories or nominal data.

21.4.1.1 Selection of Data Type

The choice of GIS data types either raster or vector varies across the various stages of GIS project discussed earlier such as the data collection, entry, editing, analysis, and output generation. For example, secondary data source of wetland map can be either in raster form prepared using digital classification or in vector form interpreted visually from Remote Sensing data (Fig. 21.14). Suppose the collected wetland

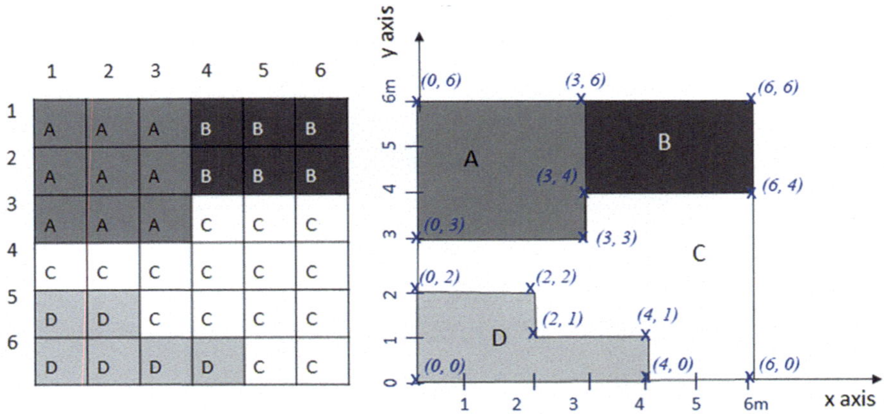

Fig. 21.13 Representation of four different land cover features in raster format with 1 m pixel size (left) and vector format (right) represented by connected points with coordinates

map is in raster form, the GIS analyst may like to convert into vector format for editing or analysing the data with other maps. Finally the output of the analysed data may be represented in vector or raster format depending on the requirements. In general it is advisable to handle the continuous spatial data (e.g., soil or water quality, rainfall, temperature) in raster format and discrete features/data (e.g., wetland classes, administrative boundaries) in vector format as discussed in the beginning of this section.

21.4.1.2 GIS/Spatial Data

In the course of GIS development, many forms of spatial data are used from secondary source and one should make sure the spatial data is from an authentic source. For example, Survey of India (SOI) topographical maps[8] are used as reference maps for Reconnaissance or preliminary survey of any ecological study. SOI maps are mostly available in 1:250,000 (small scale), 1: 50,000, and 1:25,000 (large scale) and the details provided are in *vector* form. Other maps such as village map, cadastral map, and tourist maps are also used depending on the need. Data acquired in raster or image format such as aerial photographs, remote sensing satellite images are used as primary data source. As we could see from the technological development, we have authorized/reference maps and time series global images from different sources in digital plat form (computers and mobile) such as Google Earth in *raster* form, Google map provides many features including transportation, land cover features, and infrastructure in *vector* format, most of the others like Open street maps, Bing map, Wikimaps, Open layers, etc. provide in *vector*

[8]SOI Maps are series of maps covering the entire country providing details about general geographic and topographic features with proper projection coordinates and scale along with elevation contours and necessary Bench Marks and Control points (www.soinakshe.uk.gov.in).

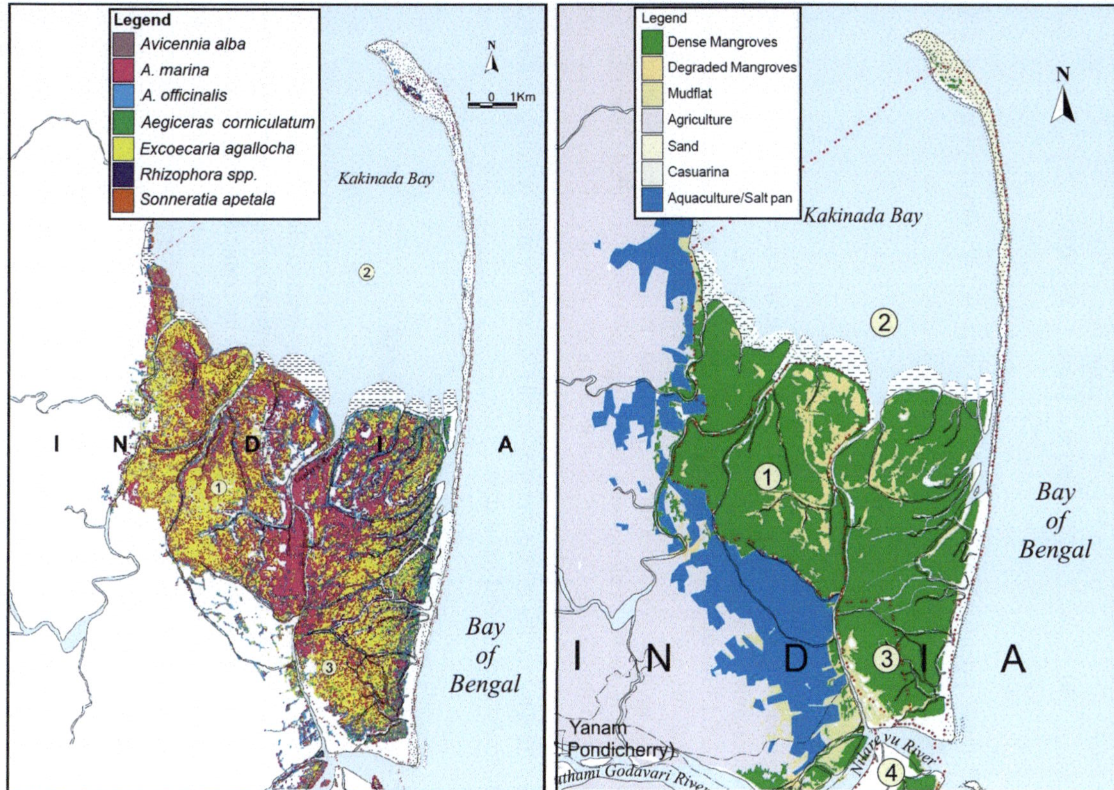

Fig. 21.14 Godavari mangroves: Mangrove community map prepared in raster format using digital classification of satellite image (left) and mangrove density mapped in vector format using visual interpretation of satellite (right). (Source: Ravishankar et al. 2004)

Table 21.2 General characteristics/Attributes of sample wetland features

Id	Wetland class	Average salinity (ppt)	pH	Area (sq.m)	Suitability	Average Elevation from Mean Sea Level (m)
A	Mangroves	28	7.0	200	High	1.0
B	Mudflat	32	6.0	340	Moderate	0.8
C	Sand dune	9	6.5	289	Low	2.3
D	Water body	30	7.3	443	Low	0.0

format. All these open data makes the initial step of GIS development of any ecological survey a hassle-free process.

21.4.2 Attribute Data

Attributes are nothing but the properties and characters of a particular spatial feature or object represented in GIS. For example, A, B, C, and D (Identifier or symbol) given in Fig. 21.13 (a) may have an attribute of wetland classes, namely mangroves, mudflat, sand dune, and water body. Other attributes may include its properties or characteristics, such as average salinity, pH, area, perimeter of each feature, etc. Collection of such attributes along with the polygonal feature is called Geospatial or GIS database (Table 21.2, General characteristics/Attributes of sample wetland features). Spatial and attribute data can be either a (i) Primary data which is collected or prepared directly by the GIS developer or (ii) Secondary data which is collected from other sources as second hand data.

21.4.3 Classification Scheme and Database

As we had seen a few of the attributes of mangrove ecosystem in Sect. 21.4.2, it is important to know the basis behind categorizing the wetland classes. The map given in Fig. 21.13 represents four different land cover features in raster format with 1 m pixel size (left) and vector format (right) represented by connected points having the coordinates(a) has four wetland classes where each class is exclusive and has its unique name (Table 21.2) and will have a clear definition which could form a part of wetland land classification scheme (Please refer to the well-structured hierarchical classification scheme defined for wetland classes in Sect. 21.4.3). Defining or adopting such a classification scheme will be the first step in any map preparation before survey or data collection. Other attributes (except suitability) about the classes given in Table 21.2 General characteristics/Attributes of sample wetland features are numeric values which are directly entered in the database. Such type of numeric data can be classified based on either (i) equal interval of values or (ii) natural break among values or having (iii) equal number of records depending the analytical or output requirements. Once, classification scheme is finalized, spatial database (collection and compilation of spatial and attribute data in an organized or structured format) is designed to make the GIS input, editing and analytical processes in a systematic manner.

21.4.4 Data Input and Editing

From the above description, GIS data possess two components. (i) *Spatial data* which containing unique geographic coordinates or other spatial identifiers and (ii) *Aspatial* or *Attribute data* which explains the properties of the spatial information. Spatial data is entered either by on-screen digitization (vector format) of points (x, y); line (x1:y1, x2:y2, ...xn:yn) and polygon (x1:y1, x2:y2,. . . .,xn:yn, x1:y1) or secondary source images (scanning/photographs/satellite images). The data may be either 2 dimensional (x, y) or 3 dimensional (x, y, z). Input of data includes necessary geometric correction as detailed in Remote sensing (Sect. 21.3.6.2.1) to match the entered data to a preferred coordinate system either geographic or projected (see Sect. 21.2.1.2).

Attributes can be entered either in tabular form using different data types, namely *Nominal*/Characters (Id and Wetland class), *Ratio* ("0" has no value—Area), *Interval* ("0" has a value—Elevation) (*integer*—Area; *float*—pH), and *Ordinal* (Suitability) as given in Table 21.2. Similar data can also be prepared as text format or in Excel/Office work sheet separately and then integrated with the GIS Database of vector data.

21.4.5 Analysis

Handling more than one layer being the major advantage of GIS, one can make a query or analyse one data or combination of multiple data. For example, an area up to 500 m from either side of the river (Fig. 21.15b) can be estimated using buffer analysis. Similarly, the extent of mangroves, mudflat, and other categories falling

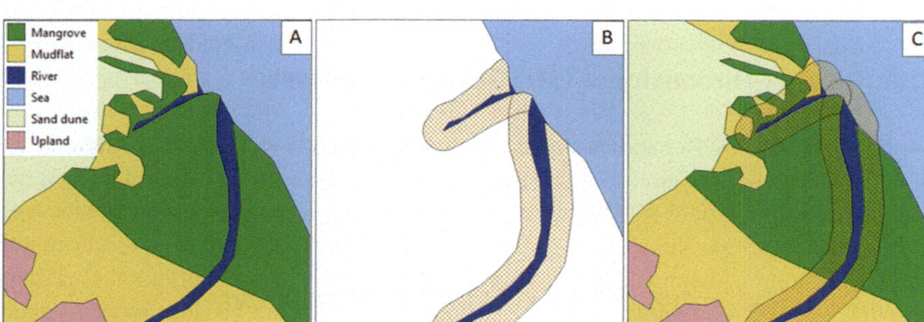

Fig. 21.15 (**a**) Wetland map, (**b**) river (blue) map with buffer zone of 500 m (light red), (**c**) overlay of buffer zone of river on wetland to estimate the area of the wetland feature under the buffer zone

under the buffer zone (Fig. 21.15c) can be extracted using overlay analysis of both A and B maps as highlighted by red around blue over the wetland features in Fig. 21.15c. Varied types of analyses can be performed to derive useful information from the GIS which includes

i. Overlay analysis: To know and estimate the extent of features from one map which are falling in or out of the features of another map. Similar to the second example given above (Fig. 21.15c)
ii. Buffer analysis: To find out and measure the area extending for a particular distance from a point or line or polygon feature as given in the first example given above (Fig. 21.15b).
iii. Neighbourhood analysis: Extension of the previous analysis. For example, to know how many soil samples are taken within the buffer zone if samples are collected evenly throughout the wetland features.
iv. Topographic analysis: We can combine any number of features along with elevation data to carryout 3D analysis. For example, if elevation profile of the region (Fig. 21.15a) is available, the extent of mangroves inundated by tidal flux can be estimated.
v. Interpolation: Different algorithms are used to generate maps of continuous data (raster) such as elevation, soil salinity, organic carbon, pH, temperature, rainfall, etc. using data collected from distributed sample locations.
vi. Network analysis: Analysing and optimizing the travelling path and the object or resource or people or any commodity travelling through a network such as road, river, canals, electrical, telecommunication network, etc. If the drainage network of a mangrove environment is available as GIS layer, we will be able to map the region getting inundated through the drainage canals. Addition of topographic data will fetch more accurate results.
vii. Visualization: Interactive way of visualizing the map individually or in combination with other maps in a two-dimensional or three-dimensional representations. Tremendous development in computer hardware and software technology paves way to visualize any ecosystem or environment virtually in it otherwise called as Virtual reality in 3 Dimension.

21.4.6 Three-Dimensional (3D) Data and Analysis

As specified earlier GIS handles spatial data either in 2- or 3-dimensional mode (x, y, z). Mostly 3D spatial data are represented in raster format and is widely known as Digital Elevation Model (DEM) which is nothing but the raster image wherein the pixel values represent the altitude of that pixel location. For example, if the wetland map/LULC map is displayed along with its topography i.e., the altitudinal is added with the spatial data we call it as 2.5D data and not a 3D data because wetland data does not vary with altitude for a particular location (x, y). However, DEM is

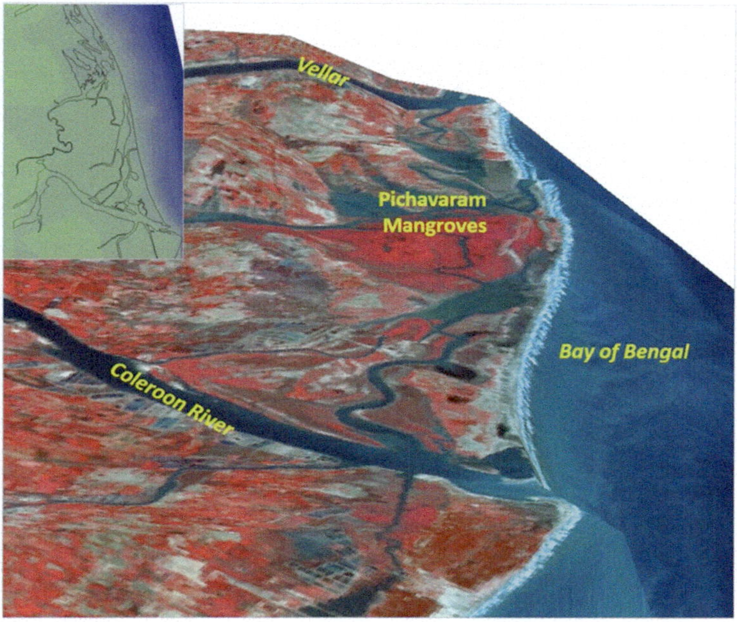

Fig. 21.16 False Colour Composite of Pichavaram mangrove wetland displayed over the integrated terrestrial elevation and sea/water depth (bathymetry) data derived from GPS and bathymetric chart (Insert). (Source: Gnanappazham L 2008)

generally accepted as 3D GIS when integrated along with other thematic[9] maps and helps to model the area of our interest. Basically there are three ways to generate topography (i) Modelling a region using 3D will add more value to the GIS developed and provides crucial details to solve various decision making problems where terrain topography/microtopography plays major role (Fig. 21.16)

21.4.7 Spatial Modelling

Geospatial modelling in the context of GIS can be defined as a system to simulate the real-world scenario for a particular time or over a period of time. It can vary from simple evaluation to prediction of some feature or phenomenon of interest (Goodchild 2005).

GIS modelling can be categorized into (i) Simple data modelling is a descriptive representation of real-world patterns in a database schema, (ii) Process modelling is the simulation of processes in the real world either static (input and output are of

[9]Thematic map is specifically designed to show a particular theme namely land use or water or geology of a particular region.

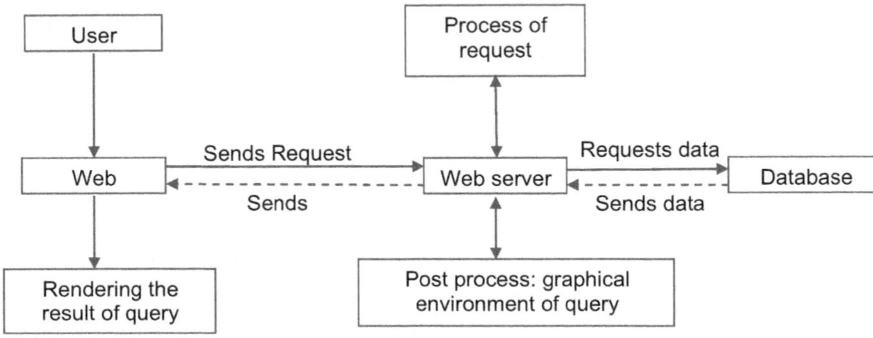

Fig. 21.17 Information flow from User—WebServer—User through WebGIS query

same time—for example extent of inundation of mangrove wetland due to tidal flux in a day) or dynamic models (output time is different from input time, for example, the future extent of mangroves if the tidal flux is going to reduce over a period of time) (https://clarklabs.org/terrset/land-change-modeler/), (iii) Space modelling which is a conceptualization of geographic space in 3 dimension using DEM/Triangulated Irregular Network/LiDAR.

21.4.8 Web GIS

As on when there is a development in computer engineering and technology, it had been incorporated simultaneously in GIS field also. Hence, the development of World Wide Web/Internet in sharing the information across the globe in no waiting time allowed the field of GIS also to share the geospatial data and the results of spatial analytics across the stakeholders from different parts of the globe. Today's world greatly depends on WebGIS and its application in every hour of human life as it has become an integral part of mobile. For example, transportation assistance, routing facility, home delivery of commodities, tracking the status, etc. completely rely on the position of the user and supplier/delivery person by his/her mobile's location.

A standalone GIS is accessible to the person handling the data from the computer on which he or she is handling. If the same is accessed or analysed through a wired network within an organization it becomes an "Enterprise GIS". If the same is performed through World Wide Web/Internet we call it as "WebGIS". Simply it is a combination of integrated or independent GIS data server and web server and the client get information out of it using a web browser installed in a computer or mobile (Fig. 21.17). The main focus of WebGIS is the dissemination of spatial data and the functionalities to the end-user with easy go approach, and its major applications include in e-governance, public utility services, real-time analysis, and mitigation measures during disasters and hazards, crowd sourcing (data input from localized resources) and reporting, etc. In general WebGIS portals can view derived information, a few can perform spatial analysis and few else can access the original data depending on the level of permission assigned. Open data portals which provide satellite data and its derived products are discussed in Sect. 21.5.

21.4.9 Spatial Data Quality

Quality of the data is the fitness of the data for the intended application. Hence, the measure of the data quality varies according to the requirements of the application and is measured in terms of Accuracy, Precision, Error, and Uncertainty. *Accuracy* is the degree to which the data matches with reality. Analysing country wide population data may need an accuracy of ten thousand scales whereas on district level would need the actual figure. Similarly the position of a city may be accurate to a kilometre in a country map but on a district map it must be between 10 and 50 m. However, there exists a linear relation between accuracy and the cost of data or data collection. *Precision* is another measure of geospatial data quality specifying the level of details and intricacies. For example, the temperature of a place is given as 27.55 °C is more precise than 27 °C. *Error* is the inverse of accuracy while *Uncertainty* is the degree to which the accuracy of data is not known.

In general the following components are defined to assess the quality of geospatial data.

1. *Positional accuracy (PA)*: The degree to which the position or coordinate of the spatial data either a pixel in an image or a point in vector data is matching with real coordinates on the ground. Hence, the accuracy is measured in terms of the degree of mismatch between coordinates of data or map point to coordinate on ground point (such points are called as Ground Control Points—GCPs) also called as Root Mean Squared Error (RMSE).

$$RMSE_x = \sqrt{\frac{\left(x_m - x_g\right)^2}{N}}, \quad RMSE_y = \sqrt{\frac{\left(y_m - y_g\right)^2}{N}}, \quad RMSE_z = \sqrt{\frac{\left(z_m - z_g\right)^2}{N}}$$

x_m, y_m, and z_m are the coordinates on map and x_g, y_g, and z_g are coordinates measured on Ground or GCPs. The planimetric accuracy is given by $\sqrt{RMSE_x{}^2 + RMSE_y{}^2}$. $RMSE_z$ is elevation accuracy.

As we have seen the accuracy depends on the intended application. Hence, the PA is given by 0.5 mm × scale. For example, PA of 1:50000 scale is 25 m (0.5 mm × 1/50000) and 1:10000 scale is 5 m (0.5 mm × 1/10000). As now all geospatial data are handled in digital mode, with the maximum possible enlargement, 0.5 mm accuracy threshold is raised to 0.25 mm and accordingly accuracy is estimated.

2. *Attribute Accuracy (AA)*: Geospatial data is encompassed with attributes of the spatial data represented in raster or vector format. And we had seen that attributes can be entered in numerical and categorical formats. Accuracy of numerical attributes is assessed using RMSE similar to positional where instead of position, the attribute values are compared with ground reality or sample data collected from field. For example, the accuracy of satellite product showing temperature of a region can be assessed by collecting representative sample points from the field. Categorical attributes given in text form can be assessed using error matrix or confusion matrix as discussed in classification accuracy (Sect. 21.3.6.2.2).
3. Other than these two measures, the spatial data should be verified for the following major components which are inevitable for the potentiality of any geospatial data.
   a. *Metadata:* This is nothing but data about the data collection and preparation process such as (i) survey method, (ii) accuracy of instruments, (iii) projection and transformation parameters, (iv) time of acquisition.
   b. *Completeness* of the data to certify if the data generated has any miss outs.
   c. *Consistency of spatial data* in its data types, positional accuracy and attribute accuracy of a map and list of maps stored in GIS. If village is chosen to be a point data, across the map and spatial database, it has to be point. For PA, scale of maps used or prepared has to be same and AA should be maintained across the map(s).
   d. *Semantic accuracy:* This deals with the naming convention of features or objects. In general, location or region specific names are assigned to the features. Sometimes according to the mapmaker. For example, one person could identify the marshy region of coastal wetland as "Mudflat" while other can name it as "Tidal flat". Such discrepancies must be avoided across the geospatial database from an organization level to national or global level.

21.5 Open Geospatial Data and Software Availability

Availability of archived and concurrent satellite and associated spatial data, and processing methods are quintessential for effective spatio-temporal monitoring of the earth's environment. In the last two decades, the trend of making the satellite data and geospatial data available for the end-user by different governmental and research agencies is widely seen, and the way in which geospatial data are collected, processed, analysed, and visualized are changed. Collaboratively contributed, authoritative and scientific geospatial data are now available for registered non-commercial users through collaborative effort of global geospatial community. Open scientific geospatial data tends to follow the FAIR (findable, accessible, interoperable, and reusable) principles in the collection, development, and dissemination of spatial and satellite data with global standards as framed by international standards development organizations such as the International Hydrographic

Organization (IHO), the International Organization for Standardization (ISO), and the Open Geospatial Consortium (OGC) (Coetzee et al. 2020).

Open source geospatial software includes diverse code libraries, algorithms, tools, applications, and platforms developed and made available under Open Source Initiative (OSI) license. Open Source Geospatial Foundation (OSGeo) software ecosystem (www.osgeo.org), founded as a not-for-profit organization with the vision of collaborative development of open geospatial technologies, data, and education for widespread use. The OSGeo projects acknowledge the principles of inclusiveness, fostering, openness, and responsibility in all their projects with an objective of open source geospatial software for community focusing on earth monitoring. OSGeo project family includes geospatial libraries (*Actinia, GeoTools, Orfeo Toolbox, GDAL, OWSlib*), web mapping GIS (*Map Server, GeoWebCache, PyWPS, GeoServer*), spatial database (*PostGIS, Open Data Cube*), and desktop GIS (*QGIS, gvSIG, GRASS GIS, Marble*). Other than OSGeo projects, contemporary platforms such as R (*R spatial tools*), Python (*GeoPython*), Javascript (*Leaflet, Map Box*), and Blender for GIS have developed open source geospatial tools (Coetzee et al. 2020).

Open geospatial data involves the collection of geospatial data and satellite images, and disseminating the data through modern technologies such as the Internet, the Internet of Things (IoT) through hosting portals. Another kind of open data follows the principle of sharing the information to user community by the authorities through dedicated data portals. Since the last decade, the recent trend of Open Spatial Data Infrastructure (SDI) emerged with the standard of transparency and collaboration with government and research organizations. Google Maps, Wikimapia, and Open Street Map (OSM) are examples of collaboratively contributed open geospatial data that follows the principle of citizen science through crowd sourcing—paid or volunteered, collaboratively maintained, and continuously updated through the Internet.

Authoritative open geospatial data includes the geospatial vector data such as administrative boundaries, point data, street lines, and associated attributes collected and maintained by government agencies are made public to user community with minimum to no restriction. Satellite data collected have also been made public through open license policy, for example, Bhuvan Data Discovery and Metadata Portal of Indian Space Research Organization (ISRO) provides thematic data and satellite imagery of global coverage. Similar portals include Copernicus: Earth Observation Program of European Union and USGS Earth Explorer and Earth Data of NASA. Cloud based platforms such as Google Earth Engine maintain the catalogue of earth observation satellite imagery, geospatial database of global coverage at multiple scales, and also JAVA based programming portal encompassing geospatial analytical tools enabling free access to scientists, researchers, and developers for a variety of analyses including monitoring the changes, trends, and quantify Earth's surface phenomenon at varied scales.

21.6 Applications of Earth Observation Techniques for Mangrove Ecosystem Monitoring

During the last four decades, monitoring of the inhospitable mangrove environment using field survey has been replaced by multi-source, multi-platform, multi-scale, multi-temporal, and multi-resolution remote sensing data. Aerial photographs, multispectral, hyperspectral, and microwave remote sensing data, LiDAR, and field spectrometry data are considered feasible, cost-effective, and time-efficient alternatives for systematic mangrove management. Using these datasets, various characteristics of mangroves such as spatial distribution, change detection, health status monitoring, field survey planning, biomass estimation, tree crown delineation, invasive species identification, anthropogenic impact assessment, ecosystem evaluation, monitoring, and management of conservation and restoration activities could be studied. In this section, we discuss different aspects of applications of remote sensing and techniques involved for effective mangrove management.

21.6.1 Mapping and Monitoring of Mangrove Wetlands

Since the development of aerial/satellite remote sensing multispectral data, they have been widely used for mapping and monitoring any feature of interest on earth. The following describes the general procedures followed while using multispectral remote sensing data as a source for an environmental or ecological survey. There are typically two modes of analysing remote sensing satellite data resulting into useful maps and statistics, namely (i) visual analysis, which completely relies on the interpreter's knowledge on the region and the field of study and (ii) digital analysis involves series of steps in converting raw satellite data in the form of pixel values to outputs. However, in practice, the digital analysis requires visual interpretation skills to validate the result, thus the hybrid of both analyses is used without which the output will be unrealistic.

21.6.1.1 Wetland Mapping Scheme

Whatever be the method of mapping one adopts, it is very much important to decide what features are to be mapped before start mapping. Generally a classification scheme is designed to identify or map any discrete or categorical features. For example, forests can be categorized into (i) Evergreen, (ii) Deciduous, (iii) Scrub, (iv) Littoral, (v) swamp, (vi) moist deciduous, etc. based on the vegetation types. Similarly, there exists different classification schemes for Landuse/Land cover (LULC),[10] wetlands, agriculture, soil, geology, etc. Tables 21.3 and 21.4 show the classification schemes of LULC and wetlands adopted in India. Visually interpreted

[10]Landuse is the nature of use the particular land is put into by human for example agriculture or industry while the land cover is the natural feature/phenomenon that covers the land for example water or vegetation or snow.

Table 21.3 LULC 1:50,000 Classification scheme LI—8, LII 3 & LIII—54 classes (NRSC 2017)

Level I	Level II	Level III
Built Up	Urban Built up	Built up -Compact (Continuous)
		Built up -Sparse (Discontinuous)
		Vegetated/Open Area
	Rural	Rural
	Industrial	Industrial area
		Ash/Cooling Pond/effluent and other waste
	Mining/Quarry	Mining—Active
		Mining—Abandoned
		Quarry
Agricultural land	Cropland	Kharif
		Rabi
		Zaid
		Cropped in 2 seasons
		Cropped in >2 seasons
	Fallow land	Fallow land
	Agriculture Plantation	Agriculture Plantation
	Aquaculture	Aquaculture
Forest	Evergreen/Semi evergreen	Dense/Closed
		Open
	Deciduous (Dry/Moist/Thorn)	Dense/Closed
		Open
	Forest Plantation	Forest Plantation
	Scrub Forest	Scrub Forest
	Swamp/Mangroves	Dense/Closed
		Open
	Tree Clad Area	Dense/Closed
		Open
Grass/Grazing	Alpine/Sub-Alpine	Alpine/Sub-Alpine
	Temperate/Sub-Tropical	Temperate/Sub-Tropical
	Tropical/Desertic	Tropical/Desertic
Wetlands	Inland	Natural (Ox-bow lake, cut-off meander, waterlogged, etc.)
		Man-made (Water logged, saltpans, etc.)
	Coastal	Lagoon, creeks, mud flats, etc.
		Saltpans
Water bodies	River	Perennial
		Non Perennial
	Canal/drain	Canal/drain
	Lake/Ponds	Permanent
		Seasonal
	Reservoir/Tank	Permanent
		Seasonal

(continued)

Table 21.3 (continued)

Level I	Level II	Level III
Wastelands	Salt Affected Land	Salt Affected Land
	Gullied/Ravinous land	Gullied
		Ravinous
	Scrub land	Dense/closed
		Open
	Sandy area	Desertic
		Coastal
		Riverine
	Barren rocky	Barren rocky

maps of Pichavaram mangroves in a large scale using high resolution (IKONOS MS + PAN; 1 m) and small scale using coarse resolution (IRS L3; 23 m) satellite data are given in Fig. 21.18 for reference.

21.6.1.2 Limitations

Even though remote sensing has enabled the user to study the natural environment in multiple ranges of spatio-temporal scale, there are few decisive factors, including cloud cover, sensor specifications, the difference in mapping scale, lack of real-time data, confusion in selecting appropriate datasets, choice of image processing, and classification techniques involved while mapping mangroves at higher floristic hierarchical level (Green et al. 1998). Recent developments in remote sensing technology have made it possible to overcome a few of these obstacles except the following.

- Since mangroves thrive in the inter-tidal region, the image scene would have been highly influenced by seasonal and diurnal tides, which makes the radiometric correction a challenging task.
- Also, the spectral reflectance obtained from the image pixel covering the fringing and open mangroves would be a mixture of vegetation, soil and water. Selection of high resolution data or suitable spectral unmixing algorithms would be needed to overcome misclassification.
- Due to the heterogeneous distribution of mangrove species in many parts of the ecosystem, it is difficult to obtain a pure pixel for unique species or community (Kuenzer et al. 2011).

21.6.2 Floristic Discrimination of Mangroves

As mentioned earlier, the distribution of mangroves is highly heterogeneous in a dense mangrove ecosystem. Conservation and management practices in such mangrove forests require baseline information of the spatial distribution of species in the forest. Traditionally the field survey methods, for example, the Point Centred Quadrant (PCQ) method records the distribution of species and the density of

Table 21.4 Classification system for mapping mangrove community zones (Nayak and Bahuguna 2001)

Level I	Level II	Level III	Level IV
Onshore areas	Beach	Muddy Sandy	Fringe tidal mangroves No mangroves
	Estuary	Inter-tidal mudflat (sandy clay)	(i) *Rhizophora, Sonneratia, Avicennia, Bruguiera* (pure/mixed communities) (ii) Saline blanks
		Inter-tidal mudflat (silty clay)	(i) *Kandelia, Avicennia, Rhizophora, Aegiceras, Sonneratia* (pure/mixed communities) (ii) Salt marsh vegetation (iii) Saline blanks
		High-tidal mudflat	(i) Salt marsh vegetation (ii) Saline blanks
		Transitional areas	Grass
	Deltaic complexes	Seaward margin of inter-tidal mudflats	*A. marina, A. alba, A. officinalis* (pure)
		Inter-tidal mudflats	(i) Pure communities of *Rhizophora, Sonneratia, Avicennia, Bruguiera,* etc. (ii) Mixed mangrove (iii) Salt marsh vegetation (iv) Saline blanks
		High-tidal mudflats	(i) Salt marsh vegetation (ii) Saline blanks
		River mouths	*Sonneratia, Bruguiera, Excoecaria, Heritiera, Lumnitzera* (pure/mixed communities)
		Along creeks (upper end)	(i) *Sonneratia, Bruguiera, Excoecaria, Heritiera, Lumnitzera* (pure/mixed communities) (ii) Freshwater species
		Transitional areas	(i) Saline blanks (ii) Grassy banks (iii) Trees/shrubs
	Gulf region	Inter-tidal mudflat (sandy clay)	(i) Pure/mixed communities of *Rhizophora, Sonneratia, Avicennia, Bruguiera*, etc.
		High-tidal mudflats	(i) *Kandelia, Avicennia, Rhizophora, Aegiceras, Sonneratia* (pure/mixed communities) (ii) Salt marsh vegetation (iii) Saline blanks
		Transitional area	Grass
Offshore areas	Islands	Sub-tidal area	Algae/seaweeds
		Inter-tidal	(i) Mangroves (ii) Sand vegetation
		Coral reefs	(i) Algae/seaweeds/seagrass (ii) Mangroves

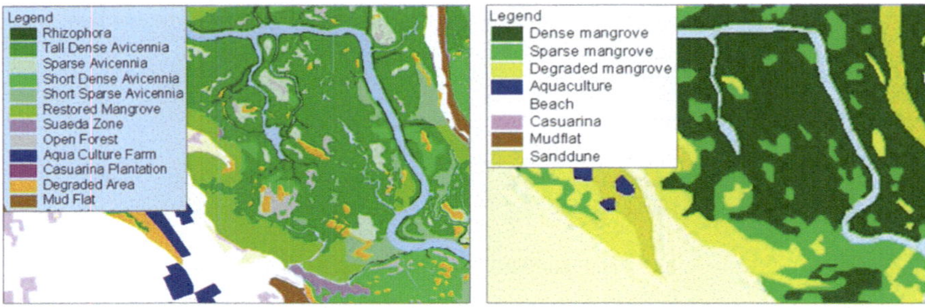

Fig. 21.18 Mangrove wetland maps of Pichavaram: Level 4 classification—large scale map prepared using high resolution data IKONOS MS + PAN data with intricacies in wetland (Left) and Level 2 classification small scale map using coarse resolution satellite data IRS L3 with less details (Right). (Source: Gnanappazham and Selvam 2011; Gnanappazham L 2008)

plant distribution by recording the number of trees in different species found at each quadrant of the ground plot of fixed dimension. This method, similar to the other field survey method requires more time, labour and financial support, which is often difficult when planning for a temporal study with short time intervals. At this instance, archived and real-time multi-source remote sensing could be an asset for spatio-temporal mapping of species distribution at multiple scales (Annexure 3).

21.6.2.1 Traditional Remote Sensing Methods

At the times of development of aerial photography, they were used for the mangrove discrimination using visual interpretation techniques. Secondary information from topographic sheets, field surveys, and colour infrared video imagery were also used for discrimination of mangroves from other associated land use/land cover. These aerial imageries were collected as grey scale image (panchromatic) or colour image with a limited number of bands often in low spatial and temporal resolutions. The visual interpretation of these products abled the user only to discriminate mangrove from other land uses and to assess damaged or loss of mangroves rather than vegetation community discrimination (Fig. 21.19).

21.6.2.2 Image Processing of Multispectral, Hyperspectral and Microwave Data

Considering mangroves, scientists used multispectral images for community level discrimination using different approaches and vegetation indices are the foremost among them. Several vegetation indices have been developed using different spectral bands of which a few of them, for example NDVI and NDWVI, are correlated with the plant canopy closure and health status of the vegetation, therefore popularly used along with other indices such as NDWI, SAVI, and SR (Jensen et al. 1991; Gupta et al. 2018). Next to VI, classification methods are widely used to map the extent. For example, unsupervised classification algorithms use only the spectral reflectance recorded in the image pixels, while the other category, supervised classification algorithms requires the visual interpretation skill of the analyst to

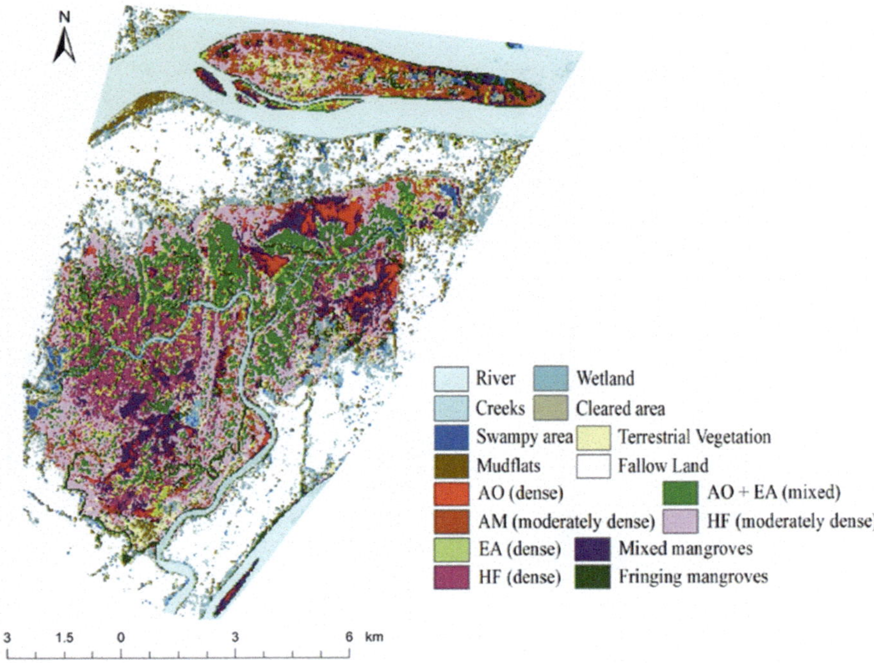

Fig. 21.19 Community level map of Bhitarkanika National Park, Odisha, India derived from the supervised classification of IRS P6 LISS-III data

extract samples to train the classifier for image classification (Figs. 21.8 and 21.9). In recent times, with the development of high spatial resolution optical imaging sensors, earth observation data at sub-metre level become available. *Object-Based Image Analysis* (OBIA) on such high resolution satellite and airborne data (Quickbird, Geoeye, IKONOS, Worldview-2, Cartosat, etc. (Please refer Annexure 1 for sensor specifications) have refined the classification accuracy of mangroves further. OBIA, unlike pixel-based classification, the images are first segmented into "objects" representing groups of pixels belonging ground patches, entities or their elements (primitives) which can be then classified into categories of interest by unsupervised, supervised or rule-based algorithms (Fig. 21.20). Thakur et al. (2020) made a systematic review on the application of multispectral remote sensing satellite data for the floristic discrimination and change detection of mangrove ecosystem.

At present, the availability of temporally dense, global coverage imaging spectrometry data is limited. There are a few space borne *hyperspectral sensors* that provide data at a moderate spatial resolution. Few airborne hyperspectral sensors are currently operational which could be flown at a lower altitude and provides data at a higher spatial resolution, but lacks global coverage. Specifications of prominent and globally used hyper spectral sensors are given in Annexure 1.

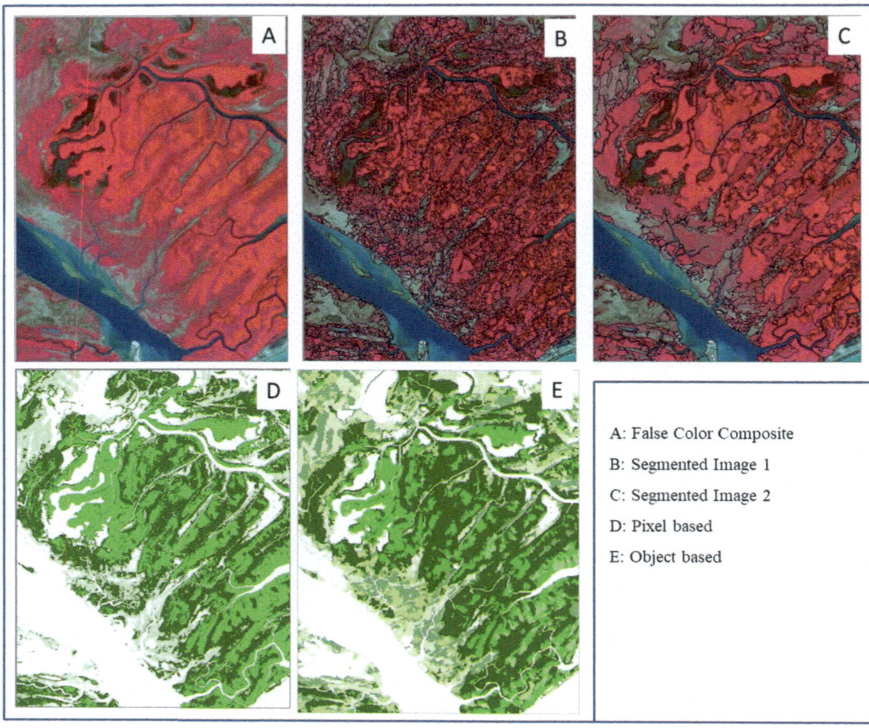

Fig. 21.20 False Colour Composite of mangroves of Krishna Delta (A); B and C show two types of segmented images. Vegetation density classes of mangroves using Pixel and Object-based classification are shown in D and E

Ground truth data is vital information for any type of satellite remote sensing for its accuracy. While categorical and qualitative ground truth data is used for multispectral data analysis, the accuracy levels of hyperspectral image analysis are increased by in-situ spectral data collected using spectroradiometer for the discrimination of features having a similar response to the light energy (spectral characteristics) such as minerals, vegetation types, soil, and water quality parameters. *Field spectrometry* is the technique which is used to quantify the radiance, irradiance, transmission/reflectance from various earth surface features in field condition (Jackson et al. 1980; ASD 2001). It is being actively used in the last two decades in forestry and vegetation sciences for species identification, classification, health status monitoring, nutrient intake estimation, invasive species monitoring etc. Several site-specific spectral libraries were collected so far for various species, including, non-native species (Underwood 2003; Aneece and Epstein 2017), wetland species (Zomer et al. 2009), Mediterranean species (Manakos et al. 2010), shrubland species (Jiménez and Díaz-Delgado 2015), coral reefs (Kutser et al.

2006), agricultural crops (Datt et al. 2003; Rao et al. 2007) and Indian mangrove species (Prasad et al. 2015).

21.6.3 Biophysical Characterization of Mangroves

Tropical forests play a critical role in ensuring the stability of the climate change through their efficiency in capturing greenhouse gases (including atmospheric carbon) and sequestrate in their biomass (Fig. 21.21).

(i). Biomass is one of the structural properties of the vegetation and its estimation is one of the essential baseline information needed for climate change mitigation programme. The change in biomass is an indicator of the stress in the vegetation induced by natural and anthropogenic disturbances. This leads to the reduction of its carbon sequestration capability and so it needs to be monitored temporally (Klemas 2013). The accurate estimation of the above ground biomass and carbon stored in the vegetation is one of the important objectives in the resolutions of the Bali Action Plan (2007) approved by the United Nations Framework Convention on Climate Change (UNFCCC) and the Kyoto Protocol (2005). The traditional field survey method was extensively used for collecting forest biophysical parameters which is usually time consuming, cost and labour intensive. Field measurement of above ground biomass (AGB) is obtained through the destructive sampling of all the plants within the ground plots of fixed dimension, for example 5 m $\times$ 5 m. The collected plant materials are then segregated into different components (leaves, stem, branches, roots, etc.) and weighed separately to determine their wet weight. These components are then dried and weighed to obtain respective dry weight to compute biomass in grams per square metre.

(ii). Plant LAI is the measure of canopy foliage content which estimates the total area of one side of photosynthetic leaf tissue per unit area of ground surface (m^2/m^2). LAI is an important parameter to study land-surface interaction processes, plant productivity, plant-atmosphere interaction, and parameterizations in climate models. Considering LAI measurements, hemispherical photographs and quantum sensors are being used for field measurement. LAI could be estimated indirectly from the remotely sensed images through image derivatives and indices. LAI estimated from optical remote sensing data could serve as a key parameter in calculating above ground biomass of vegetation (Bréda 2008; Zheng and Moskal 2009). Furthermore, field survey for biophysical parameter collection in mangrove forest is practically much difficult because of its hostile and swampy environment. Remote sensing is found to be a cost-effective and potential alternative for spatio-temporal monitoring of mangrove biophysical parameters. Multi-source, multi-sensor, multi-resolution, multi-scale, archived and real-time remote sensing data from airborne and EO satellites provides a strong database for the retrieval of biophysical properties such as biomass and Leaf Area Index

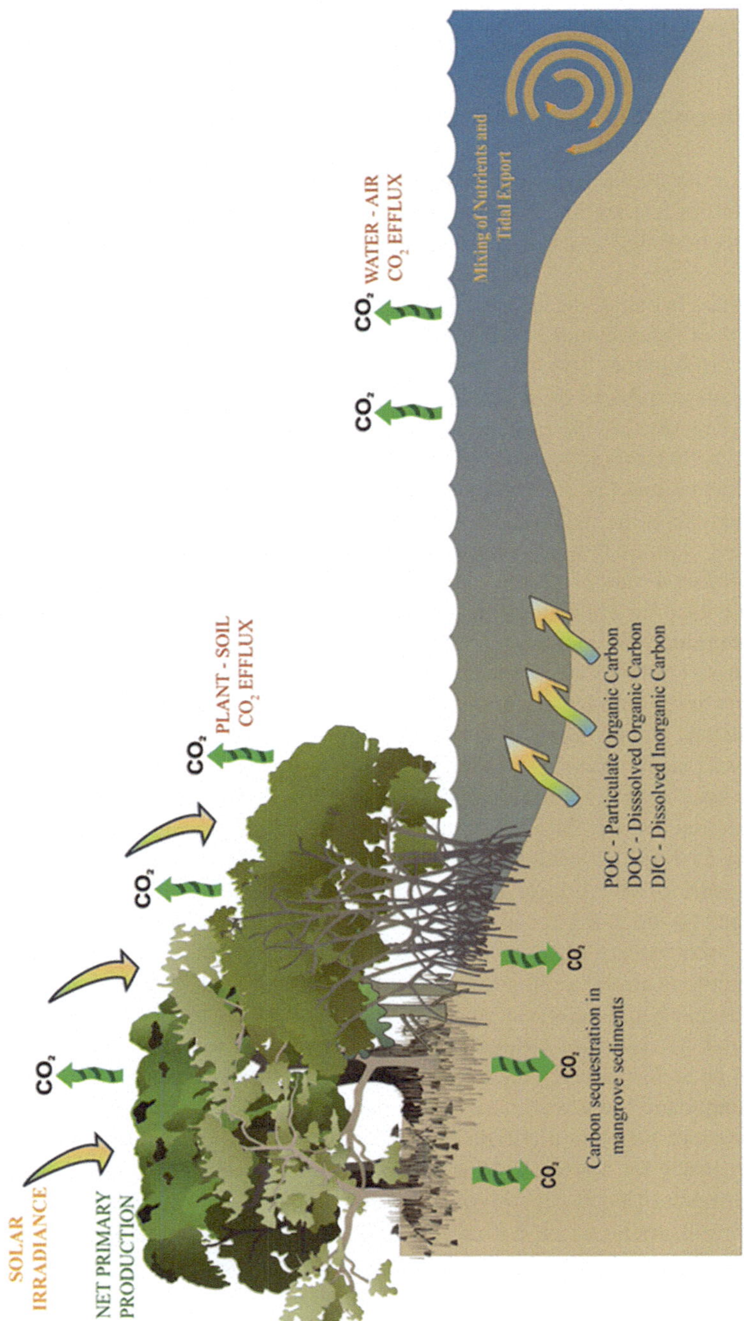

Fig. 21.21 Carbon flux pathways in mangrove ecosystem

(LAI), biochemical characteristics of mangroves and other wetland vegetation. Henceforth, the estimation of mangrove biomass by combining field survey and remote sensing is recommended for accurate biomass estimation (Heumann 2011).

(iii). Remote sensing based biomass estimation usually involves field data collection about biophysical properties such as diameter at breast height (DBH), tree height, number of individuals of a particular species, leaf area index (LAI), etc. The application of remote sensing data for biomass and productivity estimation in different ecosystems has been reviewed in detail (Kale et al. 2002; Klemas 2013; Lu et al. 2014). They mentioned that the unique characteristic of plants is displayed by its reflectance in the red and infrared regions of electromagnetic radiation and have a strong relationship with the biophysical parameters of plants. Empirical and statistical models and advanced data analytics of high resolution, multispectral, hyperspectral imagers, LIDAR and microwave sensors using allometric equations provide plethora of data for biophysical characterization of natural vegetation.

(iv). High resolution multispectral images are often used for the modelling of AGB and LAI for mangrove forests at a large scale. Vegetation indices such as NDVI, a simple ratio of multispectral bands, and specific image transformation methods such as Fourier-based Textural Ordination (FOTO), Grey Level Co-occurrence Matrices (GLCM), Artificial Neural Network (ANN) methods supported with in-situ plot biomass measurements have been proven to be successful in modelling AGB and LAI of natural vegetation (Annexure 4) VIs derived from WorldView-2 images are more efficient than other bands in predicting the biomass of high-density mangrove forests. However, using hyperspectral remote sensing data, the vegetation indices derived out of this data are found to be outperforming multispectral vegetation indices (Marshall and Thenkabail 2015). Additionally, mangrove species distribution mapped using hyperspectral image could be utilized to improve the biomass assessment (Marshall and Thenkabail 2015; Anand et al. 2020).

(v). Recent studies have proven that *active remote sensing* techniques such as SAR and LiDAR sensing have distinct advantages over optical remote sensing for mangrove biophysical parameters retrieval. The time independent and weather independent operational capability of active SAR sensors is suitable for sensing of tropical coasts including mangroves where cloud condition is prevalent. Moreover, active SAR sensing has the potential to record canopy structure and biomass (Sect. 21.6.3) while passive microwave remote sensing data can be used to indirectly infer these characteristics from their spectral response. From recent studies, it is understood that the biomass dependence of radar backscatter varies based on radar wavelength, polarization and incidence angle. Comparatively, the longer wavelength is found to be more sensitive than shorter wavelengths, and *cross polarization*[11] (HV) is more sensitive than *like polarizations*[11] HH and VV towards plant biophysical parameters. The reason is that the sensitivity of radar backscatter intensity to variations of biomass saturates after a certain level of biomass is reached, and the saturation level is higher for longer wavelengths. Availability of global coverage open source high resolution SAR data opens a new opportunity to estimate the biophysical structure parameters of mangrove ecosystems (Mougin et al. 1999; Fatoyinbo and Armstrong 2010; Pham et al. 2019). Furthermore, the advent of high performance, parallel and cloud computing resources facilitates the GIS based geospatial analysis for the global estimation of biomass and carbon sequestration of mangrove ecosystems in an efficient and cost-effective manner (Tang et al. 2018).

21.6.4 Characterizing Foliar Biochemistry of Mangroves

Assessing the biochemical and biophysical properties of wetland vegetation is vital to monitor plant primary productivity, carbon cycling, and nutrient allocation within the ecosystem (Mutanga and Skidmore 2004; Adam et al. 2009). The varying levels of the biochemical constituents of the vegetation such as carbon, hydrogen, oxygen, nitrogen, magnesium, etc. in the form of plant cells, tissues and pigments regulate the photosynthesis and consequently the plant primary productivity(Muñoz-Huerta et al. 2013; Al-Naimi et al. 2016). Some vegetation biochemical constituents are detectable at canopy level through spectral reflectance. Visible region of the spectrum (400–700 nm) is highly sensitive to leaf pigments such as chlorophyll, while reflectance at infrared region (700–2500 nm) is sensitive to non-pigment constituents such as proteins (Kokaly et al. 2009). However, spectral analysis for biochemical constituents of fresh leaves at the canopy scale would be a complex process as the absorption features of leaf water obscures the absorption features of the biochemical constituents (Kokaly and Clark 1999). Moreover, the supplementary effects of vegetation canopy structure, solar illumination effects, atmospheric influence, signal-to-noise ratio, and the reflectance anomalies from understorey vegetation, soil, roots, and branches must be considered while using remote sensing data for biochemical characterization (Asner and Martin 2008). RS based biochemical studies on Indian mangroves are very limited (Annexure 4) and have great scope further research.

[11]Cross Polarization in which Transmitted and received signals are of different polarization (HV or VH) and in Like Polarization they are of same polarization (HH or VV).

21.7 GIS in Mangrove Monitoring and Management

Since the inception of remote sensing technology, GIS became an integral component of remote sensing. Later, satellite remote sensing data and their products became the source for GIS inputs and analysis. In this section, we will see the application of different components of GIS in mangrove mapping, monitoring and its potential applications in mangrove management in the context of Indian mangroves.

21.7.1 Developing GIS of Mangrove Ecosystem

Developing a versatile GIS database will be very beneficial for the effective management of mangrove ecosystem. GIS should be developed so that, the system will help to identify the source of the problem, find the optimum solution leading to an efficient decision making process (Fig. 21.22).

21.7.1.1 Define the Study Region

Identify the boundary of the region, including mangroves and associated land cover features. It may be either reserved forest boundary or Revenue village boundary encompassing the complete mangroves under consideration from the coast line. This will help 70% of the decision making process to finalize the scale of a GIS database to be developed. In general, the preferred scale for mangrove management is 1:50,000. However, depending upon the intended activities, the larger scales from 1:25,000 to 1:5000 are adopted.

21.7.1.2 Shortlisting the Required Layers/Maps

Independent of the objectives of the study, a critical map is the *base map* of the study area which could be from either topographic maps by Survey of India (SOI) or village map from revenue village office which are authenticated source of base map details. Such details include High tide/Low tide line, reserved forest boundary, village boundary (if needed), nearby settlements, water bodies such as creeks, lagoons, rivers, streams and man-made canals, any reference points or benchmarks, etc. along with their essential attributes. More than one base map layers can be included in GIS depending upon the requirements.

Layers other than base map details generated are decided based on the objective of the study. For example, following list may be needed to include in geospatial data for Mangroves.

(i) Mangrove wetland map for the period(s) of study considered. So, depending on the requirements, one or more wetland maps, mostly derived using satellite remote sensing data are included.
(ii) Vegetation health status: NDVI would also become a supplementary layer of GIS for the selected period(s) of study.
(iii) Corresponding satellite images for regular reference.

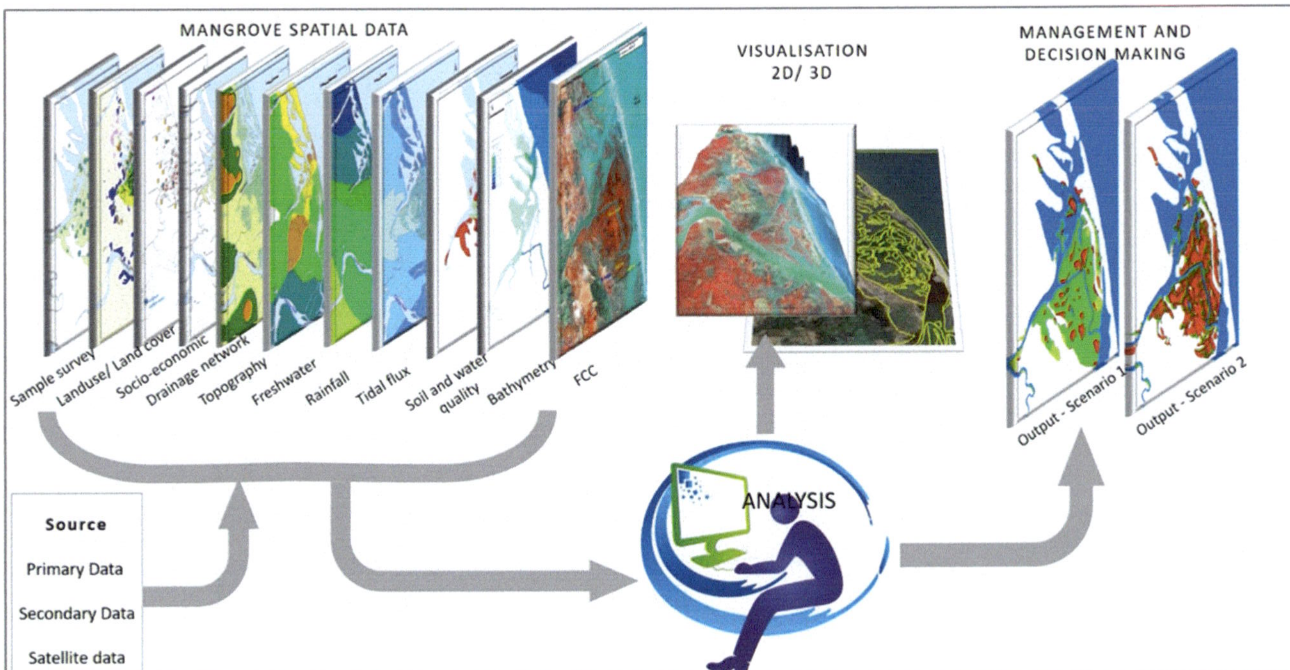

Fig. 21.22 Representation of GIS database of Pichavaram Mangrove ecosystem. (Image Source: Gnanappazham 2008)

(iv) Species composition map (derived using satellite data and verified using ground truth samples).
(v) Locations of sample data on mangroves, its diversity, health, environment including soil, water, and atmosphere, etc. collected from the field as per the requirement of the study.
(vi) Digital Elevation Model of the study region: Mangroves lie along the coast within an elevation of approximately 10 m above mean sea level depending on the tidal amplitude of the particular region. Hence, mapping the microtopography in the range of 0–10 m can be achieved by very high resolution stereo images or LiDAR from spaceborne or airborne sensors. Field surveys, including DGPS, are of limited use in a marshy environment where mangroves are very dense.

21.7.1.3 GIS Input

Categorize the GIS input data types as (i) Primary or Secondary, (ii) point or line or polygon or raster data, (iii) If raster input, then is it thematic or continuous or satellite image. Accordingly the spatial data are entered in GIS software along with attributes. Spatial and attribute data need to be prepared in the format as per the specification of the software through the process of editing and quality checking.

21.7.1.4 Analysis

There are varied types of spatial analysis carried out for mangrove related studies including mangrove change detection, monitoring, shoreline changes, time series analysis, and hydrological and environmental modelling.

Change Detection and Time Series Analysis

Application of satellite data for mangrove studies triggered by Blasco et al. (1986) for assessing the extent of Pichavaram mangroves and followed by many studies to map and monitor the mangroves of Indian coast using Landsat and IRS series of satellite data. Change detection involves overlay analysis of mangrove wetland maps of two time period under consideration and deriving the extent of (i) unchanged categories, (ii) change that had happened, especially the extent of mangrove vegetation converted to other land cover types, and (iii) from other land cover features to mangroves. Numerous change detection studies had been conducted for most of the mangroves of India and Fig. 21.23 depicts one such change detection map of mangroves of Pichavaram (Tamil Nadu). GIS analysis becomes an integral component of RS based analysis and relevant literature are included in detail (Annexure 2) marked with[a].

Shoreline Changes

Generally, the difference in positions between High Tide Line (HTL) or High Water Line (HWL) or Wet/dry line (otherwise called as shoreline proxies) is mapped between two timestamps to estimate the shoreline changes. However, there is no standard method available using remote sensing data due to the variability between tidal range and satellite data acquisition and inequalities in data resolution. ArcGIS, a

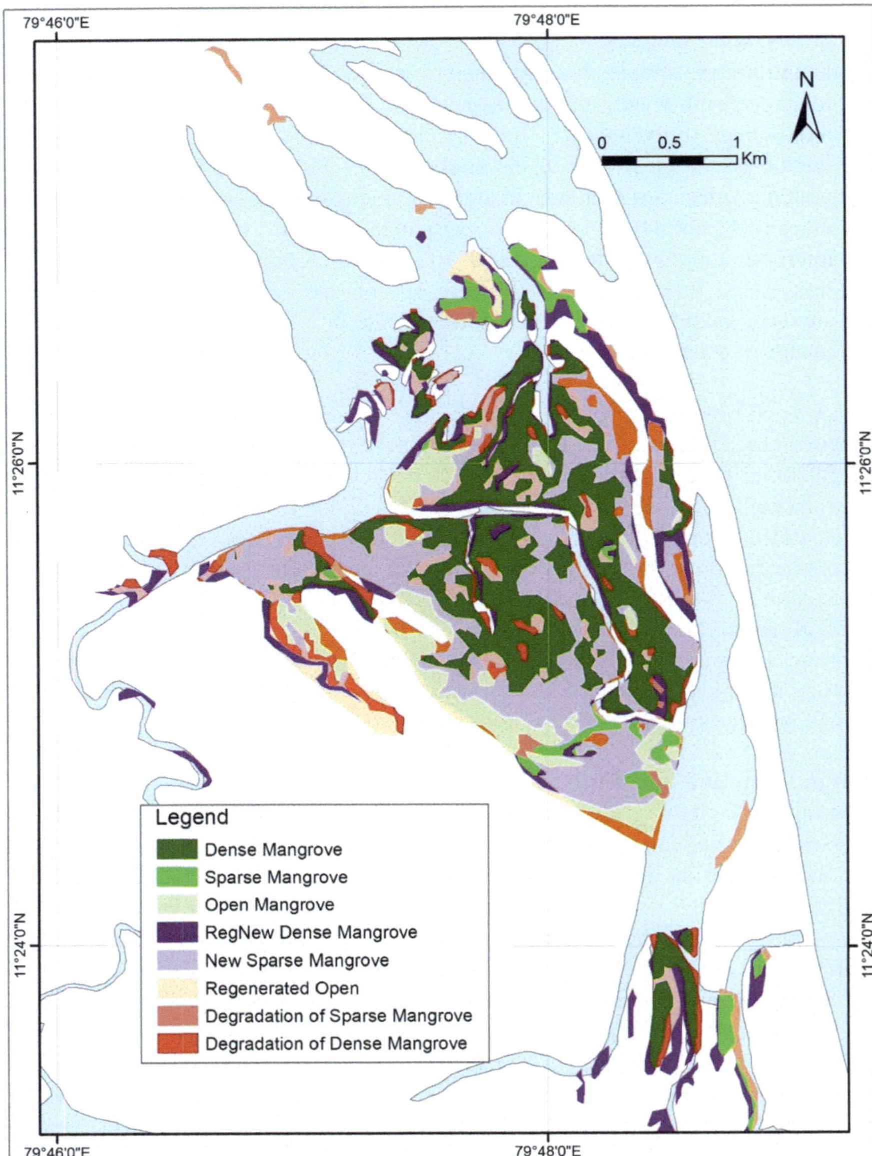

Fig. 21.23 Change in mangroves of Pichavaram between 1994 and 2002

popular commercial software has an extension developed by Unites States Geological Survey (USGS) called "Digital Shoreline Analysis System (DSAS)" that helps the user to estimate the rate of change of erosion and accretion using automated transects along the multi-temporal shorelines provided by the user (https://www.

usgs.gov/centers/whcmsc/science/digital-shoreline-analysis-system-dsas?qt-science_center_objects=0#qt-science_center_objects). For Indian coasts, Integrated Coastal and Marine Area Management (ICMAM) Programme developed the shoreline proxies depending on the varied and complex shoreline such as Wet/dry line for sandy coast, vegetative boundary along shoreline, seaward boundary of any artificial structures along the coast and the shore edge of the cliff's base. Depending on the nature of mangroves along the coast, the erosion and accretion process has either a huge impact (Fig. 21.24; Ramasubramanian et al. 2006) or no impact (Das 2020) on the mangroves and their extent (Annexure 5).

21.7.2 Environmental Parameters of Mangrove Ecosystem

Overlay analysis of more than one map helps in the decision making process and increase the efficiency of mangrove management activities as well. Researchers have worked on assessing the health of mangroves either deriving vegetation indices, environmental (soil and hydrology), and climatic parameters from time series multi-source satellite data or by integrating them with field samples on vegetation, environmental, and climate parameters and the details on literature are explained in Annexure 6. Doyle and Girod (1997) developed a high end simulation model at the landscape level, SELVA and at stand level, MANGRO to predict changes in the mangroves for unit ha and for individual trees for sea level rise and inundation levels. However, hydrological and environmental modelling of mangroves of India is very less. Modelling the impact of driving parameters on mangroves and ecosystem and vice versa, including modelling the wave dynamics has great scope in the context of Indian mangroves (Amma and Bhaskaran 2020). Nevertheless, the integration of environmental, hydrological, and climatological data as a Mangrove spatial database is recognized as the most crucial source of information for mangrove management, which is taken by the authorized organizations which are discussed in Sects. 21.7.3 and 21.7.4.

21.7.3 Mangrove Conservation and Coastal Regulation Zone (CRZ) Mapping

Indian mangroves are categorized into Reserved/Protected forests (RF/PF), National Parks (NP), Community reserves, Wild Life Sanctuary (WLS), and also as Marine and Coastal Protected Areas (MCPA) based on their ecological importance declared by Indian Forest Conservation Act, 1980 & the Wildlife (Protection) Act, 1972 (DasGupta and Shaw 2013). MCPA is further categorized into four, namely Category: I (C:I) as complete area under inter-tidal zones of NPs and WLS; C:II as a marine ecosystem of the islands; C:III A and B as sandy beaches beyond seawater with occasional interaction with evergreen forests (MoEF 2008). CRZ had the initial set up to protect the 500 m from the high tide line (HTL) from development activities when implemented in 1981 and later in 1991 the CRZ notification was implemented

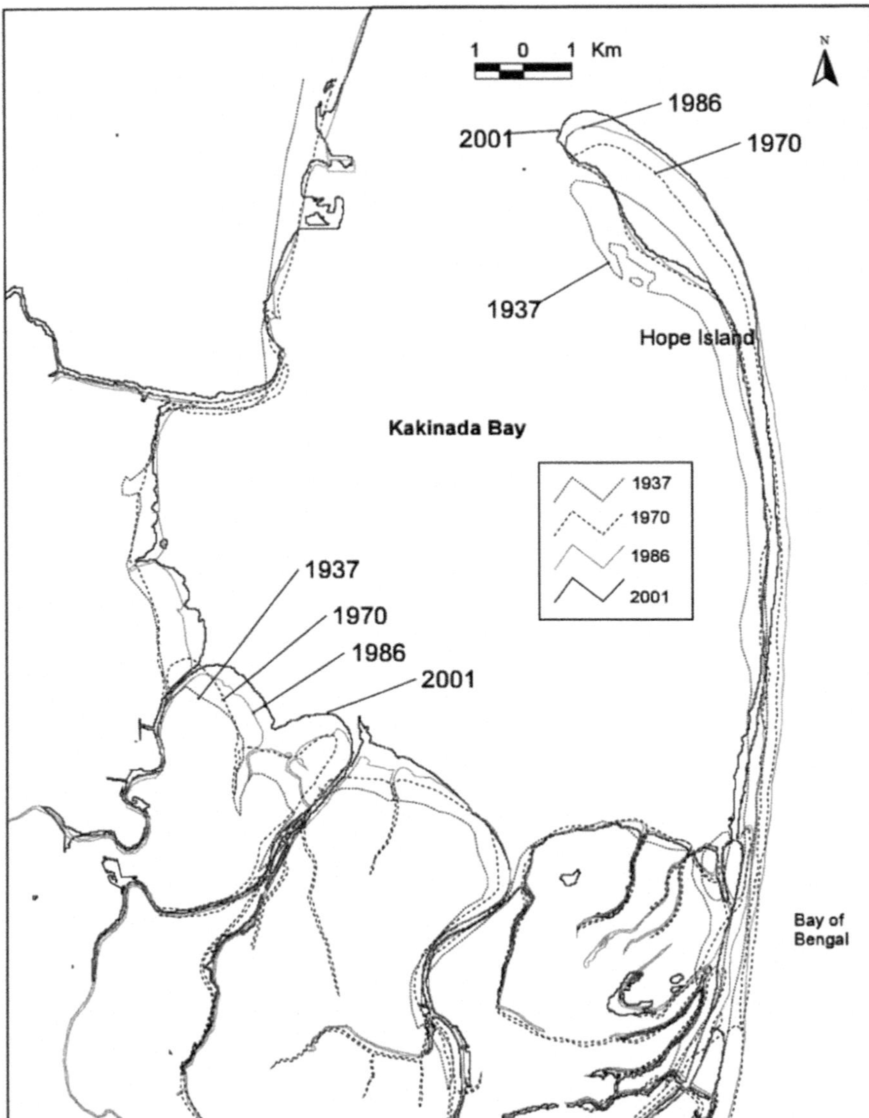

Fig. 21.24 Shoreline change along Kakinada Bay endowed by mangroves from 1935 to 2001 (Source: Ramasubramanian et al. 2006)

along with the Environmental Protection Act. Recent CRZ notification has identified 4 CRZ zones with additional measures to regulate harmful activities for coastal community and the environment (Chandramohan 2019). Since the beginning of the coastal regulation processes, Mangroves have been found among the first prioritized coastal habitats and are covered under CRZ 1A as they are identified as Ecologically Sensitive areas. An association of national and regional organizations and academic institutions is authorized to identify and prepare the maps of all four CRZ zones. Preparation of Hazard line, one of the inputs of CRZ mapping itself is the composite of the extents of the flooding on the land area due to water level fluctuations, sea level rise, and shoreline changes (erosion or accretion) occurring over a period of time which has been prepared by Survey of India and further classification of CRZ and mapping at 1:25,000 scale involve the development of exhaustive GIS layers and various kinds of spatial analysis discussed in Sect. 21.7.1.4 (NCSM 2020).

21.7.4 Integrated Coastal Zone Management

In 1979, the Indian National Mangrove Committee recommended areas of research and development in mangrove mapping, quantitative survey of seasonal variations environmental parameters, vegetation growth climatological variability for management of the mangroves which initiated national level application of remote sensing derived maps and field survey data in an integrated manner (Anonymous 2020; FSI 2017). Later the concept of ICZM was evolved during the Earth Summit of Rio de Janeiro in 1992 aiming to achieve the sustainable management of the coastal zone to ensure optimal utilization of coastal resources, maintaining the biodiversity and conserving the critical habitat in an integrated manner regardless of the ecological and political boundaries (FAO 2020). In India, the resource assessment activity of ICZM was carried out for the entire coast by developing a geospatial database using remote sensing and GIS of field surveyed data and is being used to monitor the conservation and restoration of coastal resources including mangroves (https://www.ncscm.res.in/). National level GIS based Critical Habitat Information System (CHIS) of the coastal wetlands including mangrove wetlands was developed to design appropriate management solutions to address the coastal problems (ICMAM 2000).

ENVironment Information System (ENVIS) was the initiative by MoEF in 1982–83 to provide scientific, technical, and semi-technical information on the environment of national natural ecosystems. There are about 10 ENVIS centres providing data on various aspects of ecological systems and environments including mangrove ecosystem, its flora and fauna. The major centres that deal with mangrove ecosystems are the one attached to the Institute of Ocean Management, Anna University (http://www.iomenvis.in/) and the other to Centre for Advanced Studies in Marine Biology (http://www.casmbenvis.nic.in/).

21.8 Streamlining the Geospatial Techniques for Mangrove

A large variation in assessment of the area of each mangrove forest in different publications for the same period can be noticed in the literature. Such variation in assessments is mainly due to, (i) consideration of reserved or protected forest boundary in some studies, (ii) the varied extent of study area boundary depending on the area of interest of researchers (i.e., bounding box defined for the study area), and (iii) from RS technology perspective, the resolution or scale of the data, i.e., satellite data or aerial photograph used for mapping. Thus standardization of procedures with a possible variation due to resolution of data and minimum mapping unit needs to be undertaken. Standardization needs to be extended to the classification scheme, scale of mapping, species mapping (inclusion or exclusion of associated mangroves, etc.). It is also important to define standard protocols for ground truth, field survey, sampling, and validation process. Few recent global studies have helped in the adoption of such an approach and would be a strong base for the sustainable conservation and management of mangroves and also become the base for future research activities.

21.9 Conclusions

Mangroves are an important class of natural vegetation serving diverse ecological and economical functions at the interface of ocean and land. Given the large spread and diversity as well as being not so amenable for assessment and monitoring, form a focused target for its study by remote sensing technology. This chapter presents the basic concepts of maps, survey methods, and space-based observations and Geographic Information System for understanding the various aspects of mangroves by various Geospatial tools and techniques. A list of annexures provides the various types of remote sensors for land observation and a compilation of literature on studying the various aspects of mangroves using geospatial techniques. Previous studies on Indian mangroves have been summarized for their data source, objectives, study area, etc. The remote sensing techniques and capabilities are even advancing, which allows for improved mapping, monitoring, and characterization. Use of microwave and hyperspectral data for assessing different biophysical and biochemical characterization has been discussed and studies focusing on such fields are listed. Actions on conservation and management of mangroves require integration of space-based maps with field observations as well as integrated inputs from various stakeholders. Such a spatial integration is ideally realized in a GIS framework and basic details, as well as application of GIS is presented. The chapter also brings out the crucial requirements for streamlining the Geospatial techniques for its effective utilization for the better management of mangrove forests in general and specific to Indian scenario.

Annexure 1: List of Land Observation Remote Sensors

SN	Sensor	Agency	Platform	Altitude (km)	Spectral Range (nm)	No. of spectral bands	Spatial Resolution (m)
	Low/Medium Multispectral Remote Sensors						
1	Multispectral Scanner (MSS) (1972–1983)	NASA	Landsat 1–3	915	500–1100	4	60
2	Thematic Mapper (TM) (1982–2013)	NASA	Landsat 4–5	705	450–2350	7	30
3	LISS (Linear Imagine Self Scanning System)-2 (1988–1997)	ISRO-India	IRS-1A, IRS-1B & IRS-P2	904	460–860	4	36.25
4	Enhanced Thematic Mapper (ETM+) (1999)	NASA	Landsat 7	705	450–2350	7	30
5	Advanced Spaceborne Thermal Emission And Reflection Radiometer (ASTER) (1999)	NASA-JPL	TERRA Satellite	705	520–11650	14	15–90
6	ASTER & MERIS (2002)	ESA	ENVISAT Satellite	783	555–12000 & 412–900	7, 15	1000, 300
7	Advanced Visible and Near-Infrared Radiometer (AVNIR-2) (2006)	JAXA	ALOS Satellite	702	420–890	4	10
8	Operational Land Imager (2013)[a]	NASA	Landsat 8	705	430–2290	9	30
9	Multispectral Imager (MSI) (2015)[a]	ESA	Sentinel-2A	786	443–2190	12	10–60
10	LISS-3 (2016)[a]	ISRO-India	IRS-1C, IRS-1D & RESOURCESAT-2A	817	520–1750	4	23.5
	High Resolution Multispectral Remote Sensors						
11	Optical Sensor Assembly (OSA) (1999)	DigitalGlobe	IKONOS	681	450–900	4	4
12	Quickbird (2001)	DigitalGlobe	Quickbird	450	450–900	4	2.44
13	Kompsat-MSC (2006)[a]	KARI-Korea	KOMPSAT-2	685	450–900	4	4
14	REIS (2008)	Planet-USA	RapidEye	630	440–850	5	5

(continued)

SN	Sensor	Agency	Platform	Altitude (km)	Spectral Range (nm)	No. of spectral bands	Spatial Resolution (m)
15	WV110 (2009) & WV-3 MSS (2014)[a]	DigitalGlobe	Worldview-2 & 3	770	450–1040	8	1.84
16	High Resolution Optical Imager (HiRI) (2012)[a]	Airbus Defence and Space	PLEIADES 1A/1B	694	450–890	5	2.8
17	AEISS-A (Advanced Earth Imaging Sensor System-A) (2015)[a]	KARI-Korea	KOMPSAT-3A	528	450–900	5	2.8
18	LISS-IV) (2016)[a]	ISRO-India	RESOURCESAT 2A	817	520–860	3	5.8
19	SpaceView 110 Imaging System (2016)	DigitalGlobe	Worldview-4	617	450–920	4	1.24
20	High Resolution Multi-Spectral (HRMX) (2016–2018)[a], 2019[a]	ISRO-India	Cartosat-2 Series (2C, 2D, 2E, 2F) & 3	505 & 509	470–830 & 450–860	4	2 & 1.14
21	SkySat (2018)[a]	Planet-USA	SKYSAT 14–15	505	450–900	4	1

[a]Currently operational

Microwave Remote Sensors

SN	Sensor	Agency	Temporal resolution (Days)	Band frequency information	Spatial resolution
1	ERS-1 SAR (1991), ERS-2 SAR (1995), ENVISAT ASAR (2002)	ESA	3, 35, 176	C Band: 5.3 GHz	30 m
2	JERS-1 SAR (1992), ALOS PALSAR (2006) & PALSAR 2 (2014)[a]	JAXA	44, 46 & 6–12	L Band: 1.275 GHz	18 m, 10 m & 3–10 m
3	RADARSAT-1 (1995) & 2 (2007)[a]	CSA-Canada	24	C Band: 5.3 & 5.405 GHz	10–100 m & 3–100 m
4	Radar Imaging Satellite-2 (RISAT)-2 (2009)[a], RISAT-1 (2012) & RISAT-2B (2019)[a]	ISRO-India	14 & 25	X Band: 9.59 GHz & C Band: 5.350 GHz	1–8 m, 1–50 m & 1–8 m
5	TerraSAR-X (2007)[a] & TanDEM-X (2010)[a]	DLR	2 to 4	X Band: 9.65 GHz	1–18 m
6	Aquarius SAC-D (2011)	NASA	7	Active: 1.26 GHz Passive: 1.413 GHz	76–156 km
7	Sentinel- 1A (2014) and Sentinel 1B (2016)[a]	ESA	6 to 12	C Band: 5.405 GHz	5–100 m

[a]Currently operational

Hyperspectral Remote Sensors (Imaging Spectrometers)

SN	Sensor	Agency	Platform	Altitude (km)	Spectral range (nm)	Number of spectral bands	Spectral Resolution (nm)	Spatial Resolution (m)
8	AVIRIS (1986)[a]	NASA—JPL	Airborne	20	400–2500	224	10	4 to 20
9	Hyperion (2000)	NASA	EO-1 Satellite	705	400–2500	220	10	30
10	Compact High Resolution Imaging Spectrometer (CHRIS) (2001)[a]	ESA	Proba-1 Satellite	556	415–1050	63	1.3–12	18–36
11	HySIS (2018)[a]	ISRO-India	IMS-2	636.6	400–2400	326	10	30
12	PRISMA (2019)[a]	ASI-Italy	Satellite	614	400–2505	238	10	5–30

(continued)

13	HSI (2020)	DLR-ESA	EnMAP Satellite (Upcoming)	653	420–2450	232	6.5–10	30
14	HISUI (2020)	JAXA	Satellite (Upcoming)	618	400–2505	185	10–12.5	30
15	HypsIRI (2022)	NASA—JPL	Satellite (Upcoming)	626	380–2500; 7500–12000	217	4–12	60

[a]Currently operational

Annexure 2: Mapping and Monitoring of Mangrove Wetlands Using Remote Sensing and GIS

Sl No	Study location	Datasets	Techniques	Research findings	References
1	India Coast	IRS 1C L3	K-means clustering	Mangroves classified as dense, shrub and fringe mangroves	Nayak et al. (1996)
2	India coast	IRS-1A/1B LISS-I, IRS-1C/1D L3	Vis. Int and digital analysis (1986–1993)	Mangroves area (1:250,000, 1:50,000 and 1:25,000 maps)	Nayak and Bahuguna (2001)
3	Andaman Islands,	SOI Toposheet, AP, and Landsat TM	Supervised (MLC)	Loss of 21 sq.km mangrove lost	Roy et al. (1991)
4	Tamil Nadu and Andaman Islands,	Landsat TM, SPOT, and IRS L2	Vis. Int & Fuzzy Class	Mangrove loss (1989–1996): ~ 94 sq. km.	Ramachandran et al. (1998)
5	Pichavaram, India[a]	SOI, Landsat 5 TM, and IRS 1D L3	Vis. Int & OnS Dig	Degraded mangrove forest cover has increased by about 90%	Selvam et al. (2003)
6	Mumbai[a]	SPOT-2, IRS IC and 1D	Vis. Int & OnS Dig	92.94 sq. km. (1990); 56.40 sq. km. (2001)	Vijay et al. (2005)
7	Pichavaram	IRS-1D LISS-III, and ERS-2 SAR	Multiplicative, Brovey, IHS and PC transforms	Brovey transform outperforms other methods	Shanmugam et al. (2005)
8	Godavari Estuary, India[a]	SOI, Landsat 5 TM, IRS 1C/1D L3	Vis. Int	368 ha—mangrove increased	Ramasubramanian et al. (2006)
9	Goa Coast	SPOT-1, IRS-1C/1D	MLC	MLC (OA—88.43%)	Mani Murali et al. (2006)
10	Balasore, India[a]	Landsat MSS, IRS P6 L3, SOI	Vis. Int & overlay analysis in GIS	Mangrove lost—330 ha	Reddy and Pattanaik (2007)
11	Sagar Island, India	IRS IC L3	Supervised.	2.1 sq. km. (1998); 1.3 sq. km. (1999)	Dinesh Kumar et al. (2007)
12	Sundarbans, Bangladesh and India	Geocover Landsat Mosaic	MLC	588,695.5 ha (1970's); 581,642.2 ha (2000's)	Giri et al. (2007)

(continued)

13	Godavari, AP	IRS 1B L2 and IRS P6 L3	MLC	1,250 ha of mangroves lost to aquaculture, and deforestation	Satapathy et al. (2007)
14	Gulf of Mannar, and Andaman	Resourcesat I L4, IRS IC/ID L3	GMLC	OA—85.53% (pre-tsunami) and 88% (post-tsunami)	Chatterjee et al. (2008)
15	Godavari Delta	Landsat MSS/TM/ETM+, and IRS P6 L3	MLC	194.8 sq. km. (1977); 186.0 sq. km. (2005)	Reddy and Roy (2008)
16	Global Mangroves	Global Land Survey (GLS) 2000 mosaic; Landsat	Digital analysis (mosaic)	Total mangrove—137,760 sq.km	Giri et al. (2011)
17	Mahanadi delta[a]	Landsat MSS/TM, and IRS P6 L3	Vis. Int & OnS Dig	2606 ha mangrove area loss	Pattanaik and Narendra Prasad (2011)
18	Pichavaram[a]	Landsat MSS, TM, ETM+, IRS L3, SOI toposheet	Vis. Int & Dig. Analysis	471 ha lost (1970–1991); 531 ha gain (1991–2011)	Gnanappazham and Selvam (2011)
19	Pichavaram, TN	Landsat TM, Resourcesat-1 L4	ISODATA	2.51 sq. km. area increase (1991–2006)	Srinivasa Kumar et al. (2012)
20	India Coast	Resourcesat-1 L3/IV	ISODATA and supervised	Total area—495,620 ha.	Ajai et al. (2012)
21	Kachchh coast, Gujarat[a]	IRS Resourcesat-1 LISS-IV	Vis. Int, GIS overlay	Inter-tidal mudflat area—1797.91 sq. km.	Mahapatra et al. (2013)
22	Piram Island, Gujarat [a]	IRS-1D L4, IRS-P6 LISS-III	Vis. Int & OnS Dig	Mangrove area loss—0.74 sq. km. to 0.23 sq km. (2007–2013)	Bhavsar et al. (2014)
23	Gujarat Coast[a]	IRS-P6 L3	Vis. Int & OnS Dig	Mangrove area—996.26 sq. km	Ajay et al. (2014)
24	Andaman	IRS P6 L3	Unsupervised; Supervised	Mangrove area lost—188.02 ha	Sachithanandam et al. (2014)
25	Thuraikkadu RF, TN	EO-1 Hyperion data	SAVI, SVM	OA—80.89% (healthy); 86.59% (degraded); 76.28% (sparse)	Vidhya et al. (2014)
26	Kachchh[a]	IRS-P6 L3; IRS-RS2 L4 (2014)	Vis. Int & OnS Dig	9,894 ha (2005); 11,703 ha (2011)	Upadhyay et al. (2015)
27	Sundarbans, Bangladesh and India	Corona KH, Landsat TM/ETM/OLI	Vis. Int & ISODATA	6588 sq. km. (1776);1852 sq. km. (2014)	Ghosh et al. (2015)
28	Indian Mangroves	Landsat TM/ETM+	Supervised classification	406 sq. km. increase (1998–2013)	Jayanthi et al. (2018)
29	Pichavaram	Landsat ETM+ / OLI	MLC	11.41% mangrove area increased	Vani and Rama Chandra Prasad (2018)

OA—Overall Accuracy; Vis. Int.—Visual Interpretation; OnS Dig: On screen Digitization; MLC—Maximum Likelihood Classification; ISODATA—Iterative Self-Organizing DATA; SVM—Support Vector Machine; SAVI—Soil Adjusted Vegetation Index; IHS—Intensitu Hue Saturation
[a]RS integrated GIS analysis

Annexure 3: Floristic Discrimination of Indian Mangroves

Sl. No.	Study location	Datasets	Techniques	Research findings (OA/No. of Species identified)	References
30	Bhitarkanika	IRS P6 L3 (2004)	ISODATA, MLC	12 mangrove communities	Reddy et al. (2008)
31	Lothian Island	IRS P6 L4 (2005)	Unsupervised classification	8 mangrove classes (73.44%)	Debashis and Karmaker (2010)
32	Sundarbans	IRS 1D LISS-III	MLC, ISODATA	5 mangrove classes identified	Nandy and Kushwaha (2011)
33	Sundarbans	Field spectral signatures	k-means, ANOVA, SDA, and Factor analysis	17 mangrove species	Manjunath et al. (2013)
34	Sundarbans	EO-1 Hyperion data	N-FINDR, ATGP, and LSU	7 mangrove species (74.07%)	Chakravortty and Choudhury (2013)
35	Bhitarkanika	EO-1 Hyperion data	MD, SVM, SAM	5 mangrove species (96.85%)	Kumar et al. (2013)
36	Sundarbans	Landsat TM/ETM+, EO-1 Hyperion	MLC	6 mangrove species	Giri et al. (2014)
37	Pichavaram	EO-1 Hyperion data	JM, SAM	2 mangrove classes	Padma and Sanjeevi (2014)
38	Uttara Kannada	Landsat TM/ETM+, IRS-P6 L4 MX	GMLC	3 to 5 mangrove species	Mesta et al. (2014)
39	Bhitarkanika	Field and Laboratory spectra	PCA, SDA, JM distance	8 species of *Rhizophoraceae*	Prasad and Gnanappazham (2016)
40	Pichavaram	EO-1 Hyperion data	SAM, SFF, SID, LSU	10 species (SAM + LSU)	Salghuna and Pillutla (2017)
41	Sundarbans	EO-1 Hyperion data	MD, SAM, SVM	5 mangrove classes (99.08%)	Kumar et al. (2019)
42	Lothian Island	EO-1 Hyperion, IRS Resourcesat-2 LISS-IV	OBIA	7 mangrove classes	Mondal et al. (2019)
43	Lothian Island and Bhitarkanika	AVIRIS-NG	SAM	Lothian Island (15 species); Bhitarkanika (7 species)	Chaube et al. (2019)
44	Bhitarkanika	EO-1 Hyperion data	SAM	10 species (84%)	Anand et al. (2020)

OA—Overall Accuracy; SFF—Spectral Feature Fitting; MD—Minimum Distance; MLC—Maximum Likelihood Classification; SDA—Spectral Distance Analysis; SVM—Support Vector Machine; SAM—Spectral Angle Mapper; LSU—Linear Spectral Unmixing; JM—Jeffries–Matusita distance; PCA—Principal Component Analysis; OBIA—Object-Based Image Analysis; *—RS integrated GIS analysis

Annexure 4: Studies on Biophysical and Biochemical Characterization of Indian Mangroves

Sl. No	Study location	Datasets	Techniques	Research findings	References
45	Coringa	IRS-1C L3	NDVI, MLC, ANOVA	Strong relationship between VI and basal area	Satyanarayana et al. (2001)
46	Sundarbans	IRS-Resourcesat 2 L4	NDVI, OSAVI, and TDVI; Plot biomass, CHN analysis	Total AGB—236 metric tons	Manna et al. (2014)
47	Sundarbans	Landsat MSS/TM/ ETM+	MLC, AGB—IPCC	Potential CO_2 emission is 1567.98 ± 551.69 Gg (1975–2013)	Akhand et al. (2017)
48	Bhitarkanika	Worldview-2,Plot Biomass	GLCM texture, VI's	Texture of band ratio highly correlates with plot biomass	Prasad and Gnanappazham (2018)
49	Andaman	EO-1 Hyperion; Plot Biomass	Correlates in-situ LAI and Hyperspectral VI's	Identified bands sensitive to LAI	George et al. (2018)
50	Bhitarkanika	EO-1 Hyperion data	Five ML models applied on NDVI and EVI	Total AG Carbon 459.82 kt. C. (EVI); 514.47 kt. C. (NDVI)	Anand et al. (2020)
51	India Coast incl. Andaman	SPOT-VGT (March to May) (1999–2008)	NDVI Maximum Value Compositing (MVC)	38% (very healthy) and 27% (healthy)	Chellamani et al. (2014)
52	Middle Andaman Island	EO-1 Hyperion data	Hyperspectral VI's—SR, MSR, NDVI, and NLI	Sensitive wavelengths—549 nm, 559 nm (green), 702 nm, 722 nm, 742 nm, and 763 nm (red-edge)	George et al. (2019)
53	Lothian Island	AVIRIS-NG	9 Vegetation Indices, Red-Edge Analysis	56% of the mangrove increase in stress	Hati et al. (2020)

GLCM—Grey Level Co-occurrence Matrix; SR—Simple Ratio; NLI—Non-Linear Index; MSR—Modified Simple Ratio; AGB—Above Ground Biomass; LAI—Leaf Area Index; TDVI—Transformed Difference Vegetation Index; (Please refer Muhdoni et al. 2018)

Annexure 5: Shoreline Changes

Sl. No.	Study location	Datasets	Techniques	Research findings	References
1	Godavari Estuary	SOI Toposheet, Landsat 5 TM, IRS 1C/1D L3	Vis. Int and OnS Dig	Sand spit of Hope Island has grown nearly 2.6 km. (1937–2001)	Ramasubramanian et al. (2006)
2	India Coast	Landsat TM, ETM, AWiFS and L4; CSIZ Database,	Change detection (1989–91 to 2004–06) map on 1:25,000 scale	45.5%—erosion, 35.7%—accretion, and 18.79%—stable; Net loss = 73 sq. km	SAC (2017) https://vedas.sac.gov.in/vedas/node/61
3	Karnataka Coast	Landsat MSS/TM/ETM +	DSAS in ArcGIS	70%—stable or accretion, 30%—erosion (varying magnitude)	ChenthamilSelvan et al. (2014)
4	Sundarbans	IRS 1D L3, IRS P6 AWiFS, Resourcesat 1/2 L4	Vis. Int and OnS Dig	Hooghly island—eroding Thakuran Char, and Lothian islands—accreting	Raha et al. (2014)
5	Andhra Pradesh coast	Landsat TM/ETM+, IRS-P5/P6 L3/IV, Cartosat 1	DSAS in ArcGIS	275 km—erosion; 417 km—accretion; 153 km—stable.	Kankara et al. (2015)
6	Karnataka Coast	Landsat TM/ETM+/ OLI	DSAS in ArcGIS (1991–2014)	Ankola & Karwar—accretion; Honnavar—erosion	Hegde and Akshaya (2015)
7	Tamil Nadu Coast	Landsat MSS, TM ETM+, OLI	DSAS in ArcGIS (1978–2014)	Max. erosion: Ennore 26.4 m/yr; Max. accretion: South of Pulicat 34.3 m/yr	Natesan et al. (2015)
8	India Coast	Landsat TM/ETM+, Cartosat 1 (PAN), Resourcesat—L3, L4	Long-term shore line change (1990–2016) (1:25000 scale)	33%—erosion, 29%-accretion, and 38% stable	Kankara et al. (2018) http://www.nccr.gov.in
9	Globally applicable	ArcGIS 10.4 or higher	Rate-of-change statistics for a time series of shoreline vector data	Developing add-on tools for shoreline change detection DSAS v 5.0	Himmelstoss et al. (2018) https://code.usgs.gov/cch/dsas

(continued)

10	Andhra Pradesh coast	Landsat MSS/TM/ETM +	DSAS in ArcGIS (1973–2015)	Maximum accretion: Vishakhapatnam and Srikakulam coast; Maximum Erosion: Southern part of Krishna delta	Tyagi and Rai (2020)
11	Gujarat coast	Landsat TM, Resourcesat 2—L3	GIS Overlay analysis	Kori Creek—increased erosion; Gulf of Kachchh—high accretion	Das (2020)

DSAS—Digital Shoreline Analysis System; Vis. Int.—Visual Interpretation; OnS Dig: On-screen Digitization

Annexure 6: Coastal Zone Management

Sl. No.	Study location	Datasets	Techniques	Research findings	References
12	Global: Thailand, Sundarbans	Landsat MSS, SPOT, SIR-A	Mapping, sample plots and overlay analysis	Assessment, planning, conservation and management	FAO (1994)
13	Indian mangroves	IRS L3 TM and Environmental parameters—soil, water quality, phyto, zooplankton, etc.	GIS and hydrological modelling	Mapping, Monitoring, environment impact assessment,	NCCR (1999–2002) https://www.nccr.gov.in/?q=publication/technical-reports-
14	East coast mangroves	Landsat TM, IRS L3	Change detection, Overlay analysis of environmental parameters	Identification of conservation and protection zones, Joint Mangrove Management	Selvam et al. (2002)
15	Gulf of Kutchch	Hydrodynamic, meteorological, and spatio-temporal behaviour of oil mass.	Numerical modelling and GIS based Environmental sensitivity analysis and risk assessment	Priority index due to risk of Oil spill identified in the order of coral reefs, mangroves, mudflats and rocky coast	Kankara and Subramanian (2007)

(continued)

16	Pichavaram	Landsat TM, IRS L3 (1987, 1994, 1998 and 2004) and field sample data	Visual interpretation and Change detection	22 ha of mangroves converted to aquaculture and reconversion of 15 ha of aqua farms to agriculture	Jayanthi et al. (2007)
17	Godavari	Landsat MSS, TM, IRS L3	MCL, change detection	Extent of encroachment	Reddy and Roy (2008)
18	Chennai	RTK DGPS data on run up level, IRS L3 and L4	Spatial Interpolation and Overlay analysis	Inundation Vulnerability assessment (Tsunami 2004)	Satheesh Kumar et al. (2008)
19	Indian Mangroves	IRS L3 and L4, field data from sample survey	Image processing, Weighted overlay analysis	Normalized health index of mangroves to identify hotspots	Ajai et al. (2013)
20	Gulf of Kutchch	LULC, geomorphology, wave height and tidal range	Overlay analysis	Suitable sites for mangrove plantation	Mahapatra et al. (2013)
21	Mumbai		Allometry, CHN analysis and GIS analysis	Estimated C = 34.14769 T/ha	Patil et al. (2014)
22	Pichavaram,	RS data, tidal amplitude, time series freshwater data	GIS overlay, proximity analysis	Freshwater and Tidal dynamics and impacts on mangroves	Gnanappazham and Selvam (2014)
23	Pichavaram	RS data, field samples	Numerical tidal simulation model	Categories of mangroves based on tidal influence	Sathyanathan et al. (2014)
24	Gujarat coast	Coastal geomorphology, slope, SLC rate, Tidal range and wave height	Coastal Vulnerability Index (CVI) for anticipated sea level rise	High to very high risk category: 85 km (45.67%), Moderate to low risk: 934 km (54.33%)	Mahapatra et al. (2015)
25	Gulf of Kutchch, Gujarat	IRS L3 and L4	GIS Overlay analysis	Block level mangrove change—Restoration	Upadhyay et al. (2015)
26	Krishna estuary	Resourcesat L3, Landsat TM, ETM	MS4W (MapServer) and PostgreSQL	Web GIS of mangroves wetland maps, changes and	Jayakumar K (2019)
27	Goa and Andaman	FSI maps	IRS 1A, 1B	Identification of sites for conservation and plantation	Kumar R (2020)

References

Adam E, Mutanga O, Rugege D (2009) Multispectral and hyperspectral remote sensing for identification and mapping of wetland vegetation: a review. Wetl Ecol Manag 18:281–296. https://doi.org/10.1007/s11273-009-9169-z

Ajai, Bahuguna A, Chauhan HB, et al (2012) Mangrove Inventory of India at Community Level. Natl Acad Sci Lett 36:67–77. https://doi.org/10.1007/s40009-012-0087-x

Ajai, Bahuguna A, Chauhan, H.B., Sarma, K.S., Bhattacharya S, Ashutosh S, Pandey, C.N., Thangaradjou T, Gnanppazham L, Selvam, V. and Nayak, S.R. (2013). Mangrove inventory of India at community level, National Academy Science Letters, 36(1):67–77.

Ajay P, Vijay S, Mehmood K, et al (2014) Mapping and Monitoring of Mangroves in the Coastal Districts of Gujarat State using Remote Sensing and Geo-informatics. Asian J Geoinformatics 14:15–26.

Akhand A, Mukhopadhyay A, Chanda A, et al (2017) Potential CO2Emission Due to Loss of Above Ground Biomass from the Indian Sundarban Mangroves During the Last Four Decades. J Indian Soc Remote Sens 45:147–154. https://doi.org/10.1007/s12524-016-0567-4

Al-Naimi N, Al-Ghouti MA, Balakrishnan P (2016) Investigating chlorophyll and nitrogen levels of mangroves at Al-Khor, Qatar: an integrated chemical analysis and remote sensing approach. Environ Monit Assess 188: https://doi.org/10.1007/s10661-016-5269-4

Alongi DM (2014) Carbon cycling and storage in mangrove forests. Ann Rev Mar Sci 6:195–219. https://doi.org/10.1146/annurev-marine-010213-135020

Alongi DM (2002) Present state and future of the world's mangrove forests. Environ Conserv 29:331–349. https://doi.org/10.1017/S0376892902000231

Amma PKG, Bhaskaran PK (2020) Role of mangroves in wind-wave climate modeling—A review. J Coast Conserv 24:1–14. https://doi.org/10.1007/s11852-020-00740-0

Anand A, Pandey PC, Petropoulos GP, et al (2020) Use of hyperion for mangrove forest carbon stock assessment in bhitarkanika forest reserve: A contribution towards blue carbon initiative. Remote Sens 12:. https://doi.org/10.3390/rs12040597

Aneece I, Epstein H (2017) Identifying invasive plant species using field spectroscopy in the VNIR region in successional systems of north-central Virginia. Int J Remote Sens 38:100–122. https://doi.org/10.1080/01431161.2016.1259682

Anonymous (2020) Mangrove Forest in India, http://old.cwc.gov.in/CPDAC-Website/Paper_Research_Work

ASD (2001) Field spectrometry: Techniques and instrumentation

Asner GP, Martin RE (2008) Spectral and chemical analysis of tropical forests: Scaling from leaf to canopy levels. Remote Sens Environ 112:3958–3970. https://doi.org/10.1016/j.rse.2008.07.003

Awange JL, Kyalo Kiema JB (2013) Fundamentals of Remote Sensing. In: Awange JL, Kyalo Kiema JB (eds) Environmental Geoinfoirmatics: Monitoring and Management. Springer, pp 111–118

Bhavsar DO, Jasrai YT, Pandya HA, et al (2014) Monitoring Mangrove Status using Remote Sensing and Geo-informatics in Piram Island, Gulf of Khambhat, Gujarat State, India. Int J Sci Eng Res 5:999–1005

Blasco F, Lavenu F, Baraza J (1986) Remote sensing data applied to mangrove of Kenya coast. In: Proceedings of the 20th International Symposium on Remote Sensing of the Environment Programme 3, pp 1465–1480.

Bosire JO, Dahdouh-Guebas F, Walton M, et al (2008) Functionality of restored mangroves: A review. Aquat Bot 89:251–259. https://doi.org/10.1016/j.aquabot.2008.03.010

Bréda NJJ (2008) Leaf Area Index. In: Encyclopedia of Ecology. Academic Press. pp. 2148–2154

Bunting P, Rosenqvist A, Lucas RM, et al (2018) The global mangrove watch—A new 2010 global baseline of mangrove extent. Remote Sens 10:. https://doi.org/10.3390/rs10101669

Chakravortty S, Choudhury AS (2013) Determining spatial location of sub pixels in hyperspectral data for mangrove species identification. In: 2013 International Conference on Signal Processing, Image Processing & Pattern Recognition. IEEE, pp 39–43

Chandramohan P (2019) Quick Reference for CZ for Indian Coastline—CRZ classification 2019. www.indomer.com

Chatterjee B, Porwal MC, Hussin YA (2008) Assessment of Tsunami Damage To Mangrove in India Using Remote Sensing and Gis. In: The International Archives of the Photogrammetry, Remote Sensing and Spatial Information Sciences. pp 283–288

Chaube NR, Lele N, Misra A, et al (2019) Mangrove species discrimination and health assessment using AVIRIS-NG hyperspectral data. Curr Sci 116:1136–1142. https://doi.org/10.18520/cs/v116/i7/1136-1142

Chellamani P, Singh CP, Panigrahy S (2014) Assessment of the health status of Indian mangrove ecosystems using multi temporal remote sensing data. Trop Ecol 55:245–253

ChenthamilSelvan S, Kankara RS, Rajan B (2014) Assessment of shoreline changes along Karnataka coast, India using GIS & remote sensing techniques. Indian J Geo-Marine Sci 43:1286–1291

Chow J (2018) Mangrove management for climate change adaptation and sustainable development in coastal zones. J Sustain For 37:139–156. https://doi.org/10.1080/10549811.2017.1339615

Coetzee S, Ivánová I, Mitasova H, Brovelli MA (2020) Open geospatial software and data: A review of the current state and a perspective into the future. ISPRS Int J Geo-Information 9:1–30. https://doi.org/10.3390/ijgi9020090

Dalrymple RW, Zaitlin BA, Boyd R (1992) Estuarine facies models: conceptual basis and stratigraphic implications. J Sediment Petrol 62:1130–1146. https://doi.org/10.1306/D4267A69-2B26-11D7-8648000102C1865D

Das S (2020) Does mangrove plantation reduce coastal erosion? Assessment from the west coast of India. Reg Environ Chang 20. https://doi.org/10.1007/s10113-020-01637-2

DasGupta R, Shaw R (2013) Changing perspectives of mangrove management in India—An analytical overview. Ocean Coast Manag 80:107–118. https://doi.org/10.1016/j.ocecoaman.2013.04.010

Datt B, Mcvicar TR, Niel TG Van, et al (2003) Preprocessing EO-1 Hyperion Hyperspectral Data to Support the Application of Agricultural Indexes. IEEE Trans Geosci Remote Sens 41:1246–1259

Debashis M, Karmaker S (2010) Mangrove Classification in Sundarban using High Resolution Multi-Spectral Remote Sensing Data and GIS. Asian J Environ Disaster Manag 02:197–207. https://doi.org/10.3850/s179392402010000268

Dinesh Kumar PK, Gopinath G, Laluraj CM, et al (2007) Change detection studies of Sagar Island, India, using Indian remote sensing satellite 1C linear imaging self-scan sensor III data. J Coast Res 23:1498–1502. https://doi.org/10.2112/05-0599.1

Doyle TW, Girod GF (1997) The Frequency and Intensity of Atlantic Hurricanes and Their Influence on the Structure of South Florida Mangrove Communities. In: Diaz HF, Pulwarty RS (eds) Hurricanes. Springer Berlin Heidelberg, pp 109–120

Dronova I (2015) Object-Based Image Analysis in Wetland Research: A Review. Remote Sens 7:6380–6413. https://doi.org/10.3390/rs70506380

Duke N, Meynecke J, Dittmann S (2007) A world without mangroves? Science (80) 317:41–43

Ellenbogen K (2012) Coastal Ecosystems: Why sound management of these key natural carbon sinks matter for greenhouse gas emissions and climate change? The Blue Carbon Initiative FAQs. https://www.car-spaw-rac.org/IMG/pdf/BC_FAQ_UNFCCC-2.pdf

Ellison JC (2018) Biogeomorphology of mangroves. Coast Wetl An Integr Ecosyst Approach 687–715. https://doi.org/10.1016/B978-0-444-63893-9.00020-4

FAO (1994) Mangrove forest management guidelines. FAO For Pap 117. Rome. pp 319

FAO (2007) The world's mangroves 1980–2005. FAO For Pap 153:89. https://doi.org/978-92-5-105856-5

FAO (2020) http://www.fao.org/forestry/mangrove/3942/en/

Fatoyinbo TE, Armstrong AH (2010) Remote Characterization of Biomass Measurements: Case Study of Mangrove Forests. In: Momba MNB, Bux F (eds) Biomass. Sciyo, pp 65–78

Friess DA, Rogers K, Lovelock CE, et al (2019) The State of the World's Mangrove Forests: Past, Present, and Future. Annu Rev Environ Resour 44:1–27

FSI (2017) Mangrove Cover in India: India State of Forest Report 2017 by Forest Survey of India

FSI (2019) Mangrove Cover in India: India State of Forest Report 2019 by Forest Survey of India

Fu W, Ma J, Chen P, Chen F (2020) Remote Sensing Satellites for Digital Earth. In: Guo H, Goodchild MF, Annoni A (eds) Manual of Digital Earth. Springer Singapore, pp 55–123

George R, Padalia H, Sinha SK, Kumar AS (2018) Evaluation of the use of hyperspectral vegetation indices for estimating mangrove leaf area index in middle Andaman Island, India. Remote Sens Lett 9:1099–1108. https://doi.org/10.1080/2150704X.2018.1508910

George R, Padalia H, Sinha SK, Kumar AS (2019) Evaluating sensitivity of hyperspectral indices for estimating mangrove chlorophyll in Middle Andaman Island, India. Environ Monit Assess 191. https://doi.org/10.1007/s10661-019-7679-6

Ghosh A, Schmidt S, Fickert T, Nüsser M (2015) The Indian Sundarban mangrove forests: History, utilization, conservation strategies and local perception. Diversity 7:149–169. https://doi.org/10.3390/d7020149

Giri C, Ochieng E, Tieszen LL, et al (2011) Status and distribution of mangrove forests of the world using earth observation satellite data. Glob Ecol Biogeogr 20:154–159. https://doi.org/10.1111/j.1466-8238.2010.00584.x

Giri C, Pengra B, Zhu Z, et al (2007) Monitoring mangrove forest dynamics of the Sundarbans in Bangladesh and India using multi-temporal satellite data from 1973 to 2000. Estuar Coast Shelf Sci 73:91–100. https://doi.org/10.1016/j.ecss.2006.12.019

Giri S, Mukhopadhyay A, Hazra S, et al (2014) A study on abundance and distribution of mangrove species in Indian Sundarban using remote sensing technique. J Coast Conserv 18:359–367. https://doi.org/10.1007/s11852-014-0322-3

Gnanappazham L (2008) A remote sensing and GIS based decision support system for effective management of Pichavaram mangrove wetland, South India. Dissertation, University of Madras

Gnanappazham L, Selvam V (2011) The dynamics in the distribution of mangrove forests in Pichavaram, South India—perception by user community and remote sensing. Geocarto Int 26:475–490. https://doi.org/10.1080/10106049.2011.591943

Gnanappazham L and Selvam V (2014) Response of mangroves to the change in tidal and fresh water flow–A case study in Pichavaram, South India. Ocean and Coast Manage 102, 131–138.

Goodchild M (2005) GIS and Modeling Overview. In: Maguire DJ, Batty M, Goodchild MF (eds) GIS, Spatial Analysis, and Modeling. ESRI Press, pp 1–17.

Green EP, Clark CD, Mumby PJ, et al (1998) Remote sensing techniques for mangrove mapping. Int J Remote Sens 19:935–956. https://doi.org/10.1080/014311698215801

Guo D, Onstein E (2020) State-of-the-art geospatial information processing in NoSQL databases. ISPRS Int J Geo-Information 9. https://doi.org/10.3390/ijgi9050331

Gupta K, Mukhopadhyay A, Giri S, et al (2018) An index for discrimination of mangroves from non-mangroves using LANDSAT 8 OLI imagery. MethodsX 5:1129–1139. https://doi.org/10.1016/j.mex.2018.09.011

Hamilton SE, Casey D (2016) Creation of a high spatio-temporal resolution global database of continuous mangrove forest cover for the 21st century (CGMFC-21). Glob Ecol Biogeogr 25:729–738. https://doi.org/10.1111/geb.12449

Hati JP, Goswami S, Samanta S, et al (2020) Estimation of vegetation stress in the mangrove forest using AVIRIS-NG airborne hyperspectral data. Model Earth Syst Environ. https://doi.org/10.1007/s40808-020-00916-5

Hegde AV, Akshaya BJ (2015) Shoreline Transformation Study of Karnataka Coast: Geospatial Approach. Aquat Procedia 4:151–156. https://doi.org/10.1016/j.aqpro.2015.02.021

Heumann BW (2011) Satellite remote sensing of mangrove forests: Recent advances and future opportunities. Prog Phys Geogr 35:87–108. https://doi.org/10.1177/0309133310385371

Himmelstoss EA, Henderson RE, Kratzmann MG, Farris AS (2018) Digital Shoreline Analysis System (DSAS) Version 5.0 User Guide

ICMAM (2000) Report on "Critical habitat Information System using GIS for Pichavaram Mangrove South India. Available via http://www.icmam.gov.in. Accessed on August, 2020
Jackson RD, Pinter PJ, Reginato RJ, Idso SB (1980) Hand-held radiometry. In: Agriciltural Reviews and Manuals ARM—W—19/October 1980. U.S. Deaprtment of Agriculture Report. p 66
Jayakumar K (2019) Managing Mangrove Forests Using Open Source-Based WebGIS, Coastal Management, Global Challenges and Innovations. Academic Press, pp 301–321.
Jayanthi M, Gnanappzhaam L and Ramachandran S (2007) Assessment of Impact of Aquaculture on Mangrove Environments in the South East Coast of India Using Remote Sensing and Geographical Information System (GIS). Asian Fisheries Science 20:325–338
Jayanthi M, Thirumurthy S, Nagaraj G, et al (2018) Spatial and temporal changes in mangrove cover across the protected and unprotected forests of India. Estuar Coast Shelf Sci 213:81–91. https://doi.org/10.1016/j.ecss.2018.08.016
Jensen JR (2014) Remote Sensing of the Environment: an Earth Resource Perspective, Second Edition. Pearson Education Limited
Jensen JR (2016) Introductory Digital Image Processing: A Remote Sensing Perspective. Pearson Prentice Hall, Pearson Eductaion Inc., New Jersey, USA
Jensen JR, Lin H, Yang X, et al (1991) The measurement of mangrove characteristics in Southwest Florida using SPOT multispectral data. Geocarto Int 6:13–21. https://doi.org/10.1080/10106049109354302
Jiménez M, Díaz-Delgado R (2015) Towards a Standard Plant Species Spectral Library Protocol for Vegetation Mapping: A Case Study in the Shrubland of Doñana National Park. ISPRS Int J Geo-Information 4:2472–2495. https://doi.org/10.3390/ijgi4042472
Kale MP, Singh S, Roy PS (2002) Biomass and productivity estimation using aerospace data and geographic information system. Trop Ecol 43:123–136
Kankara RS and Subramanian BR. 2007. Oil Spill Sensitivity Analysis and Risk Assessment for Gulf of Kachchh, India, using Integrated Modeling. Journal of Coastal Research. 23 (5) 1251–1258.
Kankara RS, Murthy MVR, Rajeevan M (2018) National Assessment of Shoreline changes along Indian Coast: Status Report for 26 years (1990–2016). Chennai
Kankara RS, Selvan SC, Markose VJ, et al (2015) Estimation of long and short term shoreline changes along Andhra Pradesh coast using remote sensing and GIS techniques. Procedia Eng 116:855–862. https://doi.org/10.1016/j.proeng.2015.08.374
Kathiresan K (2018) Mangrove forests of India. Curr Sci 114:976–981. https://doi.org/10.18520/cs/v114/i05/976-981
Klemas V (2013) Remote Sensing of Coastal Wetland Biomass: An Overview. J Coast Res 290:1016–1028. https://doi.org/10.2112/JCOASTRES-D-12-00237.1
Kokaly R, Clark R (1999) Spectroscopic determination of leaf biochemistry using band-depth analysis of absorption features and stepwise multiple linear regression. Remote Sens Environ 4257:267–287
Kokaly RF, Asner GP, Ollinger S V., et al (2009) Characterizing canopy biochemistry from imaging spectroscopy and its application to ecosystem studies. Remote Sens Environ 113: S78–S91. https://doi.org/10.1016/j.rse.2008.10.018
Konecny G (2014) Geoinformation: Remote Sensing, Photogrammetry, and Geographic Information Systems, Second Edi. CRC Press, Boka Raton
Kopp S, Becker P, Doshi A, et al (2019) Achieving the full vision of earth observation data cubes. Data 4:. https://doi.org/10.3390/data4030094
Kuenzer C, Bluemel A, Gebhardt S, et al (2011) Remote Sensing of Mangrove Ecosystems: A Review. Remote Sens 3:878–928. https://doi.org/10.3390/rs3050878
Kuenzer C, Ottinger M, Wegmann M, Guo H (2014) Earth observation satellite sensors for biodiversity monitoring: potentials and bottlenecks. Int J Remote Sens 35:6599–6647. https://doi.org/10.1080/01431161.2014.964349

Kumar R (2020) Conservation and management of mangroves in India, with special reference to the State of Goa and the Middle Andaman Islands, http://www.fao.org/3/x8080e/x8080e07.htm

Kumar T, Mandal A, Dutta D, et al (2019) Discrimination and classification of mangrove forests using EO-1 Hyperion data: a case study of Indian Sundarbans. Geocarto Int 34:415–442. https://doi.org/10.1080/10106049.2017.1408699

Kumar T, Panigrahy S, Kumar P, Parihar JS (2013) Classification of floristic composition of mangrove forests using hyperspectral data: case study of Bhitarkanika National Park, India. J Coast Conserv Plan Manag 17:121–132. https://doi.org/10.1007/s11852-012-0223-2

Kutser T, Miller I, Jupp DLB (2006) Mapping coral reef benthic substrates using hyperspectral space-borne images and spectral libraries. Estuar Coast Shelf Sci 70:449–460. https://doi.org/10.1016/j.ecss.2006.06.026

Lillesand TM, Kiefer RW, Chipman JW (2004) Remote Sensing and Image Interpretation, Fifth Edit. John Wiley and Sons Ltd, Hoboken, NJ, USA

Lu D, Chen Q, Wang G, et al (2014) A survey of remote sensing-based aboveground biomass estimation methods in forest ecosystems. Int J Digit Earth 1–43. https://doi.org/10.1080/17538947.2014.990526

Lugo AE, Snedaker SC (1974) The Ecology of Mangroves. Annu Rev Ecol Syst 5:39–64

Mahapatra M, Ramakrishnan R, Rajawat AS (2015) Coastal vulnerability assessment of Gujarat coast to sea level rise using GIS techniques: a preliminary study. J Coast Conserv 19:241–256. https://doi.org/10.1007/s11852-015-0384-x

Mahapatra M, Ratheesh R, Rajawat AS (2013) Potential Site Selection for Mangrove Plantation Along the Kachchh District, Gujarat, India Using Remote Sensing and Gis Techniques. Int J Geol Earth Environ Sci 3:18–23.

Manakos I, Manevski K, Petropoulos GP (2010) Development of a spectral library for mediterranean land cover types. Proc 30th EARSeL Symp Remote Sens Sci Educ Nat Cult Herit 663–668

Mani Murali R, Vethamony P, Saran AK, Jayakumar S (2006) Change detection studies in coastal zone features of Goa, India by remote sensing. Curr Sci 91:816–820

Manjunath K, Kumar T, Kundu N, Panigrahy S (2013) Discrimination of mangrove species and mudflat classes using in situ hyperspectral data: a case study of Indian Sundarbans. GIScience Remote Sens 50:400–417

Manna S, Nandy S, Chanda A, et al (2014) Estimating aboveground biomass in Avicennia marina plantation in Indian Sundarbans using high-resolution satellite data. J Appl Remote Sens 8:083638. https://doi.org/10.1117/1.jrs.8.083638

Marshall M, Thenkabail P (2015) Advantage of hyperspectral EO-1 Hyperion over multispectral IKONOS, GeoEye-1, WorldView-2, Landsat ETM+, and MODIS vegetation indices in crop biomass estimation. ISPRS J Photogramm Remote Sens 108:205–218. https://doi.org/10.1016/j.isprsjprs.2015.08.001

McGlone JC (2004) Manual of Photogrammetry, Fifth Edit. American Society for Photogrammetry and Remote Sensing.

Menéndez P, Losada IJ, Torres-Ortega S, et al (2020) The Global Flood Protection Benefits of Mangroves. Sci Rep 10:1–11. https://doi.org/10.1038/s41598-020-61136-6

Mesta PN, Setturu B, Chandran MDS, et al (2014) Inventorying, Mapping and Monitoring of Mangroves towards Sustainable Management of West Coast, India. J Geophys Remote Sens 3:1000130. https://doi.org/10.4172/2169-0049.1000130

Mitra A (2020) Ecosystem Services of Mangroves: An Overview. In: Mitra A (ed) Mangrove Forests in India: Exploring Ecosystem Services. Springer, Cham, pp 1–32

MoEF, GoI (2008) Mangroves for the Future: National Strategy and Action Plan, India (Revised Draft).

Mondal B, Saha AK, Roy A (2019) Mapping mangroves using LISS-IV and Hyperion data in part of the Indian Sundarban. Int J Remote Sens 40:9380–9400. https://doi.org/10.1080/01431161.2019.1630784

Mougin E, Proisy C, Marty G, et al (1999) Multifrequency and multipolarization radar backscattering from mangrove forests. IEEE Trans Geosci Remote Sens 37:94–102. https://doi.org/10.1109/36.739128

Muhdoni, FF, Sambah AB, Mahmudi M, Wiadnya DGR (2018) Comparison of Different Vegetation Indices for assessing Mangrove Density using Sentinel-2 Imagery. Int. J. Geomate, 14, 42–51. https://doi.org/10.21660/2018.45.7177.

Muñoz-Huerta RF, Guevara-Gonzalez RG, Contreras-Medina LM, et al (2013) A review of methods for sensing the nitrogen status in plants: Advantages, disadvantages and recent advances. Sensors 13:10823–10843. https://doi.org/10.3390/s130810823

Murdiyarso D, Purbopuspito J, Kauffman JB, et al (2015) The potential of Indonesian mangrove forests for global climate change mitigation. Nat Clim Chang 8–11. https://doi.org/10.1038/nclimate2734

Mutanga O, Skidmore AK (2004) Hyperspectral band depth analysis for a better estimation of grass biomass (Cenchrus ciliaris) measured under controlled laboratory conditions. Int J Appl Earth Obs Geoinf 5:87–96. https://doi.org/10.1016/j.jag.2004.01.001

Nandy S, Kushwaha SPS (2011) Study on the utility of IRS 1D LISS-III data and the classification techniques for mapping of Sunderban mangroves. J Coast Conserv 15:123–137. https://doi.org/10.1007/s11852-010-0126-z

Natesan U, Parthasarathy A, Vishnunath R, et al (2015) Monitoring Longterm Shoreline Changes along Tamil Nadu, India Using Geospatial Techniques. Aquat Procedia 4:325–332. https://doi.org/10.1016/j.aqpro.2015.02.044

Nayak S, Bahuguna A (2001) Application of remote sensing data to monitor mangroves and other coastal vegetation of India. Indian J Mar Sci 30:195–213

Nayak S, Chauhan P, Chauhan HB, et al (1996) IR7S-1C applications for coastal zone management. Curr. Sci. 70:614–618

Nazrul-Islam AKM (1993) Environment and vegetation of Sundarban mangrove forest. In: Lieth H, Masoom A (eds) Towards the rational use ofhigh salinity tolerant plants. Kluwer. pp 81–88

Nellemann C, Corcoran E, Duarte CM, et al (2009) Blue carbon: A Rapid Response Assessment. United Nations Environmental Programme, GRID-Arendal, www.grida.no

NCSM (2020) https://ncscm.res.in/pdf_docs/crz-2019.pdf

NRSC (2017) Natural Resource Census—Land Use Land Cover Database. Technical Report v.1

Padma S, Sanjeevi S (2014) Jeffries Matusita based mixed-measure for improved spectral matching in hyperspectral image analysis. Int J Appl Earth Obs Geoinf 32:138–151. https://doi.org/10.1016/j.jag.2014.04.001

Patil V, Singh A, Naik N, Unnikrishnan S (2014) Estimation of carbon stocks in avicennia marina stand using allometry, CHN analysis, and GIS methods. Wetlands 34:379–391. https://doi.org/10.1007/s13157-013-0505-y

Pattanaik C, Narendra Prasad S (2011) Assessment of aquaculture impact on mangroves of Mahanadi delta (Orissa), East coast of India using remote sensing and GIS. Ocean Coast Manag 54:789–795. https://doi.org/10.1016/j.ocecoaman.2011.07.013

Pham TD, Yokoya N, Bui DT, et al (2019) Remote sensing approaches for monitoring mangrove species, structure, and biomass: Opportunities and challenges. Remote Sens 11:1–24. https://doi.org/10.3390/rs11030230

Prasad KA, Gnanappazham L (2016) Multiple statistical approaches for the discrimination of mangrove species of Rhizophoraceae using transformed field and laboratory hyperspectral data. Geocarto Int 31:891–912. https://doi.org/10.1080/10106049.2015.1094521

Prasad KA, Gnanappazham L (2018) Estimation of Above Ground Biomass using High Resolution Multispectral Worldview 2 image. In: Indian Cartographer. pp 569–577

Prasad KA, Gnanappazham L, Selvam V, et al (2015) Developing a spectral library of mangrove species of Indian East Coast using field spectroscopy. Geocarto Int 30:580–599

Raha AK, Mishra A, Bhattacharya S, et al (2014) Sea Level Rise and Submergence of Sundarban Islands: A Time Series Study of Estuarine Dynamics. J Ecol Environ Sci 5:114–123

Ramachandran S, Sundaramoorthy S, Krishnamoorthy R, et al (1998) Application of remote sensing and GIS to coastal wetland ecology of Tamil Nadu and Andaman and Nicobar group of islands with special reference to mangroves. Curr Sci 75:236–244

Ramasubramanian R, Gnanappazham L, Ravishankar T, Navamuniyammal M (2006) Mangroves of Godavari—Analysis Through Remote Sensing Approach. Wetl Ecol Manag 14:29–37. https://doi.org/10.1007/s11273-005-2175-x

Rao NR, Garg PK, Ghosh SK (2007) Development of an agricultural crops spectral library and classification of crops at cultivator level using hyperspectral data. Precis Agric 8:173–185. https://doi.org/10.1007/s11119-007-9037-x

Ravishankar T, Gnanappazham L, Ramasubramanian R, et al (2004) Atlas of Mangrove Wetlands of India: Part 2 - Andhra Pradesh. M.S. Swaminathan Research Foundation, Chennai, India

Reddy CS, Pattanaik C (2007) Mangrove vegetation assessment and monitoring in Balasore district, Orissa using remote sensing and GIS. Natl Acad Sci Lett 30:377–381

Reddy CS, Pattanaik C, Murthy MSR (2008) Community zonation of mangroves in Bhitarkanika Wildlife Sanctury,Orissa, India using IRSP6LISS III data. Proc Natl Acad Sci India Sect B—Biol Sci 78:246–252

Reddy CS, Roy A (2008) Assessment of Three Decade Vegetation Dynamics in Mangroves of Godavari Delta, India Using Multi-Temporal Satellite Data and GIS. Res. J. Environ. Sci. 2:108–115

Reddy MA (2008) Text Book of Remote sensing and Geographical Information Systems, Third Edit. BS Publications, Hyderabad, India

Rees WG (2013) Physical Principles of Remote Sensing: Third Edition. Cambridge University Press.

Reichardt M (2017) Open Geospatial Consortium standards. Int Encycl Geogr People, Earth, Environ Technol 1–8. https://doi.org/10.1002/9781118786352.wbieg0348

Richards JA, Jia X (2005) Remote Sensing Digital Image Analysis: An Introduction, 4th Edition. Springer-Verlag, Berlin

Roy PS, Ranganath BK, Diwakar PG, et al (1991) Tropical forest typo mapping and monitoring using remote sensing. Int J Remote Sens 12:2205–2225. https://doi.org/10.1080/01431169108955253

SAC (2017) Shoreline Change Atlas of the Indian Coast Shoreline Change Atlas (1989–91 to 2004–06). Ahmedabad

Sachithanandam V, Mageswaran T, Ragavan P, et al (2014) Mangrove regeneration in tsunami affected area of north and south Andaman using insitu and remote sensing techniques. Indian J Geo-Marine Sci 43:1055–1061

Sahu SC, Suresh HS, Murthy IK, Ravindranath NH (2015) Mangrove Area Assessment in India: Implications of Loss of Mangroves. J Earth Sci Clim Change 06:280. https://doi.org/10.4172/2157-7617.1000280

Salghuna NN, Pillutla RCP (2017) Mapping Mangrove Species Using Hyperspectral Data: A Case Study of Pichavaram Mangrove Ecosystem, Tamil Nadu. Earth Syst Environ 1:24. https://doi.org/10.1007/s41748-017-0024-8

Satapathy DR, Krupadam RJ, Kumar LP, Wate SR (2007) The application of satellite data for the quantification of mangrove loss and coastal management in the Godavari estuary, East Coast of India. Environ Monit Assess 134:453–469. https://doi.org/10.1007/s10661-007-9636-z

Satheesh Kumar, P. Arul Murugan, R. R. Krishnamurthy, B. Prabhu Doss Batvari, M. V. Ramanamurthy, T. Usha, and Y. Pari (2008). Inundation mapping—a study based on December 2004 Tsunami Hazard along Chennai coast, Southeast India. Nat. Hazards Earth Syst. Sci., 8, 617–626

Sathyanathan R, Thattai D, Selvam V (2014) The Coleroon river flow and its effect on the Pichavaram mangrove ecosystem. J Coast Conserv 18:309–322. https://doi.org/10.1007/s11852-014-0313-4

Satyanarayana B, Thierry B, Seen D Lo, et al (2001) Remote sensing in mangrove research-relationship between vegetation indices and dendrometric parameters: A case for Coringa, east coast of India. In: 22nd Asian Conference on Remote Sensing. Singapore, pp 5–9

Schölkopf B, Smola AJ (2001) Learning with kernel: Support Vector Machines, Regularization, Optimization and Beyond. The MIT Press, Cambridge, Massachusetts, London, England, Cambridge, Massachusetts, London, England

Selvam V, Gnanappazham L, Navamuniyammal M et al (2002) Atlas of mangrove wetlands of India: part 1 Tamil Nadu, MS Swaminathan Research Foundation, Chennai, p 100.

Selvam V, Ravichandran KK, Gnanappazham L, Navamuniyammal M (2003) Assessment of community-based restoration of Pichavaram mangrove wetland using remote sensing data. Curr Sci 85:794–798

Shanmugam P, Manjunath AS, Ahn YH, et al (2005) Application of multisensor fusion techniques in remote sensing of coastal mangrove wetlands. Int J Geoinformatics 1:1–17

Snyder JP, Maling DH (1993) Flattening the Earth. Nature 366:522–522

Spadling M, Kainuma M, Collins L (2010) World Atlas of Mangroves. A collaborative project of ITTO, ISME, FAO, UNEP-WCMC, UNESCO-MAB, UNU-INWEH and TNC. Earthscan Publications, Abingdon, Oxon OX144RN, UK, Abingdon, Oxon OX144RN, UK

Srinivasa Kumar T, Mahendra RS, Nayak S, et al (2012) Identification of hot spots and well managed areas of Pichavaram mangrove using Landsat TM and Resourcesat-1 LISS IV: An example of coastal resource conservation along Tamil Nadu Coast, India. J Coast Conserv 16:1–12. https://doi.org/10.1007/s11852-011-0162-3

Tang W, Zheng M, Zhao X, et al (2018) Big Geospatial Data Analytics for Global Mangrove Biomass and Carbon Estimation. Sustainability 10:472. https://doi.org/10.3390/su10020472

Thakur S, Mondal I, Ghosh PB, et al (2020) A review of the application of multispectral remote sensing in the study of mangrove ecosystems with special emphasis on image processing techniques. Spat Inf Res 28:39–51. https://doi.org/10.1007/s41324-019-00268-y

Tomlinson PB (1994) The Botany of Mangroves. Cambridge University Press, Cambridge, UK

Townshend JRG, Justice CO (1986) Analysis of the dynamics of African vegetation using the normalized difference vegetation index. Int J Remote Sens 7:1435–1445. https://doi.org/10.1080/01431168608948946

Tso B, Mather PM (2009) Classification Methods for Remotely Sensed Data, Second Edi. CRC Press, Boka Raton, Boca Raton

Tucker CJ, Sellers PJ (1986) Satellite remote sensing of primary production. Int J Remote Sens 7:1395–1416. https://doi.org/10.1080/01431168608948944

Tyagi S, Rai SC (2020) Monitoring shoreline changes along Andhra coast of India using remote sensing and geographic information system. Indian J Geo-Marine Sci 49:218–224

Underwood E (2003) Mapping nonnative plants using hyperspectral imagery. Remote Sens Environ 86:150–161. https://doi.org/10.1016/S0034-4257(03)00096-8

Upadhyay R, Joshi N, Sampat AC, et al (2015) Mangrove Restoration and Regeneration Monitoring in Gulf of Kachchh, Gujarat State, India, Using Remote Sensing and Geo-Informatics. Int J Geosci 06:299–310. https://doi.org/10.4236/ijg.2015.64023

Vani M, Rama Chandra Prasad P (2018) Geospatial assessment of spatio-temporal changes in mangrove vegetation of pichavaram region, Tamil nadu, India. In: Threats to Mangrove Forests. Springer. pp 89–102

Vidhya R, Vijayasekaran D, Farook MA, et al (2014) Improved classification of mangroves health status using hyperspectral remote sensing data. Int Arch Photogramm Remote Sens Spat Inf Sci—ISPRS Arch XL–8:667–670. https://doi.org/10.5194/isprsarchives-XL-8-667-2014

Vijay V, Biradar RS, Inamdar AB, et al (2005) Mangrove mapping and change detection around Mumbai (Bombay) using remotely sensed data. Indian J Mar Sci 34:310–315

Woodroffe C (1992) Mangrove Sediments and Geomorphology. In: Robertson AI, Alongi DM (eds) Tropical Mangrove Ecosystems. American Geophysical Union, Washington, DC
Zheng G, Moskal LM (2009) Retrieving Leaf Area Index (LAI) Using Remote Sensing: Theories, Methods and Sensors. Sensors 9:2719–2745. https://doi.org/10.3390/s90402719
Zhou G (2001) Architecture of Future Intelligent Earth Observing Satellites (FIEOS) in 2010 and Beyond: Technical report (June 01, 2001–November 31, 2001)
Zomer R, Trabucco A, Ustin S (2009) Building spectral libraries for wetlands land cover classification and hyperspectral remote sensing. J Environ Manage 90:2170–2177. https://doi.org/10.1016/j.jenvman.2007.06.028.

The manufacturer's authorised representative in the EU is Springer Nature Customer Service Centre GmbH, Europaplatz 3, 69115 Heidelberg, Germany. If you have any concerns regarding our products, please contact ProductSafety@springernature.com

Printed and bound by CPI Group (UK) Ltd, Croydon, CR0 4YY
30/09/2026
02225240-0013